Lecture Notes in Computer Science   3967

Commenced Publication in 1973
Founding and Former Series Editors:
Gerhard Goos, Juris Hartmanis, and Jan van Leeuwen

Lecture Notes in Computer Science

Dima Grigoriev   John Harrison
Edward A. Hirsch (Eds.)

Computer Science – Theory and Applications

First International Computer Science
Symposium in Russia, CSR 2006
St. Petersburg, Russia, June 8-12, 2006
Proceedings

 Springer

Volume Editors

Dima Grigoriev
Université de Rennes, IRMAR
Campus de Beaulieu, 35042 Rennes, Cedex, France
E-mail: dimitri.grigoriev@math.univ-rennes1.fr

John Harrison
Intel Corporation
JF1-13, 2111 NE 25th Avenue, Hillsboro, OR 97124, USA
E-mail: johnh@ichips.intel.com

Edward A. Hirsch
Steklov Institute of Mathematics at St. Petersburg
27 Fontanka, St. Petersburg, 191023, Russia
E-mail: hirsch@pdmi.ras.ru

Library of Congress Control Number: 2006925113

CR Subject Classification (1998): F.1.1-2, F.2.1-2, F.4.1, I.2.6, J.3

LNCS Sublibrary: SL 1 – Theoretical Computer Science and General Issues

ISSN   0302-9743
ISBN-10   3-540-34166-8 Springer Berlin Heidelberg New York
ISBN-13   978-3-540-34166-6 Springer Berlin Heidelberg New York

Springer is a part of Springer Science+Business Media

springer.com

© Springer-Verlag Berlin Heidelberg 2006
Printed in Germany

Typesetting: Camera-ready by author, data conversion by Scientific Publishing Services, Chennai, India
Printed on acid-free paper   SPIN: 11753728   06/3142   5 4 3 2 1 0

Preface

The International Symposium on Computer Science in Russia (CSR 2006) was held on June 8–12, 2006 in St. Petersburg, Russia, hosted by the Steklov Institute of Mathematics at St. Petersburg. It was the first event in a planned series of regular international meetings that are united by their location (Russia).

The symposium was composed of two tracks: Theory and Applications/Technology. The opening lecture was given by Stephen A. Cook and ten other invited plenary lectures were given by Boaz Barak, Gerard Berry, Bob Colwell, Byron Cook, Melvin Fitting, Russell Impagliazzo, Michael Kaminski, Michael Kishinevsky, Pascal Koiran, and Omer Reingold. This volume contains the accepted papers of both tracks and also some of the abstracts of the invited speakers. The scope of the proposed topics for the symposium was quite broad and covered basically all areas of computer science and its applications. We received 279 papers in total, the contributors being from 45 countries. The Program Committee of the Theory Track selected 35 papers out of 121 submissions. The Program Committee of the Applications/Technology Track selected 29 papers out of 158 submissions.

Two workshops were co-located with CSR 2006:

- Workshop on Words and Automata (WOWA 2006);
- Tutorial on Automata-Based Programming.

The reviewing process was organized using the EasyChair conference system, thanks to Andrei Voronkov.

We are grateful to our sponsors:

- The U.S. Civilian Research & Development Foundation;
- Russian Foundation for Basic Research.

We also thank the local Organizing Committee: Dmitry Karpov, Arist Kojevnikov, Alexander Kulikov, Yury Lifshits, Sergey Nikolenko, Svetlana Obraztsova, Alexei Pastor, and, in particular, Elena Novikova.

March 2006

Dima Grigoriev
John R. Harrison
Edward A. Hirsch

Organization

Program Committee

Theory Track

Sergei Artemov	*Graduate Center CUNY, USA*
Paul Beame	*University of Washington, USA*
Michael Ben-Or	*The Hebrew University, Israel*
Andrei Bulatov	*Simon Fraser University, Canada*
Peter Bürgisser	*Paderborn, Germany*
Felipe Cucker	*City University of Hong Kong, China*
Evgeny Dantsin	*Roosevelt University, USA*
Volker Diekert	*Stuttgart, Germany*
Dima Grigoriev (Chair)	*CNRS, IRMAR, Rennes, France*
Yuri Gurevich	*Microsoft Research, USA*
Johann A. Makowsky	*Technion – Israel Institute of Technology, Israel*
Yuri Matiyasevich	*Steklov Institute/St.Petersburg, Russia*
Peter Bro Miltersen	*Aarhus, Denmark*
Grigori Mints	*Stanford, USA*
Pavel Pudlák	*Prague, Czech Republic*
Prabhakar Raghavan	*Verity Inc., USA*
Alexander Razborov	*IAS, USA and Steklov Institute/Moscow, Russia*
Michael E. Saks	*Rutgers, USA*
Alexander Shen	*LIF CNRS, France and Moscow, Russia*
Amin Shokrollahi	*EPF Lausanne, Switzerland*
Anatol Slissenko	*Paris-12, France*
Mikhail Volkov	*Ural State University, Russia*

Applications and Technology Track

Boris Babayan	*Intel, Russia*
Robert T. Bauer	*PSU, USA*
Matthias Blume	*Toyota Technological Institute, USA*
Walter Daelemans	*Antwerpen, Belgium*
Vassil Dimitrov	*Calgary, Canada*
Richard Fateman	*Berkeley, USA*
Dina Goldin	*University of Connecticut, USA*
John R. Harrison (Chair)	*Intel, USA*
John Mashey	*Techviser, USA*
Bertrand Meyer	*ETH Zurich, Switzerland and Eiffel Software, USA*
Fedor Novikov	*St. Petersburg State Polytechnical University, Russia*
Michael Parks	*Sun Microsystems, USA*

Andreas Reuter   European Media Laboratory, Germany
Mary Sheeran   Chalmers University of Technology, Sweden
Elena Troubitsyna   Åbo Akademi University, Finland
Miroslav Velev   Carnegie Mellon University, USA
Sergey Zhukov   Transas, Russia

Conference Chair

Edward A. Hirsch   Steklov Institute/St. Petersburg, Russia

Steering Committee for CSR Conferences

Anna Frid   Sobolev Institute/Novosibirsk, Russia
Edward A. Hirsch   Steklov Institute/St. Petersburg, Russia
Juhani Karhumäki   Turku, Finland
Mikhail Volkov   Ural State University, Russia

Referees

Emrah Acar   Junxiong Deng
Noga Alon   Amer Diwan
Jean-Paul Allouche   Niklas Eèn
Eugene Asarin   Robert Elsaesser
Maxim Babenko   Henning Fernau
Laurent Bartholdi   Eldar Fischer
Reuven Bar-Yehuda   Morten Fjeld
Danièle Beauquier   Lance Fortnow
Magnus Björk   Ian Foster
Johannes Borgstroem   Anna Frid
Ahmed Bouajjani   Joaquim Gabarro
Mihai Budiu   Healfdene Goguen
Eduardo Camponogara   Raphael Hauser
Olivier Carton   Rogardt Heldal
Hsun-Hsien Chang   Ulrich Hertrampf
Swarat Chaudhuri   John Hughes
Alessandra Cherubini   Piotr Indyk
David Chung   Alon Itai
Joëlle Cohen   David Janin
Richard Cole   Michael Kaminski
Stephen A. Cook   Graham Kemp
Matteo Corti   Barbara König
Maxime Crochemore   Teresa Krick
Stefan Dantchev   Manfred Kufleitner

Oliver Kullmann
Daniel Larsson
Dan Li
Yury Lifshits
Alexei Lisitsa
Anders Logg
Markus Lohrey
Martin Lotz
Sus Lundgren
Sergio Maffeis
David McAllester
Burkhard Monien
S. Muthu Muthukrishnan
Lorenz Minder
Anca Muscholl
Francesco Zappa Nardelli
Bengt Nordström
Dirk Nowotka
Matti Nykänen
Christos Papadimitriou
Holger Petersen
Rossella Petreschi
Franck Pommereau
Chung Keung Poon
Harald Raecke
Deva Ramanan

Dror Rawitz
Jan Reimann
Angelo Restificar
Søren Riis
Philip Ruemmer
Igor Rystsov
Viktor Sabelfeld
Ashish Sabharwal
Arseny Shur
Terence Sim
Jan Smith
Niklas Sörensson
Arne Storjohann
Karsten Tiemann
Michael Tiomkin
Siegbert Tiga
Paul Vitanyi
Nicolai Vorobjov
Vladimir Vovk
Vladimir Vyugin
Chao Wang
Zhenghong Wang
Alexander Wolpert
Hideki Yamasaki
Martin Ziegler
Silvano Dal Zilio Uri Zwick

Sponsors

The U.S. Civilian Research & Development Foundation.
Russian Foundation for Basic Research.

Table of Contents

Applications and Technology Track

Non-black-box Techniques in Cryptography

Boaz Barak

Princeton University

Abstract. In cryptography we typically prove the security of a scheme by reducing the task of breaking the scheme to some hard computational problem. This reduction usually done in a *black-box* fashion. By this we mean that there is an algorithm that can solve the hard problem given any black-box for breaking the scheme.

This lecture concerns exceptions to this rule: that is, schemes that are proven secure using a non-black-box reduction, that actually uses the code of a scheme-breaking attacker to construct a problem-solving algorithm. It turns out that such reductions can be used to obtain schemes with better properties that were known before. In fact, in some cases these non-black-box reductions can be obtain goals that were proven to be impossible to achieve when restricting to black-box reductions. In particular, we will present constructions of zero-knowledge protocols that are proven secure under various compositions [1, 2, 3].

We'll also discuss some of the limitations and open questions regarding non-black-box security proofs.

References

1. Barak, B.: How to go beyond the black-box simulation barrier. In: Proc. 42nd FOCS, IEEE (2001) 106–115
2. Pass, R.: Bounded-concurrent secure multi-party computation with a dishonest majority. In: Proc. 36th STOC, ACM (2004) 232–241
3. Barak, B., Sahai, A.: How to play almost any mental game over the net - concurrent composition using super-polynomial simulation. In: Proc. 46th FOCS, IEEE (2005)

D. Grigoriev, J. Harrison, and E.A. Hirsch (Eds.): CSR 2006, LNCS 3967, p. 1, 2006.
© Springer-Verlag Berlin Heidelberg 2006

Complexity of Polynomial Multiplication over Finite Fields

Michael Kaminski

Department of Computer Science,
Technion – Israel Institute of Technology,
Haifa 32000, Israel
kaminski@cs.technion.ac.il

Abstract. We prove the $\left(3 + \dfrac{(q-1)^2}{q^5 + (q-1)^3}\right) n - o(n)$ lower bound on the quadratic complexity of multiplication of two degree-n polynomials over a q-element field. The proof is based on a novel combination of two known techniques. One technique is the analysis of Hankel matrices representing bilinear forms defined by linear combinations of the coefficients of the polynomial product. The other technique is a counting argument from the coding theory.

D. Grigoriev, J. Harrison, and E.A. Hirsch (Eds.): CSR 2006, LNCS 3967, p. 2, 2006.

Synchronous Elastic Circuits

Mike Kishinevsky[1], Jordi Cortadella[2], Bill Grundmann[1],
Sava Krstić[1], and John O'Leary[1]

[1] Strategic CAD Labs, Intel Corporation, Hillsboro, Oregon, USA
[2] Universitat Politècnica de Catalunya, Barcelona, Spain

Synchronous elastic circuits (also known as latency-insensitive and latency-tolerant) behave independently of the latencies of computations and communication channels. For example, the three sequences

$$X = \langle 1, *, *, 2, *, 5, 3, \ldots \rangle \quad Y = \langle 2, *, 0, *, 1, *, 4, \ldots \rangle \quad Z = \langle *, 3, *, 2, *, *, 6, *, 7, \ldots \rangle$$

are an acceptable behavior of an elastic adder with input channels X, Y and output channel Z, where the absence of transfer on a particular channel at a given cycle is indicated by $*$. Indeed, the associated transfer subsequences (obtained by deleting the $*$'s) make up a behavior of an ordinary (non-elastic) adder:

$$X' = \langle 1, 2, 5, 3, \ldots \rangle \quad Y' = \langle 2, 0, 1, 4, \ldots \rangle \quad Z' = \langle 3, 2, 6, 7, \ldots \rangle$$

Current interest in elasticity is motivated by the difficulties with timing and communication in large synchronous designs in nanoscale technologies. The time discretization imposed by synchronicity forces to take early decisions that often complicate changes at the latest stages of the design or efficient design scaling. In modern technologies, calculating the number of cycles required to transmit data from a sender to a receiver is a problem that often cannot be solved until the final layout has been generated. Elastic circuits promise novel methods for microarchitectural design that can use variable latency components and tolerate static and dynamic changes in communication latencies, while still employing standard synchronous design tools and methods.

We will first present a simple elastic protocol, called SELF (Synchronous Elastic Flow) and describes methods for an efficient implementation of elastic systems and for the conversion of regular synchronous designs into an elastic form. Every elastic circuit $\mathcal{E}$ implements the behavior of an associated standard (non-elastic) circuit $\mathcal{C}$, as in the adder example above. For each wire X of $\mathcal{C}$, there are three in $\mathcal{E}$: the *data* wire D_X, and the single-bit control wires V_X and S_X (*valid* and *stop*). This triple of wires is a *channel* of $\mathcal{E}$. A transfer along the channel occurs when $V_X = 1$ and $S_X = 0$, thus requiring cooperation of the producer and the consumer. [CKG06] provides more details on the implementation of SELF.

We will next review theoretical foundations of SELF. Our main result states that (under favorable circumstances) "the network of elasticizations is an elasticization of the given network": if we have elastic circuits $\mathcal{E}_1, \ldots, \mathcal{E}_n$ implementing

D. Grigoriev, J. Harrison, and E.A. Hirsch (Eds.): CSR 2006, LNCS 3967, pp. 3–5, 2006.

standard circuits $C_1, \ldots, C_n$ and if C is a standard network obtained by connecting some wires of the circuits C_i, then connecting the corresponding channels (wire triples) of the elastic circuits $\mathcal{E}_i$ will produce a new elastic circuit which implements C. As a special case, we prove the characteristic property of elastic circuits: plugging an empty elastic buffer in a channel of an elastic network produces an equivalent elastic network. The details of the theory can be found in [KCKO06].

Related Work

Some researchers advocate for the modularity and efficiency of asynchronous circuits to devise a beter methodology for complex digital systems. However, asynchronous circuits require a significantly different design style and the CAD support for such circuits is still in its prehistory.

Our work addresses the following question: is there an *efficient* scheme that combines the modularity of asynchronous systems with the simplicity of synchronous implementations?

Other authors have been working towards this direction. Latency-insensitive (LI) schemes [CMSV01] were proposed to separate communication from computation and make the systems insensitive to the latencies of the computational units and channels. The implementation of LI systems is synchronous [CSV02, CN01] and uses *relay stations* at the interfaces between computational units.

In a different scenario, *synchronous interlocked pipelines* [JKB⁺02] were proposed to achieve fine-grained local handshaking at the level of stages. The implementation is conceptually similar to a discretized version of traditional asynchronous pipelines with request/acknowledge handshake signals.

A *de-synchronization* [HDGC04,BCK⁺04] approach automatically transforms synchronous specifications into asynchronous implementations by replacing the clock network with an asynchronous controller. The success of this paradigm will depend on the attitude of designers towards accepting asynchrony in their design flow.

References

[BCK⁺04] I. Blunno, J. Cortadella, A. Kondratyev, L. Lavagno, K. Lwin, and C. Sotiriou. Handshake protocols for de-synchronization. In *Proc. International Symposium on Advanced Research in Asynchronous Circuits and Systems*, pages 149–158. IEEE Computer Society Press, April 2004.

[CKG06]   J. Cortadella, M. Kishinevsky, and B. Grundmann. SELF: Specification and design of a synchronous elastic architecture for DSM systems. In *TAU'2006: Handouts of the International Workshop on Timing Issues in the Specification and Synthesis of Digital Systems*, February 2006. Available at www.lsi.upc.edu/~jordicf/gavina/BIB/reports/self_tr.pdf.

[CMSV01] L. Carloni, K.L. McMillan, and A.L. Sangiovanni-Vincentelli. Theory of latency-insensitive design. *IEEE Transactions on Computer-Aided Design*, 20(9):1059–1076, September 2001.

[CN01]   Tiberiu Chelcea and Steven M. Nowick. Robust interfaces for mixed-timing systems with application to latency-insensitive protocols. In *Proc. ACM/IEEE Design Automation Conference*, June 2001.

[CSV02]   L.P. Carloni and A.L. Sangiovanni-Vincentelli. Coping with latency in SoC design. *IEEE Micro, Special Issue on Systems on Chip*, 22(5):12, October 2002.

[HDGC04]   G.T. Hazari, M.P. Desai, A. Gupta, and S. Chakraborty. A novel technique towards eliminating the global clock in VLSI circuits. In *Int. Conf. on VLSI Design*, pages 565–570, January 2004.

[JKB⁺02]   Hans M. Jacobson, Prabhakar N. Kudva, Pradip Bose, Peter W. Cook, Stanley E. Schuster, Eric G. Mercer, and Chris J. Myers. Synchronous interlocked pipelines. In *Proc. International Symposium on Advanced Research in Asynchronous Circuits and Systems*, pages 3–12, April 2002.

[KCKO06]   S. Krstić, J. Cortadella, M. Kishinevsky, and J. O'Leary. Synchronous elastic networks. Technical Report. Available at `www.lsi.upc.edu/~jordicf/gavina/BIB/reports/elastic_nets.pdf`, 2006.

SZK Proofs for Black-Box Group Problems*

V. Arvind and Bireswar Das

Institute of Mathematical Sciences,
C.I.T Campus, Chennai 600 113, India
{arvind, bireswar}@imsc.res.in

Abstract. In this paper we classify several group-theoretic computational problems into the classes PZK and SZK (problems with perfect/statistical zero-knowledge proofs respectively). Prior to this, these problems were known to be in AM ∩ coAM. As PZK ⊆ SZK ⊆ AM ∩ coAM, we have a tighter upper bound for these problems.

1 Introduction

Motivated by cryptography, zero knowledge proof systems were introduced by Goldwasser et al [9]. These are a special kind of interactive proof systems in which the verifier gets no information other than the validity of the assertion claimed by the prover. The notion of zero knowledge is formalized by stipulating the existence of a randomized polynomial time *simulator* for a given protocol. For a given input, the simulator outputs strings following a probability distribution *indistinguishable* from the verifier's view of the interaction between prover and verifier for that input. Indistinguishability can be further qualified, leading to different notions of zero knowledge. The protocol is *perfect zero knowledge* if the simulator's distribution is identical to the verifier's view for all inputs. It is *statistical zero knowledge* if the two distributions have negligible statistical difference. The more liberal notion is *computational indistinguishability* where the two distributions cannot be distinguished by polynomial-size circuits.

Natural problems like Graph Isomorphism (GRAPH-ISO) and Quadratic Residuosity, their complements, a version of the discrete log problem are all known to have perfect zero-knowledge protocols. Some of these protocols have found cryptographic applications. For example, the Fiat-Shamir-Feige identification scheme is based on the ZK protocol for quadratic residuosity.

Our focus is complexity-theoretic in the present paper. As a complexity class SZK is intriguing. It is closed under complement and is contained in AM∩coAM. It is open if SZK is coincides with AM ∩ coAM. One approach to studying SZK is to explore for new natural problems that it contains. In [13,8], investigating SZK, it is shown that two natural promise problems, Statistical Difference (SD) and Entropy Difference (ED) are complete for SZK. We use this to exhibit several natural group-theoretic problems in SZK and PZK. These are well-studied problems and known to be in NP ∩ coAM or in AM ∩ coAM [2,5].

---

* Part of the work done was during visits to Berlin supported by a DST-DAAD project.

D. Grigoriev, J. Harrison, and E.A. Hirsch (Eds.): CSR 2006, LNCS 3967, pp. 6–17, 2006.

In this paper we put several group-theoretic problems for *permutation groups* in PZK, and for general *black-box groups* in SZK. We give a unified argument, showing that an appropriately defined group equivalence problem is reducible to Statistical Difference. One problem that requires a different technique is solvable permutation group isomorphism.

2 Preliminaries

Definition 1 (Statistical Difference). *Let X and Y be two distributions on a finite set S. The* statistical difference *between X and Y is* $\mathrm{SD}(X,Y) = \frac{1}{2}\sum_{s\in S}|Pr(X=s) - Pr(Y=s)|$.

A distribution X on S is ε-uniform if $\frac{1}{|S|}(1-\varepsilon) \le Pr[X=s] \le \frac{1}{|S|}(1+\varepsilon)$. If X is ε-uniform on S then $\mathrm{SD}(X, U_S) \le \varepsilon/2$, where U_S is the uniform distribution on S. We next define SZK.

Definition 2. *[6] An interactive proof system (P,V) for a language L is* statistical zero-knowledge *(i.e. L is in* SZK*) if for every randomized polynomial-time interactive machine V^*, there is a probabilistic polynomial-time algorithm M^* such that for $x \in L$ and all k, $M^*(x,1^k)$ outputs* fail *with probability at most $\frac{1}{2}$ and M^* has the following property: let $m^*(x,1^k)$ be the random variable for the distribution of $M^*(x,1^k)$ conditioned on $M^*(x,1^k) \neq$fail. Let $\langle P,V\rangle(x,1^k)$ be the message distribution between P and V. Then $\mathrm{SD}(m^*(x,1^k), \langle P,V\rangle(x,1^k)) \le o(\frac{1}{k^{O(1)}})$. Additionally, the protocol is* perfect zero-knowledge *($L \in$ PZK) if this statistical difference is 0 for all x and k.*

A boolean circuit $X : \{0,1\}^m \longrightarrow \{0,1\}^n$ induces a distribution on $\{0,1\}^n$ by the evaluation $X(x)$, where $x \in \{0,1\}^m$ is picked uniformly at random. We use X to denote this distribution encoded by circuit X. For $0 \le \alpha < \beta \le 1$, we define the *promise problem* $\mathrm{SD}^{\alpha,\beta} = (\mathrm{SD}_Y^{\alpha,\beta}, \mathrm{SD}_N^{\alpha,\beta})$: the input is two distributions X and Y given by circuits, and has "yes instances" $\mathrm{SD}_Y^{\alpha,\beta} = \{(X,Y) \mid \mathrm{SD}(X,Y) \le \alpha\}$ and "no instances" $\mathrm{SD}_N^{\alpha,\beta} = \{(X,Y) \mid \mathrm{SD}(X,Y) \ge \beta\}$. We recall some important results from [13, 14].

Theorem 1 (Sahai-Vadhan). *[13] The class* SZK *is closed under complement, and* $\mathrm{SD}^{1/3,2/3}$ *is complete for* SZK. *Furthermore,* $\mathrm{SD}^{0,1}$ *is in* PZK.

We recall some basic group theory. The action of a group G on a set X is defined by a map $\alpha : X \times G \longrightarrow X$ such that for all $x \in X$ (i) $\alpha(x, id) = x$, i.e., the identity $id \in G$ fixes each $x \in X$, and (ii) $\alpha(\alpha(x,g_1),g_2) = \alpha(x,g_1g_2)$ for $g_1,g_2 \in G,$. We write x^g instead of $\alpha(x,g)$ when the group action is clear from the context. The *orbit* of $x \in X$ under G action, denoted x^G, is the set, $\{y|y \in X, y = x^g \text{ for some } g \in G\}$. Notice X is *partitioned* into orbits.

Let G be a permutation group, i.e., $G \le S_n$. Each $\pi \in G$ maps $i \in [n]$ to i^π, which is the natural action of G on $[n]$. The subgroup $G^{(i)}$ of $G \le S_n$ that fixes each of $\{1,\dots,i\}$ is a *pointwise stabilizer* subgroup. Thus, we have a tower of subgroups $G = G^{(0)} \geq G^{(1)} \geq G^{(2)} \geq \cdots \geq G^{(n-1)} = \{id\}$. Notice that $[G^{(i-1)} : G^{(i)}] \leq n$. Let R_i be a set of complete and distinct coset representatives of $G^{(i)}$ in $G^{(i-1)}$ for each i. Then $\bigcup_{i=1}^{n-1} R_i$ generates G and is known as a *strong generating set* for G.

The subgroup generated by $\{xyx^{-1}y^{-1} \mid x, y \in G\}$ is the commutator subgroup G' of G. Recall that G' is the unique smallest normal subgroup of G such that G/G' is commutative. The derived series of G is $G \triangleright G' \triangleright G'' \triangleright \cdots$. We say G is *solvable* if this series terminates in $\{id\}$. There are polynomial-time algorithms to compute the derived series and to test solvability for permutation groups G given by generating sets (see e.g. [11]). A *composition series* of G is a tower of subgroups $\{id\} = G_1 \triangleleft G_2 \triangleleft \cdots \triangleleft G_m = G$ such that G_i/G_{i+1} is *simple* for each i. Recall that G is solvable iff G_i/G_{i+1} is cyclic of prime order for each i in any composition series for G.

3   Group Problems in PZK

We now show that various permutation group problems (not known to be in P) are in PZK. Examples are Coset Intersection, Double Coset Membership, Conjugate Subgroups etc. We define these problems below (see [11] for details). These are problems known to be harder than GRAPH-ISO. We show they are in PZK by a general result. We define a generic problem *Permutation Group Equivalence* PGE and show it is polynomial-time many-one reducible to $SD^{0,1}$. Since $SD^{0,1} \in$ PZK it follows that PGE $\in$ PZK. The problem PGE is generic in the sense that all considered permutation group problems (except group isomorphism) are polynomial-time many-one reducible to GE and hence are in PZK. Permutation group isomorphism requires a different approach. In fact, in this paper we show only for solvable groups that this problem is in PZK.

Definition 3. Permutation Group Equivalence PGE *has inputs of the form* (x, y, T, τ), *where* $T \subset S_n$ *and* $x, y \in \{0,1\}^m$, *for* $m = n^{O(1)}$. *Let* $G = \langle T \rangle$. *The map* $\tau : G \times S \longrightarrow S$ *is a polynomial-time computable group action of* G *on* S, *for some* $S \subseteq \{0,1\}^m$. *More precisely, given* $g \in G$ *and* $s \in S$, *the image* $s^g = \tau(g, s)$ *is polynomial-time computable. The* PGE *problem is the promise problem: given* (x, y, T, τ) *such that* $x, y \in \{0,1\}^m$ *with the promise that* τ *defines a group action of* $G = \langle T \rangle$ *on some* $S \subseteq \{0,1\}^m$ *with* $x, y \in S$, *the problem is to decide if* $\tau(g, x) = x^g = y$ *for some* $g \in G$.

Theorem 2. PGE *is polynomial-time many-one reducible to* $SD^{0,1}$.

Proof. Let (x, y, T, τ) be an input instance of PGE such that $x, y \in S$ and $S \subseteq \{0,1\}^m$. Define two circuits $X_{x,T}, X_{y,T} : \{0,1\}^k \longrightarrow \{0,1\}^m$, where k is polynomial in n to be fixed later. In the sequel we assume that it is possible to uniformly pick a random element from the set $[i]$ for each positive integer i given in unary. The circuit $X_{x,T}$ on input a random string $r \in \{0,1\}^k$ will use r to randomly sample an element from the group $G = \langle T \rangle$. This is a polynomial-time procedure based on the Schreier-Sims algorithm for computing a strong generating set $\bigcup_{i=1}^{n-1} R_i$ for G, where R_i is a complete set of distinct coset representatives of $G^{(i)}$ in $G^{(i-1)}$ for each i. Then we can sample from G uniformly at random by picking $x_i \in R_i$ uniformly at random and computing their product $g_r = x_1 x_2 \cdots x_{n-1}$. The circuit $X_{x,T}$ then outputs x^{g_r}. By construction, x^{g_r} is uniformly distributed in the G-orbit of x. Likewise the other circuit $X_{y,T}$ will output a uniformly distributed element of the G-orbit of y. Since G defines a group action on S, the two orbits are either disjoint or identical. In particular, the orbits are identical if and only if $x = y^g$ for some $g \in G$. Thus, the statistical difference between $X_{x,T}$ and $X_{y,T}$ is 0 or 1 depending on whether $x = y^g$ for some $g \in G$ or not. This proves the theorem.

We show that several permutation group problems are reducible to PGE. There is a table of reductions for permutation group problems in Luks' article [11]. It suffices to show that the following two "hardest" problems from that table are reducible to PGE (apart from permutation group isomorphism which we consider in the next section).

The *Subspace Transporter Problem* SUBSP-TRANS has input consisting of a subgroup G of S_n given by generating set T, a representation $\tau : G \longrightarrow GL(\mathbb{F}_q^m)$, and subspaces $W_1, W_2 \subseteq \mathbb{F}_q^m$ given by spanning sets. The question is whether $W_1^g = W_2$ for some $g \in G$. Here the size q of the finite field is a constant. Notice here that by W_1^g is meant the image of the subspace W_1 under the matrix $\tau(g)$.

The *Conjugacy of Groups Problem* CONJ-GROUP has inputs consisting of three permutation groups G, H_1, H_2 in S_n, given by generating sets. The question is whether there is a $g \in G$ such that $H_1^g = H_2$ (where $H_1^g = g^{-1} H_1 g$).

Lemma 1.

(a) *Let $\mathbb{F}_q$ be a fixed finite field. Given as input $X \subset \mathbb{F}_q^n$, there is a polynomial-time algorithm $\mathcal{A}$ that computes a canonical basis B of the subspace W spanned by X. The output is canonical in the following sense: if $\mathcal{A}$ is given as input any spanning set of W, the output of $\mathcal{A}$ will be B.*

(b) *Given as input $X \subset S_n$, there is a polynomial-time algorithm $\mathcal{A}$ that computes a canonical generating set B of the subgroup G generated by X. The output is canonical in the sense that $\mathcal{A}$ will output B, given any generating set X' of G as input.*

Proof. First we prove (a). Order the elements of $\mathbb{F}_q$ lexicographically. First, we search for the least i such that there is a vector $(v_1, \ldots, v_n) \in W$ with $v_i = 1$ (Notice that $v_1, \ldots v_{i-1}$ have to be zero for all elements of W). For this we can use a polynomial-time algorithm for testing feasibility of linear equations over $\mathbb{F}_q$. Having found i, we search for the least $v_{i+1} \in \mathbb{F}_q$ such that $v_i = 1$. Since q is a constant we can do this search with a constant number of similar feasibility tests. After finding the least v_{i+1} we fix it and search similarly for the least v_{i+2} and so on. Continuing thus, we can compute the lex least nonzero element $\mathbf{u_1}$ in W. Next, in order to find a basis we look for the least index $j > i$ such that there is a nonzero vector $(v_1, \ldots, v_n) \in W$ with $v_1 = v_2 = \ldots = v_{j-1} = 0$ and $v_j = 1$ again by $O(n)$ feasibility tests. After finding j, we can again pick the lex least nonzero vector $\mathbf{u_2}$ with the property that the jth index is least nonzero coordinate in $\mathbf{u_2}$. Continuing in this manner, we will clearly obtain a basis $\{\mathbf{u_1}, \mathbf{u_2}, \ldots, \mathbf{u_k}\}$ of W. By our construction this basis is canonical.

Now we prove (b). The algorithm $\mathcal{A}$ will compute a strong generating set from X for the group G using the Schreier-Sims algorithm [11]. Then using the fact that the lex least element of a coset xH (where x and H are from S_n) can be computed in polynomial time [1] we can replace each coset representative in the strong generating set by a lex least coset representative. This generating set is canonical by construction.

Theorem 3. *The problems* SUBSP-TRANS *and* CONJ-GROUP *are polynomial-time many-one reducible to* PGE.

Proof. We first consider SUBSP-TRANS. Let (T, S_1, S_2, π) be an input. Let $G = \langle T \rangle$ and $S_1, S_2 \subset \mathbb{F}_q^m$ be spanning sets of W_1 and W_2 respectively. The representation is given by $\pi : G \longrightarrow GL(\mathbb{F}_q^m)$. The reduction from SUBSP-TRANS to PGE maps (T, S_1, S_2, π) to (x, y, T, τ) where x and y are the canonical bases for W_1 and W_2 respectively, in the sense of lemma 1. The set S in Definition 3 corresponds to the set of canonical bases of all possible subspaces of $\mathbb{F}_q^m$. The group action τ is the algorithm that given $B \in S$ and $g \in G$, first computes the set of vectors $\pi(g)(B)$. Next, using the algorithm in Lemma 1, τ computes the canonical basis of subspace spanned by $\pi(g)(B)$.

The reduction is similar for CONJ-GROUP. Let (T, S_1, S_2) be an instance of CONJ-GROUP, where T, S_1 and S_2 generate G, H_1, and H_2 respectively. The reduction maps (T, S_1, S_2) to (x, y, T, τ) where x and y are the canonical strong generating sets for H_1 and H_2 respectively in the sense of Lemma 1. The set S in Definition 3 is the set of canonical strong generating sets for all subgroups of S_n. The group action τ is the algorithm that given $B \in S$ and $g \in G$, applies the algorithm $\mathcal{A}$ in lemma 1 to compute the canonical generating set for the subgroup generated by $\{g^{-1}xg \mid x \in B\}$.

Corollary 1. *The problems of Set Transporter, Coset Intersection, Double Coset Membership, Double Coset Equality, Conjugacy of Elements, Vector Transporter etc are all in* PZK *as they are polynomial time many-one reducible to* SUBSP-TRANS *or* CONJ-GROUP.

3.1   Group Nonequivalence and PZK in Liberal Sense

We now consider the complement problems. To the best of our knowledge, it is open if $\overline{\mathrm{SD}^{0,1}} \in$ PZK. However, for this part we need the following liberal definition of PZK [9], because only such PZK protocols are known for even problems like GRAPH-NONISO and Quadratic Nonresiduosity.

An interactive protocol (P, V) is *perfect zero knowledge in the liberal sense* if for every probabilistic polynomial time interactive machine V^* there exists an expected polynomial-time algorithm M^* such that for every $x \in L$ the random variable $\langle P, V^* \rangle(x)$ and $M^*(x)$ are identically distributed. Notice that in this definition [9] the simulator is required to be an *expected polynomial time* algorithm that always outputs some legal transcript. The definition we used in Section 3 is more stringent.

Similar to the proof that GRAPH-NONISO $\in$ PZK in liberal sense, we can show that Permutation Group Nonequivalence $\overline{\text{PGE}}$ is in PZK in the liberal sense. Combined with Theorem 3 we have the following.

Theorem 4. $\overline{\text{PGE}}$ *is in* PZK *in liberal sense. As a consequence, the complement of the following problems are all in* PZK *in liberal sense: Set Transporter, Coset Intersection, Double Coset Membership, Double Coset Equality, Conjugacy of Elements, Vector Transporter.*

4 Solvable Permutation Group Isomorphism is in PZK

In this section we consider *permutation group isomorphism* PERM-ISO: given two subgroups $\langle S \rangle, \langle T \rangle \leq S_n$ the problem is to test if $\langle S \rangle$ and $\langle T \rangle$ are isomorphic.

Remark. PERM-ISO is in NP $\cap$ coAM [11]. It is harder than GRAPH-ISO [11] and seems different in structure from GRAPH-ISO or PGE. Like PGE if we try to formulate PERM-ISO using group action we notice that isomorphisms between groups are not permutations on small domains (unlike PGE). Thus, we do not know how to prove certain complexity-theoretic statements for PERM-ISO that hold for GRAPH-ISO. E.g. we do not know if it is in SPP or even low for PP [10], although GRAPH-ISO is in SPP [1]. Indeed, we do not know if PERM-ISO is in SZK. However, in this section we show that PERM-ISO for solvable groups is reducible to $\text{SD}^{0,1}$ and is hence in PZK.

Definition 4. *Let X be a finite set of symbols and $FG(X)$ be the free group generated by X. A pair (X, R) is a presentation of a group G where X is a finite set of symbols and R is a set of words over $X \cup X^{-1}$ where each $w \in R$ defines the equation $w = 1$. The presentation (X, R) defines G in the sense that $G \cong FG(X)/N$, where N is the normal closure in $FG(X)$ of the subgroup generated by R. The size of (X, R) is $\|X\| + \sum_{w \in R} |w|$. Call (X, R) a short presentation of the finite group G if the size of (X, R) is $(\log |G|)^{O(1)}$.*

It is an important conjecture [4] that all finite groups have short presentations. It is known to be true for large classes of groups. In particular, it is easy to prove that solvable finite groups have short presentations.

Notice that two groups are isomorphic if and only if they have the same set of presentations. Our reduction of solvable permutation group isomorphism to $\text{SD}^{0,1}$ will use this fact. Specifically, to reduce solvable PERM-ISO to $\text{SD}^{0,1}$ we give a randomized algorithm $\mathcal{A}$ that takes as input the generating set of a solvable group $G \leq S_n$ and outputs a short presentation for G. We can consider $\mathcal{A}(G)$ as a circuit with random bits as input and a short presentation for G as output. Clearly, if $G \ncong H$ then the circuits $\mathcal{A}(G)$ and $\mathcal{A}(H)$ will output distributions with disjoint support. On the other hand, if $G \cong H$, the circuits $\mathcal{A}(G)$ and $\mathcal{A}(H)$ will compute identical probability distributions on the short presentations (for G and H).

We describe $\mathcal{A}$ in two phases. In the first phase $\mathcal{A}$ computes a random composition series for the input solvable group $G = \langle T \rangle$ following some distribution. In the second phase, $\mathcal{A}$ will deterministically compute a short presentation for G using this composition series. An ingredient for $\mathcal{A}$ is a polynomial-time sampling procedure from $L \setminus N$ where $L \leq S_n$ and $N \lhd L$ are subgroups given by generating sets. We describe this algorithm.

Lemma 2 (Sampling Lemma). *Let $L \leq S_n$ and $N \lhd L$, where both L and N are given by generating sets. There is a polynomial-time algorithm that samples from $L \setminus N$ uniformly at random (with no failure probability).*

Proof. Let $L = \langle S \rangle$ and $N = \langle T \rangle$. Recall that applying the Schreier-Sims algorithm we can compute a strong generating set for L in polynomial time. More precisely, we can compute distinct coset representatives R_i for $L^{(i)}$ in $L^{(i-1)}$ for $1 \leq i \leq n-1$, where $L^{(i)}$ is the subgroup of L that fixes each of $1, 2, \ldots, i$. Notice that $\|R_i\| \leq n$ for each i. Thus, we have the tower of subgroups $L = L^{(0)} \geq L^{(1)} \geq \ldots \geq L^{(n-1)} = 1$.

We can use the strong generating set $\bigcup R_i$ to sample uniformly at random from L as explained in proof of Theorem 2. This sampling procedure can be easily modified to sample uniformly from $L \setminus \{1\}$.

We will build on this idea, using some standard group-theoretic algorithms from [11] to sample uniformly from $L \setminus N$. Since $N \lhd L$ each set $NL^{(i)}$ is a subgroup of L. Furthermore, for each i

$$\frac{\|NL^{(i-1)}\|}{\|NL^{(i)}\|} \leq \frac{\|L^{(i-1)}\|}{\|L^{(i)}\|} \leq n - i + 1.$$

Thus, $L = NL^{(0)} \geq NL^{(1)} \geq \ldots \geq NL^{(n-1)} = N$ is also a subgroup tower with each adjacent pair of subgroups of small index. Furthermore, R_i also forms coset representatives for $NL^{(i)}$ in $NL^{(i-1)}$. However, R_i may not be all distinct coset representatives. Since we have the generating set for $NL^{(i)}$ (the union of T and the generating set for $L^{(i)}$) we can find the distinct coset representatives in polynomial time by using membership tests in $NL^{(i)}$, using the fact that $x, y \in R_i$ are not distinct coset representatives for $NL^{(i)}$ in $NL^{(i-1)}$ iff $xy^{-1} \in NL^{(i)}$. Let $S_i \subseteq R_i$ be the distinct coset representatives for each i. Let $\|S_i\| = m_i$ for each i. We can ignore the indices i for which S_i has only the identity element.

Now, each $gN \in L/N$ is uniquely expressible as $gN = (g_1 N) \cdots (g_{n-1} N) = g_1 \cdots g_{n-1} N$, $g_i \in S_i$.

Partition the nontrivial elements of L/N into sets $V_i = \{g_i \cdots g_{n-1} N \mid g_j \in S_j$ and $g_i \neq 1\}$. Clearly, $L/N \setminus \{1N\} = \biguplus_{i=1}^{n-1} V_i$. Furthermore, let $\|V_i\| = (m_i - 1) \prod_{j=i+i}^{n-1} m_j = N_i$ for each i. We can sample uniformly from V_i by uniformly picking $g_i \in S_i \setminus \{1\}$ and $g_j \in_R S_j$, $j = i + 1, \ldots, n-1$. Thus, we can sample uniformly from L/N by first picking i with probability $\frac{N_i}{\|L\|/\|N\|-1}$ and then sampling uniformly from V_i. Finally, to sample from $L \setminus N$, notice that after picking the tuple $(g_i, \ldots, g_{n-1})$ while sampling from V_i we can pick $x \in N$ (by first building a strong generating set for N). Clearly, $g = g_i \cdots g_{n-1} x$, is uniformly distributed in $L \setminus N$.

We now describe algorithm $\mathcal{A}$. Suppose S is the input to $\mathcal{A}$, where $G = \langle S \rangle$ is a solvable group. In Phase 1, $\mathcal{A}$ first computes the derived series of G (in deterministic polynomial time [11]).

Next, $\mathcal{A}(G)$ refines the derived series for G into a random composition series by inserting a chain of normal subgroups between consecutive groups of the series. It suffices to describe this refinement for $G' \lhd G$, where $G' = G_{m-1}$ and $G = G_m$. We can refine each $G_i \lhd G_{i+1}$ similarly.

Suppose $\|G/G'\| = p_1^{\alpha_1} p_2^{\alpha_2} \cdots p_l^{\alpha_l} = m$, $p_1 < p_2 < \cdots < p_l$. Using standard algorithms from [11] we can compute m in polynomial time. As m is smooth (all $p_i \leq n$) we can also factorize m in polynomial time to find the p_i. We will use the ordering of the p_i.

Let $G' = \langle T \rangle$. Since G/G' is abelian, the p_1-Sylow subgroup of G/G' is L/G' where L is generated by the union of T and $\{g^{m/p_1^{\alpha_1}} \mid g \in S\}$. Notice that $G' \lhd L \lhd G$. Applying Lemma 2, $\mathcal{A}$ can sample uniformly an $x \in L \setminus G'$. As $\|L/G'\| = p_1^{\alpha_1}$, the order of xG' is p_1^t for some $t \neq 0$. This t is easily computed by repeated powering. Clearly, $x^{p_1^{t-1}} G'$ is of order p_1. Let $x_1 = x^{p_1^{t-1}}$ and define $N_1 = \langle T \cup \{x_1\} \rangle$. Clearly, G' is normal in N_1 and $\|N_1/G'\| = p_1$. Since G/G' is abelian it follows that $G' \lhd N_1 \lhd L \lhd G$.

We now repeat the above process for the pair of groups N_1 and L. Using Lemma 2 we randomly pick $x \in L \setminus N_1$ find the order p_1^s of xN_1 in G/N_1 and set $x_2 = x^{p_1^{s-1}}$. This will give us the subgroup N_2 generated by N_1 and x_2. Thus, we get the refinement $G' \lhd N_1 \lhd N_2 \lhd L \lhd G$, where $\{N_2/N_1\} = p_1$. Continuing thus, in α_1 steps we obtain the refinement $G' \lhd N_1 \lhd N_2 \lhd \cdots \lhd N_{\alpha_1} = L \lhd G$.

Now, let M/G' be the p_2-Sylow subgroup of G/G'. We can find a generating set for M as before. Notice that $L \lhd ML \lhd G$. Thus, applying the above process we can randomly refine the series $L \lhd ML$ into a composition series where each adjacent pair of groups has index p_2. Continuing thus, $\mathcal{A}$ refines $G' \lhd G$ into a random composition series between G and G'. This process can be applied to each pair $G_i \lhd G_{i+1}$ in the derived series. To obtain a random composition series for G.

After phase 1, the computed composition series for G is described by a sequence $(x_1, x_2, \cdots, x_m)$ of elements from G, where the composition series is $id \lhd \langle x_1 \rangle \lhd \langle x_1, x_2 \rangle \lhd \cdots \lhd \langle x_1, x_2, \cdots, x_m \rangle = G$.

Observe that if $\phi : G \to H$ is an isomorphism and if $id = G_0 \lhd G_1 \lhd \cdots \lhd G_{m-1} \lhd G_m = G$ and $id = H_0 \lhd H_1 \lhd \cdots \lhd H_{m-1} \lhd H_m = H$ are the derived series of G and H respectively, then ϕ must isomorphically map G_i to H_i for each i. Furthermore, if $(x_1, x_2, \cdots, x_m)$ describes a composition series for G then $(\phi(x_1), \phi(x_2), \cdots, \phi(x_m))$ describes a composition series for H. Let X_i denote the random variable according to which x_i is picked in the above description for G. Similarly, let Y_i denote the random variable for the group H. It is easy to see that $Pr[X_1 = x_1] = Pr[Y_1 = \phi(x_1)]$. Now,

$$Pr[X_i = x_i \ 1 \leq i \leq m] = Pr[X_1 = x_1] \cdot \prod_{i=2}^{m} Pr[X_i = x_i | X_j = x_j, 1 \leq j < i].$$

Notice that to construct x_{i+1} the algorithm refines $\langle x_1, x_2, \cdots, x_i \rangle \lhd G_j$, where G_j is the appropriate group in the derived series. Now, if the algorithm finds $\phi(x_1), \phi(x_2), \cdots, \phi(x_i)$ as the first i components of the composition series for H, then the next element y_i is obtained by refining $\langle \phi(x_1), \phi(x_2), \cdots, \phi(x_i) \rangle \lhd H_j$, where $\phi : G_j \longrightarrow H_j$ is an isomorphism. Thus, it is easy to see that for $i \geq 2$ also we have

$$Pr[X_i = x_i \mid X_j = x_j \ 1 \leq j < i] = Pr[Y_i = \phi(x_i) \mid Y_j = \phi(x_j) \ 1 \leq j < i].$$

It follows that $Pr[X_i = x_i \ 1 \leq i \leq m] = Pr[Y_i = \phi(x_i) \ 1 \leq i \leq m]$.

In the second phase, the algorithm $\mathcal{A}$ computes a short presentation for G from its composition series given by $(x_1, x_2, \cdots, x_m)$. Let $p_1 = |\langle x_1 \rangle|$, $p_j = |\langle x_1, x_2, \cdots, x_j \rangle| / |\langle x_1, x_2, \cdots, x_{j-1} \rangle|$ for $j > 1$. Let the primes in this order be $p_1, p_2, \cdots, p_m$ (not necessarily distinct). Notice that each $g \in G$ can uniquely be expressed as $g = x_m^{l_m}, x_{m-1}^{l_{m-1}}, \cdots, x_1^{l_1}$, $0 \leq l_i \leq p_i - 1$.

$\mathcal{A}$ will compute the short presentation inductively. The cyclic subgroup $\langle x_1 \rangle$ has the representation (X_1, R_1) where $X_1 = \{\alpha_1\}$ and $R_1 = \{\alpha_1^{p_1}\}$. We assume inductively that $\langle x_1, x_2, \cdots, x_i \rangle$ has the presentation (X_i, R_i) where $X_i = \{\alpha_i, \alpha_2, \cdots, \alpha_i\}$. We let $X_{i+1} = X_i \cup \{\alpha_{i+1}\}$. In order to define R_{i+1} we notice that $x_{i+1} \langle x_1, \cdots, x_i \rangle = \langle x_1, \cdots, x_i \rangle x_{i+1}$ and $x_{i+1}^{p_{i+1}} \in \langle x_1, x_2, \cdots, x_i \rangle$. Thus, the new relations are: $x_{i+1}^{p_{i+1}} = u_{i+1}$, $u_{i+1} \in \langle x_1, x_2, \cdots, x_i \rangle$, and $\forall j, 1 \leq j \leq i$, $x_j x_{i+1} = x_{i+1} w_{i+1,j}$, where $w_{i+1,j} \in \langle x_1, x_2, \cdots, x_i \rangle$.

To find u_{i+1} notice that if $x \in \langle x_1, x_2, \cdots, x_i \rangle$ then x belongs to one of the cosets $x_i^j \langle x_1, x_2, \cdots, x_{i-1} \rangle$, $j = 0, \cdots, p_i - 1$. To find the exact coset $\mathcal{A}$ can do membership tests $x_i^{-j} x \in \langle x_1, x_2, \cdots, x_{i-1} \rangle$ for each j. As all the primes p_i are small, this is a polynomial-time step. By repeating the same for $\langle x_1, x_2, \cdots, x_{i-1} \rangle$, $\langle x_1, x_2, \cdots, x_{i-2} \rangle$, $\cdots$, $\langle x_1 \rangle$ the algorithm will be able to find $u_{i+1} = x_i^{l_i} \cdots x_1^{l_1}$. The corresponding relation will be $\alpha_{i+1}^{p_{i+1}} = \alpha_i^{l_i} \cdots \alpha_1^{l_1}$. The algorithm can compute $w_{i+1,j}$ and the corresponding relation similarly. Now, R_{i+1} is just R_i union the new relations. The number of relations $T(i)$ for $\langle x_1, x_2, \cdots, x_i \rangle$ follows the recurrence relation $T(i+1) = T(i) + i + 1$, $T(1) = 1$. So, the number of relation is $O(m^2)$. But $m = O(\log |G|)$. Hence the presentation is of polynomial length (more precisely it is $O(m^3)$). Suppose $\phi : G \longrightarrow H$ is an isomorphism and $\langle x_1, \cdots, x_m \rangle$ describes a composition series for G. Then $\langle \phi(x_1), \cdots, \phi(x_m) \rangle$ describes a composition series for H.

We notice that the composition series for G described by $(x_1, \cdots, x_m)$ for G and the composition series for H described by $(\phi(x_1), \cdots, \phi(x_m))$ yield the same presentation. This can be seen by observing that the process of obtaining u_{i+1} and $w_{i+1,j}$ is identical in both the cases. Thus, when $G \cong H$ it follows that the distributions produced by $\mathcal{A}(G)$ and $\mathcal{A}(H)$ will be identical. On the other hand, if $G \not\cong H$, $\mathcal{A}(G)$ and $\mathcal{A}(H)$ will have disjoint support. We have proved the following.

Theorem 5. *The problem of isomorphism testing of solvable permutation groups is polynomial time many-one reducible to* $\mathrm{SD}^{0,1}$ *and is hence in PZK.*

5   Black Box Group Problems

We next consider analogous problems over black-box groups [2, 5]. The black-box group model essentially abstracts away the internal structure of the group into a "black-box" oracle that does the group operations. In order to give uniform zero-knowledge protocols we generalize PGE to black-box groups: the *Group Equivalence Problem* GE. The key difference from the results of Section 3 is that while permutation groups can be uniformly sampled by a polynomial-time algorithm, there is no polynomial-time *uniform* sampling algorithm for black-box groups. However, the following seminal result of Babai for almost uniform sampling from black-box groups suffices to show that the considered black-box group problems are in SZK.

Theorem 6. [3] *There is a randomized algorithm that takes as input a generator set for a black-box group G and an $\epsilon > 0$ and in time polynomial in $\log |G|$ and $\log(1/\epsilon)$ it outputs a random element r of G such that for any $g \in G$, $(1 - \epsilon)/|G| \leq \text{Prob}[r = g] \leq (1 + \epsilon)/|G|$.*

As the distribution produced by the above algorithm is only ε-uniform, it turns out that we can only show that the black-box group problems are in SZK.

Theorem 7. GE *is reducible to* $\text{SD}^{\frac{1}{3},1}$ *(relative to the black box group oracle B).*

Proof. The proof is similar to Theorem 2. We reduce GE to $\text{SD}^{\epsilon_1,1}$ for some small ϵ_1. Let (q, x, y, T, τ) where elements of $\{0,1\}^q$ represents group elements, T is the set of generating elements of group G and τ is a polynomial time routine that computes the group action and has access to the group oracle B. The reduction maps (q, x, y, T, τ) to the pair of circuits (X_1, X_2), both having access to the black box group oracle B. The circuit X_1 samples $g \in G$ using Babai's algorithm. If the algorithm fails the circuit sets g to be any fixed element of G. Then it produces x^g. The circuit X_2 is similarly defined for y. As in Theorem 2, we can argue that if x and y are not in the same G-orbit the statistical difference between the two circuits will be 1. But if they are in the same orbit then we can verify that the statistical difference is less than a chosen small number ϵ_1. We can make ϵ_1 close to the ε specified by Theorem 6 by repeating Babai's algorithm and thus reducing the error introduced due to failure. As ε is inverse exponential, we can make ϵ_1 less than $\frac{1}{3}$.

Theorem 8. GE *is in* SZK^B *(where* SZK^B *stands for SZK in which both prover and verifier have access to the group oracle B).*

Proof. It suffice to observe that the proof [13] that $\text{SD}^{1/3,2/3} \in \text{SZK}$ relativizes and that $\text{SD}^{1/3,1}$ is trivially reducible to $\text{SD}^{1/3,2/3}$.

As a corollary we also get that several problems considered in [2] and some generalization of permutation group problems are in SZK^B. This partially answers an open question posed in [2] whether the considered problems are in SZK. However, we do not know if the order verification problem and group isomorphism for black-box groups are in SZK, although they are in $\text{AM} \cap \text{coAM}$.

Corollary 2. *Black box group membership testing, Disjointness of double cosets, Disjointness of subcosets, Group factorization etc are in* SZK^B.

Proof. Let (q, x, T) be an instance of black box group membership testing problem, where q is the length of the strings encoding group elements, T generates the group G. To reduce it to GE we notice that $x \in G$ if and only if some element $t \in T$ and x are in the same G-orbit where the G action is just right multiplication, i.e., $z^g = gz$.

Let (q, s, t, A, B) be an instance of double coset disjointness, where $H = \langle A \rangle$, $K = \langle B \rangle$ and the problem is to decide if HsK and HtK are disjoint. Here we notice that $HsK \cap HtK \neq \phi$ *iff* s and t are in the same $H \times K$-orbit where the action is defined by $z^{(h,k)} = h^{-1}zk$.

Disjointness of double coset and group factorization are equivalent because $Hs \cap Kt \neq \phi$ *iff* $H \cap Kts^{-1} \neq \phi$ *iff* $ts^{-1} \in KH$.

Let (q, x, A, B) be an instance of Group factorization, where $G = \langle A \rangle$, $H = \langle B \rangle$. The problem is to decide if $x \in GH$. We notice that $x \in GH$ *iff* x and the identity element e are in the same $G \times H$-orbit. The group action is defined as $z^{(g,h)} = g^{-1}zh$.

6   SZK Proof with Efficient Provers

An important question is whether we can design SZK protocols with *efficient* provers for all problems in SZK. A notion of efficient provers, considered useful for problems in SZK∩NP, is where the prover has to be a randomized algorithm that has access to an NP witness for an instance x of a language in SZK ∩ NP. This question is studied in [12] where it is shown that $\mathrm{SD}^{1/2,1}$ has such an SZK protocol. Consequently, any problem polynomial-time many-one reducible to $\mathrm{SD}^{1/2,1}$ also has such efficient provers.

As a consequence of Theorem 7 where we show that Group Equivalence for black-box groups is reducible to $\mathrm{SD}^{1/3,1}$ it follows from Corollary 2 and the above-mentioned result of [12] that all NP problems considered in Section 5 have SZK protocols with efficient provers.

Theorem 9. *Black box group membership testing, Double coset membership, Subcoset intersection, Group factorization etc are in* NP ∩ SZK^B *and have SZK protocols with efficient provers.*

7   Concluding Remarks

In this paper we show that SZK (and PZK) contains a host of natural computational black-box problems (respectively permutation group problems). As complexity classes SZK and PZK are quite intriguing. We do not known anything beyond the containment PZK ⊆ SZK ⊆ AM ∩ coAM and the closure of SZK under complement. In this context it is interesting to note that all considered permutation group problems (except solvable group isomorphism) are known to be low for PP: we can put PGE in SPP using the methods of [1, 10]. Could it be that the class PZK (or even SZK) is low for PP? We make a final remark in this context. The SZK-complete problem Entropy Difference (ED) is complete even for "nearly flat" distributions, where "flatness" is a technical measure of closeness to the uniform distribution [14]. If we consider ED with the stronger promise that the two input distributions are *uniform on their support* then we can prove that the problem is low for PP.

Acknowledgment. For comments and discussions during visits supported by a DST-DAAD project we thank Johannes Köbler.

References

1. V. ARVIND, P. P. KURUR, Graph Isomorphism is in SPP, *IEEE Foundations of Computer Science,* 743-750, 2002.
2. L. BABAI, Bounded Round Interactive Proofs in Finite Groups, *SIAM J. Discrete Math.,* 5(1): 88-111 1992.
3. L. Babai: Local Expansion of Vertex-Transitive Graphs and Random Generation in Finite Groups STOC 1991: 164-174.
4. L. BABAI, A. J. GOODMAN, W. M. KANTOR, E. M. LUKS, P. P. PLFY, Short presentations for finite groups, *Journal of Algebra,* 194 (1997), 79-112.
5. L. BABAI, E. SZEMERÉDI, On the Complexity of Matrix Group Problems I, *IEEE Foundations of Computer Science,* 229-240, 1984.
6. O. GOLDREICH, Foundations of Cryptography, Volume I, Basic Tools, *Cambridge University Press,* 2001.
7. O. GOLDREICH, S. MICALI, A. WIGDERSON, Proofs that Yield Nothing But Their Validity or All Languages in NP Have Zero-Knowledge Proof Systems, *Journal of the ACM,* 38(3): 691-729 (1991).
8. O. GOLDREICH, S. VADHAN, Comparing Entropies in Statistical Zero Knowledge with Applications to the Structure of SZK. *IEEE Conference on Computational Complexity,* 1999.
9. S. GOLDWASSER, S. MICALI, C. RACKOFF, The Knowledge Complexity of Interactive Proof Systems, *SIAM Journal of Computing,* 18(1): 186-208 (1989).
10. J. KÖBLER, U. SCHÖNING, J. TORÁN, Graph Isomorphism is Low for PP. *Computational Complexity,* 2: 301-330 (1992).
11. E. M. LUKS, Permutation groups and polynomial-time computation,in Groups and Computation, *DIMACS series in Discrete Mathematics and Theoretical Computer Science,* 139-175, 11 (1993).
12. D. MICCIANCIO, S. VADHAN, Statistical Zero-Knowledge Proofs with Efficient Provers: Lattice Problems and More, *Proceedings of the 23rd CRYPTO conference,* LNCS 2729, 282-298, 2003.
13. A. SAHAI, S. VADHAN, A Complete Promise Problem for Statistical Zero-Knowledge, *Foundations of Computer Scienceq,* 448-457, 1997.
14. S. VADHAN, *A Study of Statistical Zero-Knowledge Proofs,* Ph.D Thesis, MIT, 1999, Revised 8/00, http://www.eecs.harvard.edu/ salil/papers/phdthesis.ps.

Canonical Decomposition of a Regular Factorial Language*

S.V. Avgustinovich and A.E. Frid

Sobolev Institute of Mathematics SB RAS,
pr. Koptyuga, 4, 630090, Novosibirsk, Russia
{avgust, frid}@math.nsc.ru

Abstract. We consider decompositions of factorial languages to concatenations of factorial languages and prove that if the factorial language is regular, then so are the factors of its *canonical* decomposition.

1   Introduction and the Main Statement

In this paper we consider concatenation of languages, that is, equalities of the form $L = XY = \{xy|x \in X, y \in Y\}$, where $L, X, Y \in \Sigma^*$ for some finite alphabet Σ. In general, a language $L \in \Sigma^*$ can be decomposed to a concatenation of other languages in many ways. Even a finite language on the unary alphabet can admit several decompositions: for example,

$$(\lambda + a^2 + a^3 + a^4)(\lambda + a^2) = (\lambda + a^2 + a^3)^2.$$

As it is shown by Salomaa and Yu [5], the situation may be even more sophisticated than in this easy example. Another result demonstrating non-trivial properties of concatenation of languages has been obtained by Kunc [3] who has disproved a long-standing conjecture by Conway [2]. Conversely to the intuition, if Y is the maximal language such that $XY = YX$ for a given X, then Y can be not recursively enumerable even if X is finite.

In order to find a situation where the properties of the concatenation would be more predictable, the authors restricted themselves to considering *factorial* languages, where the word "factorial" means that the language is closed under taking a factor of an element, that is, that for all $v \in L$ the equality $v = sut$ implies $u \in L$; here $s, u, t \in \Sigma^*$ are arbitrary (possibly empty) words called a *prefix*, a *factor* and a *suffix* of v, respectively.

Note that a factorial language can also have several essentially different decompositions to factorial languages: e. g., $0^*1^* = 0^*(1^* + 0^*) = (0^* + 1^*)1^*$. However, as the authors show in [1], we can always choose a *canonical* decomposition of a factorial language, which is unique. More precisely, a decomposition $L = X_1 \cdots X_n$, where L and all X_i are factorial languages, is called *canonical* if

- Each of X_i is *indecomposable*, which means that $X_i = YZ$ implies $X_i = Y$ or $X_i = Z$;

---

* The work is supported by RFBR grants 03-01-00796 and 05-01-00364.

D. Grigoriev, J. Harrison, and E.A. Hirsch (Eds.): CSR 2006, LNCS 3967, pp. 18–22, 2006.
© Springer-Verlag Berlin Heidelberg 2006

– The decomposition is *minimal*, that is, for each i and for all $X_i' \subset X_i$ we have $L \neq X_1 \cdots X_{i-1} X_i' X_{i+1} \cdots X_n$.

Theorem 1 ([1]). *A canonical decomposition of a factorial language exists and is unique.*

The theorem is proved in a non-constructive way: in particular, we just assume that we can check if a language is decomposable. So, the methods of the proof could not be used for solving the following problem, first stated by Yu. L. Ershov:

Suppose that L is regular. Are all factors of its canonical decomposition also regular?

Regular factorial languages have been studied, e. g., by Shur [6]. Note that in general, some factors of a decomposition of a factorial language can be not regular.

Example 1. The regular language $0^*1^*2^*$ admits a decomposition

$$0^*1^*2^* = F(\{0^n 1^n 2^n\}) \cdot 2^*.$$

Here

$$F(\{0^n 1^n 2^n\}) = 0^*1^* + 1^*2^* + \{0^k 1^n 2^m | n, k, m \in \mathbb{N}, k, m \leq n\}$$

is the factorial closure of the language $\{0^n 1^n 2^n | n \in \mathbb{N}\}$; clearly, it is not regular.

After Conway's conjecture had been disproved, we could not bring ourselves to forecast the answer to Ershov's question. But fortunately, the answer turns out to be positive and not difficult to prove. So, in this note we prove the following

Theorem 2. *All factors of the canonical decomposition of a regular factorial language are regular.*

2 Proof

The main part of the proof is contained in the following

Lemma 1. *Let $L, X_1, \ldots, X_n$ be factorial languages, where L is regular and $L = X_1 \cdots X_n$. Then there exist regular factorial languages $Y_1, \ldots, Y_n$ such that $Y_i \subseteq X_i$ for $i = 1, \ldots, n$ and $L = Y_1 \cdots Y_n$.*

PROOF. Let $A = A(L)$ be an automaton recognizing L. Without loss of generality and following the notation of e. g. [4], we assume that all transitions of $A(L)$ are labelled with symbols of the alphabet Σ of the language L. Also, since L is factorial, we may assume that all the states of $A(L)$ are initial and terminal, so that $A(L) = < Q, E, \Sigma, Q, Q >$, where Q is a finite set of states and $E \subseteq Q \times Q \times \Sigma$; here $e = (p, r, a) \in E$ is a transition with the source p and the destination r, labelled with a. A *computation* in A is a sequence of transitions

$e_1, \ldots, e_k$ such that the sourse of e_i is the destination of e_{i-1} for all $i > 1$. Since all states in A are initial and terminal, all computations in it are successful, which means that their labels are always words of L. In its turn, L is the set of labels of all (successful) computations in A.

For each $i = 1, \ldots, n$ let us define the subset $E_i \subseteq E$ as follows: $e \in E_i$ if and only if all computations in $A(L)$ whose last transition is e are labelled with words from $X_1 \cdots X_i$ but among them there is a transition labelled with a word not belonging to $X_1 \cdots X_{i-1}$. Note that each transition of E belongs to exactly one of the sets E_i, so $E = E_1 \cup E_2 \cup \ldots \cup E_n$ is a partition of E.

Let Y_i be the language recognized by the automaton $A_i = < Q, E_i, \Sigma, Q, Q >$. By the construction, each of the languages Y_i is regular and factorial. We should only show that $L = Y_1 \cdots Y_n$ and $Y_i \subseteq X_i$ for all i.

Let us consider a computation $e_1, \ldots, e_m$ in A labelled with a word $a_1 \cdots a_m \in L$ and prove that if $e_i \in E_k$ and $e_j \in E_l$ for $i < j$, then $k \leq l$. Indeed, $e_i \in E_k$ means in particular that there exists a computation $f_1, \ldots, f_t, e_i$ in A labelled with a word $b_1 \cdots b_t a_i$ which does not belong to $X_1 \cdots X_{k-1}$. But since the language $X_1 \cdots X_{k-1}$ is factorial, the label $b_1 \cdots b_t a_i \cdots a_j$ of the computation $f_1, \ldots, f_t, e_i, \ldots, e_j$, which is clearly a computation in A, also does not belong to $X_1 \cdots X_{k-1}$. We see that $l \geq k$, which was to be proved.

So, in the computation $e_1, \ldots, e_m$ we observe a (possibly empty) group of transitions from E_1 labelled with a word from Y_1, followed by a (possibly empty) group of transitions from E_2 labelled with a word from Y_2, etc., so $a_1 \cdots a_m \in Y_1 \cdots Y_n$. Since the word $a_1 \cdots a_m \in L$ was chosen arbitrarily, we have $L \subseteq Y_1 \cdots Y_n$.

Now let us consider an arbitrary computation $g_1, \ldots, g_k$ in A_i labelled with a word $c_1 \cdots c_k \in Y_i$ and prove that $c_1 \cdots c_k \in X_i$. This will mean that $Y_i \subseteq X_i$ for all i. Indeed, $g_1 \in E_i$ implies that some computation $h_1, \ldots, h_l, g_1$ in A is labelled with a word $d_1 \cdots d_l c_1 \in (X_1 \cdots X_i) \backslash (X_1 \cdots X_{i-1})$. Let $d_1 \cdots d_j$ be its longest prefix from $X_1 \cdots X_{i-1}$; here $0 \leq j \leq l$. Now let us consider the computation $h_1, \ldots, h_l, g_1, \ldots g_k$, which is also a computation in A, and its label $d_1 \cdots d_l c_1 \cdots c_k \in X_1 \cdots X_i$ since $g_k \in E_i$. The longest prefix of $d_1 \cdots d_l c_1 \cdots c_k$ which belongs to $X_1 \cdots X_{i-1}$ is still $d_1 \cdots d_j$ since $X_1 \cdots X_{i-1}$ is a factorial language. Hence $d_{j+1} \cdots d_m c_1 \cdots c_k \in X_i$; since X_i is factorial, the suffix $c_1 \cdots c_k$ also lies in it. We have proved that $Y_i \subseteq X_i$.

Thus, $Y_1 \cdots Y_n \subseteq X_1 \cdots X_n = L$. Together with the inclusion $L \subseteq Y_1 \cdots Y_m$ proved above, this gives $L = Y_1 \cdots Y_n$ which proves the lemma. □

PROOF OF THEOREM 2. Let us apply Lemma 1 to the canonical decomposition $L = X_1 \cdots X_n$ of a regular language L. If we had $Y_i \subset X_i$ for some i, this would contradict to the minimality of the decomposition $L = X_1 \cdots X_n$. So, $Y_i = X_i$ for all i, and thus all languages X_i are regular, which was to be proved. □

3   Discussion and Examples

Note that Lemma 1 itself does not necessarily give a minimal (and all the more canonical) decomposition.

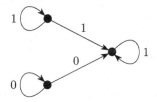

Fig. 1. A (non-minimal) automaton A recognizing the language 0^*1^*

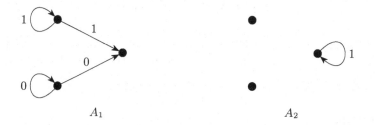

A_1 A_2

Fig. 2. Automata A_1 and A_2

Example 2. The automaton A from Fig. 1 recognizes the language $L = 0^*1^*$. However, if we decompose it starting from the decomposition $L = (0^* + 1^*)1^*$, this decomposition will not be reduced, and we will have $Y_1 = 0^* + 1^*$, $Y_2 = 1^*$ (see Fig. 2).

Here and below, we do not mark initial and terminal states of automata since we presume that *all* states are initial and terminal.

In the above proof, we walked the transitions from left to right. Symmetrically, we could walk from right to left and define the sets $E_i' \subseteq E$ as the set of all transitions e such that all computations in $A(L)$ whose first transition is e are labelled with words from $X_i \cdots X_n$, but among them there is a transition labelled with a word not belonging to $X_{i+1} \cdots X_n$. These two proofs are equivalent, but they may lead to different automata $A_1, \dots, A_n$ and even to different decompositions.

Example 3. If we consider the language L from the previous example and start with the automaton from Fig. 1 and the decomposition $L = (0^* + 1^*)1^*$, but follow the "right to left" proof, then we get the automata from Fig. 3. They give $Y_1' = 0^*$ and $Y_2 = 1^*$, which correspond to the canonical decomposition of L.

It also follows from the proof of the theorem that if a regular factorial language L is decomposable, then the factors of its canonical decompositions can be recognized by sub-automata of any automaton A recognizing L. The study of all possible partitions of A to transition-disjoint sub-automata will necessarily lead to the canonical decomposition of L, so finding a canonical decomposition of a regular factorial language is decidable.

Moreover, the non-deterministic state complexity of each of the factors of the canonical decomposition thus clearly cannot exceed the non-deterministic state complexity of the initial language. We can also note that for factorial languages,

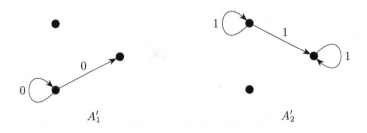

Fig. 3. Automata A_1' and A_2'

deterministic automata do not seem to be the most convenient tool for a study. It is more natural to consider automata with all the states being initial and terminal, as we did here.

References

1. S. V. Avgustinovich, A. E. Frid, *A unique decomposition theorem for factorial languages.* Internat. J. Algebra Comput. **15** (2005), 149–160.
2. J. H. Conway, *Regular Algebra and Finite Machines*, Chapman and Hall, 1971.
3. M. Kunc, *The power of commuting with finite sets of words.* STACS'05, Lecture Notes in Comput. Sci. 3404, 569–580.
4. J. Sakarovitch, *Elements de theorie des automates*, Vuibert, 2003.
5. A. Salomaa, S. Yu, *On the decomposition of finite languages*, in: Developments in Language Theory. Foundations, Applications, Perspectives, World Scientific, 2000, 22–31.
6. A. M. Shur, Subword complexity of rational languages, Discrete Analisys and Operation Research 12, no. 2 (2005), 7899 (in Russian).

Acyclic Bidirected and Skew-Symmetric Graphs: Algorithms and Structure

Maxim A. Babenko[*]

Dept. of Mechanics and Mathematics, Moscow State University,
Vorob'yovy Gory, 119899 Moscow, Russia
mab@shade.msu.ru

Abstract. *Bidirected graphs* (a sort of nonstandard graphs introduced by Edmonds and Johnson) provide a natural generalization to the notions of directed and undirected graphs. By a *weakly acyclic* bidirected graph we mean such a graph having no simple cycles. We call a bidirected graph *strongly acyclic* if it has no cycles (even non-simple). We present (generalizing results of Gabow, Kaplan, and Tarjan) a modification of the depth-first search algorithm that checks (in linear time) if a given bidirected graph is weakly acyclic (in case of negative answer a simple cycle is constructed). We use the notion of *skew-symmetric graphs* (the latter give another, somewhat more convenient graph language which is essentially equivalent to the language of bidirected graphs). We also give structural results for the class of weakly acyclic bidirected and skew-symmetric graphs explaining how one can construct any such graph starting from strongly acyclic instances and, vice versa, how one can decompose a weakly acyclic graph into strongly acyclic "parts". Finally, we extend acyclicity test to build (in linear time) such a decomposition.

1 Introduction

The notion of *bidirected graphs* was introduced by Edmonds and Johnson [3] in connection with one important class of integer linear programs generalizing problems on flows and matchings; for a survey, see also [9].

Recall that in a *bidirected* graph G three types of edges are allowed: (i) a usual directed edge, or an *arc*, that leaves one node and enters another one; (ii) an edge *from both* of its ends; or (iii) an edge *to both* of its ends. When both ends of edge coincide, the edge becomes a loop.

In what follows we use notation V_G (resp. E_G) to denote the set of nodes (resp. edges) of an undirected or bidirected graph G. When G is directed we speak of arcs rather than edges and write A_G in place of E_G.

A *walk* in a bidirected graph G is an alternating sequence $P = (s = v_0, e_1, v_1, \ldots, e_k, v_k = t)$ of nodes and edges such that each edge e_i connects nodes v_{i-1} and v_i, and for $i = 1, \ldots, k - 1$, the edges e_i, e_{i+1} form a *transit pair* at

---

[*] Supported by RFBR grants 03-01-00475 and NSh 358.2003.1.

D. Grigoriev, J. Harrison, and E.A. Hirsch (Eds.): CSR 2006, LNCS 3967, pp. 23–34, 2006.

v_i, which means that one of e_i, e_{i+1} enters and the other leaves v_i. Note that e_1 may enter s and e_k may leave t; nevertheless, we refer to P as a walk from s to t, or an s–t *walk*. P is a *cycle* if $v_0 = v_k$ and the pair e_1, e_k is transit at v_0; a cycle is usually considered up to cyclic shifts. Observe that an s–s walk is not necessarily a cycle.

If $v_i \neq v_j$ for all $1 \leq i < j < k$ and $1 < i < j \leq k$, then walk P is called *node-simple* (note that the endpoints of a node-simple walk need not be distinct). A walk is called *edge-simple* if all its edges are different.

Let X be an arbitrary subset of nodes of G. One can modify G as follows: for each node $v \in X$ and each edge e incident with v, reverse the direction of e at v. This transformation preserves the set of walks in G and thus does not change the graph in essence. We call two bidirected graphs G_1, G_2 *equivalent* if one can obtain G_2 from G_1 by applying a number of described transformations.

A bidirected graph is called *weakly (node- or edge-) acyclic* if it has no (node- or edge-) simple cycles. These two notions of acyclicity are closely connected. Given a bidirected graph G one can do the following: (i) replace each node $v \in V_G$ by a pair of nodes v_1, v_2; (ii) for each node $v \in V_G$ add an edge leaving v_1 and entering v_2; (iii) for each edge $e \in E_G$ connecting nodes $u, v \in V_G$ add an edge connecting u_i and v_j, where $i = 1$ if e enters u; $i = 2$ otherwise; similarly for j and v. This procedure yields a weakly edge-acyclic graph iff the original graph is weakly node-acyclic. The converse reduction from edge-acyclicity to node-acyclicity is also possible: (i) replace each node $v \in V_G$ by a pair of nodes v_1, v_2; (ii) for each edge $e \in E_G$ connecting nodes $u, v \in V_G$ add a node w_e and four edges connecting u_i, v_i with w_e ($i = 1, 2$); edges $u_i w_e$ should enter w_e; edges $w_e v_i$ should leave w_e; the direction of these edges at u_i (resp. v_i) should coincide with the direction of e at u (resp. v).

In what follows we shall only study the notion of weak edge-acyclicity. Hence, we drop the prefix "edge" for brevity when speaking of weakly acyclic graphs. If a bidirected graph has no (even non-simple) cycles we call it *strongly acyclic*.

One possible application of the weak acyclicity test is described in [4]. Let G be an undirected graph and M be a *perfect matching* in G (that is, a set of edges such that: (i) no two edges in M share a common node; (ii) for each node v there is a matching edge incident with v). The problem is to check if M is the unique perfect matching in G. To this aim we transform G into the bidirected graph $\overline{G}$ by assigning directions to edges as follows: every edge $e \in M$ leaves both its endpoints, every edge $e \in E_G \setminus M$ enters both its endpoints. One easily checks that the definition of matching implies that every edge-simple cycle in $\overline{G}$ is also node-simple. Moreover, each such simple cycle in $\overline{G}$ gives rise to an *alternating circuit* in G with respect to M (a circuit of even length consisting of an alternating sequence of edges belonging to M and $E_G \setminus M$). And conversely, every alternating circuit in G with respect to M generates a node-simple cycle in $\overline{G}$. It is well known (see [8]) that M is unique iff there is no alternating circuit with respect to it. Hence, the required reduction follows.

2   Skew-Symmetric Graphs

This section contains terminology and some basic facts concerning skew-symmetric graphs and explains the correspondence between these and bidirected graphs. For a more detailed survey on skew-symmetric graphs, see, e.g., [10, 6, 7, 1].

A *skew-symmetric graph* is a digraph G endowed with two bijections σ_V, σ_E such that: σ_V is an involution on the nodes (i.e., $\sigma_V(v) \neq v$ and $\sigma_V(\sigma_V(v)) = v$ for each node v), σ_A is an involution on the arcs, and for each arc e from u to v, $\sigma_E(e)$ is an arc from $\sigma_V(v)$ to $\sigma_V(u)$. For brevity, we combine the mappings σ_V, σ_A into one mapping σ on $V_G \cup A_G$ and call σ the *symmetry* (rather than skew-symmetry) of G. For a node (arc) x, its symmetric node (arc) $\sigma(x)$ is also called the *mate* of x, and we will often use notation with primes for mates, denoting $\sigma(x)$ by x'.

Observe that if G contains an arc e from a node v to its mate v', then e' is also an arc from v to v' (so the number of arcs of G from v to v' is even and these parallel arcs are partitioned into pairs of mates).

By a path (circuit) in G we mean a node-simple directed walk (cycle), unless explicitly stated otherwise. The symmetry σ is extended in a natural way to walks, cycles, paths, circuits, and other objects in G. In particular, two walks or cycles are symmetric to each other if the elements of one of them are symmetric to those of the other and go in the reverse order: for a walk $P = (v_0, a_1, v_1, \ldots, a_k, v_k)$, the symmetric walk $\sigma(P)$ is $(v'_k, a'_k, v'_{k-1}, \ldots, a'_1, v'_0)$. One easily shows that G cannot contain self-symmetric circuits (cf. [7]). We call a set of nodes X *self-symmetric* if $X' = X$.

Following terminology in [6], an arc-simple walk in G is called *regular* if it contains no pair of symmetric arcs (while symmetric nodes in it are allowed). Hence, we may speak of regular paths and regular circuits.

Next we explain the correspondence between skew-symmetric and bidirected graphs (cf. [7, Sec. 2], [1]). For sets X, A, B, we use notation $X = A \sqcup B$ when $X = A \cup B$ and $A \cap B = \emptyset$. Given a skew-symmetric graph G, choose an arbitrary partition $\pi = \{V_1, V_2\}$ of V_G such that V_2 is symmetric to V_1. Then G and π determine the bidirected graph $\overline{G}$ with node set V_1 whose edges correspond to the pairs of symmetric arcs in G. More precisely, arc mates a, a' of G generate one edge e of $\overline{G}$ connecting nodes $u, v \in V_1$ such that: (i) e goes from u to v if one of a, a' goes from u to v (and the other goes from v' to u' in V_2); (ii) e leaves both u, v if one of a, a' goes from u to v' (and the other from v to u'); (iii) e enters both u, v if one of a, a' goes from u' to v (and the other from v' to u). In particular, e is a loop if a, a' connect a pair of symmetric nodes.

Conversely, a bidirected graph $\overline{G}$ with node set $\overline{V}$ determines a skew-symmetric graph G with symmetry σ as follows. Take a copy $\sigma(v)$ of each element v of $\overline{V}$, forming the set $\overline{V}' := \{\sigma(v) \mid v \in \overline{V}\}$. Now set $V_G := \overline{V} \sqcup \overline{V}'$. For each edge e of $\overline{G}$ connecting nodes u and v, assign two "symmetric" arcs a, a' in G so as to satisfy (i)-(iii) above (where $u' = \sigma(u)$ and $v' = \sigma(v)$).

Also there is a correspondence between walks in $\overline{G}$ and pairs of symmetric walks in G. More precisely, let τ be the natural mapping of $V_G \cup A_G$ to $V_{\overline{G}} \cup E_{\overline{G}}$ (obtained by identifying the pairs of symmetric nodes and arcs). Each walk $P =$

$(v_0, a_1, v_1, \ldots, a_k, v_k)$ in G induces the sequence $\tau(P) := (\tau(v_0), \tau(a_1), \tau(v_1), \ldots, \tau(a_k), \tau(v_k))$ of nodes and edges in $\overline{G}$. One can easily check that $\tau(P)$ is a walk in $\overline{G}$ and $\tau(P) = \tau(P')$. Moreover, for any walk $\overline{P}$ in $\overline{G}$ there are exactly two preimages $\tau^{-1}(\overline{P})$ — these are certain symmetric walks P, P' in G satisfying $\tau(P) = \tau(P') = \overline{P}$.

Let us call a skew-symmetric graph *strongly acyclic* if it has no directed cycles. Each cycle in $\overline{G}$ generates a pair of symmetric cycles in G and vice versa. To obtain a similar result for the notion of weak acyclicity in bidirected graphs, suppose $\overline{G}$ is not weakly acyclic and consider an edge-simple cycle $\overline{C}$ in $\overline{G}$ having the smallest number of edges. Then $\overline{C}$ generates a pair of symmetric cycles C, C' in G (as described above). Cycles C, C' are circuits since otherwise one can shortcut them and obtain (by applying τ) a shorter edge-simple cycle in $\overline{G}$. Moreover, C and C' are regular (or, equivalently, arc-disjoint). Indeed, suppose C contains both arcs a and a' for some $a \in A_G$. Hence $\overline{C}$ traverses the edge $\tau(a)$ at least twice, contradicting the assumption. Conversely, let C be a regular circuit in G. Trivially $\overline{C} := \tau(C)$ is an edge-simple cycle in $\overline{G}$. These observations motivate the following definition: we call a skew-symmetric graph *weakly acyclic* if is has no regular circuits.

For a given set of nodes X in a directed graph G we use notation $G[X]$ to denote the directed subgraph induced by X. In case G is skew-symmetric and $X' = X$ the symmetry on G induces the symmetry on $G[X]$.

An easy part of our task is to describe the set of strongly acyclic skew-symmetric graphs. The following theorem gives the complete characterization of such graphs. (Due to lack of space we do not include proofs here; these proofs will be given in the full version of the paper.)

Theorem 1. *A skew-symmetric graph G is strongly acyclic iff there exists a partition $Z \sqcup Z'$ of V_G, such that the induced (standard directed) subgraphs $G[Z]$, $G[Z']$ are acyclic and no arc goes from Z to Z'.*

Corollary 1. *A bidirected graph G is strongly acyclic iff G is equivalent to a bidirected graph that only has directed edges forming an acyclic graph and edges leaving both endpoints.*

3   Separators and Decompositions

In this section we try to answer the following question: given a skew-symmetric weakly acyclic graph what kind of a natural certificate can be given to prove the absence of regular circuits (or, equivalently, regular cycles) in it?

Our first answer is as follows. Let G be a skew-symmetric graph. Suppose V_G is partitioned into four sets A, B, Z, Z' such that: (i) A and B are self-symmetric and nonempty; (ii) exactly one pair of symmetric arcs connects A and B; (iii) $G[A]$ and $G[B]$ are weakly acyclic; (iv) no arc leaves Z, no arc enters Z'. If these properties are satisfied we call (A, B, Z) a *weak separator* for G (see Fig. 1(a)).

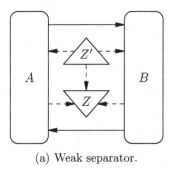
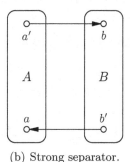

(a) Weak separator.   (b) Strong separator.

Fig. 1. Separators. Solid arcs should occur exactly once, dashed arcs may occur arbitrary number of times (including zero).

Theorem 2. *Every weakly acyclic skew-symmetric graph G is either strongly acyclic or admits a weak separator (A, B, Z). Conversely, if (A, B, Z) is a weak separator for G, then G is weakly acyclic.*

Thus, given a weakly acyclic graph G one can apply Theorem 2 to split V_G into four parts. The subgraphs $G[A]$, $G[B]$ are again weakly acyclic, so we can apply the same argument to them, etc. This recursive process (which produces two subgraphs on each steps) stops when current subgraph becomes strongly acyclic. In such case, Theorem 1 provides us with the required certificate.

Motivated by this observation we introduce the notion of a *weak acyclic decomposition* of G. By this decomposition we mean a binary tree D constructed as follows. The nodes of D correspond to self-symmetric subsets of V_G (in what follows, we make no distinction between nodes in D and these subsets). The root of D is the whole node set V_G. Any leaf X in D is a self-symmetric subset that induces a strongly acyclic subgraph $G[X]$; we attach a partition $X = Z \sqcup Z'$ as in Theorem 1 to X. Consider any non-leaf node X in D. It induces the subgraph $G[X]$ that is not strongly acyclic. Applying Theorem 2 we get a partition of X into subsets A, B, Z, Z' and attach it to X; the children of X are defined to be A and B.

An appealing special case arises when we restrict our attention to the class of strongly connected (in standard sense) skew-symmetric graphs, that is, graphs where each two nodes are connected by a (not necessarily regular) path. We need to introduce two additional definitions. Given a skew-symmetric graph H and a node s in it we call H *s-connected* if every node in H lies on a (not necessarily regular) s–s' path. Suppose the node set of a skew-symmetric graph G admits a partition (A, B) such that: (i) A and B are self-symmetric; (ii) exactly one pair of symmetric arcs $\{a'b, b'a\}$ connects A and B ($a, a' \in A$, $b, b' \in B$); (iii) $G[A]$ is weakly acyclic and a-connected, $G[B]$ is weakly acyclic and b-connected. Then we call (A, B) a *strong separator* for G (see Fig. 1(b) for an example).

Now we describe a decomposition of an arbitrary weakly acyclic skew-symmetric graph in terms of strongly connected components (hence, providing another answer to the question posed at the beginning of the section).

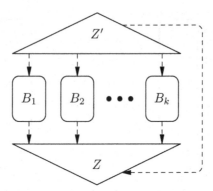

Fig. 2. Decomposition of a weakly acyclic skew-symmetric graph G. Dashed arcs may occur arbitrary number of times (including zero). Subgraphs $G[Z]$, $G[Z']$ are acyclic, subgraphs $G[B_i]$ are strongly connected and weakly acyclic.

Theorem 3. *A skew-symmetric graph G is weakly acyclic iff there exists a partition of V_G into sets $Z, Z', B_1, \ldots, B_k$ such that: (i) (standard directed) subgraphs $G[Z]$, $G[Z']$ are acyclic; (ii) sets B_i are self-symmetric, subgraphs $G[B_i]$ are strongly connected and weakly acyclic; (iii) no arc connects distinct sets B_i and B_j; (iv) no arc leaves Z, no arc enters Z'.*

Theorem 4. *A skew-symmetric graph B is strongly connected and weakly acyclic iff it admits a strong separator (A, B).*

An example of such decomposition is presented in Fig. 2. For $k = 0$ the decomposition in Theorem 3 coincides with such in Theorem 1.

Consider an arbitrary weakly acyclic skew-symmetric graph G. Add auxiliary nodes $\{s, s'\}$ and arcs $\{sv, v's'\}$, $v \in V_G \setminus \{s, s'\}$ thus making G s-connected. Similarly to its weak counterpart, a *strong acyclic decomposition* of G is a tree D constructed as follows. The nodes of D correspond to self-symmetric subsets of V_G. Each such subset A induces the a-connected graph $G[A]$ for some $a \in A$. The root of D is the whole node set V_G. Consider a node A of D. Applying Theorem 3 one gets a partition of A into subsets $Z, Z', B_1, \ldots, B_k$ and attaches it to A. Each of B_i is strongly connected and thus Theorem 4 applies. Hence, we can further decompose each of B_i into $X_i \sqcup Y_i$ ($X_i' = X_i$, $Y_i' = Y_i$) with the only pair of symmetric arcs $\{x_i'y_i, y_i'x_i\}$ ($x_i \in X_i$, $y_i \in Y_i$) connecting X_i and Y_i. The induced subgraphs $G[X_i]$ (resp. $G[Y_i]$) are x_i-connected (resp. y_i-connected). We define the children of A to be $X_1, Y_1, \ldots, X_k, Y_k$. Clearly, leaf nodes of D correspond to certain strongly acyclic subgraphs.

4   Algorithms

We need some additional notation. For a set of nodes X denote the set of arcs entering (resp. leaving) X by $\delta^{\mathrm{in}}(X)$ (resp. $\delta^{\mathrm{out}}(X)$). Denote the set of arcs having both endpoints in X by $\gamma(X)$.

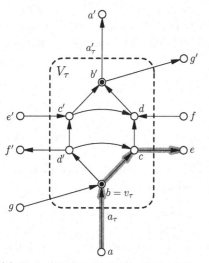
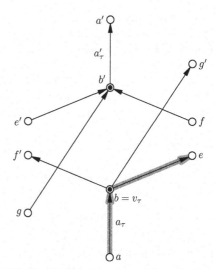

(a) Bud τ in graph G together with a shaded path $\overline{P}$.

(b) Graph G/τ together with a shaded path P.

Fig. 3. Buds, trimming, and path restoration. Base and antibase nodes b, b' are marked. Path $\overline{P}$ is a preimage of P.

Let V_τ be a symmetric set of nodes in a skew-symmetric graph G; $a_\tau \in \delta^{\mathrm{in}}(V_\tau)$. Let v_τ denote the head of a_τ. Suppose every node in V_τ is reachable from v_τ by a regular path in $G[V_\tau]$. Then we call $\tau = (V_\tau, a_\tau)$ a *bud*. (Note that our definition of bud is weaker than the corresponding one in [6].) The arc a_τ (resp. node v_τ) is called the *base arc* (resp. *base node*) of τ, arc a'_τ (resp. node v'_τ) is called the *antibase arc* (resp. *the antibase node*) of τ. For an arbitrary bud τ we denote its set of nodes by V_τ, base arc by a_τ, and base node by v_τ. An example of bud is given in Fig. 3(a).

Consider an arbitrary bud τ in a skew-symmetric graph G. By *trimming* τ we mean the following transformation of G: (i) all nodes in $V_\tau \setminus \{v_\tau, v'_\tau\}$ and arcs in $\gamma(V_\tau)$ are removed from G; (ii) all arcs in $\delta^{\mathrm{in}}(V_\tau) \setminus \{a_\tau\}$ are transformed into arcs entering v'_τ (the tails of these arcs are not changed); (iii) all arcs in $\delta^{\mathrm{out}}(V_\tau) \setminus \{a'_\tau\}$ are transformed into arcs leaving v_τ (the heads of these arcs are not changed). The resulting graph (which is obviously skew-symmetric) is denoted by G/τ. Thus, each arc of the original graph G not belonging to $\gamma(V_\tau)$ has its *image* in the trimmed graph G/τ. Fig. 3 gives an example of bud trimming.

Let P be a regular path in G/τ. One can lift this path to G as follows: if P does not contain neither a_τ, nor a'_τ leave P as it is. Otherwise, consider the case when P contains a_τ (the symmetric case is analogous). Split P into two parts: the part P_1 from the beginning of P to v_τ and the part P_2 from v_τ to the end of P. Let a be the first arc of P_2. The arc a leaves v_τ in G/τ and thus corresponds to some arc $\overline{a}$ leaving V_τ in G ($\overline{a} \neq a'_\tau$). Let $u \in V_\tau$ be the tail of a in G and Q be a regular v_τ–u path in $G[V_\tau]$ (existence of Q follows from the definition of bud). Consider the path $\overline{P} := P_1 \circ Q \circ P_2$ (here $U \circ V$ denotes the path obtained by concatenating U and V). One can easily show that $\overline{P}$ is regular. We call $\overline{P}$ a *preimage of* P (under trimming G by τ). Clearly, $\overline{P}$ is not unique. An example of such path restoration is shown in Fig. 3: the shaded path $\overline{P}$ on the left picture corresponds to the shaded path P on the right picture.

Given a skew-symmetric graph G we check if it is weakly acyclic as follows (we refer to this algorithm as ACYCLICITY-TEST). For technical reasons we require G to obey the following two properties:

(i) *Degree property*: for any node v in G at most one arc enters v or at most one arc leaves v.

(ii) *Loop property*: G must not contain parallel arcs connecting symmetric nodes (these arcs correspond to loops in bidirected graphs).

Degree property implies that a regular walk in G cannot contain a pair of symmetric nodes (loosely speaking, the notions of node- and arc-regularity coincide for G). This property can be satisfied by applying the reductions described in Section 1. It can be easily shown that degree and loop properties are preserved by trimmings.

Our algorithm adopts ideas from [4] to the case of skew-symmetric graphs. The algorithm is a variation of both depth-first-search (DFS) procedure (see [2]) and regular reachability algorithm (see [6]). It has, however, two essential differences. Firstly, unlike standard DFS, which is carried out in a static graph, our algorithm changes G by trimming some buds. Secondly, unlike regular reachability algorithm, we do not trim a bud as soon as we discover it. Rather, trimming is postponed up to the moment when it can be done "safely".

Let H be a current graph. Each pair of symmetric nodes in G is mapped to a certain pair of symmetric nodes in H. This mapping is defined by induction on the number of trimmings performed so far. Initially this mapping is identity. When a bud τ is trimmed and nodes $V_\tau \setminus \{v_\tau, v'_\tau\}$ are removed, the mapping is changed so as to send the pairs of removed nodes to $\{v_\tau, v'_\tau\}$. Given this mapping, we may also speak of the *preimage* $\overline{X}$ of any self-symmetric node set X in H.

The algorithm recursively grows a directed forest F. At every moment this forest has no symmetric nodes (or, equivalently, does not intersect the symmetric forest F'). Thus, every path in such forest is regular. The algorithm assigns *colors* to nodes. There are five possible colors: white, gray, black, antigray, and antiblack. White color assigned to v means that v is not yet discovered. Since the algorithm processes nodes in pairs, if v is white then so is v'. Other four colors also occur in pairs: if v is gray then v' is antigray, if v is black then v' is antiblack (and vice versa). All nodes outside both F and F' are white, nodes in F are black or gray, nodes in F' are antiblack or antigray.

At any given moment the algorithm implicitly maintains a regular path starting from a root of F. As in usual DFS, this path can be extracted by examining the recursion stack. The nodes on this path are gray, the symmetric nodes are antigray. No other node is gray or antigray. Black color denotes nodes which are already completely processed by the algorithm; the mates of such nodes are antiblack.

The core of the algorithm is the following recursive procedure. It has two arguments — a node u and optionally an arc q entering u (q may be omitted when u is a root node for a new tree in F). Firstly, the procedure marks u as gray and adds u to F (together with q if q is given). Secondly, it scans all arcs leaving u. Let a be such arc, v be its head. Several cases are possible (if no case applies, then a is skipped and next arc is fetched and examined):

(i) *Circuit case:* If v is gray, then there exists a regular circuit in the current graph (it can be obtained by adding the arc a to the gray v–u path in F). The procedure halts reporting the existence of a regular circuit in G (which is constructed from C in a postprocessing stage, see below).

(ii) *Recursion case:* If v is white, the recursive call with parameters (v, uv) is made.

(iii) *Trimming case:* If v is antiblack, the procedure constructs a certain bud in the current graph and trims it as follows. One can show that each time trimming case occurs the node v' is an ancestor of u in F. Let P denote the corresponding u–v' path. Let a_τ be the (unique) arc of F entering u (u has at least two outgoing arcs and hence cannot the a root of F, see below). Let H denote the current graph. Finally, let V_τ be the union of node sets of P and P'. One can easily show that $\tau = (V_\tau, a_\tau)$ is a bud in H (buds formed by a pair of symmetric regular paths are called *elementary* in [6]). The procedure trims τ and replaces H by H/τ. The forest F is updated by removing nodes in $V_\tau \setminus \{u, u'\}$ and arcs in $\gamma(V_\tau)$. All other arcs of F are replaced by their images under trimming by τ. Since a_τ belongs to F, it follows that the structure of forest is preserved. Note that trimming can produce new (previously unexisting) arcs leaving u.

When all arcs leaving u are fetched and processed the procedure marks u as black, u' as antiblack and exits.

ACYCLICITY-TEST initially makes all nodes white. Then, it looks for symmetric pairs of white nodes in G. Consider such a pair $\{v, v'\}$ and assume, without loss of generality, that out-degree of v is at most 1. Invoke the above-described procedure at v (passing no arc) and proceed to the next pair.

If all recursive calls complete normally, we claim that the initial graph is weakly acyclic. Otherwise, some recursive call halts yielding a regular circuit C in a current graph. During the postprocessing stage we consider the sequence of the trimmed buds in the reverse order and undo the corresponding trimmings. Each time we undo a trimming of a certain bud τ we also replace C by its preimage (as described in Section 3). At each such step the regularity of C is preserved, thus at the end of postprocessing we obtain a regular circuit in the original graph, as required.

It can be shown that this algorithm is correct and can be implemented to run in linear time.

Now we address the problem of building a weak acyclic decomposition. We solve it by a modified version of ACYCLICITY-TEST which we call DECOMPOSE. Let G be a skew-symmetric graph with a designated node s. Suppose we are

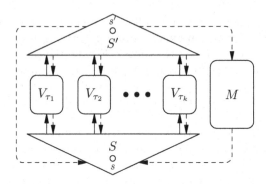

Fig. 4. A barrier. Solid arcs should occur exactly once, dashed arcs may occur arbitrary number of times (including zero).

given a collection of buds $\tau_1, \ldots, \tau_k$ in G together with node sets S and M. Additionally, suppose the following properties hold: (i) $\{S, S', M, V_{\tau_1}, \ldots, V_{\tau_k}\}$ is a partition of V_G with $s \in S$; (ii) no arc goes from S to $S' \cup M$; (iii) no arc connects distinct sets V_{τ_i} and V_{τ_j}; (iv) no arc connects V_{τ_i} and M; (v) the arc e_{τ_i} is the only one going from S to V_{τ_i}. Then we call the tuple $\mathcal{B} = (S, M; \tau_1, \ldots, \tau_k)$ an s–s' *barrier* ([6], see Fig. 4 for an example).

Let us introduce one more weak acyclicity certificate (which is needed for technical reasons) and show how to construct a weak decomposition from it. Let $\mathcal{B} = (S, M; \tau_1, \ldots, \tau_k)$ be a barrier in G. Put $\widetilde{G} := G/\tau_1/\ldots/\tau_k$, $W := S \cup \{v_{\tau_1}, \ldots, v_{\tau_k}\}$. We call $\mathcal{B}$ *acyclic* if the following conditions are satisfied: (i) subgraphs $G[M], G[V_{\tau_1}], \ldots, G[V_{\tau_k}]$ are weakly acyclic. (ii) the (standard directed) subgraph $\widetilde{G}[W]$ is acyclic.

Suppose we are given an acyclic barrier $\mathcal{B}$ of G with $M = \emptyset$. Additionally, suppose that weak acyclic decompositions of $G[V_{\tau_i}]$ are also given. A weak acyclic decomposition of G can be obtained as follows. Consider the graph $\widetilde{G}$ and the set W as in definition of an acyclic barrier. Order the nodes in W topologically: $W = \{w_1, \ldots, w_n\}$; for $i > j$ no arc in $\widetilde{G}$ goes from w_i to w_j. Also, assume that buds τ_i are numbered according to the ordering of the corresponding base nodes v_{τ_i} in W. Let these base nodes separate the sequence $w_1, \ldots, w_n$ into parts $Z_1, \ldots, Z_{k+1}$ (some of them may be empty). In other words, let $\{Z_i\}$ be the collection of sequences of nodes such that $w_1, \ldots, w_n = Z_1, v_{\tau_1}, Z_2, \ldots, Z_k, v_{\tau_k}, Z_{k+1}$. Additionally, put $A_i := (Z_1 \cup Z_1') \cup V_{\tau_1} \cup \ldots \cup V_{\tau_{i-1}} \cup (Z_i \cup Z_i')$. Obviously, sets A_i are self-symmetric, $A_{k+1} = V_G$. The graph $G[A_1]$ is strongly acyclic (this readily follows from Theorem 1 by putting $Z := Z_1$). One can show that for each $i \geq 2$ the triple $(A_{i-1}, V_{\tau_{i-1}}, Z_i)$ is a weak separator for $G[A_i]$. Using known decompositions of $G[V_{\tau_i}]$ these separators can be combined into a decomposition of G. An example is depicted in Fig. 5.

Buds that are trimmed by the algorithm are identified in a current graph but can also be regarded as buds in the original graph G. Namely, let H be a current graph and τ be a bud in H. One can see that $(\overline{V}_\tau, \overline{a}_\tau)$, where $\overline{a}_\tau$ (resp. $\overline{V}_\tau$) is the preimage of a_τ (resp. V_τ), is a bud in G. This bud will be denoted by $\overline{\tau}$.

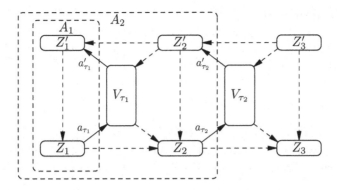

Fig. 5. Constructing a weak decomposition from an acyclic barrier. Solid arcs should occur exactly once, dashed arcs may occur arbitrary number of times (including zero). Not all possible dashed arcs are shown.

Observe that the node sets of preimages $\overline{\tau}$ of buds τ trimmed by ACYCLICITY-TEST are distinct sets forming a laminar family in V_G. At any moment the current graph H can be obtained from G by trimming the set of inclusion-wise maximal buds (which were discovered up to that moment). For each such bud $\overline{\tau}$ we maintain an acyclic $v'_{\overline{\tau}}$-barrier $\mathcal{B}^{\overline{\tau}}$ with the empty M-part.

Nodes in H can be of two possible kinds: *simple* and *complex*. Simple nodes are nodes that were not touched by trimmings, that is, they do not belong to any of $V_{\overline{\tau}}$ sets for all trimmed buds τ. Complex nodes are base and antibase nodes of maximal trimmed buds.

The following key properties of ACYCLICITY-TEST can be shown: (i) the (standard directed) subgraph induced by the set of black nodes is acyclic; (ii) no arc goes from black node to gray, white or antiblack node.

DECOMPOSE consists of two phases: *traversal* and *postprocessing*. During the first phase we invoke ACYCLICITY-TEST modified as follows. Suppose the algorithm trims a bud τ in H. First, suppose that the node v_τ was simple prior to that trimming. We construct $\mathcal{B}^{\overline{\tau}}$ as follows. Let B be the set of black simple nodes in V_τ, $\overline{\tau}_1, \ldots, \overline{\tau}_k$ be the preimages (in G) of trimmed buds corresponding to base nodes in V_τ. Putting $\mathcal{B}^{\overline{\tau}} := (B \cup \{v'_\tau\}, \emptyset; \overline{\tau}_1, \ldots, \overline{\tau}_k)$ we obtain a required acyclic barrier for τ.

Situation gets more involved when v_τ is a complex node (hence, the algorithm performs several trimmings at this node). Define B as above. Let $\overline{\phi}$ be the already trimmed inclusion-wise maximal bud at v_τ. Consider a barrier $\mathcal{B}^{\overline{\phi}} = (Q, \emptyset; \overline{\phi}_1, \ldots, \overline{\phi}_l)$. We put $\mathcal{B}^{\overline{\tau}} := (Q \cup B; \emptyset; \overline{\phi}_1, \ldots, \overline{\phi}_l, \overline{\tau}_1, \ldots, \overline{\tau}_k)$. It can be shown that $\mathcal{B}^{\overline{\tau}}$ is a required acyclic barrier for $\overline{\tau}$.

When traversal of G is complete the algorithm builds a final acyclic barrier in G. Observe that at that moment all nodes are black or antiblack. The set of simple black nodes B^* in the final graph and the inclusion-wise maximal trimmed buds $\overline{\tau}_1^*, \ldots, \overline{\tau}_k^*$ induce the acyclic barrier $\mathcal{B}^* := (B^*, \emptyset; \overline{\tau}_1^*, \ldots, \overline{\tau}_k^*)$ in G. During the postprocessing phase the algorithm constructs the desired decomposition of G from acyclic barriers recursively as indicated above.

Acknowledgements

The author is thankful to Alexander Karzanov for constant attention, collaboration, and many insightful discussions.

References

1. Maxim A. Babenko and Alexander V. Karzanov. Free multiflows in bidirected and skew-symmetric graphs. 2005. Submitted to a special issue of *DAM*.
2. T. Cormen, C. Leiserson, and R. Rivest. *Introduction to Algorithms*. MIT Press, 1990.
3. J. Edmonds and E. L. Johnson. Matching, a well-solved class of integer linear programs. *Combinatorial Structures and Their Applications*, pages 89–92, 1970.
4. Harold N. Gabow, Haim Kaplan, and Robert E. Tarjan. Unique maximum matching algorithms. pages 70–78, 1999.
5. Harold N. Gabow and Robert E. Tarjan. A linear-time algorithm for a special case of disjoint set union. *J. Comp. and Syst. Sci.*, 30:209–221, 1986.
6. Andrew V. Goldberg and Alexander V. Karzanov. Path problems in skew-symmetric graphs. *Combinatorica*, 16(3):353–382, 1996.
7. Andrew V. Goldberg and Alexander V. Karzanov. Maximum skew-symmetric flows and matchings. *Mathematical Programming*, 100(3):537–568, 2004.
8. L. Lovász and M. D. Plummer. *Matching Theory*. North-Holland, NY, 1986.
9. A. Schrijver. *Combinatorial Optimization*, volume A. Springer, Berlin, 2003.
10. W. T. Tutte. Antisymmetrical digraphs. *Canadian J. Math.*, 19:1101–1117, 1967.

Inductive Type Schemas as Functors

Freiric Barral[1] and Sergei Soloviev[2]

[1] LMU München, Institut für Informatik,
Oettingenstraße 67, 80 538 München, Germany
barral@tcs.informatik.uni-muenchen.de
[2] IRIT, Université Paul Sabatier,
118 route de Narbonne, 31062 Toulouse Cedex 4, France
soloviev@irit.fr

Abstract. Parametric inductive types can be seen as functions taking type parameters as arguments and returning the instantiated inductive types. Given functions between parameters one can construct a function between the instantiated inductive types representing the change of parameters along these functions. It is well known that it is not a functor w.r.t. intensional equality based on standard reductions. We investigate a simple type system with inductive types and iteration and show by modular rewriting techniques that new reductions can be safely added to make this construction a functor, while the decidability of the internal conversion relation based on the strong normalization and confluence properties is preserved. Possible applications: new categorical and computational structures on λ-calculus, certified computation.

1 Introduction

This paper is part of a larger project where we consider how some new computational rules may be incorporated in a λ-calculus with inductive types.

One of the main difficulties in applications of computer assisted reasoning based on λ-calculus is that the representation of real computations is very indirect, it is in fact complex coding, satisfactory for theoretical results but lacking the directness and transparency required for efficient applications. Extensions of typed systems using "real-life" inductive types like natural numbers, lists, and trees, with corresponding constructors and iteration/recursion operators are helpful but not sufficient.

Symbolic computation, for example, often includes the transformations of symbolic expressions that were never studied from the point of view of properties of the corresponding rewriting system. The importance of the problem of certified computation, symbolic or numerical (i.e., computation together with the proof of its correctness) was emphasized several years ago (cf. [2]) but since it was studied in a very limited number of cases.

The possibility used in most proof-assistants is to obtain a proof-term representing the proof of equality of two terms representing computations. This term should be carried everywhere, and this turns out to be very cumbersome and inefficient.

D. Grigoriev, J. Harrison, and E.A. Hirsch (Eds.): CSR 2006, LNCS 3967, pp. 35–45, 2006.
© Springer-Verlag Berlin Heidelberg 2006

One of the reasons is that the reduction system incorporated in the underlying typed λ-calculus is very restrictive. Thus even very simple equalities used routinely very often require the proof-term corresponding to this equality to be carried with it. It may require quite complex manipulations if the equality is used within another computation.

Our approach is based on extensions of reduction systems preserving good properties of the system as a whole, such as strong normalization (SN) and Church-Rosser property (CR).

In this paper we address the problem of rules representing the func toriality of the schemas of inductive types. We show that the corresponding extensions of the λ-calculus remain SN and CR. As result, the categorical computations using functoriality can be safely incorporated in "an intensional way" into proof-assistants. This will considerably lighten the proofs used in certified computations.

Notice, that there are well known categorical structures defined on certain systems of simply-typed λ-calculus, for example, cartesian closed structure on the calculus with surjective pairing and terminal object. These structures have numerous applications. Our approach will permit to "lift" them to corresponding classes of parametrized inductive types and obtain new categorical models and computational structures.

Details of proofs can be found on the web: `http://www.tcs.informatik.uni-muenchen.de/ barral/doc/Proofs.pdf`

2   Systems of Inductive Types

Given functions between parameters of inductive types with the same schema, one can construct by iteration the function between these types representing the change of parameters along these functions (it can be seen as a generalized Map function for arbitrary inductive types). The "minimal" system to study this "generalized Map" is the system of inductive types equipped with iteration. We shall consider only inductive types satisfying the condition of strict positivity.

2.1   Types and Schemas

We will use the appropriate vector notation for finite lists of syntactic expressions. Within more complex type and term expressions it is to be unfolded as follows:

- $\overrightarrow{\rho} \to \sigma ::= \sigma \mid \overrightarrow{\rho} \to \rho \to \sigma$.
- $\lambda \overrightarrow{x}.t ::= t \mid \lambda x \lambda \overrightarrow{x}.t$.
- $t\,\overrightarrow{s} ::= t \mid (ts)\,\overrightarrow{s}$
- $f^{\sigma \to \tau} \circ g^{\overrightarrow{\rho} \to \sigma} ::= \lambda \overrightarrow{x^\rho}.f(g\,\overrightarrow{x})$

We assume that the countable sets TVar of type variables and Const of constructors are given.

Definition 1 (Types). *We define simultaneously*

- *the grammar of* types:

$$\mathsf{Ty} \ni \rho, \sigma ::= \rho \to \sigma \mid \mu\alpha\,(\overrightarrow{c : \kappa_{\overrightarrow{\rho},\overrightarrow{\sigma}}(\alpha)}).$$

where $\alpha \in \mathsf{TVar}$, $\overrightarrow{c} \subseteq \mathsf{Const}$, $\overrightarrow{\kappa_{\overrightarrow{\rho},\overrightarrow{\sigma}}(\alpha)}$ *are constructor types over* α.
- *and the set* $\mathsf{KT}_{\overrightarrow{\rho},\overrightarrow{\sigma}}(\alpha)$ *of* constructor types *over* α *with type parameters* $\overrightarrow{\rho}, \overrightarrow{\sigma}$ *(we assume* $\alpha \notin FV(\overrightarrow{\rho}, \overrightarrow{\sigma})$*):*

$$\mathsf{KT}_{\overrightarrow{\rho},\overrightarrow{\sigma}}(\alpha) \ni \kappa_{\overrightarrow{\rho},\overrightarrow{\sigma}}(\alpha) = \overrightarrow{\rho} \to \overrightarrow{(\overrightarrow{\sigma}_i \to \alpha)}_{1 \leqslant i \leqslant n} \to \alpha$$

with $\overrightarrow{\sigma_1} :: \cdots :: \overrightarrow{\sigma_n} = \overrightarrow{\sigma}$.

The types with $\to$ as main symbol are called arrow or functional types, those with the binding symbol μ are called inductive types.

In the above definition $\overrightarrow{\rho}$ and $\overrightarrow{(\overrightarrow{\sigma}_i \to \alpha)}$ stand for the types of the arguments of a constructor. The types of the form ρ are called parametric operators, those of the form $\overrightarrow{\sigma} \to \alpha$ are called recursive operators, (0-recursive if $\overrightarrow{\sigma}$ is empty and 1-recursive otherwise)

We assume that the constructors are uniquely determined by their inductive type and that the constructors within an inductive type are different.

The types of constructors defined above verify the so-called strict positivity condition.

Note, that we have fixed a particular order of arguments of a constructor (first parametric and then recursive), It doesn't influence significantly the expressivity of the system and simplifies the presentation.

A constructor type $\kappa(\alpha)$ has always the form $\overrightarrow{\tau} \to \alpha$. We shall write $\kappa^-(\alpha)$ for the list of types $\overrightarrow{\tau}$. For $\mu = \mu\alpha\,(\overrightarrow{c : \kappa_{\overrightarrow{\rho},\overrightarrow{\sigma}}(\alpha)})$, we will write $c_k : \kappa_{\overrightarrow{\rho},\overrightarrow{\sigma}}(\alpha) \in \mu$ if $c_k : \kappa_{\overrightarrow{\rho},\overrightarrow{\sigma}}(\alpha) \in \overrightarrow{c : \kappa_{\overrightarrow{\rho},\overrightarrow{\sigma}}(\alpha)}$

Definition 2 (Schema of inductive type). *Given a list of variables* $\overrightarrow{k}$ *for inductive type constructors, type variables* $\overrightarrow{\pi}, \overrightarrow{\theta}, \alpha$ *(with* $\alpha \notin \overrightarrow{\pi} \cup \overrightarrow{\theta}$*), and a constructor type* $\kappa_{\overrightarrow{\pi},\overrightarrow{\theta}}(\alpha)$, *we define the* schema of inductive type $\mathbf{S}$ *by*

$$\mathbf{S}_{\overrightarrow{\pi},\overrightarrow{\theta}}(\overrightarrow{k}) ::= \mu\alpha(\overrightarrow{k : \kappa_{\overrightarrow{\pi},\overrightarrow{\theta}}(\alpha)})$$

Each inductive type $\mu\alpha\,(\overrightarrow{c : \kappa_{\overrightarrow{\rho},\overrightarrow{\sigma}}(\alpha)})$ is obtained by instantiation of the constructor variables $\overrightarrow{k}$ and type parameters $\overrightarrow{\pi}, \overrightarrow{\theta}$ of a schema of the inductive type $\mu\alpha(\overrightarrow{k : \kappa_{\overrightarrow{\pi},\overrightarrow{\theta}}(\alpha)})$.

The schemas considered in the following example represent all types of constructors of inductive types relevant to our study. The same inductive types will be used later to illustrate main technical ideas of this paper.

Example 1. $\mathbf{N} = \mu\alpha(k_1 : \alpha, k_2 : \alpha \to \alpha)$ *(schema of the type of natural numbers)*;$\mathbf{L}_\pi = \mu\alpha(k_1 : \alpha, k_2 : \pi \to \alpha \to \alpha)$ *(schema of lists over ρ)*;

$\mathbf{T}_{\pi,\theta} = \mu\alpha(k_1 : \alpha, k_2 : \pi \to (\theta \to \alpha) \to \alpha)$ *(θ-branching tree).*

Instantiations of these schemas may be $\mathsf{N} = \mathbf{N}(0, \mathsf{s}) = \mu\alpha(0 : \alpha, \mathsf{s} : \alpha \to \alpha$ *(natural numbers),* $\mathsf{N}' = \mathbf{N}(0', \mathsf{s}')$ *(a "copy" of N),* $\mathsf{L}(\mathsf{N}') = \mathbf{L}_{\mathsf{N}'}(\mathsf{nil}, \mathsf{cons})$ *(lists over N' with standard names of constructors),* $\mathsf{T}(\mathsf{N}, \mathsf{N}) = \mathbf{T}_{\mathsf{N},\mathsf{N}}(\mathsf{leaf}, \mathsf{node})$ *(infinitely branching tree over N),* $\mathbf{T}_{\mathsf{N},\mathsf{N}'}(\mathsf{leaf}', \mathsf{node}')$ *etc.*

Sometimes, when the names of parameters are not relevant, we may omit them altogether.

2.2   Terms

The terms of our systems are those of the simply typed λ-calculus extended by constructor constants from Const and iteration operators (iterators) $(\!|\overrightarrow{t}|\!)^{\mu,\tau}$ for all inductive types μ and τ (μ stands for the source and τ for the target type).

Definition 3 (Terms). *The set of terms Λ is generated by the following grammar:*

$$\Lambda \ni t ::= x \mid \lambda x^\tau t \mid (t\ t) \mid c_k \mid (\!|\overrightarrow{t}|\!)^{\mu,\tau} \ ,$$

with $x \in$ Var, $c_k \in$ Const and $\tau, \mu \subseteq$ Ty.

Definition 4 (Typing). *the typing relation is defined by*

$$\frac{(x,\rho) \in \Gamma}{\Gamma \vdash x : \rho} \qquad \frac{\Gamma, x : \rho \vdash r : \sigma}{\Gamma \vdash \lambda x^\rho.r : \rho \to \sigma} \qquad \frac{\Gamma \vdash s : \rho \qquad \Gamma \vdash r : \rho \to \sigma}{\Gamma \vdash rs : \sigma}$$

$$\frac{(c_k : \kappa_{\overrightarrow{\rho},\overrightarrow{\sigma}}(\alpha) \in \mu) \qquad \Gamma \vdash \overrightarrow{r} : \kappa_{\overrightarrow{\rho},\overrightarrow{\sigma}}^-(\mu)}{\Gamma \vdash c_k \overrightarrow{r} : \mu}$$

$$\frac{\mu\alpha\,\overline{(c : \kappa_{\overrightarrow{\rho},\overrightarrow{\sigma}}(\alpha))} = \mu \qquad \overrightarrow{\Gamma \vdash t : \kappa_{\overrightarrow{\rho},\overrightarrow{\sigma}}(\tau)}}{\Gamma \vdash (\!|\overrightarrow{t}|\!)^{\mu,\tau} : \mu \to \tau}$$

Example 2. *Let $\mu = \mathsf{N}, \mathsf{L}(\rho), \mathsf{T}(\rho, \sigma)$, and τ be the "target-type" in the last rule. The types of iterator terms $\overrightarrow{t}$ (step types) must be:*

- *τ and $\tau \to \tau$ in case of N;*
- *τ and $\rho \to \tau \to \tau$ in case of lists $\mathsf{L}(\rho)$;*
- *τ and $\rho \to (\sigma \to \tau) \to \tau$ in case of trees $\mathsf{T}(\rho, \sigma)$.*

Their particularly simple form is due to the use of iteration (as opposed to primitive recursion where the step type should contain type for the argument of the constructor as well).

2.3   Reductions

Simultaneous substitution of terms $\overrightarrow{s}$ to variables $\overrightarrow{y}$ in a term t, $t\{\overrightarrow{s}/\overrightarrow{y}\}$ is defined by structural induction as usual:

$$x\{\overrightarrow{s}/\overrightarrow{y}\} \qquad ::= \begin{cases} s_i \text{ if } x = y_i \text{ with } i \text{ the smallest index s.t. } y_i \in \overrightarrow{y} \\ x \text{ otherwise} \end{cases}$$

$$(\lambda x^\tau t)\{\overrightarrow{s}/\overrightarrow{y}\} ::= \lambda x^\tau t\{\overrightarrow{s}/\overrightarrow{y}\} \text{ where } x \notin \overrightarrow{y} \text{ and } x \notin \overrightarrow{FV(s)}$$

$$(t\ t)\{\overrightarrow{s}/\overrightarrow{y}\} \quad ::= t\ \{\overrightarrow{s}/\overrightarrow{y}\}t\{\overrightarrow{s}/\overrightarrow{y}\}$$

$$(\!|\overrightarrow{t}|\!)^{\mu,\tau}\{\overrightarrow{s}/\overrightarrow{y}\} ::= (\!|t\{\overrightarrow{s}/\overrightarrow{y}\}|\!)^{\mu,\tau}$$

Note that we can always ensure the condition of the second clause by α-conversion (variable condition).

Definition 5 (β-reduction). *We define the relation of β-reduction by the following rule:*

$$(\beta) \quad (\lambda x^\tau \cdot t)\ u \longmapsto_\beta t\{^u/_x\}$$

Definition 6 (η-expansion). *We define the relation of η-expansion by the following rule:*

$$(\eta) \quad t \quad \longmapsto_\eta \quad \lambda x^\tau \cdot t\ x$$

where $t : \tau \to \upsilon$, t is not an abstraction, $x \notin \mathsf{FV}(t)$.

Usual restriction concerning applicative position is incorporated in the definition of one-step reduction below.

We define two ι-reductions ; first the traditional one:

Definition 7 (ι-reduction). *Let $\mu \equiv \mu\alpha\,(\overrightarrow{\mathsf{c} : \kappa_{\overrightarrow{p},\overrightarrow{\sigma}}(\alpha)})$, $\mathsf{c}_k : \kappa_{\overrightarrow{p},\overrightarrow{\sigma}}(\alpha) \in \mu$, and $\kappa \equiv \overrightarrow{\rho} \to (\overrightarrow{\sigma}_i \to \alpha)_{1 \leqslant i \leqslant n} \to \alpha$ over α in μ. Given a term $\mathsf{c}_k\,\overrightarrow{p}\,\overrightarrow{r}$, where $\overrightarrow{p}$ (with p_i of type ρ_i) denotes parameter arguments and $\overrightarrow{r}$ (with r_i of type $\overrightarrow{\sigma}_i \to \mu$) n recursive arguments, and the terms $\overrightarrow{t}$ of step type $\kappa_{\overrightarrow{p},\overrightarrow{\sigma}}(\tau)$, the ι-reduction is defined by:*

$$(\iota) \quad (\!|\overrightarrow{t}|\!)^{\mu,\tau}(\mathsf{c}_k{}^\mu\,\overrightarrow{p}\,\overrightarrow{r}) \longmapsto_\iota t_k\,\overrightarrow{p}\,\overrightarrow{((\!|\overrightarrow{t}|\!)^{\mu,\tau} \circ r_i)} = t_k\,\overrightarrow{p}\,\overrightarrow{(\lambda\overrightarrow{x}\,(\!|\overrightarrow{t}|\!)^{\mu,\tau}(r_i\,\overrightarrow{x}))}\ .$$

This reduction may create β-redexes. Obvious redexes may appear due to the composition at the right part of the term and if iteration terms are abstractions. If abstracted variables corresponding to 1-recursive arguments are in applicative position inside this also may produce subsequent β-redexes. Good news is that this "cascade" of β-reductions will stop short because the types of arguments of r_i and of arguments of variables corresponding to 1-recursive arguments inside iteration terms are always inductive, not arrow types.

Since our system is equipped with η-expansion, one can always expand 1-recursive variables inside iteration terms as a pre-condition and then define a modified ι-reduction carrying out all these administrative β-reductions in one step.

Specialization of iteration terms and modified substitution:

Definition 8. *Let $\overrightarrow{y} = y_1^{\overrightarrow{\sigma_1} \to \tau}, ..., y_n^{\overrightarrow{\sigma_n} \to \tau}$ we define inductively the set of terms $It(\overrightarrow{y})$ where these variables always appear applied to maximal number of arguments*

$$It(\overrightarrow{y}) \ni t ::= (y_i\,\overrightarrow{t})^\tau \mid x \mid \lambda z.t \mid tt \mid (\!|\overrightarrow{t}|\!) \mid \mathsf{c}_k\,\overrightarrow{t}$$

Definition 9. *Modified simultaneous substitution of composition of $\overrightarrow{u}$ and $\overrightarrow{r}$ (with $u_i : \mu_i \to \tau, r_i : \overrightarrow{\sigma_i} \to \mu_i$) in t, $t\langle \overrightarrow{u \bullet r}/\overrightarrow{y}\rangle$, is defined recursively on $\mathcal{I}t(\overrightarrow{y})$:*

$$
\begin{aligned}
y_i \overrightarrow{t}\langle \overrightarrow{u \bullet r}/\overrightarrow{y}\rangle &::= u(r\overrightarrow{t\langle \overrightarrow{u \bullet r}/\overrightarrow{y}\rangle}) \; \textit{essential case}\\
x\langle \overrightarrow{u \bullet r}/\overrightarrow{y}\rangle &::= x\\
(\lambda z.t)\langle \overrightarrow{u \bullet r}/\overrightarrow{y}\rangle &::= \lambda z.t\langle \overrightarrow{u \bullet r}/\overrightarrow{y}\rangle \; \textit{where } z \notin \overrightarrow{y} \textit{ and } z \notin \overrightarrow{FV(u \bullet r)}\\
(tt)\langle \overrightarrow{u \bullet r}/\overrightarrow{y}\rangle &::= t\langle \overrightarrow{u \bullet r}/\overrightarrow{y}\rangle t\langle \overrightarrow{u \bullet r}/\overrightarrow{y}\rangle\\
(\!|\overrightarrow{t}|\!)\langle \overrightarrow{u \bullet r}/\overrightarrow{y}\rangle &::= (\!|\overrightarrow{t\langle \overrightarrow{u \bullet r}/\overrightarrow{y}\rangle}|\!)\\
\mathsf{c_k}\overrightarrow{t}\langle \overrightarrow{u \bullet r}/\overrightarrow{y}\rangle &::= \mathsf{c_k}\overrightarrow{t\langle \overrightarrow{u \bullet r}/\overrightarrow{y}\rangle}
\end{aligned}
$$

(At the third line α-conversion may be needed.)

Definition 10 ($\iota 2$-reduction). *Let $\mu \equiv \mu\alpha\,(\overrightarrow{\mathsf{c} : \kappa_{\overrightarrow{\rho}, \overrightarrow{\sigma}}(\alpha)})$, $\mathsf{c_k} : \kappa_{\overrightarrow{\rho}, \overrightarrow{\sigma}}(\alpha) \in \mu$, and $\kappa \equiv \overrightarrow{\rho} \to (\overrightarrow{\sigma}_i \to \alpha)_{1\leqslant i \leqslant n} \to \alpha$ over α in μ. Let a term $\mathsf{c_k}\overrightarrow{p}\,\overrightarrow{r}$ be given, where $\overrightarrow{p}$ (p_i of type ρ_i) denote parameter arguments and $\overrightarrow{r}$ (r_i of type $\overrightarrow{\sigma}_i \to \mu$) the n recursive arguments of $\mathsf{c_k}$ respectively. Let the terms $\overrightarrow{t}$ be the terms of step-type (iteration terms) with t_k be fully η-expanded externally, i.e., $t_k = \lambda\overrightarrow{x^\rho}y^{\overrightarrow{\sigma}_i \to \tau}.s_k$, and, moreover $s_k \in \mathcal{I}t(y_i^{\overrightarrow{\sigma_i} \to \tau})$. Under these conditions the $\iota 2$-reduction is defined by:*

$$(\!|\dots, \lambda\overrightarrow{x}\,\overrightarrow{y}.s_k, \dots|\!)(\mathsf{c_k}\,\overrightarrow{p}\,\overrightarrow{r}) \longmapsto_\iota s_k\{\overrightarrow{p}/\overrightarrow{x}\}\langle \overrightarrow{(\!|t|\!)\bullet r}/\overrightarrow{y}\rangle$$

Example 3 (multiplication by 2). *Although primitive recursion is encodable in our system, for sake of simplicity we present here a function directly encodable using iteration, the multiplication by 2 of natural numbers:*

$$\times 2 \equiv (\!|0, \lambda x.\mathsf{s}(\mathsf{s}x)|\!)$$

the associated $\iota 2$-reduction for the term $\mathsf{s}\,t$ is:

$$(\times 2)\mathsf{s}t \longrightarrow_{\iota 2} \mathsf{s}(\mathsf{s}((\times 2)t))$$

the selection of even branches in a tree can be defined as:

$$\mathsf{sel2} \equiv (\!|\mathsf{leaf}, \lambda xy.\mathsf{node}\,x\,(\lambda z.y((\times 2)z))|\!)$$

the associated $\iota 2$-reduction for the term $\mathsf{node}\,t\,f$ is:

$$\mathsf{sel2}(\mathsf{node}\,t\,f) \longrightarrow_{\iota 2} \mathsf{node}\,t\,(\lambda z.\mathsf{sel2}(f((\times 2)z)))$$

The "true" reduction relation of our system is defined now via contextual closure. The usual restriction on applicative position in η-expansion is incorporated here.

Definition 11 (One-step Reduction). *The One-step reduction* $\longrightarrow_R$ *is defined as the smallest relation such that:*

$$\frac{r \longmapsto_R r'}{r \longrightarrow_R r'} \qquad \frac{s \longrightarrow_R s'}{rs \longrightarrow_R rs'} \qquad \frac{r \longrightarrow_R r'}{\lambda x.r \longrightarrow_R \lambda x.r'} \qquad \frac{r \longrightarrow_R r' \quad r \not\longmapsto_\eta r'}{rs \longrightarrow_R r's}$$

$$\frac{t \longrightarrow_R t'}{(\![\overrightarrow{r}, t, \overrightarrow{s}]\!) \longrightarrow_R (\![\overrightarrow{r}, t', \overrightarrow{s}]\!)}$$

R can be for example β, η, ι, $\iota 2$.

The transitive, resp. transitive symmetric, closure of $\longrightarrow_R$ will be written $\longrightarrow_R^+$, resp. $\longrightarrow_R^*$. The *R-derivations* (sequences of terms such that two successive terms are in a one step reduction relation $\longrightarrow_R$) will be denoted by $d, e\ldots$. The expression $t \xrightarrow{\infty}_R$ will denote an infinite derivation beginning at t.

2.4   General Results

Theorem 1. *$\beta\eta\iota$ is convergent (cf [3]).*

The alternative ι-reduction is proved convergent using the fact that it is a (particularly simple) embedding (i.e., a reduction preserving encoding of a system within another).

Theorem 2.   *1. $\beta\eta\iota 2$ is embeddable in $\beta\eta\iota$,*
2. $\beta\eta\iota 2$ is convergent.

2.5   Inductive Type Schemas as Functors

We define the category $\mathcal{I}$ whose class of objects $\mathcal{I}_0$ is the set of Inductive Types μ and the class of arrows $\mathcal{I}_1$ is the set of terms A of types $\mu \to \mu'$ (where μ and μ' are inductive types) defined inductively as follows:

$$\mathcal{I}_1 \ni a, a' ::= \lambda x^\mu . x \mid (\![\overrightarrow{t}]\!) \mid a \circ a'$$

Definition 12 (Instantiation of schemas). *We define the function* $\mathbf{Cp_S^0}$ *for each schema of inductive type taking as arguments the constructor names $\overrightarrow{c}$ and inductive types $\overrightarrow{\rho}, \overrightarrow{\sigma} \in \mathcal{I}_0$ and returning the corresponding instantiated inductive type* $\mathbf{S}\{\overrightarrow{c}, \overrightarrow{\rho}, \overrightarrow{\sigma} / \overrightarrow{k}, \overrightarrow{\pi}, \overrightarrow{\theta}\}$

$$\mathbf{Cp_S^0} : \mathrm{Const}^{|\overrightarrow{k}|} \times \mathcal{I}^{|\overrightarrow{\rho}| + |\overrightarrow{\sigma}|} \to \mathcal{I}$$
$$(\overrightarrow{c}, \overrightarrow{\rho}, \overrightarrow{\sigma}) \qquad \mapsto \varphi = \mathbf{S}\{\overrightarrow{c}, \overrightarrow{\rho}, \overrightarrow{\sigma} / \overrightarrow{k}, \overrightarrow{\pi}, \overrightarrow{\theta}\}$$

If moreover, given inductive types $\overrightarrow{\rho'}, \overrightarrow{\sigma'}$ there exists terms in $\mathcal{I}_1$, $\overrightarrow{f} : \overrightarrow{\rho} \to \overrightarrow{\rho'}$, and $\overrightarrow{f'} : \overrightarrow{\sigma'} \to \overrightarrow{\sigma}$, we define the function $\mathbf{Cp_S^1}$ taking these terms and the relabeling function $l : \overrightarrow{c} \mapsto \overrightarrow{c'}$ as arguments by:

$$\mathbf{Cp_S^1}(l, \overrightarrow{f}, \overrightarrow{f'}) := (\![\overrightarrow{t}]\!) : \mathbf{Cp_S^0}(\overrightarrow{c}, \overrightarrow{\rho}, \overrightarrow{\sigma}) \to \mathbf{Cp_S^0}(\overrightarrow{l(c)}, \overrightarrow{\rho'}, \overrightarrow{\sigma'})$$

where $\vec{t} = t_1, \ldots, t_n$, $l(c_k) = c'_k$, *and*

$$t_k = \lambda \overrightarrow{x^\rho} \overrightarrow{y^{\sigma_i \to \varphi'}} \cdot c'_k \overrightarrow{\circ_x(f)} \lambda \vec{z}.y \circ_z \overrightarrow{(f')} \ .$$

where the function $\circ_x(f_k)$ *which returns a* β*-reduced form of* $f_k x$ *is defined recursively:*

- $\circ_x(\lambda x^\mu.x) = x$
- $\circ_x(\langle\!|\, \vec{t}\, \rangle\!|) = \langle\!|\, \vec{t}\, \rangle\!| x$
- $\circ_x(a \circ a') = \circ_x(a)\{\circ_x(a')/x\}$

With the notation of the definition above and given constructor names $\overrightarrow{c''}$, types $\overrightarrow{\rho''}, \overrightarrow{\sigma''}$, relabelling functions $l' : \overrightarrow{c'} \to \overrightarrow{c''}$, and terms $g : \overrightarrow{\rho} \to \overrightarrow{\rho''} \in \mathcal{I}_1$ and $g' : \overrightarrow{\sigma''} \to \overrightarrow{\sigma} \in \mathcal{I}_1$, the following equalities are provable:

$$\mathbf{Cp_S^1}(l, \vec{g}, \vec{g'}) \circ \mathbf{Cp_S^1}(l, \vec{f}, \vec{f'}) = \mathbf{Cp_S^1}(l' \circ l, \overrightarrow{g \circ f}, \overrightarrow{f' \circ g'})$$
$$\mathbf{Cp_S^1}(\mathrm{id}, \vec{\mathrm{id}}, \vec{\mathrm{id}}) = \mathrm{id}_{\mathbf{Cp_S^0}}$$

This means that with respect to an extensional model the pair $(\mathbf{Cp_S^0}, \mathbf{Cp_S^1})$ defines a functor. This result is well known, and a categorical proof (of a generalization of this result) can be found for example in Varmo Vene's doctoral thesis ([4]). However, these equalities do not hold w.r.t. the conversion relation. We shall extend the reduction relation in order to obtain a functor w.r.t. the conversion relation while preserving confluence and strong normalization of the underlying rewrite system.

In the following we will lighten the notation and omit all unnecessary material. In spirit of category theory we will write $\mathbf{Cp_S}$ instead of $\mathbf{Cp_S^0}$, $\mathbf{Cp_S^1}$ and often just $\mathbf{Cp}$ when $\mathbf{Cp}$ will be clear from context. In the same way we will not write the relabelling functions which we will not consider as part of the calculus.

Definition 13 (χ-reductions).

$$\mathbf{Cp}(\vec{g}, \vec{g'})(\mathbf{Cp}(\vec{f}, \vec{f'})t) \longmapsto_{\chi_\circ} \mathbf{Cp}(\overrightarrow{g \circ f}, \overrightarrow{f' \circ g'})t$$
$$\mathbf{Cp}(\vec{\mathrm{id}}, \vec{\mathrm{id}})t \longmapsto_{\chi_{\mathrm{id}}} t$$

Example 4. *The function* sel2 *of the previous example can be written as* $\mathbf{Cp_T}(\mathrm{id}, \times 2)$. *The* $\chi_\circ$*-reduction states that selecting even branches from a tree where one as already selected even branches should reduce to selecting the even branches of the even branches of this tree.*

$$\mathbf{Cp_T}(\mathrm{id}, \times 2)(\mathbf{Cp_T}(\mathrm{id}, \times 2)t) \longrightarrow_{\chi_\circ} \mathbf{Cp_T}(\mathrm{id}, \times 2 \circ \times 2)t$$

And now, to prove that $\mathbf{Cp_T}(\mathrm{id}, \times 2)(\mathbf{Cp_T}(\mathrm{id}, \times 2)t) = \mathbf{Cp_T}(\mathrm{id}, \times 4)t$, *one only has to prove* $\times 2 \circ \times 2 = \times 4$.

3   Main Theorems

3.1   Adjournment

Definition 14 (Adjournment). *Given two reduction relations S and R, we say that S is adjournable w.r.t. R in a derivation d, if*

$$d = t \longrightarrow_S \longrightarrow_R \overset{\infty}{\longrightarrow}_{RS} \Rightarrow \exists\, e = t \longrightarrow_R \overset{\infty}{\longrightarrow}_{RS}$$

If S is adjournable w.r.t. to R for derivation d, then we say that S is adjournable w.r.t. to R (cf. [1]).

Remark 1. *S is adjournable w.r.t. R in particular in the case: $S; R \subseteq R; R \cup S$. The notion of adjournability is traditionally expressed with this weaker condition (where d is not taken into account).*

Lemma 1 (Adjournment). *If R and S are strongly normalizing and S is adjournable w.r.t. to R then RS is strongly normalizing.*

3.2   Convergence of $\beta\eta\iota\chi_\circ$

Theorem 3 (Strong normalization of $\beta\eta\iota\chi_\circ$). *The $\chi_\circ$-reduction is strongly normalizing and adjournable with respect to $\beta\eta\iota$-reduction.*

By Newman lemma, a strongly normalizing and locally confluent system is confluent, so we need only to check local confluence.

Theorem 4 (Confluence). *The λ-calculus with $\beta\eta\iota\chi_\circ$-reduction is locally confluent.*

3.3   Pre-adjusted Adjournment

Definition 15 (insertability). *Given two reduction relation R, T, with $T \subset R$, T is said to be* insertable *in R if there exists a relation S on the support of R with $T \subseteq S$ and the two following conditions hold:*

$$S^{-1}; (R \setminus T) \subseteq R^+; S^{-1} \qquad\qquad S^{-1}; R \subseteq T^*; S^{-1}$$

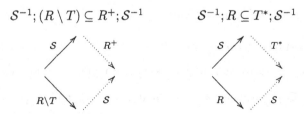

Lemma 2 (insertion). *Given two reduction relations R, T such that T is insertable in R and T is strongly normalizing. If there exists an infinite derivation d from t and an object t' with $t \rightarrow_T t'$, then there exists an infinite derivation d' from t'.*

Definition 16 (Conditional Adjournment). *Let R, S be reduction relations, an infinite derivation $d = t \longrightarrow_S \longrightarrow_R \overset{\infty}{\longrightarrow}_{RS}$ beginning with t and $\mathcal{P}$ a predicate on the objects. Then S is adjournable w.r.t. R in d under condition $\mathcal{P}$, if*

$$d = t \longrightarrow_S \longrightarrow_R \overset{\infty}{\longrightarrow}_{RS} \wedge \mathcal{P}(t) \Rightarrow \exists e = t \longrightarrow_R \overset{\infty}{\longrightarrow}_{RS}.$$

S is adjournable w.r.t. R under condition $\mathcal{P}$, if S is adjournable w.r.t. R in d under condition $\mathcal{P}$ for all d.

Definition 17 (realization). *Let T be a reduction relation and $\mathcal{P}$ a predicate on the objects. T realizes $\mathcal{P}$ for t if $\exists t', t \rightarrow_T^* t' \wedge \mathcal{P}(t')$. T realizes $\mathcal{P}$ if T realizes $\mathcal{P}$ for all objects.*

Lemma 3 (pre-adjusted adjournment). *Given reductions relations R, S, T with $S, T \subseteq R$, S is adjournable with respect to R under condition $\mathcal{P}$ and T is insertable in R, strongly normalizing and realizes $\mathcal{P}$, then S is adjournable w.r.t. R.*

Definition 18 (unrestricted η-expansion $\overline{\eta}$). *we define the rewrite rule for unrestricted η-expansion $\overline{\eta}$ by:*

$$t \longmapsto_{\overline{\eta}} \lambda x^\rho . tx \ if \ t : \rho \to \sigma$$

The one step reduction reduction relation $\longrightarrow_{\overline{\eta}}$ is defined as the contextual closure of $\longmapsto_{\overline{\eta}}$

Lemma 4 (weak condition for insertability). *In the definition of the insertability, if the relation S is the transitive reflexive closure of a reduction relation T', we establish some sufficient condition for T to be insertable. Given reduction relations R, T, T', if the relation T' verifies*

$$T'^{-1}; R \setminus T \subseteq R^*; R \setminus T; R^*; (T'^{-1})^* \wedge T'^{-1}; T \subseteq R^*; (T'^{-1})^*$$

then it T is insertable:

$$(T'^{-1})^*; R \setminus T; R^+; (T'^{-1})^* \wedge (T'^{-1})^*; R; T^*; (T'^{-1})^*$$

Proof. cf [3]

Lemma 5. *η-expansion is insertable.*

3.4   Convergence of $\beta\eta\iota\chi$

The difficult case is an overlap with ι-reduction, trying to adjourn directly:

$$\mathbf{Cp}_{\overrightarrow{\mathsf{id}}, \overrightarrow{\mathsf{id}}}(\mathsf{node}\, p\, x) \longrightarrow_{\chi_{\mathsf{id}}} \mathsf{node}\, p\, x \longrightarrow_{\beta\eta\iota} \mathsf{node}\, p'\, x$$

results in first applying a ι-contraction and then a χ_{id}-contraction:

$$
\begin{aligned}
\mathbf{Cp}_{\overrightarrow{\mathsf{id}}, \overrightarrow{\mathsf{id}}}(\mathsf{node}\, p\, x) &\longrightarrow_{\iota} & \mathsf{node}\, x\, \lambda z.yz\{^p/_x\}\langle^{\mathbf{Cp}_{\overrightarrow{\mathsf{id}}, \overrightarrow{\mathsf{id}}} \bullet^x}/_y\rangle \\
&\equiv & \mathsf{node}\, p\, \lambda z.\mathbf{Cp}_{\overrightarrow{\mathsf{id}}, \overrightarrow{\mathsf{id}}}(xz) \\
&\longrightarrow_{\chi_{\mathsf{id}}} & \mathsf{node}\, p\, \lambda z.xz
\end{aligned}
$$

And there is no way to close the fork with the initial sequence. We need to incorporate some η-expansion in the first sequence before being able to apply the adjournment lemma.

Theorem 5 (Strong Normalization).

1. χ_{id}-*reduction is strongly normalizing,*
2. χ_{id}-*reduction is adjournable with respect to $\beta\eta\iota\chi_0$ under the condition that 1-recursive arguments $\overrightarrow{r} = r_1 \ldots r_n$ of a constructor $c\,\overrightarrow{p}\,\overrightarrow{r}$ of type μ are fully eta-expansed externally, i.e. $r_i = \overrightarrow{r_i'} = \lambda\overrightarrow{x}.(r_i'\,\overrightarrow{x})^\mu$.*

Theorem 6 (Confluence). $\beta\eta\iota\chi$ *is locally confluent.*

4 Conclusion

We have designed a system where the reduction relation ensures the functorial laws w.r.t. intensional equality for certain classes of categories of inductive types.

The extension of this result to primitive recursion or to more general inductive types satisfying the monotonicity condition seems to be more difficult, but feasible.

An interesting task is to handle more categorical properties as studied by Vene [4] or Wadler [5].

Another motivating goal is to incorporate directly (and in a more efficient way) certain computations into proof-assistants based on type theory.

Acknowledgements. We would like to thank David Chemouil for fruitful discussions and Ralph Sobek, Andreas Abel and anonymous referees for useful remarks that helped to improve this paper.

References

1. L. Bachmair and N. Dershowitz. Commutation, transformation, and termination. In Jörg H. Siekmann, editor, *Proceedings of the 8th International Conference on Automated Deduction*, volume 230 of *LNCS*, pages 5–20, Oxford, UK, July 1986. Springer.
2. Henk Barendregt and Erik Barendsen. Autarkic computations in formal proofs. *J. Autom. Reasoning*, 28(3):321–336, 2002.
3. David Chemouil. *Types inductifs, isomorphismes et récriture extensionnelle.* Thèse de doctorat, Université Paul Sabatier, Toulouse, septembre 2004.
4. Varmo Vene. *Categorical Programming with Inductive and Coinductive Types.* PhD thesis (Diss. Math. Univ. Tartuensis 23), Dept. of Computer Science, Univ. of Tartu, August 2000.
5. P. Wadler. Theorems for free! In *Functional Programming Languages and Computer Architecture.* Springer Verlag, 1989.

Unfolding Synthesis of Asynchronous Automata[*]

Nicolas Baudru and Rémi Morin

Laboratoire d'Informatique Fondamentale de Marseille
39 rue Frédéric Joliot-Curie, F-13453 Marseille Cedex 13, France

Abstract. Zielonka's theorem shows that each regular set of Mazurkiewicz traces can be implemented as a system of synchronized processes provided with some distributed control structure called an asynchronous automaton. This paper gives a new algorithm for the synthesis of a non-deterministic asynchronous automaton from a regular Mazurkiewicz trace language. Our approach is based on an unfolding procedure that improves the complexity of Zielonka's and Pighizzini's techniques: Our construction is polynomial in terms of the number of states but still double-exponential in the size of the alphabet. As opposed to Métivier's work, our algorithm does not restrict to acyclic dependence alphabets.

1 Introduction

One of the major contributions in the theory of Mazurkiewicz traces [5] characterizes regular languages by means of asynchronous automata [17] which are devices with a distributed control structure. So far all known constructions of asynchronous automata from regular trace languages are quite involved and yield an exponential explosion of the number of states [7, 12]. Furthermore conversions of non-deterministic asynchronous automata into deterministic ones rely on Zielonka's time-stamping function [8, 13] and suffer from the same state-explosion problem. Interestingly heuristics to build small deterministic asynchronous automata were proposed in [15].

Zielonka's theorem and related techniques are fundamental tools in concurrency theory. For instance they are useful to compare the expressive power of classical models of concurrency such as Petri nets, asynchronous systems, and concurrent automata [10, 16]. These methods have been adapted already to the construction of communicating finite-state machines from collections of message sequence charts [1, 6, 11].

In this paper we give a new construction of a non-deterministic asynchronous automaton. Our algorithm starts from the specification of a regular trace language in the form of a possibly non-deterministic automaton. The latter is unfolded inductively on the alphabet into an automaton that enjoys several structural properties (Section 4). Next this unfolding automaton is used as the common skeleton of all local processes of an asynchronous automaton (Section 3). Due to the structural properties of the unfolding this asynchronous automaton accepts precisely the specified regular trace language.

We show that the number of local states built is polynomial in the number of states in the specification and double-exponential in the size of the alphabet (Subsection 4.4). Therefore *our approach subsumes the complexity of Zielonka's and Pighizzini's constructions* (Subsection 2.3).

---

[*] Supported by the ANR project SOAPDC.

D. Grigoriev, J. Harrison, and E.A. Hirsch (Eds.): CSR 2006, LNCS 3967, pp. 46–57, 2006.
© Springer-Verlag Berlin Heidelberg 2006

2   Background and Main Result

In this paper we fix a finite alphabet Σ provided with a total order $\sqsubseteq$. An automaton over a subset $T \subseteq \Sigma$ is a structure $\mathcal{A} = (Q, \imath, T, \longrightarrow, F)$ where Q is a *finite* set of states, $\imath \in Q$ is an initial state, $\longrightarrow \subseteq Q \times T \times Q$ is a set of transitions, and $F \subseteq Q$ is a subset of final states. We write $q \xrightarrow{a} q'$ to denote $(q, a, q') \in \longrightarrow$. An automaton $\mathcal{A}$ is called *deterministic* if we have $q \xrightarrow{a} q' \wedge q \xrightarrow{a} q'' \Rightarrow q' = q''$. For any word $u = a_1...a_n \in \Sigma^*$, we write $q \xrightarrow{u} q'$ if there are some states $q_0, q_1, ..., q_n \in Q$ such that $q = q_0 \xrightarrow{a_1} q_1...q_{n-1} \xrightarrow{a_n} q_n = q'$. The language $L(\mathcal{A})$ accepted by some automaton $\mathcal{A}$ consists of all words $u \in \Sigma^*$ such that $\imath \xrightarrow{u} q$ for some $q \in F$. A subset of words $L \subseteq \Sigma^*$ is *regular* if it is accepted by some automaton.

2.1   Mazurkiewicz Traces

We fix an *independence relation* $\parallel$ over Σ, that is, a binary relation $\parallel \subseteq \Sigma \times \Sigma$ which is irreflexive and symmetric. For any subset of actions $T \subseteq \Sigma$, the *dependence graph* of T is the undirected graph (V, E) whose set of vertices is $V = T$ and whose edges denote dependence, i.e. $\{a, b\} \in E \Leftrightarrow a \nparallel b$.

The *trace equivalence* $\sim$ associated with the independence alphabet $(\Sigma, \parallel)$ is the least congruence over Σ^* such that $ab \sim ba$ for all pairs of independent actions $a \parallel b$. For a word $u \in \Sigma^*$, the *trace* $[u] = \{v \in \Sigma^* \mid v \sim u\}$ collects all words that are equivalent to u. We extend this notation from words to sets of words in a natural way: For all $L \subseteq \Sigma^*$, we put $[L] = \{v \in \Sigma^* \mid \exists u \in L, v \sim u\}$.

A *trace language* is a subset of words $L \subseteq \Sigma^*$ that is closed for trace equivalence: $u \in L \wedge v \sim u \Rightarrow v \in L$. Equivalently we require that $L = [L]$. As usual a trace language L is called *regular* if it is accepted by some automaton.

2.2   Asynchronous Systems vs. Asynchronous Automata

Two classical automata-based models are known to correspond to regular trace languages. Let us first recall the basic notion of an asynchronous system [3].

DEFINITION 2.1.   *An automaton $\mathcal{A} = (Q, \imath, \Sigma, \longrightarrow, F)$ over the alphabet Σ is called* an asynchronous system over $(\Sigma, \parallel)$ *if we have*

ID: $q_1 \xrightarrow{a} q_2 \wedge q_2 \xrightarrow{b} q_3 \wedge a \parallel b$ *implies* $q_1 \xrightarrow{b} q_4 \wedge q_4 \xrightarrow{a} q_3$ *for some* $q_4 \in Q$.

The Independent Diamond property ID ensures that the language $L(\mathcal{A})$ of any asynchronous system is closed for the commutation of independent adjacent actions. Thus it is a regular trace language. Conversely it is easy to observe that *any regular trace language is the language of some deterministic asynchronous system.*

We recall now a more involved model of communicating processes known as asynchronous automata [17]. A finite family $\delta = (\Sigma_k)_{k \in K}$ of subsets of Σ is called a *distribution of* $(\Sigma, \parallel)$ if we have $a \nparallel b \Leftrightarrow \exists k \in K, \{a, b\} \subseteq \Sigma_k$ for all actions $a, b \in \Sigma$. Note that each subset Σ_k is a clique of the dependence graph $(\Sigma, \nparallel)$ and a distribution δ is simply a clique covering of $(\Sigma, \nparallel)$. We fix an arbitrary distribution $\delta = (\Sigma_k)_{k \in K}$ in the rest of this paper. We call *processes* the elements of K. The *location* $\mathrm{Loc}(a)$ of an action $a \in \Sigma$ consists of all processes $k \in K$ such that $a \in \Sigma_k$: $\mathrm{Loc}(a) = \{k \in K \mid a \in \Sigma_k\}$.

DEFINITION 2.2. *An asynchronous automaton* over the distribution $(\Sigma_k)_{k\in K}$ *consists of a family of finite sets of states* $(Q_k)_{k\in K}$, *a family of initial local states* $(\imath_k)_{k\in K}$ *with* $\imath_k \in Q_k$, *a subset of final global states* $F \subseteq \prod_{k\in K} Q_k$, *and a transition relation* $\partial_a \subseteq \prod_{k\in \text{Loc}(a)} Q_k \times \prod_{k\in \text{Loc}(a)} Q_k$ *for each action* $a \in \Sigma$.

The set of *global states* $Q = \prod_{k\in K} Q_k$ can be provided with a set of global transitions $\longrightarrow$ in such a way that an asynchronous automaton is viewed as a particular automaton. Given an action $a \in \Sigma$ and two global states $q = (q_k)_{k\in K}$ and $r = (r_k)_{k\in K}$, we put $q \xrightarrow{a} r$ if $((q_k)_{k\in \text{Loc}(a)}, (r_k)_{k\in \text{Loc}(a)}) \in \partial_a$ and $q_k = r_k$ for all $k \in K \setminus \text{Loc}(a)$. The initial global state $\imath$ consists of the collection of initial local states: $\imath = (\imath_k)_{k\in K}$. Then the *global automaton* $\mathcal{A} = (Q, \imath, \Sigma, \longrightarrow, F)$ satisfies Property ID of Def. 2.1. Thus it is an asynchronous system over $(\Sigma, \|)$ and $L(\mathcal{A})$ is a regular trace language. An asynchronous automaton is *deterministic* if its global automaton is deterministic, i.e. the local transition relations ∂_a are partial functions.

2.3   Main Result and Comparisons to Related Works

Although deterministic asynchronous automata appear as a restricted subclass of deterministic asynchronous systems, Zielonka's theorem asserts that any regular trace language can be implemented in the form of a deterministic asynchronous automaton.

THEOREM 2.3. *[17] For any regular trace language* L *there exists a deterministic asynchronous automaton whose global automaton* $\mathcal{A}$ *satisfies* $L = L(\mathcal{A})$.

In [12] a complexity analysis of Zielonka's construction is detailed. Let $|Q|$ be the number of states of the minimal deterministic automaton that accepts L and $|K|$ be the number of processes. Then the number of local states built by Zielonka's technique in each process $k \in K$ is $|Q_k| \leqslant 2^{O(2^{|K|} \cdot |Q| \log |Q|)}$. The simplified construction by Cori et al. in [4] also suffers from this exponential state-explosion [5].

Another construction proposed by Pighizzini [14] builds a non-deterministic asynchronous automaton from particular rational expressions. This simpler approach proceeds inductively on the structure of the rational expression. Each step can easily be shown to be polynomial. In particular the number of local states in each process is (at least) *doubled* by each restricted iteration. Consequently in some cases the number of local states in each process is *exponential* in the length of the rational expression.

In the present paper we give a new construction that is *polynomial in* $|Q|$ (Th. 5.9): It produces $|Q_k| \leqslant (3.|\Sigma|.|Q|)^d$ local states for each process, where $d = 2^{|\Sigma|}$, $|\Sigma|$ is the size of Σ, and $|Q|$ is the number of states of some (possibly non-deterministic) asynchronous system that accepts L.

With the help of two simple examples we present our new approach in the next section. It consists basically in two steps: A *naive construction* applied on an *unfolded* automaton. Comparisons with known techniques is hard since this twofold approach has no similarity with previous methods. On the other hand we have applied recently our unfolding strategy in the framework of Message Sequence Charts [2]. We believe also that our approach can be strengthened in order to build *deterministic* asynchronous automata from deterministic asynchronous systems with a similar complexity cost.

3   Twofold Strategy

In this section we fix a (possibly non-deterministic) automaton $\mathcal{A} = (Q, \imath, \Sigma, \longrightarrow, F)$ over the alphabet Σ. We fix also a distribution $\delta = (\Sigma_k)_{k \in K}$ of $(\Sigma, \|)$. We introduce a basic construction of a *projected asynchronous automaton* $\widehat{\mathcal{A}}$ associated with $\mathcal{A}$. In general $L(\widehat{\mathcal{A}}) \neq L(\mathcal{A})$ even if we assume that $\mathcal{A}$ satisfies Axiom ID of Def. 2.1. Our strategy will appear as a method to unfold an *asynchronous system* $\mathcal{A}$ into a larger automaton $\mathcal{A}_{\mathrm{Unf}}$ that represents the same language: $[L(\mathcal{A}_{\mathrm{Unf}})] = L(\mathcal{A})$ and such that the projected asynchronous automaton of the unfolding $\mathcal{A}_{\mathrm{Unf}}$ yields a correct implementation: $L(\widehat{\mathcal{A}_{\mathrm{Unf}}}) = [L(\mathcal{A}_{\mathrm{Unf}})] = L(\mathcal{A})$. Note that $\mathcal{A}_{\mathrm{Unf}}$ will not fulfill ID in general.

The construction of the *projected asynchronous automaton* $\widehat{\mathcal{A}}$ over δ from the automaton $\mathcal{A}$ proceeds as follows. First the local states are copies of states of $\mathcal{A}$: We put $Q_k = Q$ for each process $k \in K$. The initial state $(\imath, ..., \imath)$ consists of $|K|$ copies of the initial state of $\mathcal{A}$. Moreover for each $a \in \Sigma$, the pair $((q_k)_{k \in \mathrm{Loc}(a)}, (r_k)_{k \in \mathrm{Loc}(a)})$ belongs to the transition relation ∂_a if there exist two states $q, r \in Q$ and a transition $q \xrightarrow{a} r$ in $\mathcal{A}$ such that the two following conditions are satisfied:

- for all $k \in \mathrm{Loc}(a)$, $q_k \xrightarrow{u} q$ in $\mathcal{A}$ for some word $u \in (\Sigma \setminus \Sigma_k)^\star$;
- for all $k \in \mathrm{Loc}(a)$, $r_k = r$; in particular all r_k are equal.

To conclude this definition, a global state $(q_k)_{k \in K}$ is *final* if there exists a final state $q \in F$ such that for all $k \in K$ there exists a path $q_k \xrightarrow{u} q$ in $\mathcal{A}$ for some word $u \in (\Sigma \setminus \Sigma_k)^\star$. The next result can be proved straightforwardly.

PROPOSITION 3.1. *We have* $L(\mathcal{A}) \subseteq [L(\mathcal{A})] \subseteq L(\widehat{\mathcal{A}})$.

EXAMPLE 3.2. We consider the independence alphabet $(\Sigma, \|)$ where $\Sigma = \{a, b, c\}$, $a \| b$ but $a \nparallel c \nparallel b$. Let $\mathcal{A}$ be the asynchronous system depicted in Fig. 1 and δ be the distribution with two processes $\Sigma_a = \{a, c\}$ and $\Sigma_b = \{b, c\}$. We assume here that all states of $\mathcal{A}$ are final. Then we get $L(\widehat{\mathcal{A}}) = \Sigma^\star$ whereas the word cc does not belong to $L(\mathcal{A})$. Consider now the asynchronous system $\mathcal{A}'$ depicted in Fig. 2. We can check that $L(\mathcal{A}') = L(\mathcal{A})$ and $L(\widehat{\mathcal{A}'}) = L(\mathcal{A})$.

This example shows that for some automata $\mathcal{A}$ the naive construction of the projected asynchronous automaton does not provide a correct implementation. However it is possible to unfold the automaton $\mathcal{A}$ to get a larger automaton $\mathcal{A}'$ for which the naive construction is correct. The aim of this paper is to show that this unfolding process is feasible with a polynomial cost for any asynchronous system $\mathcal{A}$.

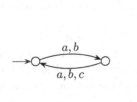

Fig. 1. Asynchronous system $\mathcal{A}$

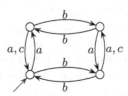

Fig. 2. Asynchronous system $\mathcal{A}'$

4   Unfolding Algorithm

In the rest of the paper we fix some asynchronous system $\mathcal{A} = (Q, \imath, \Sigma, \longrightarrow, F)$ that is possibly non-deterministic. The aim of this section is to associate $\mathcal{A}$ with a family of automata called *boxes* and *triangles* which are defined inductively. The last box built by this construction is called the *unfolding* of $\mathcal{A}$ (Def. 4.1).

Boxes and triangles are related to $\mathcal{A}$ by means of morphisms which are defined as follows. Let $\mathcal{A}_1 = (Q_1, \imath_1, T, \longrightarrow_1, F_1)$ and $\mathcal{A}_2 = (Q_2, \imath_2, T, \longrightarrow_2, F_2)$ be two automata over a subset of actions $T \subseteq \Sigma$. A *morphism* $\sigma : \mathcal{A}_1 \to \mathcal{A}_2$ *from* $\mathcal{A}_1$ *to* $\mathcal{A}_2$ is a mapping $\sigma : Q_1 \to Q_2$ from Q_1 to Q_2 such that $\sigma(\imath_1) = \imath_2$, $\sigma(F_1) \subseteq F_2$, and $q_1 \xrightarrow{a}_1 q_1'$ implies $\sigma(q_1) \xrightarrow{a}_2 \sigma(q_1')$. In particular, we have then $L(\mathcal{A}_1) \subseteq L(\mathcal{A}_2)$.

Now boxes and triangles are associated with an initial state that may not correspond to the initial state of $\mathcal{A}$. They are associated also with a subset of actions $T \subseteq \Sigma$. For these reasons, for any state $q \in Q$ and any subset of actions $T \subseteq \Sigma$, we let $\mathcal{A}_{T,q}$ denote the automaton $(Q, q, T, \longrightarrow_T, F)$ where $\longrightarrow_T$ is the restriction of $\longrightarrow$ to the transitions labeled by actions in T: $\longrightarrow_T = \longrightarrow \cap (Q \times T \times Q)$.

In this section we shall define the box $\square_{T,q}$ for all states $q \in Q$ and all subsets of actions $T \subseteq \Sigma$. The box $\square_{T,q}$ is a pair $(\mathcal{B}_{T,q}, \beta_{T,q})$ where $\mathcal{B}_{T,q}$ is an automaton over T and $\beta_{T,q} : \mathcal{B}_{T,q} \to \mathcal{A}_{T,q}$ is a morphism. Similarly, we shall define the triangle $\triangle_{T,q}$ for all states q and all *non-empty* subsets of actions T. The triangle $\triangle_{T,q}$ is a pair $(\mathcal{T}_{T,q}, \tau_{T,q})$ where $\mathcal{T}_{T,q}$ is an automaton over T and $\tau_{T,q} : \mathcal{T}_{T,q} \to \mathcal{A}_{T,q}$ is a morphism.

The *height* of a box $\square_{T,q}$ or a triangle $\triangle_{T,q}$ is the cardinality of T. Boxes and triangles are defined inductively on the height. We first define the box $\square_{\emptyset,q}$ for all states $q \in Q$. Next triangles of height h are built upon boxes of height $g < h$ and boxes of height h are built upon either triangles of height h or boxes of height $g < h$, whether the dependence graph $(T, \not\parallel)$ is connected or not.

The base case deals with the boxes of height 0. For all states $q \in Q$, the box $\square_{\emptyset,q}$ consists of the morphism $\beta_{\emptyset,q} : \{q\} \to Q$ that maps q to itself together with the automaton $\mathcal{B}_{\emptyset,q} = (\{q\}, q, \emptyset, \emptyset, F_{\emptyset,q})$ where $F_{\emptyset,q} = \{q\}$ if $q \in F$ and $F_{\emptyset,q} = \emptyset$ otherwise. In general a state of a box or a triangle is final if it is associated with a final state of $\mathcal{A}$.

DEFINITION 4.1.   *The* unfolding $\mathcal{A}_{\mathrm{Unf}}$ *of* $\mathcal{A}$ *is the box* $\mathcal{B}_{\Sigma,\imath}$.

4.1   Building Triangles from Boxes

Triangles are made of boxes of lower height. Boxes are inserted into a triangle recursively on the height along a tree-like structure and several copies of the same box may appear within a triangle. We want to keep track of this structure in order to prove properties of triangles (and boxes) inductively: This enables us to allow for different copies of the same box within a triangle.

To do this, each state of a triangle is associated with a *rank* $k \in \mathbb{N}$ such that all states with the same rank come from the same copy of the same box. It is also important to keep track of the height each state comes from, because boxes of a triangle are inserted recursively on the height. For these reasons, a state of a triangle $\triangle_{T^\circ, q^\circ} = (\mathcal{T}_{T^\circ, q^\circ}, \tau_{T^\circ, q^\circ})$ is encoded as a quadruple $v = (w, T, q, k)$ such that w is a state from the box $\square_{T,q}$ with height $h = |T|$ and v is added to the triangle within the

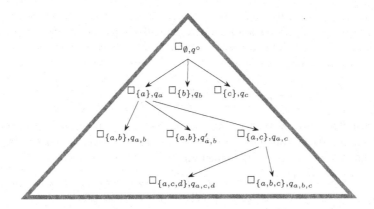

Fig. 3. Tree structure of triangles $\triangle_{T^\circ,q^\circ}$ with $T^\circ = \{a,b,c,d\}$

k-th box inserted into the triangle. Moreover this box is a copy of $\square_{T,q}$. In that case the state v maps to $\tau_{T^\circ,q^\circ}(v) = \beta_{T,q}(w)$, that is, the insertion of boxes preserves the correspondance to the states of $\mathcal{A}$. Moreover the morphism τ_{T°,q° of a triangle $\triangle_{T^\circ,q^\circ}$ is encoded in the data structure of its states.

We denote by $\mathcal{B}' = \text{MARK}(\mathcal{B}, T, q, k)$ the generic process that creates a copy $\mathcal{B}'$ of an automaton $\mathcal{B}$ by replacing each state w of $\mathcal{B}$ by $v = (w, T, q, k)$. The construction of the triangle $\triangle_{T^\circ,q^\circ}$ starts with using this marking procedure and building a copy $\text{MARK}(\square_{\emptyset,q^\circ}, \emptyset, q^\circ, 1)$ of the base box $\square_{\emptyset,q^\circ}$ which gets rank $k = 1$ and whose marked initial state $(\imath_{\square,\emptyset,q^\circ}, \emptyset, q^\circ, 1)$ becomes the initial state of $\triangle_{T^\circ,q^\circ}$. Along the construction of this triangle, an integer variable k counts the number of boxes already inserted in the triangle to make sure that all copies inserted get distinct ranks. The construction of the triangle $\triangle_{T^\circ,q^\circ}$ proceeds by successive insertions of copies of boxes according to the single following rule.

RULE 4.2. *A new copy of the box $\square_{T',q'}$ is inserted into the triangle $\triangle_{T^\circ,q^\circ}$ in construction if there exists a state $v = (w, T, q, l)$ in the triangle in construction and an action $a \in \Sigma$ such that*

T_1: *$\beta_{T,q}(w) \xrightarrow{a} q'$ in the automaton $\mathcal{A}_{T^\circ,q^\circ}$;*
T_2: *$T' = T \cup \{a\}$ and $T \subset T' \subset T^\circ$;*
T_3: *no a-transition relates sofar v to the initial state of some copy of the box $\square_{T',q'}$ in the triangle in construction.*

In that case some a-transition is added in the triangle in construction from v to the initial state of the new copy of the box $\square_{T',q'}$.

Note here that Condition T_2 ensures that inserted boxes have height at most $|T^\circ| - 1$. By construction all copies of boxes inserted in a triangle are related in a tree-like structure built along the application of the above rules. It is easy to implement the construction of a triangle from boxes as specified by the insertion rules above by means of a list of inserted boxes whose possible successors have not been investigated, in a depth-first-search or breadth-first-search way. Condition T_2 ensures also that if a new copy of the box $\Box_{T',q'}$ is inserted and connected from $v = (w, T, q, l)$ then $T \subset T' \subset T^\circ$. This shows that this insertion process eventually stops and the resulting tree has depth at most $|T^\circ - 1|$. Moreover, since we start from the empty box and transitions in boxes $\Box_{T,q}$ carry actions from T, we get the next obvious property.

LEMMA 4.3. *If a word* $u \in \Sigma^*$ *leads in the triangle* $\triangle_{T^\circ, q^\circ}$ *from its initial state to some state* $v = (w, T, q, l)$ *then* $u \in T^*$ *and all actions from* T *appear in* u.

Note also that it is easy to check that the mapping τ_{T°, q° induced by the data structure builds a morphism from $\mathcal{T}_{T^\circ, q^\circ}$ to $\mathcal{A}_{T^\circ, q^\circ}$. For latter purposes we define the list of missing transitions to state $q' \in Q$ in the triangle $\triangle_{T^\circ, q^\circ}$ as follows.

DEFINITION 4.4. *Let* $T^\circ \subseteq \Sigma$ *be a subset of actions and* q°, q' *be two states of* $\mathcal{A}$. *The set of missing transitions* MISSING(T°, q°, q') *consists of all pairs* (v, a) *where* $v = (w, T, q, l)$ *is a state of* $\triangle_{T^\circ, q^\circ}$ *and* a *is an action such that*

- $\beta_{T,q}(w) \xrightarrow{a} q'$ *in the automaton* $\mathcal{A}_{T^\circ, q^\circ}$;
- $T \subset T \cup \{a\} = T^\circ$.

Note here that the insertion rule T_2 for triangles forbids to insert a copy of the box $\mathcal{B}_{T^\circ, q'}$ and to connect its initial state with a transition labeled by a from state v. Note also that $|$MISSING$(T^\circ, q^\circ, q')|$ is less than the number of states in $\triangle_{T^\circ, q^\circ}$.

4.2   Building Boxes from Triangles

As announced in the introduction of this section the construction of the box $\Box_{T^\circ, q^\circ}$ depends on the connectivity of the dependence graph of T°. Assume first that $T^\circ \subseteq \Sigma$ is not connected. Let T_1 denote the connected component of $(T^\circ, \not\parallel)$ that contains the least action $a \in T^\circ$ w.r.t. the total order $\sqsubseteq$ over Σ. We put $T_2 = T^\circ \setminus T_1$. The construction of the box $\Box_{T^\circ, q^\circ}$ starts with building a copy of the box $\Box_{T_2, q^\circ}$. Next for each state w of $\Box_{T_2, q^\circ}$ and each transition $\beta_{T_2, q^\circ}(w) \xrightarrow{a} q$ in $\mathcal{A}_{T^\circ, q^\circ}$ with $a \in T_1$, the algorithm adds some a-transition from the copy of w to the initial state of a new copy of $\Box_{T_1, q}$.

We come now to the definition of boxes associated with a *connected* set of actions. This part is more subtle than the two previous constructions which have a tree-structure and create no new loop. Let $T^\circ \subseteq \Sigma$ be a connected (non-empty) subset of actions. Basically the box $\Box_{T^\circ, q^\circ}$ collects all triangles $\triangle_{T^\circ, q}$ for all states $q \in Q$. Each triangle is replicated a fixed number of times and copies of triangles are connected in some very specific way. We adopt a data structure similar to triangles (and unconnected boxes). A node w of a box $\Box_{T^\circ, q^\circ}$ is a quadruple (v, T°, q, k) where v is a node of the triangle $\triangle_{T^\circ, q}$ and $k \in \mathbb{N}$. The rank k will allow us to distinguish between different copies of the same triangle within a box.

The construction of the box $\Box_{T^\circ, q^\circ}$ consists in two steps. First m copies of each triangle $\triangle_{T^\circ, q}$ are inserted in the box and the first copy of $\triangle_{T^\circ, q^\circ}$ gets rank 1; moreover the first copy of its initial state is the initial state of the box. The value of m will be discussed below. In a second step some transitions are added to connect these triangles to each other according to the single following rule.

RULE 4.5. *For each triangle* $\triangle_{T^\circ, q}$, *for each state* $q' \in Q$, *and for each missing transition* $(v, a) \in$ MISSING(T°, q, q') *we add some* a-*transition from each copy of state* v *to*

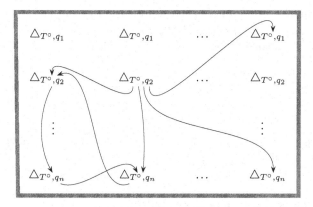

Fig. 4. Square structure of a box $\Box_{T^\circ,q^\circ}$ with T° connected

the initial state of some copy of triangle $\triangle_{T^\circ,q'}$. In this process of connecting triangles we obey to the two following requirements:

C_1: *No added transition connects two states from the same copy of a triangle.*
C_2: *At most one transition connects one copy of $\triangle_{T^\circ,q}$ to one copy of $\triangle_{T^\circ,q'}$.*

Condition C_1 requires that there is no added transition from state (v, T°, q, l) with rank l to the (initial) state $(\imath_{\triangle,T^\circ,q}, T^\circ, q, l)$. To do so it is sufficient to have two copies of each triangle. Condition C_2 ensures that if we add from a copy of $\triangle_{T^\circ,q}$ of rank l some transition $(v_1, T^\circ, q, l) \xrightarrow{a_1} (\imath_{\triangle,T^\circ,q'}, T^\circ, q', l')$ and some transition $(v_2, T^\circ, q, l) \xrightarrow{a_2} (\imath_{\triangle,T^\circ,q'}, T^\circ, q', l')$ to the same copy of $\triangle_{T^\circ,q'}$ then $v_1 = v_2$ and $a_1 = a_2$. Recall that the number of added transitions from a fixed copy of $\triangle_{T^\circ,q}$ to copies of $\triangle_{T^\circ,q'}$ is $|\mathrm{MISSING}(T^\circ, q, q')|$. Altogether it is sufficient to take

$$m = \max_{q,q' \in Q} |\mathrm{MISSING}(T^\circ, q, q')| + 1 \tag{1}$$

From the definition of missing transitions (Def. 4.4) it follows that the data-structure defines a morphism from the box $\Box_{T^\circ,q^\circ}$ to $\mathcal{A}_{T^\circ,q^\circ}$. Furthermore Definition 4.4 and Lemma 4.3 yield easily the following useful property.

LEMMA 4.6. *Within a box $\Box_{T^\circ,q^\circ}$ associated with a connected set of actions T°, if a non-empty word $u \in \Sigma^\star$ leads from the initial state of a triangle to the initial state of a triangle then the alphabet of u is precisely T°.*

4.3   Some Notations and a Useful Observation

First, for each path $s = q \xrightarrow{u} q'$ in some automaton $\mathcal{G}$ over Σ and for each action $a \in \Sigma$ we denote by $s|a$ the sequence of transitions labeled by a that appear along s.

Let T be a non-empty subset of Σ. Let v be a state from the triangle $\mathcal{T}_{T,q}$. By construction of $\mathcal{T}_{T,q}$, v is a quadruple (w, T', q', k') such that w is a state from the box $\Box_{T',q'}$ and $k' \in \mathbb{N}$. Then we say that the *box location* of v is $l^\Box(v) = (T', q', k')$. We define the *sequence of boxes* $\mathbb{L}^\Box(s)$ visited along a path $s = v \xrightarrow{u} v'$ in $\mathcal{T}_{T,q}$ as follows:

- If the length of s is 0 then s corresponds to a state v of $\mathcal{T}_{T,q}$ and $\mathbb{L}^{\square}(s) = l^{\square}(v)$.
- If s is a product $s = s' \cdot t$ where t is the transition $v'' \xrightarrow{a} v'$ then $\mathbb{L}^{\square}(s) = \mathbb{L}^{\square}(s')$ if $l^{\square}(v'') = l^{\square}(v')$ and $\mathbb{L}^{\square}(s) = \mathbb{L}^{\square}(s').l^{\square}(v')$ otherwise.

Similarly we define the sequence of boxes $\mathbb{L}^{\square}(s)$ visited along a path s in a box $\mathcal{B}_{T,q}$ where T is an *unconnected* set of actions and the sequence of triangles $\mathbb{L}^{\triangle}(s)$ visited along a path s in a box $\mathcal{B}_{T,q}$ where T is a non-empty *connected* set of actions.

By means of Lemma 4.6 the next fact is easy to show.

LEMMA 4.7. *Let T be a non-empty connected set of actions. Let $a \in T$ be some action. Let $s_1 = v \xrightarrow{u_1} v'$ and $s_2 = v \xrightarrow{u_2} v'$ be two paths from v to v' in a box $\mathcal{B}_{T,q}$. If $s_1|a = s_2|a$ then $\mathbb{L}^{\triangle}(s_1) = \mathbb{L}^{\triangle}(s_2)$.*

4.4 Complexity of This Unfolding Construction

For all naturals $n \geqslant 0$ we denote by $\mathfrak{B}_n$ the maximal number of states in a box $\mathcal{B}_{T,q}$ with $|T| = n$ and $q \in Q$. Similarly for all naturals $n \geqslant 1$ we denote by $\mathfrak{T}_n$ the maximal number of states in a triangle $\mathcal{T}_{T,q}$ with $|T| = n$ and $q \in Q$. Noteworthy $\mathfrak{B}_0 = 1$ and $\mathfrak{T}_1 = 1$. Moreover $\mathfrak{T}_n$ is non-decreasing because the triangle $\triangle_{T',q}$ is a subautomaton of the triangle $\triangle_{T,q}$ as soon as $T' \subseteq T$. In the following we assume $2 \leqslant n \leqslant |\Sigma|$. Consider some subset $T \subseteq \Sigma$ with $|T| = n$. Each triangle $\mathcal{T}_{T,q}$ is built inductively upon boxes of height $h \leqslant n - 1$. We distinguish two kinds of boxes. First boxes of height $h < n - 1$ are inserted. Each of these boxes appears also in some triangle $\mathcal{T}_{T',q}$ with $T' \subset T$ and $|T'| = n - 1$. Each of these triangles is a subautomaton of $\mathcal{T}_{T,q}$ with at most $\mathfrak{T}_{n-1}$ states. Moreover there are only n such triangles which give rise to at most $n.\mathfrak{T}_{n-1}$ states built along this first step. Second, boxes of height $n - 1$ are inserted and connected to states inserted at height $n - 2$. Each of these states belongs to some box $\square_{T',q'}$ with $|T'| = n - 2$; it gives rise to at most $2.|Q|$ boxes at height $n - 1$ because $|T \setminus T'| = 2$: This produces at most $2.|Q|.\mathfrak{B}_{n-1}$ new states. Altogether we get

$$\mathfrak{T}_n \leqslant n.\mathfrak{T}_{n-1}.(1 + 2.|Q|.\mathfrak{B}_{n-1}) \leqslant 3.|\Sigma|.|Q|.\mathfrak{T}_{n-1}.\mathfrak{B}_{n-1} \tag{2}$$

Assume now $1 \leqslant n \leqslant |\Sigma|$ and consider a connected subset $T \subseteq \Sigma$ with $|T| = n$. Then each box $\mathcal{B}_{T,q}$ is built upon all triangles $\mathcal{T}_{T,q'}$ of height n. It follows from (1) that $m \leqslant \mathfrak{T}_n + 1 \leqslant 2.\mathfrak{T}_n$. Therefore the box $\mathcal{B}_{T,q}$ contains at most $2.\mathfrak{T}_n$ copies of each triangle $\mathcal{T}_{T,q'}$. It follows that we have $(*) \ |\mathcal{B}_{T,q}| \leqslant 2.|Q|.\mathfrak{T}_n^2$. Consider now a *non-connected* subset $T \subseteq \Sigma$ with $|T| = n$. Then $\mathcal{B}_{T,q}$ consists of at most $1 + (n-1).Q.\mathfrak{B}_{n-1}$ boxes of height at most $n - 1$. Therefore we have also $(**) \ |\mathcal{B}_{T,q}| \leqslant |\Sigma|.|Q|.\mathfrak{B}_{n-1}^2$. From (2), $(*)$, and $(**)$ we get the next result by an immediate induction.

LEMMA 4.8. *If $1 \leqslant n \leqslant |\Sigma|$ then $\mathfrak{T}_n \leqslant (3.|\Sigma|.Q)^{2^n-1}$ and $\mathfrak{B}_n \leqslant (3.|\Sigma|.Q)^{2^n-1}$.*

As a consequence the unfolding automaton $\mathcal{A}_{\mathrm{Unf}}$ has at most $(3.|\Sigma|.Q)^{2^{|\Sigma|}-1}$ states.

5 Properties of the Unfolding Construction

In this section we fix a regular trace language L over the independence alphabet $(\Sigma, \|)$. We assume that the possibly non-deterministic automaton $\mathcal{A}$ fulfills Property ID of Def. 2.1 and satisfies $L(\mathcal{A}) = L$.

5.1 Arched Executions for Boxes and Triangles

Let $\mathcal{G}$ be some automaton over Σ and $\widehat{\mathcal{G}}$ be its projected asynchronous automaton. For each global state q of $\widehat{\mathcal{G}}$ we denote by $q{\downarrow}k$ the local state of process k in q. Let q_1 and q_2 be two global states of $\widehat{\mathcal{G}}$. A *true step* $q_1 \xrightarrow{a} q_2$ of action a from q_1 to q_2 in $\widehat{\mathcal{G}}$ consists of a transition $q \xrightarrow{a} r$ in $\mathcal{G}$ such that $q_1{\downarrow}k = q_2{\downarrow}k$ for all $k \notin \mathrm{Loc}(a)$, $q_1{\downarrow}k = q$ and $q_2{\downarrow}k = r$ for all $k \in \mathrm{Loc}(a)$. If $q_1{\downarrow}j = q_2{\downarrow}j$ for all processes $j \neq k$, $q_1{\downarrow}k \xrightarrow{a} q_2{\downarrow}k$, and $k \notin \mathrm{Loc}(a)$ then $q_1 \xrightarrow{\varepsilon} q_2$ is called a ε-*step* of process k from q_1 to q_2.

DEFINITION 5.1. *An execution of $u \in \Sigma^\star$ from q to q' in $\widehat{\mathcal{G}}$ is a sequence of n true or ε-steps $q_{i-1} \xrightarrow{x_i} q_i$ such that $q_0 = q$, $q_n = q'$, and $u = x_1...x_n$ with $x_i \in \Sigma \cup \{\varepsilon\}$.*

For each execution s of $u \in \Sigma^\star$ and each process $k \in K$ we denote by $s{\downarrow}k$ the path of $\mathcal{G}$ followed by process k along s. For each state q of $\mathcal{G}$ we denote by $\widehat{q}$ the global state of $\widehat{\mathcal{G}}$ such that each process is at state q. A global state is *coherent* if it is equal to some $\widehat{q}$. Notice that the initial state and all final states of $\widehat{\mathcal{G}}$ are coherent. An execution from q_1 to q_2 in $\widehat{\mathcal{G}}$ is called *arched* if both q_1 and q_2 are coherent. The next observation shows how arched executions are related to the language of $\widehat{\mathcal{G}}$.

PROPOSITION 5.2. *For all words $u \in \Sigma^\star$ we have $u \in L(\widehat{\mathcal{G}})$ if and only if there exists an arched execution of u from the initial state of $\widehat{\mathcal{G}}$ to some of its final states.*

The following result expresses a main property of boxes: It asserts that active processes visit the same sequence of triangles along an arched execution within a box.

PROPOSITION 5.3. *Let $\mathcal{B}_{T,q}$ be a box with T a connected set of actions and s be an arched execution in $\widehat{\mathcal{B}_{T,q}}$. Then $\mathbb{L}^\triangle(s{\downarrow}k) = \mathbb{L}^\triangle(s{\downarrow}k')$ for all $k, k' \in \mathrm{Loc}(T)$.*

Proof. Since T is connected it is sufficient to show that for all actions $a \in T$ and for all processes $k, k' \in \mathrm{Loc}(a)$ we have $\mathbb{L}^\triangle(s{\downarrow}k) = \mathbb{L}^\triangle(s{\downarrow}k')$. So we fix an action $a \in T$ and two processes $k, k' \in \mathrm{Loc}(a)$. Since k and k' synchronize on the same transition at each occurrence of a, we have $(s{\downarrow}k)|a = (s{\downarrow}k')|a$. It follows by Lemma 4.7 that $\mathbb{L}^\triangle(s{\downarrow}k) = \mathbb{L}^\triangle(s{\downarrow}k')$. ∎

Since processes in $K \backslash \mathrm{Loc}(T)$ are involved in ε-steps only, we can change their behavior within an execution of u without affecting the resulting word u and we get:

PROPOSITION 5.4. *Let $\mathcal{B}_{T,q}$ be a box with T some connected set of actions and s be an arched execution of u in $\widehat{\mathcal{B}_{T,q}}$ from $\widehat{w}$ to $\widehat{w}'$. Then there exists some arched execution s° of u in $\widehat{\mathcal{B}_{T,q}}$ from $\widehat{w}$ to $\widehat{w}'$ such that $\mathbb{L}^\triangle(s^\circ{\downarrow}k) = \mathbb{L}^\triangle(s^\circ{\downarrow}j)$ for all $j, k \in K$.*

Noteworthy it is easy to adapt these remarks to unconnected boxes and triangles due to their tree-like structure.

5.2 A Technical Lemma and a Key Property

We come now to the main technical lemma of this paper. It completes Proposition 5.4 and asserts that we can split any arched execution s associated with a box $\mathcal{B}_{T,q}$ with a connected set of actions T into an equivalent series of arched executions $s_0 \cdot t_1 \cdot s_1 \cdot$

$t_2 \cdot \ldots \cdot t_n \cdot s_n$ where each s_i is an arched execution within a component triangle and each t_i corresponds to the unique added transition from a triangle to another triangle. This decomposition will allow us to reason about arched executions inductively on the construction of the unfolding.

LEMMA 5.5. *Let $\mathcal{B}_{T,q}$ be a box with T some connected set of actions and $l_1, \ldots, l_n$ be a sequence of triangle locations within $\mathcal{B}_{T,q}$ with $n \geqslant 2$. Let $s : \widehat{w} \xrightarrow{u} \widehat{w}'$ be an arched execution of u in $\widehat{\mathcal{B}_{T,q}}$ from $\widehat{w}$ to $\widehat{w}'$ such that $\mathbb{L}^{\triangle}(s° \downarrow k) = l_1 \ldots l_n$ for all $k \in K$. Then there exist a transition $w_2 \xrightarrow{a} w_3$ in $\mathcal{B}_{T,q}$ and three arched executions $s_1 : \widehat{w} \xrightarrow{u_1} \widehat{w}_2$, $s_2 : \widehat{w}_2 \xrightarrow{u_2} \widehat{w}_3$, and $s_3 : \widehat{w}_3 \xrightarrow{u_3} \widehat{w}'$ such that*

- $\mathbb{L}^{\triangle}(s_1 \downarrow k) = l_1$ *for all $k \in K$;*
- $s_2 \downarrow k = w_2 \xrightarrow{a} w_3$ *for all $k \in K$;*
- $\mathbb{L}^{\triangle}(s_3 \downarrow k) = l_2 \ldots l_n$ *for all $k \in K$;*
- $(s_1 \cdot s_2 \cdot s_3 \downarrow k) = (s \downarrow k)$ *for all $k \in K$;*
- $s_1 \cdot s_2 \cdot s_3$ *is an execution of $u_1.u_2.u_3$ in $\widehat{\mathcal{B}_{T,q}}$ and $u_1.u_2.u_3 \sim u$.*

Intuitively we require that all processes leave together the first triangle along the unique transition $w_2 \xrightarrow{a} w_3$ that leads from l_1 to l_2. This result relies on Lemma 4.6 and the two properties C_1 and C_2 of Rule 4.5.

Observe now that the tree-structure of triangles and boxes associated to unconnected sets of actions ensures that we can state a similar result for all triangles and all boxes.

LEMMA 5.6. *Let $T \subseteq \Sigma$ be a non-empty subset of actions. If s is an arched execution of u from $\widehat{w}_1$ to $\widehat{w}_2$ in $\widehat{\mathcal{B}_{T,q}}$ (resp. $\widehat{\mathcal{T}_{T,q}}$) then there exists some path $w_1 \xrightarrow{u'} w_2$ in $\mathcal{B}_{T,q}$ (resp. $\mathcal{T}_{T,q}$) such that $u' \sim u$.*

Proof. Observe first that this property holds also trivially for the empty boxes $\mathcal{B}_{\emptyset,q}$. We proceed now by induction on the size of T along the construction of triangles and boxes. Let $n = |T|$. Assume that the property holds for all triangles $\mathcal{T}_{T^\dagger,q}$ with $|T^\dagger| \leqslant n$. Assume also that T is connected. By Lemma 5.5 we can split the execution s into an equivalent series of arched executions $s_0 \cdot t_1 \cdot s_1 \cdot t_2 \cdot \ldots \cdot t_n \cdot s_n$ where each s_i is an arched execution within a component triangle and each t_i corresponds to the unique added transition from a triangle to another triangle. By induction hypothesis each s_i corresponds to a path in the corresponding triangle. In that way we get a path for the sequence s. The case where T is not connected is similar due to the tree-structure of these boxes. The case of triangles is also similar.   ∎

Recall now that arched executions are closely related to the langage of the projected asynchronous automaton (Prop. 5.2). As an immediate corollary we get the following key result.

PROPOSITION 5.7. *We have $L(\widehat{\mathcal{A}_{\mathrm{Unf}}}) \subseteq [L(\mathcal{A}_{\mathrm{Unf}})]$.*

5.3 Main Result

Due to the morphisms from boxes and triangles to asynchronous systems $\mathcal{A}_{T,q}$ we have the inclusion relation $[L(\mathcal{B}_{T,q})] \subseteq L(\mathcal{A}_{T,q})$ for each box $\mathcal{B}_{T,q}$ and similarly $[L(\mathcal{T}_{T,q})] \subseteq L(\mathcal{A}_{T,q})$ for each triangle $\mathcal{T}_{T,q}$. We can check by an easy induction that boxes satisfy the converse inclusion relation, which leads us to the next statement.

PROPOSITION 5.8. *We have* $[L(\mathcal{A}_{\mathrm{Unf}})] = L(\mathcal{A})$.

We come to the main statement of this paper.

THEOREM 5.9. *The asynchronous automaton* $\widehat{\mathcal{A}_{\mathrm{Unf}}}$ *satisfies* $L(\widehat{\mathcal{A}_{\mathrm{Unf}}}) = L(\mathcal{A})$. *Moreover the number of states in each process is* $|Q_k| \leqslant (3.|\Sigma|.|Q|)^d$ *where* $d = 2^{|\Sigma|}$.

Proof. By Proposition 5.8 we have $[L(\mathcal{A}_{\mathrm{Unf}})] = L(\mathcal{A})$. By Proposition 3.1 we have also $[L(\mathcal{A}_{\mathrm{Unf}})] \subseteq L(\widehat{\mathcal{A}_{\mathrm{Unf}}})$ hence $L(\mathcal{A}) \subseteq L(\widehat{\mathcal{A}_{\mathrm{Unf}}})$. Now Proposition 5.7 shows that $L(\widehat{\mathcal{A}_{\mathrm{Unf}}}) \subseteq [L(\mathcal{A}_{\mathrm{Unf}})] = L(\mathcal{A})$. The complexity result follows from Lemma 4.8. ∎

References

1. Baudru N. and Morin R.: *Safe Implementability of Regular Message Sequence Charts Specifications*. Proc. of the ACIS 4th Int. Conf. SNDP (2003) 210–217
2. Baudru N. and Morin R.: *The Synthesis Problem of Netcharts*. (2006) – Submitted
3. Bednarczyk M.A.: *Categories of Asynchronous Systems*. PhD thesis in Computer Science (University of Sussex, 1988)
4. Cori R., Métivier Y. and Zielonka W.: *Asynchronous mappings and asynchronous cellular automata*. Inform. and Comput. **106** (1993) 159–202
5. Diekert V. and Rozenberg G.: *The Book of Traces*. (World Scientific, 1995)
6. Genest B., Muscholl A. and Kuske D.: *A Kleene Theorem for a Class of Communicating Automata with Effective Algorithms*. DLT, LNCS **3340** (2004) 30–48
7. Genest B. and Muscholl A.: *Constructing Exponential-size Deterministic Zielonka Automata*. Technical report (2006) – 12 pages
8. Klarlund N., Mukund M. and Sohoni M.: *Determinizing Asynchronous Automata*. ICALP, LNCS **820** (1994) 130–141
9. Métivier Y.: *An algorithm for computing asynchronous automata in the case of acyclic non-commutation graph*. ICALP, LNCS **267** (1987) 226–236
10. Morin R.: *Concurrent Automata vs. Asynchronous Systems*. LNCS **3618** (2005) 686–698
11. Mukund M., Narayan Kumar K. and Sohoni M.: *Synthesizing distributed finite-state systems from MSCs*. CONCUR, LNCS **1877** (2000) 521–535
12. Mukund M. and Sohoni M.: *Gossiping, Asynchronous Automata and Zielonka's Theorem*. Report TCS-94-2, SPIC Science Foundation (Madras, India, 1994)
13. Muscholl A.: *On the complementation of Büchi asynchronous cellular automata*. ICALP, LNCS **820** (1994) 142–153
14. Pighizzini G.: *Synthesis of Nondeterministic Asynchronous Automata*. Algebra, Logic and Applications, vol. **5** (1993) 109–126
15. Ştefănescu A., Esparza J. and Muscholl A.: *Synthesis of distributed algorithms using asynchronous automata*. CONCUR, LNCS **2761** (2003) 20–34
16. Thiagarajan P.S.: *Regular Event Structures and Finite Petri Nets: A Conjecture*. Formal and Natural Computing, LNCS **2300** (2002) 244–256
17. Zielonka W.: *Notes on finite asynchronous automata*. RAIRO, Theoretical Informatics and Applications **21** (Gauthiers-Villars, 1987) 99–135

Conjugacy and Equivalence of Weighted Automata and Functional Transducers

Marie-Pierre Béal[1], Sylvain Lombardy[2], and Jacques Sakarovitch[3]

[1] Institut Gaspard-Monge, Université Marne-la-Vallée
[2] LIAFA, Université Paris 7
[3] LTCI, CNRS / Ecole Nationale Supérieure des Télécommunications (UMR 5141)
`beal@univ-mlv.fr`, `lombardy@liafa.jussieu.fr`, `sakarovitch@enst.fr`

Abstract. We show that two equivalent $\mathbb{K}$-automata are conjugate to a third one, when $\mathbb{K}$ is equal to $\mathbb{B}$, $\mathbb{N}$, $\mathbb{Z}$, or any (skew) field and that the same holds true for functional tranducers as well.

EXTENDED ABSTRACT

1 Presentation of the Results

In a recent paper ([1]), we have studied the equivalence of $\mathbb{Z}$-automata. This equivalence is known to be decidable (with polynomial complexity) for more than forty years but we showed there two results that give more *structural information* on two equivalent $\mathbb{Z}$-automata. We first proved that two equivalent $\mathbb{Z}$-automata are related by a series of *three* conjugacies — we shall define conjugacy later in the paper — and then that every conjugacy relation can be decomposed into a sequence of three operations: state (out-)splitting (also known as *covering*), circulation of coefficients, and state (in-)merging (also known as *co-covering*). Altogether, we reached a decomposition of any equivalence between $\mathbb{Z}$-automata as the one described at Figure 1 [Conjugacy is represented by double-line arrows, coverings by simple solid arrows, co-coverings by simple dashed arrows, and circulation by simple dotted arrows].

At the end of our ICALP paper we mentioned two problems open by the gap between these results and those that were formerly known. First, whether *three* conjugacies are necessary (in general), and, if yes, whether it is decidable when *two* conjugacies suffice. Second, whether, in the case of $\mathbb{N}$-automata, the whole chain of conjugacies could be always realized with transfer matrices in $\mathbb{N}$ and, if not, whether it is decidable when this property holds.

We answer these two questions here. By means of techniques different from the ones that where developed in [1], we show that *two conjugacies* always suffice and that this property holds not only for $\mathbb{Z}$-automata but also for $\mathbb{N}$-automata and other families of automata as stated by the following.

Theorem 1. *Let $\mathbb{K}$ be $\mathbb{B}$, $\mathbb{N}$, $\mathbb{Z}$, or any (skew) field. Two $\mathbb{K}$-automata are equivalent if and only if there exists a third $\mathbb{K}$-automaton that is conjugate to both of them.*

D. Grigoriev, J. Harrison, and E.A. Hirsch (Eds.): CSR 2006, LNCS 3967, pp. 58–69, 2006.
© Springer-Verlag Berlin Heidelberg 2006

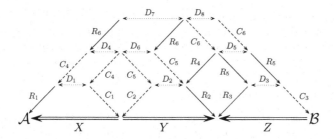

Fig. 1. Structural decomposition of the equivalence of two $\mathbb{Z}$-automata

Moreover, an analoguous result holds for functional transducers as well.

Theorem 2. *Two functional transducers are equivalent if and only if there exists a third functional transducer that is conjugate to both of them*

Together with these results on conjugacy, we extend the decomposition of conjugacy by means of covering, co-covering and "circulation" as follow (we shall define covering and co-covering more precisely at Section 3). We state the first one for sake of completeness.

Theorem 3 ([1]). *Let $\mathbb{K}$ be a field $\mathbb{F}$ or the ring $\mathbb{Z}$ and let $\mathcal{A}$ and $\mathcal{B}$ be two $\mathbb{K}$-automata. We have $\mathcal{A} \xrightarrow{X} \mathcal{B}$ if and only if there exists two $\mathbb{K}$-automata $\mathcal{C}$ and $\mathcal{D}$ and a circulation matrix D such that $\mathcal{C}$ is a co-$\mathbb{K}$-covering of $\mathcal{A}$, $\mathcal{D}$ a $\mathbb{K}$-covering of $\mathcal{B}$ and $\mathcal{C} \xrightarrow{D} \mathcal{D}$.*

Theorem 4. *Let $\mathbb{K}$ be the semiring $\mathbb{N}$ or the Boolean semiring $\mathbb{B}$ and let $\mathcal{A}$ and $\mathcal{B}$ be two trim $\mathbb{K}$-automata. We have $\mathcal{A} \xrightarrow{X} \mathcal{B}$ if and only if there exists a $\mathbb{K}$-automaton $\mathcal{C}$ that is a co-$\mathbb{K}$-covering of $\mathcal{A}$ and a $\mathbb{K}$-covering of $\mathcal{B}$.*

Theorem 5. *Let $\mathcal{A}$ and $\mathcal{B}$ be two trim functional transducers. We have $\mathcal{A} \xrightarrow{X} \mathcal{B}$ if and only if there exists two (functional) transducers $\mathcal{C}$ and $\mathcal{D}$ and a diagonal matrix of words D such that $\mathcal{C}$ is a co-covering of $\mathcal{A}$, $\mathcal{D}$ a covering of $\mathcal{B}$ and $\mathcal{C} \xrightarrow{D} \mathcal{D}$.*

In other words, Figure 1 can be replaced by Figure 2 where $\mathcal{A}$ and $\mathcal{B}$ are taken in any family considered in Theorems 1 and 2.

The present result on conjugacy is both *stronger* and *broader* than the preceeding ones. Stronger as the number of conjugacies is reduced from *three* to *two*, broader as the result apply not only to $\mathbb{Z}$-automata (indeed to automata with

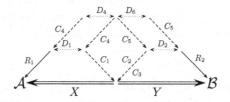

Fig. 2. Structural decomposition of the equivalence of two $\mathbb{K}$-automata

multiplicity in an Euclidean domain) but to a much larger family of automata. It answers in particular to what was a long standing problem for the authors: is it possible to transform an N-automaton into any other equivalent one using only state splitting and state merging? The answer is thus positive, and the chain of operations is rather short. The benefit brought by the change from $\mathbb{Z}$ into $\mathbb{N}$ is well illustrated by the following consequence.

Theorem 6. *If two regular languages have the same generating function (i.e. the numbers of words of every length is the same in both languages) then there exists a letter-to-letter rational function that realizes a bijection between the two languages.*

2   Conjugacy and Covering of Automata

A *finite* automaton $\mathcal{A}$ over an alphabet A with multiplicity in a semiring $\mathbb{K}$, or $\mathbb{K}$-automaton for short, can be written in a compact way as $\mathcal{A} = \langle I, E, T \rangle$ where E is a square matrix of finite dimension Q whose entries are linear combinations (with coefficients in $\mathbb{K}$) of letters in A and where I and T are two vectors — respectively row vector and column vector — with entries in $\mathbb{K}$ as well. We can view each entry $E_{p,q}$ as the label of a unique arc which goes from state p to state q in the graph whose set of vertices is Q (if $E_{p,q} = 0_{\mathbb{K}}$, we consider that there is *no* arc from p and q).

The *behaviour* of $\mathcal{A}$, denoted $|\mathcal{A}|$, is the series such that the coefficient of a word w is the coefficient of w in $I\,E^{|w|}\,T$. It is part of Kleene-Schützenberger Theorem that every $\mathbb{K}$-rational series is the behaviour of a $\mathbb{K}$-automaton of the form we have just defined. For missing definitions, we refer to [4, 2, 10].

2.1   Conjugacy

Definition 1. *A $\mathbb{K}$-automaton $\mathcal{A} = \langle I, E, T \rangle$ is conjugate to a $\mathbb{K}$-automaton $\mathcal{B} = \langle J, F, U \rangle$ if there exists a matrix X with entries in $\mathbb{K}$ such that*
$$I X = J, \qquad E X = X F, \quad and \quad T = X U.$$
The matrix X is the transfer matrix *of the conjugacy and we write $\mathcal{A} \xrightarrow{X} \mathcal{B}$.*

Remark that in spite of the idea conveyed by the terminology, the conjugacy relation *is not an equivalence* but a *preorder* relation. Suppose that $\mathcal{A} \xrightarrow{X} \mathcal{C}$ holds; if $\mathcal{C} \xrightarrow{Y} \mathcal{B}$ then $\mathcal{A} \xrightarrow{XY} \mathcal{B}$, but if $\mathcal{B} \xrightarrow{Y} \mathcal{C}$ then $\mathcal{A}$ is not necessarily conjugate to $\mathcal{B}$, and we write $\mathcal{A} \xrightarrow{X} \mathcal{C} \xleftarrow{Y} \mathcal{B}$ or even $\mathcal{A} \xrightarrow{X}\xleftarrow{Y} \mathcal{B}$.

This being well understood, we shall speak of "conjugate automata" when the orientation does not matter. For instance, we state that, obviously, two conjugate automata are equivalent (*i.e.* have the same behaviour).

2.2   Covering

The standard notion of morphisms of automata — which consists in merging states and does not tell enough on transitions — is not well-suited to $\mathbb{K}$-automata. Hence the definitions of $\mathbb{K}$-coverings and co-$\mathbb{K}$-coverings. These have probably stated independently a number of times. We describe them here in terms of conjugacy. A definition closer to the classical morphisms could be given and then the definitions below become propositions (*cf.* [1, 10]).

Let $\varphi\colon Q \to R$ be a surjective map and H_φ the $Q \times R$-matrix where the (q, r) entry is 1 if $\varphi(q) = r$, 0 otherwise. Since φ is a map, each row of H_φ contains exactly one 1 and since φ is surjective, each column of H_φ contains at least one 1. Such a matrix is called an *amalgamation matrix* ([6, Def. 8.2.4]).

Let $\mathcal{A}$ and $\mathcal{B}$ be two $\mathbb{K}$-automata of dimension Q and R respectively. We say that $\mathcal{B}$ is *a* $\mathbb{K}$-*quotient* of $\mathcal{A}$ and conversely that $\mathcal{A}$ is *a* $\mathbb{K}$-*covering* of $\mathcal{B}$ if there exists a surjective map $\varphi\colon Q \to R$ such that $\mathcal{A}$ is conjugate to $\mathcal{B}$ by H_φ

The notion of $\mathbb{K}$-quotient is *lateralized* since the conjugacy relation is not symmetric. Somehow, it is the price we pay for extending the notion of morphism to $\mathbb{K}$-automata. Therefore the *dual* notions co-$\mathbb{K}$-*quotient* and co-$\mathbb{K}$-*covering* are defined in a natural way. We say that $\mathcal{B}$ is *a* co-$\mathbb{K}$-*quotient* of $\mathcal{A}$ and conversely that $\mathcal{A}$ is *a* co-$\mathbb{K}$-*covering* of $\mathcal{B}$ if there exists a surjective map $\varphi\colon Q \to R$ such that $\mathcal{B}$ is conjugate to $\mathcal{A}$ by ${}^t H_\varphi$.

We also write $\varphi\colon \mathcal{A} \to \mathcal{B}$ and call φ, by way of metonymy, *a* $\mathbb{K}$-*covering, or a* co-$\mathbb{K}$-*covering from* $\mathcal{A}$ *onto* $\mathcal{B}$.

3   The Joint Reduction

The proof of Theorems 1 and 2 relies on the idea of *joint reduction* which is defined by means of the notion of *representation*.

An automaton $\mathcal{A} = \langle I, E, T \rangle$ of dimension Q can equivalently be described as the representation $\mathcal{A} = (I, \mu, T)$ where $\mu\colon A^* \to \mathbb{K}^{Q \times Q}$ is the morphism defined by the equality

$$E = \sum_{a \in A} \mu(a)\, a \ .$$

This equality makes sense since the entries of E are assumed to be linear combinations of letters of A with coefficients in $\mathbb{K}$. And the coefficient of any word w in the series $|\mathcal{A}|$ is $I\,\mu(w)\,T$.

The set of vectors $\{I\,\mu(w) \mid w \in A^*\}$ (row vectors of dimension Q), that is, the *phase space* of $\mathcal{A}$, plays a key role in the study of $\mathcal{A}$, as exemplifyed by the following two contrasting cases.

If $\mathcal{A}$ is a Boolean automaton, this set of vectors (each vector represents a subset of the dimension Q) is finite and makes up the states of the *determinized automaton* $\mathcal{D}$ of $\mathcal{A}$ (by the subset construction). Moreover, if we form the matrix X whose rows are the states of $\mathcal{D}$, then $\mathcal{D}$ is conjugate to $\mathcal{A}$ by X.

If $\mathcal{A}$ is a $\mathbb{K}$-automaton with $\mathbb{K}$ a field, the *left reduction* of $\mathcal{A}$ — recalled with more detail below — consists in choosing a prefix-closed set P of words such that the vectors $\{I\,\mu(p) \mid p \in P\}$ is a basis of the vector space generated by $\{I\,\mu(w) \mid w \in A^*\}$ (*cf.* [2]). Moreover the (left-)reduced automaton is conjugate to $\mathcal{A}$ by the matrix X whose rows are the vectors $\{I\,\mu(p) \mid p \in P\}$.

Let now $\mathcal{A} = (I, \mu, T)$ and $\mathcal{B} = (I', \kappa, T')$ be two $\mathbb{K}$-automata of dimension Q and R respectively. We consider the *union* of $\mathcal{A}$ and $\mathcal{B}$ and thus the vectors $[I\,\mu(w)|I'\,\kappa(w)]$ of dimension $Q \cup R$. These vectors, for w in A^*, generate a $\mathbb{K}$-module W. The *(left) joint reduction* of $\mathcal{A}$ and $\mathcal{B}$ consists in computing — when it is possible — a finite set G of vectors $[x|y]$ which generate the same $\mathbb{K}$-module W. Then the matrix M whose *rows* are these vectors $[x|y]$ provides in some sense a $\mathbb{K}$-automaton $\mathcal{C}$ which is conjugate to both $\mathcal{A}$ and $\mathcal{B}$ with the transfer matrices X and Y respectively, where X and Y are the 'left' and 'right' parts of the matrix M respectively.

In every case listed in the above Theorems 1 and 2, and which we consider now, the finite set G is effectively computable.

3.1   Joint Reduction in Fields

Let $\mathbb{K}$ be a field and let $\mathcal{A} = (I, \mu, T)$ be a $\mathbb{K}$-automaton of dimension n.

The reduction algorithm for $\mathbb{K}$-automata is split into two dual parts. The first part consists in computing a prefix-closed subset P of A^* such that the set $G = \{I\,\mu(w) \mid w \in P\}$ is free and, for every letter a, and every word in P, $I\,\mu(wa)$ is lineary dependant from G. The set G has at most n elements and an automaton $\mathcal{C} = (J, \kappa, U)$, whose states are the elements of G, is defined by:

$$J_x = \begin{cases} 1 & \text{if } x = I \ , \\ 0 & \text{otherwise} \ , \end{cases} \qquad \forall x \in G, \quad U_x = xT \ ,$$

$$\forall a, \quad \exists!\kappa(a), \qquad \forall x \in G, \ x\mu(a) = \sum_{y \in G} \kappa(a)_{x,y} y \ .$$

This can be viewed as a change of basis: the set G generates the smallest subspace of $\mathbb{K}^n$ that contains every $I\,\mu(w)$ and if G is completed into a basis B, after changing the canonical basis by B and projection, one gets the automaton $\mathcal{C}$. Finally, if M is the matrix whose *rows* are the elements of G, it holds $\mathcal{C} \xRightarrow{M} \mathcal{A}$.

The second part is similar and consists in computing a basis of the subspace of $\mathbb{K}^{|G|}$ generated by the vectors $\kappa(w)\,U$. It is a nice result (by Schützenberger) that after these two semi-reductions, the outcome is a $\mathbb{K}$-automaton of smallest dimension that is equivalent to $\mathcal{A}$.

We focus here on the first part which we call *left reduction*. Let $\mathcal{A} = (I, \mu, T)$ and $\mathcal{B} = (I', \mu', T')$ be two equivalent $\mathbb{K}$-automata and let $\mathcal{C}_0 = (J, \kappa, U)$ be the automaton obtained by left reduction of $\mathcal{A} + \mathcal{B}$. The automaton $\mathcal{A} + \mathcal{B}$ has a representation equal to $([I|I'], \mathrm{diag}(\mu, \mu'), [T|T'])$, where $[I|I']$ is obtained by horizontally joining the row vectors I and I', $[T|T']$ by vertically stacking the column vectors T and T', and for every letter a, $[\mathrm{diag}(\mu|\mu')](a)$ is the matrix whose diagonal blocks are $\mu(a)$ and $\mu'(a)$.

The automaton $\mathcal{C}_0$ is conjugate to $\mathcal{A} + \mathcal{B}$ by the matrix $[X|Y]$, in which every row has the form $[I\,\mu(w)|I'\,\mu'(w)]$ where w is a word. It holds:

$$J\,[X|Y] = [I|I'], \quad \forall a, \ \kappa(a)\,[X|Y] = [X|Y]\,(\mathrm{diag}(\mu|\mu'))(a), \quad U = [X|Y]\,[T|T']$$

As $\mathcal{A}$ and $\mathcal{B}$ are equivalent, $XT = YT'$ and thus $U = 2XT = 2YT'$. Let $\mathcal{C} = \langle J, \kappa, U/2 \rangle$; it immediatly comes $\mathcal{C} \xRightarrow{X} \mathcal{A}$ and $\mathcal{C} \xRightarrow{Y} \mathcal{B}$.

3.2   Joint Reduction in $\mathbb{Z}$

The result and the algorithm are basically the same as the previous ones if the multiplicity semiring is $\mathbb{Z}$. As in vector spaces, there is a dimension theory in the free $\mathbb{Z}$-modules and it is still possible to compute a basis G of the submodule of $\mathbb{Z}^n$ generated by the vectors $I\mu(w)$. However, this basis does not correspond any more to a prefix-closed set of words. and the algorithm to compute it is explained in [1].

3.3   Joint Reduction in $\mathbb{N}$

There is no dimension theory in the $\mathbb{N}$-modules and thus no reduction algorithm for $\mathbb{N}$-automata similar to the previous ones.

However, given $\mathcal{A}+\mathcal{B}$ our aim is not the reduction itself but the computation of a set G of vectors with the 3 properties: for every $z = [x|y]$ in G, $xT = yT'$ holds, the $\mathbb{N}$-module $\langle G \rangle$ generated by G is closed under multiplication by $(\mathrm{diag}(\mu|\mu'))(a)$, for every letter a (which is important to effectively build the automaton $\mathcal{C}$), and finally G is finite. It can be noted that in the preceeding algorithms, the freeness of the generating set G is used only to garantee finiteness. An algorithm that compute such a G for $\mathbb{N}$-automata can be roughly sketched as follows.

Start from $G = \{[I|I']\}$. While $\langle G \rangle$ is not closed under $(\mathrm{diag}(\mu|\mu'))(a)$, take $z = [x|y]$ in $G(\mathrm{diag}(\mu|\mu'))(a) \setminus \langle G \rangle$ add z to G, and *reduce* G. The reduction goes as follow: while G contains z and z' such that $z < z'$ (in the product order of $\mathbb{N}^{Q \cup R}$) replace z' by $z' - z$. This algorithm ends since at every step, either the size of vectors of G decreases or the size of G increases. The size of vectors cannot decrease infinitely and as vectors of G are pairwise incomparable (after the reduction step), G has only a finite number of elements.

The outcome of this algorithm is not canonically associated to $\mathcal{A}$ and $\mathcal{B}$ and even its size (in contrast to what happens with fields) may depend on the order in which comparable pairs are considered during the reduction step. Yet, an automaton $\mathcal{C}$ whose states are the elements of G is built as the previous cases.

3.4   Joint Reduction in $\mathbb{B}$

In $\mathbb{B}$, as in many semirings that cannot be embedded in rings, there is no subtraction. Therefore it is quite difficult to *reduce* vectors $[I\mu(w)|I'\mu'(w)]$ to find a "minimal" set of generators. As $\mathbb{B}$ is finite, the simplest way is to keep all the vectors $[I\mu(w)|I'\mu'(w)]$. The automaton $\mathcal{C}$ obtained from this set is nothing else that the *determinised* automaton of $\mathcal{A} \cup \mathcal{B}$. For the same reason as above, this automaton is conjugate both to $\mathcal{A}$ and $\mathcal{B}$.

3.5   Joint Reduction of Functional Transducers

With transducers, difficulties of automata with multiplicities and Boolean automata meet. On the one hand, if $\mathcal{T} = (I, \mu, T)$, the set $\{I\mu(w) \mid w \in A^*\}$ may be infinite and, in the other hand, as in the Boolean case, the substraction is not allowed in the semiring of multiplicities that can be associated to them.

If the transducers $\mathcal{A}$ and $\mathcal{B}$ were sequentialisable, it would be sufficient to consider the sequentialised transducer of their union that would be conjugate to each of them. The idea of the sequentialisation (*cf.* [3, 7]) is to compute a (finite) set of vectors of words, each vector being the information that can not be output and that is necessary for further computation.

On general functional transducers, this algorithm does not always end. We present now a *pseudo-sequentialisation*, that stops on any functional transducer. This algorithm allows to split vectors of words when their components are different enough, which induces non deterministic transitions.

We describe first this algorithm on one functional transducer and then explain how to use it for the joint reduction.

Definition 2. *Let k be a positive integer and X be a se of words. Two words u and v of A^* are k-related in X, if there exists a finite sequence $w_0, ..., w_n$ of words such that $u = w_0$, $v = w_n$, for every i in $[1;n]$, $d(w_{i-1}, w_i) \leqslant k$ and there exists i in $[1;n]$ such that w_i is a prefix of u and v. The set X is k-related if every pair of its elements is k-related in X.*

The k-relation is an equivalence on X.

Definition 3. *Let α be a vector of words. The k-decomposition of α is the smallest set of vectors $D_k(\alpha)$ such that, for every $\beta \in D_k(\alpha)$ the set of components of β is k-related and $\alpha = \sum_{\beta \in D_k(\alpha)} \beta$.*

Obviously, the vectors of $D_k(\alpha)$ have disjoint supports. We shall applies this decomposition to vectors of words and then reduce them with respect to their greatest common prefix; this second step is exactly the same as in the classical sequentialisation algorithm.

Definition 4. *Let α be a vector of words. We denote $\overset{\circ}{\alpha}$ the greatest common prefix of non zero components of α and $\alpha^\sharp = \overset{\circ}{\alpha}{}^{-1} \alpha$.*

Definition 5. *Let $\mathcal{T} = (I, \mu, T)$ be a functional transducer and let k be non negative integer The k-pseudo-sequentialised transducer $\mathcal{S}$ of $\mathcal{T}$ is defined by:*
- *for every β in $D_k(I)$, $\beta^\sharp$ is an initial state with initial weight $\overset{\circ}{\beta}$;*
- *for every state α, for every letter a, for every β in $D_k(\alpha\mu(a))$, there is a transition labeled by a with output $\overset{\circ}{\beta}$ from α to $\beta^\sharp$.*
- *for every state α, α is final with output $w = \alpha T$ if w is non zero.*

Proposition 1. *For k, the k-pseudo-sequentialised transducer $\mathcal{S}$ of a functional transducer $\mathcal{T}$ is a finite transducer that is conjugate to $\mathcal{T}$.*

The transducer $\mathcal{S}$ is finite since the components of its states (that are vectors) are bounded by k.

If the k-pseudo-sequentialisation is applied to the union of two equivalent functional transducers $\mathcal{A} = (I, \mu, T)$ and $\mathcal{B} = (I', \mu', T')$, it gives a transducer $\mathcal{C}$ which is conjugate to $\mathcal{A} \cup \mathcal{B}$ with a matrix $M = [X|Y]$, but, in general, this transducer is not conjugate to $\mathcal{A}$ with X and to $\mathcal{B}$ with Y. Actually, if k is too small, there may be rows $[x|y]$ of M such that $xT \neq yT'$.

Let k be equal to $n^2 L$, where n is the maximum of dimensions of $\mathcal{A}$ and $\mathcal{B}$ and L is the longest output of transitions or terminal functions of $\mathcal{A}$ and $\mathcal{B}$. In this case, the k-pseudo-sequentialised transducer is unambiguous, which implies that $xT = yT'$ for every state $[x|y]$ of $\mathcal{C}$. Therefore, the transducer $\mathcal{C}$ is conjugate both to $\mathcal{A}$ and $\mathcal{B}$.

Example 1. Figure 3 shows the transducer $\mathcal{T}_1$ and its k-pseudo-sequentialised $\mathcal{S}_1$ (the result is the same with any positive k), where $\mathcal{T}_1$ is the (left) transducer that replaces systematically factors abb by baa when reading words *from right to left*; $\mathcal{T}_1$ is thus co-sequential, that is, input co-deterministic (*cf.* [10]). The transducer $\mathcal{S}_1$ is conjugate to $\mathcal{T}_1$ with the transfer matrix M:

$$M = \begin{bmatrix} bb & b & 1 \\ b & 1 & 0 \\ 1 & 0 & 0 \\ 0 & b & 1 \\ 0 & 0 & 1 \end{bmatrix}.$$

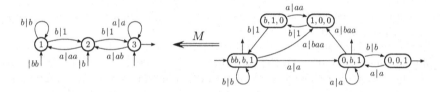

Fig. 3. The transducers $\mathcal{T}_1$ and $\mathcal{S}_1$

The above list may lead to think that a joint reduction procedure may be found for any semiring. This is certainly not the case and the tropical semirings for instance, or the non functional transducers, are not likely to admit a joint reduction procedure.

4   From Conjugacy to Coverings

It remains to show Theorems 3, 4 and 5.

4.1   The Case of Fields and Integers

We have proved Theorem 3 in [1]. Actually, every matrix M can be decomposed in a product HDK, where ${}^t H$ and K are amalgamation matrices and D is a diagonal matrix whose entries are invertible. If $\mathbb{K}$ is a field, the dimension of D is the number of non zero entries of M, and if $\mathbb{K} = \mathbb{Z}$, as the only invertible elements are 1 and -1, every non zero element has to be decomposed in a sum of ± 1 and the dimension of D is the sum of the absolute values of the entries of M.

The proof consists then in proving that there exist automata $\mathcal{C}$ and $\mathcal{D}$ such that $\mathcal{A} \overset{H}{\Longrightarrow} \mathcal{C} \overset{D}{\Longrightarrow} \mathcal{D} \overset{K}{\Longrightarrow} \mathcal{B}$. The construction of $\mathcal{C}$ and $\mathcal{D}$ amounts to fill in blocks of their transition matrix knowing the sum of the rows and the columns.

For natural integers, the proof is exactly the same. The unique invertible element of $\mathbb{N}$ is 1, thus D is the identity matrix. However, to get the expected form, the matrix M must have no zero row or column.[1] This is ensured by the assumption that $\mathcal{A}$ and $\mathcal{B}$ are trim.

4.2   The Boolean Case

Let $\mathcal{A} = (I, \mu, T)$ and $\mathcal{B} = (J, \kappa, U)$ be two trim automata such that there exists a $n \times m$ Boolean matrix X that verifies $\mathcal{A} \overset{X}{\Longrightarrow} \mathcal{B}$.

Let k be the number of non zero entries of matrix X. We define $\varphi \colon [1;k] \to [1;n]$ and $\psi \colon [1;k] \to [1;m]$, such that $x_{\varphi(i),\psi(i)}$ is the i-th non zero entry of X. Let H_φ and H_ψ be the matrices associated to these applications. It holds $X = {}^t H_\varphi \, H_\psi$. We define $\mathcal{C} = (K, \zeta, V)$ with dimension k by:

$$K = I \, {}^t H_\varphi \quad , \quad V = H_\psi U \quad ,$$
$$\forall (p,q) \in [1;k]^2, \qquad \zeta(a)_{p,q} = \mu(a)_{\varphi(p),\varphi(q)} \wedge \kappa(a)_{\psi(p),\psi(q)} \quad .$$

It is then easy to check that $\mathcal{C} \overset{{}^t H_\varphi}{\Longrightarrow} \mathcal{A}$ and $\mathcal{C} \overset{H_\psi}{\Longrightarrow} \mathcal{B}$, which means that $\mathcal{C}$ is a co-$\mathbb{B}$-covering of $\mathcal{A}$ and a $\mathbb{B}$-covering of $\mathcal{B}$.

In the case were $\mathcal{A}$ is the determinised automaton of $\mathcal{B}$ (which arises if one applies the algorithm given in the previous section), the automaton built in this way is the Schützenberger covering of $\mathcal{B}$, a construction that appears naturally in a number of problems for automata with multiplicity (cf. [5, 9, 10]).

4.3   The Functional Transducer Case

Let $\mathcal{A} = (I, \mu, T)$ and $\mathcal{B} = (J, \kappa, U)$ be two trim functional transducers and let X be a $n \times m$ matrix of words such that $\mathcal{A} \overset{X}{\Longrightarrow} \mathcal{B}$. Let k be the number of non zero entries of X. The matrix X can be decomposed into HDK, where H and K are Boolean matrices and D is a diagonal matrix of words of dimension k.

This diagonal matrix corresponds to a circulation of words. Actually, in the framework of transducers, the circulation of words is a well-known operation that is needed for instance in the minimisation of sequential transducers. This operation can be related to the circulation of invertible elements for fields if we consider words as elements of the free group.

We want to prove that there exists $\mathcal{A}' = (I', \mu', T')$ and $\mathcal{B}' = (J', \kappa', U')$ such that $\mathcal{A} \overset{H}{\Longrightarrow} \mathcal{A}' \overset{D}{\Longrightarrow} \mathcal{B}' \overset{K}{\Longrightarrow} \mathcal{B}$. We set $I' = IH$, $J' = I'D$, $U' = KU$ and $T' = DU'$.

For every letter a, there exists a matrix $\zeta(a)$ such that $H\zeta(a) = \mu(a)HD$ and $\zeta(a)K = DK\kappa(a)$. As H and K are Boolean matrices, $\zeta(a)$ can be factorised in $\mu'(a)D$ and $D\kappa'(a)$, which gives the solutions.

---

[1] In the previous case, this technical item is handled by considering that $0 = 1 + (-1)$.

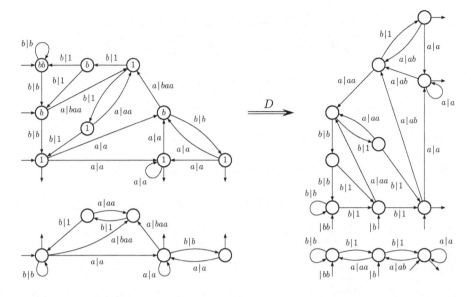

Fig. 4. The instance of Theorem 5 for S_1 and T_1

5   An Application

Theorem 6 is a striking consequence of the strengthening of our conjugacy result of [1] and answers a question on automatic structures.

Let A and B be two (Boolean) *unambiguous* automata the languages L and K respectively and suppose that L and K have the same generating functions. It amounts to say that if we forget the labels in A and B (and replace them all by the same letter x) we have two equivalent $\mathbb{N}$-automata A' and B': the coefficient of x^n in $|A'|$ and thus in $|B'|$ is the number of words of length n in L and thus in K.

By Theorem 1, A' and B' are both conjugate to a same $\mathbb{N}$-automaton C' (on x^*). By Theorem 4 there exist D' and E' such that D' is a co-$\mathbb{N}$-covering of C' and a $\mathbb{N}$-covering of A' and E' is a co-$\mathbb{N}$-covering of C' and a $\mathbb{N}$-covering of B'. By a diamond lemma ([1, Proposition 6]) there exists a $\mathbb{N}$-automaton T' (on x^*) which is a co-$\mathbb{N}$-covering of D' and of E'.

Every transition of T' is mapped, via the co-$\mathbb{N}$-coverings and the $\mathbb{N}$-coverings onto a transition of A' and onto a transition of B'. But these are transitions of A and B and every transition of T' may thus be labelled by a pair of letters (one coming from A and one coming from B) and hence turned into a letter-to-letter transducer T. As the projection on each component gives an unambiguous automaton, T realises a bijective function.

Remark 1. Theorem 6 bears some similarity with an old result by Maurer and Nivat (*cf.* [8]) on rational bijections. It is indeed completely different: it is more restricted in the sense it applies only to languages with the same generating functions whereas Maurer and Nivat considered bijections between languages

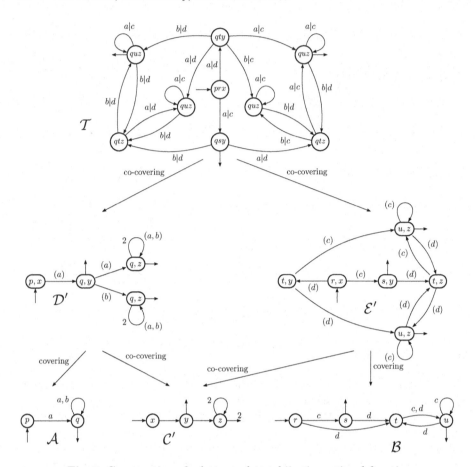

Fig. 5. Construction of a letter-to-letter bijective rational function

with 'comparable' growth functions, and it is much more precise in the sense that the transducer which realizes the bijection is letter-to-letter. It is this last property that makes the result interesting for the study of automatic structures.

Figure 5 shows the construction for the two languages $L = a(a + b)^*$ and $K = (c + dc + dd)^* \setminus cc(c + d)^*$ recognized by their minimal deterministic (and thus unambiguous) automata $\mathcal{A}$ and $\mathcal{B}$.

References

1. BÉAL, M.-P., LOMBARDY, S. AND SAKAROVITCH, J. On the equivalence of $\mathbb{Z}$-automata. *Proc. ICALP'05*, LNCS 3580, Springer (2005) 397–409.
2. BERSTEL, J., AND REUTENAUER, CH. *Rational Series and their Languages.* Springer, 1988.
3. CHOFFRUT, CH. Une caractrisation des fonctions squentielles et des fonctions sous-squentielles en tant que relations rationnelles. *Theoret. Comput. Sci. 5* (1977), 325–337.

4. EILENBERG, S. *Automata, Languages, and Machines. Vol. A.* Academic Press, 1974.

5. KLIMANN, I., LOMBARDY, S., MAIRESSE J., AND PRIEUR, CH. Deciding unambiguity and sequentiality from a finitely ambiguous max-plus automaton. *Theoret. Comput. Sci. 327* (2004), 349–373.

6. LIND, D., AND MARCUS, B. *An Introduction to Symbolic Dynamics and Coding.* Cambridge University Press, 1995.

7. LOMBARDY, S. AND SAKAROVITCH, J. Sequential ?. *Theoret. Comput. Sci.*, to appear.

8. MAURER, H., AND NIVAT, M. Rational Bijection of Rational Sets. *Acta Informatica 13* (1980) 365–378.

9. SAKAROVITCH, J., A construction on automata that has remained hidden. *Theoret. Computer Sci. 204* (1998), 205–231.

10. SAKAROVITCH, J., *Eléments de théorie des automates*, Vuibert, 2003. English translation, Cambridge Universit Press, to appear.

Applications of the Linear Matroid Parity Algorithm to Approximating Steiner Trees

Piotr Berman, Martin Fürer[1,*], and Alexander Zelikovsky[2,**]

[1] Department of Computer Science and Engineering,
Pennsylvania State University
{berman, furer}@cse.psu.edu
http://www.cse.psu.edu/{~berman, ~furer}
[2] Computer Science Department,
Georgia State University
alexz@cs.gsu.edu
http://www.cs.gsu.edu/~cscazz

Abstract. The Steiner tree problem in unweighted graphs requires to find a minimum size connected subgraph containing a given subset of nodes (terminals). In this paper we investigate applications of the linear matroid parity algorithm to the Steiner tree problem for two classes of graphs: where the terminals form a vertex cover and where terminals form a dominating set. As all these problems are MAX-SNP-hard, the issue is what approximation can be obtained in polynomial time. The previously best approximation ratio for the first class of graphs (also known as unweighted quasi-bipartite graphs) is ≈ 1.217 (Gröpl et al. [4]) is reduced in this paper to $8/7 - 1/160 \approx 1.137$. For the case of graphs where terminals form a dominating set, an approximation ratio of $4/3$ is achieved.

Keywords: Steiner trees, matroid, parity matroid problem, approximation ratio.

1 Introduction

One of the strongest results in matroid theory is the polynomial time solution to the parity problem in linear matroids. Briefly stated, we have a collection of $2 \times n$ matrices, and our goal is to find a maximum size sub-collection such that they can be stacked into a single matrix of maximum rank (i.e., its rank is equal to the number of rows). This problem was solved by Lovász [6], and much more efficient algorithms were later found by Orlin and Vande Vate [7] and Gabow and Stallmann [3] for graphic matroids.

Numerous graph problems can be expressed in the language of linear matroids, therefore one should expect that a number of them would be solved using an algorithm for matroid parity. Known examples are finding minimum feedback

---

* Supported in part by NSF Grant CCR-0209099.
** Supported in part by CRDF Award #MOM2-3049-CS-03.

D. Grigoriev, J. Harrison, and E.A. Hirsch (Eds.): CSR 2006, LNCS 3967, pp. 70–79, 2006.

vertex sets in cubic graphs, equivalently, finding maximum size acyclic induced subgraphs and finding maximum planar subgraphs [2]. However, until now, very few such applications were found. We hope that a technique described in this paper will allow to find many more such applications.

We will show approximation algorithms for two MAX-SNP hard versions of the unweighted Steiner tree problem. In each, we are given an undirected graph $G = (V, E)$ with a set $P \subseteq V$ of n terminals. The goal is to find a subset of edges that forms a tree containing all the terminals. The cost of the tree is defined as the number of edges it contains. We measure the quality of our algorithm by the *performance ratio*, i.e., the worst case ratio of the cost of the obtained solution to the cost of an optimum solution.

In these two problems we restrict ourselves to graphs where P forms a vertex cover or a dominating set, respectively. Our approximation algorithms for these problems are based on solving a general *base cover* problem for matroids using the algorithm for matroid parity.

The graphs where terminals form a vertex cover are also known as quasi-bipartite graphs. The best approximation ratio for general quasi-bipartite graphs is ≈ 1.279 [8] and for the subclass of uniform quasi-bipartite graphs (i.e., those where all edges incident to the same Steiner point have the same weight) is ≈ 1.217 [4]. In this paper we give an approximation ratio of ≈ 1.137 for the even narrower subclass of unweighted quasi-bipartite graphs. This class is especially interesting since it is one of the narrowest classes of graphs for which the Steiner tree problem is MAX SNP complete.[1] Note that the tight example for the 1.21-approximation algorithm of [4] used unweighted quasi-bipartite graphs, thus showing that matroid parity essentially improves Steiner tree approximations.

In the next section we review notations and terminology. Section 3 shows how to approximate the base cover problem with the performance ratio $8/7 - 1/160$. Section 4 gives a 4/3-approximation algorithm for the case when the terminals dominate all vertices.

2   Notation and Terminology

In this paper we use the following notation and terminology:

- $P \subseteq V$ is the set of terminals.
- $\#A$ is the cardinality of A.
- $N(a, B)$ is the set of nodes of B that are adjacent to a.
- If A is a set of objects and a is an object, $A + a$ and $A - a$ are shorthands for $A \cup \{a\}$ and $A - \{a\}$.
- $union(X)$ is the union of sets that belong to family X.

In a graph, *collapsing a set of nodes* means viewing them as a new single node. Afterwards, internal edges of the set are disregarded, and edges incident on any element of the set will be viewed as incident on the new node.

---

[1] The reduction is from the vertex cover problem restricted to graphs of degree at most 4.

A *Steiner tree* is a tree (a connected acyclic subgraph) containing P, the set of *terminals*. This tree is a union of maximal subtrees in which all terminals are leaves. We call such subtrees *full components*. An internal node of a Steiner tree is called a *Steiner point* if its degree is at least 3.

A *matroid* $M = (X, I)$ is a finite set X with a family I of *independent* subsets of X such that

- if $A \in I$ and $B \subseteq A$ then $B \in I$ (hereditary property)
- if $A, B \in I$, $|A| > |B|$, then there exists $a \in A - B$ such that $B + a \in I$ (exchange property)

Any maximal independent set of M (i.e., a *base* of M) has the same size which is called the rank of M, $rank(M)$. For any subset E of X, $rank(E)$ is the size of a maximal independent subset of E. The span of E, $span(E)$ is the maximum superset of E with the rank equal $rank(E)$. If E is a set of subsets of X, then we use $span(E)$ for $span(union(E))$.

An undirected graph $G = (V, E)$ is associated with a *graphic matroid* $M = (E, I)$, where a subset of edges A is independent, i.e., $A \in I$, if A does not contain cycles. A graphic matroid is a *linear* matroid, i.e., it can be represented as a linear space where each independent set is a linearly independent set of elements. Any base of M corresponding to a connected graph G is a spanning tree of G.

Given a partition of X into pairs (e_i, e_{i+1}), the *matroid parity problem* asks for a maximum number of pairs whose union is independent. Lovász [5] showed that the parity matroid problem can be solved efficiently for linear matroids.

3   The Base Cover Problem

When the terminals form a vertex cover in an unweighted graph $G = (V, E)$, then all full components are stars, i.e., subgraphs consisting of a Steiner node connected to some subset of terminals. Then each Steiner tree chooses a set of stars spanning all terminals. This section is devoted to the problem in matroids corresponding to this Steiner tree problem.

We first formulate this problem as a base cover problem (BCP) in matroids and give necessary notations. Then we prove two main properties of the solutions for BCP and give the approximation algorithms based on solution of the matroid parity problem. Finally, we prove a performance ratio of ≈ 1.137 for the proposed algorithms.

Let C be a hereditary family of subsets of X spanning a matroid $H = (X, I)$, i.e., $rank(union(C)) = rank(H)$. A subfamily $D = \{d_1, \ldots, d_m\}$ of disjoint sets from C is called *valid* if $union(D) \in H$. The cost of D is defined as $cost(D) = \sum_{i=1}^{m}(1 + |d_i|)$.

The Base Cover Problem. Given a hereditary family of subsets C spanning a matroid H, find the minimum cost valid subfamily $D \subseteq C$.

We will approximate the base cover problem for the special case of the graphic matroid H of the complete graph $G_P = (P, E_P)$ whose vertices are the terminals P of a Steiner tree problem. A subset c of edges of G_P is in C if all the endpoint of edges in c are leaves of the same full component. Therefore, such a subset c corresponds to a star in G (if the terminals form a dominating set).

Note that the rank of H, $rank(H) = n$, equals to the number of terminals minus 1, and the rank of any edge set A is the number $n - k$, where k is the number of connected components induced by A in G_P. The cost of a Steiner tree in an unweighted graph $G = (V, E)$ equals to the number of terminals minus 1 (which is the rank of the matroid) plus the number of stars used to connect them. Therefore, the cost of D equals the cost of the corresponding Steiner tree.

In two cases there exists a polynomial time exact algorithm. If C consists of singleton sets, every solution costs $2n$. If C is a collection of singletons and pairs, this is equivalent to the linear matroid parity problem, as we are maximizing the number of times when we can purchase two elements for the price of 3 (thus minimizing the number of times when we purchase elements individually, for the price of 2.)

We need some more notations:

$D_i = \{d \in D \mid \#d \geq i\}$
$valid(D) \equiv D$ is valid
$valid_i(D) \equiv valid(D)$ and $D = D_i$
$cost_1(D) = \min\{cost(E) \mid rank(E) = n$ and $valid(E)$ and $E_2 = D\}$
       (used if $valid_2(D)$, it is the cost of a solution obtained from D
       by adding singleton sets, equals $cost(D) + 2(n - rank(D))$)
$rank_2(D) = \max\{rank(E) \mid$ and $valid_2(E)$ and $E_3 = D\}$
       (used if $valid_3(D)$, it is the largest rank of a valid solution
       obtained from D by inserting pairs)
$cost_2(D) = \min\{cost_1(E) \mid valid(E)$ and $E_3 = D\}$
       (used if $valid_3(D)$, it is the least cost of a valid solution
       obtained from D by inserting pairs and singleton sets
$rank_3(D) = \max\{rank_2(E) \mid valid_3(E)$ and $D \subseteq E\}$
       (used if $valid_3(D)$, it is the largest rank of a valid solution
       obtained from D by inserting non-singleton sets)
$r_{max} = \max\{rank(D) \mid valid_2(D)\} = rank_3(\emptyset)$
$c_{min} = \min\{cost(D) \mid valid(D)\}$
H/I is defined for a subspace I of the linear space H, it is a result of a
       linear mapping l such that $l^{-1}(0) = I$

Our algorithm requires that we can compute $rank(D)$ in polynomial time. If $valid_2(D)$, then $cost_1(D) = 2n - (\#union(D) - \#D)$. Assuming $valid_3(D)$, we can find the largest collection D' of pairs of elements such that $valid_2(D \cup D')$ by solving the linear matroid parity problem in $H/span(D)$; if $\#D' = k$, then $rank_2(D) = rank(D) + 2k$ and $cost_2(D) = cost_1(D) - k$. Thus to minimize $cost(D)$ such that $valid(D)$ it suffices to minimize $cost_2(D)$ such that $valid_3(D)$.

The next two lemmas convey the crux of our method. Lemma 1 says that we can restrict our search to partial solutions satisfying $rank_3(D) = r_{max}$. In other words, we prove that any maximum rank partial solution containing only sets of size 2 or more will have the same singletons as some optimal solution. Lemma 2 allows to compute $rank_3$ and to perform the analysis of our (almost) greedy algorithm.

Lemma 1. *Assume $valid_2(D)$ and $rank(D) = r_{max}$. Then there exists E such that $span(E) = span(D)$ and $cost_1(E) = c_{min}$.*

Proof. Let $conform(D, E)$ be the union of set intersections $d \cap e$ such that $d \in D$, $e \in E$ and $\#(d \cap e) \geq 2$. Among the sets E such that $valid_2(E)$ and $cost_1(E) = c_{min}$ we can choose one that maximizes the size of $conform(D, E)$. It suffices to show that $span(D) \subseteq span(E)$: because $rank(D) = r_{max} \geq rank(E)$, $span(E)$ cannot properly contain $span(D)$. Thus this inclusion implies equality.

Suppose, by the way of contradiction, that $span(D) \not\subseteq span(E)$. Then $union(D) \not\subseteq span(E)$ and for some $u \in d \in D$ we have $u \notin span(E)$. Because $valid_2(D)$ holds, d has another element say v. Vector v must satisfy one of the following conditions.

Case 1: $v \notin union(E)$ and $rank(union(E) + u + v) = rank(E) + 2$. Then $E' = E + \{u, v\}$ satisfies $valid_2(E')$ and $cost_1(E')$ is by 1 smaller than $cost_1(E) = c_{min}$, a contradiction.

Case 2: $v \notin union(E)$ and $rank(union(E) + u + v) = rank(E) + 1$. Because $union(D)$ and $union(E) + u$ are independent sets of elements, and $union(E) + u + v$ is not, there exists a $w \in union(E) - union(D)$ such that the set $union(E) + u + v - w$ is independent. Let e be the set of w in E. One can see that $E' = (E - e + (e - w) + \{u, v\})_2$ satisfies $valid_2(E')$, $cost_1(E') = cost_1(E)$ and $conform(D, E') = conform(D, E) + u + v$, a contradiction.

Case 3: $v \in union(E) - conform(D, E)$. Let e be the set of v in E. One can see that $E' = E - e + (e - v) + \{u, v\}$ gives the same contradiction as in Case 2.

Case 4: $v \in conform(D, E)$, i.e., for some $e \in E$, $v \in d \cap e$ and $\#(d \cap e) \geq 2$. If $e - d$ has at most one element, we obtain a contradiction with $E' = E - e + ((e \cap d) + u)$, as $valid_2(E')$ holds, $cost_1(E') \leq cost_1(E)$ and $conform(D, E') = conform(D, E) + u$. If $e - d$ has at least two elements, we obtain the same contradiction with $E' = E - e + (e - d) + ((e \cap d) + u)$. $\square$

Lemma 2. *Assume $valid_3(D)$ and $rank_2(D) < r_{max}$. Then there exists E such that $valid_3(E)$, $rank_2(E) = rank_2(D) + 1$ and either*
(1) $E = (D - d + (d - w))_3$ for some $w \in d \in D$, or
(2) $E = D + d$ for some triple d of elements.

Proof. Let F be such that $valid_2(F)$ and $rank(F) = r_{max}$. Among D' such that $valid_2(D')$, $D'_3 = D$ and $rank(D') = rank_2(D)$ choose one that maximizes the size of $conform(D', F)$. Because $rank(F) > rank_2(D) = rank(D')$, for some $u \in f \in F$ we have $u \notin span(D')$. Let v be another element of f; this vector must satisfy one of the following cases.

Case 1: $v \notin union(D')$ and $rank(union(D') + u + v) = rank(D') + 2$. Then $valid_2(D' + \{u, v\})$, $(D' + \{u, v\})_3 = D$ and $rank(D' + \{u, v\}) = rank(D') + 2$, a contradiction to our choice of D'.

Case 2: $v \notin union(D')$ and $rank(union(D') + u + v) = rank(D') + 1$. As in Case 2 of the previous proof, for some $w \in union(D') - union(F)$, the set $union(D') + u + v - w$ is independent. Let d be the set of w in D', i.e., $d \in w \in D'$.

Case 2.1: $\#d = 2$. Then we can replace d in D' by $\{u, v\}$, and this will preserve the property $valid_2(D')$, and the values of D'_3 and $rank(D')$. However the size of $conform(D', F)$ will increase, a contradiction to our choice of D'.

Case 2.2: $\#d > 2$. Then for $D'' = D' - d + (d - w) + \{u, v\}$ we have $valid_2(D'')$ while $rank(D'') = rank(D') + 1 = rank_2(D) + 1$. Therefore we have found $E = D''_3$ such that $rank_2(E) = rank_2(D) + 1$. Note that $E = (D - d + (d - w))_3$, which satisfies Case (1) of the claim.

Case 3: $v \in union(D') - conform(D', F)$. We can define $w = v$ and proceed exactly as in Case 2.

Case 4: $v \in conform(D', F)$. Let d' be the set of v in D'. If $\#d' > 2$, we can proceed as in Case 2.2, with $d = d'$ and $w = v$. If $\#d' = 2$, then $d' \subseteq f \in F$, consequently for $d = d' + u$ we have $valid_2(D' - d' + d)$ and $rank(D' - d' + d) = rank(D') + 1$. Note that $E = (D' - d' + d)_3$ satisfies Case (2) of the claim. $\square$

One consequence of Lemma 2 is that the following algorithm, on input D satisfying $valid_3(D)$, finds E such that $D \subseteq E$, $valid_3(E)$ and $rank_2(E) = rank_3(D)$, in particular, the algorithm computes $rank_3(D)$:

Algorithm 1:

    $E \leftarrow D$
    while there exists a triple d of elements such that
            $valid_3(E + d)$ and $rank_2(E + d) > rank_2(E)$ do
      $E \leftarrow E + d$
    return E

To see that this algorithm is correct, it suffices to analyze it for $D = \emptyset$. Let us apply Lemma 2, starting from the empty collection of vector sets. Initially, we have $rank_2$ equal to some $2m$. After k iterations consistent with Case (2), we obtain $rank_2$ equal to $2m + k$. Suppose that now we can apply Case (1) of the lemma. Then we obtain E consisting of $k - 1$ triples with $rank_2(E) = 2m + k + 1$. This means that a sequence of $k - 1$ triple insertions increases $rank_2$ by $k + 1$, hence one of these insertions increases $rank_2$ by at least 2, which is impossible. Because $rank_2$ is computable in polynomial time, this algorithm computes $rank_3$ in polynomial time.

Now we have all the tools to analyze the following algorithm for approximating the base cover problem.

Algorithm 2:

(* quick stage *)
$A \leftarrow$ empty collection of vector sets
while there exists d such that $valid(A + d)$ and $\#d \geq 8$
$\quad A \leftarrow A + d$
(* careful stage *)
for $i \leftarrow 7$ down to 3
$\quad$ while there exists d such that $valid_i(A + d)$ and $rank_3(A + d) = rank_3(A)$
$\quad\quad A \leftarrow A + d$
(* finishing stage *)
$\quad$ while there exists d such that $valid_2(A + d)$ and $rank_2(A + d) = rank_2(D)$
$\quad\quad A \leftarrow A + d$
$\quad$ while there exists d such that $valid(A + d)$
$\quad\quad A \leftarrow A + d$
return A

Theorem 1. *In polynomial time, Algorithm 2 produces a base cover with cost at most $1 + 1/7 - 1/160$ times the optimum cost.*

Proof. Say that a collection of vector sets B forms a best solution. We want to find the worst case ratio $cost(A)/cost(B)$. Each time the algorithm adds a set d to the partial solution A, the remaining problem is to find the least cost base cover for $H/span(A)$. In the analysis, we modify B after each selection. Each selection d reduces the rank of $H/span(A)$ by $\#d$, hence we have to remove $\#d$ elements from the sets of B. At the same time, the cost of A increases by $1 + \#d$.

During the quick stage the ratio between the increase of $cost(A)$ and the decrease of the remaining $cost(B)$ is at most $9/8$. Later we cannot argue in the same manner, but we have to average the cost increases of the careful stage with the mistake-free finishing stage. We achieve this goal as follows. First, at the beginning of the careful stage we modify B so that it contains a minimum number of singleton sets. By Lemma 1 this can be done without changing the solution cost. Afterwards, during the careful stage, we monitor B_3, while preserving the invariant that $rank_3(B_3)$ is maximal for the linear space $H/span(A)$. During this process, B_3 consists of the "original" sets (with some elements removed) and some "new" triples.

After selecting some d during the careful stage, we remove $\#d$ elements from the sets of B. If one of these sets becomes empty, then the remaining solution has $\#d = i$ fewer elements and 1 fewer set, so its costs decreases by $i + 1$, the same as the increase of the cost of A. Otherwise, we need to account for the difference between these two quantities, i.e., 1. We split 1 into i equal parts, and we charge them to the elements of B_3 that had elements removed.

Now, the size of elements of B_3 drops for two reasons. First, we remove some elements to maintain the invariant that $union(B)$ is the base of $H/span(A)$. Second, if as a result some sets drop in size to 1, we have to reduce the number of singleton sets in B to the minimum to maintain the second invariant. We do it using the modifications described in Lemma 2. If even one such modification is of kind (2), it effectively replaces in B a singleton set and a pair by a triple which decreases the cost of B, so there are no charges. A modification of kind (1) removes one vector from a set of B_3, while also replacing a singleton set in B by a pair. As a result, we exchange a size decrease to 1 with a size decrease to at least 2.

Summarizing, a selection of d during the careful stage makes i or fewer decreases of the sets in B_3; if there are fewer than i decreases, this is a step without a charge, otherwise there are exactly i decreases, each associated with a charge of $1/i$. Because we always select the largest possible set, a charge of $1/i$ can be applied only to a set of size at most i, otherwise this set would be selected rather than d. Once a set drops in size to 2, it does not accept further decreases (unless there are no charges).

Now we can see what the possible ratios are between the cost of a set in B and the charges it receives. Singleton sets and pairs do not accept any charges. A triple can receive one charge, at most $1/3$, and the resulting ratio is at most $(1/3)/4 = 1/12$. A quadruple can receive a most two charges, the first at most $1/4$, when it is reduced in size to a triple, and then at most $1/3$, for the resulting ratio $(1/4 + 1/3)/5 = 7/60$. For the larger sets we get the ratios $(1/5 + 1/4 + 1/3)/6 = 47/360$, $(1/6 + 1/5 + 1/4 + 1/3)/7 = (1 - 1/20)/7 = 1/7 - 1/140$ and $(1/7 + 1/6 + 1/5 + 1/4 + 1/3)/8 = (8/7 - 1/20)/8 = 1/7 - 1/160$. □

4   Terminals Forming a Dominating Set

Now we consider the case when all edges have length 1 and the set P of terminals forms a dominating set (i.e., each node is in P or is adjacent to P). To assure an approximation ratio of $4/3$ we may use a version of Algorithm 2. In the quick stage, we first sequentially contract edges connecting the terminals, and then i-stars for $i \geq 4$; one can easily see that if we spend x in the quick stage, then opt, the cost of the minimum Steiner tree, is decreased by at least $3x/4$. Therefore we may assume that we deal with a graph G that is an output of the quick stage, and consequently P is an independent set (and a dominating set), and each node in $V - P$ is adjacent to at most 3 nodes from P.

The next 4 lemmas contain the necessary graph-theoretic observations. The proof of Lemma 5 includes a description of the algorithm needed (a variation of the careful and finishing stages) and shows an approximation ratio of $4/3$.

For the sake of analysis, let G' be obtained from G by deleting all edges from the graph that are not incident on any one of the terminals. We will say that connected components of G' are V-components, while an intersection of a V-component with P is a P-component.

We define an edge weighted graph $G_P = (P, E_P)$ where edges $\{u, v\}$ in G_P are identified with simple paths between u and v within a full component of G. The edge weights are the corresponding path lengths.

Because of space constraints, we refer to a planned journal version or the web page of the second author, for the proof of the next three lemmas.

Lemma 3. *Any minimum spanning tree (MST) of G_P can be formed as follows: points of each P-component are connected with paths of length 2, and then P components are connected with paths of length 3.*

We say that a Steiner tree is *normal* if every internal node u is adjacent in the tree to a terminal; in proofs we will use $p(u)$ to denote a (unique) selected terminal adjacent in the tree to u. As before, *opt* is the minimum cost of a Steiner tree.

Lemma 4. *There exists a normal minimum tree with cost opt.*

We say that a Steiner tree is *special* if (1) it is normal, (2) its full components have at most 4 terminals each, (3) its full components have at most one Steiner point, and (4) the union of minimum spanning trees of its full components forms an MST of G_P. For *any* Steiner tree T we will use a_T and b_T to denote the number of full components with 3 and 4 terminals respectively, and exactly one Steiner point. As before, *mst* denotes the cost of an MST of G_P.

Lemma 5. *The cost of a special Steiner tree T equals $mst - a_T - 2b_T$.*

Lemma 6. *There exists a special Steiner tree T with cost at most $\frac{4}{3}opt - \frac{1}{3}b_T$.*

Proof. Consider a normal Steiner tree T of P with cost *opt* (it exists by Lemma 4). In the first part of the proof, we will transform it into a Steiner tree satisfying conditions (1), (2) and (3). At an intermediate stage of the transformation let x be the cost of the full components of T that satisfy conditions (2) and (3), y be the cost of the remaining components. At each stage we will decrease y, while satisfying the invariant $x + \frac{4}{3}y + \frac{1}{3}b_T \le \frac{4}{3}opt$. Observe that this invariant holds initially, because the cost of each component counted by a_T has cost at least 4.

Assume then that $y > 0$ and consider a full component C that violates (2) or (3). Suppose that a Steiner point u in C is adjacent (via edges or paths of length 2 that do not go through a Steiner point) to some four terminals, $p(u), q, r$ and s. Because no node is adjacent to more than 3 terminals, we may assume that the path from u to s has length 2. We may remove the paths from u to q, r and s from C, thus decreasing its cost by some $z \ge 4$, and create a new component consisting of these paths and the edge $\{u, p\}$. In such a stage $\frac{4}{3}y$ decreases by $\frac{4}{3}z \ge z + 1 + \frac{1}{3}$, while x increases by 1 and $\frac{1}{3}b_T$ increases by $\frac{1}{3}$. If we cannot perform a stage like that, we may infer that all full component with exactly one Steiner node satisfy (2) and (3). Suppose then that C contains more than one Steiner point; one of them, say u, is adjacent to exactly one other, say v. We can accomplish a stage by removing the Steiner point u from C together with the incident paths, and creating a new component consisting of these paths and the edge $\{v, p(v)\}$. If the new component contains 3 terminals, we decreased y by some $z \ge 3$, and increased x by $z + 1$, and if the new component contains 4 terminals, we can repeat the previous calculation.

Now we may assume that $y = 0$ and $x + \frac{1}{3}b_T \le \frac{4}{3}opt$. Suppose that condition (4) is violated, i.e., the union U of MSTs of the full components of T has cost larger than *mst* (it is clear that U is a spanning tree of G_P). Let M be an MST of G_P. By Lemma 3, M consists of paths of length 2 and 3, and because T is normal, U also consists of such paths. Therefore we can insert into U a path of length 2 from M, and from the resulting cycle remove a path of length 3, that belongs to MST of a full component C. If C is just a path, we can replace C by the new path of length 2, decreasing the cost of T. If C contains a Steiner point u, then we removed from U a path of the form $(p(u), u, v)$, and so we can remove path (u, v) from C; because this path has length 2, x remains unchanged, while b_T could only decrease. Therefore we can decrease the cost of U without increasing $x + \frac{1}{3}b_T$. $\qquad\square$

Lemma 7. *Assume that there exists a special Steiner tree T with cost c. Then we can find a special Steiner tree with cost $c + \frac{1}{3}b_T$ in polynomial time.*

Proof. (Sketch) By Lemma 5, $c = mst - a_T - 2b_T$. Therefore it suffices to find a special Steiner tree T' such that $a_{T'} + 2b_{T'} \geq a_T + 2b_T - \frac{1}{3}b_T$.

We will use Algorithm 2 as follows. Our linear space will be over the field $\mathbf{Z_2}$; the base elements will be nodes and P-components. Elements in the problem presented to Algorithm 2 will correspond to paths that can be used in an MST of G_P according to Lemma 3. Assume that such a path connects the terminals p_1 and p_2, which in turn belong to the P-components P_1 and P_2. If $P_1 = P_2$ (and the length of this path is 2), we map this path into the vector $p_1 + p_2$, otherwise (the case of a path of length 3), we map it into $P_1 + P_2$. Finally, we identify a 3-star or 4-star with a set of elements that correspond to the sets of paths that form MSTs of these stars.

It is quite easy to see that Algorithm 2 results in $b_{T'} \geq \frac{1}{3}b_T$. Moreover, Lemma 2 assures that $2a_{T'} + 3b_{T'} \geq 2a_T + 3b_T$. We obtain the desired result by dividing the above inequalities by 2 and adding them together. $\qquad\square$

Combining the previous Lemmas, we get:

Theorem 2. *If the terminals form a dominating set, then the minimum Steiner tree problem can be approximated in polynomial time with a performance ratio of $4/3$.* $\qquad\square$

References

1. M. Bern and P. Plassmann. *The Steiner tree problem with edge lengths 1 and 2.* Inform. Process. Lett. 32: 171–176, 1989.
2. G. Calinescu, C.G. Fernandes, U. Finkler and H. Karloff. *A better approximation algorithm for finding planar subgraphs.* Proc. 7th SODA: 16–25, 1996.
3. H.N. Gabow and M. Stallmann. *Efficient algorithms for graphic matroid intersection and parity.* Proc. ICALP'85. Lecture Notes in Computer Science, 194: 210–220, 1985.
4. C. Gröpl, S. Hougardy, T.Nierhoff and H.J.Prömel. *Steiner trees in uniformly quasi-bipartite graphs.* Information Processing Letters, 83:195–200, 2002.
5. L. Lovász, *The matroid matching problem.* Algebraic Methods in Graph Theory, vol. 2, North Holland, 495–518, 1981.
6. L. Lovász and M.D. Plummer. *Matching Theory.* Elsevier Science, Amsterdam, 1986.
7. Orlin and Vande Vate, *Solving the Linear Matroid Parity Problem as a Sequence of Matroid Intersection Problems.* Mathematical Programming 47 (Series A), 81–106, 1990.
8. G. Robins and A. Zelikovsky. *Tighter Bounds for Graph Steiner Tree Approximation.* SIAM Journal on Discrete Mathematics, 19:122-134, 2005.

Tuples of Disjoint NP-Sets

(Extended Abstract)

Olaf Beyersdorff

Institut für Informatik, Humboldt-Universität zu Berlin, 10099 Berlin, Germany
beyersdo@informatik.hu-berlin.de

Abstract. Disjoint NP-pairs are a well studied complexity theoretic concept with important applications in cryptography and propositional proof complexity. In this paper we introduce a natural generalization of the notion of disjoint NP-pairs to disjoint k-tuples of NP-sets for $k \geq 2$. We define subclasses of the class of all disjoint k-tuples of NP-sets. These subclasses are associated with a propositional proof system and possess complete tuples which are defined from the proof system.

In our main result we show that complete disjoint NP-pairs exist if and only if complete disjoint k-tuples of NP-sets exist for all $k \geq 2$. Further, this is equivalent to the existence of a propositional proof system in which the disjointness of all k-tuples is shortly provable. We also show that a strengthening of this conditions characterizes the existence of optimal proof systems.

1 Introduction

During the last years the theory of disjoint NP-pairs has been intensively studied. This interest stems mainly from the applications of disjoint NP-pairs in the field of cryptography [9, 16] and propositional proof complexity [18, 13]. In this paper we investigate a natural generalization of disjoint NP-pairs: instead of pairs we consider k-tuples of pairwise disjoint NP-sets. Concepts such as reductions and separators are smoothly generalized from pairs to k-tuples.

One of the major open problems in the field of disjoint NP-pairs is the question, posed by Razborov [19], whether there exist disjoint NP-pairs that are complete for the class of all pairs under suitable reductions. Glaßer et al. [6] gave a characterization in terms of uniform enumerations of disjoint NP-pairs and also proved that the answer to the problem does not depend on the reductions used, i.e. there are reductions for pairs which vary in strength but are equivalent with respect to the existence of complete pairs.

The close relation between propositional proof systems and disjoint NP-pairs provides a partial answer to the question of the existence of complete pairs. Namely, the existence of optimal propositional proof systems is a sufficient condition for the existence of complete disjoint NP-pairs. This result is already implicitly contained in [19]. However, Glaßer et al. [7] construct an oracle relative to which there exist complete pairs but optimal proof systems do not exist.

D. Grigoriev, J. Harrison, and E.A. Hirsch (Eds.): CSR 2006, LNCS 3967, pp. 80–91, 2006.

Hence, the problems on the existence of optimal proof systems and of complete disjoint NP-pairs appear to be of different strength.

Our main contribution in this paper is the characterization of these two problems in terms of disjoint k-tuples of NP-sets. In particular we address the question whether there exist complete disjoint k-tuples under different reductions. Considering this problem it is easy to see that the existence of complete k-tuples implies the existence of complete l-tuples for $l \leq k$: the first l components of a complete k-tuple are complete for all l-tuples. Conversely, it is a priori not clear how to construct a complete k-tuple from a complete l-tuple for $l < k$. Therefore it might be tempting to conjecture that the existence of complete k-tuples forms a hierarchy of assumptions of increasing strength for greater k. However, we show that this does not happen: there exist complete disjoint NP-pairs if and only if there exist complete disjoint k-tuples of NP-sets for all $k \geq 2$, and this is even true under reductions of different strength. Further, we prove that this is equivalent to the existence of a propositional proof system in which the disjointness of all k-tuples with respect to suitable propositional representations of these tuples is provable with short proofs. We also characterize the existence of optimal proof systems with a similar but apparently stronger condition.

We achieve this by extending the connection between proof systems and NP-pairs to k-tuples. In particular we define representations for disjoint k-tuples of NP-sets. This can be done on a propositional level with sequences of tautologies but also with first-order formulas in arithmetic theories. To any propositional proof system P we associate a subclass $\mathsf{DNPP}_k(P)$ of the class of all disjoint k-tuples of NP-sets. This subclass contains those k-tuples for which the disjointness is provable with short P-proofs. We show that the classes $\mathsf{DNPP}_k(P)$ possess complete tuples which are defined from the proof system P. Somewhat surprisingly, under suitable conditions on P these non-uniform classes $\mathsf{DNPP}_k(P)$ equal their uniform versions which are defined via arithmetic representations. This enables us to further characterize the existence of complete disjoint k-tuples by a condition on arithmetic theories.

The paper is organized as follows. In Sect. 2 we recall some relevant definitions concerning propositional proof systems and disjoint NP-pairs. We also give a very brief description of the correspondence between propositional proof systems and arithmetic theories. This reference to bounded arithmetic, however, only plays a role in Sect. 5 where we analyse arithmetic representations. The rest of the paper and in particular the main results in Sect. 6 are fully presented on the propositional level.

In Sect. 3 we define the basic concepts such as reductions and separators that we need for the investigation of disjoint k-tuples of NP-sets.

In Sect. 4 we define propositional representations for k-tuples and introduce the complexity classes $\mathsf{DNPP}_k(P)$ of all disjoint k-tuples of NP-sets that are representable in the system P. We show that these classes are closed under our reductions for k-tuples. Further, we define k-tuples from propositional proof systems which serve as hard languages for $\mathsf{DNPP}_k(P)$. In particular we generalize the interpolation pair from [18] and demonstrate that even these generalized variants still capture the feasible interpolation property of the proof system.

In Sect. 5 we define first-order variants of the propositional representations from Sect. 4. We utilize the correspondence between proof systems and bounded arithmetic to show that a k-tuple of NP-sets is representable in P if and only if it is representable in the arithmetic theory associated with P. This equivalence allows easy proofs for the representability of the canonical k-tuples associated with P, thereby improving the hardness results for $\mathsf{DNPP}_k(P)$ from Sect. 4 to completeness results for proof systems corresponding to arithmetic theories.

The main results on the connections between complete NP-pairs, complete k-tuples and optimal proof systems follow in Sect. 6.

Due to space limitations we only sketch proofs or omit them in this extended abstract. The complete paper is available as a technical report [2].

2   Preliminaries

Propositional Proof Systems. Propositional proof systems were defined in a very general way by Cook and Reckhow in [5] as polynomial time functions P which have as its range the set of all tautologies. A string π with $P(\pi) = \varphi$ is called a P-proof of the tautology φ. By $P \vdash_{\leq m} \varphi$ we indicate that there is a P-proof of φ of length $\leq m$. If Φ is a set of propositional formulas we write $P \vdash_* \Phi$ if there is a polynomial p such that $P \vdash_{\leq p(|\varphi|)} \varphi$ for all $\varphi \in \Phi$. If $\Phi = \{\varphi_n \mid n \geq 0\}$ is a sequence of formulas we also write $P \vdash_* \varphi_n$ instead of $P \vdash_* \Phi$.

Proof systems are compared according to their strength by simulations introduced in [5] and [14]. Given two proof systems P and S we say that S *simulates* P (denoted by $P \leq S$) if there exists a polynomial p such that for all tautologies φ and P-proofs π of φ there is a S-proof π' of φ with $|\pi'| \leq p(|\pi|)$. If such a proof π' can even be computed from π in polynomial time we say that S *p-simulates* P and denote this by $P \leq_p S$. A proof system is called *(p-)optimal* if it (p-)simulates all proof systems. Whether or not optimal proof systems exist is an open problem posed by Krajíček and Pudlák [14].

In [3] we investigated several natural properties of propositional proof systems. We will just define those which we need in this paper. We say that a propositional proof system P is *closed under substitutions by constants* if there exists a polynomial q such that $P \vdash_{\leq n} \varphi(\bar{x}, \bar{y})$ implies $P \vdash_{\leq q(n)} \varphi(\bar{a}, \bar{y})$ for all formulas $\varphi(\bar{x}, \bar{y})$ and constants $\bar{a} \in \{0,1\}^{|\bar{x}|}$. We call P *efficiently closed under substitutions by constants* if we can transform any P-proof of a formula $\varphi(\bar{x}, \bar{y})$ in polynomial time to a P-proof of $\varphi(\bar{a}, \bar{y})$. A system P is *closed under disjunctions* if there is a polynomial q such that $P \vdash_{\leq m} \varphi$ implies $P \vdash_{\leq q(m+|\psi|)} \varphi \vee \psi$ for arbitrary formulas ψ. Similarly, we say that a proof system P is *closed under conjunctions* if there is a polynomial q such that $P \vdash_{\leq m} \varphi \wedge \psi$ implies $P \vdash_{\leq q(m)} \varphi$ and $P \vdash_{\leq q(m)} \psi$, and likewise $P \vdash_{\leq m} \varphi$ and $P \vdash_{\leq n} \psi$ imply $P \vdash_{\leq q(m+n)} \varphi \wedge \psi$ for all formulas φ and ψ. As with closure under substitutions by constants we also consider efficient versions of closure under disjunctions and conjunctions.

Propositional Proof Systems and Arithmetic Theories. In Sect. 5 we will use the correspondence of propositional proof systems to theories of bounded arithmetic. Here we will just briefly introduce some notation and otherwise refer to the monograph [11]. To explain the correspondence we have to translate first-order arithmetic formulas into propositional formulas. An arithmetic formula in prenex normal form with only bounded existential quantifiers is called a Σ_1^b-formula. These formulas describe NP-predicates. Likewise, Π_1^b-formulas only have bounded universal quantifiers and describe coNP-predicates. A Σ_1^b- or Π_1^b-formula $\varphi(x)$ is translated into a sequence $\|\varphi(x)\|^n$ of propositional formulas containing one formula per input length for the number x. We use $\|\varphi(x)\|$ to denote the set $\{\|\varphi(x)\|^n \mid n \geq 1\}$.

The *reflection principle* for a propositional proof system P states a strong form of the consistency of the proof system P. It is formalized by the $\forall\Pi_1^b$-formula

$$\mathrm{RFN}(P) = (\forall\pi)(\forall\varphi)\mathrm{Prf}_P(\pi, \varphi) \to \mathrm{Taut}(\varphi)$$

where Prf_P and Taut are suitable arithmetic formulas describing P-proofs and tautologies, respectively. A proof system P has the *reflection property* if $P \vdash_* \|\mathrm{RFN}(P)\|^n$ holds.

In [15] a general correspondence between arithmetic theories T and propositional proof systems P is introduced. Pairs (T, P) from this correspondence possess in particular the following two properties:

1. Let $\varphi(x)$ be a Π_1^b-formula such that $T \vdash (\forall x)\varphi(x)$. Then there exists a polynomial time computable function f that on input 1^n outputs a P-proof of $\|\varphi(x)\|^n$.
2. $T \vdash \mathrm{RFN}(P)$ and if $T \vdash \mathrm{RFN}(Q)$ for some proof system Q, then $Q \leq_p P$.

We call a proof system P *regular* if there exists an arithmetic theory T such that properties 1 and 2 are fulfilled for (T, P). Probably the most important example of a regular proof system is the extended Frege system EF that corresponds to the theory S_2^1. This correspondence was established in [4] and [15].

Disjoint NP-Pairs. A pair (A, B) is called a *disjoint* NP-*pair* if $A, B \in$ NP and $A \cap B = \emptyset$. The pair (A, B) is called *p-separable* if there exists a polynomial time computable set C such that $A \subseteq C$ and $B \cap C = \emptyset$. Grollmann and Selman [9] defined the following reduction between disjoint NP-pairs (A, B) and (C, D): $((A, B) \leq_p (C, D))$ if there exists a polynomial time computable function f such that $f(A) \subseteq C$ and $f(B) \subseteq D$. This variant of a many-one reduction for pairs was strengthened by Köbler et al. [10] to: $(A, B) \leq_s (C, D)$ if there exists a function $f \in$ FP such that $f^{-1}(C) = A$ and $f^{-1}(D) = B$.

The link between disjoint NP-pairs and propositional proof systems was established by Razborov [19], who associated a canonical disjoint NP-pair with a proof system. This canonical pair is linked to the automatizablility and the reflection property of the proof system. Pudlák [18] introduced an *interpolation pair* for a proof system P which is p-separable if and only if the proof system P

has the feasible interpolation property [12]. In [1] we analysed a variant of the interpolation pair. More information on the connection between disjoint NP-pairs and propositional proof systems can be found in [18, 1, 3, 8].

3   Basic Definitions and Properties

Definition 1. *Let $k \geq 2$ be a number. A tupel $(A_1, \ldots, A_k)$ is a* disjoint k-tuple *of* NP-sets *if all components $A_1, \ldots, A_k$ are nonempty languages in* NP *which are pairwise disjoint.*

We generalize the notion of a separator of a disjoint NP-pair as follows:

Definition 2. *A function $f : \{0,1\}^* \to \{1, \ldots, k\}$ is a* separator *for a disjoint k-tuple $(A_1, \ldots, A_k)$ if $a \in A_i$ implies $\underline{f(a) = i}$ for $i = 1, \ldots, k$ and all $a \in \{0,1\}^*$. For inputs from the complement $\overline{A_1 \cup \cdots \cup A_k}$ the function f may answer arbitrarily. If $(A_1, \ldots, A_k)$ is a disjoint k-tuple of* NP-sets *that has a polynomial time computable separator we call the tuple* p-separable, *otherwise* p-inseparable.

Whether there exist p-inseparable disjoint k-tuples of NP-sets is certainly a hard problem that cannot be answered with our current techniques. At least we can show that this question is not harder than the previously studied question whether there exist p-inseparable disjoint NP-pairs.

Theorem 3. *The following are equivalent:*

1. *For all numbers $k \geq 2$ there exist p-inseparable disjoint k-tuples of* NP-sets.
2. *There exists a number $k \geq 2$ such that there exist p-inseparable disjoint k-tuples of* NP-sets.
3. *There exist p-inseparable disjoint* NP-pairs.

Proof. The implications $1 \Rightarrow 2$ and $3 \Rightarrow 1$ are immediate. To prove $2 \Rightarrow 3$ let us assume that all disjoint NP-pairs are p-separable. To separate a k-tuple $(A_1, \ldots, A_k)$ for some $k \geq 2$ we evaluate all separators $f_{i,j}$ for all disjoint NP-pairs (A_i, A_j) and output the number i such that we received 1 at all evaluations $f_{i,j}$. If no such i exists, then we know that the input is outside $A_1 \cup \cdots \cup A_k$, and we can answer arbitrarily. □

Let us pause to give an example of a disjoint k-tuple of NP-sets that is derived from the Clique-Colouring pair (cf. [18]). The tuple $(C_1, \ldots, C_k)$ has components C_i that contain all $i + 1$-colourable graphs with a clique of size i. Clearly, the components C_i are NP-sets which are pairwise disjoint. This tuple is also p-separable, but to devise a separator for $(C_1, \ldots, C_k)$ is considerably simpler than to separate the Clique-Colouring pair: given a graph G we output the maximal number i between 1 and k such that G contains a clique of size i. For graphs with n vertices this number i can be computed in time $O(n^k)$.

Candidates for p-inseparable tuples arise from one-way functions. Let $\Sigma = \{a_1, \ldots, a_k\}$ be an alphabet of size $k \geq 2$. To an injective one-way function

$f : \Sigma^* \to \Sigma^*$ we assign a disjoint k-tuple $(A_1(f), \ldots, A_k(f))$ of NP-sets with components

$$A_i(f) = \{(y, j) \mid (\exists x) f(x) = y \text{ and } x_j = a_i\}$$

where x_j is the j-th letter of x. This tuple is p-inseparable if f has indeed the one-way property.

Next we define reductions for k-tuples. We will only consider variants of many-one reductions which are easily obtained from the reductions $\leq_p$ and $\leq_s$ for pairs.

Definition 4. *A k-tupel $(A_1, \ldots, A_k)$ is polynomially reducible to a k-tupel $(B_1, \ldots, B_k)$, denoted by $(A_1, \ldots, A_k) \leq_p (B_1, \ldots, B_k)$, if there exists a polynomial time computable function f such that $f(A_i) \subseteq B_i$ for $i = 1, \ldots, k$. If additionally $f(\overline{A_1 \cup \cdots \cup A_k}) \subseteq \overline{B_1 \cup \cdots \cup B_k}$ holds, then we call the reduction performed by f strong. Strong reductions are denoted by $\leq_s$.*

From $\leq_p$ and $\leq_s$ we define equivalence relations $\equiv_p$ and $\equiv_s$ and call their equivalence classes degrees.

Following common terminology we call a disjoint k-tuple of NP-sets $\leq_p$-complete if every disjoint k-tuple of NP-sets $\leq_p$-reduces to it. Similarly, we speak of $\leq_s$-complete tuples.

In the next theorem we separate the reductions $\leq_p$ and $\leq_s$ on the domain of all p-separable disjoint k-tuples of NP-sets:

Theorem 5. *For all numbers $k \geq 2$ the following holds:*

1. *All p-separable disjoint k-tuples of NP-sets are $\leq_p$-equivalent. They form the minimal $\leq_p$-degree of disjoint k-tuples of NP-sets.*
2. *If $\mathsf{P} \neq \mathsf{NP}$, then there exist infinitely many $\leq_s$-degrees of p-separable disjoint k-tuples of NP-sets.*

Proof. Part 1 is easy. For part 2 we use the result of Ladner [17] that there exist infinitely many different $\leq_m^p$-degrees of NP-sets assuming $\mathsf{P} \neq \mathsf{NP}$. Therefore Ladner's theorem together with the following claim imply part 2.

Claim: Let $(A_1, \ldots, A_k)$ and $(B_1, \ldots, B_k)$ be p-separable disjoint k-tuple of NP-sets . Let further $\overline{B_1 \cup \cdots \cup B_k} \neq \emptyset$. Then $(A_1, \ldots, A_k) \leq_s (B_1, \ldots, B_k)$ if and only if $A_i \leq_m^p B_i$ for all $i = 1, \ldots, k$. $\square$

4   Disjoint k-Tuples from Propositional Proof Systems

In [3] we defined propositional representations for NP-sets as follows:

Definition 6. *Let A be a NP-set over the alphabet $\{0, 1\}$. A propositional representation for A is a sequence of propositional formulas $\varphi_n(\bar{x}, \bar{y})$ such that:*

1. *$\varphi_n(\bar{x}, \bar{y})$ has propositional variables $\bar{x}$ and $\bar{y}$ such that $\bar{x}$ is a vector of n propositional variables.*
2. *There exists a polynomial time algorithm that on input 1^n outputs $\varphi_n(\bar{x}, \bar{y})$.*
3. *Let $\bar{a} \in \{0, 1\}^n$. Then $\bar{a} \in A$ if and only if $\varphi_n(\bar{a}, \bar{y})$ is satisfiable.*

Once we have propositional descriptions of NP-sets we can now represent disjoint k-tuples of NP-sets in propositional proof systems.

Definition 7. *Let P be a propositional proof system. A k-tuple $(A_1, \ldots, A_k)$ of NP-sets is representable in P if there exist propositional representations $\varphi_n^i(\bar{x}, \bar{y}^i)$ of A_i for $i = 1, \ldots, k$ such that for each $1 \leq i < j \leq k$ the formulas $\varphi_n^i(\bar{x}, \bar{y}^i)$ and $\varphi_n^j(\bar{x}, \bar{y}^j)$ have only the variables $\bar{x}$ in common, and further*

$$P \vdash_* \bigwedge_{1 \leq i < j \leq k} \neg \varphi_n^i(\bar{x}, \bar{y}^i) \vee \neg \varphi_n^j(\bar{x}, \bar{y}^j) \ .$$

By $\mathsf{DNPP}_k(P)$ we denote the class of all disjoint k-tuples of NP-sets which are representable in P.

For $\mathsf{DNPP}_2(P)$ we will also write $\mathsf{DNPP}(P)$. In [3] we have analysed this class for some standard proof systems. As the classes $\mathsf{DNPP}_k(P)$ provide natural generalizations of $\mathsf{DNPP}(P)$ we have chosen the same notation for the classes of k-tuples. The next proposition shows that these classes are closed under the reductions $\leq_p$ and $\leq_s$.

Proposition 8. *Let P be a proof system that is closed under conjunctions and disjunctions and that simulates resolution. Then for all numbers $k \geq 2$ the class $\mathsf{DNPP}_k(P)$ is closed under $\leq_p$.*

Now we want to associate tuples of NP-sets with proof systems. It is not clear how the canonical pair could be modified for k-tuples but the interpolation pair [18] can be expanded to a k-tuple $(I_1(P), \ldots, I_k(P))$ by

$$I_i(P) = \{(\varphi_1, \ldots, \varphi_k, \pi) \mid \mathrm{Var}(\varphi_j) \cap \mathrm{Var}(\varphi_l) = \emptyset \text{ for all } 1 \leq j < l \leq k,$$
$$\neg \varphi_i \in \mathrm{SAT} \text{ and } P(\pi) = \bigwedge_{1 \leq j < l \leq k} \varphi_j \vee \varphi_l\}$$

for $i = 1, \ldots, k$, where $\mathrm{Var}(\varphi)$ denotes the set of propositional variables occurring in φ. This tuple still captures the feasible interpolation property of the proof system P as the next theorem shows.

Theorem 9. *Let P be a propositional proof system that is efficiently closed under substitutions by constants and conjunctions. Then $(I_1(P), \ldots, I_k(P))$ is p-separable if and only if P has the feasible interpolation property.*

Searching for canonical candidates for hard tuples for the classes $\mathsf{DNPP}_k(P)$ we modify the interpolation tuple to the following tuple $(U_1(P), \ldots, U_k(P))$ with

$$U_i(P) = \{(\varphi_1, \ldots, \varphi_k, 1^m) \mid \mathrm{Var}(\varphi_j) \cap \mathrm{Var}(\varphi_l) = \emptyset \text{ for all } 1 \leq j < l \leq k,$$
$$\neg \varphi_i \in \mathrm{SAT} \text{ and } P \vdash_{\leq m} \bigwedge_{1 \leq j < l \leq k} \varphi_j \vee \varphi_l\}$$

for $i = 1, \ldots, k$. The next theorem shows that for all reasonable proof systems P these tuples are hard for the classes $\mathsf{DNPP}_k(P)$.

Theorem 10. *Let P be a proof system that is closed under substitutions by constants. Then $(U_1(P), \ldots, U_k(P))$ is $\leq_s$-hard for $\mathsf{DNPP}_k(P)$ for all $k \geq 2$.*

Proof. Let $(A_1, \ldots, A_k)$ be a disjoint k-tuple of NP-sets and let $\varphi_n^i(\bar{x}, \bar{y}^i)$ be propositional representations of A_i for $i = 1, \ldots, k$ such that we have polynomial size P-proofs of

$$\bigwedge_{1 \leq i < j \leq k} \neg\varphi_n^i(\bar{x}, \bar{y}^i) \vee \neg\varphi_n^j(\bar{x}, \bar{y}^j) \ .$$

Then the $\leq_s$-reduction from $(A_1, \ldots, A_k)$ to $(U_1(P), \ldots, U_k(P))$ is performed by

$$a \ \mapsto \ (\neg\varphi_{|a|}^1(\bar{a}, \bar{y}^1), \ldots, \neg\varphi_{|a|}^k(\bar{a}, \bar{y}^k), 1^{p(|a|)})$$

for some suitable polynomial p.   □

5 Arithmetic Representations

In [19] and [1] arithmetic representations of disjoint NP-pairs were investigated. These form a uniform first-order counterpart to the propositional representations introduced in the previous section. We now generalize the notion of arithmetic representations to disjoint k-tuples of NP-sets.

Definition 11. *A Σ_1^b-formula φ is an* arithmetic representation *of an NP-set A if for all natural numbers a we have $\mathbb{N} \models \varphi(a)$ if and only if $a \in A$.*
A disjoint k-tuple $(A_1, \ldots, A_k)$ of NP-sets is representable *in an arithmetic theory T if there are Σ_1^b-formulas $\varphi_1(x), \ldots, \varphi_k(x)$ representing $A_1, \ldots, A_k$ such that $T \vdash (\forall x) \bigwedge_{1 \leq i < j \leq k} \neg\varphi_i(x) \vee \neg\varphi_j(x)$. The class $\mathsf{DNPP}_k(T)$ contains all disjoint k-tuples of NP-sets that are representable in T.*

We now show that the classes $\mathsf{DNPP}_k(T)$ and $\mathsf{DNPP}_k(P)$ coincide for regular proof systems P corresponding to the theory T.

Theorem 12. *Let $P \geq EF$ be a regular proof system which is closed under substitutions by constants and conjunctions and let $T \supseteq S_2^1$ be a theory corresponding to T. Then we have $\mathsf{DNPP}_k(P) = \mathsf{DNPP}_k(T)$ for all $k \geq 2$.*

At first sight Theorem 12 might come as a surprise as it states that the non-uniform and uniform concepts equal when representing disjoint k-tuples of NP-sets in regular proof systems. Uniform representations of k-tuples are translated via $\|.\|$ to non-uniform representations in a straightforward manner. For the transformation of propositional representations into first-order formulas it is, however, necessary to essentially change the representations of the components.
We now observe that the k-tuples that we associated with a proof system P are representable in P if the system is regular.

Lemma 13. *Let P be a regular proof system. Then for all numbers $k \geq 2$ the k-tuples $(I_1(P), \ldots, I_k(P))$ and $(U_1(P), \ldots, U_k(P))$ are representable in P.*

Proof. We choose straightforward arithmetic representations for the components $U_i(P)$ and $I_i(P)$. Using the reflection principle of P we can prove the disjointness of the components of the U- and I-tuples in the theory T associated with P, from which the lemma follows by Theorem 12.   □

With this lemma we can improve the hardness result of Theorem 10 to a completeness result for regular proof systems. Additionally, we can show the $\leq_s$-completeness of the interpolation tuple for $\mathsf{DNPP}_k(P)$:

Theorem 14. *Let $P \geq EF$ be a regular proof system that is efficiently closed under substitutions by constants. Then for all $k \geq 2$ the tuples $(U_1(P), \ldots, U_k(P))$ and $(I_1(P), \ldots, I_k(P))$ are $\leq_s$-complete for $\mathsf{DNPP}_k(P)$. In particular we have $(U_1(P), \ldots, U_k(P)) \equiv_s (I_1(P), \ldots, I_k(P))$.*

This theorem is true for EF as well as for all extensions $EF + \|\Phi\|$ of the extended Frege system for polynomial time sets Φ of true Π_1^b-formulas. The equivalence of the interpolation tuple and the U-tuple for strong systems as stated in Theorem 14 might come unexpected as the first idea for a reduction from the U-tuple to the I-tuple probably is to generate proofs for $\bigwedge_{1 \leq j < l \leq k} \varphi_j \vee \varphi_l$ at input $(\varphi_1, \ldots, \varphi_k, 1^m)$. This, however, is not possible for extensions of EF, because a reduction from $(U_1(P), \ldots, U_k(P))$ to $(I_1(P), \ldots, I_k(P))$ of the form $(\varphi_1, \ldots, \varphi_k, 1^m) \mapsto (\varphi_1, \ldots, \varphi_k, \pi)$ implies the automatizability of the system P. But it is known that automatizability fails for strong systems $P \geq EF$ under cryptographic assumptions [16, 18].

6   On Complete Disjoint k-Tuples of NP-Sets

In this section we will study the question whether there exist complete disjoint k-tuples of NP-sets under the reductions $\leq_p$ and $\leq_s$. We will not be able to answer this question but we will relate it to the previously studied questions whether there exist complete disjoint NP-pairs or optimal propositional proof systems. The following is the main theorem of this section:

Theorem 15. *The following conditions are equivalent:*

1. *For all numbers $k \geq 2$ there exists a $\leq_s$-complete disjoint k-tuple of NP-sets.*
2. *For all numbers $k \geq 2$ there exists a $\leq_p$-complete disjoint k-tuple of NP-sets.*
3. *There exists a $\leq_p$-complete disjoint NP-pair.*
4. *There exists a number $k \geq 2$ such that there exists a $\leq_p$-complete disjoint k-tuple of NP-sets.*
5. *There exists a propositional proof system P such that for all numbers $k \geq 2$ all disjoint k-tuples of NP-sets are representable in P.*
6. *There exists a propositional proof system P such that all disjoint NP-pairs are representable in P.*

Proof. (Sketch) The proof is structured as follows: $1 \Rightarrow 2 \Rightarrow 3 \Rightarrow 6 \Rightarrow 1$ and $3 \Leftrightarrow 4$, $5 \Leftrightarrow 6$. Apparently, items 1 to 4 are conditions of decreasing strength.

For the implication $3 \Rightarrow 6$ assume that (A, B) is a $\leq_p$-complete pair. We choose some representations φ_n and ψ_n for A and B, respectively. Using Proposition 8 we can show that all disjoint NP-pairs are representable in the proof system $EF + \{\neg\varphi_n \vee \neg\psi_n \mid n \geq 0\}$.

The most interesting part of the proof is the implication $6 \Rightarrow 1$. Assuming that all pairs are representable in the proof system P we first choose a system $Q \geq P$ with sufficient closure properties. Then for each disjoint k-tuple $(A_1, \ldots, A_k)$ all pairs (A_i, A_j) are representable in Q. However, we might need different representations for the sets A_i to prove the disjointness of all these pairs. For example proving $A_1 \cap A_2 = \emptyset$ and $A_1 \cap A_3 = \emptyset$ might require two different representations for A_1. For this reason we cannot simply reduce $(A_1, \ldots, A_k)$ to $(U_1(Q), \ldots, U_k(Q))$. But we can reduce $(A_1, \ldots, A_k)$ to a suitable modification of the U-tuple of Q, thereby showing the $\leq_s$-completeness of this tuple.   □

Using Theorem 12 we can also characterize the existence of complete disjoint k-tuples of NP-sets by a condition on arithmetic theories, thereby extending the list of characterizations from Theorem 15 by the following item:

Theorem 16. *The conditions 1 to 6 of Theorem 15 are equivalent to the existence of a finitely axiomatized arithmetic theory in which all disjoint k-tuples of NP-sets are representable for all $k \geq 2$.*

In Theorem 15 we stated that the existence of complete disjoint NP-pairs is equivalent to the existence of a proof system P in which every NP-pair is representable. By definition this condition means that for all disjoint NP-pairs there exists a representation for which the disjointness of the pair is provable with short P-proofs. If we strengthen this condition by requiring that this is possible for all disjoint NP-pairs and all representations we arrive at a condition which is strong enough to characterize the existence of optimal proof systems.

Theorem 17. *The following conditions are equivalent:*

1. *There exists an optimal propositional proof system.*
2. *There exists a propositional proof system P such that for all $k \geq 2$ the system P proves the disjointness of all disjoint k-tuples of NP-sets with respect to all representations, i.e. for all disjoint k-tuples $(A_1, \ldots, A_k)$ of NP-sets and all representations $\varphi_n^1, \ldots, \varphi_n^k$ of $A_1, \ldots, A_k$ we have $P \vdash_* \bigwedge_{1 \leq i < j \leq k} \neg\varphi_n^i \vee \neg\varphi_n^j$.*
3. *There exists a propositional proof system P that proves the disjointness of all disjoint NP-pairs with respect to all representations.*

Proof. (Sketch) For the implication $1 \Rightarrow 2$ let P be an optimal proof system. For all choices of representations of k-tuples the sequence of tautologies expressing the disjointness of the tuple can be generated in polynomial time. Therefore these sequences have polynomial size P-proofs.

For $3 \Rightarrow 1$ we use the following fact: if optimal proof systems do not exist, then every proof system P admits hard sequences of tautologies, i.e. the sequence can be generated in polynomial time but does not have polynomial size P-proofs. Given a proof system P and an NP-pair (A, B) we code these hard tautologies into propositional representations of A and B and obtain representations for which P does not prove the disjointness of (A, B).   □

As an immediate corollary to Theorems 15 and 17 we get a strengthening of a theorem of Köbler, Messner and Torán [10], stating that the existence of optimal proof systems implies the existence of $\leq_s$-complete disjoint NP-pairs:

Corollary 18. *If there exist optimal propositional proof systems, then there exist $\leq_s$-complete disjoint k-tuples of NP-sets for all numbers $k \geq 2$.*

Acknowledgements. For helpful conversations and suggestions on this work I am very grateful to Johannes Köbler, Jan Krajíček, Pavel Pudlák, and Zenon Sadowski.

References

1. O. Beyersdorff. Representable disjoint NP-pairs. In *Proc. 24th Conference on Foundations of Software Technology and Theoretical Computer Science*, pages 122–134, 2004.
2. O. Beyersdorff. Tuples of disjoint NP-sets. Technical Report TR05-123, Electronic Colloquium on Computational Complexity, 2005.
3. O. Beyersdorff. Disjoint NP-pairs from propositional proof systems. In *Proc. 3rd Conference on Theory and Applications of Models of Computation*, 2006.
4. S. R. Buss. *Bounded Arithmetic*. Bibliopolis, Napoli, 1986.
5. S. A. Cook and R. A. Reckhow. The relative efficiency of propositional proof systems. *The Journal of Symbolic Logic*, 44:36–50, 1979.
6. C. Glaßer, A. L. Selman, and S. Sengupta. Reductions between disjoint NP-pairs. In *Proc. 19th Annual IEEE Conference on Computational Complexity*, pages 42–53, 2004.
7. C. Glaßer, A. L. Selman, S. Sengupta, and L. Zhang. Disjoint NP-pairs. *SIAM Journal on Computing*, 33(6):1369–1416, 2004.
8. C. Glaßer, A. L. Selman, and L. Zhang. Survey of disjoint NP-pairs and relations to propositional proof systems. Technical Report TR05-072, Electronic Colloquium on Computational Complexity, 2005.
9. J. Grollmann and A. L. Selman. Complexity measures for public-key cryptosystems. *SIAM Journal on Computing*, 17(2):309–335, 1988.
10. J. Köbler, J. Messner, and J. Torán. Optimal proof systems imply complete sets for promise classes. *Information and Computation*, 184:71–92, 2003.
11. J. Krajíček. *Bounded Arithmetic, Propositional Logic, and Complexity Theory*, volume 60 of *Encyclopedia of Mathematics and Its Applications*. Cambridge University Press, Cambridge, 1995.
12. J. Krajíček. Interpolation theorems, lower bounds for proof systems and independence results for bounded arithmetic. *The Journal of Symbolic Logic*, 62(2): 457–486, 1997.
13. J. Krajíček. Dual weak pigeonhole principle, pseudo-surjective functions, and provability of circuit lower bounds. *The Journal of Symbolic Logic*, 69(1):265–286, 2004.
14. J. Krajíček and P. Pudlák. Propositional proof systems, the consistency of first order theories and the complexity of computations. *The Journal of Symbolic Logic*, 54:1963–1079, 1989.

15. J. Krajíček and P. Pudlák. Quantified propositional calculi and fragments of bounded arithmetic. *Zeitschrift für mathematische Logik und Grundlagen der Mathematik*, 36:29–46, 1990.

16. J. Krajíček and P. Pudlák. Some consequences of cryptographical conjectures for S_2^1 and *EF*. *Information and Computation*, 140(1):82–94, 1998.

17. R. E. Ladner. On the structure of polynomial-time reducibility. *Journal of the ACM*, 22:155–171, 1975.

18. P. Pudlák. On reducibility and symmetry of disjoint NP-pairs. *Theoretical Computer Science*, 295:323–339, 2003.

19. A. A. Razborov. On provably disjoint NP-pairs. Technical Report TR94-006, Electronic Colloquium on Computational Complexity, 1994.

Constructive Equivalence Relations on Computable Probability Measures

Laurent Bienvenu

Laboratoire d'Informatique Fondamentale, Marseille, France

Abstract. We study the equivalence relations on probability measures corresponding respectively to having the same Martin-Löf random reals, having the same Kolmogorov-Loveland random reals, and having the same computably random reals. In particular, we show that, when restricted to the class of strongly positive generalized Bernoulli measures, they all coincide with the classical equivalence, which requires that two measures have the same nullsets.

1   Introduction

Since the first attempt made in 1919 by R. von Mises to define what it means for an infinite sequence of zeros and ones to be random, many definitions of randomness have been proposed. The most satisfactory so far was given in 1966 by P. Martin-Löf (and is now called Martin-Löf randomness), but some other proposals have also received a lot of attention, such as Mises-Wald-Churh stochasticity, Kolmogorov-Loveland stochasticity, Kolmogorov-Loveland randomness, Schnorr randomness, Kurtz randomness, computable randomness, etc. (for an excellent and detailed survey, see [1]). Although they were originally meant to describe randomness relative to the uniform measure, their definition can often be extended to other (computable) measures. It is for example the case for all the above notions, except for stochasticity (it relies on the law of large numbers, which does not hold for all computable measures). Relations between the different notions have been extensively studied. In this paper, we propose a different approach, as we look at these notions from the measure point of view. In classical probability theory, two probability measures are said to be *equivalent* if they have the same nullsets, or, in other words, if they have the same sets of measure 1, which means that they are in some sense quite similar. Since defining a notion of randomness means choosing for each computable measure μ a particular set of μ-measure 1 and calling its elements *random*, it is natural to define at the same time a constructive equivalence relation, saying that two measures are similar if they have the same random elements.

This is what we do here, focusing on three particular notions of randomness: Martin-Löf randomness, Kolmogorov-Loveland randomness, and computable randomness. In Section 3, we discuss the particular case of generalized Bernoulli measures. In classical probability theory, Kakutani's theorem provides a very simple characterization of equivalence for generalized Bernoulli measures. As

D. Grigoriev, J. Harrison, and E.A. Hirsch (Eds.): CSR 2006, LNCS 3967, pp. 92–103, 2006.
© Springer-Verlag Berlin Heidelberg 2006

Vovk did with Martin-Löf randomness, we prove an analogue of Kakutani's theorem in terms of Kolmogorov-Loveland randomness and computable randomness (we in fact prove the second half, the first having been done by Muchnik et al. in [5]).

In Section 4, we study these three equivalence relations for arbitrary computable measures. Theorem 23 is a first step in comparing them; we show in particular that, for two computable measures, having the same Martin-Löf random elements is a stronger condition than classical equivalence.

2   Definitions and Concepts

2.1   Measures and Semimeasures on the Cantor Space

The Cantor space is the set $\{0,1\}^\omega$ (which we abbreviate by 2^ω) of infinite binary sequences (also called reals) endowed with the product topology. For all $u \in \{0,1\}^*$, we denote by $\mathcal{O}_u$ the open set $\{\alpha \in 2^\omega : \forall i < |u| \; \alpha_i = u_i\}$. The set $\{\mathcal{O}_u : u \in \{0,1\}^*\}$ is a base for the product topology, and gives us a handy way to describe measures on 2^ω:

Theorem 1 (Caratheodory's extension theorem). *Let f be a real function, taking its values in $[0,1]$, and such that $f(2^\omega) = 1$ and for all $u \in \{0,1\}^*$, $f(\mathcal{O}_u) = f(\mathcal{O}_{u0}) + f(\mathcal{O}_{u1})$. There exists a unique measure μ on 2^ω which extends f.*

Hence, from now on we can identify a measure with its restriction to the open sets $\mathcal{O}_u$'s. The canonical measure on 2^ω is the Lebesgue measure λ, defined by $\lambda(\mathcal{O}_u) = 2^{-|u|}$ for all u (it is of course computable).

The notion of measure can be extended to the notion of semi-measure:

Definition 2. *A semimeasure is a real function μ defined on $\{\mathcal{O}_u : u \in \{0,1\}^*\}$, and taking its values in $[0,1]$, such that for all $u \in \{0,1\}^*$: $\mu(\mathcal{O}_u) \geqslant \mu(\mathcal{O}_{u0}) + \mu(\mathcal{O}_{u1})$.*

In the remaining of this paper, we often abbreviate $\mu(\mathcal{O}_u)$ by $\mu(u)$.

Definition 3. *We say that a (semi)measure μ is computable if the function $u \mapsto \mu(\mathcal{O}_u)$ is computable.*

We say that a semimeasure μ is enumerable if there exists a computable function $h : \{0,1\}^* \times \mathbb{N} \to \mathbb{R}$ such that for all $u \in \{0,1\}^*$, $n \mapsto h(u,n)$ is non-decreasing and $\lim_{n \to +\infty} h(u,n) = \mu(\mathcal{O}_u)$.

We finally recall the classical definition of equivalence:

Definition 4. *Two measures μ and ν are equivalent if for all $X \subseteq 2^\omega$:*

$$\mu(X) = 0 \leftrightarrow \nu(X) = 0$$

They are said to be inconsistent if there exists a set $Y \subseteq 2^\omega$ whose measure is 1 for either μ or ν, and 0 for the other measure.

2.2 Martin-Löf Randomness

Definition 5. *An open set* $\mathcal{V}$ *is said to be computably enumerable (c.e.) if there exists a computably enumerable* $A \subset \{0, 1\}^*$ *such that* $\mathcal{V} = \bigcup_{u \in A} \mathcal{O}_u$.

A collection $\{\mathcal{V}_n : n \in \mathbb{N}\}$ *of c.e. open sets is said to be computable if there exists a computable function* $f : \mathbb{N}^2 \to \{0, 1\}^*$ *such that for all* n, $\mathcal{V}_n = \bigcup_{k \in \mathbb{N}} \mathcal{O}_{f(n,k)}$.

A μ-*Martin-Löf test is a computable collection of c.e. sets* $\{\mathcal{V}_n\}_n$ *such that for all* n, $\mu(\mathcal{V}_n) \leqslant 2^{-n}$.

$\alpha \in 2^\omega$ *is said to pass the* μ-*Martin-Löf test* $\{\mathcal{V}_n\}_n$ *if* $\alpha \notin \bigcap_n \mathcal{V}_n$.

$\alpha \in 2^\omega$ *is said to be* μ-*Martin-Löf random (*μ-*ML-random for short) if it passes all* μ-*Martin-Löf tests. We denote by* $\mu\mathbf{MLR}$ *the set of* μ-*ML-random infinite sequences.*

For every μ-Martin-Löf test $\{\mathcal{V}_n\}_n$, it is obvious that $\mu(\bigcap_n \mathcal{V}_n) = 0$. Since there are only countably many μ-Martin-Löf tests, it immediately follows that:

Proposition 6. *For every computable measure* μ: $\mu(\mu\mathbf{MLR}) = 1$

We will use the following fundamental theorem, which gives a pure measure-theoretic characterization of ML randomness:

Theorem 7 (Levin). *(a) There exists a universal enumerable semi-measure, that is there exists an enumerable semimeasure* $\mathbf{M}$ *such that for every enumerable semimeasure* μ *there exists a real constant* $c > 0$ *such that* $\mathbf{M} \geqslant c\mu$.

(b) Let μ *be a computable measure, and* $\alpha \in 2^\omega$. $\alpha \in \mu\mathbf{MLR}$ *if and only if* $\left\{ \frac{\mathbf{M}(\alpha_{[0,n]})}{\mu(\alpha_{[0,n]})} : n \in \mathbb{N} \right\}$ *is bounded. Equivalently,* $\alpha \in \mu\mathbf{MLR}$ *if and only if for every enumerable semimeasure* ν, $\left\{ \frac{\nu(\alpha_{[0,n]})}{\mu(\alpha_{[0,n]})} : n \in \mathbb{N} \right\}$ *is bounded.*

2.3 Kolmogorov-Loveland Randomness and Computable Randomness

C. Schnorr argued (in [6] and [7]) that the notion of Martin-Löf randomness is not fully satisfactory as a notion of effective randomness, since the definition of a Martin-Löf test involves open sets which are only *computably enumerable*, and not computable. He then proposed two alternative (and weaker) notions, which are now known as Schnorr randomness (which we will not discuss here) and computable randomness. The latter relies on the "unpredictability principle": we want to define a sequence α as being random if there is no computable strategy which asymptotically wins an infinite amount of money by making bets on the values of α's bits. In [5], Muchnik et al. defined another (stronger than these last two) notion of randomness, based on the same principle, which, following Merkle et al. in [4], we call Kolmogorov-Loveland randomness. Intuitively, it is based on the following infinite game, which depends on a measure μ. Suppose that all the bits of an infinite sequence α are initially hidden. A player, whose initial capital is 1, tries to guess its bits. At the n-th move, the player chooses a bit which has not been revealed yet and predicts its value (0 or 1) by betting an amount of money which does not exceed his capital. The bit is then revealed to the player. If his guess was wrong, the player loses his stake. If it was correct, the player wins an amount of money which is equal to his stake multiplied by $\frac{\mu(bet\ is\ incorrect|history)}{\mu(bet\ is\ correct|history)}$. This factor may seem strange at first, but it ensures that the game is fair, i.e. that the player's expectancy is 0 at every move. Suppose we know the bit we are about to bet on has μ-probability $\frac{9}{10}$ to be 1, and we predict its value to be 1. Since we're not taking a huge risk, if we turn out to be correct, our reward will be smaller than our stake (in this particular example, it will be equal to $\frac{1}{9}$ times our stake). We now make this more formal:

Definition 8. *A finite assignment (f.a.) is a sequence*

$$x = (k_0, v_0)...(k_{n-1}, v_{n-1}) \in (\mathbb{N} \times \{0, 1\})^*$$

where the k_i's are pairwise distinct. The set $\{k_0, k_1, ..., k_{n-1}\}$ is called the domain of x and is denoted by $dom(x)$. We denote by $x_1.x_2$ the concatenation of two consistent finite assignments. The set of all finite assignments is denoted by FA.

For all $x \in FA$, $x = (k_0, v_0)...(k_{n-1}, v_{n-1})$, we define $\mathcal{O}_x = \{\alpha \in 2^\omega : \forall i \in [0, n)\ \alpha_{k_i} = v_i\}$, and abbreviate $\mu(\mathcal{O}_x)$ by $\mu(x)$.

Definition 9. *A strategy is a (total) function $S : FA \to \mathbb{N} \times \{0, 1\} \times [0, 1]$ such that for all x, the first component of $S(x)$ is not in $dom(x)$. We call an element of $\mathbb{N} \times \{0, 1\} \times [0, 1]$ a bet.*

S is said to be monotonic if for all $x \in FA$, the first component of $S(x)$ is greater than every $k \in dom(x)$.

This definition means that, having already bet on bits $k_0, ..., k_{n-1}$ and having read the corresponding values $v_0, ..., v_{n-1}$, the player bets a fraction ρ of his capital on the fact that $\alpha_{k_n} = v_n$, where $S((k_0, v_0), ..., (k_{n-1}, v_{n-1})) = (k_n, v_n, \rho)$.

We now see how to run the strategy S on the sequence α. As the game depends on the reference measure μ, we call it the μ-game of S against α. We let x_0 be the empty finite assignment, and we set $V_0(\alpha, S, \mu) = 1$. We define by induction:

- $S(x_n) = (k_n, v_n, \rho_n)$
- $x_{n+1} = x_n.(k_n, \alpha_n)$
- If $v_n \neq \alpha_{k_n}$, $V_{n+1}(\alpha, S, \mu) = V_n(\alpha, S, \mu)(1 - \rho_n)$. And if $v_n = \alpha_{k_n}$, $V_{n+1}(\alpha, S, \mu) = V_n(\alpha, S, \mu)(1 + \rho_n \frac{\mu(x_n.(k_n, \bar{v}_n))}{\mu(x_n.(k_n, v_n))})$ where $\bar{v}_n$ is 0 if v_n is 1, and 1 if v_n is 0. By convention, if $\mu(x_n.(k_n, v_n)) = 0$ and $\rho_n \neq 0$, we set $V_{n+1}(\alpha, S, \mu) = +\infty$

x_n represents the history of the game before the n-th move (by convention, there is a 0-th move), and $V_n(\alpha, S, \mu)$ represents the player's capital before the n-th move.

We say that S succeeds on α in the μ-game if $\limsup_n V_n(\alpha, S, \mu) = +\infty$

Definition 10. *An infinite sequence α is said to be μ-Kolmogorov-Loveland random (respectively μ-computably random) if no computable strategy (respectively computable monotonic strategy) succeeds on it in the μ-game. We denote by μKLR the set of μ-Kolmogorov-Loveland random sequences, and by μCR the set of μ-computably random sequences.*

Proposition 11 (Muchnik et al. [5]). *For all computable measures μ, we have:*

$$\mu\mathbf{MLR} \subseteq \mu\mathbf{KLR} \subseteq \mu\mathbf{CR}$$

If μ is taken to be the uniform measure, the second inclusion is strict. One of the most important open questions in the field of algorithmic randomness is whether or not, in this case, the first inclusion is strict as well. It is not even known whether there exists some computable measure μ such that the first inclusion is strict.

3   Generalized Bernoulli Measures

We start our discussion with the class of generalized Bernoulli measures. Intuitively, a generalized Bernoulli measure corresponds to choosing an infinite sequence in $\alpha \in 2^\omega$, where the bits are chosen independently by biased coin tosses, such that the probability of α_i to be 1 is p_i (that is, depends on i, and only on i).

Definition 12. *Let $\{p_i\}_{i \in \mathbb{N}}$ be a sequence of real numbers such that $p_i \in [0,1]$ for all i. The generalized Bernoulli measure μ of parameter $\{p_i\}_i$ is defined by*

$$\mu(\mathcal{O}_u) = \prod_{i<|u|,\ \alpha_i=0} (1-p_i) \prod_{i<|u|,\ \alpha_i=1} p_i$$

It is said to be strongly positive if there exists $\varepsilon > 0$ such that for all i, $p_i \in [\varepsilon, 1-\varepsilon]$.

Remark: The generalized Bernoulli measure μ of parameter $\{p_i\}_i$ is computable if and only if $\{p_i\}_i$ is a computable sequence of real numbers.

The class of generalized Bernoulli measures is of high importance in the field of algorithmic randomness. It is indeed one of the simplest extensions of Lebesgue measure and has interesting applications. For example, as we will see below, it has been used by Shen' to distinguish between two notions of randomness, namely Martin-Löf randomness and Kolmogorov-Loveland stochasticity (for a definition of the latter, see for example [4]), the equivalence of which was left as an open question by Kolmogorov.

In 1948, Kakutani gave a criterion for two generalized Bernoulli measures to be equivalent.

Theorem 13 (Kakutani [2]). *Let μ and ν be two strongly positive generalized Bernoulli measures, respectively of parameter $\{p_i\}_i$ and $\{q_i\}_i$.*
(a) If $\sum_i (p_i - q_i)^2 < +\infty$, then $\mu \sim \nu$.
(b) If $\sum_i (p_i - q_i)^2 = +\infty$, then μ and ν are inconsistent.

Vovk proved an analogue of this theorem in terms of ML-randomness:

Theorem 14 (Vovk [10]). *Let μ and ν be two computable strongly positive generalized Bernoulli measures, respectively of parameter $\{p_i\}_i$ and $\{q_i\}_i$. We have:*
*(a) If $\sum_i (p_i - q_i)^2 < +\infty$, then μ**MLR** $= \nu$**MLR***
*(b) If $\sum_i (p_i - q_i)^2 = +\infty$, then μ**MLR** $\cap\ \nu$**MLR** $= \emptyset$*

This last theorem is quite fundamental, as it is one of the main ingredients used in [8] to prove that KL-stochasticity is not equivalent to Martin-Löf randomness with respect to the uniform measure:

Theorem 15 (Shen' [8] - van Lambalgen [9]). *(i) If μ is a computable generalized Bernoulli measure of parameter $\{p_i\}_i$ such that $\lim p_i = \frac{1}{2}$, then μ**MLR** $\subseteq$ **Stoch** (**Stoch** denotes the set of KL-stochastic sequences).*
*(ii) There exists a computable measure μ such that $\mu($**Stoch**$) = 1$ whereas $\mu(\lambda$**MLR**$) = 0$ (which obviously implies that these two sets are distinct).*

(to get (ii) from (i), it suffices to take $p_i = \frac{1}{2} + \frac{1}{\sqrt{i+4}}$ and apply Theorem 14)

In turn, Muchnik et al. strengthened the part (b) of Theorem 14 as follows:

Theorem 16 (Muchnik et al [5]). *Let μ and ν be two computable strongly positive generalized Bernoulli measures, respectively of parameter $\{p_i\}_i$ and $\{q_i\}_i$. If $\sum_i (p_i - q_i)^2 = +\infty$, then μ**CR** $\cap\ \nu$**CR** $= \emptyset$ (a fortiori, μ**KLR** $\cap\ \nu$**KLR** $= \emptyset$ and μ**MLR** $\cap\ \nu$**MLR** $= \emptyset$).*

Looking at Theorem 14 and Theorem 16, it is natural to ask whether Theorem 14.a holds if one replaces **MLR** by **CR** or **KLR**. This is indeed the case, and we will see later on that this in fact strengthens Theorem 14.a.

Theorem 17. *Let μ and ν be two computable strongly positive generalized Bernoulli measures, respectively of parameter $\{p_i\}_i$ and $\{q_i\}_i$. If $\sum_i (p_i - q_i)^2 < +\infty$, then μ**CR** $= \nu$**CR** and μ**KLR** $= \nu$**KLR**.*

We will prove the following proposition. It immediately implies Theorem 17 since by a well known result (see for example [7]), if there is a computable strategy S such that $\limsup_n V_n(\alpha, S, \mu) = +\infty$, there exists a computable strategy S' such that $\lim_n V_n(\alpha, S', \mu) = +\infty$

Proposition 18. *Let μ and ν be two computable strongly positive generalized Bernoulli measures, respectively of parameter $\{p_i\}_i$ and $\{q_i\}_i$, with $\sum_i (p_i-q_i)^2 < +\infty$. Let $\alpha \in 2^\omega$, and suppose there exists a computable strategy S_1 such that $\lim V_n(\alpha, S_1, \mu) = +\infty$. Then, there exists a computable strategy S_2 such that for all n (up to an additive constant): $V_n(\alpha, S_2, \nu) \geqslant \ln V_n(\alpha, S_1, \mu)$. Moreover, if S_1 is monotonic, S_2 can be taken to be monotonic as well.*

Proof: Suppose there exists such a strategy S_1. Let S_2 be the strategy which simulates S_1 and, at the n-th move, when S_1 makes a bet (i_n, v_n, ρ_n), makes a bet $(i_n, v_n, \frac{\rho_n}{V_n(\alpha, S_2, \nu)})$ Intuitively, this means that when S_1 bets a fraction ρ of its capital, S_2 bets the amount ρ, independently of its capital. Notice that S_2 might not have enough money to afford this bet; we will discuss this at the end of the proof.

During the n-th move, there are three cases:

	$\left\vert V_{n+1}(\alpha, S_1, \mu)/V_n(\alpha, S_1, \mu) \right.$	$\left. V_{n+1}(\alpha, S_2, \nu) - V_n(\alpha, S_2, \nu) \right\vert$
P loses	$1 - \rho_n$	$-\rho_n$
P wins and $\alpha_{i_n} = 0$	$1 + \rho_n \frac{p_{i_n}}{1 - p_{i_n}}$	$\rho_n \frac{q_{i_n}}{1 - q_{i_n}}$
P wins and $\alpha_{i_n} = 1$	$1 + \rho_n \frac{1 - p_{i_n}}{p_{i_n}}$	$\rho_n \frac{1 - q_{i_n}}{q_{i_n}}$

Let x_n be either $-\rho_n$ or $\rho_n \frac{p_{i_n}}{1 - p_{i_n}}$ or $\rho_n \frac{1 - p_{i_n}}{p_{i_n}}$, depending on the result of the bet. With this notation, we have:

	$\left\vert V_{n+1}(\alpha, S_1, \mu)/V_n(\alpha, S_1, \mu) \right.$	$\left. V_{n+1}(\alpha, S_2, \nu) - V_n(\alpha, S_2, \nu) \right\vert$
P loses	$1 + x_n$	x_n
P wins and $\alpha_{i_n} = 0$	$1 + x_n$	$x_n(1 + \frac{q_{i_n} - p_{i_n}}{p_{i_n}(1 - q_{i_n})})$
P wins and $\alpha_{i_n} = 1$	$1 + x_n$	$x_n(1 + \frac{p_{i_n} - q_{i_n}}{q_{i_n}(1 - p_{i_n})})$

It follows by induction:

$$V_n(\alpha, S_1, \mu) = \prod_{k=0}^{n-1}(1 + x_k)$$

By the hypothesis of strong positivity, let $\varepsilon > 0$ be such that for all i, $p_i \in [\varepsilon, 1-\varepsilon]$, and $q_i \in [\varepsilon, 1-\varepsilon]$. By definition of x_n, we have for all n: $-1 \leqslant x_n \leqslant \varepsilon^{-1}$. Let C, C' and C'' be three positive constants such that:

- $C = \sqrt{\sum_{i=0}^{\infty} |p_i - q_i|^2}$.
- for all $t \in [-1, \varepsilon^{-1}]$: $ln(1 + t) \leqslant t - C't^2$.
- for all $t \geqslant 0$: $\varepsilon^{-2}C\sqrt{t} \leqslant C't + C''$.

We then have:

$$ln\, V_n(\alpha, S_1, \mu) = \sum_{k=0}^{n-1} ln(1 + x_k) \leqslant \sum_{k=0}^{n-1}(x_k - C'x_k^2)$$

Concerning the strategy S_2, we have in the three above cases:

$$V_{n+1}(\alpha, S_2, \nu) - V_n(\alpha, S_2, \nu) \geqslant x_n - \varepsilon^{-2}|x_n||p_{i_n} - q_{i_n}|$$

Hence, by induction, for all n:

$$V_n(\alpha, S_2, \nu) \geqslant \sum_{k=0}^{n-1}(x_k - \varepsilon^{-2}|x_k||p_{i_k} - q_{i_k}|)$$

By the Cauchy-Schwarz inequality:

$$V_n(\alpha, S_2, \nu) \geqslant \sum_{k=0}^{n-1} x_k - \varepsilon^{-2} \sqrt{\sum_{k=0}^{n-1} |p_{i_k} - q_{i_k}|^2} \sqrt{\sum_{k=0}^{n-1} x_k^2}$$

By definition of C, C', and C'' :

$$V_n(\alpha, S_2, \nu) \geqslant \sum_{k=0}^{n-1} (x_k - C' x_k^2) - C''$$

and thus

$$V_n(\alpha, S_2, \nu) \geqslant ln \, V_n(\alpha, S_1, \mu) - C''$$

Recall that we assumed $\lim_n V_n(\alpha, S_1, \mu) = +\infty$. It follows that $\lim_n V_n(\alpha, S_2, \nu) = +\infty$. Hence, without loss of generality (up to modifying S_2 for a finite number of bets), we can assume that for all n, $V_n(\alpha, S_2, \nu) \geqslant 1$, and thus, S_2 is always allowed at the n-th move to make the bet $(k_n, \mathrm{v}_n, \frac{p_n}{V_n(\alpha, S_2, \nu)})$. Thus, S_2 wins against α in the ν-game. Moreover, by construction, if S_1 is monotonic, S_2 is monotonic as well. This means that for all $\alpha \in 2^\omega$: if $\alpha \notin \mu\mathbf{KLR}$, then $\alpha \notin \nu\mathbf{KLR}$ and if $\alpha \notin \mu\mathbf{CR}$, then $\alpha \notin \nu\mathbf{CR}$. By symmetry, the proposition is proven. ☐

Remark: Although we do not discuss these notions here, we can apply Proposition 18 to get Theorem 17 for other notions of randomness which can be defined in terms of strategies, like Schnorr randomness or Kurtz randomness (see for example [1]).

The hypothesis of strong positivity of the above theorems cannot be removed, as asserted by the following proposition, which is a simple effective version of the Borel-Cantelli lemma.

Proposition 19. *Let $\{p_i\}_i$ be a computable sequence taking its values in $(0,1)$, converging to 0. Let μ be the generalized Bernoulli measure of parameter $\{p_i\}_i$.*

(a) If $\sum_i p_i < +\infty$, then $\mu\mathbf{MLR} = \mu\mathbf{KLR} = \mu\mathbf{CR} = \{0,1\}^* 0^\omega$
(b) If $\sum_i p_i = +\infty$, then $\mu\mathbf{CR} \cap \{0,1\}^* 0^\omega = \emptyset$ (a fortiori $\mu\mathbf{KLR} \cap \{0,1\}^* 0^\omega = \emptyset$, and $\mu\mathbf{MLR} \cap \{0,1\}^* 0^\omega = \emptyset$)

(We omit the proof of this fact, which is rather simple).
Thus, although $\sum_i (\frac{1}{i+1} - \frac{1}{(i+1)^2})^2 < +\infty$, the measures μ and ν of respective parameter $\{\frac{1}{i+1}\}_i$ and $\{\frac{1}{(i+1)^2}\}_i$ do not satisfy $\mu\mathbf{MLR} = \nu\mathbf{MLR}$ nor $\mu\mathbf{KLR} = \nu\mathbf{KLR}$.

4 Arbitrary Computable Measures

We now turn our attention to arbitrary computable measures. We first show that, similarly to Theorem 7, computable randomness has a purely measure-theoretical characterization:

Proposition 20. *Let μ be a computable measure, and $\alpha \in 2^\omega$. Then $\alpha \in \mu\mathbf{CR}$ if and only if for all computable measure ν, $\{\frac{\nu(\alpha_{[0,n]})}{\mu(\alpha_{[0,n]})} : n \in \mathbb{N}\}$ is bounded.*

Proof: Suppose there exists a measure ν such that $\sup\{\frac{\nu(\alpha_{[0,n]})}{\mu(\alpha_{[0,n]})} : n \in \mathbb{N}\} = +\infty$.
We use a remark which can be found in [5], stating that the quantity $\frac{\nu(\alpha_{[0,n]})}{\mu(\alpha_{[0,n]})}$ can be interpreted as the capital after the n-th move of some strategy playing the μ-game against α. Let S be the strategy which at the n-th move:

- if $\frac{\mu(\alpha_{[0,n-1]})\nu(\alpha_{[0,n-1]}1)}{\mu(\alpha_{[0,n-1]}1)\nu(\alpha_{[0,n-1]})} \leqslant 1$, S bets $(n, 0, 1 - \frac{\mu(\alpha_{[0,n-1]})\nu(\alpha_{[0,n-1]}1)}{\mu(\alpha_{[0,n-1]}1)\nu(\alpha_{[0,n-1]})})$
- else, S bets $(n, 1, 1 - \frac{\mu(\alpha_{[0,n-1]})\nu(\alpha_{[0,n-1]}0)}{\mu(\alpha_{[0,n-1]}0)\nu(\alpha_{[0,n-1]})})$

By a simple calculation, in the μ-game, for all n: $V_n(\alpha, S, \mu) = \frac{\nu(\alpha_{[0,n-1]})}{\mu(\alpha_{[0,n-1]})}$, and hence, by the hypothesis, $\alpha \notin \mu\mathbf{CR}$.

Conversely, suppose that there exists a monotonic strategy S such that in the μ-game: $\limsup_n V_n(\alpha, S, \mu) = +\infty$. Without loss of generality, we can assume that for every $\beta \in 2^\omega$, S bets on every bit in order (up to betting 0 on the bits S skips). We then define the measure ν by

$$\nu(u) = V_{|u|}(u0^\omega, S)\mu(u)$$

(since by the fairness of the μ-game, for all u, $V_{|u|}(u0^\omega, S, \mu) = V_{|u0|}(u00^\omega, S, \mu)\mu(u0) + V_{|u1|}(u10^\omega, S)\mu(u1)$, this does define a measure). S being a winning strategy in the μ-game against α, if follows that $\limsup_n \frac{\nu(\alpha_{[0,n]})}{\mu(\alpha_{[0,n]})} = +\infty$. $\qquad \square$

From the remark we mentioned in the above proof, Muchnik et al. derived the following important proposition:

Proposition 21 (Muchnik [5]). *Let μ and ν be two computable measures. For all $\alpha \in 2^\omega$, if $\alpha \in \mu\mathbf{MLR} \setminus \nu\mathbf{MLR}$, then $\alpha \notin \nu\mathbf{CR}$*

Proof: Since $\alpha \in \mu\mathbf{MLR}$, there exists a $C > 0$ such that for all n: $\frac{\mathbf{M}(\alpha_{[0,n]})}{\mu(\alpha_{[0,n]})} \leqslant C$ (and hence $\frac{\mu(\alpha_{[0,n]})}{\mathbf{M}(\alpha_{[0,n]})} \geqslant C^{-1}$).

Since $\alpha \notin \nu\mathbf{MLR}$, $\limsup_n \frac{\mathbf{M}(\alpha_{[0,n]})}{\nu(\alpha_{[0,n]})} = +\infty$. Combining the two, we get

$$\limsup_n \frac{\mu(\alpha_{[0,n]})}{\mathbf{M}(\alpha_{[0,n]})} \frac{\mathbf{M}(\alpha_{[0,n]})}{\nu(\alpha_{[0,n]})}) = +\infty \quad i.e. \quad \limsup_n \frac{\mu(\alpha_{[0,n]})}{\nu(\alpha_{[0,n]})} = +\infty$$

Applying Proposition 20, we get $\alpha \notin \nu\mathbf{CR}$. $\qquad \square$

This last proposition has a quite negative consequence: it is not possible to find a computable measure which, similarly to Theorem 15, would separate the notions of μ-ML randomness and μ-KL randomness. In fact, it is even impossible to find a computable measure wich separates μ-ML randomness and μ-computable randomness:

Proposition 22. *For all computable measures μ and ν: $\nu(\mu\mathbf{CR} \setminus \mu\mathbf{MLR}) = 0$*

Proof: By Proposition 6, $\nu(\nu\mathbf{MLR}) = 1$, so: $\nu(\mu\mathbf{CR} \setminus \mu\mathbf{MLR}) = \nu(\nu\mathbf{MLR} \cap \mu\mathbf{CR} \setminus \mu\mathbf{MLR})$. But by Proposition 21, $\nu\mathbf{MLR} \cap \mu\mathbf{CR} \setminus \mu\mathbf{MLR} = \emptyset$, hence the result. □

Proposition 23. *For all computable measures μ and ν, we have the following implications:*

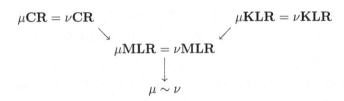

Proof: The first two implications are a direct consequence of Proposition 21. Suppose $\mu\mathbf{CR} = \nu\mathbf{CR}$. Let $\alpha \in \mu\mathbf{MLR}$. If $\alpha \notin \nu\mathbf{MLR}$, then by Proposition 21, $\alpha \notin \nu\mathbf{CR}$, hence by the hypothesis, $\alpha \notin \mu\mathbf{CR}$ and thus $\alpha \notin \mu\mathbf{MLR}$, a contradiction. This proves $\mu\mathbf{MLR} \subseteq \nu\mathbf{MLR}$. By symmetry: $\mu\mathbf{MLR} = \nu\mathbf{MLR}$. The other implication is entirely similar.

Let us now suppose that $\mu \nsim \nu$ and let us prove $\mu\mathbf{MLR} \neq \nu\mathbf{MLR}$. Without loss of generality, we can assume that there exists $X \subseteq 2^\omega$ such that $\mu(X) > 0$ and $\nu(X) = 0$. Let q be a positive computable real number such that $\mu(X) \geqslant q$. Since $\nu(X) = inf\{\nu(\mathcal{O}) : \mathcal{O}\ open\ and\ X \subseteq \mathcal{O}\}$, for all k, there is an open set $\mathcal{O}$ such that $X \subseteq \mathcal{O}$ (and hence $\mu(\mathcal{O}) \geqslant q$) and $\nu(\mathcal{O}) \leqslant 2^{-k}$. Since $\mathcal{O}$ can be written as $\mathcal{O} = \bigcup_{u \in A} \mathcal{O}_u$ for some subset A of $\subseteq \{0,1\}^*$, there exists a finite subset B of A such that $\mu(\bigcup_{u \in B} \mathcal{O}_u) \geqslant \mu(\bigcup_{u \in A} \mathcal{O}_u) - 2^{-k}$. Hence, $\mu(\bigcup_{u \in B} \mathcal{O}_u) \geqslant q - 2^{-k}$, but since it is a subset of $\mathcal{O}$, we also have $\nu(\bigcup_{u \in B} \mathcal{O}_u) \leqslant 2^{-k}$.

Given a $k \in \mathbb{N}$, it is possible to effectively find a finite subset C_k of $\{0,1\}^*$ such that $\mu(\bigcup_{u \in C_k}) \geqslant q - 2^{-k}$ and $\nu(\bigcup_{u \in C_k}) \leqslant 2^{-k}$: it suffices to enumerate the finite subsets $\{0,1\}^*$ until we find one satisfying the two conditions (by the above discussion, there exists such a set). Thus, the family $\{\mathcal{V}_n\}_n$ defined by $\mathcal{V}_n = \bigcup_{k \geqslant n} \bigcup_{u \in C_k} \mathcal{O}_u$ is a computable family of c.e. open sets. Moreover, we have, for all n: $\mu(\mathcal{V}_n) \geqslant q$ and $\nu(\mathcal{V}_n) \leqslant 2^{-n+1}$. In other words, $\{\mathcal{V}_{n+1}\}_n$ is a ν-Martin-Löf test, and since $\{\mathcal{V}_n\}_n$ is decreasing, we have: $\mu(\bigcap_{n=0}^{\infty} \mathcal{V}_n) \geqslant q$ and $\nu(\bigcap_{n=0}^{\infty} \mathcal{V}_n) = 0$.

Therefore, by definition of ML-randomness, $\nu\mathbf{MLR} \cap \bigcap_{n=0}^{\infty} \mathcal{V}_n = \emptyset$. But the μ-measure of $\mu\mathbf{MLR} \cap \bigcap_{n=0}^{\infty} \mathcal{V}_n$ is at least q (by Proposition 6), and thus it is non-empty. It follows that $\mu\mathbf{MLR} \neq \nu\mathbf{MLR}$. □

Remark: Combining Proposition 23 with Theorem 17, we obtain Theorem 14.a as a corollary.

Back to generalized Bernoulli measures, we can make a synthesis of all the above results:

Theorem 24. *Let μ and ν be two computable strongly positive generalized Bernoulli measures, of respective parameter $\{p_i\}_i$ and $\{q_i\}_i$.*

(a) *The following are equivalent :*
  (i) *μ and ν are inconsistent*
  (ii) *$\sum_i (p_i - q_i)^2 = +\infty$*
  (iii) *$\mu\mathbf{CR} \cap \nu\mathbf{CR} = \emptyset$*
  (iv) *$\mu\mathbf{KLR} \cap \nu\mathbf{KLR} = \emptyset$*
  (v) *$\mu\mathbf{MLR} \cap \nu\mathbf{MLR} = \emptyset$*

(b) *The following are equivalent :*
  (i) *$\mu \sim \nu$*
  (ii) *$\sum_i (p_i - q_i)^2 < +\infty$*
  (iii) *$\mu\mathbf{CR} = \nu\mathbf{CR}$*
  (iv) *$\mu\mathbf{KLR} = \nu\mathbf{KLR}$*
  (v) *$\mu\mathbf{MLR} = \nu\mathbf{MLR}$*

Proof: (a) By Theorem 13, we get: (i) $\rightarrow$ (ii). By Theorem 16: (ii) $\rightarrow$ (iii). (iii) $\rightarrow$ (iv) and (iv) $\rightarrow$ (v) are trivial. And finally, by Proposition 6: (v) $\rightarrow$ (i).

(b) Here too, we get (i) $\rightarrow$ (ii) by Theorem 13. (ii) $\rightarrow$ (iii) and (ii) $\rightarrow$ (iv) is exactly Theorem 17. Finally, (iii) $\rightarrow$ (v), (iv) $\rightarrow$ (v), and (v) $\rightarrow$ (i) is exactly Theorem 23.   □

5   Open Questions

As a conclusion, we give the main questions this paper leaves open.

Open question: Is any other implication than the ones given in 23 hold ? Is is true in particular that two computable measures are consistent if and only if they have the same ML-random reals ?

  Also, we have a fully satisfactory result for the class of strongly positive generalized Bernoulli measures, but it would be nicer to have one for the whole class of generalized Bernoulli measures.

Open question: For the class of computable generalized Bernoulli measures (not necessarily strongly positive), does the equivalence (iii) $\leftrightarrow$ (iv) $\leftrightarrow$ (v) of Theorem 24.b still hold ?

  Proposition 19 suggests that the answer might be yes.

Acknowledgments. We would like to thank Bruno Durand, Serge Grigorieff, and Alexander Shen' for helpful discussions. We also thank an anonymous referee for a very detailed and relevant review.

References

1. R. DOWNEY, D. HIRSCHFELDT. *Algorithmic Randomness and Complexity.* Book in preparation. See http://www.mcs.vuw.ac.nz/~downey
2. S. KAKUTANI. *On the equivalence of infinite product measures.* Ann. Math., 49: 214-224, 1948.
3. P. MARTIN-LÖF. *The definition of random sequences.* Information and Control, 9(6):602-619, 1966.

4. W. MERKLE, J.S. MILLER, A. NIES, J. REIMANN, F. STEPHAN. *Kolmogorov-Loveland Randomness and Stochasticity.* Annals of Pure and Applied Logic, 138 (1-3):183-210, 2006.

5. A.A. MUCHNIK, A.L. SEMENOV, V.A. USPENSKY. *Mathematical metaphysics of randomness.* Theor. Comput. Sci., 207(2):263-317, 1998.

6. C. P. SCHNORR. *A unified approach to the definition of random sequences.* Math. Systems Theory, 5:246-258, 1971.

7. C. P. SCHNORR. *Zufälligkeit und Wahrscheinlichkeit.* Springer-Verlag Notes in Mathematics, 218, 1971.

8. A. KH. SHEN'. *On relations between different algorithmic definitions of randomness.* Soviet Mathematical Doklady, 38:316-319, 1988.

9. M. VAN LAMBALGEN. *Random sequences.* Doctoral dissertation, University of Amsterdam, 1987.

10. V.G. VOVK. *On a randomness criterion.* Soviet Mathematics Doklady, 35:656-660, 1987.

Planar Dimer Tilings

Olivier Bodini and Thomas Fernique

LIRMM CNRS-UMR 5506 and Université Montpellier II,
161 rue Ada 34392 Montpellier, Cedex 5 - France
{bodini, fernique}@lirmm.fr

Abstract. Domino tilings of finite domains of the plane are used to model dimer systems in statistical physics. In this paper, we study *dimer tilings*, which generalize domino tilings and are indeed equivalent to perfect matchings of planar graphs. We use *height functions*, a notion previously introduced by Thurston in [10] for domino tilings, to prove that a dimer tiling of a given domain can be computed using any Single-Source-Shortest-Paths algorithm on a planar graph. We also endow the set of dimers tilings of a given domain with a structure of distributive lattice and show that it can be effectively visited by a simple algorithmical operation called flip.

1 Dimer Tilings

A *cell* is a polygonal closed set of $\mathbb{R}^2$ and a *domain* is a finite set of cells with disjoint interiors. Two cells of a domain are said *adjacent* if they share at least one boundary edge. A domain $\mathcal{C}$ is said *tileable* if its cells can be grouped two by two, two grouped cells being adjacent. If it exists, such a grouping is called a *dimer tiling* of $\mathcal{C}$. Notice that a tileable domain can admit many dimer tilings: we denote by $\Delta(\mathcal{C})$ the set of dimer tilings of $\mathcal{C}$. Fig. 1 illustrates these notions.

Dimer tilings arise for example in statistical physics to model the behavior of dimer systems: the cells are squares and the dimer are rectangles, called *dominoes* (see e.g. [5,6]). In this context, it is particularly interesting to endow the set of dimer tilings of a domain with a structure suitable for performing random sampling.

Dimer tilings are also connected with *perfect matchings* of planar graphs. Indeed, we can associate to a set of cells $\mathcal{C}$ an undirected planar graph denoted by $A(\mathcal{C})$: to each cell of $\mathcal{C}$ corresponds a vertex of $A(\mathcal{C})$, two of them being connected by an edge if the corresponding cells are adjacent. Then, there is a natural bijection between the perfect matchings of $A(\mathcal{C})$ and the dimer tilings of $\mathcal{C}$ (see Fig. 2).

In this paper, we mainly focus on the two following problems: the first want to compute a dimer tiling, and the second want to provide a way to move on the set of dimer tilings of a fixed domain. This paper is organized as follows. In Section 2, we associate to a domain a weighted directed graph and define *height function* over the vertices of this graph. Such an approach has been firstly used in [10] to compute in linear time a dimer tiling of a simply connected domain

D. Grigoriev, J. Harrison, and E.A. Hirsch (Eds.): CSR 2006, LNCS 3967, pp. 104–113, 2006.
© Springer-Verlag Berlin Heidelberg 2006

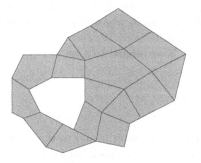
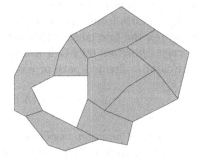

Fig. 1. A domain of 16 cells (left) and a dimer tiling of it, the grouped cells being represented by a single polygon (right)

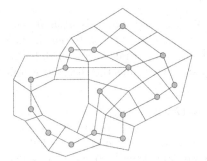
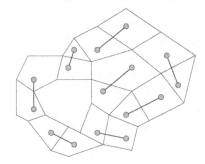

Fig. 2. Left, the planar graph corresponding to the set of cells of Fig. 1. Right, the perfect matching of this graph corresponding to the dimer tiling of Fig. 1.

made of square cells, and then extended in [1] to regular cells (all with the same number of edges). We extend here these results to domain with holes made of non-regular cells. First, we show in Section 3 and 4 that the computation of a dimer tiling of a bipartite domain (or, equivalently, of a perfect matching of a bipartite planar graph) can be reduced to a single-source-shortest-paths problem on a planer graph. This yields in particular a $\mathcal{O}(n\ln(n)^3)$-algorithm to compute a dimer tiling of a domain which is, contrarily to [10, 1], neither necessarily simply connected nor made of regular cells. Second, we endow in Section 5 the set of dimer tilings of a domain with a structure of distributive lattice and defines a simple effective operation, called • ••, which allows to visit it. This can be used for example to perform random sampling on the set of dimer tilings.

2   General Settings

2.1   Weight and Height Function

A directed graph is denoted by $G = (V, E)$, V (resp. E) being the set of vertices (resp. edges) of G.

A *directed path* is a sequence $e_1, \ldots, e_p$ of edges such that e_i points to the vertex e_{i+1} starts from, and a *circuit* is a directed path whose last edge points to the vertex the first edge of this path start from.

A *weight function* over G is a map $w : E \to \mathbb{R}$, extended to a set of edges X (in particular a directed path or a circuit) by:

$$w(X) = \sum_{e \in X} w(e).$$

Then, the *height function* associated to the weight function w and to the vertex $v^* \in V$ is the map from V to $\mathbb{R} \cup \{-\infty\}$ denoted by h_{w,v^*} and defined by:

$$\forall v \in V, \quad h_{w,v^*}(v) = \inf\{w(p) \mid p \text{ is a path from } v^* \text{ to } v\}.$$

2.2   The Graph of a Bipartite Domain

Let $\mathcal{C}$ be a domain as defined in Section 1. We suppose that $\mathcal{C}$ is *bipartite*: we split it into two sets $\mathcal{C}_b$ and $\mathcal{C}_w$ (resp. the *black* cells and the *white* ones), such that two adjacent cells never belong to the same set (that is, are of different colors). Then, given an orientation of the plane, we define the directed graph $G(\mathcal{C})$ as follows (see Fig. 3):

- to each vertex of $\mathcal{C}$ corresponds a vertex of $G(\mathcal{C})$;
- to an edge shared by two adjacent cells of $\mathcal{C}$ (and thus of different colors) corresponds in $G(\mathcal{C})$ an edge directed so that the black cell is on its left;
- to an edge on the boundary of the domain (and thus belongs to only one cell) correspond in $G(\mathcal{C})$ a bidirected edge.

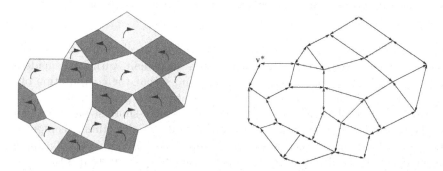

Fig. 3. The orientations of black and white cells of the bipartite domain of Fig. 1 (left). The corresponding directed graph, whose bidirected edges correspond to the boundaries of the domain, the vertex v^* being on the outer boundary (right).

In particular, to each cell of $\mathcal{C}$ naturally corresponds a circuit of $G(\mathcal{C})$, called *cell-circuit* in all what follows.

> From now on and up to the end of the paper, $\mathcal{C}$ stands for a set of cells whose union is a bipartite connected domain, and v^* stands for a fixed vertex of the outer boundary of $G(\mathcal{C})$.

Once v^* is fixed, we will simply denote by h_w the height function h_{w,v^*} associated to the weight function w. Then, the idea of the paper is to define particular weight functions on the graph $G(\mathcal{C})$, such that height functions can be used to compute the ones which correspond to dimer tilings of $\mathcal{C}$.

3   Counters and Dimer Tilings

We define here *counters* and use them to give a characterization of tileable domains. The results provided here are then used in the next section to compute effectively a dimer tiling.

Definition 1. *A* counter *over $G(\mathcal{C})$ is a weight function δ such that $\delta(e) = 0$ for any bidirected edge e and $\delta(c) = 1$ for any cell-circuit c. A counter is moreover said* binary *if $\delta(e) \in \{0,1\}$ for any edge e.*

Clearly, a binary counter weights exactly one edge of a cell-circuit by 1, the other ones having weight 0. Hence, grouping two cells which share an edge of weight 1 yields a dimer tiling (see Fig. 4). Conversely, it is straightforward to similarly derive a binary counter from a dimer tiling. Thus, we use indifferently the terms *dimer tiling* or *binary counter* in all that follows.

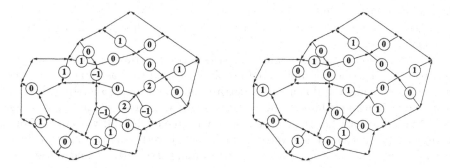

Fig. 4. A counter (left) and the binary counter corresponding to the dimer tiling of Fig. 1 (right). For the sake of clarity, the weights 0 of bidirected edges are not represented.

Let us now consider two counters δ and δ', and let c be a circuit of $G(\mathcal{C})$. One easily proves by induction on the size of c that $\delta(c) = \delta'(c)$. In particular, if $\mathcal{C}$ is tileable then one can consider δ' to be a *binary* counter: one has $\delta'(c) \in \mathbb{N}$, and this yields $\delta(c) \in \mathbb{N}$. Thus:

Proposition 1. *If $\mathcal{C}$ is tileable, then $\delta(c) \in \mathbb{N}$ for any counter δ.*

Conversely, suppose that for any counter δ' and any circuit c, $\delta'(c) \in \mathbb{N}$. In such a case, we can define the notion of δ-shortest path from a vertex v to a vertex v': it is a path p (not necessarily unique) which satisfies

$$\delta(p) = \min\{\delta(p') \mid p' \text{is a path from } v \text{ to } v'\}.$$

Then one has the following properties:

Proposition 2. *If δ and δ' are counters over $G(\mathcal{C})$, then any δ-shortest path is also a δ'-shortest path.*

Proposition 3. *If δ and δ' are counters over $G(\mathcal{C})$, then $h_\delta = h'_\delta$ yields $\delta = \delta'$.*

We then prove:

Theorem 1. *Let δ be a counter over $G(\mathcal{C})$ and set for an edge e from v to v':*

$$\delta_\perp(e) = \delta(e) - (h_\delta(v') - h_\delta(v)).$$

Then, $\delta_\perp$ is the binary counter such that $h_{\delta_\perp}(v) = 0$ for any vertex v.

Thus, Th. 1 and Prop. 1 yield that $\mathcal{C}$ is tileable if and only $\delta(c) \in \mathbb{N}$ for any counter δ over $G(\mathcal{C})$ and any circuit c of $G(\mathcal{C})$. This provide a characterization of tileable bipartite domains that we use in the following section.

4   Computing a Binary Counter

The previous section has defined counters and binary counters. We are especially interested in binary counters since they correspond to dimer tilings. Here, we first show how to compute a counter in linear time, and we then use Th. 1 to derive a binary counter from it.

The first step to compute counter consists in constructing a particular weighted tree. Let $A(\mathcal{C})$ be the undirected graph associated to $\mathcal{C}$ as explained in the introduction. Since the domain is bipartite and connected, so is $A(\mathcal{C})$. Let us assign color black or white to the vertices of $A(\mathcal{C})$, so that two linked vertices have different colors. Let then T be a spanning tree of $A(\mathcal{C})$. If we remove from T an edge e between a black vertex b and a white vertex w, this splits T into two trees: we denote by $T_{e,b}$ the one which contains the vertex b and we set:

$$d_T(e) = \#\{\text{black vertices in } T_{e,b}\} - \#\{\text{white vertices in } T_{e,b}\}.$$

It defines a function d_T from the edges of T to $\mathbb{Z}$ (see Fig. 5). One checks:

Proposition 4. *Let v be a vertex of T and $e_1, \ldots, e_k$ be the edges of T containing v. If the domain has as much black as white cells, then one has:*

$$\sum_{i=1}^{k} d_T(e_i) = 1.$$

Notice that a tileable bipartite domain has necessarily as much black as white cells since each black cell is grouped with a white cell. Simple examples show that the converse is however false.

The second step to compute a counter consists in deriving from the function d_T a counter δ_T over $G(\mathcal{C})$. We proceed as follows. We set $\delta_T(e) = 0$ for any

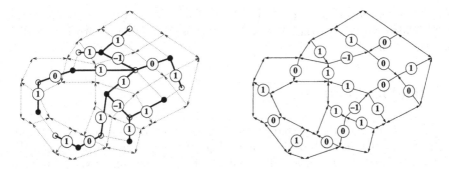

Fig. 5. A spanning tree T of $A(\mathcal{C})$ and the weights given by d_T (left). The corresponding weight function δ_T over $G(\mathcal{C})$ is a counter (right).

bidirected edge (that is, an edge of the boundaries of the domain). Otherwise, a directed edge e corresponds to an edge shared by two cells of $\mathcal{C}$, say C_b and C_w, which correspond in the graph $A(\mathcal{C})$ respectively to two vertices b and w, connected by an undirected edge e' of $A(\mathcal{C})$; we set $\delta_T(e) = d_T(e')$ if e' belongs to the tree T, $\delta_T(e) = 0$ else (see Fig. 5). Prop. 4 then yields that $\delta_T(c) = 1$ for any cell-circuit c. Thus, δ_T is a counter.

Let us study the complexity of the construction of this counter. Let n be the number of cells of the domain $\mathcal{C}$, or equivalently, the number of vertices of $A(\mathcal{C})$. Constructing a spanning tree T can be done in linear time by a greedy algorithm. Then, the weight function d_T can be computed recursively in linear time, starting from the leaves of T. Deriving δ_T from d_T can be performed in linear time since, the graph $G(\mathcal{C})$ being planar, it has $\mathcal{O}(n)$ edges. Thus, the counter δ_T can be computed in linear time.

Then, Th. 1 allows to derive in linear time the binary counter $\delta_\perp$ from the height function of δ_T. This height function can be computed by any single-source-shortest-paths algorithm on the planar graph $G(\mathcal{C})$ weighted by δ_T, the source being the vertex v^*. In particular, [3] provides a $\mathcal{O}(n \ln(n)^3)$-algorithm to do this. Thus, summing up all what preceeds leads to the following theorem:

Theorem 2. *A dimer tiling of a bipartite tileable domain can be constructed by a $\mathcal{O}(n \ln(n)^3)$-algorithm.*

The previous algorithm can also be used to detect the case no dimer tiling exists:

- if the bipartite domain is not balanced (hence not tileable), then the construction of the weight function d_T leads to a vertex such that the sum of the weights of its adjacent edges is not equal to 1;
- otherwise, δ_T is a counter, and since $\mathcal{C}$ is not tileable, there exists a circuit c such that $\delta_T(c) \notin \mathbb{N}$; more precisely $\delta_T(c) < 0$ since $\delta_T(c) \in \mathbb{Z}$ by construction. Thus, the shortest paths starting from v^* are not defined, and the algorithm of [3] detects it (as most of the shortest paths algorithms).

Notice that our algorithm has a complexity similar to the $\mathcal{O}(n \ln(n))$ algorithm of [9], which deals with the case of square cells and domain with a bounded number of holes.

5   Random Sampling

In this section, we suppose that there exists at least one binary counter over $G(\mathcal{C})$, that is, the set $\Delta(\mathcal{C})$ of the dimer tilings of the domain $\mathcal{C}$ is not empty. We endow $\Delta(\mathcal{C})$ with a structure of distributive lattice which can be visited using a simple operation called *flip*.

5.1   Flips

Definition 2. *Let δ be a binary counter over $G(\mathcal{C})$. A δ-nodule is a maximal[1] set of vertices of $G(\mathcal{C})$ such that any two of them are linked by a directed path p which satisfies $\delta(p) = 0$.*

Notice that if δ is a binary counter, then a path p such that $\delta(p) = 0$ is always a shortest path for δ. Thus, Prop. 2 yields that two binary counters define the same nodules: we thus simply call *nodule* a δ-nodule. Moreover, it is worth noticing that h_δ is always constant over the vertices of a nodule: intuitively, a nodule can be seen as an expanded vertex.

  We call *incoming* (resp. *outcoming*) edge of a nodule a directed edge of $G(\mathcal{C})$ which links a vertex outside the nodule to a vertex inside of it (resp. inside to outside). We also denote by A^* the nodule which contains the vertex v^*, and we then define the following operation:

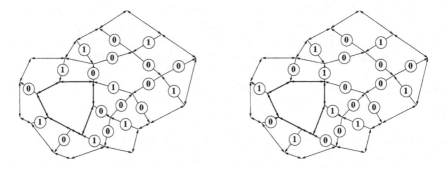

Fig. 6. The seven vertices which belong to the bidirected edges around the hole form a nodule. A (decreasing) flip on this nodule transforms the binary counter on the left into the one on the right. It corresponds on Fig. 7 to the flip from the upper dimer tiling to the one immediatly below.

---

[1] For inclusion.

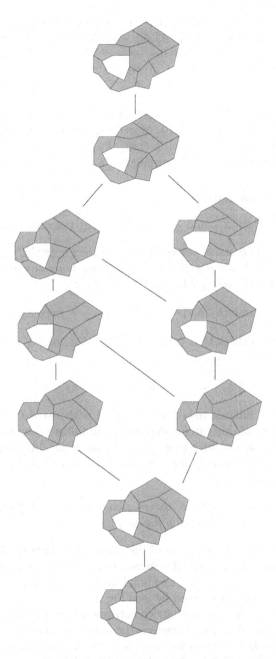

Fig. 7. The distributive lattice of all the dimer tilings of a domain. The dimer tiling $\delta_\perp$ defined in Th. 1 is located at the bottom of this lattice. The flips allow to move between connected tilings. Notice that, excepted the upper flip which is performed on the nodule depicted in Fig. 6, all the other flips are performed on nodules reduceed to a single vertex.

Definition 3. *Let A be a nodule, $A \neq A^*$. Suppose that δ is a binary counter such that $\delta(e)$ is equal to 1 (resp. 0) for each incoming edge of A, and $\delta(e)$ is equal to 0 (resp. 1) for each outcoming edge of A. We call* decreasing flip *(resp.* increasing flip*) the operation which exchanges the weights of the incoming and outcoming edges of A.*

One easily checks that a flip on a nodule A transforms a binary counter δ into a binary counter, say δ' (see Fig. 6). Moreover, if it is an increasing flip (resp. decreasing), then $h_{\delta'}(v) = h_\delta(v) + 1$ (resp. $h_{\delta'}(v) = h_\delta(v) - 1$) if $v \in A$ and $h_{\delta'}(v) = h_\delta(v)$ otherwise: flips act on heights in a very simple way.

5.2   A Distributive Lattice

We define two operations $\vee$ and $\wedge$ over the set $\Delta(\mathcal{C})$ of dimer tilings of $\mathcal{C}$:

Proposition 5. *Let δ and δ' be two binary counters over $G(\mathcal{C})$. Then the height functions $\min(h_\delta, h_{\delta'})$ and $\max(h_\delta, h_{\delta'})$ are height functions of binary counters over $G(\mathcal{C})$, respectively denoted by $\delta \wedge \delta'$ and $\delta \vee \delta'$:*

$$h_{\delta \wedge \delta'} = \min(h_\delta, h_{\delta'}) \qquad and \qquad h_{\delta \vee \delta'} = \max(h_\delta, h_{\delta'}).$$

It is then easy to check that $(\Delta(\mathcal{C}), \wedge, \vee)$ is a distributive lattice. We denote by $\preceq$ the associated partial order:

$$\delta \preceq \delta' \quad \Leftrightarrow \quad \delta = \delta \wedge \delta' \quad \Leftrightarrow \quad h_\delta \leq h_{\delta'}.$$

Notice that it is not difficult to endow a finite set with a structure of distributive lattice. The interest of this specific definition will follows from the way the flips act on this lattice. Recall first that, given δ and δ' in $\Delta(\mathcal{C})$, one says that δ' *covers* δ for the partial order $\preceq$ if $\delta \prec \delta'$ and if, for any $\delta'' \in \Delta(\mathcal{C})$, $\delta \prec \delta''$ yields $\delta' \preceq \delta''$. Then one has:

Theorem 3. *A binary counter δ' covers a binary counter δ if and only if δ can be obtained performing a decreasing flip on δ'.*

In other words, the Hasse's diagram of the distributive lattice $(\Delta(\mathcal{C}), \wedge, \vee)$ (two elements are linked if and only if one covers the other) is isomorphic to the undirected graph whose vertices correspond to dimer tilings, each of them being connected to the ones it is covered by (see Fig. 7).

This structure of distributive lattice corresponds to the one given in [7] in terms of orientations of graphs. However, the weight functions we use here allow to perform *effectively* flips (it suffices to check the weights of the incoming and outcoming edges of a nodule). Thus, the results of [8] concerning random sampling over distributive lattice can be effectively applied to generate randomly a dimer tiling of a given domain.

In [4], we also use this structure together with the flips to generate *all* the $|\Delta(\mathcal{C})|$ dimer tilings of a domain $\mathcal{C}$, performing less than $2|\Delta(\mathcal{C})|$ flips. Since a flip can be performed in linear time, this leads to an algorithm in $\mathcal{O}(n|\Delta(\mathcal{C})|)$.

It thus improves the $\mathcal{O}(n^2)$ algorithm of [2] which is moreover restricted to the case of square cells.

Acknowledgements. We would like to thank Valérie Berthé for corrections and anonymous referees for many useful comments.

References

1. O. Bodini, M. Latapy, Generalized Tilings with Height Functions. Morfismos **7** (2003).
2. S. Desreux, M. Matamala, I. Rapaport, E. Remila, Domino tiling and related models: space of configurations of domains with holes. Theoret. Comput. Sci. **319** (2004), 83–101.
3. J. Fakcharoenphol, S. Rao, Planar graphs, negative weight edges, shortest paths, and near linear time. FOCS 2001, 232–241.
4. T. Fernique, Pavages d'une polycellule. LIRMM Research Report 04002 (2004), available at http://www.lirmm.fr/~fernique/info/memoire_mim3.ps.gz
5. P. W. Kasteleyn, The statistics of dimers on a lattice. I. The number of dimer arrangements on a quadratic lattice. Physica **27** (1961), 1209–1225.
6. R. Kenyon, The planar dimer model with boundary: a survey. Directions in mathematical quasicrystals, M. Baake and R. Moody, eds. CRM monograph series (AMS, Providence, RI, 2000).
7. J. Propp, Lattice structure of orientations of graphs. Preprint (1993), available at http://www.math.wisc.edu/ propp/orient.html.
8. J. Propp, Generating random elements of finite distributive lattices. Preprint (1997), available at http://www.math.wisc.edu/~propp/wilf.ps.gz.
9. N. Thiant, An $\mathcal{O}(n \log n)$-algorithm for finding a domino tiling of a plane picture whose number of holes is bounded. Theoret. Comput. Sci. **303** (2003), 353–374.
10. W. P. Thurston, Conway's tiling group. American Mathematical Monthly, **97** (1990), 757–773.

The Complexity of Equality
Constraint Languages

Manuel Bodirsky[1] and Jan Kára[2,*]

[1] Algorithms and Complexity Department, Humboldt University, Berlin
bodirsky@informatik.hu-berlin.de
[2] Department of Applied Mathematics, Charles University, Prague
kara@kam.mff.cuni.cz

Abstract. We apply the algebraic approach to infinite-valued constraint satisfaction to classify the computational complexity of all constraint satisfaction problems with templates that have a highly transitive automorphism group. A relational structure has such an automorphism group if and only if all the constraint types are Boolean combinations of the equality relation, and we call the corresponding constraint languages *equality constraint languages*. We show that an equality constraint language is tractable if it admits a constant unary or an injective binary polymorphism, and is NP-complete otherwise.

Keywords: Constraint Satisfaction, Logic in Computer Science, Computational Complexity, Clones on Infinite Domains.

1 Introduction

In a constraint satisfaction problem we are given a set of variables and a set of constraints on those variables, and want to find an assignment of values to the variables such that all the constraints are satisfied. The computational complexity of the constraint satisfaction problem depends on the constraint language that we are allowed to use in the instances of the constraint satisfaction problem, and attracted a lot of interest in recent years; see e.g. [6] for an introduction to the state-of-the-art of the techniques used to study the computational complexity of constraint satisfaction problems.

Formally, we can define constraint satisfaction problems (CSPs) as *homomorphism problems* for relational structures. Let Γ be a (not necessarily finite) structure with a relational signature τ. Then the constraint satisfaction problem CSP(Γ) is a computational problem where we are given a *finite* τ-structure S and want to know whether there is a homomorphism from S to Γ; for the detailed definitions, see Section 2. We show two examples.

---

* The second author has been supported by a Marie Curie fellowship of the graduate program "Combinatorics, Geometry, and Computation", HPMT-CT-2001-00282. Also supported by project 1M0021620808 of the Ministry of Education of the Czech Republic.

D. Grigoriev, J. Harrison, and E.A. Hirsch (Eds.): CSR 2006, LNCS 3967, pp. 114–126, 2006.

Example 1. Let Γ be the relational structure $(\mathbb{N}; =, \neq)$. Then $\mathrm{CSP}(\Gamma)$ is the computational problem to determine for a given set of equality or inequality constraints on a finite set of variables whether the variables can be mapped to the natural numbers such that variables x, y with a constraint $x = y$ are mapped to the same value and variables x, y with a constraint $x \neq y$ are mapped to distinct values.

This problem is tractable: for this, we consider the undirected graph on the variables of an instance S of $\mathrm{CSP}(\Gamma)$, where two variables x and y are joined iff there is a constraint $x = y$ in S. Then it is easy to see that S does not have a solution if and only if it contains an inequality-constraint $x \neq y$ such that y is reachable from x in the graph defined above. Clearly, such a reachability test can be performed in polynomial time.

Example 2. Let Γ be the relational structure $(\mathbb{N}; S)$, where S is the ternary relation $S := \{ (x_1, x_2, x_3) \in \mathbb{N}^3 \mid (x_1 = x_2 \wedge x_2 \neq x_3) \vee (x_1 \neq x_2 \wedge x_2 = x_3) \}$. Here the problem $\mathrm{CSP}(\Gamma)$ turns out to be NP-complete (see Section 5).

In this paper we consider constraint satisfaction problems where the infinite template $\Gamma = (D; R_1, \ldots, R_k)$ has a *highly transitive* automorphism group, i.e., if every permutation of D is an automorphism of Γ. That is, we study the constraint satisfaction problems for templates with the highest possible degree of symmetry. We will see in Section 2 that Γ has a highly transitive automorphism group if and only if all relations $R_1, \ldots, R_k$ can be defined with a Boolean combination of atoms of the form $x = y$. (A Boolean combination is a formula built from atomic formulas with the usual connectives of conjunction, disjunction, and negation.) We say that such a relational structure defines an *equality constraint language*. Later, we also discuss the case where the template has infinitely many relation symbols $R_1, R_2, \ldots$ Note that Example 1 and 2 are both equality constraint languages.

The main result of this paper is a full classification of the computational complexity of equality constraint languages. They are either tractable, or NP-complete. The containment in NP is easy to see: a nondeterministic algorithm can guess which variables in an instance S denote the same element in Γ and can verify whether this gives rise to a solution for S. To prove that certain equality constraint languages are NP-hard (Section 5) we apply the algebraic approach to constraint satisfaction, which was previously mainly applied to constraint satisfaction with finite templates.

Some equality constraint languages are tractable. These languages are described by certain closure properties. The most interesting languages here are those that are *closed under an injective binary operation*. The polynomial-time algorithm for such languages, which is presented in Section 6, is an instantiation of the relational consistency algorithm as introduced in [8]. Our contribution here is the proof that this algorithm is *complete* for equality constraint languages that are closed under an injective binary polymorphism, this is, the algorithm rejects an instance if and only if the instance does not have a solution.

2   Fundamental Concepts for the Algebraic Approach

We introduce classical concepts that are fundamental for the algebraic approach to constraint satisfaction. A general introduction to these concepts is [12]; for clones and polymorphisms we refer to [17].

Structures. A *relational language* τ is a (here always at most countable) set of *relation symbols* R_i, each associated with a finite *arity* k_i. A *(relational) structure* Γ over the *(relational) language* τ (also called τ-*structure*) is a countable set D_Γ (the *domain*) together with a relation $R_i \subseteq D_\Gamma^{k_i}$ for each relation symbol of arity k_i from τ. For simplicity, we use the same symbol for a relation symbol and the corresponding relation. If necessary, we write R^Γ to indicate that we are talking about the relation R belonging to the structure Γ. For a τ-structure Γ and $R \in \tau$ it will also be convenient to say that $R(u_1, \ldots, u_k)$ *holds in* Γ iff $(u_1, \ldots, u_k) \in R$. We sometimes write $\bar{u}$ for a tuple $(u_1, \ldots, u_k)$ of some length k. If we add relations to a given structure Γ, we call the resulting structure Γ' an *expansion* of Γ, and Γ is called a *reduct* of Γ'.

Homomorphisms. Let Γ and Γ' be τ-structures. A *homomorphism* from Γ to Γ' is a function f from D_Γ to $D_{\Gamma'}$ such that for each n-ary relation symbol R in τ and each n-tuple $(a_1, \ldots, a_n)$, if $(a_1, \ldots, a_n) \in R^\Gamma$, then $(f(a_1), \ldots, f(a_n)) \in R^{\Gamma'}$. In this case we say that the map f *preserves* the relation R. Isomorphisms from Γ to Γ are called *automorphisms*, and homomorphisms from Γ to Γ are called *endomorphisms*. The set of all automorphisms of a structure Γ is a group, and the set of all endomorphisms of a structure Γ is a monoid with respect to composition.

Polymorphisms. Let D be a countable set, and O be the set of *finitary operations* on D, i.e., functions from D^k to D for finite k. We say that a k-ary operation $f \in O$ *preserves* an m-ary relation $R \subseteq D^m$ if whenever $R(x_1^i, \ldots, x_m^i)$ holds in Γ for all $1 \leq i \leq k$, then $R\big(f(x_1^1, \ldots, x_1^k), \ldots, f(x_m^1, \ldots, x_m^k)\big)$ holds in Γ. If f preserves all relations of a relational τ-structure Γ, we say that f is a *polymorphism* of Γ. In other words, f is a homomorphism from $\Gamma^k = \Gamma \times \ldots \times \Gamma$ to Γ, where $\Gamma_1 \times \Gamma_2$ is the *(categorical- or cross-) product* of the two relational τ-structures Γ_1 and Γ_2. Hence, the unary polymorphisms of Γ are the endomorphisms of Γ, and the unary bijective polymorphisms are the automorphisms of Γ.

Clones. An operation π is a *projection* if for all n-tuples, $\pi(x_1, \ldots, x_n) = x_i$ for some fixed $i \in \{1, \ldots, n\}$. The *composition* of a k-ary operation f and k operations $g_1, \ldots, g_k$ of arity n is an n-ary operation defined by

$$f(g_1, \ldots, g_k)(x_1, \ldots, x_n) = f\big(g_1(x_1, \ldots, x_n), \ldots, g_k(x_1, \ldots, x_n)\big) \ .$$

A *clone* F is a set of operations from O that is closed under compositions and that contains all projections. We write D_F for the *domain* D of the clone F. It is easy to verify that the set $Pol(\Gamma)$ of all polymorphisms of Γ is a clone with the domain D_Γ. Moreover, $Pol(\Gamma)$ is also closed under interpolations: we say that an operation $f \in O$ is *interpolated* by a set $F \subseteq O$ if for every finite subset B of D there is some operation $g \in F$ such that $f|_B = g|_B$ (f restricted to B equals g restricted to B, i.e., $f(\overline{s}) = g(\overline{s})$ for every $\overline{s} \in B^k$). The set of operations that are interpolated by F is called the *local closure* of F; if F equals its local closure, we say that F is *locally closed*. The following is a well-known fact:

Proposition 1 (see e.g. [16]). *A set $F \subseteq O$ of operations is locally closed if and only if F is the set of polymorphisms of Γ for some relational structure Γ.*

An operation is called *essentially unary* iff there is a unary operation f_0 such that $f(x_1, \ldots, x_k) = f_0(x_i)$ for some fixed $i \in \{1, \ldots, k\}$. We say that a k-ary operation f *depends on argument i* iff there is no $k-1$-ary operation f' such that $f(x_1, \ldots, x_k) = f'(x_1, \ldots, x_{i-1}, x_{i+1}, \ldots, x_k)$. Hence, an essentially unary operation is an operation that depends on one argument only. We can equivalently characterize k-ary operations that depend on the i-th argument by requiring that there are elements $x_1, \ldots, x_k$ and x_i' such that $f(x_1, \ldots, x_k) \neq f(x_1, \ldots, x_{i-1}, x_i', x_{i+1}, \ldots, x_k)$. We refer to [16] and [17] for a general introduction to clones.

3   The Algebraic Approach

A τ-formula is called *primitive positive*, if it has the form $\exists x_1 \ldots x_k . \psi_1 \wedge \cdots \wedge \psi_l$, where ψ_i is an atomic τ-formula that might contain free variables and existentially quantified variables from $x_1, \ldots, x_k$. The atomic formula ψ_i might also be of the form $x = y$. A formula is called *existential positive*, if it is a disjunctive combination of primitive positive formulas (equivalently, if it is a first-order formula without universal quantifiers and negations). Every formula with k free variables defines on a structure Γ a k-ary relation. Primitive positive definability of relations is an important concept in constraint satisfaction because primitive positive definable relations can be 'simulated' by the constraint satisfaction problem. The following is frequently used in hardness proofs for constraint satisfaction problems; see e.g. [13].

Lemma 1. *Let Γ be a relational structure and let R be a relation that has a primitive positive definition in Γ. Then the constraint satisfaction problems of Γ and of the expansion of Γ by R have the same computational complexity.*

The algebraic approach to constraint satisfaction (see e.g. [4,5,13]) is based on the following preservation statements that characterize syntactic restrictions of first-order definability.

Theorem 1 (from [3,10,14]). *Let Γ be a finite relational structure. Then*

1. *A relation R has a first-order definition in Γ if and only if it is preserved by all automorphisms of Γ;*
2. *A relation R has an existential positive definition in Γ if and only if it is preserved by all endomorphisms of Γ;*

3. *A relation R has a primitive positive definition in Γ if and only if it is preserved by all polymorphisms of Γ.*

These statements do not hold for infinite structures in general. However, we have the following.

Theorem 2 (from [2, 1]). *Let Γ be a countably infinite relational structure. Then Statement 1 of Theorem 1 holds if and only if Γ is ω-categorical, i.e., if the first-order theory of Γ has only one countable model up to isomorphism. For ω-categorical Γ, Statements 2 and 3 hold as well.*

Let G be a permutation group on a countable infinite set D. An *orbit of k-tuples in Γ* is a largest set O of k-tuples in Γ such that for all $\bar{s}, \bar{t} \in O$ there is a permutation α of Γ such that $(\alpha(s_1), \ldots, \alpha(s_k)) = (t_1, \ldots, t_k)$. A permutation group G on a countably infinite set D is called *oligomorphic*, if it has only finitely many orbits of k-tuples from D, for all $k \geq 1$; see [7]. The next theorem can be seen as a reformulation of the theorem of Ryll-Nardzewski, Engeler, and Svenonius (see [12]), and is also closely related to the first part of Theorem 2.

Theorem 3 (See [7]). *Let Γ be a relational structure. Then the following are equivalent.*

- *Γ is ω-categorical;*
- *the automorphism group of Γ is oligomorphic;*
- *every k-ary first-order definable relation in Γ is the union of a finite number of orbits of k-tuples of the automorphism group of Γ.*

Now it is easy to see that a relational structure $\Gamma = (V; R_1, R_2, \ldots)$ has a highly transitive automorphism group if and only if all relations can be defined with Boolean combinations of the equality relation. Clearly, such relations are preserved by all permutations of V. On the other hand, if Γ has a highly transitive automorphism group, it is in particular ω-categorical. Hence, every k-ary relation R from Γ is the union of a finite number of orbits of k-tuples of the automorphism group of Γ. It is easy to see that the orbits of k-tuples of a highly transitive permutation group can be described by a conjunction of equality and inequality relations.

4   Representations of Relations

From now on, unless stated otherwise, $\Gamma = (D; R_1, R_2, \ldots)$ is a relational structure on a countably infinite domain D where every relation R_i can be defined by a Boolean combination of atoms of the form $x = y$. Note that the automorphism group of Γ is the full symmetric group on D, which is clearly oligomorphic.

Both the hardness results and the algorithm for equality constraint languages use a special representation of the relations in Γ, which we are now going to describe. Theorem 3 implies that every k-ary relation in Γ is a union of orbits of k-tuples of the automorphism group of Γ. Let $\bar{s}$ be a k-tuple from one of these orbits. We define the equivalence relation ρ on the set $\{1, \ldots, k\}$ that contains those pairs $\{i, j\}$ where $s_i = s_j$. Clearly, all tuples in the orbit lead to the same equivalence relation ρ. Hence, every k-ary relation R in Γ corresponds uniquely to a set of equivalence relations on $\{1, \ldots, k\}$, which we call the *representation* of R. Sometimes we identify a relation R from Γ with its representation and for example freely write $\rho \in R$ if ρ is an equivalence relation from the representation of R. Let $|R|$ denote the number of orbits of k-tuples contained in R. Hence, $|R|$ also denotes the number of equivalence relations in the representation of R.

Definition 1. *Let ρ and ρ' be equivalence relations on a set X. We say that ρ is finer than ρ', and write $\rho \subseteq \rho'$, if $\rho(x, y)$ implies $\rho'(x, y)$ for each $x, y \in X$. We also say that in this case ρ' is coarser than ρ. The intersection of two equivalence relations ρ and ρ', denoted by $\rho \cap \rho'$, is the equivalence relation σ such that $\sigma(x, y)$ if and only if $\rho(x, y)$ and $\rho'(x, y)$. Finally, let $c(\rho)$ denote the number of equivalence classes in ρ.*

Lemma 2. *For a k-ary relation R in an equality constraint language on a countable set D the following are equivalent.*

1. *R is preserved by every injection of D^2 into D;*
2. *R is preserved by an injective binary operation on D;*
3. *R is preserved by a binary operation f and there are two k-element subsets S_1, S_2 of the domain such that f restricted to $S_1 \times S_2$ is injective;*
4. *The representation of R is closed under intersections, i.e., $\rho \cap \rho' \in R$ for all equivalence relations $\rho, \rho' \in R$;*

Proof. The implication from (1) to (2) and from (2) to (3) is immediate. Let ρ and ρ' be two equivalence relations from the representation of R. Pick two k-tuples $\bar{s}$ and $\bar{s}'$ in R that lie in the orbits that are described by ρ and ρ'. Now, let f be a binary operation of D that is injective on its restriction to $S_1 \times S_2$ for two k-element subsets S_1, S_2. Let α_1 and α_2 be permutations of D that map the entries of the k-tuples $\bar{s}$ and $\bar{s}'$ to S_1 and S_2, respectively. Then by injectivity of f the k-tuple $\bar{s}'' := (f(\alpha_1(s_1), \alpha_2(s_1')), \ldots, f(\alpha_1(s_k), \alpha_2(s_k')))$ satisfies $s_i'' = s_j''$ if and only if $\rho(i, j)$ and $\rho'(i, j)$. Hence, we found a tuple in R that lies in the orbit that is described by $\rho \cap \rho'$, which is therefore also contained in the representation of R, and therefore (3) implies (4). Every injection of D^2 into D preserves every relation with an intersection-closed representation, because it maps two tuples that correspond to equivalence relations ρ and ρ' to a tuple that corresponds to $\rho \cap \rho'$. We thus proved that (4) implies (1). $\square$

If a relation R has a representation that is closed under intersections, we also write that R is $\cap$-*closed*.

Corollary 1. *An operation f on a countable set D and the permutations on D locally generate an injective binary operation g if and only if every equality constraint relation that is preserved by f is $\cap$-closed.*

Proof. If f and the permutations locally generate an injective binary operation g, then every relation R that is preserved by f is also preserved by g, and Lemma 2 shows that R is $\cap$-closed. Conversely, if every equality constraint relation R preserved by f is $\cap$-closed, we claim that f and the permutations locally generate all injective binary operations. Suppose the contrary. Then there is a relation R that is preserved by f but not by an injective binary operation g. An application of Lemma 2 in the other direction shows that R cannot be $\cap$-closed, contradicting the assumption.   □

5   A Generic Hardness Proof

In this section we prove that every equality constraint language without a constant unary or an injective binary polymorphism is NP-hard. Let us start with a fundamental lemma on non-injective endomorphisms.

Lemma 3. *If Γ has a non-injective endomorphism f, then Γ also has a constant endomorphism.*

Proof. Let f be an endomorphism of Γ such that $f(x) = f(y)$ for two distinct points x, y from D. Let $a_1, a_2, \ldots$ be an enumeration of D. We construct an infinite sequence of endomorphisms $e_1, e_2, \ldots$ where e_i is an endomorphism that maps the points $a_1, \ldots, a_i$ to a_1. This suffices, since by local closure the mapping defined by $e(x) = a_1$ for all x is an endomorphism of Γ.

For e_1 we take the identity map, which clearly is an endomorphism with the desired properties. To define e_i for $i \geq 2$ let α be an automorphism of Γ that maps $a_1 = e_{i-1}(a_1) = \cdots = e_{i-1}(a_{i-1})$ to x, and $e_{i-1}(a_i)$ to y. Then the endomorphism $f(\alpha(e_{i-1}))$ is constant on $a_1, \ldots, a_i$. There is also an automorphism α' that maps $f(\alpha(e_{i-1}(a_1)))$ to a_1. Then $e_i := \alpha'(f(\alpha(e_{i-1})))$ is an endomorphism with the desired properties.   □

Lemma 4. *If Γ does not have a constant endomorphism, then there is a primitive positive definition of the relation $x \neq y$ in Γ.*

Proof. Suppose Γ has a k-ary polymorphism f that does not preserve $\neq$, i.e., there are k-tuples $\bar{u}$ and $\bar{v}$ such that $u_i \neq v_i$ for all $i \in \{1, \ldots, k\}$, but $f(\bar{u}) \neq f(\bar{v})$. Let $\alpha_2, \ldots, \alpha_k$ be permutations of D that map u_1 to u_i and v_1 to v_i. Then the endomorphism $g(x) := f(x, \alpha_2(x), \ldots, \alpha_k(x))$ is not injective, because $g(u_1) = f(u_1, \ldots, u_k) = f(v_1, \ldots, v_k) = g(v_1)$, and by Lemma 3 locally generates a constant, in contradiction to the assumptions. Hence, every polymorphism of Γ preserves $\neq$, and by Theorem 2 the relation $\neq$ has a primitive positive definition.   □

Due to the following lemma we can focus on binary operations in some later proofs.

Lemma 5. *Every essentially at least binary operation together with all permutations locally generates a binary operation that depends on both arguments.*

Proof. Let k be a k-ary operation, where $k > 2$, that depends on all arguments. In particular, f depends on the first argument, and hence there are two k-tuples $(a_1, \ldots, a_k)$ and $(a'_1, a_2, \ldots, a_k)$ with $f(a_1, \ldots, a_k) \neq f(a'_1, a_2, \ldots, a_k)$. Suppose first that there are $b_1, \ldots, b_k$ such that $b_i \neq a_i$ for $i \geq 2$ and $f(b_1, b_2, \ldots, b_k) \neq f(b_1, a_2, \ldots, a_k)$. We can then define permutations α_i of D for $i \geq 3$, such that a_2 is sent to a_i and d_2 is sent to d_i. The binary operation g defined by $g(x, y) = f(x, y, \alpha_3(y), \ldots, \alpha_k(y))$ depends on both arguments, as $g(a_1, a_2) \neq g(a'_1, a_2)$ and $g(b_1, b_2) \neq g(b_1, a_2)$, and hence we are done in this case.

So suppose that for every b_1 and every $b_2, \ldots, b_k$ such that $b_i \neq a_i$ for $i \in \{2, \ldots, k\}$ it holds that $f(b_1, b_2, \ldots, b_k) = f(b_1, a_2, \ldots, a_k)$. Since f depends on the second coordinate, there are elements $c_1, c_2, \ldots, c_k$ and c'_2 with $f(c_1, \ldots, c_k) \neq f(c_1, c'_2, c_3, \ldots, c_k)$. The value $f(c_1, a_2, \ldots, a_k)$ can be equal to either $f(c_1, \ldots, c_k)$ or to $f(c_1, c'_2, c_3, \ldots, c_k)$, but not to both. We can assume without loss of generality that $f(c_1, \ldots, c_k) \neq f(c_1, a_2, \ldots, a_k)$. Let us choose $d_2, \ldots, d_k$ such that $d_i \neq a_i$ and $d_i \neq c_i$ for $i \in \{2, \ldots, k\}$. Since c_i and d_i are distinct for all $2 \leq i \leq k$, we can define permutations α_i of D for $i \geq 3$ such that b_2 is sent to b_i and c_2 is sent to c_i.

We claim that the operation g defined by $g(x, y) := f(x, y, \alpha_3(y), \ldots, \alpha_k(y))$ depends on both arguments. Indeed, from the beginning of the previous paragraphs we know that $g(a_1, d_2) = f(a_1, d_2, \ldots, d_k) = f(a_1, a_2, \ldots, a_k)$, and that $g(a'_1, d_2) = f(a'_1, d_2, \ldots, d_k) = f(a'_1, a_2, \ldots, a_k)$. By the choice of the values $a_1, \ldots, a_k$ and a'_1 these two values are distinct, and we have shown that g depends on the first argument. For the second argument, note that $g(c_1, d_2) = f(c_1, d_2, \ldots, d_k) = f(c_1, a_2, \ldots, a_k)$ and that $g(c_1, c_2) = f(c_1, c_2, \ldots, c_k)$. But in the previous paragraph we also saw that these two values are distinct, and hence g also depends on the second argument. $\square$

Now comes the central argument.

Theorem 4. *Let f be a binary operation that depends on both arguments. Then f together with all permutations locally generates either a constant unary operation or a binary injective operation.*

Proof. Suppose that f does not locally generate a constant operation. We want to use Corollary 1 and show that every equality constraint relation R that is preserved by f is $\cap$-closed, which implies that f locally generates a binary injective polymorphism. Suppose for contradiction that R is an n-ary equality constraint relation, $n \geq 2$, that is closed under f but not $\cap$-closed, i.e., there are two equivalence relations ρ and ρ' in R such that $\rho \cap \rho'$ is not in R. Choose ρ and ρ' such that $(c(\rho), c(\rho'))$ is lexicographically maximal. Let $\bar{s} := (s_1, \ldots, s_n)$ and $\bar{t} := (t_1, \ldots, t_n)$ be n-tuples of D that have the equivalence relations ρ and ρ'. Because ρ is not finer than ρ' we can find indices p and q such that $s_p = s_q$, $t_p \neq t_q$. Let r be the number of equivalence classes of ρ that are contained in the equivalence class of p in ρ'. Choose p and q such that r is minimal.

Consider $2n-1$ distinct elements $a_1, \ldots, a_{2n-1}$ from D. By the infinite pigeonhole principle, there is an infinite subset S_1 of D such that $f(a_1, b) = f(a_1, b')$ for all $b, b' \in S_1$, or $f(a_1, b) \neq f(a_1, b')$ for all $b, b' \in S_1$. We apply the same argument to a_2 instead of a_1, and S_1 instead of D, and obtain an infinite subset S_2 of S_1. The argument can be iterated to obtain an infinite subset S_{2n-1} such that for all $a \in \{a_1, \ldots, a_{2n-1}\}$ we either have $f(a, b) \neq f(a, b')$ for all $b, b' \in B$, or $f(a, b) = f(a, b')$ for all $b, b' \in B$. Then there is also an n-element subset A of $\{a_1, \ldots, a_{2n-1}\}$ and an n-element subset B of S_{2n-1} such that either $f(a, b) \neq f(a, b')$ for all $a \in A$ and $b, b' \in B$, or $f(a, b) = f(a, b')$ for all $a \in A$ and $b, b' \in B$. Not-e that in the latter case $f(a, b) \neq f(a', b)$ for all distinct elements $a, a' \in A$, and $b \in B$. Otherwise, if $f(a, b) = f(a', b)$, then f does not preserve the inequality relation, because there is a $b' \in B$ such that $b' \neq b$ and $f(a, b) = f(a, b')$, and hence $f(a, b) = f(a', b')$, but $a \neq a'$ and $b \neq b'$. But this is impossible, because Lemma 3 shows that in this case f locally generates a constant operation. Therefore, we found two n-element sets A and B such that either $f(a, b) \neq f(a', b)$ for all $a, a' \in A$ and $b \in B$, or $f(a, b) \neq f(a, b')$ for all $a \in A$ and $b, b' \in B$. Without loss of generality we assume that the first case applies.

Since f cannot only depend on the first argument, there are elements u, v_1, and v_2 in D such that $v_1 \neq v_2$ and $f(u, v_1) \neq f(u, v_2)$. We can assume that v_2 is from B: For this, consider any element v' of B. If $f(u, v') \neq f(u, v_1)$, we choose v' instead of v_2 and are done. If $f(u, v') = f(u, v_1)$, then $f(u, v') \neq f(u, v_2)$, and we choose v' instead of v_2 and v_2 instead of v_1. We can also assume that $f(u, v) \neq f(u', v)$ for all $u' \in A$, $v \in B$: The reason here is that if there are elements $a \in A$ and $b_1, b_2 \in B$ such that $f(a, b_1) \neq f(a, b_2)$ we choose $u = a$, $v_1 = b_1$, and $v_2 = b_2$. Otherwise, we know that $f(a, b_1) = f(a, b_2)$ for all $a \in A$ and $b_1, b_2 \in B$. But then, $f(u, v) = f(u', v)$ is impossible for all $u' \in A$ and $v \in B$ due to Lemma 3.

Let α_1 be a permutation of D that maps $s_p = s_q$ to u and the other entries in $\bar{s}$ to A. Let α_2 be a permutation of D that maps t_p to v_1, t_q to v_2, and the other entries in $\bar{t}$ to B. Consider the equivalence relation σ of the tuple $(f(\alpha_1(s_1), \alpha_2(t_1)), \ldots, f(\alpha_1(s_n), \alpha_2(t_n)))$. Because f preserves R, we know that σ is contained in R. If $r = 0$, then due to the way we apply the operation f to $\alpha_1(\bar{s})$ and $\alpha_2(\bar{t})$ it is easy to see that σ has more equivalence classes than ρ, contradicting the maximal choice of ρ. If $r \geq 1$, then σ has more equivalence classes than ρ', for the following reason. Every equivalence class C of ρ' either consists of a union of equivalence classes from ρ, or contains an element from an equivalence class in ρ that is not contained in C. But also in the latter case, by the choice of p and q such that r is minimal, we can infer that C contains some equivalence class from ρ. Hence, in both cases we can associate in that way one equivalence class from ρ to every class in ρ'. Due to the way we apply the operation f to $\alpha_1(\bar{s})$ and $\alpha_2(\bar{t})$, all these equivalence classes correspond to distinct equivalence classes in σ. Moreover, $f(\alpha_1(s_q), \alpha_2(t_q))$ will lie in yet another equivalence class of σ. Thus, σ has more equivalence classes than ρ'. Since σ is not coarser than ρ, the existence of the relations ρ and σ then contradicts the choice of ρ and ρ' where $(c(\rho), c(\rho'))$ was lexicographically maximal.   □

Hence, if the template is not preserved by a constant unary or an injective binary operation, we have a primitive positive definition of every first-order definable relation, in particular for the relation S that was defined in Example 2 in the introduction.

Lemma 6. *If the relation S has a primitive positive definition in Γ, then $CSP(\Gamma)$ is NP-hard.*

Proof. First observe that by identification of arguments x and y, if S has a primitive positive definition in Γ, then the inequality relation has a primitive positive definition in Γ as well. We prove the NP-hardness by reduction from the NP-hard problem 3-COLORING [9]. Let $G = (V, E)$ be a graph that is an instance of 3-COLORING. We construct an instance of $CSP(\Gamma)$ that has a polynomial size in $|V|$ and $|E|$ and is satisfiable if and only if G has a proper 3-coloring. Lemma 1 asserts we can use inequality constraints and the relation S to formulate this instance. The set of variables in this instance is $V \cup V' \cup \{c_1, c_2, c_3\}$, where V' is a copy of V, and c_1, c_2, c_3 are three new variables representing colors. We impose inequality constraints on each pair in c_1, c_2, c_3 and on each pair (u, v) for $uv \in E$. We impose the constraint S on (c_1, v', c_2) for each $v' \in V'$, and on (v', v, c_3) for each $v \in V$ where v' is the copy of v in V'. By construction, a solution to these constraints induces a proper 3-coloring of G. Conversely, a simple case analysis shows that any proper 3-coloring can be extended in a way that satisfies these constraints.   □

As we already mentioned in the introduction, the constraint satisfaction problem for equality constraint languages is always contained in NP. By combining the results obtained in this section and using Theorem 2 and Lemma 1 we therefore proved the following main result of this section.

Theorem 5. *If Γ has no constant unary and no injective binary polymorphism, then $CSP(\Gamma)$ is NP-complete.*

6   Algorithmic Results

The case that Γ contains a constant unary polymorphism gives rise to trivially tractable constraint satisfaction problems: If an instance of such a constraint satisfaction problem has a solution, then there is also a solution that maps all variables to a single point. In this case an instance of $CSP(\Gamma)$ is satisfiable if and only if it does not contain a constraint R where R denotes the empty relation in Γ. Clearly, this can be tested efficiently. To finish the classification of the complexity of equality constraint languages we are left with the case that Γ has a binary injective polymorphism.

Lemma 7. *Let Γ be closed under a binary injective polymorphism, and let R be a k-ary relation from Γ. Then for every equivalence relation ρ on $\{1, \ldots, k\}$ (note, that ρ need not be in R) either there is no $\sigma \in R$ that is coarser than ρ, or there exists an equivalence relation $\sigma \in R$ such that σ is coarser than ρ and σ is finer than any $\sigma' \in R$ coarser than ρ. Furthermore, σ can be computed in time $O(k^2|R|)$.*

Proof. First we compute the set R' of equivalence relations of R that are coarser than ρ. The set R' can be computed straightforwardly in time $O(k^2|R|)$ by checking each equivalence relation in R. If R' is empty we are done. Otherwise, because R is closed under intersections, we know that $\sigma = \cap_{\sigma' \in R'} \sigma'$ is in R. It is even in R', since if two equivalence relations are both coarser than another, then so is their intersection. We can find σ with the following procedure.

– We start with an arbitrary equivalence relation τ in R'.
– For each $\sigma' \in R'$, if σ' is finer than τ, then set τ to be σ'.

The procedure clearly runs in time $O(k^2|R|)$. ☐

Theorem 6. *Let Γ be closed under a binary injective polymorphism, and let S be an instance of CSP(Γ) with n variables and q constraints. Let k be the maximal arity of the constraints, and let m be the maximal number of equivalence relations in the representations for the constraints. Then there is an algorithm that decides the satisfiability of S in time $O(qm(qmk^2 + n))$.*

Proof. We start by assigning each variable a unique value. Then we check whether each constraint is satisfied. If we find an unsatisfied l-ary constraint R, let $x_1, \ldots, x_l$ be the variables of that constraint. Let ρ be the equivalence relation on the elements $\{1, \ldots, l\}$ that contains all pairs $\{i, j\}$ where x_i got the same value as x_j. Using the algorithm from Lemma 7 we either find that there is no $\sigma \in R$ coarser than ρ, in which case we answer that the problem does not have a solution. Otherwise we find the unique finest equivalence relation $\sigma \in R$. In this case we reassign the values to the variables in the following way: If $\sigma(i, j)$, we assume without loss of generality that $i < j$, and change the value of all variables with the value of x_j to the value of x_i. Finally we restart the procedure with the new assignment for the variables. If all the constraints are satisfied we have computed a solution.

To show the correctness of this algorithm we prove by induction that each of the introduced equalities holds in every solution of the problem. In the beginning we introduced no equality (all the values were mutually different). We introduce an equality only if we find an unsatisfied constraint. In that case we have computed the set of equalities (an equivalence relation) that is contained in every other set of equalities acceptable for the constraint. Because the constraint must be satisfied in every solution we introduce only the equalities that hold in every solution.

Because the set of acceptable equivalence relations is made smaller each time the constraints are not yet satisfied, we have to recompute the assignment at most qm times. Finding the unsatisfied constraint can take $O(qmk^2)$ and changing the assignment can take $O(n)$. Putting the terms together yields the claimed bound on the time complexity. ☐

Note that the asymptotic running time of the algorithm can be substantially improved by using better data structures.

In the standard case that the signature of Γ is finite, the algorithm clearly establishes the tractability of CSP(Γ) for injective binary polymorphisms, since in this case k and m are bounded by constants that only depend on Γ. If Γ has a countable signature, there are various possibilities to define tractability of CSP(Γ). We refer to the discussion in [6]. The definition of tractability chosen there is to require that for every reduct Γ' of Γ with a finite signature the problem CSP(Γ') is tractable. If Γ has an injective binary polymorphism, this requirement is clearly fulfilled, because we can again use the above algorithm with the same argument. If we allow that the instances contain arbitrary relations from the signature, we have to discuss how to represent the constraints in the instance. For equality constraint languages, one natural candidate to represent the constraints in the instance is the representation that we already used in the formulation of the algorithm: a constraint is represented by a list of equivalence relations on its arguments. Now, the detailed complexity analysis given above shows that we even obtain tractability in the stronger sense where instances might contain arbitrary constraints in the above representation.

7 Conclusion and Remarks

We combine the results of Section 5 and Section 6 and obtain the following.

Theorem 7. *An equality constraint language with template Γ is tractable if Γ has a constant unary or an injective binary polymorphism. Otherwise it is NP-complete.*

In other words, unless P=NP, an equality constraint language with template Γ is tractable if and only if every relation in Γ contains all tuples of the form $(a, \ldots, a)$ for all $a \in \Gamma$, or if all relations are $\cap$-closed.

We would like to conclude with a remark on the relationship of the presented results with questions from universal algebra. The lattice of clones that contain all the permutations is a recent research focus in universal algebra [15,11], and a full classification seems to be out of reach. However, the lattice of *locally closed* clones that contain the set of all permutations S_ω is considerably simpler. The lattice has a smallest element, the clone that is locally generated by S_ω. Above this clone the lattice has exactly two minimal clones that correspond to the maximally tractable equality constraint languages. Is it possible to give a full description of the locally closed clones that contain all the permutations?

Acknowledgements. We thank the anonymous referees.

References

1. M. Bodirsky. Constraint satisfaction with infinite domains. PhD thesis, Humboldt-Universitat zu Berlin, 2004.
2. M. Bodirsky and J. Nešetřil. Constraint satisfaction with countable homogeneous templates. In *Proceedings of CSL'03*, pages 44–57, 2003.
3. V. G. Bodnarčuk, L. A. Kalužnin, V. N. Kotov, and B. A. Romov. Galois theory for post algebras, part I and II. *Cybernetics*, 5:243–539, 1969.

4. A. Bulatov. Tractable conservative constraint satisfaction problems. In *Proceedings of LICS'03*, pages 321–330, 2003.
5. A. Bulatov, A. Krokhin, and P. Jeavons. The complexity of maximal constraint languages. In *Proceedings of STOC'01*, pages 667–674, 2001.
6. A. Bulatov, A. Krokhin, and P. G. Jeavons. Classifying the complexity of constraints using finite algebras. *SIAM Journal on Computing*, 34:720–742, 2005.
7. P. J. Cameron. *Oligomorphic Permutation Groups*. Cambridge University Press, 1990.
8. R. Dechter and P. van Beek. Local and global relational consistency. *TCS*, 173(1):283–308, 1997.
9. Garey and Johnson. *A Guide to NP-completeness*. CSLI Press, Stanford, 1978.
10. D. Geiger. Closed systems of functions and predicates. *Pacific Journal of Mathematics*, 27:95–100, 1968.
11. L. Heindorf. The maximal clones on countable sets that include all permutations. *Algebra univers.*, 48:209–222, 2002.
12. W. Hodges. *A shorter model theory*. Cambridge University Press, 1997.
13. P. Jeavons, D. Cohen, and M. Gyssens. Closure properties of constraints. *Journal of the ACM*, 44(4):527–548, 1997.
14. M. Krasner. Généralisation et analogues de la théorie de Galois. *Congrés de la Victoire de l'Ass. France avancement des sciences*, pages 54–58, 1945.
15. M. Pinsker. The number of unary clones containing the permutations on an infinite set. *Acta Sci. Math. (Szeged)*, 2005. To appear.
16. R. Pöschel and L. A. Kalužnin. *Funktionen- und Relationenalgebren*. Deutscher Verlag der Wissenschaften, 1979.
17. A. Szendrei. *Clones in universal Algebra*. Seminaire de mathematiques superieures. Les Presses de L'Universite de Montreal, 1986.

Window Subsequence Problems for Compressed Texts*

Patrick Cégielski[1], Irène Guessarian[2,**], Yury Lifshits[3], and Yuri Matiyasevich[3]

[1] LACL, UMR-FRE 2673, Université Paris 12, Route forestière Hurtault,
F-77300 Fontainebleau, France
`cegielski@univ-paris12.fr`
[2] LIAFA, UMR 7089 and Université Paris 6, 2 Place Jussieu,
75254 Paris Cedex 5, France
`ig@liafa.jussieu.fr`
[3] Steklov Institute of Mathematics, Fontanka 27, St. Petersburg, Russia
`yura@logic.pdmi.ras.ru, yumat@pdmi.ras.ru`

Abstract. Given two strings (a text t of length n and a pattern p) and a natural number w, *window subsequence problems* consist in deciding whether p occurs as a subsequence of t and/or finding the number of size (at most) w windows of text t which contain pattern p as a subsequence, *i.e.* the letters of pattern p occur in the text window, in the same order as in p, but not necessarily consecutively (they may be interleaved with other letters). We are searching for subsequences in a text which is compressed using Lempel-Ziv-like compression algorithms, without decompressing the text, and we would like our algorithms to be almost optimal, in the sense that they run in time $O(m)$ where m is the size of the compressed text. The pattern is uncompressed (because the compression algorithms are evolutive: various occurrences of a same pattern look different in the text).

1 Introduction

We are concerned with searching information in a compressed text *without decompressing* the text. We will search to decide whether a pattern occurs as a *subsequence* of a text: pattern $p = p_1 \ldots p_k$ is said to be a subsequence of text t if $p_1, \ldots, p_k$ occur in t, in the same order as in p, but not necessarily consecutively (they may be interleaved with other letters). It is also demanded that the subsequences consisting of p be contained in text windows of (at most) a fixed size w. Pattern matching in compressed texts has already been studied in *e.g.* [R99, GKPR96]. Subsequence matching within windows of size w at most is a more difficult problem, which emerged due to its applications in knowledge discovery and datamining [M02], and as a first step for solving problems in molecular biology. One quite important use of subsequence matching consists

---

* Support by grants INTAS–04-77-7173 and NSh–2203-2003-1 is gratefully acknowledged.
** Corresponding author.

D. Grigoriev, J. Harrison, and E.A. Hirsch (Eds.): CSR 2006, LNCS 3967, pp. 127–136, 2006.
© Springer-Verlag Berlin Heidelberg 2006

in recognizing frequent patterns in large sequences of data. Knowledge of frequent patterns is then used to determine association rules in databases and to predict the behavior of large data. Consider for instance a text t consisting of a university WWW-server logfile containing requests to see WWW pages, and suppose we want to see how often, within a time window of at most 10 units of time, the sequence of events $e_1e_2e_3e_4$ has occurred, where: $e_1 = $ 'Computer Science Department homepage', $e_2 = $ 'Graduate Course Descriptions', $e_3 = $ 'CS586 homepage', $e_4 = $ 'homework'. This will be achieved by counting the number of 10-windows of t containing $p = e_1e_2e_3e_4$ as a subsequence.

Most efficient compression algorithms are evolutive, in the sense that the text represented by each compression symbol is determined dynamically, hence the encoding of a subword is different for different occurrences of this subword in the text. It is thus less useful for the search to encode the pattern. Moreover, pattern sizes are usually smaller than text sizes by several orders of magnitude. We will thus search for a plain (not encoded) pattern in an encoded (compressed) text.

We address several *window subsequence* problems in three models of compression: Lempel-Ziv (in short LZ [LZ77]), Lempel-Ziv-Welch (in short LZW [LZ78, W84]), and straight-line-programs (in short SLP [R03]). We show that, for all three models, as soon as there is a significant (say quadratic) difference in size between the compressed text and the original text, searching directly in the compressed text is more efficient than the naive decompress-then-search approach.

The paper is organised as follows: in Section 2, we recall the compression models, in Section 3 we define five window subsequence problems, in Section 4 we describe auxiliary data structures, and show how to compute them, yielding algorithms for the window subsequence problems.

Related Results

Different versions of pattern-matching and subsequence problems have been considered. Subsequence problems are different from pattern-matching problems in two respects: 1. the letters of the pattern need not be consecutive in the text, and 2. the size of the text window where the pattern occurs is bounded. Some related problems are as follows.

It was shown in [GKPR96] that the pattern matching problem can be solved in polynomial time, even if *both* text and pattern are given in LZ compressed form.

It was shown in [L05, LL05] that problem 1 below (Section 3), is both NP and co-NP-hard if *both* text and pattern are given in LZ compressed form.

It was shown in [ABF95] that finding the first occurrence of the pattern (pattern matching with compressed text and uncompressed pattern) in an LZW-style compressed text can be done in time $O(m + k^2)$ or $O(m \log k + k)$.

Next, we should mention [BKLPR02], where *Compressed Pattern Matching* problems were extended to the two-dimensional case: it was shown that complexity increases in this setting. *Compressed Pattern Matching* is NP-complete while *Fully Compressed Pattern Matching* is Σ_2^P-complete.

Recently, applications for algorithms on compressed texts in analysis of message sequence charts were found, see [GM02].

Besides pattern matching the membership of a compressed text in a formal language was studied. In [GKPR96], the authors presented a polynomial algorithm for deciding membership in a regular language. Recently, [MS04] showed that this problem is P-complete. On the other hand, it was shown in [Loh04] that deciding membership in a context-free language is PSPACE-complete.

2   Compression Algorithms

2.1   Notations

An *alphabet* is a finite non-empty set $A = \{a_1 \ldots, a_i\}$. We will also use an extra letter a_0. A *word* t on A is a sequence $t[1]t[2]\cdots t[n]$ of letters from A (also denoted by $t_1 t_2 \cdots t_n$ and called the *text*). The number n is called the *length* of t and will be denoted by $|t|$. The only length zero word is the *empty word*, denoted by ε. Given integers $k \leq n$, $i < j \leq n$ and t a length n word, let $t[k]$ (resp. $t[-k]$, $t[i..j]$) denote the kth leftmost letter of t (resp. the kth rightmost letter, the subword $t[i]t[i+1]\cdots t[j]$ of t).

Let t be a word from A^*.

2.2   Lempel-Ziv-Welch Algorithm

Compression

1. Let $T_0 = \varepsilon$
2. Assume words $T_0, T_1, \ldots, T_{k-1}$ were defined, and

$$ta_0 = T_0 T_1 \ldots T_{k-1} s \tag{1}$$

with s non-empty. Let T_k be the shortest prefix of s which is not among $T_0, T_1, \ldots, T_{k-1}$; there exists a unique pair consisting of a number r_k and a letter $c_k \in A \cup \{a_0\}$ such that $r_k < k$ and $T_k = T_{r_k} c_k$.

$$ta_0 = \overbrace{* \cdots * c_1}^{T_1} \overbrace{* \cdots * c_2}^{T_2} * \cdots * \overbrace{* \cdots * c_m}^{T_m} \tag{2}$$
$$\underbrace{}_{T_{r_1}} \underbrace{}_{T_{r_2}} \underbrace{}_{T_{r_m}}$$

The *LZW–compression* of t is the sequence of elements $r_1, c_1, r_2, c_2, \ldots, r_m$, c_m from $A \cup \mathbb{N}$ where m is defined by the condition $c_m = a_0$.

Decompression

1. Let $T_0 = \varepsilon$
2. Repeat $T_k = T_{r_k} c_k$ until $c_k = a_0$
3. Let $t = T_0 T_1 \ldots T_{m-1} T_m$

2.3 Lempel-Ziv Algorithm

Compression

1. Let $T_0 = \varepsilon$
2. Assume words $T_0, T_1, \ldots, T_{k-1}$ were defined, and

$$t = T_0 T_1 \ldots T_{k-1} s \tag{3}$$

with s non-empty. Let T_k be the longest prefix of s which is a subword of $T_0 T_1 \ldots T_{k-1}$, if such a prefix exists, otherwise let $T_k = a_j$ where a_j is the first letter of S; in the former case there exists a unique pair of numbers q_k and r_k such that $1 \leq q_k < r_k \leq |T_0 T_1 \ldots T_{k-1}|$ and $T_k = t[q_k..r_k]$; in the latter case, we define formally $q_k = r_k = -j$.

$$t = \overbrace{* \cdots *}^{T_1} * \ldots * \underbrace{t[q_k] \cdots t[r_k]}_{T_k} * \ldots * \overbrace{* \cdots *}^{T_{k-1}} \overbrace{* \cdots *}^{T_k} * \ldots * \tag{4}$$

The *LZ–compression* of t is the sequence of numbers $q_1, r_1, q_2, r_2, \ldots, q_m, r_m$ where m is such that $t = T_1 \ldots T_m$.

Decompression

1. Let $t = \varepsilon$
2. For $k = 1$ to m do : if $q_k < 0$ then let $t := t\, a_{-q_k}$ else let $t := t\, t[q_k..r_k]$

2.4 Straight-Line Programs

A *straight–line program* compression (in short SLP) $\mathcal{P}$ of **size** m is a sequence of assignments: $X_i := exp_i$ for $i = 1, \ldots, m$, where each X_i is a **non-terminal** and each expression exp_i is either $exp_i = a$ with $a \in A$, or $exp_i = X_j X_k$ with $k, j < i$. A straight–line program can be viewed as a context-free grammar with initial symbol X_m generating a single word $val(\mathcal{P}) = val(X_m)$ which is the decompression of the text represented in compressed form by the SLP.

2.5 Comparison of Compression Models

The Lempel–Ziv–Welch (resp. Lempel–Ziv) algorithm is usually called LZ78 (resp. LZ77). According to [R99] "LZ78 is less interesting [than LZ77] from the theoretical point of view, but much more interesting from the practical point of view". The size of the LZ compression of a text is smaller than the size of its LZW compression. The drawback is that the LZ compression is harder to compute.

More specifically, LZ–decompression can yield an exponential blow-up, while LZW–decompression is bounded by a quadratic growth of text size. Given an LZW–compressed text of length m, we can easily construct in time $O(m)$ an SLP of size $O(m)$ generating the decompression. Given an LZ–compressed text of length m, and of original length n, we can construct in time $O(m \log n)$ an SLP of size $O(m \log n)$ generating the same decompressed text [R03].

3   The Problems

Let $t = t_1 t_2 \cdots t_n \in A^*$ be the *text* and $P = p_1 p_2 \cdots p_k$ be the *pattern* also in A^*. A size w *window* of t, in short w-*window*, is a size w subword $t_{i+1} t_{i+2} \cdots t_{i+w}$ of t; words corresponding to different values of i are considered to be different windows, even if they are equal as words; thus, there are $n - w + 1$ such windows in t. The word p is a *subsequence* of t iff there exist integers $1 \leq i_1 < i_2 < \cdots < i_k \leq n$ such that $t_{i_j} = p_j$ for $1 \leq j \leq k$. If moreover, $i_k - i_1 < w$, p is a *subsequence* of t *in a w-window*. A window containing p as a subsequence is said to be *minimal* if neither $t_{i+2} \cdots t_{i+w}$ nor $t_{i+1} t_{i+2} \cdots t_{i+w-1}$ contain p.

Example 1. If $t =$ "dans ville il y a vie" (a French advertisement), then "vie" is a subword and hence a subsequence of t. "vile" is neither a subword, nor a subsequence of t in a 4-window, but it is a subsequence of t in a 5-window. "ville" and "vie" are two minimal windows containing the pattern "vie". See figure 1.   □

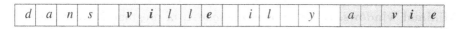

Fig. 1. A text with two 5-windows containing "vie" (in gray), and a single 5-window containing "vile"

Given an alphabet A, a text t on A^* and a pattern P, we consider five window problems:

- **Problem 1.** Given a compression of t and a pattern P, to decide whether pattern P is a subsequence of text t.
- **Problem 2.** Given a compression of t and a pattern P, to compute the number of minimal windows of t containing pattern P as a subsequence.
- **Problem 3.** Given a compression of t and a pattern P, to decide whether pattern P is a subsequence of a w-window of text t.
- **Problem 4.** Given a compression of t, a pattern P, and a number w, to compute the number of w-windows of t containing pattern P as a subsequence.
- **Problem 5.** Given a compression of t, a pattern P, and a number w, to compute the number of minimal windows of t which are of size at most w and which contain pattern P as a subsequence.

4   The Window Subsequence Algorithm

4.1   Auxiliary Data Structures We Are Using

From now on we will consider a text t compressed by an SLP $\mathcal{P}$ of size m. Let $|P| = k$, and let $P_1, \ldots, P_l$ (by convention $P_1 = P$) be all the different subwords of pattern P. We may note that $l = 1 + 2 + \cdots + k = k(k+1)/2 \leq k^2$.

We introduce two basic and three problem-oriented data structures.

The basic structures are two $m \times l$ arrays.

Left inclusion array. For every non-terminal X_i of program $\mathcal{P}$ and every sub-word P_j of pattern P, denote by $L_{i,j}$ the length of the shortest **prefix** of $val(X_i)$ containing P_j. If there is no such prefix we set $L_{i,j} = \infty$.

$$
\overbrace{val(X_i) = * \ldots * \; p_\alpha * \ldots * p_{\alpha+1} * \ldots * p_{\alpha+l_j}}^{L_{i,j} \text{ symbols}} * \ldots *
\tag{5}
$$

Fig. 2. $L_{i,j}$ for $P_j = p_\alpha p_{\alpha+1} \ldots p_{\alpha+l_j}$

Right inclusion array. For every non-terminal X_i of program $\mathcal{P}$ and every subword P_j of pattern P, denote by $R_{i,j}$ the length of the shortest **suffix** of $val(X_i)$ containing P_j. If there is no such prefix we set $R_{i,j} = \infty$.

The data structures we will use to solve the problems are the following three one-dimensional integer arrays:

Minimal windows. For every non-terminal X_i of program $\mathcal{P}$, we denote by MW_i the number of minimal windows of $val(X_i)$ containing P.

Windows of constant size. For every non-terminal X_i of program $\mathcal{P}$, we denote by FW_i the number of w-windows of $val(X_i)$ containing P.

Bounded minimal windows. For every non-terminal X_i of program $\mathcal{P}$, we denote by BMW_i the number of minimal windows of $val(X_i)$ which contain P and have size at most w.

4.2   Efficient Computation of These Data Structures

Let us show how to efficiently compute the above five arrays.

Left inclusion array. We use structural induction over non-terminals of the SLP in order to compute the left inclusions array; the algorithm is as follows:
Basis. If $exp_i = a$, then

$$
L_{i,j} = \begin{cases} 1 & \text{if } P_j = a, \\ \infty & \text{otherwise.} \end{cases}
$$

Induction. If $exp_i = X_p X_q$, two cases can occur:

(i) either P_j is contained in $val(X_p)$, i.e. $L_{p,j} \neq \infty$; in that case we have $L_{i,j} = L_{p,j}$,

(ii) otherwise, let P_u be the longest prefix of P_j such that $L_{p,u} < \infty$; in this case $L_{i,j} = |val(X_p)| + L_{q,v}$ where P_v is such that $P_j = P_u P_v$.

$$
\underbrace{\overbrace{*.. * \; p_{i_u} * .. * p_{i_u+1} * .. * p_{i_u+l_u}}^{P_u \text{ subsequence}} *..*}_{val(X_p)} \underbrace{*.. * \; \overbrace{p_{i_v} * .. * p_{i_v+1} * .. * p_{i_v+l_v}}^{P_v \text{ subsequence}} *..*}_{val(X_q)}
\tag{6}
$$

Using binary search to find P_u, the complexity of one inductive step will be $O(\log k)$, and the overall complexity will be $O(ml \log k) \leq O(mk^2 \log k)$.

Right inclusion array. Analogous to the left inclusions.

Minimal windows. We will use left and right inclusion arrays together with structural induction on the SLP structure. Let us first describe the intuitive idea. Minimal windows in $val(X_i)$ for $X_i := X_p X_q$ are of one of three types: (i) either they are entirely inside $val(X_p)$, (ii) or they are entirely inside $val(X_q)$, (iii) or they are overlapping on both $val(X_p)$ and $val(X_q)$. Type (iii) minimal windows will be called *boundary* windows (see Figure 3). To count the number of minimal windows in X_i we add the already counted numbers for X_p and X_q together with the number $B_{p,q}$ of *boundary* windows. Notice that for every decomposition $P = P_u P_v$ there is at most one boundary minimal window in which P_u is inside $val(X_p)$ and P_v is inside $val(X_q)$. Using left and right inclusion arrays we can determine decompositions of P for which such a boundary minimal window exists. However, counting must be done carefully: the same boundary window may correspond to several decompositions of P. So we run over all decompositions from $|P_u| = k-1$ to $|P_u| = 1$ and update our counter only when the following two conditions hold: 1) P_u (resp. P_v) is embedded in X_p (resp. X_q) and 2) the window is shifted from the previous successful embedding. To check these conditions, we will use a marker α in the program computing $B_{p,q}$: α will be set to 1 if we know that the next-to-be-studied window cannot be minimal. For the first and last boundary windows, we must also check that they do not contain a minimal window of type (i) or (ii), and this is also taken care of by marker α.

$$
\underbrace{\overbrace{\underbrace{*..* \; p_1 *..* p_2 * ... * p_l}_{P_u \text{ subsequence}} *..*}^{R_{p,u} \text{ symbols}}}_{val(X_p)} \; \underbrace{\overbrace{*..* \; \underbrace{p_{l+1} *..* p_{l+2} * .. * p_k}_{P_v \text{ subsequence}} * ..*}^{L_{q,v} \text{ symbols}}}_{val(X_q)} \tag{7}
$$

Fig. 3. A boundary window of length $R_{p,u} + L_{q,v}$ for $P = P_u P_v$, $P_u = p_1 \ldots p_l$, $P_v = p_{l+1} \ldots p_k$

The algorithm is as follows.
Basis. If $exp_i = a$, then

$$
MW_i = \begin{cases} 1 & \text{if } P = a, \\ 0 & \text{otherwise.} \end{cases}
$$

Induction. If $exp_i = X_p X_q$, then $MW_i = MW_p + MW_q + B_{p,q}$.

$B_{p,q}$ is determined by the following FOR loop (by convention "advance" is a shorthand for $u := u'$; $v := v'$; and we write u (resp v) instead of P_u (resp. P_v)):

$B := 0$; $\alpha := 0$; $u := p_1 \ldots p_{k-1}$; $v := p_k$;

IF $(R_{p,u} = R_{p,P} < \infty)$ THEN $\alpha := 1$; ENDIF
FOR $l = k - 1$ TO 1 DO
    $l := l - 1$; $u' := p_1 \ldots p_l$; $v' := p_{l+1} \ldots p_k$;
    IF $(L_{q,v'} = L_{q,v} \wedge \infty > R_{p,u} \geq R_{p,u'})$ THEN $\alpha := 0$; advance; ENDIF
    IF $(\infty > L_{q,v'} > L_{q,v} \wedge R_{p,u} = R_{p,u'})$ THEN
        IF $\alpha \neq 1$ THEN $B := B + 1$; $\alpha := 1$; ENDIF advance; ENDIF
    IF $(\infty > L_{q,v'} > L_{q,v} \wedge \infty > R_{p,u} > R_{p,u'})$ THEN
        IF $\alpha \neq 1$ THEN $B := B + 1$; ENDIF $\alpha := 0$; advance; ENDIF
ENDFOR
IF $(L_{q,P} > L_{q,v'} \wedge \alpha \neq 1)$ THEN $B := B + 1$; ENDIF
$B_{p,q} = B$;
    Thus the complexity of computing each $B_{p,q}$ is $O(k)$ and the overall complexity of computing the MW structure is $O(mk)$.

Minimal windows of size bounded by w. Computing this structure is the same as computing minimal windows. We just ignore boundary minimal windows of size more than w (*i.e.* increment B only if $(R_{p,u} + L_{q,v}) \leq w$).

Windows of constant size w. The main observation is that any w-window containing P also contains a *minimal window* containing P. Again w-windows of $val(X_i)$ (with $exp_i = X_p X_q$) containing P are (i) either entirely inside $val(X_p)$, (ii) or entirely inside $val(X_q)$, (iii) or overlapping on both $val(X_p)$ and $val(X_q)$. Thus we only need to explain how to count boundary windows. In the same way as in the previous section we run over all decompositions of P, starting from P entirely contained in $val(X_p)$ to finish with P entirely contained in $val(X_q)$. For every decomposition, using information from left and right inclusion arrays, we find a minimal window corresponding to this decomposition. In counting the number of boundary w-windows, we have to be careful, because several minimal windows can be included in the same w-window containing P as a subsequence; hence we cannot just count the number of w-windows containing a minimal window: we have to only count the *new* w-windows contributed by the current minimal window.
    The number $FB_{p,q}$ of boundary w-windows is determined by a FOR loop quite similar to the previous one; we replace the statement $B := B + 1$; by a subprogram called "update" which is defined by:

IF $(R_{min} + L_{q,v}) \leq w$
    THEN $B := B + R_{min} - R_{p,u}$;
    ELSE IF $(R_{p,u} + L_{q,v}) \leq w$ THEN $B := B + w - (R_{p,u} + L_{q,v}) + 1$; ENDIF
ENDIF
$R_{min} := R_{p,u}$;

The number $FB_{p,q}$ boundary w-windows is defined by the following FOR loop:

$B := 0$; $l := k$; $u := P$; $R_{min} := R_{p,P}$;   // w-windows with P in $val(X_p)$
IF $(R_{min} < w)$ THEN $B := B + w - R_{min}$; ENDIF
$\alpha := 0$; $u := p_1 \ldots p_{k-1}$; $v := p_k$;
IF $(R_{p,u} = R_{p,P} < \infty)$ THEN $\alpha := 1$; ENDIF

FOR $l = k - 1$ TO 1 DO $\qquad$ // w-windows with P in val(X_p) and val(X_q)

$\quad l := l - 1$; $u' := p_1 \ldots p_l$; $v' := p_{l+1} \ldots p_k$;

$\quad$ IF $(L_{q,v'} = L_{q,v} \wedge \infty > R_{p,u} \geq R_{p,u'})$ THEN $\alpha := 0$; advance; ENDIF

$\quad$ IF $(\infty > L_{q,v'} > L_{q,v} \wedge R_{p,u} = R_{p,u'})$ THEN

$\qquad\qquad$ IF $\alpha \neq 1$ THEN update; $\alpha := 1$; ENDIF advance; ENDIF

$\quad$ IF $(\infty > L_{q,v'} > L_{q,v} \wedge \infty > R_{p,u} > R_{p,u'})$ THEN

$\qquad\qquad$ IF $\alpha \neq 1$ THEN update; ENDIF $\alpha := 0$; advance; ENDIF

ENDFOR

IF $(L_{q,P} > L_{q,v'} \wedge \alpha \neq 1)$ THEN update; ENDIF

IF $(R_{min} + L_{q,P}) \leq w$ $\qquad\qquad\qquad$ // w-windows with P in val(X_q)

$\quad$ THEN $B := B + R_{min} - 1$;

$\quad$ ELSE IF $L_{q,P} \leq w$ THEN $B := B + w - L_{q,P}$; ENDIF

ENDIF

$FB_{p,q} = B$;

So we can estimate the complexity of this step by $O(k)$ and and the overall complexity of computing the FW structure is $O(mk)$.

4.3 Final Algorithm and Its Complexity

Our structures contain answers to all five problems:

1. Pattern P is a subsequence of text t iff $L_{m,1} \neq \infty$ (letting $P_1 = P$),
2. The number of minimal windows of t which contain P is equal to MW_m,
3. Pattern P is a subsequence of some w-window iff $FW_m \neq 0$,
4. The number of w-windows containing P is equal to FW_m,
5. The number of minimal windows of size at most w and which contain P is equal to BMW_m.

So the final complexity of our algorithm in the case of compression by straight-line program is $O(mk^2 \log k)$, where m is the size of the compressed text and k is the pattern size.

Since LZW is easily converted to SLP, for LZW compression the complexity of our algorithm is $O(mk^2 \log k)$, where m is now the size of the LZW-compressed text.

For LZ compression we also can convert it to SLP. That gives as complexity $O(mk^2 \log k \log n)$. Here m is the size of the LZ-compressed text, n is the original text size and k is the pattern size.

5   Conclusions

We introduced in the present paper a new algorithm for a series of window subsequence problems. We showed that for SLP and LZW compression our algorithm is *linear* in the size of the *compressed* text. In the case of LZ compression it is only $\log n$ times worse than linear. These results show that all subsequence search problems can be done efficiently for compressed texts without unpacking.

An open question we have is the following. Is it possible to reduce the k-dependant factor in our algorithm complexity?

References

[ABF95]   A. Amir, G. Benson, M. Farach, Let sleeping files lie: pattern matching in
            Z–compressed files, J. Comput. Syst. Sci., Vol. 52 (2) (1996), pp. 299–307.

[BKLPR02]   P. Berman, M. Karpinski, L. Larmore, W. Plandowski, W. Rytter,
            On the Complexity of Pattern Matching for Highly Compressed Two-
            Dimensional Texts, Journal of Computer and Systems Science, Vol. 65
            (2), (2002), pp. 332–350.

[C88]   M. Crochemore, String-matching with constraints, Proc. MFCS'88, LNCS
            324, Springer-Verlag, Berlin (1988), pp. 44–58.

[GKPR96]   L. Gasieniec, M. Karpinski, W. Plandowski and W. Rytter. Efficient Algo-
            rithms for Lempel-Ziv Encoding (Extended Abstract), Proceedings of the
            5th Scandinavian Workshop on Algorithm Theory (SWAT 1996), LNCS
            1097, Springer-Verlag, Berlin (1996), pp. 392–403.

[GM02]   B. Genest, A. Muscholl, Pattern Matching and Membership for Hierar-
            chical Message Sequence Charts, In Proceedings of the 5th Latin Amer-
            ican Symposium on Theoretical Informatics (LATIN 2002), LNCS 2286,
            Springer-Verlag, Berlin (2002), pp. 326–340.

[LZ77]   G. Ziv, A. Lempel, A universal algorithm for sequential data compress-
            sion, IEEE Transactions on Information Theory, Vol. 23 (3), (1977), pp.
            337–343.

[LZ78]   G. Ziv, A. Lempel, Compresssion of individual sequences via variable-rate
            coding, IEEE Transactions on Information Theory, Vol. 24, (1978), pp.
            530–536.

[L05]   Yu. Lifshits, On the computational complexity of embedding of com-
            pressed texts, St.Petersburg State University Diploma thesis, (2005);
            http://logic.pdmi.ras.ru/~yura/en/diplomen.pdf.

[LL05]   Yu. Lifshits, M. Lohrey, Querying and Embedding Compressed Texts, to
            appear (2005).

[Loh04]   M. Lohrey, Word problems on compressed word, ICALP 2004, Springer-
            Verlag, LNCS 3142, Berlin (2004), pp. 906–918.

[M02]   H. Mannila, Local and Global Methods in Data Mining: Basic Techniques
            and open Problems, Proc. ICALP 2002, LNCS 2380, Springer-Verlag,
            Berlin (2002), pp. 57–68.

[MS04]   N. Markey, P. Schnoebelen, A PTIME-complete matching problem for
            SLP-compressed words, Information Processing Letters, Vol. 90 (1),
            (2004), pp. 3–6.

[Ma71]   Yu. Matiyasevich, Real-time recognition of the inclusion relation, Za-
            piski Nauchnykh Leningradskovo Otdeleniya Mat. Inst. Steklova Akad.
            Nauk SSSR, Vol. 20, (1971), pp. 104–114. Translated into English,
            Journal of Soviet Mathematics, Vol. 1, (1973), pp. 64–70; http://
            logic.pdmi.ras.ru/~yumat/Journal.

[R99]   W. Rytter, Algorithms on compressed strings and arrays, Proc. SOF-
            SEM'99, LNCS 1725, Springer-Verlag, Berlin (1999), pp. 48–65.

[R03]   W. Rytter, Application of Lempel-Ziv factorization to the approximation
            of grammar-based compression, TCS 1-3(299) (2003), pp. 763–774.

[S71]   A. Slissenko, String-matching in real time, LNCS 64, Springer-Verlag,
            Berlin (1978), pp. 493–496.

[W84]   T. Welch, A technique for high performance data compresssion, Com-
            puter, (June 1984), pp. 8–19.

Efficient Algorithms in Zero-Characteristic for a New Model of Representation of Algebraic Varieties

Alexander L. Chistov

St. Petersburg Department of Steklov Mathematical Institute,
Fontanka 27, St. Petersburg 191011, Russia

Abstract. We suggest a model of representation of algebraic varieties based on representative systems of points of its irreducible components. Deterministic polynomial–time algorithms to substantiate this model are described in zero–characteristic. The main result here is a construction of the intersection of algebraic varieties. As applications we get efficient algorithms for constructing the smooth stratification and smooth cover of an algebraic variety introduced by the author earlier.

The present work concludes the series of papers [1], [2], [3], [4], [5], [6], [11], [7], [8] (the correction of Lemma 2 [8], see in [9]), [9], [10], where the polynomial–time algorithms for to algebraic varieties in zero–characteristic are suggested (we do not use the results of [10] in the present paper; but the particular case of [10] from [7] is necessary here). Before formulating our results we describe how to give a quasiprojective algebraic variety using a representative system of points of its irreducible components. The model of representation of algebraic varieties suggested here slightly generalizes the one outlined in [6], see the remarks below. In [6] the description of the algorithms for this representation was postponed. It hase become possible only using the results of four more papers [4], [7], [8], [9].

Let k be a field of zero–characteristic with algebraic closure $\overline{k}$. Let $X_0, X_1, \ldots$ be independent variables over k. Denote by $\mathbb{P}^n(\overline{k})$, $n \geq 0$, the projective space over the field $\overline{k}$ with coordinates $X_0, \ldots, X_n$. We shall suppose that $\mathbb{P}^n(\overline{k})$ is defined over k (the structure of defined over k algebraic variety on $\mathbb{P}^n(\overline{k})$ is given here, e.g., by the homogeneous ring $k[X_0, \ldots, X_n]$ defined over k). For arbitrary homogeneous polynomials $g_1, \ldots, g_m \in \overline{k}[X_0, \ldots, X_n]$ we shall denote by $\mathcal{Z}(g_1, \ldots, g_m)$ the set of all common zeroes of polynomials $g_1, \ldots, g_m$ in $\mathbb{P}^n(\overline{k})$. The similar notations will be used for the sets of zeroes of ideals and polynomials with other fields of coefficients in affine and projective spaces (this will be seen from the context).

Let W be a quasiprojective algebraic variety in $\mathbb{P}^n(\overline{k})$ and W is defined over k. Then we represent

$$W = \bigcup_{1 \leq i \leq b} W^{(i)} \setminus \bigcup_{b+1 \leq i \leq a} W^{(i)}, \tag{1}$$

D. Grigoriev, J. Harrison, and E.A. Hirsch (Eds.): CSR 2006, LNCS 3967, pp. 137–146, 2006.
© Springer-Verlag Berlin Heidelberg 2006

where $1 \leq b \leq a$ are integers, and all $W^{(i)}$, $1 \leq i \leq a$, are projective algebraic varieties in $\mathbb{P}^n(\overline{k})$ defined over k. Each algebraic variety $W^{(i)}$, $1 \leq i \leq a$, is a union of some irreducible components of the variety $V^{(i)} = \mathcal{Z}(f_1^{(i)}, \ldots, f_{m(i)}^{(i)}) \subset \mathbb{P}^n(\overline{k})$ where homogeneous polynomials $f_1^{(i)}, \ldots, f_{m(i)}^{(i)} \in k[X_0, \ldots, X_n]$ are given, $m(i) \geq 1$. For every $0 \leq s \leq n$ denote by $V^{(i,s)}$ (respectively $W^{(i,s)}$) the union of all irreducible components of dimension $n - s$ of $\mathcal{Z}(f_1^{(i)}, \ldots, f_{m(i)}^{(i)})$ (respectively $W^{(i)}$). Therefore $W^{(i,s)}$ is a union of some irreducible components of $V^{(i,s)}$. For every $0 \leq s \leq n$ the family of linear forms $L_{s+1}^{(i,s)}, \ldots, L_n^{(i,s)} \in k[X_0, \ldots, X_n]$ is given such that the number of points

$$\#V^{(i,s)} \cap \mathcal{Z}(L_{s+1}^{(i,s)}, \ldots, L_n^{(i,s)}) < +\infty \tag{2}$$

is finite, every point $\xi \in V^{(i,s)} \cap \mathcal{Z}(L_{s+1}^{(i,s)}, \ldots, L_n^{(i,s)})$ is a smooth point of the algebraic variety $\mathcal{Z}(f_1^{(i)}, \ldots, f_{m(i)}^{(i)})$, and the intersection of the tangent spaces in the point ξ of $V_s^{(i)}$ and $\mathcal{Z}(L_{s+1}^{(i,s)}, \ldots, L_n^{(i,s)})$ is transversal, i.e.,

$$T_{\xi, V_s^{(i)}} \cap \mathcal{Z}(L_{s+1}^{(i,s)}, \ldots, L_n^{(i,s)}) = \{\xi\} \tag{3}$$

(we consider the tangent space $T_{\xi, V_s^{(i)}}$ as a subspace of $\mathbb{P}^n(\overline{k})$). The set of points

$$\Xi^{(i,s)} = W^{(i,s)} \cap \mathcal{Z}(L_{s+1}^{(i,s)}, \ldots, L_n^{(i,s)})$$

is given. Each point from $\Xi^{(i,s)}$ is represented in form (11), see below. Hence the following property holds. Let $\xi \in V^{(i,s)} \cap \mathcal{Z}(L_{s+1}^{(i,s)}, \ldots, L_n^{(i,s)})$ and E be the uniquely defined irreducible over k component of the algebraic variety $V^{(i,s)}$ such that $\xi \in E$. Then $\xi \in \Xi^{(i,s)}$ if and only if E is a component of $W^{(i)}$. In what follows, unless we state otherwise, we assume that the degrees $\deg_{X_0, \ldots, X_n} f_j^{(i)} < d$ for all i, j.

Thus, formally the suggested in this paper representation of W is a quadruple

$$(f, L, \Xi, b), \tag{4}$$

where f is a family of polynomials

$$f_j^{(i)}, \quad 1 \leq j \leq m(i), 1 \leq i \leq a, \tag{5}$$

L is a family of linear forms

$$L_w^{(i,s)}, \quad s+1 \leq w \leq n, 0 \leq s \leq n, 1 \leq i \leq a, \tag{6}$$

and Ξ is a family of finite sets of points

$$\Xi^{(i,s)}, \quad 0 \leq s \leq n, 1 \leq i \leq a. \tag{7}$$

Denote also

$$\Xi^{(i)} = \bigcup_{0 \leq s \leq n} \Xi^{(i,s)}. \tag{8}$$

In [6] only the case $b = 1$ is considered. In the present paper in comparison with [6] we replace the lower index s by the upper one to avoid an ambiguity of the notation when one considers more than one algebraic variety, see below.

Notice that (2) and (3) can be always verified using the algorithms from [3], see the Introduction of [6] for details. Note also that for a given point $y \in \mathbb{P}^n(\overline{k})$ one can decide whether $y \in W^{(i,s)}$ using the algorithms from [3], see the Introduction of [6]. Thus, for a given point $y \in \mathbb{P}^n(\overline{k})$ one can decide also within the polynomial time whether $y \in W^{(i)}$, $1 \le i \le a$, and whether $y \in W$.

Let $W = \bigcup_{i \in I} W_i$ be the decomposition of W into the union of defined over k and irreducible over k (respectively irreducible over $\overline{k}$) components and representation (4) is given. Then using Theorem 1 [7] (or more strong Theorem 3 [9]) and Theorem 2 [7] one can construct for every $i \in I$ the representation $(f, L_i, \Xi_i, 1)$ of the irreducible component W_i. The working time of this algorithm is polynomial in d^n and the size of input, see [7] (and also the proof of Theorem 1 for the partial case $\nu = 1$). Notice that the case when W_i is irreducible over $\overline{k}$ is reduced to the one of irreducible over k components. Namely, we construct the minimal field of definition k_i of W_i containing k and replace the ground field k by k_i in the representation of W_i, see [7] for details. So in what follows we consider only the decomposition into defined over k components.

Further, see Theorem 3 below, let W_1, W_2 be two quasiprojective algebraic varieties which are similar to W and are given in the similar way (with the same bound d for degrees of the polynomials). Then one can decide whether $W_1 = W_2$ within the time in d^n and the size of input. In [6] this is proved only when W_1 and W_2 are projective algebraic varieties. The general case of quasiprojective algebraic varieties is difficult.

For the proof of Theorem 2 (and hence also of Theorem 4) we need at first describe an algorithm for constructing an intersection of ν quasiprojective algebraic varieties given in model under consideration within the time in $d^{n\nu}$ and the size of input, see Theorem 1 below. This algorithm uses the reduction to the diagonal. Here one needs to apply Theorem 1 [9]. The last theorem has a long proof. It is based on [8], [7], [6] and other our papers, see the Introduction of [9]. Besides in Theorem 1 indices of intersection of algebraic varieties are computed when they are defined. Notice here that at the output of the algorithm from Theorem 1 the intersection of quasiprojective algebraic varieties is not given in the model under consideration. The irreducible components of the intersections not always can be given using representative systems of points of irreducible components of an algebraic variety $\mathcal{V}$ with good upper bounds for degrees of polynomials giving $\mathcal{V}$. Still at the output of the algorithm from Theorem 1 we get all the information about the intersection. One can consider the representation of the intersection of algebraic varieties from assertion (a) of Theorem 1 as a generalization of representation (4) to the case of intersections of algebraic varieties given in form (4).

Denote $m = m(1)$, and $f_i = f_i^{(1)}$, $1 \le i \le m$. Let $V = \mathcal{Z}(f_1, \ldots, f_m) \subset \mathbb{P}^n(\overline{k})$ be an algebraic variety. Recall the following definition.

Definition 1. *Smooth cover of the algebraic variety V is a finite family*

$$V_\alpha, \quad \alpha \in A, \tag{9}$$

of quasiprojective smooth algebraic varieties $V_\alpha \subset \mathbb{P}^n(\overline{k})$, $\alpha \in A$ such that V is represented as a union $V = \cup_{\alpha \in A} V_\alpha$. Further, we shall require that all irreducible components of V_α have the same dimension (which depends only on α). Smooth stratification of the algebraic variety V is a smooth cover V_α, $\alpha \in A$, of V such that additionally for any two $\alpha_1, \alpha_2 \in A$ if $\alpha_1 \neq \alpha_2$ then $V_{\alpha_1} \cap V_{\alpha_2} = \emptyset$.

In this paper we assume that the degree of an arbitrary projective algebraic variety is the sum of the degrees of all its irreducible components (of different dimensions). The degree of a quasiprojective algebraic variety $\mathcal{V}$ is by definition the degree of its closure in the corresponding projective space.

In [6] using the construction of local parameters from [4] we prove the existence of smooth cover as well as smooth stratification (9) of the algebraic variety V with the bound for the degrees of strata $2^{2^{n^C}} d^n$, and the number of strata $2^{2^{n^C}} d^n$ (respectively the number of strata $2^{2^{n^C}} d^{n(n+1)/2}$) for an absolute constant $0 < C \in \mathbb{R}$, see Theorem 2 [6]. The constructions of [6] are quite explicit. It turns out that it is sufficient to use additionally only Theorem 1 and Theorem 2 (the last one only in the case $(\nu, nn_1) = (3, 2)$) to obtain the algorithms for constructing the smooth cover and smooth stratification from Theorem 2 [6] within the time polynomial in $2^{2^{n^C}} d^{n^2}$ and the size of input, see Theorem 4 below.

Now we proceed to the precise statements. Let the integers a, b; $m(i), 1 \leq i \leq b$, the homogeneous polynomials $f_j^{(i)} \in k[X_0, \ldots, X_n]$, $1 \leq j \leq m(i)$, $1 \leq i \leq a$, the algebraic varieties $V^{(i)}$, $W^{(i)}$, and W be as above.

Let the field $k = \mathbb{Q}(t_1, \ldots, t_l, \theta)$ where $t_1, \ldots, t_l$ are algebraically independent over the field $\mathbb{Q}$ and θ is algebraic over $\mathbb{Q}(t_1, \ldots, t_l)$ with the minimal polynomial $F \in \mathbb{Q}[t_1, \ldots, t_l, Z]$ and leading coefficient $\mathrm{lc}_Z F$ of F is equal to 1. We shall represent each polynomial $f = f_j^{(i)}$ in the form

$$f = \frac{1}{a_0} \sum_{i_0, \ldots, i_n} \sum_{0 \leq j < \deg_Z F} a_{i_0, \ldots, i_n, j} \theta^j X_0^{i_0} \cdots X_n^{i_n},$$

where $a_0, a_{i_0, \ldots, i_n, j} \in \mathbb{Z}[t_1, \ldots, t_l]$, $\mathrm{G\,C\,D}_{i_0, \ldots, i_n, j}(a_0, a_{i_0, \ldots, i_n, j}) = 1$. Define the length $\mathrm{l}(a)$ of an integer a by the formula $\mathrm{l}(a) = \min\{s \in \mathbb{Z} : |a| < 2^{s-1}\}$. The length of coefficients $\mathrm{l}(f)$ of the polynomial f is defined to be the maximum of lengths of coefficients from $\mathbb{Z}$ of polynomials $a_0, a_{i_0, \ldots, i_n, j}$ and the degree

$$\deg_{t_\gamma}(f) = \max_{i_0, \ldots, i_n, j} \{\deg_{t_\gamma}(a_0), \deg_{t_\gamma}(a_{i_0, \ldots, i_n, j})\},$$

where $1 \leq \gamma \leq l$. In the similar way we shall define degrees and lengths of integer coefficients of other polynomials, in particular $\deg_{t_\gamma} F$ and $\mathrm{l}(F)$ are defined. We shall suppose that we have the following bounds

$$\deg_{X_0, \ldots, X_n}(f_j^{(i)}) < d, \ \deg_{t_\gamma}(f_j^{(i)}) < d_2, \ \mathrm{l}(f_j^{(i)}) < M, \tag{10}$$
$$\deg_Z(F) < d_1, \ \deg_{t_\gamma}(F) < d_1, \ \mathrm{l}(F) < M_1.$$

for all $1 \leq j \leq m(i)$, $1 \leq i \leq a$, $1 \leq \gamma \leq l$. The size $L(f)$ of the polynomial f is defined to be the product of $l(f)$ to the number of all the coefficients from $\mathbb{Z}$ of f in the dense representation. We have

$$L(f_j^{(i)}) < (\binom{d+n}{n} d_1 + 1) d_2^l M$$

Similarly $L(F) < d_1^{l+1} M_1$.

Remark 1. *Unless we state otherwise, in what follows we suppose l to be fixed. The working time of the algorithms from Theorem 1, Theorem 2 and Theorem 4, see below, is essentially the same as for solving systems of polynomial equations with a finite set of solutions in the projective space. So this theorems can be formulated also in the case when l is not fixed. Notice that the constants $O(\ldots)$, see Theorem 1, Theorem 2 and Theorem 4 below, in the estimate of the lengths of integer coefficients of linear forms L'_j, L_j are absolute; they does not depend on l.*

We shall represent a point $z \in V$ with coordinates from a finite extension of k as follows. An index $0 \leq i_0 \leq n$ is known such that $X_{i_0}(z) \neq 0$ and an isomorphism of fields

$$k(z) = k((X_1/X_{i_0})(z), \ldots, (X_n/X_{i_0})(z)) = k[\eta] \simeq k[Z]/(\Phi) \qquad (11)$$

is given where $\eta = \sum_{0 \leq i \leq n} c_i (X_i/X_{i_0})(z)$, the coefficients $c_i \in \mathbb{Z}$ are given and $\Phi \in k[Z]$ is minimal polynomial of η over k with leading coefficient $\mathrm{lc}_Z \Phi = 1$ (so the point z is defined up to a conjugation over k).

Let $g \in k[\eta]$ be an arbitrary element. Then $g = G(\eta)$ for the uniquely defined polynomial $A \in k[Z]$ such that $\deg_Z G < \deg_Z \Phi$. The length of integer coefficients $l(g)$, the size $L(g)$ and the degrees $\deg_{t_\alpha} g$, $1 \leq \alpha \leq l$, of g are defined by the formulas

$$l(g) = l(G), \quad L(g) = L(G), \quad \deg_{t_\alpha} g = \deg_{t_\alpha} G.$$

Let us define the size of the point z to be $L(\Phi) + \sum_{0 \leq i \leq n} L(X_i/X_{i_0})$. Now let $\varXi$ be an arbitrary finite set of points defined over k, and every point from $\varXi$ is given in form (11). Hence $\varXi$ gives a zero–dimensional algebraic varieties defined over k. Put the size $L(\varXi)$ of $\varXi$ to be the sum of sizes of its irreducible over k components.

Recall that in [9] we give the definition of transversality of intersection of algebraic varieties. Now we give the analogous natural definition related to the proper intersections. Let $W_1, \ldots, W_\nu \subset \mathbb{P}^n(\overline{k})$, $\nu \geq 1$, be ν quasiprojective algebraic varieties defined over k. Let E be an arbitrary defined over k and irreducible over k component of $W_1 \cap \ldots \cap W_\nu$. We shall say that the intersection $W_1 \cap \ldots \cap W_\nu$ is proper at E (in the ambient space $\mathbb{P}^n(\overline{k})$) if and only if for every defined over k and irreducible over k component E_i, $1 \leq i \leq \nu$, of W_i such that $E_i \supset W$ the equality $\sum_{1 \leq i \leq \nu} (n - \dim E_i) = n - \dim E$ holds (and hence $\dim E_i$ depends only on i). The intersection of $W_1, \ldots, W_\nu$ is proper (in the ambient space $\mathbb{P}^n(\overline{k})$) if and only if it is proper at every its defined over k and irreducible over k component.

If $\nu = 2$ then the index of intersection $i(W_1, W_2; E) = i(W_1, W_2; E')$ where E' is an arbitrary irreducible over $\overline{k}$ component of E, and the index of intersection $i(W_1, W_2; E')$ is defined in the usual way.

For an arbitrary $\nu > 2$ we define the index of intersection $i(W_1, \ldots, W_\nu; E)$ of the algebraic varieties $W_1, \ldots, W_\nu$ at E recursively by the formula $i(W_1, \ldots, W_\nu; E) = \sum_{E''} i(W_1, \ldots, W_{\nu-1}; E'')i(E'', W_\nu; E)$ where E'' runs over all the irreducible over k components of $W_1 \cap \ldots \cap W_{\nu-1}$ such that $E'' \supset E$. For $\nu = 1$ it is natural to put $i(W_1; E) = 1$.

Here all the indices of intersection are considered in $\mathbb{P}^n(\overline{k})$. To specify this we denote $i_{\mathbb{P}^n(\overline{k})}(W_1, \ldots, W_\nu; E) = i(W_1, \ldots, W_\nu; E)$. Assume that all W_j, $1 \le j \le \nu$, are subvarieties of an affine space $\mathbb{A}^n(\overline{k})$. One can identify $\mathbb{A}^n(\overline{k}) = \mathbb{P}^n(\overline{k}) \setminus \mathcal{Z}(X_0)$. We shall denote also in this case $i_{\mathbb{A}^n(\overline{k})}(W_1, \ldots, W_\nu; E) = i(W_1, \ldots, W_\nu; E)$ when the last index of intersection is defined.

We shall use the reduction to diagonal for indices of intersection. Namely, let $\nu \ge 1$ be an integer. Let us identify the affine space $\mathbb{A}^{n\nu}(\overline{k}) = (\mathbb{A}^n(\overline{k}))^\nu$. Put $\Delta = \{(x, x, \ldots, x) \in \mathbb{A}^{n\nu}(\overline{k}) : x \in \mathbb{A}^n(\overline{k})\}$ to be the diagonal subvariety. Now we identify

$$\mathbb{A}^n(\overline{k}) = (\mathbb{A}^n(\overline{k}))^\nu \cap \Delta. \tag{12}$$

Let $W_1, \ldots, W_\nu$ be affine algebraic varieties in $\mathbb{A}^n(\overline{k})$ and E be a defined over k and irreducible over k component of $W_1 \cap \ldots \cap W_\nu$ such that the last intersection is proper at E. Then by (12) the variety E is a defined over k and irreducible over k component of $(W_1 \times \ldots \times W_\nu) \cap \Delta$. Obviously the last intersection is proper at E. We have

$$i_{\mathbb{A}^n(\overline{k})}(W_1, \ldots, W_\nu; E) = i_{\mathbb{A}^{n\nu}(\overline{k})}(W_1 \times \ldots \times W_\nu, \Delta; E). \tag{13}$$

This formula of reduction to diagonal is well known for $\nu = 2$. For an arbitrary ν it is proved by the induction on ν using the general properties of indices of intersection (we leave the details to the reader).

Let $\nu \ge 1$ be an integer. For every integer $1 \le \alpha \le \nu$ let $m_\alpha(i)$, a_α, b_α, $f_{\alpha,j}^{(i)}$, W_α, V_α, $W_\alpha^{(i)}$, $V_\alpha^{(i)}$, $W_\alpha^{(i)}$, $V_\alpha^{(i,s)}$, $W_\alpha^{(i,s)}$, $L_{\alpha,\beta}^{(i,s)}$, $\Xi_\alpha^{(i,s)}$, $(f_\alpha, L_\alpha, \Xi_\alpha, b_\alpha) = \rho_\alpha$ are similar to the introduced above $m(i)$, a, b, $f_j^{(i)}$, W, V, $W^{(i)}$, $V^{(i)}$, $W^{(i)}$, $V^{(i,s)}$, $W^{(i,s)}$, $L_\beta^{(i,s)}$, $\Xi^{(i,s)}$, $(f, L, \Xi, b) = \rho$ respectively. In what follows in this paper we suppose that inequalities (10) with $f_{\alpha,j}^{(i)}$ in place of $f_j^{(i)}$ hold for every $1 \le \alpha \le \nu$.

Theorem 1. *Assume that for every $1 \le \alpha \le \nu$ a representation $(f_\alpha, L_\alpha, \Xi_\alpha, b_\alpha)$ of a quasiprojective algebraic variety W_α is given. Then one can construct linear forms $L_0, \ldots, L_{n+1} \in k[X_0, \ldots, X_n]$ with integer coefficients of length $O(n\nu \log d + \sum_{1 \le i \le \nu} \log b_i + \log(\sum_{1 \le i \le \nu} a_i))$, the finite set of indices J, and for every $j \in J$ the finite set Ξ_j of points from $W_1 \cap \ldots \cap W_\nu$ (each point is represented in form (11)) such that the following assertions hold.*

(a) For every $j \in J$ there is the unique defined over k and irreducible over k component of $W_1 \cap \ldots \cap W_\nu$ (denote it by E_j) such that $\dim E_j = n - s(j)$ and $\Xi_j = E_j \cap \mathcal{Z}(L_{s(j)+1}, \ldots, L_n)$ in $\mathbb{P}^n(\overline{k})$. Conversely, for every defined over k and irreducible over k component E' of $W_1 \cap \ldots \cap W_\nu$ there is $j' \in J$ such that $E' = E_{j'}$. Hence $W_1 \cap \ldots \cap W_\nu = \bigcup_{j \in J} E_j$ is the decomposition of $W_1 \cap \ldots \cap W_\nu$ into the union of irreducible over k components. Let $\overline{E}_j$ be the closure of E_j with respect to the Zariski topology in $\mathbb{P}^n(\overline{k})$. Then for every $j \in J$

$$\Xi_j \cap \mathcal{Z}(L_0) = \emptyset, \quad \#\Xi_j = \#(L_{n+1}/L_0)(\Xi_j) = \deg \overline{E}_j \qquad (14)$$

(here and below $\#(.)$ denotes the number of elements of a set), and all points of Ξ_j are smooth points of $W_1 \cap \ldots \cap W_\nu$. The variety

$$E_j = \overline{E}_j \setminus \left(\bigcup_{1 \le \alpha \le \nu} \bigcup_{b_\alpha + 1 \le i \le a_\alpha} W_\alpha^{(i)} \right).$$

(b) If the intersection of $W_1, \ldots, W_\nu$ is proper at E_j then one can compute the index of intersection $i_{\mathbb{P}^n(\overline{k})}(W_1, \ldots, W_\nu; E_j)$, and for every point $\xi \in \Xi_j$ the equality $i_{\mathbb{P}^n(\overline{k})}(W_1, \ldots, W_\nu; E_j) = i_{\mathbb{P}^n(\overline{k})}(W_1, \ldots, W_\nu, \mathcal{Z}(L_{s(j)+1}, \ldots, L_n); \xi)$ holds.

(c) For an arbitrary point $z \in \mathbb{P}^n(\overline{k})$ (given in form (11)) one can decide whether $z \in E_j$, and more than that, compute the multiplicity $\mu(z, E_j)$ of the point z at E_j.

(d) Let us identify the set of all $(n+2)$-tuples of linear forms from $\overline{k}[X_0, \ldots, X_n]$ with the affine space $\mathbb{A}^{(n+1)(n+2)}(\overline{k})$. For every $j \in J$ and an arbitrary $\lambda^* = (L_0^*, \ldots, L_{n+1}^*)$, where all $L_j^* \in \overline{k}[X_0, \ldots, X_n]$ are linear forms, put $\Xi_j^* = E_j \cap \mathcal{Z}(L_{s(j)+1}^*, \ldots, L_n^*) \subset \mathbb{P}^n(\overline{k})$. Let $\mathfrak{l} \in \mathbb{A}^{(n+1)(n+2)}(\overline{k})$ be a line defined over a field k' such that $\overline{\lambda} = (L_0, \ldots, L_{n+1}) \in \mathfrak{l}$. Then for all $\lambda^* \in \mathfrak{l}(k')$ (here $\mathfrak{l}(k')$ is the set of all k'-points of $\mathfrak{l}$), except at most a polynomial in $d^{n\nu}(\sum_{1 \le i \le \nu} a_i) \prod_{1 \le \alpha \le \nu} b_\alpha$ number, assertion (a) holds with $\lambda^*, \Xi_j^*, j \in J$, in place of $\overline{\lambda}, \Xi_j, j \in J$. For every element $\lambda^* \in \mathfrak{l}(k')$ one can decide whether assertions (a) hold with $\lambda^*, \Xi_j^*, j \in J$, in place of $\overline{\lambda}, \Xi_j, j \in J$.

The working time of the algorithm for constructing linear forms $L_0, \ldots, L_n$ and the family of finite sets Ξ_j, $j \in J$, satisfying (a) and also of the algorithm from assertion (b) is polynomial in $d^{n\nu}$, $\prod_{1 \le \alpha \le \nu} b_\alpha$, and the sum of sizes $\sum_{1 \le \alpha \le \nu} L((f_\alpha, L_\alpha, \Xi_\alpha, b_\alpha))$. More precisely, this working time is polynomial in $d^{n\nu}$, $\prod_{1 \le \alpha \le \nu} b_\alpha$, $\sum_{1 \le \alpha \le \nu} a_\alpha$, M, M_1, d_1, d_2, $\sum_{1 \le \alpha \le \nu, 1 \le i \le a_\alpha} m(\alpha, i)$, and

$$\sum_{0 \le s \le n, 1 \le \alpha \le \nu, 1 \le i \le a_\alpha} L(\Xi_\alpha^{(i,s)}), \qquad \sum_{0 \le s \le n, s+1 \le w \le n, 1 \le \alpha \le \nu, 1 \le i \le a_\alpha} L(L_{\alpha,w}^{(i,s)}).$$

The working time of the algorithm from (c) (respectively (d)) is polynomial in the same values and the size $L(z)$ of the point z (respectively the size $L(\lambda^*)$ of the element λ^*).

Theorem 2. *Suppose that the assumptions of Theorem 1 hold, i.e., for every $1 \leq \alpha \leq \nu$ a representation $(f_\alpha, L_\alpha, \Xi_\alpha, b_\alpha)$ of a quasiprojective algebraic variety W_α is given. Let ν_1 be an integer such that $1 \leq \nu_1 \leq \nu$. Then one can construct linear forms $L_0, \ldots, L_{n+1} \in k[X_0, \ldots, X_n]$ with integer coefficients of length $O(n\nu \log d + \sum_{1 \leq i \leq \nu} \log b_i + \log(\sum_{1 \leq i \leq \nu} a_i))$, the finite set of indices $J^{(1)}$ (respectively $J^{(2)}$), and for every $j \in J^{(1)}$ (respectively $j \in J^{(2)}$) the finite set Ξ_j of points from $W_1 \cap \ldots \cap W_{\nu_1}$ (respectively $W_{\nu_1+1} \cap \ldots \cap W_\nu$) such that the following assertions hold.*

(a) *For every $j \in J^{(1)}$ (respectively $j \in J^{(2)}$) there is the unique defined over k and irreducible over k component E of $W_1 \cap \ldots \cap W_{\nu_1}$ (respectively $W_{\nu_1+1} \cap \ldots \cap W_\nu$) such that $\dim E = n - s(j)$ and $\Xi_j = E \cap \mathcal{Z}(L_{s(j)+1}, \ldots, L_n)$ in $\mathbb{P}^n(\overline{k})$. Denote $E = E_j$. Conversely, for every defined over k and irreducible over k component E' of $W_1 \cap \ldots \cap W_{\nu_1}$ (respectively $W_{\nu_1+1} \cap \ldots \cap W_\nu$) there is $j' \in J^{(1)}$ (respectively $j' \in J^{(2)}$) such that $E' = E_{j'}$. Hence*

$$W_1 \cap \ldots \cap W_{\nu_1} = \bigcup_{j \in J^{(1)}} E_j, \qquad W_{\nu_1+1} \cap \ldots \cap W_\nu = \bigcup_{j \in J^{(2)}} E_j$$

are the decompositions of $W_1 \cap \ldots \cap W_{\nu_1}$ and $W_{\nu_1+1} \cap \ldots \cap W_\nu$ into the unions of irreducible over k components. Let $\overline{E}_j$ be the closure of E_j with respect to the Zariski topology in $\mathbb{P}^n(\overline{k})$. Then for every $j \in J^{(1)}$ (respectively $j \in J^{(2)}$)

$$\Xi_j \cap \mathcal{Z}(L_0) = \emptyset, \quad \#\Xi_j = \#(L_{n+1}/L_0)(\Xi_j) = \deg \overline{E}_j, \tag{15}$$

and all the points of Ξ_j are smooth points of $W_1 \cap \ldots \cap W_{\nu_1}$ (respectively $W_{\nu_1+1} \cap \ldots \cap W_\nu$). Put $A^{(1)} = \{1, \ldots, \nu_1\}$, $A^{(2)} = \{\nu_1 + 1, \ldots, \nu\}$. Then for every $j \in J^{(i)}$, $i = 1, 2$ the variety

$$E_j = \overline{E}_j \setminus \left(\bigcup_{\alpha \in A^{(i)}} \bigcup_{b_\alpha + 1 \leq i \leq a_\alpha} W_\alpha^{(i)} \right).$$

(b) *One can decide for every $j_1 \in J^{(1)}$ for every $j_2 \in J^{(2)}$ whether $E_{j_1} \subset \overline{E}_{j_2}$. More precisely, for the constructed linear forms $L_0, \ldots, L_{n+1}$ the inclusion $E_{j_1} \subset \overline{E}_{j_2}$ holds if and only if $\Xi_{j_1} \subset \overline{E}_{j_2}$.*

(c) *One can decide for every $j_1 \in J^{(1)}$ for every $j_2 \in J^{(2)}$ whether $E_{j_1} \subset E_{j_2}$.*

The working time of each of the algorithms from this theorem is polynomial in $d^{n\nu}$, $\prod_{1 \leq \alpha \leq \nu} b_\alpha$, and the sum of sizes $\sum_{1 \leq \alpha \leq \nu} \mathrm{L}((f_\alpha, L_\alpha, \Xi_\alpha, b_\alpha))$. More precisely, the considered working time is polynomial in $d^{n\nu}$, $\prod_{1 \leq \alpha \leq \nu} b_\alpha$, $\sum_{1 \leq \alpha \leq \nu} a_\alpha$, M, M_1, d_1, d_2, $\sum_{1 \leq \alpha \leq \nu,\, 1 \leq i \leq a_\alpha} m(\alpha, i)$, and

$$\sum_{0 \leq s \leq n,\, 1 \leq \alpha \leq \nu,\, 1 \leq i \leq a_\alpha} \mathrm{L}(\Xi_\alpha^{(i,s)}), \qquad \sum_{0 \leq s \leq n,\, s+1 \leq w \leq n,\, 1 \leq \alpha \leq \nu,\, 1 \leq i \leq a_\alpha} \mathrm{L}(L_{\alpha,w}^{(i,s)}).$$

As an immediate consequence of Theorem 2 we get the following result.

Theorem 3. *Assume that representations $(f_\alpha, L_\alpha, \Xi_\alpha, b_\alpha)$, $\alpha = 1, 2$ of two quasiprojective algebraic variety W_α are given. Then one can decide whether*

$W_1 \subset W_2$. Hence one can decide also whether $W_2 \subset W_1$ and whether $W_1 = W_2$. The working time of this algorithm is polynomial in d^n and the sizes $L((f_\alpha, L_\alpha, \Xi_\alpha, b_\alpha))$ of the given representations of W_α, $\alpha = 1, 2$.

Theorem 4. *One can construct a smooth cover (respectively a smooth stratification) V_α, $\alpha \in A$, of the algebraic variety V such that every quasiprojective algebraic variety V_α is defined over k, irreducible over $\overline{k}$, and represented in the accepted way. Let $\dim V_\alpha = n - s$ where $0 \le s \le n$, and $s = s(\alpha)$ depends on α. Let (4) be the constructed representation of V_α (it depends on α). Denote $h_{\alpha,j} = f_j^{(1)}$, $1 \le j \le m(1)$, and $\Delta_\alpha = f_1^{(2)}$ if $a \ge 2$, see (5). Then the constructed representation of V_α satisfies the following properties.*

 (i) *The integer $b = 1$. For the case of smooth cover $a = 2$ if $s < n$, and $a = 1$ if $s = n$. For the case of smooth stratification $a \le 2^{2^{n^C}} d^{n(n+1)/2}$ for an absolute constant $0 < C \in \mathbb{R}$.*

 (ii) *$m(1) = s$ and if $a \ge 2$ then $m(2) = 1$. Hence if $a \ge 2$ then V_α is an irreducible component of the algebraic variety $\mathcal{Z}(h_{\alpha,1}, \ldots, h_{\alpha,s}) \setminus \mathcal{Z}(\Delta_\alpha)$ in the case of the smooth cover (respectively an open in the Zariski topology subset of an irreducible component of the latter algebraic variety in the case of smooth stratification).*

 (iii) *There are linearly independent linear forms $Y_0, \ldots, Y_n \in k[X_0, \ldots, X_n]$ such that $X_i = \sum_{0 \le j \le n} x_{i,j} Y_j$, all the coefficients $x_{i,j} \in k$, all $x_{i,j}$ are integers with lengths $O(2^{n^C} + n \log d)$ for an absolute constant $C > 0$, and*

$$\Delta_\alpha = \det(\partial h_{\alpha,i}/\partial Y_j)_{1 \le i,j \le s} = \det(\sum_{0 \le v \le n} x_{v,j} \partial h_{\alpha,i}/\partial X_v)_{1 \le i,j \le s}.$$

Hence V_α is a smooth algebraic variety by the implicit function theorem. Besides that, in the case of smooth cover one can take $Y_i = X_{\sigma(i)}$ for some permutation σ of the set $0, \ldots, n$.

 (iv) *The lengths of integer coefficients of all linear forms from the family L is $O(2^{n^C} + n \log d)$ for an absolute constant $0 < C \in \mathbb{R}$.*

 (v) *For all $\alpha \in A$, $1 \le j \le s(\alpha)$ degrees $\deg_{X_0, \ldots, X_n} h_{\alpha,j}$ are less than $n^{2^{s(\alpha)^C}} d$ for an absolute constant $0 < C \in \mathbb{R}$. In the case of smooth stratification for all $i > 2$, j degrees $\deg_{X_0, \ldots, X_n} f_j^{(i)}$ are less than $2^{2^{n^C}} d$.*

 (vi) *For all $\alpha \in A$, $1 \le j \le s(\alpha)$ lengths of coefficients of polynomials $h_{\alpha,j}$ are bounded from above by a polynomial in $n^{2^{s(\alpha)^C}} d^{ns(\alpha)}$, d_1, d_2, M, M_1, m for an absolute constant $0 < C \in \mathbb{R}$. Further, in the case of smooth stratification lengths of coefficients of all polynomials from the family f are bounded from above by a polynomial in $2^{2^{n^C}} d^{n^2}$, d_1, d_2, M, M_1, m for an absolute constant $0 < C \in \mathbb{R}$. The similar estimation holds for the size $L(\xi)$ of each point $\xi \in \Xi^{(i,s)}$ for all i, s, see (7).*

 (vii) *The number of elements $\#A$ of A is bounded from above by $2^{2^{n^C}} d^n$ for the case of smooth cover (respectively $2^{2^{n^C}} d^{n(n+1)/2}$ for the case of smooth stratification) for an absolute constant $0 < C \in \mathbb{R}$.*

The working time of the algorithm for constructing the smooth cover (respectively the smooth stratification) in the size of the output and $2^{2^{n^C}} d^{n^2}$, d_1, d_2, M, M_1, m, *where* $0 < C \in \mathbb{R}$ *is an absolute constant.*

References

1. **Chistov A. L.**: *"Polynomial–Time Computation of the Dimension of Algebraic Varieties in Zero–Characteristic"*, Journal of Symbolic Computation, 22 # 1 (1996) 1–25.
2. **Chistov A. L.**: *"Polynomial–time computation of the dimensions of components of algebraic varieties in zero–characteristic"*, Journal of Pure and Applied Algebra, 117 & 118 (1997) 145–175.
3. **Chistov A. L.**: *"Polynomial–time computation of the degrees of algebraic varieties in zero–characteristic and its applications"*, Zap. Nauchn. Semin. St-Petersburg. Otdel. Mat. Inst. Steklov (POMI) v. 258 (1999), p. 7–59 (in Russian) [English transl.: Journal of Mathematical Sciences 108 (6), (2002) p. 897–933] [1]
4. **Chistov A. L.**: *"Strong version of the basic deciding algorithm for the existential theory of real fields"*, Zap. Nauchn. Semin. St-Petersburg. Otdel. Mat. Inst. Steklov (POMI) 256 (1999) p. 168–211 (in Russian). [English transl.: J. of Mathematical Sciences v.107 No.5 p.4265–4295]
5. **Chistov A. L.**: *"Efficient Construction of Local Parameters of Irreducible Components of an Algebraic Variety"*, Proc. of the St.–Petersburg Mathematical Society, 1999, v. 7, p. 230–266 (in Russian) [English transl. in: Proceedings of the St.Petersburg Mathematical Society, American Math. Soc. Publisher v. VII, 2001].
6. **Chistov A. L.**: *"Efficient Smooth Stratification of an Algebraic Variety in Zero–Characteristic and its Applications"*, Zap. Nauchn. Semin. St-Petersburg. Otdel. Mat. Inst. Steklov (POMI) v. 266 (2000) p.254–311 (in Russian) [English transl.: Journal of Mathematical Sciences v. 113 (5), (2003), p 689–717].
7. **Chistov A. L.**: *"Monodromy and irreducibility criteria with algorithmic applications in zero–characteristic"*, Zap. Nauchn. Semin. St-Petersburg. Otdel. Mat. Inst. Steklov (POMI) 292 (2002) p. 130–152 (in Russian) [in English, see Preprint of St.Petersburg Mathematical Society (2004)].
8. **Chistov A. L.**: *"Polynomial–time computation of the degree of a dominant morphism in zero–characteristic I"*, Zap. Nauchn. Semin. St-Petersburg. Otdel. Mat. Inst. Steklov (POMI) 307 (2004) p. 189–235 (in Russian) [in English, see Preprint of St.Petersburg Mathematical Society (2004)].
9. **Chistov A. L.**: *"Polynomial–time computation of the degree of a dominant morphism in zero–characteristic II"*, Preprint of St.Petersburg Mathematical Society (2004), http://www.MathSoc.spb.ru
10. **Chistov A. L.**: *"A deterministic polynomial–time algorithm for the first Bertini theorem"*, Preprint of St.Petersburg Mathematical Society (2004), http://www.MathSoc.spb.ru
11. **Chistov A. L.**: *"A correction in the statement of my theorem on the efficient smooth cover and smooth stratification of an algebraic variety."*, Preprint of the St.–Petersburg Mathematical Society, 2004, #13, http://www.mathsoc.spb.ru/preprint/2004/index.html#13

---

[1] Preliminary versions of [3], [4], [5] and [6] in English can be found as Preprints (1999), http://www.MathSoc.spb.ru

Relativisation Provides Natural Separations for Resolution-Based Proof Systems[*]

Stefan Dantchev

Department Computer Science, Durham University,
Science Laboratories, South Road, Durham, DH1 3LE, UK
s.s.dantchev@durham.ac.uk

Abstract. We prove a number of simplified and improved separations between pairs of Resolution with bounded conjunction proof systems, Res(d), as well as their tree-like versions, Res* (d). The tautologies, which we use, are natural combinatorial principles: the Least Number Principle (LNP_n) and a variant of the Induction Principle (IP_n).

LNP_n is known to be easy for resolution. We prove that its relativisation is hard for resolution, and, more generally, the relativisation of LNP_n iterated d times provides a natural separation between Res (d) and Res ($d + 1$). We prove the same result for the iterated relativisation of IP_n if the tree-like proof system Res* (d) is considered instead of Res (d).

1 Introduction

We study the power of relativisation in Propositional Proof Complexity, i.e. we are interested in the following question: Given a propositional proof system is there a *first-order (FO) sentence which is easy but whose relativisation is hard* (within the system)?

The main motivation for studying relativisation comes from a work of Krajicek, [7]. He defines a combinatorics of first order (FO) structure and a relation of covering between FO structures and propositional proof systems. The combinatorics contains all the sentences, easy for the proof system. On the other hand, as defined in [7], it is closed under relativisation. Thus the existence of a sentence, which is easy, but whose relativisation is hard for the underlying proof system, would imply that it is impossible to capture the class of "easy" sentences by a combinatorics.

The proof system we consider is *Resolution with bounded conjunction*, denoted by Res(d). It is an extension of Resolution in which conjunctions of up to d literals are allowed instead of single literals. The Tree-like Res(d) is usually denoted by Res* (d). Krajicek proved that Tree-like Resolution, and even Res* (d), have combinatorics associated with it. This follows also from Riis' complexity gap theorem for Tree-like Resolution [10], and shows that the sentences, easy for Tree-like Resolution, remain easy after having been relativised.

The next natural system to look at is Resolution. It is stronger than Res* (d) for any d, $1 \leq d \leq n$ (equivalent to Res* (n), in fact), and yet weak enough so that one could expect that it can easily prove some property of the whole universe, but cannot

---

[*] This research was funded by EPSRC under grant EP/C526120/1.

D. Grigoriev, J. Harrison, and E.A. Hirsch (Eds.): CSR 2006, LNCS 3967, pp. 147–158, 2006.

prove it for an arbitrary subset. As we show in the paper, this is indeed the case. The example is very natural, *Least Number Principle* (LNP_n), saying that *a finite partially ordered n-element set* has a *minimal element*. It is easy to see that LNP_n is easy for Resolution, and we prove that its relativisation $Rel(LNP_n)$ is hard. A more general result has been proven in [3]; however the lower bound there is weaker. We also consider *iterated relativisation*, Rel^d (d is the number of iterations, a constant), and show that $Rel^d(LNP_n)$ is hard for $Res(d)$, but easy for $Res(d+1)$.

We finally consider the relativisation question for $Res^*(d)$, where the FO language is enriched with a *built-in order*. The *complexity gap theorem does not hold* in this setting, and we are able to show that relativisation makes some sentences harder. There is a variant of *Induction Principle* (IP_n), saying that there is a property which: holds for the minimal element; if it holds for a particular element, there is a bigger one for which the property holds, too; therefore the property holds for the maximal element. We prove that $Rel^d(IP_n)$ is *easy for $Res^*(d+1)$*, but *hard for $Res^*(d)$*.

More precisely, our results are the following:

1. Any Resolution proof of the relativised Minimum Element Principle, $Rel(LNP_n)$ is of size $2^{\Omega(n)}$. Firstly, this answers positively to the Krajicek's question. Secondly, observing that $Rel(LNP_n)$ has an $O(n^3)$-size Res(2), we get an exponential separation between Resolution and Res(2). A similar result was proven in [12] (see also [1] for a weaker, quasi-polynomial, separation). Our proof is rather simple, especially when compared to the proof in [12], which is a corollary of a more general result,
2. $Rel^d(LNP_n)$ has an $O(dn^3)$-size Res$(d+1)$ proof, but requires $2^{\Omega(n^\varepsilon)}$-size Res$(d)$ proof. This holds for any constant d (ε is a constant, dependent on d). These separations were first proven in [12]. As a matter of fact, we use their method but our tautologies are more natural, and our proof is a bit simpler.
3. $Rel^d(IP_n)$ has an $O(dn^2)$-size Res$^*(d+1)$ proof, but requires Res$^*(d)$ proofs of size $2^{\Omega(\frac{n}{d})}$. This holds for any d, $0 \le d \le n$. A similar result was proven in [4]. Again, our tautologies are more natural, while the proof is simpler.

The rest of the paper is organised as follows. In the section 2 we define Resolution and its extension Res(d), and outline general methods for proving lower bounds for these proof systems. In the sections 3 and 4 we prove the results about Resolution complexity of Minimal Element Principle and Tree-like Resolution complexity of Induction Principle respectively.

2   Preliminaries

2.1   Denotations and Conventions

We use the notation $[k] = \{1, 2, \ldots k\}$. We denote by $\top$ and $\bot$ the boolean values "true" and "false", respectively. A *literal* is either a propositional variable or a negated variable. A *d-conjunction* (*d-disjunction*) is a conjunction (disjunction) of at most d literals. A *term* (*d-term*) is either a conjunction (*d-conjunction*) or a constant, $\top$ or $\bot$. A *d-DNF* (or *d-clause*) is a disjunction of (unbounded number of) *d-conjunctions*.

Sometimes, when clear from the context, we will say "clause" instead of "d-clause", even though, formally speaking, a clause is 1-clause.

As we are interested in translating FO sentences into sets of clauses, we assume that a finite n-element universe $\mathbb{U}$ is given. The elements of $\mathbb{U}$ are the first n positive natural numbers, i.e. $\mathbb{U} = [n]$. When we say "element" we always assume an element from the universe. We will not explain the translation itself; the details can be found in [11] or [7].

2.2 Resolution and Extensions

We shall describe the proof system $\mathrm{Res}(d)$ which is an extension of Resolution, first introduced by Krajicek [6]. It is used to prove inconsistency of a given set of d-clauses. We start off with the given clauses and derive new ones. The goal is to derive the empty clause.

There are four derivation rules. The $\wedge$-*introduction rule* is

$$\frac{\mathcal{C}_1 \vee \bigwedge_{j \in \mathbb{J}_1} L_j \quad \mathcal{C}_2 \vee \bigwedge_{j \in \mathbb{J}_2} L_j}{\mathcal{C}_1 \vee \mathcal{C}_2 \vee \bigwedge_{j \in \mathbb{J}_1 \cup \mathbb{J}_2} L_j}.$$

The *cut (or resolution) rule* is

$$\frac{\mathcal{C}_1 \vee \bigvee_{j \in \mathbb{J}} L_j \quad \mathcal{C}_2 \vee \bigwedge_{j \in \mathbb{J}} \neg L_j}{\mathcal{C}_1 \vee \mathcal{C}_2}.$$

The two *weakening rules* are

$$\frac{\mathcal{C}}{\mathcal{C} \vee \bigwedge_{j \in \mathbb{J}} L_j} \quad \text{and} \quad \frac{\mathcal{C} \vee \bigwedge_{j \in \mathbb{J}_1 \cup \mathbb{J}_2} L_j}{\mathcal{C} \vee \bigwedge_{j \in \mathbb{J}_1} L_j}.$$

Here $\mathcal{C}$'s are d-clauses, L's are literals, and we have assumed that $|\mathbb{J}_1 \cup \mathbb{J}_2| \le d$, $|\mathbb{J}| \le d$.

A $\mathrm{Res}(d)$-proof can be considered as a directed acyclic graph (DAG), whose sources are the initial clauses, called also axioms, and whose only sink is the empty clause. We will measure the size of a proof as the number of the internal nodes of the graph, i.e. the number of applications of a derivation rule.

Whenever we say "we refute a FO sentence in $\mathrm{Res}(d)$", we mean is that we first translate the sentence into a set of clauses defined on a finite universe of size n, and then refute it with $\mathrm{Res}(d)$. The size of the refutation is then a function in n. We will often use "prove" and "refute" as synonyms.

In the denotation "$\mathrm{Res}(d)$", d is, in fact, a function in the number of propositional variables. Important special cases are $\mathrm{Res}(\log)$ as well as $\mathrm{Res}(\mathrm{const})$.

Clearly $\mathrm{Res}(1)$ is (ordinary) Resolution. In this case, we have only usual clauses, i.e. disjunctions of literals. The cut rule becomes the usual resolution rule, and only the first weakening rule is meaningful.

2.3 Proving Lower Bounds for Resolution and $\mathrm{Res}(d)$

We will first describe the *search problem*, associated to an *inconsistent set of clauses*, as defined in [6]: Given a truth assignment, find a clause, falsified under the assignment.

We can use a refutation of the given clause to solve the search problem as follows. We first turn around all the edges of the graph of the proof. The contradiction now becomes the only root (source) of the new graph, and the axioms and the initial formulae become the leaves (sinks). We perform a search in the new graph, starting from the root, which is falsified by any assignment, and always going to a vertex which is falsified under the given assignment. Such a vertex always exists as the inference rules are sound. We end up at a leaf, which is one of the initial clauses.

Thus, if we want to prove *existence of a particular kind of clauses in any proof*, we can use an *adversary argument in solving the search problem*. The argument is particularly nice for Resolution as developed by Pudlak in [8]. There are two players, named *Prover* and *Adversary*. An unsatisfiable set of clauses is given. Adversary claims wrongly that there is a satisfying assignment. *Prover* holds a Resolution refutation, and uses it to solve the search problem. A *position* in the game is a partial assignment of the propositional variables. The positions can be viewed as conjunctions of literals. All the possible positions in the game are exactly negations of all the clauses in the Prover's refutation. The game start from the empty position (which corresponds to $\top$, the negation of the empty clause). Prover has two kind of moves:

1. She queries a variable, whose value is unknown in the current position. Adversary answers, and the position then is extended with the answer.
2. She forgets a value of a variable, which is known. The current position is then reduced, i.e., the variable value becomes unknown.

The game is over, when the current partial assignment falsifies one of the clauses. Prover then wins, having shown a contradiction.

We will be interested in *deterministic Adversary's strategies* which allows to prove the existence of certain kind of clauses in a Resolution refutation.

In order to prove *lower bounds on Resolution proofs*, we will use the known technique, "bottleneck counting". It has been introduced by Haken in his seminal paper [5] (for the modern treatment see [2]). We first define the concept of *big clause*. We then design *random restrictions*, so that they "kill" (i.e. evaluate to $\top$) any big clause with high probability (whp). By the union bound, If there are few big clause, there is a restriction which kills them all. We now consider the *restricted set of clauses*, and using *Prover-Adversary game*, show that there has to be *at least one big clause in the restricted proof*, which is a contradiction and completes the argument.

The case of Res(d) is not so easy. A general method for proving lower bounds is developed in [12]. We first *hit the refutation by random restrictions*, such that all the d-clauses in the proof, under the restrictions, *can be represented by short boolean decision trees whp*. We then use the fact, proven in [12], that such a proof *can be transformed into small width Resolution proof*. Finally we consider the *restricted set of clauses*, and using *Prover-Adversary game*, show that there has to be *at least one big clause in the Resolution proof*. This gives the desired contradiction to the assumption that the initial Res(d) proof contains small number of d-clauses.

The case of Tree-like proofs, either Resolution or Res(d), is much simpler as a *tree-like proof* of a given set of clause *is equivalent* to a *decision tree, solving the search problem*. We can use pretty straightforward adversary argument against a decision tree, in order to show that it has to have many nodes.

3   Relativised Minimal Element Principle

3.1   Minimal Element Principle (LNP_n) and Relativisation

The Minimal Element Principle, LNP_n, states that a (partial) order, defined on a finite set of n elements, has a minimal element. Its negation can be expressed as the following FO sentence:

$$\exists L \,((\forall x \,\neg L\,(x,x)) \wedge$$
$$(\forall x,y,z \,\,(L\,(x,y) \wedge L\,(y,z)) \rightarrow L\,(x,z)) \wedge \qquad (1)$$
$$(\forall x \exists y \,L\,(y,x)))\,.$$

Here $L\,(x,y)$ stands for $x < y$.

The encoding of LNP_n as a set of clauses is as follows. The purely universal part (the first two lines of (1)) translate straightforwardly into

$$\neg L_{ii} \quad i \in [n]$$
$$\neg L_{ij} \vee \neg L_{jk} \vee L_{ik} \quad i,j,k \in [n]\,,$$

while the last line translates into

$$\bigvee_{j \in [n]} s_{ij} \quad i \in [n]$$
$$\neg s_{ij} \vee L_{ji} \quad i,j \in [n]\,,$$

where s is the Skolem relation, witnessing the existential variable, i.e. for each i $s_{ij} = \top$ implies that the j-th element is smaller than the i-th one.

The d-relativised Minimal Element Principle, $Rel^d\,(LNP_n)$, is as follows. Let R^p, $1 \leq p \leq d$, be the unary predicates which we relativise by, and let us denote by $\mathcal{R}\,(x)$ the conjunction $\bigwedge_{p \in [d]} R^p\,(x)$. $Rel^d\,(LNP_n)$ is the following sentence:

$$\exists L \,((\forall x \,\mathcal{R}\,(x) \rightarrow \neg L\,(x,x)) \wedge$$
$$(\forall x,y,z \,\mathcal{R}\,(x) \wedge \mathcal{R}\,(y) \wedge \mathcal{R}\,(z) \rightarrow$$
$$(L\,(x,y) \wedge L\,(y,z)) \rightarrow L\,(x,z)) \wedge$$
$$(\forall x \exists y \,\mathcal{R}\,(x) \rightarrow (\mathcal{R}\,(y) \wedge L\,(y,x))) \wedge$$
$$(\exists x \,\mathcal{R}\,(x)))\,.$$

The corresponding translation into clauses gives

$$\neg \mathcal{R}_i \vee \neg L_{ii} \quad i \in [n]$$
$$\neg \mathcal{R}_i \vee \neg \mathcal{R}_j \vee \neg \mathcal{R}_k \vee \neg L_{ij} \vee \neg L_{jk} \vee L_{ik} \quad i,j,k \in [n]\,,$$

as well as

$$\bigvee_{j \in [n]} s_{ij} \quad i \in [n]$$
$$\neg s_{ij} \vee \neg \mathcal{R}_i \vee R_j^q \quad i,j \in [n]\,,\ q \in [d] \qquad (2)$$
$$\neg s_{ij} \vee \neg \mathcal{R}_i \vee L_{ji} \quad i,j \in [n]\,,$$

which say that if the element i is in the set $\mathcal{R}$ and j is a witness of i (i.e. $s_{ij} = \top$) then j is in the set $\mathcal{R}$, too, and moreover j is smaller than i. Finally we add

$$\bigvee_{j \in [n]} t_j$$

$$\neg t_i \vee R_i^p \quad i \in [n], \; p \in [d], \tag{3}$$

which say that $\mathcal{R}$, the intersection of the sets R^p, $p \in [d]$, is nonempty. Note that in the above a propositional variable of the form R_i^p corresponds to the formula $R_p(x)$, and therefore $\neg \mathcal{R}_i$ is the disjunction $\bigvee_{p \in [d]} \neg R_i^p$.

3.2   The Upper Bound

We will first prove that $Rel^d (LNP_n)$ is easy for $Res\,(d+1)$.

Proposition 1. *There is an $O\left(dn^3\right) Res(d+1)$ proof of $Rel^d\left(LNP_n\right)$.*

Proof. The proof is an easy adaptation of the well known polynomial-size Resolution refutation of LNP_n It can be found in the full version of the paper. $\qquad\square$

3.3   An Optimal Lower Bound

We will prove that $Rel\left(LNP_n\right)$ is *exponentially hard* for Resolution.

Proposition 2. *Any Resolution proof of $Rel\left(LNP_n\right)$ is of size $2^{\Omega(n)}$.*

Proof. The idea is to randomly divide the universe $\mathbb{U}$ into two approximately equal parts. One of them, $\mathbb{R}$, will represent the predicate R; all the variables within it will remain unset. The rest, $\mathbb{C}$, will be the "chaotic" part; all the variables within $\mathbb{C}$ and most of the variables between $\mathbb{C}$ and $\mathbb{R}$ will be set at random. It is now intuitively clear that while $\mathbb{C}$ kills with positive probability a certain number of "big" clauses, $\mathbb{R}$ allows to show, via an adversary argument, that at least one such clause must be present in any resolution refutation, after it has been hit by the random restrictions. Therefore a huge number of "big" clauses must have been presented in the original refutation.

Let us first observe that the variables t_i just ensure that the predicate R is non-empty. W.l.o.g. we can eliminate them by setting $t_n = R_n = \top$ and for all $i \neq n \; t_i = \bot$.
    The *random restrictions* are as follows.

1. We first set all the variables R_i, $i \in [n-1]$, to either $\top$ or $\bot$ independently at random with equal probabilities, $1/2$. Let us denote the set of variables with $R_i = \top$ by $\mathbb{R}$, and the set of variables with $R_i = \bot$ by $\mathbb{C}$, $\mathbb{C} = \mathbb{U} \setminus \mathbb{R}$.
2. We now set all the variables L_{ij} with at least one endpoint in $\mathbb{C}$, i.e. $\{i,j\} \cap \mathbb{C} \neq \emptyset$, to either $\top$ or $\bot$ independently at random with equal probabilities, $1/2$.
3. For each $i \in \mathbb{C}$, $j \neq i$ we set s_{ij} to either $\top$ or $\bot$ independently at random with equal probabilities $1/2$. Note that it is possible to set all the s_{ij} to $\bot$, thus violating an axiom. It however happens with small probability, $1/2^{n-1}$ for a fixed i.
4. We finally set all the variables s_{ij} with $i \in \mathbb{R}$, $j \in \mathbb{C}$ to $\bot$.

Note the unset variables define exactly the non-relativised principle on $\mathbb{R}$, $LNP_{|\mathbb{R}|}$.

By the Chernoff's bound the probability that $\mathbb{R}$ contains less than $n/4$ elements is exponentially small, and we therefore have:

Claim. The probability that the random restrictions are inconsistent (i.e. violate an axiom) or $|\mathbb{R}| \leq n/4$ is at most $(n-1)\, 2^{-(n-1)} + e^{-n/16}$.

A *big clause* is one which contains at least $n/8$ busy elements. The element i is *busy* in the clause $\mathcal{C}$ iff $\mathcal{C}$ contains one of the following variables, either positively or negatively: $R_i, L_{ij}, L_{ji}, s_{ij}$ for some $j, j \neq i$.

It can be easily seen that

Claim. A clause, containing k busy elements, does not evaluate to $\top$ under the random restrictions with probability at most $(3/4)^{k/2}$.

Indeed, let us consider the different cases of a busy variable i in the clause $\mathcal{C}$:

1. The variable R_i is present in $\mathcal{C}$: The probability that the corresponding literals does not evaluate to $\top$ is $1/2$.
2. The variable s_{ij} for some $j \neq i$ is present in $\mathcal{C}$: The corresponding literal does not evaluate to $\top$ if either $i \in \mathbb{R}$, i.e. it remains unset, or $i \in \mathbb{C}$, but it evaluates to $\perp$. The probability of this is $3/4$.
3. Either the variable L_{ij} or the variable L_{ji} for some $j \neq i$ is present in $\mathcal{C}$. Let us denote the set of all such elements by $\mathbb{V}$, $|\mathbb{V}| = l$, and the corresponding sub-clause by $\mathcal{E}$. Construct the graph $\mathbb{G}$ with the vertex set $\mathbb{V}$ and the edge set $\mathbb{E}$ determined by the variables L_{ij}, i.e. $\mathbb{E} = \{\{i,j\} \mid L_{ij} \text{ is present in } \mathcal{C}\}$. Consider any spanning forest of $\mathbb{G}$. Assume that all the roots are in $\mathbb{R}$ as this only increases the probability that $\mathcal{E}$ does not evaluate to $\top$. Going from the root to the leaves in each tree, we see that the probability that the corresponding edge does not evaluate to $\top$ is $3/4$ (the same reason as in the 2nd case). Moreover all the edge variables are independent from each other, and also there are at most $l/2$ roots (exactly $l/2$ iff the forest consists of trees having a root and a single leaf only). Therefore the probability that the sub-clause $\mathcal{E}$ does not evaluate to $\top$ is at most $(3/4)^{l/2}$.

As the events from 1, 2 and 3 are independent for different elements from $\mathbb{U}$, we have completed the argument for observation (3.3)

We can now present the main argument in the proof. Assume there is a resolution refutation of $Rel\,(LNP_n)$ which contains less than $(4/3)^{n/16}$ big clauses. From the observations 3.3 and 3.3, using the union-bound on probabilities, we can conclude that there is a restriction which is consistent, "kills" all the big clauses (evaluating them to $\top$), and leaves $\mathbb{R}$ big enough ($|\mathbb{R}| \geq n/4$). Recall that the restricted refutation is nothing, but a resolution refutation of $LNP_{|\mathbb{R}|}$ on $\mathbb{R}$. What remains to show is that any such refutation must contain a big clause which would contradict to the assumption there were "few" big clauses in the original refutation.

We will the Prover-Adversary game for $LNP_{|\mathbb{R}|}$. At any time $\mathbb{R}$ is represented as a disjoint union of three sets, $\mathbb{R} = \mathbb{B} \uplus \mathbb{W} \uplus \mathbb{F}$. $\mathbb{B}$ is the set of all the elements busy in the current clause. The elements of $\mathbb{B}$ are always totally ordered. $\mathbb{W}$ is the set of witnesses for some elements in $\mathbb{B}$, i.e. for each $i \in \mathbb{B}$ there is an element $j \in \mathbb{W} \uplus \mathbb{B}$ such that $s_{ij} = \top$. We assume that, at any time, any element of $\mathbb{W}$ is smaller than all the elements of $\mathbb{B}$. $\mathbb{F}$ is the set of all remaining elements which we call "free". It is obvious how Adversary maintains these sets in the Prover-Adversary game. When a variable, which makes an element $i \in \mathbb{W} \uplus \mathbb{F}$ busy, is queried he adds i at the bottom of the totally ordered set $\mathbb{B}$, choose some $j \in \mathbb{F}$ and move it to $\mathbb{W}$ setting $s_{ij} = \top$. When all the variables, which kept an element $\mathbb{B}$ busy, are forgotten, Adversary removes i from $\mathbb{B}$ and remove the corresponding witness j from $\mathbb{W}$ if it is there (note that it may be in $\mathbb{B}$, too, in which case it is not removed). In this way Adversary can maintain the partial assignment consistent as far as $\mathbb{B} \neq \emptyset$. Note also that $|\mathbb{B}| \geq |\mathbb{W}|$. Therefore at the moment a contradiction is reached we have $|\mathbb{B}| \geq |\mathbb{R}| / 2 \geq n/8$ as claimed.   □

3.4   General Lower Bounds: Rel^d (LNP_n) Is *Sub-exponentially* Hard for *Res* (d)

We first give the necessary background from [12].

Definition 1. *(Definition 3.1, [12]) A decision tree is a rooted binary tree in which every internal node queries a propositional variable, and the leaves are labelled by either $\top$ or $\bot$.*

  Thus every path from the root to a leaf may be viewed as a partial assignment. Let us denote by $Br_v(\mathfrak{T})$, for $v \in \{\top, \bot\}$, the set of paths (partial assignments) in the decision tree $\mathfrak{T}$ which lead from the root to a leaf labelled by v.

  A decision tree $\mathfrak{T}$ strongly represents a DNF $\mathcal{F}$ iff for every $\pi \in Br_v(\mathfrak{T})$, $\mathcal{F}\!\restriction_\pi = v$.

  The representation height of $\mathcal{F}$, $h(\mathcal{F})$, is the minimum height of a decision tree strongly representing $\mathcal{F}$.

Definition 2. *(Definition 3.2, [12])Let $\mathcal{F}$ be a DNF, and $\mathbb{S}$ be a set of variables. We say that $\mathbb{S}$ is a cover of $\mathcal{F}$ iff every conjunction of $\mathcal{F}$ contains a variable from $\mathbb{S}$. The covering number of $\mathcal{F}$, $c(\mathcal{F})$, is the minimum size of a cover of $\mathcal{F}$.*

Lemma 1. *(Corollary 3, [12]; see also [9] for an improved version) Let $d \geq 1$, $\alpha > 0$, $1 \geq \beta, \gamma > 0$, $s > 0$, and let $\mathfrak{D}$ be a distribution on partial assignments such that for every d-DNF $\mathcal{G}$, $Pr_{\rho \in \mathfrak{D}}\,[\mathcal{G}\!\restriction_\rho \neq \top] \leq \alpha 2^{-\beta(c(\mathcal{G}))^\gamma}$. Then for every d-DNF $\mathcal{F}$:*

$$Pr_{\rho \in \mathfrak{D}}\,[h(\mathcal{F}\!\restriction_\rho) \geq s] \leq \alpha d2^{-2(\beta/4)^d(s/2)^{\gamma^d}}.$$

Lemma 2. *(Theorem 10, [12]) Let $\mathfrak{C}$ be a set of clauses of width at most w. If $\mathfrak{C}$ has a Res (d) refutation so that for each line $\mathcal{F}$ of the refutation, of Γ, $h(\mathcal{F}) \leq w$, then $\mathfrak{C}$ has a Resolution refutation of width at most dw.*

We also need the following construction which is only mentioned in [12], but whose proof can be found in the full version of the same paper.

Lemma 3. *(Sub-sections 8.3 and 8.4 in the full version of [12]) There is an undirected graph $\mathbb{G}([n], \mathbb{E})$ on n vertices and max-degree $\theta(\ln n)$ such that any Resolution refutation of LNP_n, restricted on $\mathbb{G}$, is of width $\Omega(n)$.*

  LNP_n, restricted on $\mathbb{G}$, means that for each element i the witness j has to be a neighbour of i in $\mathbb{G}$, i.e. we set $s_{ij} = \bot$ whenever $\{i, j\} \notin \mathbb{E}$.

We can now prove the desired result:

Proposition 3. *For every constant $d \geq 1$ there is a constant $\varepsilon_d \in (0, 1]$ such that any $Res(d)$ refutation of $Rel^d (LNP_n)$ is of size $2^{\Omega(n^{\varepsilon_d})}$.*

Proof. It is a fairly straightforward, though rather technical, application of Lemma 1. It can be found in the full version of the paper.   □

4   Relativised Induction Principle

In this section we consider a variant of the Induction Principle, denoted further by IP_n. It can be encoded as a FO sentence if a built-in predicate, defining a total order on the universe, is added to the language. It is easy to show that IP_n is easy for Tree-like Resolution, and so is $Rel^d (IP_n)$, but for Res* $(d + 1)$. Finally we prove that $Rel^d (IP_n)$ is hard for Res* (d).

4.1   Induction Principle

The negation of the Induction Principle, we consider, is the following simple statement: Given an ordered universe, there is a property P, such that

1. The property holds for the smallest element.
2. If $P(x)$ hold for some x, then there is y, bigger than x, and such that $P(y)$ holds.
3. The property does not hold for the biggest element.

The universe $\mathbb{U}$ can now be considered as the set of first n natural numbers. In our language we can use the relation symbol < with its usual meaning. We can also use any constant c as well as $n - c$ (note that in the language n denotes the maximal element of $\mathbb{U}$, while 1 denotes the minimal one). The Induction Principle, we have just described, can be written as

$$P(1) \wedge \forall x \exists y ((x < y) \wedge (P(x) \rightarrow P(y))) \wedge \neg P(\mathbf{n}).$$

The translation into propositional logic gives the following set of clauses

$$P_1, \ \neg P_n$$
$$\bigvee_{j=i+1}^{n} s_{ij} \ \ 1 \leq i \leq n - 1$$
$$\neg s_{ij} \vee \neg P_i \vee P_j \ \ 1 \leq i < j \leq n.$$

The relativised version translation is

$$P_1, \ \neg P_n$$
$$R_1^p, \ R_n^p \ \ p \in [d]$$
$$\bigvee_{j=i+1}^{n} s_{ij} \ \ 1 \leq i \leq n - 1 \tag{4}$$
$$\neg s_{ij} \vee \neg \mathcal{R}_i \vee R_j^p \ \ 1 \leq i < j \leq n, \ p \in [d] \tag{5}$$
$$\neg s_{ij} \vee \neg \mathcal{R}_i \vee \neg P_i \vee P_j \ \ 1 \leq i < j \leq n. \tag{6}$$

4.2 The Upper Bound

We will first show that $Rel^d (IP_n)$ is easy for Res* $(d + 1)$

Proposition 4. *There is an* $O\left(dn^2\right)$ *Res** $(d + 1)$ *proof of* $Rel^d (IP_n)$.

Proof. It is rather straightforward, and can be found in the full version of the paper. $\square$

4.3 The Lower Bound

We will first prove it for $d = 1$, i.e. that $Rel (IP_n)$ is exponentially hard for Tree-like Resolution. We will then generalise it to any d.

Proposition 5. *Any Tree-like Resolution proof of* $Rel (IP_n)$ *is of size* $2^{\Omega(n)}$.

Proof. We will use an adversary strategy against the decision tree solving the search problem.

We say that the variables P_i, R_i and s_{ij} for $j > i$ are *associated* to the i-the element. When one of these has been queried for the first time by Prover, Adversary fixes all of them, so that the i-th element becomes *busy*. Initially only the maximum element is busy as the singleton clauses $\neg P_n$ and R_n force the values of the corresponding variables. For technical reasons only, we assume that the $(n - 1)$-th element is busy too, by setting $s_{n-1\,n} = \top$, $R_{n-1} = \top$ and $P_{n-1} = \bot$. The elements that are not busy we call "*free*", with a single exception, *the source*. The source is the biggest element j, such that $R_j = P_j = \top$. Initially the source is the first element. It is important to note that no contradiction can be found as far as there is at least one free element bigger than the source. All the variables associated to the elements smaller that the source are set (consistently with the axioms) in the current partial assignment. Thus there are free elements only between the source and the maximal element. Informally speaking, Prover's strategy is moving the source towards the end of the universe, the $n - 2$-nd element.

We will prove that at any stage in the Prover-Adversary game, the number of free elements can be used to lower bound the subtree, rooted at the current node of the tree. More precisely, if $S(k)$ is the size of the subtree rooted at a node, where there are k such elements, we will show that $S(k) \geq \varphi_k$. Here φ_k is the k-th Fibonacci number, defined by

$$\varphi_0 = \varphi_1 = 1$$
$$\varphi_k = \varphi_{k-1} + \varphi_{k-2} \quad \text{for } k \geq 2.$$

Initially, we have $n - 3$ free elements bigger than the source, therefore the inequality we claim, together with the known asymptotic $\varphi_k \sim \frac{1}{\sqrt{5}} \left(\frac{1+\sqrt{5}}{2}\right)^k$, implies the desired lower bound.

What remains is to prove $S(k) \geq \varphi_k$. We use induction on k. The basis cases $k = 0$ or $k = 1$ are trivial. To prove the induction step, we consider all the possibilities for a Prover's query:

It is about either a busy element. As already explained, the value of such a variable is already known in the current partial assignment. Adversary answers; the value of k does not change.

The query is about a free element i, and recall that it is bigger than the source. If the variable queried is either R_i or P_i, Adversary first sets $s_{in} = \top$, $s_{ij} = \bot$ for all j, $i < j < n$, and then chooses between the two possibilities: either $R_i = \top$, $P_i = \bot$ or $R_i = \bot$, $P_i = \top$. If the variable queried is s_{ij} for some $j > i$ Adversary first sets $R_i = P_i = \bot$ $s_{il} = \top$ for all $l \neq j$, and then chooses either $s_{ij} = \top$ or $s_{ij} = \bot$. In any of the above cases, Adversary was free to choose the value of the variable queried, i.e. to force a branch of Prover's decision tree, while keeping the current partial assignment consistent. The number of free elements, k, decreases by one. Therefore we have

$$S(k) \geq 2S(k-1).$$

By the induction hypothesis $S(k-1) > \varphi_{k-1}$, and then $S(k) \geq 2\varphi_{k-1} \geq \varphi_k$.

The query is about the source, i.e. the variable queried is s_{ij}, where i is the source's index. If the j-th element is busy, Adversary answers $\bot$. If the j-th element is free, but far away from the source, that is there are at least two free elements between the source and the j-th element, Adversary answers $\bot$, too. Neither the position of the source nor the value of k changes. The only remaining case is when the j-th element is both free and near to the source, that is one of the two smallest free elements, bigger than the source. Adversary is now free to move the source to any of these two elements, by giving the corresponding answer: $\top$ - source is moved to the j-th element or $\bot$ - source is moved to the other nearest element. In one of these choices k decreases by one, and in the other it decreases by two. Therefore we have

$$S(k) \geq S(k-1) + S(k-2).$$

The induction hypothesis gives $S(k-1) \geq \varphi_{k-1}$ and $S(k-2) \geq \varphi_{k-2}$. Thus

$$S(k) \geq \varphi_{k-1} + \varphi_{k-2} = \varphi_k.$$

This completes the proof. □

We will show how to modify the proof in order to prove the following more general proposition.

Proposition 6. *Any* $Res^*(d)$ *proof of* $Rel^d(IP_n)$ *is of size* $2^{\Omega(\frac{n}{d})}$.

Proof. The difference from the previous argument is that the number of free elements may be reduced by at most d whenever a query (which is now a d-conjunction) has been made. This gives the following recurrence

$$S(k) \geq S(k-d) + S(k-d-1)$$

whose solution gives the lower bound given in the statement. All the details can be found in the full version of the paper. □

Acknowledgements. I would like to thank Jan Krajíček for asking the question which led to writing this paper. Many thanks to Søren Riis for the helpful discussions, I have had with him.

References

1. A. Atserias and M. Bonet. On the automatizability of resolution and related propositional proof systems. In *16th Annual Conference of the European Association for Computer Science Logic*, 2002.
2. P. Beame and T. Pitassi. Simplified and improved resolution lower bounds. In *Proceedings of the 37th annual IEEE symposium on Foundation Of Computer Science*, pages 274–282, 1996.
3. S. Dantchev and S. Riis. On relativisation and complexity gap for resolution-based proof systems. In *The 17th Annual Conference of the EACSL, Computer Science Logic*, volume 2803 of *LNCS*, pages 142–154. Springer, August 2003.
4. J.L. Esteban, N. Galesi, and J. Mesner. On the complexity of resolution with bounded conjunctions. In *Proceedings of the 29th International Colloquium on Automata, Languages and Programming*, 2002.
5. A. Haken. The intractability of resolution. *Theoretical Computer Science*, 39:297–308, 1985.
6. J. Krajíček. *Bounded Arithmetic, Propositional Logic, and Complexity Theory*. Cambridge University Press, 1995.
7. J. Krajicek. Combinatorics of first order structures and propositional proof systems. *Archive for Mathematical Logic*, 43(4), 2004.
8. P. Pudlák. Proofs as games. *American Mathematical Monthly*, pages 541–550, June-July 2000.
9. A. Razborov. Pseudorandom generators hard for k – dnf resolution and polynomial calculus resolution. 2002 – 2003.
10. S. Riis. A complexity gap for tree-resolution. *Computational Complexity*, 10:179–209, 2001.
11. S.M. Riis and M. Sitharam. Generating hard tautologies using predicate logic and the symmetric group. *Logic Journal of the IGPL*, 8(6):787–795, 2000.
12. N. Segerlind, S. Buss, and R. Impagliazzo. A switching lemma for small restrictions and lower bounds for k-dnf resolution. In *Proceedings of the 43rd annual symposium on Foundations Of Computer Science*. IEEE, November 2002.

Bounded-Degree Forbidden Patterns Problems Are Constraint Satisfaction Problems

Stefan Dantchev and Florent Madelaine

Department of Computer Science, University of Durham,
Science Labs, South Road, Durham DH1 3LE, UK
{s.s.dantchev, f.r.madelaine}@durham.ac.uk

Abstract. Forbidden Patterns problem (FPP) is a proper generalisation of Constraint Satisfaction Problem (CSP). FPP was introduced in [1] as a combinatorial counterpart of MMSNP, a logic which was in turn introduced in relation to CSP by Feder and Vardi [2]. We prove that *Forbidden Patterns Problems are Constraint Satisfaction Problems* when restricted to *graphs of bounded degree*. This is a generalisation of a result by Häggkvist and Hell who showed that F-moteness of bounded-degree graphs is a CSP (that is, for a given graph F there exists a graph H so that the class of bounded-degree graphs that do not admit a homomorphism from F is exactly the same as the class of bounded-degree graphs that are homomorphic to H). Forbidden-pattern property is a strict generalisation of F-moteness (in fact of F-moteness combined with a CSP) as it involves both vertex- and edge-colourings of the graph F, and thus allows to express $\mathcal{NP}$-complete problems (while F-moteness is always in $\mathcal{P}$). We finally extend our result to arbitrary relational structures, and prove that every problem in MMSNP, restricted to connected inputs of bounded (hyper-graph) degree, is in fact in CSP.

Keywords: Logic in Computer Science, Constraint Satisfaction, Graph Homomorphism, Duality, Monadic Second Order Logic.

1 Introduction

Graph Homomorphisms and related problems have received considerable attention in the recent years not only as a topic in Combinatorics and Graph Theory but also in relation with Constraint Satisfaction. A lot of possible directions of research in the area have been opened by different motivations, and have been explored (the very recent monograph [3] serves as a good survey of the area).

The present work was motivated mainly by Constraint Satisfaction Problem (CSP) and its generalisations, Forbidden Patterns Problem (FPP), introduced in [1]. Our results fall into the category of so-called (restricted) duality theorems, and have the same flavour as the results in [4].

To better explain the motivation behind our work, we need to start with MMSNP, a fragment of monadic second order logic, that was introduced in

D. Grigoriev, J. Harrison, and E.A. Hirsch (Eds.): CSR 2006, LNCS 3967, pp. 159–170, 2006.

[2], and was motivated by the search of a logical characterisation of Constraint Satisfaction Problem. Though computationally very closely related to CSP, the logic MMSNP is too strong and captures problems which are not constraint satisfaction problems [2, 5] (or, in a more general setting, at least not CSPs with a finite template [6]). The combinatorial counterpart of MMSNP is the class of Forbidden Patterns Problems. FPPs were studied at length in [7] where an exact characterisation of FPPs that are not CSPs was proved. This characterisation subsumes the characterisation of duality pairs obtained by Tardif and Nešetřil in [4].

In [8], Häggkvist and Hell showed a new kind of (restricted) duality result for Graph Homomorphism (or CSP). They built a universal graphs H for the class of F-mote graphs, i.e. the class of graphs that do not admit a homomorphism from F, of bounded degree b. Thus, any degree-d graph G is F-mote if, and only if, there is a homomorphism from G to H. The result can be seen as a restricted form of *duality* and provides a "good characterisation" (in the sense of Edmonds) for F-moteness. In the last decade, a series of papers built upon the Häggkvist and Hell's technique. In particular, a number of results by Nešetřil and various coauthors are relevant to our work. First, the complexity of H-colouring of bounded-degree graphs was investigated in more details in [9]. It was conjectured that if H is a triangle-free graph with chromatic number three then the H-colouring problem for 3-bounded graphs is $\mathcal{NP}$-complete. This conjectured was disproved in a very recent paper [10], where the authors investigated more generally the existence of universal graphs for the class of graphs that are simultaneously F-mote and H-colorable.

The main contribution of the present paper is a proof that Forbidden Patterns Problems, restricted to inputs of bounded-degree, become Constraint Satisfaction Problems. More precisely, given a forbidden patterns problem Ω and a bound b on the degree of the input graph, we explicitly construct a graph H, such that any graph G of bounded degree b is a yes-instance of Ω if, and only if, G is homomorphic to H. We say that H is a universal graph for Ω (and b) as it does not depend on the input G.

This result is a significant jump in the (descriptive) complexity from the result of Häggkvist and Hell as their result corresponds to problems, which are captured by fragments of first-order logic, and therefore are in $\mathcal{P}$. In our case we retain problems which are $\mathcal{NP}$-complete even on bounded-degree graphs.

Finally, we investigate the fragment of existential MSO logic that corresponds to the original FPPs as defines in [1], Feder and Vardi's logic MMSNP. It is perhaps worth mentioning that in the present paper we assume a definition of FPP that is more general than the original one in that the new definition allows colourings not only of the vertices but also of the edges of the input graph.

The rest of paper is organised as follows. In section 2 we give the necessary background as well as some motivating examples. Then we prove the main result in section 3. Finally, in section 4, we briefly discuss the logic MMSNP and extend our result to arbitrary relational structures. In 5, we conclude by a few questions and open problems.

2   Preliminaries

2.1   Definitions

To simplify the exposition, we shall assume mostly that the input is a graph G. We use standard graph notation. We denote by $V(G)$ the vertex set of G and by $E(G)$ the edge set of G. Let G and H be two graphs. A *graph homomorphism* is a vertex mapping $h : V(G) \rightarrow V(H)$ such that for every vertices x and y in $V(G)$, if (x, y) belongs to $E(G)$ then $(h(x), h(y))$ belongs to $E(H)$. Let F and G be two graphs. We say that G is F-*mote* if, and only if, there is no homomorphism from F to G. The F-*moteness* is the property of being F-mote, a term that we also use to refer to the corresponding decision problem.

Constraint Satisfaction Problems. The *constraint satisfaction problem* with *template H* is the decision problem with,

- input: a graph G; and,
- question: does there exist a homomorphism from G to H?

Example 1. The constraint satisfaction problem with template K_3 (the clique with three elements, *i.e.* a triangle) is nothing else than the 3 colorability problem from graph theory. Hence, the restriction to (undirected) graphs is also known as the H-*colouring problem*.

We denote by CSP the class of constraint satisfaction problems (not necessarily graph problems).

Edge and Vertex Coloured Forbidden Patterns Problems. Let $\mathcal{V}$ (respectively, $\mathcal{E}$) be a finite set of *vertex colours* (respectively, *edge colours*). A *pattern* is a connected graph whose vertices are coloured with $\mathcal{V}$ and whose edges are coloured with $\mathcal{E}$. Formally, it is a triple $(F, f^{\mathcal{V}}, f^{\mathcal{E}})$, where F is a finite graph, $f^{\mathcal{V}}$ is a mapping from $V(F)$ to $\mathcal{V}$ and $f^{\mathcal{E}}$ is a mapping from $E(F)$ to $\mathcal{E}$.

Let G be a graph, $g^{\mathcal{V}} : V(G) \rightarrow \mathcal{V}$ and $g^{\mathcal{E}} : E(G) \rightarrow \mathcal{E}$. Let H be a graph, $h^{\mathcal{V}} : V(H) \rightarrow \mathcal{V}$ and $h^{\mathcal{E}} : E(H) \rightarrow \mathcal{E}$. We say that a homomorphism h from G to H *preserve the colours* (given by $g^{\mathcal{V}}, g^{\mathcal{E}}, h^{\mathcal{V}}$ and $h^{\mathcal{E}}$) if, for every x in $V(G)$, $g^{\mathcal{V}}(x) = h^{\mathcal{V}} \circ h(x)$ holds; and, for every $e = (x, y)$ in $E(G)$, $g^{\mathcal{E}}(e) = h^{\mathcal{E}}(e')$ holds, where $e' = (h(x), h(y))$ belongs to $E(H)$. When the colours are clear from the context, we simply write that h *preserves colours*. Note that the composition of two homomorphism that preserve colours is also a homomorphism that preserve colours.

In this paper, patterns are used to model constraints in a negative fashion and consequently, we refer to them as *forbidden* patterns. Let $\mathfrak{F}$ be a finite set of forbidden patterns. Let G be a graph, $g^{\mathcal{V}} : V(G) \rightarrow \mathcal{V}$ and $g^{\mathcal{E}} : E(G) \rightarrow \mathcal{E}$. We say that $(G, g^{\mathcal{V}}, g^{\mathcal{E}})$ is *valid* with respect to $\mathfrak{F}$ if, and only if, for any pattern $(F, f^{\mathcal{V}}, f^{\mathcal{E}})$ there does not exist any homomorphism h from F to G that preserves these colours

The *problem with forbidden patterns $\mathfrak{F}$* is the decision problem with,

- input: a graph G
- question: does there exist any mapping $g^{\mathcal{V}} : V(G) \to \mathcal{V}$ and $g^{\mathcal{E}} : E(G) \to \mathcal{E}$ such that $(G, g^{\mathcal{V}}, g^{\mathcal{E}})$ is valid with respect to $\mathfrak{F}$.

Remark 2. The class of forbidden patterns problems with colours over the vertices only, corresponds to the problems that can be expressed by a formula in Feder and Vardi's MMSNP (Monotone Monadic SNP without inequalities, see [2, 1]). Note that allowing colours over the edges does not amount to drop the hypothesis of monadicity. Rather, it corresponds to a logic, let's call it MMSNP$_2$, which is similar to MMSNP but allows *first-order* variables over edges (just like Courcelle's MSO (Monadic Second Order logic) and MSO$_2$, see [11]).

2.2   Motivating Examples

In this section, we motivate this paper through the study of the following concrete graph problems.

1. VERTEX-NO-MONO-TRI: consists of the graphs for which there exists a partition of the vertex set in two sets such that no triangle has its three vertices occurring in a single partition. It was proved in [2, 5] that this problem is not in CSP and in [12] that it is $\mathcal{NP}$-complete.
2. TRI-FREE-TRI: consists of the graphs that are both three colourable (tripartite) and in which there is no triangle. It was proved in [5] that this problem is not in CSP.
3. EDGE-NO-MONO-TRI: consists of the graphs for which there exists a partition of the edge set in two sets such that no triangle has its three edges occurring in a single partition. It is known to be $\mathcal{NP}$-complete (see [13]).

The above examples can be formulated as Forbidden Patterns Problems. The corresponding set of forbidden patterns are depicted on Figure 1. In the case of EDGE-NO-MONO-TRI, the two type of colours for edges are depicted with dashed and full line respectively. In [8], Häggvist and Hell proved a result that can be rephrased as follows.

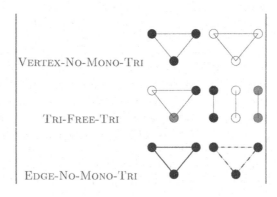

Fig. 1. Some forbidden patterns problems

Theorem 3. *Let F be a connected graph and b a positive integer. There exists a graph U such that there is no homomorphism from F to U; and, for every graph G of bounded degree b, there is no homomorphism from F to G if, and only if, there is a homomorphism from G to U.*

Using this theorem, we prove the following.

Corollary 4. *Let b be a positive integer. The problem* VERTEX-NO-MONO-TRI, *restricted to graph of bounded degree b, is a constraint satisfaction problem with input restricted to graph of bounded degree b.*

Proof. Let b be a positive integer. Let U be the universal graph for the class of graph of bounded degree b given by Theorem 3 for $F := K_3$. Let U' be the graph that consists of two disjoint copies U_1 and U_2, together with all the edges (x_1, x_2), where $x_1 \in V(U_1)$ and $x_2 \in V(U_2)$. Let G be a graph of bounded degree b. We model VERTEX-NO-MONO-TRI as a forbidden patterns problem with $\mathcal{V} := \{1, 2\}$ and two forbidden patterns (K_3, c_1) and (K_3, c_2) where c_1 is the constant 1 and c_2 the constant 2. We drop altogether the notation for edge colours since there is a single edge colour. Assume that G is a yes-instance of VERTEX-NO-MONO-TRI. Let $g^\mathcal{V} : V(G) \to \mathcal{V}$ such that $(G, g^\mathcal{V})$ is valid. Let G_1, respectively G_2, be the subgraph of G induced by the vertices of colour 1, respectively 2. It follows that there is neither a homomorphism from K_3 to G_1, nor from K_3 to G_2. Thus, there exist a homomorphism h_1 from G_1 to U_1 and a homomorphism h_2 from G_2 to U_2. Let h be the mapping induced by the union of h_1 and h_2. It is a homomorphism from G to U' by construction of U'. Conversely, if there is a h homomorphism from G to U', it induces a colour map $g^\mathcal{V}$ from G to $\mathcal{V}$, according to which copy of U a vertex of G is mapped to. Similarly to the above, it follows that $(G, g^\mathcal{V})$ is valid. Thus, VERTEX-NO-MONO-TRI, restricted to graph of bounded degree b, is the constraint satisfaction problem with template U' and input restricted to graph of bounded degree b.

It does not seem to be straightforward to come up with a similar graph construction to prove the same for the problem TRI-FREE-TRI. However, Theorem 3 has been generalised in [10] and can be reformulated as follows. Here, we give a shorter proof than in [10].

Theorem 5. *Let F and H be connected graphs, b a positive integer. Then there exists a graph U with the following properties:*

1. *there exists a homomorphism u from U to H;*
2. *For every graph G of bounded degree b such that there is no homomorphism from F to G and there exists a homomorphism g from G to H, there exist a homomorphism $G \xrightarrow{a} U$ such that $g = u \circ a$; and,*
3. *there is no homomorphism from F to U.*

Proof. By Häggvist an Hell's theorem (Theorem 3), there exists a finite graph $U_{F,b}$ such that for any graph G of bounded degree b, there is no homomorphism from F to G if, and only if, there is a homomorphism from G to $U_{F,b}$.

Choose for U the (categorical) product of $U_{F,b}$ and H and denote by π_1 and π_2 the two projections (see [3] for details). Choose for u the projection π_2.

Let G be a graph of bounded degree b such that there is no homomorphism from F to G and there is a homomorphism g from G to H. It follows that there exists a homomorphism h from G to $U_{F,b}$. We choose for a the homomorphism (h,g) and by definition of the product we have $g = u \circ a$.

Finally there can not be a homomorphism f from F to U as otherwise, by composition with π_1, there would be a homomorphism from F to $U_{F,b}$ in contradiction with Theorem 3.

The above result involves problems that corresponds to the intersection of a constraint satisfaction problem with template H, with a very simple forbidden patterns problem that has a single vertex colour, a single edge colour and a single forbidden pattern F. Applying this theorem with $F = H = K_3$, we get the problem TRI-FREE-TRI (intersection of 3-colorability with triangle free, see previous examples) and the following result.

Corollary 6. *Let b be a positive integer. The problem* TRI-FREE-TRI, *restricted to graph of bounded degree b, is a constraint satisfaction problem with input restricted to graph of bounded degree b.*

Note that, it is unlikely that the problems corresponding to Theorem 5 are as general as FPP, even when restricted to colours over vertices only.

Proposition 7. *If $\mathcal{P} \neq \mathcal{NP}$ then* VERTEX-NO-MONO-TRI *can not be represented as a problem of the form of Theorem 5.*

Proof. Let F and H be two graphs such that VERTEX-NO-MONO-TRI is the intersection of a constraint satisfaction problem with template H with the very simple forbidden patterns problem that has a single vertex colour, a single edge colour and a single forbidden pattern F. By Proposition 13 in [14], H must have a loop and a graph G is in VERTEX-NO-MONO-TRI if, and only if, there is no homomorphism from F to G. This last property is expressible in first-order logic. However, it was proved in [12] that this problem is $\mathcal{NP}$-complete. Since checking property expressible in first-order logic can be done in polynomial time, the result follows.

We now turn our attention to the problem EDGE-NO-MONO-TRI. It seems to be difficult to prove the same result for this problem as a direct corollary from Theorem 5 or Theorem 3. We shall deal with this example as a corollary of our main result: this result is a generalisation of Theorem 5 to the class of all (edge and vertex coloured) forbidden patterns problems.

3   Bounded Degree Forbidden Patterns Problems

In this section, we prove that forbidden patterns problems over graphs, when restricted to input of bounded degree, become constraint satisfaction problems. In graph parlance, we show that there is a universal graph for the (yes-instance of) bounded degree forbidden patterns problems.

Theorem 8. *Let b be a positive integer. Let Ω be the problem with forbidden patterns $\mathfrak{F}$. There exists a universal graph U for the graphs of bounded degree b in Ω. That is,*

1. *There exist $u^{\mathcal{V}} : V(U) \to \mathcal{V}$ and $u^{\mathcal{E}} : E(U) \to \mathcal{E}$ such that $(U, u^{\mathcal{V}}, u^{\mathcal{E}})$ is valid w.r.t. $\mathfrak{F}$;*
2. *For every graph G of bounded degree b such that there exist $g^{\mathcal{V}} : V(G) \to \mathcal{V}$ and $g^{\mathcal{E}} : E(G) \to \mathcal{E}$ such that $(G, g^{\mathcal{V}}, g^{\mathcal{E}})$ is valid w.r.t. $\mathfrak{F}$, there exist a homomorphism a from G to U that preserves these colours.*
3. *Conversely, for every graph G of bounded degree b, if there exists a homomorphism $G \xrightarrow{a} U$ then a induces colour maps $g^{\mathcal{V}}$ and $g^{\mathcal{E}}$ such that $(G, g^{\mathcal{V}}, g^{\mathcal{E}})$ is valid w.r.t. $\mathfrak{F}$.*

Remark 9. Using the standard notion of product of graphs, our new proof of Theorem 5 makes this result a direct corollary of Häggvist and Hell's theorem. However the proof in [10] present an idea that is key to our paper: the fact that Hell and Häggvist general technique can be extended to take *coloured* graphs into account. Hence, apart from the fact that we allow also colours on the edges, one could see Theorem 8 as the true statement behind Dreyer, Malon and Nešetřil's proof.

We shall first explain the intuition and ideas behind the proof by giving an informal outline. Let us denote the largest diameter of a forbidden patterns by m. This parameter, together with the degree-bound b, are absolute constants, i.e. they are not part of the input of the FPP. The universal graph U has to contain all possible *small graphs* that are *yes-instances of the FPP*, "small" meaning "of diameter $m + 1$" here and thereafter. The intuition then is that each graph G, which is a yes-instance of the FPP, however big, can be homomorphically mapped to U as such a mapping should be only locally consistent. Given that the degree of G is bounded by b, we need to distinguish among no more than $X \leq b^{m+2}$ vertices in a small neighbourhood, so we can use X many different labels in constructing the vertex set of U. On the other hand, in order to define the adjacency relation of the universal graph, *i.e.* to correctly "glue" all the possible small neighbourhoods, any vertex of U should carry information not only about its label, but also about its neighbourhood. In other words, any such vertex will represent a small graph together with a vertex which is the "centre" (or the root) of the small graph. Thus the vertex set of the universal graph will consists of all such rooted small graphs vertex- and edge-coloured in all possible ways that are yes-instances of the FPP. Two vertices will be adjacent in U if, and only if, the graphs they represent "agree", i.e. have most of their vertices with the same labels and colours and the induced subgraphs of these vertices coincide including the edge colours; for the precise definition of what "agree" means, one should see the formal proof below. It is now intuitively clear why a yes-instance G of the FPP should be homomorphic to the universal graph U: the vertices of G can be labelled so that any two adjacent vertices get different labels, then one can choose a good vertex- and edge-colouring of G, and because of the construction of U now every vertex u in G can be mapped to the vertex of

U that represents the small neighbourhood of u rooted at u. It is straightforward to see that the mapping preserves edges.

The universal graph U has a very useful property, namely every small neighbourhood of U rooted at a vertex v is homomorphic to the small graph represented by the vertex u (see Lemma 11 below). This property immediately implies that a no-instance of the FPP cannot be homomorphic to U. Indeed, suppose for the sake of contradiction, there is a no-instance G that is homomorphic to U. Fix the colouring induced by the homomorphism and observe that G there is a homomorphism from the FPP graph F into G. The composition of the two homomorphisms gives a homomorphism from F into U, and by the property above, by another composition, we get a homomorphism from the FPP graph F to some small graph represented by a vertex of U. This gives a contradiction with our construction as we have taken small no-instances only to be represented by the vertices of the universal graph.

We illustrated the construction of Theorem 8 for our final example EDGE-NO-MONO-TRI on Figure 2 for input of bounded degree $b \geq 3$. Any input G can be labelled by elements from $\{1, 2, \ldots, 10, \ldots, X\}$ such that for every vertex x in $V(G)$, the vertices at distance at most 2 (the diameter of a triangle plus one) of x have a different label.

Corollary 10. *Let b be a positive integer. The problem* EDGE-NO-MONO-TRI, *restricted to graph of bounded degree b, is a constraint satisfaction problem with input restricted to graph of bounded degree b.*

The remaining of this section is devoted to the formal proof of Theorem 8.

Let b be a positive integer and $\mathfrak{F}$ be a set of forbidden patterns. We write $\delta(G)$ to denote the *diameter* of a graph G: that is, the minimum for all vertices x in $V(G)$ of the maximum distance from x to any other vertex y in $V(G)$. Let $m := \max\{\delta(F)$ such that $(F, f^{\mathcal{V}}, f^{\mathcal{E}}) \in \mathfrak{F}\}$. Let $X := 1 + \Sigma_{j=0}^{m} b(b-1)^j$.

Construction of U. Let $\mathcal{S}$ be the set of connected graphs $(S, s^{\mathcal{V}}, s^{\mathcal{E}})$ that are valid w.r.t. $\mathfrak{F}$, such that $V(S)$ is a subset of $\{1, 2, \ldots, X\}$. Let U be the graph with:

- vertices $(v, (S, s^{\mathcal{V}}, s^{\mathcal{E}}))$, where $(S, s^{\mathcal{V}}, s^{\mathcal{E}}) \in \mathcal{S}$ and $v \in S$; and,
- such that $(v, (S, s^{\mathcal{V}}, s^{\mathcal{E}}))$ is adjacent to $(v', (S', s'^{\mathcal{V}}, s'^{\mathcal{E}}))$ if, and only if, the following holds:
  (i) (v, v') belongs to both $E(S)$ and $E(S')$; and,
  (ii) the induced coloured subgraph of S induced by every vertex at distance at most m of v (respectively, v') is *identical* to the induced coloured subgraph of S' induced by every vertex at distance at most m of v (respectively, v').

U Is a Yes-Instance. We first define *"universal colours"* for U as follows.

- $u^{\mathcal{V}}(v, (S, s^{\mathcal{V}}, s^{\mathcal{E}})) := s^{\mathcal{V}}(v)$, for every vertex $(S, s^{\mathcal{V}}, s^{\mathcal{E}})$ in $V(U)$; and,
- $u^{\mathcal{E}}((v, (S, s^{\mathcal{V}}, s^{\mathcal{E}}))(v', (S', s'^{\mathcal{V}}, s'^{\mathcal{E}}))) := s^{\mathcal{E}}(v, v')$, for every edge $((v, (S, s^{\mathcal{V}}, s^{\mathcal{E}}))(v', (S', s'^{\mathcal{V}}, s'^{\mathcal{E}})))$ in $E(U)$.

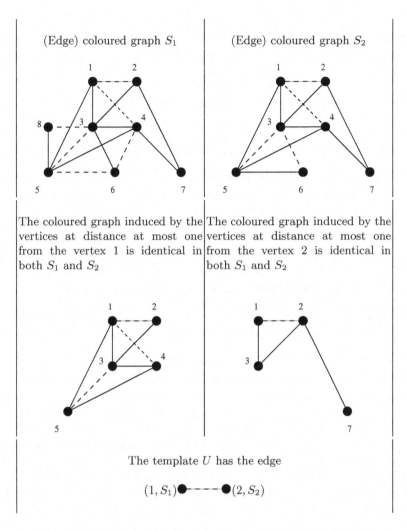

The coloured graph induced by the vertices at distance at most one from the vertex 1 is identical in both S_1 and S_2

The coloured graph induced by the vertices at distance at most one from the vertex 2 is identical in both S_1 and S_2

The template U has the edge

$$(1, S_1) \bullet\!-\!-\!-\!-\!\bullet (2, S_2)$$

Fig. 2. Illustration of the construction for EDGE-NO-MONO-TRI

We need the following lemma.

Lemma 11. *Let $(v, (S, s^{\mathcal{V}}, s^{\mathcal{E}}))$ in $V(U)$. Let $(B, c^{\mathcal{B}}, s^{\mathcal{E}})$ be the coloured graph induced by all vertices at distance at most m from $(v, (S, s^{\mathcal{V}}, s^{\mathcal{E}}))$. Then, there exists a homomorphism from $(B, c^{\mathcal{B}}, s^{\mathcal{E}})$ to $(S, s^{\mathcal{V}}, s^{\mathcal{E}})$ that preserves the colours.*

Proof. Let $(v', (S', s'^{\mathcal{V}}, s'^{\mathcal{E}}))$ in $V(B)$. We set $h(v', (S', s'^{\mathcal{V}}, s'^{\mathcal{E}})) := v'$. By induction on the distance from $(v, (S, s^{\mathcal{V}}, s^{\mathcal{E}}))$ to $(v', (S', s'^{\mathcal{V}}, s'^{\mathcal{E}}))$, using (i), it follows that the vertex v' belongs to $V(S)$ and that h is homomorphism. Similarly, using (ii) it follows that h preserves colours.

We now prove that $(U, u^{\mathcal{V}}, u^{\mathcal{E}})$ is valid with respect to the forbidden patterns $\mathfrak{F}$. Assume for contradiction that $(U, u^{\mathcal{V}}, u^{\mathcal{E}})$ is not valid and that there exists some forbidden pattern $(F, f^\mathcal{V}, f^\mathcal{E})$ in $\mathfrak{F}$ and some homomorphism f' from F to U that preserves these colours. Since the diameter of F is at most m, there exists a vertex x in $V(F)$ such that every vertex of F is at distance at most m. Hence, it is also the case for $f'(x)$ in the homomorphic image of F via f'. By lemma 11, there exists a homomorphism h from F to S, where $f'(F) = (v, (S, s^\mathcal{V}, s^\mathcal{E}))$. Both homomorphisms are colour preserving, thus composing these two homomorphisms, we get that $h \circ f'$ is a colour-preserving homomorphism from F to S. However, by definition of S, the coloured graph $(S, s^\mathcal{V}, s^\mathcal{E})$ is valid with respect to the forbidden patterns $\mathfrak{F}$. We reach a contradiction and the result follows.

Yes-Instances Are Homomorphic to U. Let G be a graph of bounded degree b for which there exist $g^\mathcal{V} : V(G) \to \mathcal{V}$ and $g^\mathcal{E} : E(G) \to \mathcal{E}$ such that $(G, g^\mathcal{V}, g^\mathcal{E})$ is valid w.r.t. $\mathfrak{F}$. Since G has bounded degree b, for every vertex x in $V(G)$, there are at most $X - 1$ vertices y at distance at most m from x. Therefore, there exists a map χ from $V(G)$ to $\{1, 2, \ldots, X\}$ such that every two distinct vertices within distance m or less take a different colour via χ. Thus, for every vertex x in G, the subgraph of G induced by the vertices at distance at most m of x can be identified (via the labelling χ) to a graph S_x with domain $\{1, 2, \ldots, X\}$. Similarly, the restriction of $g^\mathcal{V}$ and $g^\mathcal{E}$ to this subgraph induce colour maps $s_x^\mathcal{V}$ and $s_x^\mathcal{E}$ of S. We set $a(x) := (\chi(x), (S_x, s_x^\mathcal{V}, s_x^\mathcal{E}))$. It follows directly from the definition of U that a is homomorphism that preserve colours.

No-instances Are Not Homomorphic to U. Let G be a graph of bounded degree b that is a no instance of the forbidden patterns problem represented by $\mathfrak{F}$ and assume for contradiction that $G \xrightarrow{a} U$. The homomorphism a together with the universal colouring $(U, u^\mathcal{V}, u^\mathcal{E})$ induces colourings $g^\mathcal{V}$ and $g^\mathcal{E}$ as follows: For every vertex x in $V(G)$, set $g^\mathcal{V}(x) := u^\mathcal{V}(a(x))$; and, for every edge (x, y) in $E(G)$, set $g^\mathcal{E}(x, y) := u^\mathcal{E}(a(x), a(y))$. Since G is a no instance, there exists a forbidden pattern $(F, f^\mathcal{V}, f^\mathcal{E})$ in $\mathfrak{F}$ and a homomorphism f' from F to G that preserve these colours. Composing the two homomorphisms, we get that $a \circ f'$ is a homomorphism from F to U that preserves the colours of F and U. This contradicts the fact that $(U, u^\mathcal{V}, u^\mathcal{E})$ is valid w.r.t. $\mathfrak{F}$. This concludes the proof of Theorem 8.

4   Logic

In [2], a fragment of MSO was defined in the search of a logic to capture CSP by Feder and Vardi: the logic MMSNP (Monotone Monadic SNP without inequalities). They proved that though CSP is strictly included in MMSNP, for every problem Ω in MMSNP, there exists a problem Ω' in CSP such that Ω reduces to Ω' by a polynomial-time Karp reduction and Ω' reduces to Ω by a randomised polynomial-time Turing reduction. While in [1], we attached ourselves to understand the difference between CSP and MMSNP and characterised precisely the problems in MMSNP that are not in CSP, we adopt the opposite stance in the present work and show that the two classes coincide once restricted to input of bounded degree.

The proof technique of Theorem 8 generalises to arbitrary structures and since every problem in MMSNP is a disjunction of problems that are forbidden patterns problems without edge colours (*i.e.* with a single edge colour to be precise), we get the following result (the proof has been omitted due to space restriction and can be found in the online appendix [14]).

Theorem 12. *On connected input of bounded degree, MMSNP coincides with CSP.*

5 Conclusion and Open Problems

We have proved that every forbidden patterns (with colours on both edges and vertices) problem is in fact a constraint satisfaction problem, when restricted to input of bounded degree. We derived from this result that the logic MMSNP coincides with CSP on connected input of bounded degree (whereas CSP is strictly included in MMSNP for unrestricted inputs).

Theorem 12 can be easily adapted to the extension $MMSNP_2$ of MMSNP with first order quantification over edges as well as vertices. Since CSP is subsumed by MMSNP, this proves that MMSNP and $MMSNP_2$ have the same expressive power for connected input of bounded degree. This phenomenon can be related to a result due to Courcelle [11], who proved that the same properties of graphs of degree at most b, where b is a fixed integer, can be expressed by monadic second-order formulae using edge and vertex quantifications (MSO_2) as well as by monadic second-order formulae using vertex quantifications only (MSO_1). Courcelle also proved that MSO_2 is more expressive than MSO_1 for unrestricted input. This leads us to ask the following question. *Is $MMSNP_2$ more restrictive than $MMSNP_1$ for unrestricted input?* Proving, for example, that the problem EDGE-NO-MONO-TRI is not expressible in $MMSNP_1$ would answer this question affirmatively.

Courcelle has presented other restrictions over graphs and digraphs that ensure that MSO_1 and MSO_2 have the same expressive power, such as planarity or bounded tree-width. This provokes the following questions. *Is it the case that $MMSNP_2$ coincides with $MMSNP_1$ for planar graphs or graphs of bounded tree width?* and, *if it is the case, do they also collapse to CSP?*

References

1. Madelaine, F., Stewart, I.A.: Constraint satisfaction, logic and forbidden patterns. submitted. (2005)
2. Feder, T., Vardi, M.Y.: The computational structure of monotone monadic SNP and constraint satisfaction: a study through datalog and group theory. SIAM J. Comput. **28** (1999) 57–104
3. Hell, P., Nešetřil, J.: Graphs and homomorphisms. Oxford University Press (2004)
4. Nešetřil, J., Tardif, C.: Duality theorems for finite structures (characterising gaps and good characterisations). Journal of Combin. Theory Ser. B **80** (2000) 80–97

5. Madelaine, F., Stewart, I.A.: Some problems not definable using structures homomorphisms. Ars Combinatoria **LXVII** (2003)
6. Bordisky, M.:   Constraint Satisfaction with Infinite Domains.   PhD thesis, Humboldt-Universität zu Berlin (2004)
7. Madelaine, F.: Constraint satisfaction problems and related logic. PhD thesis, University of Leicester, Department of Mathematics and Computer Science (2003)
8. Häggkvist, R., Hell, P.: Universality of A-mote graphs. European Journal of Combinatorics **14** (1993) 23–27
9. Hell, P., Galluccio, A., Nešetřil, J.: The complexity of H-coloring bounded degree graphs. Discrete Mathematics **222** (2000) 101–109
10. Paul A. Dreyer Jr, Malon, C., Nešetřil, J.: Universal H-colourable graphs without a given configuration. Discrete Mathematics **250**(1-3) (2002) 245–252
11. Courcelle, B.: The monadic second order logic of graphs VI: On several representations of graphs by relational structures. Discrete Applied Mathematics **54** (1994) 117–149
12. Achlioptas, D.:   The complexity of G-free colourability.   Discrete Mathematics **165-166** (1997) Pages 21–30
13. Garey, M., Johnson, D.: Computers and intractability: a guide to NP-completeness. Freeman, San Francisco, California (1979)
14. Madelaine, F., Dantchev, S.: Online appendix of this paper. electronic form (2006) available from www.dur.ac.uk/f.r.madelaine.

Isolation and Reducibility Properties and the Collapse Result[*]

Sergey M. Dudakov

CS Dept., Tver State University, 33, Zhelyabova str., Tver, 170000, Russia
Sergey.Dudakov@tversu.ru

Abstract. Two methods are used usually for to establish the collapse result for theories. They use the isolation property and the reducibility property. Early it is shown that the reducibility implies the isolation. We prove that these methods are equivalent.

1 Introduction

Relational database is a finite set of finite table. The content of tables changes but its structure does not. Hence, a database is a finite structure of some fixed language Ω which is called also a *database scheme*.

Elements of tables are usually from some infinite *universe*. In this case databases are called *finite states* over the universe. For example such universe can be the set of natural number, the set of real number, a set of string over some alphabet etc. Such universe can have own relations which form some signature Σ. For numbers these relations can be the addition, the multiplication, the order relation. For strings the order and the concatenation can be used.

For extracting information from relational databases first order languages are usually used, i.e. *queries* to databases are first order formulas. In query we can use only database relations from Ω or database and universe relations from (Ω, Σ). It is known that some information can't be extracted using Ω-queries. For example if Ω contain alone unary predicate symbol P it is impossible to write Ω-query which will be true iff the cardinality of P is odd.

Gurevich Yu.Sh. demonstrated that if besides relations from Ω we use an universe *order relation* then the expressive power of first order languages grows. The problem is the following. Can the expressive power of FO languages grow if we add to language $(\Omega, <)$ other universe relations? Queries of the language $(\Omega, <)$ are called *restricted* and ones of the language (Ω, Σ) are called *extended*.

This problem is investigated in [2, 1, 3, 5, 7, 6]. The answer depends on an universe and relations. For example if the universe is a set of natural number ω and additional relations are $+$ and $\times$ then the expressive power grows, but if we add only $+$ then it remains the same.

It is clear that investigating the expressive power we must consider only queries expressing inner properties of a database. They must not depend on

[*] The work was sponsored by Russian Foundation of Basic Research, projects 04-01-00015 and 04-01-00565.

D. Grigoriev, J. Harrison, and E.A. Hirsch (Eds.): CSR 2006, LNCS 3967, pp. 171–177, 2006.

coding the database elements by universe elements. More formally we consider only locally generic queries. A query is *locally generic* in a universe if it is preserved by all preserving the order database isomorphisms in the universe.

If in a universe for every database scheme every locally generic extended query is equivalent over finite states to some restricted query then this universe holds the *collapse result*. Elementary equivalent universes holds or lacks the collapse result simultaneously because the collapse result can be expressed as a set of closed Σ-formulas. Hence, the collapse result is the property of theories.

There are two main methods used for establishing the collapse result for theories.

One method uses *isolation* properties. It was introduced in [1] and was improved in [6]. Isolations properties imply the collapse result.

Other methods was proposed in [3] where it is proved that the collapse result is held for universes without the independent property. In [5, 7] it is shown that the previous result is a partial case of more general theorem: the collapse result is implied by the *reducibility* property.

The discussed question is "which method is more general?" In [5, 7] it is shown that the reducibility implies the second isolation properties.

In this paper we prove that any isolation property also implies the reducibility. Hence these methods are equivalent.

In the following section we give main definitions, then adduce our construction, and in the last section we prove main theorems. In the conclusion we formulate an open question we are interested in.

2 Main Definitions

The main concepts of the model theory can be found in [4].

The notion $\phi(\bar{x})$ means the formula ϕ doesn't contain free variables except $\bar{x}$. If $\phi(\bar{x})$ is a Σ-formula, $\mathfrak{A}$ is a Σ-structure, and $\bar{\mathsf{a}} \in \mathfrak{A}$ is a tuple then the notion $\phi(\bar{\mathsf{a}})$ means the value of ϕ when the value of the variable x_i is a_i for each i.

Definition 1. *In a structure* $(\mathfrak{A}, D)$ *of language* (Σ, Q) *the set D is* pseudo-finite *if any* (Σ, Q)*-formula from* $\mathrm{Fin}(\mathfrak{A}, Q)$ *is true for* $(\mathfrak{A}, D)$.

$\mathrm{Fin}(\mathfrak{A}, Q)$ *is a set of all* (Σ, Q)*-formulas which are true for all* (Σ, Q)*-structures* $(\mathfrak{A}, D')$ *with finite D'.*

We consider a (Σ, P)-structure $(\mathfrak{A}, I)$. We suppose that the structure has a linear order relation $<$ and this symbol belongs to the language Σ.

Definition 2. (Σ, P)*-structure* $(\mathfrak{A}, I)$ *holds first (second) isolation* property *if I is indiscernible sequence in* $\mathfrak{A}$ *(in* $(\mathfrak{A}, I)$*) and for any* (Σ, P)*-structure* $(\mathfrak{B}, J) \equiv (\mathfrak{A}, I)$ *for any pseudo-finite set* $D \subseteq J$ *for any finite set* $E \subseteq \mathfrak{B}$ *and for any* $\mathsf{b} \in \mathfrak{B}$ *the type* $\mathrm{tp}(\mathsf{b}/D \cup E)$ *in* $\mathfrak{B}$ *(in* $(\mathfrak{B}, J)$*) is isolated in* $\mathfrak{B}$ *(in* $(\mathfrak{B}, J)$*) by some type* $\mathrm{tp}(\mathsf{b}/D_0 \cup E)$ *for some countable set* $D_0 \subseteq D$*, i.e.* $\mathrm{tp}(\mathsf{b}/D \cup E)$ *is the unique type over* $D \cup E$ *containing* $\mathrm{tp}(\mathsf{b}/D_0 \cup E)$.

Definition 3. *In the* (Σ, P)-*structure* $(\mathfrak{A}, I)$ *a formula* $\phi(\bar{x}, \bar{y})$ *is called* I-*reducible* to order *(or* I-*reducible) if there is a quantifier-free order formula* $\psi(\bar{x}, \bar{z})$ *and for any* $\bar{a} \in \mathfrak{A}$ *there is* $\bar{c}_{\bar{a}} \in I$ *for which*

$$\phi(\bar{x}, \bar{a}) \equiv \psi(\bar{x}, \bar{c}_{\bar{a}})$$

for any $\bar{x} \in I$. *The variables* $\bar{x}$ *we call* I-*bounded in this case.*
  The structure $(\mathfrak{A}, I)$ *is* I-*reducible if any* Σ-*formula is* I-*reducible.*

Definition 4. *Let* $\mathfrak{A}$ *be a* Σ-*structure. Let a language* Ω *be disjoint with* Σ *and contains only predicate and constant symbols. Such languages are called* database schemes. Ω-*structure* $\mathfrak{B}$ *is a* finite state *over* $\mathfrak{A}$ *if the support of* $\mathfrak{B}$ *is a finite subset of the support of* $\mathfrak{A}$. *In this case the pair* $(\mathfrak{B}, \mathfrak{A})$ *can be considered as the united* (Ω, Σ)-*structure. Two finite states* $\mathfrak{B}_1$ *and* $\mathfrak{B}_2$ *are* order isomorphic *in* $\mathfrak{A}$ *if there is an isomorphism from* $\mathfrak{B}_1$ *to* $\mathfrak{B}_2$ *preserving the order* $<$ *in* $\mathfrak{A}$.
  A query *is a first-order formula. Queries of the language* (Ω, Σ) *are called* extended. *Queries of the language* $(\Omega, <)$ *are called* restricted. *A query is* locally generic *in* $\mathfrak{A}$ *if it is simultaneously true or false for any two order isomorphic finite states.*
  Σ-*structure* $\mathfrak{A}$ *holds* the collapse result *if for any database scheme* Ω *and for any locally generic extended query* ϕ *of language* (Ω, Σ) *there is a restricted* $(\Omega, <)$-*query* ψ *which is equivalent* ϕ *for any finite state over* $\mathfrak{A}$. *A theory holds the collapse result if some its model holds it.*

Theorem 1 (see [6]). *If a structure* $(\mathfrak{A}, I)$ *holds the first or second isolation property then the theory* $\mathrm{Th}(\mathfrak{A})$ *holds the collapse result.*

Theorem 2 (see [7, 5] and [3]). *If a structure* $(\mathfrak{A}, I)$ *is* I-*reducible then there are a structure* $(\mathfrak{B}, J') \equiv (\mathfrak{A}, I)$ *and a set* $J \subseteq J'$ *such that the structure* $(\mathfrak{B}, J)$ *holds the second isolation property.*

3   Construction

We consider the structure $(\mathfrak{A}, I)$ of the language (Σ, P).

Definition 5. *Let* $\phi(\bar{x}, \bar{y})$ *be a* (Σ, P)-*formula. Let* $\bar{a} \in \mathfrak{A}$. *Let* $\bar{d}_1, \bar{d}_2 \in I$. *We call the pair* $(\bar{d}_1, \bar{d}_2)$ *a* difference pair *for the formula* $\phi(\bar{x}, \bar{a})$ *if these tuples has the same order type, there is a number* i_0 *and* $\mathrm{d}_1^{(i)} = \mathrm{d}_2^{(i)}$ *for all* $i \neq i_0$, *and* $\phi(\bar{d}_1, \bar{a}) \not\equiv \phi(\bar{d}_2, \bar{a})$.
  If $\bar{d}_1^{(i_0)} < \bar{d}_2^{(i_0)}$ *then the interval* $[\mathrm{d}_1^{(i_0)}; \mathrm{d}_2^{(i_0)}]$ *in* I *is called* difference interval *and is denoted as* $\mathrm{diff}[\bar{d}_1, \bar{d}_2]$.
  If $(\bar{d}_1, \bar{d}_2)$ *and* $(\bar{d}_3, \bar{d}_4)$ *are two difference pairs and intervals* $\mathrm{diff}[\bar{d}_1, \bar{d}_2]$ *and* $\mathrm{diff}[\bar{d}_3, \bar{d}_4]$ *are disjoint then these pairs are also called* disjoint.
  An element $\mathrm{e} \in I$ *is called* defining *for* $\phi(\bar{x}, \bar{a})$ *if for any its neighborhood* $O(\mathrm{e})$ *in* I *there is a difference pair* $(\bar{d}_1, \bar{d}_2)$ *for* $\phi(\bar{x}, \bar{a})$ *such that* $\mathrm{diff}[\bar{d}_1, \bar{d}_2] \subseteq O(\mathrm{e})$.

Theorem 3. *Let I be an indiscernible sequence in a Σ-structure $\mathfrak{A}$ and the order on I be dense and complete. Then a formula $\phi(\bar{x}, \bar{y})$ is I-reducible iff there is a natural constant $M \in \omega$ and for any $\bar{a} \in \mathfrak{A}$ a number of defining elements for $\phi(\bar{x}, \bar{a})$ is bounded by M.*

Proof. If $\phi(\bar{x}, \bar{y})$ is I-reducible by order formula $\psi(\bar{x}, \bar{z})$ then for each $\bar{a} \in \mathfrak{A}$ there is $\bar{c}_{\bar{a}} \in I$ such that

$$(\mathfrak{A}, I) \models (\forall x \in P)(\phi(\bar{x}, \bar{a}) \leftrightarrow \psi(\bar{x}, \bar{c}_{\bar{a}})).$$

Thus, all defining for $\phi(\bar{x}, \bar{a})$ elements are in $\bar{c}_{\bar{a}}$.

Suppose the number of defining elements of $\phi(\bar{x}, \bar{y})$ is bounded. It is enough to prove that if $\bar{c} \in I$ is defining elements for $\phi(\bar{x}, \bar{a})$ then for any $\bar{d}_1, \bar{d}_2 \in I$ having the same order type over $\bar{c}$

$$\phi(\bar{d}_1, \bar{a}) \equiv \phi(\bar{d}_2, \bar{a}).$$

We use the induction by the number of unequal components of $\bar{d}_1$ and $\bar{d}_2$.

The base. Let an unequal component be unique: $d_1^{(i_0)} \neq d_2^{(i_0)}$. Let us suppose

$$\phi(\bar{d}_1, \bar{a}) \not\equiv \phi(\bar{d}_2, \bar{a}).$$

Let us consider the set

$$C = \text{diff}[\bar{d}_1; \bar{d}_2] = I \cap [d_1^{(i_0)}; d_2^{(i_0)}].$$

Let for any $\mathsf{c} \in C$ the notion $\bar{d}_{\mathsf{c}}$ means the tuple obtaining from $\bar{d}_1$ by replacing $d_1^{(i_0)}$ with c. Let C_1 be a set of all $\mathsf{c} \in C$ for which

$$\phi(\bar{d}_{\mathsf{c}}, \bar{a}) \equiv \phi(\bar{d}_1, \bar{a}),$$

and C_2 be a set of all $\mathsf{c} \in C$ for which

$$\phi(\bar{d}_{\mathsf{c}}, \bar{a}) \equiv \phi(\bar{d}_2, \bar{a}).$$

As (C_1, C_2) is a partition of C and the order on C is dense complete so at least one of C_1 and C_2 contains a boundary point e. Then e is a defining element for $\phi(\bar{x}, \bar{a})$, hence, it belongs to $\bar{c}$ and $\bar{d}_1$ and $\bar{d}_2$ have difference order types over $\bar{c}$. It contradicts the assumption.

Let $\bar{d}_1$ and $\bar{d}_2$ have n unequal components, $n > 1$. For example let us consider the case when the least unequal component belongs to $\bar{d}_1$. Let this component be $d_1^{(i_0)}$. Let the tuple $\bar{d}_2'$ be obtained from $\bar{d}_2$ by replacing $d_2^{(i_0)}$ with $d_1^{(i_0)}$. Then the tuples $\bar{d}_2$ and $\bar{d}_2'$ have one unequal component and have the same order type over $\bar{c}$. Using the base we obtain that

$$\phi(\bar{d}_2, \bar{a}) \equiv \phi(\bar{d}_2', \bar{a}).$$

The tuples $\bar{d}_1$ and $\bar{d}_2'$ have $n - 1$ unequal components and by the inductive assumption

$$\phi(\bar{d}_1, \bar{a}) \equiv \phi(\bar{d}_2', \bar{a}).$$

It implies

$$\phi(\bar{d}_1, \bar{a}) \equiv \phi(\bar{d}_2, \bar{a}).$$

Theorem 4. *Let I be an indiscernible sequence in a Σ-structure $\mathfrak{A}$. Let for any Σ-formula $\phi(\bar{x}, \bar{y})$ there be a constant M such that for any $\bar{a} \in \mathfrak{A}$ there aren't more than M disjoint difference pairs for $\phi(\bar{x}, \bar{a})$.*

Then there are (Σ, P)-structure $(\mathfrak{B}, J') \equiv (\mathfrak{A}, I)$ and a set $J \subseteq J'$ such that the structure $(\mathfrak{B}, J)$ is J-reducible.

Proof. Let a structure $(\mathfrak{B}, J') \equiv (\mathfrak{A}, I)$ be $(2^\omega)^+$-saturated. Then J' contains a subset J which is order isomorphic the set of real numbers. Thus, the order on J is dense complete. Let us suppose that the structure $(\mathfrak{B}, J)$ isn't J-reducible. Then by the theorem 3 there is a formula $\phi(\bar{x}, \bar{y})$ and a sequence of tuples $\bar{a}_i \in \mathfrak{B}$, $i \in \omega$ such that there are more that i defining elements for $\phi(\bar{x}, \bar{a}_i)$. Hence for any $i \in \omega$ we can construct i disjoint difference pairs. Then the same is possible in the structure $(\mathfrak{A}, I)$ that contradicts the theorem assumption.

4   Main Result

Theorem 5. *Let a structure $(\mathfrak{A}, I)$ holds the first isolation property.*

Then there is a structure $(\mathfrak{B}, J') \equiv (\mathfrak{A}, I)$ and a set $J \subseteq J'$ such that the structure $(\mathfrak{B}, J)$ is J-reducible.

Proof. Let us suppose there is no J-reducible structure $(\mathfrak{B}, J)$ where $(\mathfrak{B}, J') \equiv (\mathfrak{A}, I)$ and $J \subseteq J'$. By the theorem 4 there is a Σ-formula $\phi(\bar{x}, \bar{y}')$ for which we can construct i disjoint difference pair for any $i \in \omega$. We suppose that in the formula $\phi(\bar{x}, \bar{y}')$ the number of non-I-bounded variables $\bar{y}'$ is the least possible. This number must be greater than 0. Let $\bar{y}' = (\bar{y}, z)$.

Let $(\bar{a}_i, b_i)_{i \in \omega}$ be a sequence of tuple such that for $\phi(\bar{x}, \bar{y}, z)$ there are at least i disjoint difference pairs: $(\bar{d}_1^{ij}, \bar{d}_2^{ij})$, $j = 1, \ldots, i$. Let for each j the tuple $\bar{d}_3^{ij}$ is obtained from $\bar{d}_1^{ij}$ and $\bar{d}_2^{ij}$ by replacing an unequal component with any element inside $\mathrm{diff}[\bar{d}_1^{ij}, \bar{d}_2^{ij}]$. As $\phi(\bar{d}_1^{ij}, \bar{a}_i, b_i) \not\equiv \phi(\bar{d}_2^{ij}, \bar{a}_i, b_i)$, so $(\bar{d}_1^{ij}, \bar{d}_3^{ij})$ or $(\bar{d}_2^{ij}, \bar{d}_3^{ij})$ isn't a difference pair.

Let D_i be a set of these tuples:

$$D_i = \{(\bar{d}_1^{ij}, \bar{d}_2^{ij}, \bar{d}_3^{ij}) : j = 1, \ldots, i\}.$$

Let us consider the structures $(\mathfrak{A}, I, D_i, \bar{a}_i, b_i)$, $i \in \omega$. Let $(\mathfrak{B}, J, D, \bar{a}, b)$ be its ultraproduct modulo any non-principal ultrafilter over ω. Then the set D is pseudo-finite and has a cardinality not less than ω^+. Let E be a set of element of $\bar{a}$.

If F is a set of some tuples then let $[F]$ be a set of all elements of these tuples. Let us consider the type $K = \mathrm{tp}(b/E \cup [D])$ in $\mathfrak{B}$. Let us suppose K can be isolated by some type $K_0 = \mathrm{tp}(b/E \cup D_0)$ for some countable set $D_0 \subseteq [D]$. As D_0 is countable and D is not so there are at least ω^+ tuples $(\bar{d}_1, \bar{d}_2, \bar{d}_3) \in D$ for which $\mathrm{diff}[\bar{d}_1, \bar{d}_2] \cap D_0 = \emptyset$. Let D' be a set of all such tuples.

Let a formula $\Phi(\bar{u}, \bar{v}, \bar{y}, z)$ mean that $(\bar{u}, \bar{v})$ is a difference pair for $\phi(\bar{x}, \bar{y}, z)$. Evidently, $\Phi(\bar{d}_1, \bar{d}_2, z)$ is in K for any $(\bar{d}_1, \bar{d}_2)$ from D'. Let us prove that a set $K_0 \cup \{\neg\Phi(\bar{d}_1, \bar{d}_2, \bar{a}, z)\}$ holds in $\mathfrak{B}$ for some $(\bar{d}_1, \bar{d}_2)$ from D'.

Let us suppose $\mathfrak{B}$ lacks the set $K_0 \cup \{\neg\Phi(\bar{d}_1, \bar{d}_2, \bar{a}, z)\}$ for all $(\bar{d}_1, \bar{d}_2)$ from D'. As all these sets are countable and $\mathfrak{B}$ is ω^+-saturated then for each $(\bar{d}_1, \bar{d}_2)$ there is finite subset $K_0^{(\bar{d}_1, \bar{d}_2)} \subseteq K_0$ which is not consistent with $\neg\Phi(\bar{d}_1, \bar{d}_2, \bar{a}, z)$. The set D' contains at least ω^+ tuples $(\bar{d}_1, \bar{d}_2, \bar{d}_3)$ and there are countable many $K_0^{(\bar{d}_1, \bar{d}_2)}$, hence, for ω^+ tuples from D' the sets $K_0^{(\bar{d}_1, \bar{d}_2)}$ are same. Let us denote this set by K_0' and the set of tuples $(\bar{d}_1, \bar{d}_2, \bar{d}_3)$ from D' inconsistent with K_0' by D''.

Let $\Psi(\bar{d}, \bar{d}_1, \bar{d}_2, \bar{a}, z)$ be a conjunction of all formulas from $K_0' \cup \{\neg\Phi(\bar{d}_1, \bar{d}_2, \bar{a}, x)\}$ where $\bar{d} \in D_0$. The count of non-I-bounded variables in the formula $(\exists z)\Psi(\bar{w}, \bar{u}, \bar{v}, \bar{y}, z)$ is less than in the formula $\phi(\bar{x}, \bar{y}, z)$. Hence, for $(\exists z)\Psi(\bar{w}, \bar{u}, \bar{v}, \bar{a}, z)$ the count of disjoint difference pairs $(\bar{c}_1^k, \bar{c}_2^k)$ is bound by some constant M. Let us consider any such pair $(\bar{c}_1^k, \bar{c}_2^k)$. If the tuple $\bar{c}_3^k$ is obtained from $\bar{c}_1^k$ and $\bar{c}_2^k$ by replacing an unequal component with any element from $\mathrm{diff}[\bar{c}_1^k, \bar{c}_2^k]$ then $(\bar{c}_1^k, \bar{c}_3^k)$ or $(\bar{c}_3^k, \bar{c}_2^k)$ is a difference pair again. As the count of difference pair is bounded then we can make intervals $\mathrm{diff}[\bar{c}_1^k, \bar{c}_2^k]$ to intersect with no more than one $\mathrm{diff}[\bar{d}_1, \bar{d}_2]$ where $(\bar{d}_1, \bar{d}_2, \bar{d}_3) \in D''$.

Let us select a tuple $(\bar{d}_1, \bar{d}_2, \bar{d}_3)$ from D'' such that $\mathrm{diff}[\bar{d}_1, \bar{d}_2]$ doesn't intersect with any $\mathrm{diff}[\bar{c}_1^k, \bar{c}_2^k]$. Let us suppose the pair $(\bar{d}_1, \bar{d}_3)$ isn't difference for $\phi(\bar{x}, \bar{a}, b)$. As

$$\mathrm{diff}[(\bar{d}, \bar{d}_1, \bar{d}_2), (\bar{d}, \bar{d}_1, \bar{d}_3)] = \mathrm{diff}[\bar{d}_2, \bar{d}_3]$$

doesn't intersect with any $\mathrm{diff}[\bar{c}_1^k, \bar{c}_2^k]$ so the formula $(\exists z)\Psi(\bar{d}, \bar{d}_1, \bar{d}_2, \bar{a}, z)$ has the same value as $(\exists z)\Psi(\bar{d}, \bar{d}_1, \bar{d}_3, \bar{a}, z)$. But $\Psi(\bar{d}, \bar{d}_1, \bar{d}_3, \bar{a}, z)$ is true when $z = b$. Hence the formula $(\exists z)\Psi(\bar{d}, \bar{d}_1, \bar{d}_2, \bar{a}, z)$ is true also. It means the set $K_0' \cup \{\neg\Phi(\bar{d}_1, \bar{d}_2, \bar{a}, z)\}$ holds in $\mathfrak{B}$. It contradicts the assumption.

We have proved that the set $K_0 \cup \{\neg\Phi(\bar{d}_1, \bar{d}_2, \bar{a}, z)\}$ holds in $\mathfrak{B}$. It can be expanded to some type K'. But evidently the types K and K' are unequal. Hence, the type K isn't isolated by K_0.

We prove that the type K isn't isolated by any type K_0 over countable $D_0 \subseteq [D]$ in $\mathfrak{B}$. As $(\mathfrak{B}, J) \equiv (\mathfrak{A}, I)$ and $[D] \subseteq J$ is pseudo-finite so the structure $(\mathfrak{A}, I)$ lacks the first isolation property.

Theorem 6. *Let a (Σ, P)-structure $(\mathfrak{A}, I)$ holds the second isolation property. Then there is a structure $(\mathfrak{B}, J') \equiv (\mathfrak{A}, I)$ and a set $J \subseteq J'$ such that the structure $(\mathfrak{B}, J)$ is J-reducible.*

Proof. Let $\mathfrak{A}' = (\mathfrak{A}, I)$. Then the structure $(\mathfrak{A}', I)$ holds the first isolation property. By the theorem 5 there is a structure $(\mathfrak{B}', J') \equiv (\mathfrak{A}', I)$ and a set $J \subseteq J'$ such that $(\mathfrak{B}', J)$ is J-reducible. But $\mathfrak{B}' = (\mathfrak{B}, J')$ where $\mathfrak{B}$ is the reduct of $\mathfrak{B}'$ to the language Σ. Thus $(\mathfrak{B}, J)$ is J-reducible also.

5   Conclusion

We have established that the method of the isolation properties and the method of the reducibility are equivalent. We are interested in the following question.

Question 1. Does a theory exist which holds the collapse result but doesn't hold the isolation properties neither the reducibility?

Acknowledgments

We thank prof. Michael A. Taitslin for discussions and comments.

References

1. Belegradek O.V., Stolboushkin A.P., Taitslin M.A. Extended order-generic queries // Annals of Pure and Applied Logic 97, 1999, pp. 85–125.
2. Benedict M., Dong G., Libkin L., Wong L. Relational expressive power of constraint query languages. // Proc. 15th ACM Symp. on Principles of Database Systems. 1996. pp.5–16.
3. Baldwin J., Benedikt M. Stability theory, permutations of indiscernibles, and embedded finite models. // Trans. Amer. Math. Soc. 2000. 352(11). pp.4937–4969.
4. C.C.Chang, H.J.Keisler. Model Theory. North Holland, 3rd edition, 1990.
5. Dudakov S.M. The collapse theorem for theories of I-reducible algebraic systems // Izvestiya RAN: Ser. Mat., 68:5, 2004, pp.67–90.
6. Taitslin M.A. Restricted pseudo-finite homogeneity ans isolation properties. // Tver state university notes, Ser. "Mathematics and Computer science", Tver, Tver state university, 2003. 2(1). pp.5—15. (in Russian)
7. Taitslin M.A. A general condition for collapse results // Annals of Pure and Applied Logic. 2001. 113. pp. 323–330.

Incremental Branching Programs

Extended Abstract

Anna Gál[1,*], Michal Koucký[2,**], and Pierre McKenzie[3,***]

[1] University of Texas at Austin
[2] Mathematical Institute, Prague, Czech Republic
[3] Université de Montréal

Abstract. We propose a new model of restricted branching programs
which we call *incremental branching programs*. We show that *syntactic* incremental branching programs capture previously studied structured models of computation for the problem GEN, namely marking machines [Co74] and Poon's extension [Po93] of jumping automata on graphs [CoRa80]. We then prove exponential size lower bounds for our syntactic incremental model, and for some other restricted branching program models as well. We further show that nondeterministic syntactic incremental branching programs are provably stronger than their deterministic counterpart when solving a natural NL-complete GEN subproblem. It remains open if syntactic incremental branching programs are as powerful as unrestricted branching programs for GEN problems.

1 Introduction

Is the complexity class L of the problems solvable in deterministic logarithmic space properly contained in the class P of problems solvable in polynomial time? This question arose in the late 1960's [Co71] and remains open today. As is well known, L is captured by polynomial size branching programs. To separate L from P, it thus suffices to identify a language in P that no polynomial size branching program can recognize.

In this paper we focus on the problem GEN, a P-complete problem having natural NL-complete and L-complete subproblems [Co74, JoLa77, BaMc91]. We consider a restriction on branching programs, that all currently known size-efficient branching programs solving GEN and its subproblems seem to possess. We call branching programs obeying this restriction *incremental branching programs*. As in other restricted branching program models, we consider syntactic and semantic versions of the model: the syntactic restriction imposes a condition

---

* Supported in part by NSF Grant CCF-0430695 and an Alfred P. Sloan Fellowship.
** Part of this work was done while being a postdoctoral fellow at McGill University, Canada and at CWI, Amsterdam, Netherlands. Supported in part by NWO vici project 2004-2009, project No. 1M0021620808 of MŠMT ČR and grant 201/05/0124 of GA ČR, Institutional Research Plan No. AV0Z10190503.
*** Supported by the NSERC of Canada and the (Québec) FQRNT

D. Grigoriev, J. Harrison, and E.A. Hirsch (Eds.): CSR 2006, LNCS 3967, pp. 178–190, 2006.
© Springer-Verlag Berlin Heidelberg 2006

on all the graph-theoretic paths in the branching program, and the semantic restriction imposes a condition only on the paths actually traversed by an input. We prove exponential size lower bounds for syntactic incremental branching programs computing GEN. We also obtain exponential lower bounds for some other branching program models for computing GEN, that can be viewed in a common framework with incremental branching programs, but do not require the incrementality restriction. We refer to this more general framework as *tight computation*. This framework is specific to the n-way branching program model of [BoCo82]. The models captured by this framework include read-once branching programs and an extension of monotone nondeterministic branching programs to n-way branching programs (called *S-monotone programs* where S is a subset of the possible values of the variables), in addition to incremental branching programs.

Why should one consider a new branching program restriction when so many restrictions have been investigated (see for example [We00])? Here are our main reasons:

1. The model that we propose offers a new perspective on the known structured lower bounds for GEN.
2. Our analysis of incremental branching programs reveals certain properties that *any* purported subexponential-size branching program solving GEN must have (Remark 1). Hence the analysis of incremental branching programs may lead to new insights into computation of unrestricted branching programs.
3. All currently known upper bounds for GEN and its various subproblems can be achieved by syntactic incremental branching programs, and it remains open if syntactic incremental branching programs are as powerful as unrestricted branching programs for GEN problems.
4. While so far we have not been able to analyze them, *semantic* incremental branching programs may provide the answer to Cook's [Co74] and Edmonds' [EPA99] requests for a computational model intermediate between marking machines and NNJAG's on the one hand, and unrestricted branching programs on the other.

In this paper we show that strong exponential size lower bounds for syntactic incremental branching programs for GEN follow from [Co74, PTC77] via our Symmetrization lemma, and slightly weaker lower bounds can be derived from monotone circuit depth lower bounds [RaMc99], revealing an informative connection. In particular, marking machine lower bounds (albeit weaker than those from [Co74, PTC77]) follow from monotone circuit depth lower bounds. Figure 1 summarizes our main bounds on the sizes of syntactic incremental branching programs and $\{1\}$-monotone nondeterministic n-way branching programs solving the $n \times n$ instances of the P-complete problem GEN, of the NL-complete GEN restriction $\text{GEN}_{(2rows)}$ and of the L-complete problem $\text{GEN}_{(1row)}$. For $\text{GEN}_{(2rows)}$, we note a super-polynomial separation between the power of deterministic and nondeterministic syntactic incremental branching programs. See Section 2 for the definition of GEN and our branching program models.

	Deterministic syntactic incremental	Nondeterministic syntactic incremental	Nondeterministic $\{1\}$-monotone
GEN	$O(n^2 2^n) \cap \Omega(2^{cn/\log n})$	$O(2^n) \cap \Omega(2^{cn/\log n})$	$O(2^n) \cap \Omega(2^{n^\delta})$
GEN$_{(2rows)}$	$n^{\Theta(\log n)}$	$O(n^2)$	$O(n^2)$
GEN$_{(1row)}$	$O(n^2)$		

Fig. 1. Main size bounds presented in this paper. Here δ and c are specific constants.

Our proofs are based on reductions of varying degrees of difficulty and they appeal to the lower bounds from [Co74, PTC77, RaMc99, EPA99]. Our work raises the following open questions: Are $f(n)$-size semantic incremental branching programs strictly more powerful than $O(f(n))$-size syntactic incremental branching programs? Can unrestricted (n-way) branching programs for GEN be simulated by semantic, or even by syntactic incremental branching programs without a significant size blowup? It should be noted that in the context of read-k-times branching programs, the semantic variant is provably much more powerful than its syntactic counterpart [Ju95, BJS01], indicating that semantic incremental branching programs may also behave quite differently from their syntactic counterparts.

Due to space constraints, we omit all proofs from this extended abstract. Proofs and more details can be found in [GKM05] or in the forthcoming journal version.

2   Definitions and Preliminaries

We write $[n]$ for $\{1, 2, \ldots, n\}$. When $T \subseteq [n] \times [n] \times [n]$ and $S \subseteq [n]$, we write $\langle S \rangle_T$ for the *closure of S under T*, defined as the smallest $S' \supseteq S$ such that the following holds for every $(i, j, k) \in [n] \times [n] \times [n]$: if $i \in S'$ and $j \in S'$ and $(i, j, k) \in T$ then $k \in S'$. We will work with the following problems:

Problem GEN: Given a function $g : [n] \times [n] \to [n]$ prescribing $T^g \subseteq [n] \times [n] \times [n]$, determine whether $n \in \langle \{1\} \rangle_{T^g}$.

Problem RELGEN: Given an n^3-length bit string prescribing a set $T \subseteq [n] \times [n] \times [n]$, determine whether $n \in \langle \{1\} \rangle_T$.

We will talk about n-GEN and n-RELGEN when only a particular value n is considered, this will be necessary since we work in nonuniform models. (The name GEN comes from "Generating Problem" and the name RELGEN comes from the *relational* version of the Generating Problem.) We will view n-RELGEN as a Boolean function of n^3 variables (n-RELGEN : $\{0, 1\}^{n^3} \to \{0, 1\}$), and n-GEN as a function over n^2 n-ary variables (n-GEN : $[n]^{n^2} \to \{0, 1\}$). Note that n-RELGEN is a monotone Boolean function while the Boolean version of n-GEN (over $n^2 \log_2 n$ variables obtained by encoding the values of the n^2 n-ary variables as $\log_2 n$-length bit strings) is not monotone. We call an instance T of GEN or RELGEN *positive* if $n \in \langle \{1\} \rangle_T$ otherwise the instance T is *negative*. It is known that (i) GEN (and thus RELGEN) is P-complete [Co74, JoLa77], (ii) $\mathrm{GEN}_{(2rows)}$, namely the restriction of GEN in which $i * j \neq 1 \Rightarrow i \leq 2$, is NL-complete [BaMc91], (iii) $\mathrm{GEN}_{(1row)}$ namely the restriction of GEN in which $i * j \neq 1 \Rightarrow i = 1$, is L-complete [BaMc91].

Fix $n > 0$. We call $[n]$ the set of n-GEN elements. Both an n-GEN instance and an n-RELGEN instance define a set of triples $(i, j, k) \in [n] \times [n] \times [n]$ which we denote $i * j = k$ (note that we retain this notation even when k is not uniquely defined from i and j). Recall that an n-GEN instance is defined by a function $g : [n] \times [n] \to [n]$, thus the corresponding set of triples T^g has the property that for each pair $(i, j) \in [n] \times [n]$ there is exactly one value $k \in [n]$ such that $(i, j, k) \in T^g$. On the other hand, n-RELGEN instances may involve arbitrary sets $T \subseteq [n] \times [n] \times [n]$, including the possibility that some $i * j$ is not assigned any value, that is $(i, j, k) \notin T$ for any k. (The name RELGEN indicates that the underlying set of triples corresponds to a relation, rather than a function.)

For $i \in [n]$, we write $\chi_n(i)$ for the n-bit string $0^{i-1}10^{n-i}$ (i.e. the characteristic vector of the singleton set $\{i\}$), and we write $\chi(i)$ for $\chi_n(i)$ when n is understood. Given an n-GEN instance g, we write $\chi_n(g)$ (or $\chi(g)$ when n is understood) for the length-n^3 n-RELGEN instance $\chi(g(1,1))\chi(g(1,2)) \cdots \chi(g(n,n))$. Note that for any n-GEN instance g, $\mathrm{GEN}(g) = \mathrm{RELGEN}(\chi_n(g))$.

An n-RELGEN instance $w \in \{0,1\}^{n^3}$ is considered as divided up into n^2 n-bit blocks denoted w_{i*j} for $(i, j) \in [n] \times [n]$ and concatenated to form w. We will refer to the Boolean variables of n-RELGEN as $w_{i,j,k}$.

A triple of the form $i * j = 1$ for some i and j is called a *trivial* triple (since the element 1 is always trivially included in the closure $\langle \{1\} \rangle_T$). For an n-RELGEN instance $w \in \{0,1\}^{n^3}$, we write $\mathrm{TRIVEXT}(w)$ (for *trivial extension*) to mean $w \vee (10^{n-1})^{n^2}$, that is, the n-RELGEN instance obtained from w by adding to the set of triples represented by w all the trivial triples, setting $w_{i,j,1} = 1$ for each $i, j \in [n]$. A block w_{i*j} is said to be *heavy* if two (or more) of its bits are 1.

2.1 Branching Programs

Since n-GEN is defined over n-ary input variables, it is convenient for us to work with n-way branching programs, first defined by Borodin and Cook [BoCo82]. We also need the nondeterministic extension of the model, defined by Borodin, Razborov and Smolensky [BRS93].

Definition 1. *[BRS93] A nondeterministic n-way branching program is a directed acyclic rooted multi-graph with a distinguished sink node labeled ACCEPT. The edges out of non-sink nodes are either unlabeled, or labeled "$x_i = j$" for some variable x_i and $j \in [n]$. Only inputs satisfying the statement on the label may follow the labeled edges, all inputs are allowed to follow the unlabeled edges. An input $x_1, \ldots, x_t$ is accepted by the program if there is at least one directed path leading from the root to the ACCEPT node, such that the input $x_1, \ldots, x_t$ is allowed to follow it. (As there may be multiple edges between*

182 A. Gál, M. Koucký, and P. McKenzie

*two nodes in a branching program by a path we always understand a sequence
of edges* $(v_1, v_2), (v_2, v_3), \ldots, (v_{m-1}, v_m)$ *rather than just a sequence of vertices*
$v_1, v_2, \ldots, v_m$.) *A nondeterministic* n-*way branching program computes a func-
tion* $f : [n]^t \rightarrow \{0, 1\}$ *if* $f(x_1, \ldots, x_t) = 1$ *if and only if* $x_1, \ldots, x_t$ *is accepted
by the program. A* deterministic n-*way branching program* must *satisfy the ad-
ditional restrictions that it has no unlabeled edges, and there are exactly* n *edges
out of each non-sink node with the* n *possible labels* $x = j$ *for* $j = 1, \ldots, n$ *for
the same variable* x.

The *size* of a branching program is the number of its nodes. If the program
contains other sink nodes in addition to the ACCEPT node, they are labeled
REJECT. Note that deterministic programs computing non-constant functions
have at least one REJECT node, but REJECT nodes may be omitted from
nondeterministic programs. Note also that the nondeterministic model defined
above is usually called a switching-and-rectifier network if the underlying graph
is not required to be acyclic.

A path is called *consistent* if for every variable x_i and for all values $j_1 \neq j_2$,
the labels $x_i = j_1$ and $x_i = j_2$ do not both appear on the path. Note that since
no input will follow an inconsistent path, the correctness of the program in itself
gives no requirements for inconsistent paths.

Definition 1 contains as a special case (deterministic or nondeterministic)
Boolean branching programs if $n = 2$ (using the values 0 and 1 instead of 1 and
2, of course). Several slightly different definitions of nondeterministic branching
programs have appeared in the literature, in particular they may involve introducing guessing nodes. The size necessary to compute a given function under
these different definitions remains polynomially related (see e.g. [Ra91]), note
however that this is not automatically inherited in various restricted versions of
the models.

For an n-ary function $f : [n]^t \rightarrow \{0, 1\}$ let $f_{bin} : \{0, 1\}^{t \log_2 n} \rightarrow \{0, 1\}$ be the
Boolean function obtained from f by encoding its variables as binary strings. If
f can be computed by n-way branching programs of size $s(n)$ then f_{bin} can be
computed by Boolean branching programs of size $n \cdot s(n)$. No size increase occurs
in the reverse direction. Thus, proving super-polynomial size lower bounds for
deterministic or nondeterministic n-way branching programs computing n-GEN
would separate P from L or NL, respectively.

2.2 Tight Computation of GEN

Given a (deterministic or nondeterministic) n-way branching program P computing n-GEN, we denote by MON(P) the nondeterministic *Boolean* branching
program obtained from P as follows: replace each edge label of the form $i * j = k$
of P by the label $w_{i,j,k} = 1$.

Since MON(P) uses only edge labels of the form $x = 1$, it computes a
monotone Boolean function, which we denote by $f_{\text{MON}(P)}$. Note that for any
n-GEN instance g, we have GEN$(g) = $ RELGEN$(\chi(g)) = f_{\text{MON}(P)}(\chi(g))$. On
the other hand, the fact that P computes n-GEN in itself does not place any
requirements on the value of $f_{\text{MON}(P)}$ over inputs $w \in \{0, 1\}^{n^3}$ that are not

of the form $\chi(g)$ for any n-GEN instance g. In particular, while we know that MON(P) computes a monotone Boolean function that agrees with RELGEN on inputs of the form $\chi(g)$, we have no reason to expect that MON(P) would actually compute n-RELGEN, or even agree with n-RELGEN on any other inputs except what is implied by the monotonicity of $f_{\text{MON}(P)}$.

It turns out that the following additional requirement on n-way branching programs computing n-GEN is sufficient to obtain exponential lower bounds. We require that in addition to inputs of the form $\chi(g)$, $f_{\text{MON}(P)}$ agrees with RELGEN also on the trivial extensions TRIVEXT($\chi(g)$). This is equivalent to just requiring that if $f_{\text{MON}(P)}(\chi(g)) = 0$ then $f_{\text{MON}(P)}(\text{TRIVEXT}(\chi(g))) = 0$ as well, since $f_{\text{MON}(P)}$ is monotone. Thus, we just require that for any n-GEN instance g that has no accepting path in P, no path of P can reach the ACCEPT node if the only edges used in addition to the edges that g could traverse are labeled by trivial triples of the form $i * j = 1$.

Definition 2. *We say that a (deterministic or nondeterministic) n-way branching program P tightly computes n-GEN if it computes n-GEN, and for any n-GEN instance g if $f_{\text{MON}(P)}(\chi(g)) = 0$ then $f_{\text{MON}(P)}(\text{TRIVEXT}(\chi(g))) = 0$.*

Definition 3. *We say that a Boolean function $f : \{0,1\}^{n^3} \to \{0,1\}$ represents n-GEN if $f(\chi(g)) = RELGEN(\chi(g))$ for any n-GEN instance g.*

We say that a Boolean function $f : \{0,1\}^{n^3} \to \{0,1\}$ tightly represents n-GEN if it represents n-GEN and $f(\text{TRIVEXT}(\chi(g))) = RELGEN(\text{TRIVEXT}(\chi(g)))$ for any n-GEN instance g.

The following is immediate from the above definitions:

Proposition 1. *An n-way branching program P computes n-GEN if and only if $f_{\text{MON}(P)}$ represents n-GEN, and an n-way branching program P tightly computes n-GEN if and only if $f_{\text{MON}(P)}$ tightly represents n-GEN.*

Note that in models where inconsistent paths are excluded, for example in deterministic read-once branching programs, tight computation is automatically guaranteed. Next we define two versions of the model that guarantee tight computation of GEN – without excluding inconsistent paths in general: {1}-monotone nondeterministic n-way branching programs and syntactic incremental branching programs.

2.3 Monotone Nondeterministic n-Way Branching Programs

In the case of nondeterministic Boolean branching programs, Definition 1 can easily be modified to define *monotone* nondeterministic Boolean branching programs, by simply forbidding labels of the form $x = 0$ (using the values 0 and 1 instead of 1 and 2, of course), see e.g. [Ra91].

We propose the following extension to nondeterministic n-way branching programs when $n > 2$. Similarly to the Boolean case, we simply forbid some of the n possible values to be used in edge labels. Just like in the case of Boolean branching programs, "monotonicity" makes more sense in the nondeterministic framework, and not every n-ary function can be computed under this restriction.

Definition 4. *Let $\emptyset \neq S \subset [n]$. A nondeterministic n-way branching program is S-monotone if edge labels $x = j$ with $j \in S$ do not appear in the program.*

Definition 5. *Let $x, y \in [n]^t$ and $\emptyset \neq S \subset [n]$. We say that $x <_S y$ if for any $i \in [t]$ either $x_i = y_i$ or $x_i \in S$ and $y_i \notin S$. A function $f : [n]^t \rightarrow \{0,1\}$ is S-monotone, if for any $x, y \in [n]^t$ such that $x <_S y$, we have $f(x) \leq f(y)$.*

Note that our definition of S-monotone functions includes both monotone and anti-monotone Boolean functions as a special case. Clearly, S-monotone branching programs can compute only S-monotone functions. Notice that GEN is a $\{1\}$-monotone function, and the variant of GEN where the problem is to determine whether n is in the closure of some fixed starting set S is an S-monotone function. We can match the current best upper bounds for various GEN subproblems by $\{1\}$-monotone nondeterministic n-way branching programs without significant increase in size. That is, so far we have no results separating the power of monotone and non-monotone nondeterministic n-way branching programs for computing GEN.

2.4 Incremental Branching Programs

It seems natural for an n-way branching program solving n-GEN to try to find the elements in the closure $\langle\{1\}\rangle_{T^g}$ "incrementally", that is by asking questions $i * j = ?$ only for elements i, j already known to be in $\langle\{1\}\rangle_{T^g}$. In fact the current best upper bounds (known to us) for various subproblems of GEN can be achieved by constructions that have this property. We formally define the incrementality property below.

Let P be a (deterministic or nondeterministic) branching program computing n-GEN. For each vertex u of P, we will define the set $A(u)$ ("available set") of elements that have already been generated along *every* path reaching u.

Given a path π from the root to some vertex u in P, let T^π be the set of triples that appear as edge labels along π. Let PATHS(u) denote the set of all graph theoretic paths in P starting from the root and reaching u. Let $A(u) = \bigcap_{\pi \in \text{PATHS}(u)} \langle\{1\}\rangle_{T^\pi}$. We obtain a potentially larger set, if we only require that its elements are generated along every path reaching u that may be followed by some GEN instance. Recall that a path in P is consistent if for every pair (i,j) and for all values $k_1 \neq k_2$, the labels $i * j = k_1$ and $i * j = k_2$ do not both appear along the path. If π is consistent then $T^\pi \subseteq T^g$ for some n-GEN instance g. On the other hand, no GEN instance can follow an inconsistent path. We denote by GENPATHS(u) the set of all consistent paths starting from the root and reaching u. Let $A_{\text{GEN}}(u) = \bigcap_{\pi \in \text{GENPATHS}(u)} \langle\{1\}\rangle_{T^\pi}$. Note that GENPATHS$(u) \subseteq$ PATHS(u), and thus $A(u) \subseteq A_{\text{GEN}}(u)$.

Definition 6. *A (deterministic or nondeterministic) n-way branching program for n-GEN is (semantic) incremental if for every edge with label $i * j = k$ directed out of a node u the condition $\{i,j\} \subseteq A_{\text{GEN}}(u)$ holds. The program is syntactic incremental if for every edge with label $i * j = k$ directed out of a node u the stronger condition $\{i,j\} \subseteq A(u)$ holds.*

Let P be a branching program for n-GEN and let $\pi = (v_0, v_1), \ldots, (v_\ell, v_{\ell+1})$ be a path in P. We allow $\ell = 0$ so the path may consist of a single edge. We say that $i \in [n]$ is *useful* for π if the last edge $(v_\ell, v_{\ell+1})$ of π is labeled by $i * j = k$ or by $j * i = k$ for some $j, k \in [n]$, and none of the edges (v_t, v_{t+1}) for $t \in \{0, \ldots, \ell-1\}$ is labeled by $k * j = i$ for any $j, k \in [n]$. For a node u of P, let $U(u)$ be the set of elements that are useful for some path π starting in u and leading to an arbitrary node of P. Notice, the program P is syntactic incremental if and only if $U(u) \subseteq A(u)$ for every node u of P. We denote by $\max U(P)$ the maximum size of $U(u)$ for any node u in P.

3   Lower Bounds Derived from Monotone Circuit Depth Bounds

3.1   Lower Bounds in Models Without Requiring Incrementality

By the definitions of the previous section, every n-way branching program P computing GEN has an associated monotone nondeterministic Boolean branching program $\mathrm{MON}(P)$ computing some function $f_{\mathrm{MON}(P)}$ that represents GEN. By standard arguments, this yields monotone formulae for $f_{\mathrm{MON}(P)}$. Thus, proving that every monotone function representing GEN requires large monotone circuit depth would be sufficient to obtain lower bounds for unrestricted (deterministic or nondeterministic) n-way branching programs computing GEN. We can even define specific monotone Boolean functions representing GEN such that monotone circuit depth lower bounds for them would imply lower bounds for every monotone function representing GEN[1]. Unfortunately, we do not know how to prove monotone circuit depth lower bounds for these functions.

However, the monotone circuit depth lower bounds of [RaMc99] for the REL-GEN function are sufficient to derive the following statement.

Theorem 1. *For some $\gamma > 0$ and any t large enough, any function $f : \{0, 1\}^t \to \{0, 1\}$ that tightly represents GEN requires monotone circuits of depth t^γ.*

Using Proposition 1 and Theorem 1 we derive the following.

Theorem 2. *For some $\epsilon > 0$ and all n large enough, any (deterministic or nondeterministic) n-way branching program that tightly computes n-GEN has size 2^{n^ϵ}.*

Remark 1. Tight computation places no restrictions on the model itself, but instead requires correctness of the computation in a slightly stronger sense: it places requirements regarding acceptance on some graph-theoretic paths that no GEN instance would follow. Correctness in the usual sense places no requirements on the computation along such paths. However, tight computation places no requirements on paths with inconsistencies that do not involve trivial triples. Hence the significance of Theorem 2 is that any purported subexponential size

---

[1] For example, the n^2-th slice function of n-RELGEN has this property. See e.g. [Be81, We87] for more on slice functions.

branching program P solving GEN would need to make critical use of trivial triples, although such triples appear oblivious to any progress.

As noted at the end of Section 2.2, tight computation is automatically guaranteed in read-once deterministic branching programs.

Corollary 1. *For some $\epsilon > 0$ and all n large enough, any read-once deterministic n-way branching program computing n-GEN has size 2^{n^ϵ}.*

Proposition 2. *Any $\{1\}$-monotone nondeterministic n-way branching program computing n-GEN computes n-GEN tightly.*

Theorem 3. *For some $\epsilon > 0$ and all n large enough, any $\{1\}$-monotone nondeterministic n-way branching program computing n-GEN has size 2^{n^ϵ}.*

3.2   Lower Bounds for Syntactic Incremental Branching Programs

Proposition 3. *Any (deterministic or nondeterministic) syntactic incremental n-way branching program computing n-GEN computes n-GEN tightly.*

The statement will follow from the following much stronger statement.

Proposition 4. *Let P be a (deterministic or nondeterministic) syntactic incremental n-way branching program computing n-GEN. Then, for every $w \in \{0,1\}^{n^3}$ such that $RELGEN(w) = 0$, $f_{MON(P)}(w) = 0$ as well.*

Corollary 2. *Any (deterministic or nondeterministic) syntactic incremental n-way branching program computing n-GEN has size 2^{n^ϵ} for some $\epsilon > 0$.*

4   Syntactic Incrementality and Pebbling

Marking machines were defined by Cook [Co74] as a model for computing GEN. Jumping automata on graphs were defined by Cook and Rackoff [CoRa80] as a computational model for solving graph s-t-connectivity; Poon [Po93] defined an extension of this model called *node-named jumping automata on graphs* (NN-JAG). We show in this section that syntactic incremental branching programs can be efficiently simulated by marking machines and NNJAG's, and vice versa.

4.1   Marking Machines

We adapt Cook's definition [Co74] to our terminology while using a slightly different set of rules that are analogous to the pebbling rules of the games on graphs introduced by Paterson and Hewitt [PaHe70]. We discuss the difference between these rules in the full version of the paper. A *marking machine M* operates on an instance $T \subseteq [n] \times [n] \times [n]$ of n-GEN. Each configuration of M is one of the subsets of $[n]$; it identifies the set of marked elements of $[n]$. The *initial configuration* of M is the empty set. At each step of a computation, M

(nondeterministically) changes its configuration C to C' in one of the following ways: M marks the element 1, i.e., $C' = C \cup \{1\}$, or M removes a mark from an arbitrary element $r \in C$, i.e., $C' = C \setminus \{r\}$, or it marks an element $z \notin C$ provided that $(x, y, z) \in T$ for some $x, y \in C$, i.e., $C' = C \cup \{z\}$. A configuration C is *accepting* if $n \in C$. M *accepts input* T iff there is a sequence of configurations $C_0, C_1, \ldots, C_m$, where C_0 is the initial configuration, C_i follows from C_{i-1} by a legal move and C_m is an accepting configuration. We say that M *accepts* T *using only ℓ markers* if there is an accepting computation of M on T in which all configurations are of size at most ℓ.

We first establish the relationship between incremental branching programs and marking machines. As the main measure of size of marking machines is the number of marks used it should not come as a surprise that this measure relates to $\max U(\cdot)$ of branching programs.

Proposition 5. *1. If P is a (deterministic or nondeterministic) syntactic incremental n-way branching program computing n-GEN then there is a marking machine M that accepts every positive instance of n-GEN using at most $\max U(P)$ markers.*

2. If M is a marking machine M that accepts every positive instance of n-GEN using at most ℓ markers then there is a nondeterministic syntactic incremental n-way branching program P of size $1 + (\sum_{i=1}^{\ell} \binom{n}{i})^2$ that computes n-GEN.

This proposition means that lower bounds on the number of markers needed by a marking machine to solve arbitrary instances of n-GEN imply lower bounds on $\max U(P)$ for syntactic incremental branching programs computing n-GEN. Such lower bounds can be further translated into lower bounds on the size of these branching programs as the following lemma indicates.

Lemma 1 (Symmetrization lemma). *Let $k, n \geq 2$ be integers. Let P be a nondeterministic syntactic incremental kn-way branching program that computes kn-GEN. Then there is a nondeterministic syntactic incremental n-way branching program P' that computes n-GEN and such that $size(P') \leq size(P)$ and $\max U(P') \leq 2 + \log_k size(P)$.*

In [Co74] Cook proves that there are instances of n-GEN that cannot be accepted by marking machines with $o(\sqrt{n})$ markers. This was later improved by Paul et al. in [PTC77]:

Proposition 6 ([PTC77]). *There is a constant $c > 0$ such that for all $n \geq 2$ there is an instance T of n-GEN that cannot be accepted by any marking machine using less than $cn/\log n$ markers.*

The previous three claims give us the following corollary.

Theorem 4. *There is a constant $c > 0$ such that for all n large enough if a nondeterministic syntactic incremental n-way branching program P solves n-GEN then it has size at least $2^{cn/\log n}$.*

4.2   Jumping Automata on Graphs

By n-STCONN we denote the sub-problem of s-t-connectivity restricted to graphs on n vertices.

As defined in [Po93], a (deterministic) NNJAG J is a finite state automaton with p distinguished pebbles, q states and a transition function δ. (p, q and δ can non-uniformly depend on n.) The input to J is an instance G of STCONN. J computes on G by moving pebbles initially placed on vertex 1 along edges of G. At each step of its computation J can also jump a pebble from its current location to a vertex occupied by another pebble. J can detect the names of vertices that are occupied by its pebbles. NNJAG *solves* n-STCONN if on every instance G of n-STCONN, J accepts G iff there is a path from vertex 1 to vertex n in G. The *size of a NNJAG* J is the number of its possible configurations, i.e., qn^p.

Proposition 7.   *1. For any deterministic syntactic incremental n^2-way branching program P that computes n^2-GEN$_{(2rows)}$ there is a NNJAG of size at most $5n^4 \cdot size(P)^2$ that solves n-STCONN.*

   2. If J is an NNJAG solving n-STCONN then there is a deterministic syntactic incremental n-way branching program P of size at most $O(n^2 + n^3(size(J))^2)$ that computes n-GEN$_{(2rows)}$.

There is a long sequence of lower bounds for various types of jumping automata on graphs. The strongest one was obtained by Edmonds et al., which we use to obtain Theorem 5.

Proposition 8.   *[EPA99] If NNJAG J solves n-STCONN then J has size at least $n^{\Omega(\log n)}$.*

Theorem 5.   *There is a constant $c > 0$ such that for all n large enough if a deterministic syntactic incremental n-way branching program P solves n-GEN$_{(2rows)}$ then it has size at least $n^{c \log n}$.*

We should note here that Edmonds et al. [EPA99] in fact prove a lower bound for probabilistic NNJAG's. Their lower bound thus implies also a lower bound for appropriately defined probabilistic syntactic incremental branching programs.

5   Upper Bounds

Our upper bounds are stated in Figure 1. We mention that the super-polynomial separation between the power of deterministic and nondeterministic syntactic incremental branching programs for GEN$_{(2rows)}$ cannot be made exponential as a consequence of the following theorem (which also holds in the semantic model).

Theorem 6.   *If P is a nondeterministic syntactic incremental branching program that computes n-GEN then there is a deterministic syntactic incremental branching program P' of size at most $size(P)^{O(\log size(P))}$ that computes n-GEN.*

References

[BaMc91]   D. BARRINGTON AND P. MCKENZIE, Oracle branching programs and Logspace versus P, *Information and Computation* **95** (1991), pp. 96–115.

[BJS01]   P. BEAME, T.S. JAYRAM AND M.E. SAKS, Time-Space Tradeoffs for Branching Programs J. Computer and Systems Science **63** (4), pp. 2001542–572

[Be81]   S.J. BERKOWITZ, *On some relationships between monotone and non-monotone circuit complexity*, manuscript, University of Toronto, Computer Science Department (1981).

[BoCo82]   A. BORODIN AND S.A. COOK, A time-space trade-off for sorting on a general sequential model of computation *SIAM J. on Computing* **11**, 2 (1982), pp. 287–297.

[BRS93]   A. BORODIN, A. RAZBOROV AND R. SMOLENSKY, On lower bounds for read-k-times branching programs. *Computational Complexity* **3** (1993) pp. 1-18.

[Co71]   S.A. COOK, Characterizations of pushdown machines in terms of time-bounded computers, *J. of the Association for Computing Machinery* **18** (1), pp. 1971.4–18

[Co74]   S.A. COOK, An observation on time-storage trade-off, J. Computer and Systems Science **9(3)** (1974), pp. 308–316

[CoRa80]   S.A. COOK AND C.W. RACKOFF, Space lower bounds for maze thread-ability on restricted machines, *SIAM J. on Computing* **9**,   (1980), pp. 636–652.

[EPA99]   J. EDMONDS, C.K. POON AND D. ACHLIOPTAS, Tight lower bounds for st-connectivity on the NNJAG model, *SIAM J. on Computing* **28**, 6 (1999), pp. 2257–2284.

[GKM05]   A. GÁL, M. KOUCKÝ AND P. MCKENZIE, Incremental branching programs, ECCC TR05-136 (2005), pp. 1-18.

[JoLa77]   N.D. JONES AND W.T. LAASER, Complete problems for deterministic polynomial time, Theoretical Computer Science **3** (1977), pp. 105–117.

[Ju95]   S. JUKNA, A note on read-k-times branching programs, *RAIRO Theoretical Informatics and Applications* 29, pp. 75-83.

[KaWi88]   M. KARCHMER AND A. WIGDERSON, Monotone circuits for connectivity require super-logarithmic depth, *Proc. of the 20th ACM Symp. on the Theory of Computing* (1988), pp. 539–550. Full version in: *SIAM J. on Disc. Math.* **3**, no. 2 (1990) pp. 255–265.

[PTC77]   W. J. PAUL, R. E. TARJAN AND J. R. CELONI, Space bounds for a game on graphs, *Mathematical Systems Theory* **10**,(1977), 239-251.

[PaHe70]   M.S. PATERSON AND C.E. HEWITT, Comparative schematology, *Record of Project MAC Conference on Concurrent Systems and Parallel Computations (June 1970)*, pp. 119-128, ACM, New Jersey, December 1970.

[Po93]   C.K. POON, Space bounds for graph connectivity problems on node-named JAGs and node-ordered JAGs *Proc. of the 34th IEEE Symp. on the Foundations of Computer Science* (1993), pp. 218–227.

[Ra91]   A. RAZBOROV, Lower bounds for deterministic and nondeterministic branching programs, *Proceedings of the 8th FCT, Lecture Notes in Computer Science*, **529** (1991) pp. 47-60.

[RaMc99]   R. RAZ AND P. MCKENZIE, Separation of the monotone NC hierarchy, *Combinatorica* **19**(3)(1999), pp. 403-435.

[Sa70]   W.J. SAVITCH, Relationships between nondeterministic and deterministic tape complexities, J. Computer and Systems Science **4(2)** (1970), pp. 177-192

[We87]   I. WEGENER, The complexity of Boolean functions, Wiley-Teubner, 1987.

[We00]   I. WEGENER, Branching programs and binary decision diagrams, SIAM Monographs on Discrete Mathematics and Applications, 2000.

Logic of Proofs for Bounded Arithmetic

Evan Goris*

The Graduate Center of the City University of New York,
365 Fifth Avenue, 10016 New York, NY, USA
evangoris@gmail.com

Abstract. The logic of proofs is known to be complete for the semantics of proofs in Peano Arithmetic PA. In this paper we present a refinement of this theorem, we will show that we can assure that all the operations on proofs can be realized by feasible, that is PTIME-computable, functions. In particular we will show that the logic of proofs is complete for the semantics of proofs in Buss' bounded arithmetic S^1_2. In view of recent applications of the Logic of Proofs in epistemology this result shows that explicit knowledge in the propositional framework can be made computationally feasible.

1 Introduction

In [1] the Logic of Proofs (LP) is shown to be arithmetically complete with respect to the semantics of proofs in Peano Arithmetic PA. The mathematical contribution of this paper is the extension of this result to the semantics of proofs in S^1_2. Although interesting in itself the extension of this theorem to S^1_2 is of fundamental value for computer science as well.

Namely, recently LP has been applied in the field of epistemic logic [2, 3]. One of the main merits of LP over the more traditional epistemic logics is that LP can express, using its proof terms, how hard it is to obtain certain knowledge. In contrast, for any theorem A, the basic epistemic logic S4 proves $\Box A$. However in general, the work involved with obtaining knowledge of A (for example by proving it, if it happens to be a theorem) does not polynomialy depend on the length of A. And thus neither is the epistemic assertion $\Box A$. This undermines the real word way of reasoning about knowledge, in which we can safely assume that certain agents do not posses certain information and or knowledge, although in principle they could. For example, according to the modal logic of knowledge everybody who knows the rules of chess also knows a winning strategy for the White. However, in the real world such a strategy is simply not feasibly obtainable.

This feature of epistemic logics like S4, called logical omniscience, is one of the major problems in the field of epistemic logic. LP is not logical omniscient in the following sense. For an assertion A, the epistemic assertion 'A is known'

---

* Research supported by CUNY Community College Collaborative Incentive Research Grant 91639-0001 "Mathematical Foundations of Knowledge Representation"

D. Grigoriev, J. Harrison, and E.A. Hirsch (Eds.): CSR 2006, LNCS 3967, pp. 191–201, 2006.

takes the form $t{:}A$ in LP. Where t is an object in a term language. By [4] t must be at least as long as a derivation of A. Thus a theorem $t{:}A$ of LP does not only say that A is knowable in principle but also honestly reflects how hard it is to obtain knowledge about A.

Looking at this from the other direction some information is still missing. Namely, the length of a proof term can only be a good indication of an upper bound for the complexity of the evidence it represents when the operations that occur in it are feasibly computable. In this paper we give a mathematical application for which LP is complete and in which each proof term represents a PTIME-computable function.

The paper is organized as follows. In Section 2 we define the logic LP. In Section 3 we discuss S_2^1, the formalization of syntax and provability in S_2^1, the interpretation of LP in S_2^1 and show that LP is complete with respect to this interpretation. The proof is a refinement of the one given in [1] for the interpretation of LP in PA.

The author would like to thank Professor Sergei Artemov and Walter Dean for useful suggestions and remarks.

2   Logic of Proofs

LP has a long history and many variations. Here we consider the version from [1]. For an extensive overview of LP we refer the reader to [5].

The language of LP consist of the following. We have countably many propositional variables $p_1, p_2, \ldots$, countably many proof variables $x_1, x_2, \ldots$ and countably many axiom constants $c_1, c_2, \ldots$ The following definitions show how we can construct more complex expressions.

Definition 2.1 (LP-terms). *We define* LP*-terms as follows.*

 - *Axiom constants and proof variables are terms,*
 - *If s and t are terms then so are $s + t$, $s \cdot t$ and $!t$.*

Definition 2.2 (LP-formulas). *We define* LP*-formulas as follows.*

 - $\perp$ *and any propositional variable is a formula,*
 - *If A and B are formulas then so is $A \to B$,*
 - *If A is a formula and t is a term then $t{:}A$ is a formula.*

Definition 2.3 (LP). *As axioms we take all instances of the following schemata.*

A0 *"Propositional logic",*
A1 $t{:}A \to A$,
A2 $s{:}(A \to B) \to t{:}A \to (s{\cdot}t){:}B$,
A3 $s{:}A \to (s + t){:}A \wedge (t + s){:}A$,
A4 $t{:}A \to !t{:}(t{:}A)$,
A5 $c{:}A$, c *an axiom constant and* A *an instance of* **A0-A4**.

The set of theorems of LP *is obtained by closing the set of axioms under modus ponens.*

Definition 2.4 (Constant specification). *A constant specification is a set of pairs* $\langle c, F \rangle$ *where* c *is a proof constant and* F *an instance of one of* **A0** − **A4**.

With $\mathsf{LP}_{\mathcal{CS}}$ we denote the fragment of LP, where **A5** is restricted to $c{:}A$ for $\langle c, A \rangle \in \mathcal{CS}$. Let us write $\bigwedge \mathcal{CS}$ for $\bigwedge \{c{:}A \mid \langle c, A \rangle \in \mathcal{CS}\}$. The following is obvious.

$$\mathsf{LP}_\emptyset \vdash \bigwedge \mathcal{CS} \to A \Leftrightarrow \mathsf{LP}_{\mathcal{CS}} \vdash A \ .$$

Let X be a finite set of formulas. Let $\mathcal{T}(X) = \{t \mid \text{for some } A, \, t{:}A \in X\}$. We say that X is *adequate* when

− X is closed under subformulas,
− X is closed under single negation,
− If $t \in \mathrm{Sub}(\mathcal{T}(X))$ and $A \in X$ then $t{:}A \in X$.

Clearly any finite set of formulas can be extended to a finite adequate set of formulas. Also notice that when X is adequate, then $\mathrm{Sub}(\mathcal{T}(X)) = \mathcal{T}(X)$.

We say that a set of formulas Γ is *inconsistent* if for some $X_1, \ldots, X_k \in \Gamma$ we have $\mathsf{LP} \vdash \neg X_1 \vee \cdots \vee \neg X_k$. A set is *consistent* when it is not inconsistent. We say that a set Γ is *maximal consistent in* X when $\Gamma \subseteq X$, Γ is consistent and if $\Gamma \subsetneq \Gamma' \subseteq X$, then Γ' is inconsistent.

The proof of the following lemma is standard.

Lemma 2.5. *If* Γ *is maximal consistent in* X *then for every* $\neg A \in X$ *we have* $A \in \Gamma$ *or* $\neg A \in \Gamma$.

The following lemma is an immediate corollary to Lemma 2.5

Lemma 2.6. *Let* X *be adequate and let* Γ *be maximally consistent in* X. *Then*

1. *If* $A, A \to B \in \Gamma$ *then* $B \in \Gamma$,
2. *If* $t{:}A \in \Gamma$ *then* $A \in \Gamma$,
3. *If* $s{:}(A \to B) \in \Gamma$, $t{:}A \in \Gamma$ *and* $s \cdot t \in \mathcal{T}(X)$ *then* $(s \cdot t){:}B \in \Gamma$,
4. *If* $s{:}A \in \Gamma$ *or* $t{:}A \in \Gamma$ *then* $s + t \in \mathcal{T}(X)$ *implies* $(s + t){:}A \in \Gamma$,
5. *If* $t{:}A \in \Gamma$ *and* $!t \in \mathcal{T}(X)$ *then* $!t{:}(t{:}A) \in \Gamma$.

Notice that in the above two lemmas Γ is not necessarily finite. If we choose X finite and Γ maximally consistent in X, then Γ is finite as well. In [1] an explicit algorithm is given that, given a formula A such that $\mathsf{LP} \nvdash A$, constructs a finite Γ that satisfies Lemma 2.6 such that $\neg A \in \Gamma$.

3   Arithmetical Interpretation

In this section we introduce the fragment of arithmetic S_2^1 [6, 7, 8], the formalization of syntax and provability in S_2^1 and the interpretation of LP in S_2^1. The discussion on S_2^1 is basically meant to fix the notation, for a detailed treatment of the subject see [6, 8].

3.1 Bounded Arithmetic

In this paper we will use the approach of [6]. In [6] the theory S_2, a conservative extension of $I\Delta_0 + \Omega_1$, and its fragments S_2^i are formulated. The theories S_2^i are first-order theories in the language $\mathcal{L} = \{+, \times, \lfloor \frac{x}{2} \rfloor, |\ |, \sharp\}$. The intended meaning of $+$ and $\times$ is as usual. $\lfloor \frac{x}{2} \rfloor$ is the bitwise shift-right operation, $|x|$ is the length of the binary representation of x and the intended meaning of $x \sharp y$ is $2^{|x||y|}$.

A quantifier in a formula is *bounded* if its of the form $\forall x \leq t$ or $\exists x \leq t$, where t is a term. We say that the quantifier is *sharply bounded* if t is of the form $|t'|$. The classes of formulas Σ_i^b are defined as follows.

- A formula is Σ_1^b if all quantifiers are bounded and all universal quantifiers are sharply bounded.
- A formula is Π_1^b if its negation is Σ_1^b.
- The class of formulas Σ_{i+1}^b is the least class that contains $\Sigma_i^b \cup \Pi_i^b$ and is closed under $\wedge$, $\vee$ and bounded existential quantification and sharply bounded universal quantification.
- The class of formulas Π_{i+1}^b is the least class that contains $\Sigma_i^b \cup \Pi_i^b$ and is closed under $\wedge$, $\vee$ and bounded universal quantification and sharply bounded existential quantification.

In addition we define

- A formula is Δ_i^b if it is both Σ_i^b and Π_i^b.

These definitions relativize to arithmetical theories in the obvious way.

All the (axiomatizations of the) theories S_2^i contain a basic set of axioms BASIC, defining the recursive properties of the functions in the language. The theories S_2^i are then axiomatized over BASIC by the polynomial induction scheme for Σ_i^b formulas ϕ:

$$\phi(0) \wedge \forall x \leq y \left(\phi \left(\lfloor \frac{x}{2} \rfloor \right) \rightarrow \phi(x) \right) \rightarrow \phi(y) \ .$$

An important relation of bounded arithmetic with complexity theory is as follows.

Theorem 3.1. *If $\sigma(x, y)$ is a Σ_1^b formula such that $S_2^1 \vdash \forall x \exists! y \sigma(x, y)$. Then $\sigma(x, y)$ defines the graph of a PTIME-computable function.*

For a proof of this theorem, and its reverse: every PTIME-computable function has a Σ_1^b-definition that is provably total in S_2^1, see [6,7].

One of the fundamental theorems of [9] projects to S_2^1 as follows (see [8]).

Theorem 3.2 (Parikh's Theorem). *Let $\sigma(x, y)$ be a Σ_1^b formula such that $S_2^1 \vdash \forall x \exists y \sigma(x, y)$. Then there exists a term t such that $S_2^1 \vdash \forall x \exists y \leq t(x) \delta(x, y)$.*

Since the exponential function majorizes any term of S_2^1, Parikh's theorem implies that exponentiation does not have a provably total Σ_1^b definition.

One further fundamental result is as follows. Let f be a new function symbol and let $\phi(x, y)$ be a Σ_1^b formula such that $S_2^1 \vdash \forall x \exists! y \phi(x, y)$. Let $\Sigma_i^b(f)$ and $S_2^1(f)$

be defined exactly as Σ_1^b and S_2^1 but in the language $\mathcal{L} \cup \{f\}$ (in particular f may be used in bounding terms and induction schemes) and $S_2^1(f)$ has an additional axiom $f(x) = y \leftrightarrow \phi(x, y)$.

Theorem 3.3. $S_2^i(f)$ *is conservative over* S_2^i *and any* $\Sigma_i^b(f)$ *formula is* $S_2^i(f)$ *equivalent to a* Σ_i^b *formula.*

In more informal terms, Σ_i^b definable functions can be freely used. A similar statement holds for Δ_1^b definable predicates. For a proof see [8].

3.2   Formalization of Syntax

First some notation. Elements from $\mathbf{N}$ are printed in boldface: $\mathbf{n}$, $\mathbf{m}$ etc. A sequence $x_1, \ldots, x_n$ is usually written as $\overline{x}$. The Gödel number of some syntactic object f is denoted by $\ulcorner f \urcorner$. For exact details on such a coding we refer the reader to [6].

A simple but important definition is the canonical representation (numeral), of an element of $\mathbf{N}$. We define

$$\underline{\mathbf{0}} = 0$$

$$\underline{\mathbf{n}} = \begin{cases} S(S(0)) \cdot \underline{\mathbf{m}} & \text{if } \mathbf{n} = 2\mathbf{m} \\ S(S(S(0)) \cdot \underline{\mathbf{m}}) & \text{if } \mathbf{n} = 2\mathbf{m} + 1 \end{cases}$$

We can define a Σ_1^b function $\text{num}(x)$ in S_2^1 such that for any $\mathbf{n} \in \mathbf{N}$ we have

$$\text{num}(\mathbf{n}) = \ulcorner \underline{\mathbf{n}} \urcorner .$$

In addition we can define a function $\text{sub}(x, y, z)$, Σ_1^b definable in S_2^1, that satisfies the following.

$$S_2^1 \vdash \text{sub}\left(\ulcorner \phi(\overline{y}, x) \urcorner, \ulcorner x \urcorner, \ulcorner t \urcorner \right) = \ulcorner \phi(\overline{y}, t) \urcorner . \tag{1}$$

Using such functions one can proof a fixed point theorem [8].

Lemma 3.4. *For any formula* $\phi(\overline{x}, y)$ *there exists a formula* $\epsilon(\overline{x})$ *such that*

$$S_2^1 \vdash \epsilon(\overline{x}) \leftrightarrow \phi(\overline{x}, \ulcorner \epsilon(\overline{x}) \urcorner) .$$

From now on we will not make distinction between numerals and numbers (that is, we systematically confuse numerals with elements of the standard model).

In what follows we let $\text{isProof}(x)$ be a Δ_1^b-formula that defines the codes of proofs. It is well known that there exists a Δ_1^b formula $\text{Proof}(x, y)$ for which we have the following.

$$S_2^1 \vdash \phi \Leftrightarrow \mathbf{N} \models \exists x \text{Proof}(x, \ulcorner \phi \urcorner) . \tag{2}$$

Moreover there exist PTIME computable functions $\oplus$, $\otimes$ and e for which the following holds.

$$\mathbf{N} \models \text{Proof}(x, \ulcorner \phi \urcorner) \wedge \text{Proof}(y, \ulcorner \phi \to \psi \urcorner) \to \text{Proof}(y \otimes x, \ulcorner \psi \urcorner) , \tag{3}$$

$$\mathbf{N} \models \text{Proof}(x, \ulcorner \phi \urcorner) \vee \text{Proof}(y, \ulcorner \phi \urcorner) \to \text{Proof}(x \oplus y, \ulcorner \phi \urcorner) , \tag{4}$$

$$\mathbf{N} \models \text{Proof}(x, \ulcorner \phi \urcorner) \to \text{Proof}(e(x), \ulcorner \text{Proof}(x, \ulcorner \phi \urcorner) \urcorner) . \tag{5}$$

3.3   Arithmetical Interpretation of LP

Any Δ_1^b formula $\mathrm{Prf}(x, y)$ that satisfies (2) will be called a *proof predicate*. If it in addition comes with functions $\oplus$, $\otimes$ and e that satisfy the conditions (3), (4) and (5) that way say that it is a *normal proof predicate*. We say that a structure

$$\langle \mathrm{Prf}(x, y), \oplus, \otimes, e, * \rangle$$

is an *arithmetical interpretation* (which we also denote by $*$) when $\mathrm{Prf}(x,y)$ is a normal proof predicate with the functions $\oplus$, $\otimes$ and e. Moreover, $*$ is a mapping from propositional variables to sentences of S_2^1 and from proof variables and proof constants to numbers.

Given an arithmetical interpretation $\langle \mathrm{Prf}(x, y), \oplus, \otimes, e, * \rangle$, we can extend the map $*$ to the full language of LP as follows.

- $(s \cdot t)^* = s^* \otimes t^*$, $(s + t)^* = s^* \oplus t^*$ and $(!t)^* = e(t^*)$,
- $(A \to B)^* = A^* \to B^*$,
- $(t{:}A)^* = \mathrm{Prf}(t^*, \ulcorner A^* \urcorner)$.

We immediate get the following.

Theorem 3.5 (Arithmetical soundness). *If* $\mathsf{LP}_{\mathcal{CS}} \vdash A$. *Then for any arithmetical interpretation* $*$ *such that* $S_2^1 \vdash \mathcal{CS}^*$ *we have* $S_2^1 \vdash A^*$.

3.4   Arithmetical Completeness

For now we focus on $\mathsf{LP}_\emptyset$. We show that $\mathsf{LP}_\emptyset$ is arithmetically complete. A more general version, for arbitrary constant specifications, will be proved as an corollary.

Apart from a formalization of the syntax of S_2^1 in S_2^1 we simultaneous assume a disjoint formalization of the syntax of LP. So for any LP formula or term θ we have a code $\ulcorner \theta \urcorner$ and from this code we can deduce (in S_2^1) that it is indeed the code of an LP object (and not an S_2^1 object).

Theorem 3.6 (Arithmetical completeness). $\mathsf{LP}_\emptyset \vdash A$ *iff for any arithmetical interpretation* $*$ *we have that* $S_2^1 \vdash A^*$

The proof of this theorem is what constitutes the rest of this section. The soundness direction is a special case of Theorem 3.5. To show completeness assume that $\mathsf{LP}_\emptyset \nvdash A$. Let X be some adequate set such that $\neg A \in X$ and let Γ be maximal consistent in X such that $\neg A \in \Gamma$. (In particular Γ is finite.) We will construct a proof predicate $\mathrm{Prf}(x, y)$, PTIME operations on codes of proofs $\tilde{\oplus}$, $\tilde{\otimes}$ and $\tilde{e}$, and a mapping $*$ from propositional variables to sentences of S_2^1, and from proof variables and proof constants to $\mathbf{N}$ such that, (when $*$ is extended to the full language of $\mathsf{LP}_\emptyset$ as explained above) $S_2^1 \vdash A^*$ for all $A \in \Gamma$.

Let us first decide on what objects should serve as 'proofs'. To begin with, all usual proofs are 'proofs'. That is all sequence of formulas, each of which is an axiom of S_2^1, or can be obtain by an inference rule from earlier formulas in the sequence. This way the left to right direction of (2) is easily satisfied. As an extra source of 'proofs' we will use the finite set of LP terms $\mathcal{T}(X)$. The theorems of a 'proof' $t \in \mathcal{T}(X)$ will be the set $\{A \mid t{:}A \in \Gamma\}$.

We now wish to formalize the contents of the last two paragraphs. Let us suppose that we have a formula $\mathrm{Prf}(x, y)$. We define a translation from LP formulas A and LP terms t to S_2^1 sentences $A^\dagger$ and numbers $t^\dagger$ as follows.

1. $p^\dagger \equiv \ulcorner p \urcorner = \ulcorner p \urcorner$, if $p \in \Gamma$ and $p^\dagger \equiv \ulcorner p \urcorner \neq \ulcorner p \urcorner$ otherwise,
2. $t^\dagger = \ulcorner t \urcorner$, for any proof term t,
3. $(A \to B)^\dagger \equiv A^\dagger \to B^\dagger$,
4. $(t{:}A)^\dagger \equiv \mathrm{Prf}(t^\dagger, \ulcorner A^\dagger \urcorner)$.

To carry out the proof as in [1], we would like to construct a function $\mathrm{tr}(p, f)$ which, given a code p of a normal proof predicate $\mathrm{Prf}(x, y)$ and a code f of an LP-formula F gives us the code of the S_2^1 sentence $F^\dagger$. There is some difficulty constructing a Σ_1^b definition of such a function over S_2^1.

Fact 1. *There exists a sequence $F_0, F_1, F_2, \ldots$ of LP formulas such that for any S_2^1 term $s(x)$, there exists $n \geq 0$ such that $\ulcorner F_n^\dagger \urcorner > s(\ulcorner F_n \urcorner)$.*

By Parikh's theorem [7], such a function cannot be shown total in S_2^1. Since we only need $\ulcorner F^\dagger \urcorner$ for only finitely many F's, (namely those $F \in X$) the following suffices. Let $\mathrm{sub}_{x,y}(p, z_1, z_2)$ be a Σ_1^b definable and provably total function which satisfies (see (1) above)

$$\mathrm{sub}_{x,y}(\ulcorner \phi(x, y) \urcorner, \ulcorner t_1 \urcorner, \ulcorner t_2 \urcorner) = \ulcorner \phi(t_1, t_2) \urcorner \ .$$

As usual, we write $\dot{\to}$ for the Σ_1^b definable and provably total function that satisfies

$$\ulcorner \phi \urcorner \dot{\to} \ulcorner \psi \urcorner = \ulcorner \phi \to \psi \urcorner \ .$$

For each $F \in X$ we define with induction on F a Δ_1^b-formula $\phi_F(p, x)$, defining the graph of $\mathrm{tr}(p, \ulcorner F \urcorner)$, as follows.

– If $F \equiv \bot$ then
$$\phi_F(p, x) \equiv x = \ulcorner 0 \neq 0 \urcorner$$

– If $F \equiv p \in \Gamma$ then
$$\phi_F(p, x) \equiv x = \ulcorner \ulcorner p \urcorner = \ulcorner p \urcorner \urcorner$$

– If $F \equiv p \notin \Gamma$ then
$$\phi_F(p, x) \equiv x = \ulcorner \ulcorner p \urcorner \neq \ulcorner p \urcorner \urcorner$$

– If $F \equiv F_0 \to F_1$ then
$$\phi_F(p, x) \equiv \exists x_0 x_1 \leq x (\phi_{F_0}(p, x_0) \wedge \phi_{F_1}(p, x_1) \wedge x = x_0 \dot{\to} x_1)$$

– If $F \equiv t{:}F'$ then
$$\phi_F(p, x) \equiv \exists x' \leq x (\phi_{F'}(p, x') \wedge x = \mathrm{sub}_{x,y}(p, \ulcorner t \urcorner, \mathrm{num}(x')))$$

Let Prf(x,y) satisfy the following fixed point equation (see Lemma 3.4, also recall that Γ is finite).

$$\mathsf{S}_2^1 \vdash \mathrm{Prf}(x,y) \leftrightarrow \mathrm{Proof}(x,y) \vee \bigvee_{t:F\in\Gamma} (\phi_F(\ulcorner \mathrm{Prf}(x,y)\urcorner, y) \wedge x = \ulcorner t \urcorner) \qquad (6)$$

For briefity we put

$$T(x) \equiv \bigvee \{x = \ulcorner t \urcorner \mid t \in T(X)\} \ ,$$

$$\phi_F(u) \equiv \phi_F(\ulcorner \mathrm{Prf}(x,y)\urcorner, u) \ .$$

With induction on F one easily shows that for all $F, G \in \Gamma$ we have

$$\mathsf{S}_2^1 \vdash \phi_F(y) \leftrightarrow y = \ulcorner F^\dagger \urcorner \qquad (7)$$

and

$$F \not\equiv G \Rightarrow F^\dagger \not\equiv G^\dagger \ . \qquad (8)$$

Lemma 3.7. *Prf(x,y) is Δ_1^b in S_2^1*

Proof. From (7) one easily derives that for each $F \in \Gamma$, ϕ_F is Δ_1^b in S_2^1. □

Lemma 3.8. *For all $F \in X$ we have $F \in \Gamma$ implies $\mathsf{S}_2^1 \vdash F^\dagger$ and $F \notin \Gamma$ implies $\mathsf{S}_2^1 \vdash \neg F^\dagger$.*

Thus we have that, for each $F \in \Gamma$, $F^\dagger$ is S_2^1 provable. Since Γ is finite, we can find one single S_2^1 proof that proves them all. So let **g** be some number such that for all $A \in \Gamma$ we have

$$\mathbf{N} \models \mathrm{Proof}(\mathbf{g}, \ulcorner A^\dagger \urcorner) \ . \qquad (9)$$

Lemma 3.9. $\mathsf{S}_2^1 \vdash \exists x \, Prf(x,y) \leftrightarrow \exists x \, Proof(x,y)$

Proof. The right to left direction is clear by (6). For the other direction reason in S_2^1 and assume that for some x we have $\mathrm{Prf}(x,y)$. In the case $\mathrm{Proof}(x,y)$ we are done at once so assume that this is not so. Then by (6) we have

$$\bigvee_{t:F\in\Gamma} (\phi_F(y) \wedge x = \ulcorner t \urcorner) \ .$$

For each $t{:}F \in \Gamma$ we have $\phi_F(y) \to y = \ulcorner F^\dagger \urcorner$ and thus

$$\bigvee_{t:F\in\Gamma} y = \ulcorner F^\dagger \urcorner \ .$$

Since we have

$$\bigwedge_{F\in\Gamma} \mathrm{Proof}(\mathbf{g}, \ulcorner F^\dagger \urcorner)$$

we conclude that $\mathrm{Proof}(\mathbf{g}, y)$. □

Let $\pi(x)$ be a function such that for all $\mathbf{n}$ and all formulas ϕ for which $\mathbf{N} \models$ Proof$(\mathbf{n}, \ulcorner \phi \urcorner)$ we have

$$\mathbf{N} \models \mathrm{Proof}(\pi(\mathbf{n}), \mathrm{Proof}(\mathbf{n}, \ulcorner \phi \urcorner) \to \mathrm{Prf}(\mathbf{n}, \ulcorner \phi \urcorner)) \ . \tag{10}$$

Let $\underline{+}$, $\underline{\times}$ and $\underline{!}$ stand for Σ_1^b definable functions that takes codes $\ulcorner t_0 \urcorner$, $\ulcorner t_1 \urcorner$ of LP terms t_0 and t_1 to codes $\ulcorner t_0 + t_1 \urcorner$, $\ulcorner t_0 \cdot t_1 \urcorner$ and $\ulcorner !t_0 \urcorner$ of the LP terms $t_0 + t_1$, $t_0 \cdot t_1$ and $!t_0$ resp.

Recall that $T(x)$ defines the codes of the terms in $T(X)$ and that isProof(x) defines the codes of 'real' proofs in S_2^1. We define the function $\tilde{\oplus}$ as follows.

$$x \tilde{\oplus} y = \begin{cases} x \oplus y & \mathrm{isProof}(x) \wedge \mathrm{isProof}(y) \\ \mathbf{g} \oplus y & T(x) \wedge \mathrm{isProof}(y) \\ x \oplus \mathbf{g} & \mathrm{isProof}(x) \wedge T(y) \\ x \underline{+} y & T(x) \wedge T(y) \wedge T(x \underline{+} y) \\ \mathbf{g} & T(x) \wedge T(y) \wedge \neg T(x \underline{+} y) \end{cases}$$

Similarly we define the function $\tilde{\otimes}$ as follows.

$$x \tilde{\otimes} y = \begin{cases} x \otimes y & \mathrm{isProof}(x) \wedge \mathrm{isProof}(y) \\ \mathbf{g} \otimes y & T(x) \wedge \mathrm{isProof}(y) \\ x \otimes \mathbf{g} & \mathrm{isProof}(x) \wedge T(y) \\ x \underline{\times} y & T(x) \wedge T(y) \wedge T(x \underline{\times} y) \\ \mathbf{g} & T(x) \wedge T(y) \wedge \neg T(x \underline{\times} y) \end{cases}$$

And finally we define the function $\tilde{e}$ as follows.

$$\tilde{e}(x) = \begin{cases} !x & T(x) \wedge T(!x) \\ \mathbf{g} & T(x) \wedge \neg T(!x) \\ \pi(x) \otimes e(x) & \mathrm{isProof}(x) \end{cases}$$

And to finish we define the mapping $*$.

1. $p^* \equiv \ulcorner p \urcorner = \ulcorner p \urcorner$ if $p \in \Gamma$,
2. $p^* \equiv \ulcorner p \urcorner \neq \ulcorner p \urcorner$ if $S \notin \Gamma$,
3. $x^* \equiv \ulcorner x \urcorner$, $a^* = \ulcorner a \urcorner$.

Lemma 3.10. $\tilde{\oplus}$, $\tilde{\otimes}$ and $\tilde{e}$ are PTIME computable

Proof. All functions and predicates occurring in their definitions are PTIME computable. □

Now we have finished the definition of our translation of LP into S_2^1. The relation with our preliminary translation is as follows.

Lemma 3.11. If $t \in T(X)$ then $t^\dagger = t^*$. If $F \in X$ then $F^\dagger = F^*$.

Corollary 3.12. *For all $A \in \Gamma$ we have $S_2^1 \vdash A^*$*

Proof. Induction on F. We only write out the case $F \equiv t{:}F'$. We have that $\phi_{F'}(\ulcorner F'^\dagger \urcorner)$ is provable. If $F \in \Gamma$ then $\mathrm{Prf}(\ulcorner t \urcorner, \ulcorner F'^\dagger \urcorner)(= (t{:}F')^\dagger)$ is provable. If $F \notin \Gamma$ then by (8) we have that $\phi_{G'}(\ulcorner F'^\dagger \urcorner)$ is provably false for any $t{:}G' \in \Gamma$. Since $\ulcorner t \urcorner$ is never the code of a 'real' proof in S_2^1 we also have that $\mathrm{Proof}(\ulcorner t \urcorner, \ulcorner F'^\dagger \urcorner)$ is provably false and thus $\mathrm{Prf}(\ulcorner t \urcorner, \ulcorner F'^\dagger \urcorner)(= (t{:}F')^\dagger)$ is provably false. □

Lemma 3.13. *$\mathrm{Prf}(x, y)$ is a normal proof predicate, that is:*

1. $\mathbf{N} \models \mathrm{Prf}(x, \ulcorner \phi \urcorner) \wedge \mathrm{Prf}(y, \ulcorner \phi \to \psi \urcorner) \to \mathrm{Prf}(y \tilde{\otimes} x, \ulcorner \psi \urcorner)$,
2. $\mathbf{N} \models \mathrm{Prf}(x, \ulcorner \phi \urcorner) \vee \mathrm{Prf}(y, \ulcorner \phi \urcorner) \to \mathrm{Prf}(x \tilde{\oplus} y, \ulcorner \phi \urcorner)$,
3. $\mathbf{N} \models \mathrm{Prf}(x, \ulcorner \phi \urcorner) \to \mathrm{Prf}(\tilde{e}(x), \ulcorner \mathrm{Prf}(x, \ulcorner \phi \urcorner) \urcorner)$.

Proof. We only write out Item 3. Suppose $\mathrm{Prf}(x, \ulcorner \phi \urcorner)$. There are three cases, corresponding to the three disjunct that make up $\tilde{e}$, to consider. *Case 1:* isProof(x). In this case

$$\tilde{e}(x) = \pi(x) \otimes e(x) \text{ and } \mathrm{Proof}(x, \ulcorner \phi \urcorner) \ .$$

We thus have

$$\mathrm{Proof}(e(x), \ulcorner \mathrm{Proof}(x, \ulcorner \phi \urcorner) \urcorner) \ ,$$

and

$$\mathrm{Proof}(\pi(x), \ulcorner \mathrm{Proof}(x, \ulcorner \phi \urcorner) \to \mathrm{Prf}(x, \ulcorner \phi \urcorner) \urcorner) \ .$$

Clearly now $\mathrm{Proof}(\pi(x) \otimes e(x), \ulcorner \mathrm{Prf}(x, \ulcorner \phi \urcorner) \urcorner)$, and thus

$$\mathrm{Prf}(\pi(x) \otimes e(x), \ulcorner \mathrm{Prf}(x, \ulcorner \phi \urcorner) \urcorner) \ .$$

Case 2: $T(x)$ and $T(!x)$. In this case

$$\tilde{e}(x) = \ !x \ .$$

And by (6) we have for some $t{:}A \in \Gamma$ that

$$x = \ulcorner t \urcorner \text{ and } \phi = A^\dagger \ .$$

But since $!t \in T(X)$ we also have that $!t{:}t{:}A \in \Gamma$, and thus $\mathrm{Prf}(\ulcorner !t \urcorner, \ulcorner (t{:}A)^\dagger \urcorner)$ is true. That is,

$$\mathrm{Prf}(!x, \ulcorner \mathrm{Prf}(x, \ulcorner \phi \urcorner) \urcorner)$$

is true.

Case 3: $T(x)$ but $\neg T(!x)$. In this case

$$\tilde{e}(x) = \mathbf{g} \ .$$

Again we have for some $t{:}A \in \Gamma$ that

$$x = \ulcorner t \urcorner \text{ and } \phi = A^\dagger \ .$$

So $(t{:}A)^\dagger = \mathrm{Prf}(x, \ulcorner \phi \urcorner)$, so by choice of $\mathbf{g}$ we have $\mathrm{Proof}(\mathbf{g}, \mathrm{Prf}(x, \ulcorner \phi \urcorner))$. □

Proof (of Theorem 3.4). Lemmata 3.7 and 3.9 show that Prf is a proof predicate and Lemma 3.13 shows that it is normal. Finally Lemma 3.8 shows that, if S_2^1 is consistent, $S_2^1 \nvdash A^*$.   □

Corollary 3.14. $\text{LP}_{\mathcal{CS}} \vdash A$ *iff for any arithmetical interpretation $*$ that meets $\mathcal{CS}$ we have $S_2^1 \vdash A^*$*

Proof. Again, soundness is an easy consequence of the definitions. Suppose now that $\text{LP}_{\mathcal{CS}} \nvdash A^*$. Then $\text{LP}_\emptyset \nvdash \bigwedge \mathcal{CS} \to A$. As we have seen above this gives us an arithmetical interpretation $*$ such that $S_2^1 \vdash (\bigwedge \mathcal{CS})^* \wedge \neg A^*$. Clearly $*$ meets $\mathcal{CS}$.   □

References

1. Artemov, S.N.: Explicit provability and constructive semantics. Bulletin of Symbolic Logic **7**(1) (2001) 1–36
2. Artemov, S., Nogina, E.: On epistemic logic with justification. In van der Meyden, R., ed.: Theoretical Aspects of Rationality and Knowledge Proceedings of the Tenth Conference (TARK 2005), June 10-12, 2005, Singapore., National University of Singapore (2005) 279–294
3. Artemov, S.: Evidence-Based Common Knowledge. Technical Report TR-2004018, CUNY Ph.D. Program in Computer Science (2004)
4. Artemov, S., Kuznets, R.: Explicit knowledge is not logically omniscient. Unpublished (2005)
5. Artemov, S.N., Beklemishev, L.D.: Provability logic. In Gabbay, D., Guenthner, F., eds.: Handbook of Philosophical Logic. Volume 13. 2nd edn. Kluwer (2004) 229–403
6. Buss, S.R.: Bounded arithmetic. Bibliopolis, Napels (1986) Revision of PhD. thesis.
7. Krajíček, J.: Bounded arithmetic, propositional logic, and complexity theory. Cambridge University Press, New York, NY, USA (1995)
8. Buss, S.R.: First-Order Theory of Arithmetic. In Buss, S.R., ed.: Handbook of Proof Theory. Studies in Logic and the Foundations of Mathematics, Vol.137. Elsevier, Amsterdam (1998) 475–546
9. Parikh, R.: Existence and feasability in arithmetic. Journal of Symbolic Logic **36** (1971) 494–508

On a Maximal NFA Without Mergible States

Igor Grunsky[1], Oleksiy Kurganskyy[1], and Igor Potapov[2]

[1] Institute of Applied Mathematics and Mechanics,
Ukrainian National Academy of Sciences,
74 R. Luxemburg St, Donetsk, Ukraine
grunsky@iamm.ac.donetsk.ua, kurgansk@gmx.de
[2] Department of Computer Science,
University of Liverpool, Chadwick Building,
Peach St, Liverpool L69 7ZF, UK
igor@csc.liv.ac.uk

Abstract. In this paper we answer an open question about the exact bound on the maximal number of non-mergible states in nondeterministic finite automaton (NFA). It is shown that the maximal possible number of non-mergible states in a NFA that accepts a given regular language L is not greater than $2^n - 1$, where n is the number of states in the minimal deterministic finite automaton that accepts L. Next we show that the bound is reachable by constructing a NFA that have exactly $2^n - 1$ non-mergible states. As a generalization of this result we show that the number of states in a NFA that does not contain a subset of k mergible states, where $k > 1$, is bounded by $(k - 1)(2^n - 1)$ and the bound is reachable.

1 Introduction

In this paper we answer an open question about the exact bound on the maximal number of non-mergible states in nondeterministic finite automaton (NFA). The notion of mergibility is closely related to the problem of finding small representations of regular languages, for which there is both practical and theoretical motivation.

It is well known that in some cases we can reduce the size of a finite automaton by merging its states without changing its accepting power. The question about reducing the number of states and transition of finite automata is well studied and has efficient solution for deterministic finite automata (DFA), but it is more complex for non-deterministic finite automata (NFA).

In many cases problems for non-deterministic automata are highly non-trivial and usually cannot be solved using classical methods for deterministic automata [6, 7]. For example the problem of identification of mergible states is quite difficult for NFA since the order of merging could lead to different non-mergible representations of the same regular language.

In this paper we work on the problem of mergible states[1] existence that can be formulated as follows:

---

[1] I.e. the states that we can merge without changing the accepted language of NFA.

D. Grigoriev, J. Harrison, and E.A. Hirsch (Eds.): CSR 2006, LNCS 3967, pp. 202–210, 2006.

Problem 1. Given a regular language L. Is there a constant E_L such that any NFA that accepts L with a number of states greater than E_L has mergible states?

According to [2] one can distinguish two main ways of merging states: (1) a *weak* method (the weak mergibility), where two states are merged by simply collapsing one into the other and consolidating all their input and output transitions, and (2) a *strong* method (the strong mergibility), where one state is merged into another one by redirecting its input transitions toward the other state and completely deleting all its output transitions.

Very recently the upper bound on non-mergible states in case of the weak mergibility was found in [1]. Then it was shown in [2] that in the case of the strong mergibility the upper bound does not exist. The proof of the upper bound for the weak mergibility provided in [1] gives an effectively computable constant which, however, is very large and involves some imbricated Stirling numbers. In this paper we provide a solution for the Problem 1 (in a context of the week mergibility) by proving the lower and upper bounds that match exactly. We show that the maximal possible number of non-mergible states in a NFA that accepts a given language L is not greater than $2^n - 1$, where n is the number of states in the minimal deterministic finite automaton that accepts L. Next we show that the bound is reachable by construction of a NFA that have exactly $2^n - 1$ non-mergible states. As a generalization of this result we show that the number of states in a NFA that does not contain a subset of k mergible states, where $k > 1$, is bounded by $(k-1)(2^n - 1)$ and the bound is reachable.

The proof is based on our recent technique that was used in [4,5] for characterization of the languages representable by the undirected graphs. In this paper, without loss of generality, we use edge labeled graphs instead of NFAs. In particular we represent a language L as a set of all paths in an edge-labeled graph that lead us to the final vertex instead of saying that L is accepted by a NFA.

On the one hand, this approach helps to study the properties of mergible and non-mergible states of a NFA in terms of edge-label graphs. On the other hand we believe that presented proofs, in the current form, are more readable and can be adapted for further research on graph representable languages [5]. The paper is organized as follows. In the first part of the paper we introduce definitions and prove several useful properties about equivalent and quasi-equivalent vertices. Then in the next section we prove our main result. The paper ends with some conclusions.

2   Preliminaries

2.1   Basic Notations and Definitions

Let X be a finite alphabet and X^* be a free monoid on the set X. We denote the empty word by e. The length of a word $w = x_1 x_2 \ldots x_k$, where $x_i \in X$, $1 \le i \le k$, is denoted by $|w|$. The concatenation of two words $u \in X^*$, $v \in X^*$ we denote as uv. We call u as a prefix of the word uv. We denote by X_e the language $X \bigcup \{e\}$.

Definition 1. *Given languages $L, L' \subseteq X^*$. The basic language operations that we use are defined as follows:*

- *the concatenation of two languages: $LL' = \{wu | w \in L, u \in L'\}$*
- *the prefix contraction of a word w from the language L: $w^{-1}L = \{u | wu \in L\}$*
- *the prefix contraction of the language L' from the language L: $(L')^{-1}L = \cap_{w \in L'} w^{-1}L$.*

Note that $(L')^{-1}L$ we can define as follows:

$$(L')^{-1}L = \{u | \forall w \in L' : wu \in L\}.$$

Definition 2. *Let $G = (S, X, E, s_0, F)$ be a simple (i.e. without multiple edges), directed edge-labeled graph, where S is a set of vertices, X is a finite set of labels, $E \subset S \times X_e \times S$ is a set of directed labeled edges, $F \subset S$ is a set of final vertices. By sxt or (s, x, t) we denote an edge from the vertex s to the vertex t with the label x.*

Note that an edge in a graph G can be labeled with the empty word e. From now on by a *graph* we understand an edge-labeled graph. Any path in a graph is defined as a finite sequence

$$p = s_1 x_1 s_2 x_2 \ldots s_{k-1} x_{k-1} s_k$$

such that $s_i x_i s_{i+1} \in E$ for $1 \leq i < k$. The path p begins in the vertex s_1 and ends in s_k. The word $x_1 x_2 \ldots x_{k-1} \in X^*$ is a label of the path p that we denote as $\mu(p)$.

By **s** we denote the set of labels of all paths from the initial vertex to the vertex s. The set of labels of all paths from a vertex s to final vertices we denote by $\mathbf{s}^{-1}$. We also say that $\mathbf{s}^{-1}$ is a language generated by the vertex s and $L(G)$ is a language generated by the initial vertex of the graph G, i.e. $L(G) = \mathbf{s_0}^{-1}$.

Let A be a NFA that accepts a regular language L. We can think about a structure of A as a finite directed edge-labeled graph G_A. In accordance with our definitions we can state that $L = L(G_A)$. Moreover, it is obvious that we can uphold the above language equality by adding/deleting any states or transitions in A and adding/deleting corresponding vertices or labeled edges at the same time in G_A.

Definition 3. *Two vertices $s, t \in S$ are equivalent, if $\mathbf{s}^{-1} = \mathbf{t}^{-1}$. The relation between s and t we denote as ϵ.*

Definition 4. *Given a graph G that generates a language L. Two vertices s and t from the graph G are quasi-equivalent if $(\mathbf{s})^{-1}L = (\mathbf{t})^{-1}L$. The relation between s and t we denote as α.*

Definition 5. *A graph G is deterministic if every vertex of G does not have any two outgoing incident edges with the same label and does not have any outgoing edges lebeled by the empty word.*

Let us consider an equivalence relation $\rho \subseteq S \times S$, i.e. a reflexive, transitive and symmetric relation. By $\rho(s)$, where $s \in S$, we denote a set $\{t | (s, t) \in \rho\}$.

Definition 6. *The graph* $G/\rho = (G', X', E', s_0', F')$ *is a factor graph of a graph* G *if* $G' = \{\rho(s) | s \in S\}$, $\rho(s)x\rho(s') \in E'$ *for all edges* $sxs' \in E$, $s_0' = \rho(s_0)$ *and* $F' = \{\rho(s) | s \in F\}$.

By the merging of two vertices we understand a weak method, where two vertices are merged by simply collapsing one into the other and consolidate all their incomming and outgoing edges. We say that the graph G/ρ can be constructed from G by merging vertices of each set $\rho(s)$, $s \in S$.

Definition 7. *A subset R of vertices S in a graph G is mergible iff the language generated by the graph after merging of these vertices does not change, i.e.* $L(G) = L(G/\rho)$, *where* $(s, t) \in \rho$ *iff* $s = t$ *or* $s, t \in R$.

Now we define a simple graph that we call *Line*. Let $w = x_1 x_2 \ldots x_{k-1}$ be a word of a length $k-1$. By $Line(w)$ we denote a directed graph that is isomorphic to the graph with k vertices $s_1, ..., s_k$, and with the set of edges $\{s_i x_i s_{i+1} | 1 \le i < k\}$. We also call the vertex s_1 as initial and the vertex s_k as final. An example of a graph $Line(w)$, where $w = abcb$ is shown on the Figure 1.

Fig. 1. An example of $Line(w)$, where $w = abcb$

2.2 Intermediate Results

In this paper we consider languages in a given finite alphabet X, and graphs with the same alphabet X for labels. Without loss of generality we assume that all vertices in graphs are reachable from initial vertices and from any vertex at least one finite vertex is reachable in G.

Lemma 1. *Let G be a directed edge-labeled graph and s, t are two vertices in G, that $x^{-1}t^{-1} \supseteq s^{-1}$, and the edge (t, x, s) does not belong to the graph G. Then adding the new edge (t, x, s) into the graph G does not the language $L(G)$.*

Proof. Let us add one edge (t, x, s) to the graph G, that is obviously can only extend a language $L(G)$. Thus, it is enough to show that for any path p which goes via the edge (t, x, s) from the initial vertex to a final one we can find a path p' from the initial vertex to a final one without passing the edge (t, x, s) with the same label, i.e. $\mu(p) = \mu(p')$.

Let us consider a path $p = s_1 x_1 s_2 x_2 \ldots s_{k-1} x_{k-1} s_k$ from the initial vertex to a final one such that passes the edge (t, x, s) in the extended graph G. Let i be the maximum number, such that $(t, x, s) = (s_i, x_i, s_{i+1})$. According to the initial condition $x^{-1}t^{-1} \supseteq s^{-1}$ there is a path $s_i y_1 t_1 y_2 t_2 \ldots t_{l-1} y_l t_l$ from s_i to a final vertex t_l with the label $y_1 y_2 \ldots y_l$ such that $y_1 y_2 \ldots y_l = x_i x_{i+1} x_{i+2} \ldots x_{k-1}$, where $x_i = x$, but which does not pass the edge (t, x, s). Thus the number of (t, x, s)-edges in the path $p' = s_1 x_1 s_2 \ldots s_i y_1 t_1 y_2 t_2 \ldots t_{l-1} y_l t_l$ is less than in p by one. From it follows that, by repeating the above reduction a finite number of times, for any path p in G from the initial vertex to a final one we can construct a path p' in G from the initial vertex to a final one, that does not contain the edge (t, x, s), and $\mu(p) = \mu(p')$.

Lemma 2. *For any (finite or infinite) graph G the identity $L(G) = L(G/\epsilon)$ holds.*

Proof. Let G be an initial edge-labeled graph and let s and t be equivalent vertices in G. Let us assume that for some vertex s' the graph G contains the edge (s, x, s') and there is no edge (t, x, s') in the graph G. In this case we meet initial condition of Lemma 1, so the language represented by the graph G would not be changed by extending G with the edge (t, x, s'). Moreover, a language generated by any vertex in the extended graph would not be changed as well.

Let us extend the graph G in the following way: for any equivalent vertices s and t and some vertex s' we add the edges (t, x, s') iff the edge (s, x, s') belongs to the graph.

From it follows that the sequence $\epsilon(s_1) x_1 \epsilon(s_2) x_2 \ldots x_{k-1} \epsilon(s_k)$ is a path from the initial vertex to a final one in the factor graph G/ϵ iff there exist a path $r_1 x_1 r_2 x_2 \ldots x_{k-1} r_k$ in the graph G, where $r_i \in \epsilon(s_i)$, $1 \leq i \leq k$. Above facts show that for any graph G the equality $L(G) = L(G/\epsilon)$ holds.

Lemma 3. *Given a graph G and a language L, then $L(G) \subseteq L$ iff $\mathbf{s}^{-1} \subseteq (\mathbf{s})^{-1} L$ for any vertex s in the graph G.*

Proof. The above lemma follows from the fact that a set of all paths via a vertex s to a final vertex can be represented by $\mathbf{ss}^{-1}$. Since $\mathbf{ss}^{-1} \subseteq L$ holds it is equivalent to $\mathbf{s}^{-1} \subseteq (\mathbf{s})^{-1} L$. The proof of the reverse part is trivial since from the statement $\mathbf{s}^{-1} \subseteq (\mathbf{s})^{-1} L$, where s is the initial vertex, follows that $L(G) \subseteq L$.

The next lemma gives us the common property for equivalent and quasi-equivalent vertices.

Lemma 4. *For any graph G the identity $L(G) = L(G/\alpha)$ holds.*

Proof. Let G be an initial edge-labeled graph. Note that the operation of merging any two vertices in a edge-labeled graph can only extend the language that it represents. Let us now assume that $W = (\mathbf{s})^{-1} L(G) = (\mathbf{t})^{-1} L(G)$, i.e. s and t are quasi-equivalent. From Lemma 3 follows that $\mathbf{s}^{-1} \subseteq W$ and $\mathbf{t}^{-1} \subseteq W$.

Let us construct an extension of a graph G by adding new vertices and edges in such a way that $\mathbf{s}^{-1} = \mathbf{t}^{-1} = W$ holds in the extended graph. In particular we can do it as follows. Let us add by direct sum all graphs $Line(w)$, $w \in W$, to the graph G and merge the vertex s with the initial vertices of $Line(w)$, $w \in W$. Next we can repeat the same construction with the vertex t. According to the Lemma 1 the language represented by a graph G after its extension has not been changed.

Since the vertices s and t are now equivalent in the extended graph G we can merge s and t without changing the language represented by this graph. Now we can delete all introduced edges and nodes to get a graph that can be constructed by just merging quasi-equivalent vertices s and t.

Corollary 1. *Quasi-equivalent vertices are mergible.*

Definition 8. *A graph G is saturated if for any vertex s the equality $(s)^{-1}L = s^{-1}$ holds, i.e. any quasi-equivalent vertices in the graph are equivalent.*

Lemma 5. *Two vertices in a saturated graph are mergible iff they are equivalent.*

Proof. It is obvious that equivalent vertices are mergible.

Assume s and t are not equivalent in a saturated graph G. Then there are two words u and w, such that either $u \in s$, $u \notin t$, $w \notin s^{-1}$, $w \in t^{-1}$ or $u \notin s$, $u \in t$, $w \in s^{-1}$, $w \notin t^{-1}$ holds. In this case the word uw does not belong to the language $L(G)$, but could belong to the language $L(G)$ if we merge the vertices s and t, i.e. s and t are not mergible.

3   Main Result

Theorem 1. *Given a regular language L and a graph G such that it does not contain any mergible vertices and $L(G) = L$. The number of vertices in the graph G is bounded by $2^n - 1$, where $n = |\{w^{-1}L | w \in L\}|$ and the bound is reachable.*

Proof. Let us consider a graph G satisfying to the above statement. Since quasi-equivalent vertices are mergible then the maximum number of non-mergible vertices is bounded by a maximal number of vertices that are not pairwise quasi-equivalent. Since a set $(W)^{-1}L$, where $W \subseteq L$, is defined as $\cap_{w \in W} w^{-1}L$ the number of different quasi-equivalent classes in G cannot exceed the number of different non-empty subsets of the set $\{w^{-1}L | w \in L\}$, which is $2^n - 1$.

Now we prove that $2^n - 1$ is a reachable bound. The Figure 2 shows a deterministic graph H with the initial vertex n and all final vertices. The vertex k, $1 \leq k \leq n$, generates the language that contains only all possible initial prefixes of the words from the language defined by the regular expression $(a_1 a_2 \ldots a_n)^*$, where $a_i \in \{0, 1\}$ for $i \in \{1, 2, \ldots, n\} - \{k\}$ and $a_k = 0$. Hence the graph H does not have any equivalent vertices.

Let us construct a saturated graph G with $2^n - 1$ vertices such that $L(H) = L(G)$. First we define a set of vertices $S = \{s | s \subseteq \{1, 2, \ldots, n\}, s \neq \emptyset\}$. The vertex $\{n\}$ is initial, and all vertices are final. For any $s \in S$, $t \in S$ the edge $(s, 0, t)$ belongs to G iff for all $x \in s - \{1\}$ we have that $x - 1 \in t$ and if $1 \in s$ then $n \in t$. For any $s \in S$, $t \in S$ the edge $(s, 1, t)$ belongs to G iff $1 \notin s$ and for all $x \in s$ we have that $x - 1 \in t$. It follows from the construction that for each vertex $s \in S$ the equality $s^{-1} = \cap_{i \in s} i^{-1}$ holds, where i^{-1} is a language generated by a vertex i from the graph H. Hence the graph G is saturated and for any vertex $s \in S$ the language s^{-1} consists of all possible prefixes of the language defined by the regular expression $(a_1 a_2 \ldots a_n)^*$, where $a_i = (0 \vee 1)$ for $i \in \{1, 2, \ldots, n\} - s$ and $a_i = 0$ for $i \in s$. From it follows that for different vertices s we have different

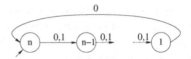

Fig. 2. Deterministic graph with the initial vertex n and all final vertices

languages s^{-1} and according to the Lemma 5 there are no mergible vertices in G. This ends our proof.

Corollary 2. *Let $k > 1$ be an integer, L be a regular language, and G be a finite graph that does not contain a subset of k mergible vertices and $L(G) = L$. The number of vertices in the graph G is bounded by $(k-1)(2^n - 1)$, where $n = |\{w^{-1}L|w \in L\}|$ and the bound is reachable.*

Proof. Let L be a regular language, G be a finite graph with a number of vertices greater than $(k-1)(2^n-1)$ and such that $L(G) = L$, where $n = |\{w^{-1}L|w \in L\}|$. The number of different quasi-equivalent classes in G is not greater than $2^n - 1$. According to the pigeonhole principle there is a subset of k quasi-equivalent and therefore mergible vertices in the graph G.

Consider the graph G and the language L from the proof of the Theorem 1. We have that $n = |\{w^{-1}L|w \in L\}|$ and G consists of $(2^n - 1)$ non-mergible vertices. Let $G_1, G_2, \ldots, G_{k-1}$ be $k - 1$ copies of G, and s_{0i} the inital vertex of G_i, where $1 < i < k$. Let H be a graph constructed by direct sum of the graphs G_i, $1 < i < k$, and such that H has the initial vertex s_{01} and for all vertices s, t in H, such that $x^{-1}t^{-1} \supseteq s^{-1}$ and the edge (t, x, s) does not belong to the graph H we construct the edge (t, x, s). does not change the langauge $L(H)$. By applying Lemma 1 we state that the above construction does not change the langauge $L(H)$. So it is obvious that $L(G) = L(H)$, all final vertices in H are reachable and it does not contain a subset of k quasi-equivalent vertices.

Corollary 3. *Let $k > 1$ be an integer, L be a regular language, and A be a NFA that does not contain a subset of k mergible states and accept L. The maximal number of states in A is bounded by $(k-1)(2^n - 1)$, where n is the number of states in the minimal deterministic finite automaton that accepts L.*

Proof. The proof is straightforward and follows from the Corollary 2. By a NFA A we understand the graph G from Corollary 2. In addition to it we use the fact that a number of states in the minimal deterministic automaton that accepts L is equal to the cardinality of a set $\{w^{-1}L|w \in L\}$.

4   Conclusion

In this paper we answered a question about the exact bound on non-mergible states in a nondeterministic finite automaton. It was shown that the tight bound on the maximal possible number of non-mergible states in NFA that accepts a given language L is $2^n - 1$, where n is the number of states in the minimaldeterministic finite automaton that accepts L. It is easy to see that the same bound holds for a case of ε-free NFA since the upper bound proofs cover a case of ε-free NFA and the low bound constructions do not contain any ε-transitions.

Another interesting aspect of this paper is the use of quasi-equivalence relation that have a number of important properties not only for finite but also for infinite graphs or automata.

Corollary 4. *Let L be a regular language and G be an infinite graph such that $L = L(G)$. The factor graph G/α is finite.*

Let us show using simple example that the above statement does not hold if we substitute the relation α by ϵ. The infinite graph G on Figure 3 have one initial vertex that is the vertex without incoming edges and one final vertex that is the vertex without outgoing edges.

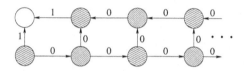

Fig. 3. A infinite graph G which is isomorphic to G/ϵ

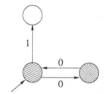

Fig. 4. The graph G/α

It is easy to see that the language generated by G corresponds to the expression $(00)^*1$. The above graph does not have any equivalent vertices since each vertex of G generates a different language. On the other hand, the vertices colored with the same pattern are quasi-equivalent and mergible, because of the fact that $(0^{2i})^{-1}L(G) = L(G)$ and $(00^{2i})^{-1}L(G) = 0^{-1}L(G)$, for any $i \geq 0$.

The factor graph G/α is shown on Figure 4.

The question about applicability of quasi-equivalence relation for the minimization of NFA that accepts a given regular language could be promising next step along the research initiated in [8].

References

1. C. Campeanu, N. Santean, S. Yu. Mergible states in large NFA. Theoretical Computer Science, 330, 23-34, (2005)
2. C. Campeanu, N. Santean, S. Yu. Large NFA without mergible states. Proceedings of 7th International Workshop on Descriptional Complexity of Formal Systems, DCFS 2005, 75-84 (2005)

3. S. Eilenberg. Automata, Languages, and Machines. Academic Press, Inc. Orlando, FL, USA (1974)
4. I.S. Grunsky, O. Kurganskyy. Properties of languages that generated by undirected vertex-labeled graphs. In proceeding of 8-th International Seminar on Discrete mathematics and its applications, Moscow State University, 267-269 (2004) [in Russian].
5. I.S. Grunsky, O. Kurganskyy, I. Potapov. Languages Representable by Vertex-labeled Graphs. MFCS, LNCS 3618, 435-446 (2005)
6. J. Hromkovic, S. Seibert, T. Wilke. Translating Regular Expressions into Small ε-Free Nondeterministic Finite Automata. J. Comput. Syst. Sci. 62(4): 565-588 (2001)
7. J. Hromkovic, G. Schnitger. NFAs With and Without ε-Transitions. ICALP 2005, 385-396 (2005)
8. L. Ilie, R. Solis-Oba, S. Yu. Reducing the Size of NFAs by Using Equivalences and Preorders. CPM 2005: 310-321 (2005)

Expressiveness of Metric Modalities for Continuous Time

Yoram Hirshfeld and Alexander Rabinovich

Sackler Faculty of Exact Sciences,
Tel Aviv Univ., Israel 69978
{joram, rabinoa}@post.tau.ac.il

Abstract. We prove a conjecture by A. Pnueli and strengthen it showing a sequence of "counting modalities" none of which is expressible in the temporal logic generated by the previous modalities, over the real line, or over the positive reals. We use this sequence to prove that over the real line there is no finite temporal logic that can express all the natural properties that arise when dealing with systems that evolve in continuous time.

1 Introduction

Temporal Logic based on the two modalities "Since" and "Until" (*TL*) is a most popular framework for reasoning about the evolving of a system in time. By Kamp's theorem [12] this logic has the same expressive power as the monadic first order predicate logic. Therefore the choice between monadic logic and temporal logic is merely a matter of personal preference.

Temporal logic and the monadic logic are equivalent whether the system evolves in discrete steps or in continuous time, But for continuous time both logics are not strong enough to express properties like: "X will happen within 1 unit of time", and we need a better version of the logics.

Following the work R. Koymans, T. Henzinger and others, [14, 3, 2, 13, 5, 17, 1, 7], and more, we introduced in [9, 11] the logic QTL (Quantitative Temporal Logic), which has besides the modalities $Until$ and $Since$ two metric modalities: $\Diamond_1(X)$ and $\overleftarrow{\Diamond}_1(X)$. The first one says that X will happen (at least once) within the next unit of time, and the second says that X happened within the last unit of time. We proved:

1. This logic consumes the different metric temporal logics that we found in the literature, like $MITL$ [2, 1, 7].
2. The validity and satisfiability problem for this logic is decidable, whether we are interested in systems with *finite variability*, or in all systems evolving in time (a system has finite variability if it changes only at finitely many points, in any finite interval of time).

An important question was not answered: is this logic expressive enough to express all the important properties about evolving systems? If not, which modalities should we add?

D. Grigoriev, J. Harrison, and E.A. Hirsch (Eds.): CSR 2006, LNCS 3967, pp. 211–220, 2006.

A. Pnueli suggested the modality $P_2(X, Y)$: "X and then Y will both occur in the next unit of time". $P_2(X, Y)$ was probably thought of as a natural strengthening of the metric logics that were presented before. It can serve as a first in a sequence of extensions of the logic, where for each natural number n, we add the modality $P_n(X_1,, \ldots, X_n)$. $P_n(X_1,, \ldots, X_n)$ says that there is an increasing sequence of points $t_1,, \ldots, t_n$ in the coming unit interval such that $X_i(t_i)$ holds for $i = 1,, \ldots, n$. It probably seemed pointless to define new modalities when you cannot prove that they can express something new. Pnueli conjectured that the modality $P_2(X, Y)$ cannot be expressed in the different (equivalent) metric logic that we defined above, but he left it at that (we were unable to locate where this conjecture was first published. It is attributed to Pnueli in later papers like [2] and [17]).

Here we prove Pnueli's conjecture: We denote by $C_2(X)$ the modality "X will be true at least twice in the next unit of time". This is a special case of $P_2(X, Y)$ where $Y = X$. We prove:

- $C_2(X)$ cannot be expressed in QTL (and the equivalent languages). Moreover:
- For every n let us define the modality $C_n(X)$ that says that X will hold at least at n points of the next unit interval. Then the modality $C_{n+1}(X)$ cannot be expressed in the logic $QTL(C_1,, \ldots,, C_n)$, which is generated by QTL, and the extra n modalities $QTL(C_1(X),, \ldots,, C_n(X))$

Therefore there is a proper hierarchy of temporal logics, and it is important to investigate how to extend the logic QTL to a full strength, yet decidable logic. Counting modalities like $C_n(X)$ are not a natural choice of a modality and it maybe suspected that a better chosen finite set of modalities together with QTL is as strong as, or even stronger than QTL with all the modalities C_n. Not so! We were able to prove:

- No finite temporal logic can express all the statements $C_n(X)$.

The last claim needs to be made exact: No finite temporal logic, whose modalities are defined in a natural monadic predicate logic, can express all the counting modalities over continuous time, extended in both directions; i.e, over the full real line. We believe that the same is true also when we consider continuous time with a first point 0, i.e, positive time line, but the proof will be more difficult.

When stated formally the result seems even stronger: Let L be *second order* monadic logic of order, together with the predicate $B(t, s)$ which says that $s = t + 1$. The modalities $C_n(X)$ are expressible in this logic, but no temporal logic with a finite *or infinite* family of modalities which are defined by formulas with bounded quantifier depth can express all the modalities $C_n(X)$.

In predicate logic the expressive power grows with the increasing of the quantifier depth. In temporal logic this is achieved by increasing the nesting depth of the modalities. Kamp showed that for the simplest logic of order iterating the modal operations can replace the complex use of quantifier. Our result, together with previous evidence (see [15]) suggests that this was a lucky peculiarity of the first-order monadic logic of linear order, and it cannot be expected to hold for strong logics.

These results leave open and emphasize even more the main question: Is the logic QTL enriched by all the modalities $P_n(X_1, , \ldots, X_n)$ the appropriate maximal decidable temporal logic? If not, what is its strength, and what is the appropriate continuous metric temporal logic?

The paper is divided as follows: In section 2 we recall the definitions and the previous results concerning the continuous time logics. In section 3 we prove Pnueli's conjecture and its generalization, that the modalities C_i create a strictly increasing family of logics. In section 4 we discuss the more general and abstract result: that no temporal logic based on modalities with finite quantifier depth can express all the modalities C_n.

2 Monadic Logic and Quantitative Temporal Logic

2.1 MLO - Monadic Logic of Order

The natural way to discuss systems that evolve in time is classical predicate logic. The language has a name for the order relation of the time line, and a supply of unary predicate names to denote a properties that the system may or may not have at any point in time. Hence:

The syntax of the monadic predicate logic of order - MLO has in its vocabulary *individual* (first order) variables $t_0, t_1, \ldots$, monadic *predicate* variables $X_0, X_1, \ldots$, and one binary relation $<$ (the order). **Atomic formulas** are of the form $X(t)$, $t_1 = t_2$ and $t_1 < t_2$. **Well formed formulas** of the monadic logic MLO are obtained from atomic formulas using Boolean connectives $\neg, \vee, \wedge, \rightarrow$ and the (first order) quantifiers $\exists t$ and $\forall t$ and the (second-order) quantifiers $\exists X$ and $\forall X$. The formulas which do not use $\exists X$ and $\forall X$ are called first-order MLO formulas ($FOMLO$). Note that $FOMLO$ formulas may contain free monadic predicate variables, and they will be assigned to particular predicates in a structure.

A structure for MLO is a tuple $M = \langle A, <, P_1, \ldots, P_n \rangle$, where A is a set linearly ordered by the relation $<$, and $P_1, \cdots, P_n$, are one-place predicates (sets) that correspond to the predicate names in the logic. We shall use the simple notation $\langle A, < \rangle$ when the particular predicates are not essential to the discussion.

The main models are: the **continuous canonical model** $\langle R^+, < \rangle$, the non-negative real line, and the **discrete canonical model** $\langle N, < \rangle$, the naturals.

As is common we will use the assigned formal names to refer to objects in the meta discussion. Thus we will write:

$$M \models \varphi[t_1, \ldots, t_k; X_1, \ldots, X_m]$$

where M is a structure, φ a formula, $t_1, \cdots, t_k$ elements of M and $X_1, \ldots, X_m$ predicates in M, instead of the correct but tedious form:

$$M, \tau_1, \ldots, \tau_k; P_1, \ldots, P_m \models_{MLO} \varphi(t_1, \ldots, t_k; X_1, \ldots, X_m),$$

where $\tau_1, \ldots, \tau_k$ and $P_1 \cdots, P_m$ are names in the metalanguage for elements and predicates in M.

2.2 Temporal Logics

Temporal logics evolved in philosophical logic and were enthusiastically embraced by a large body of computer scientists. It uses logical constructs called "modalities" to create a language that is free from variables and quantifiers. Here is the general logical framework to define temporal logics:

The syntax of the Temporal Logic $TL(O_1^{(k_1)}, \ldots, O_n^{(k_n)}, \ldots)$ has in its vocabulary *monadic predicate names* $P_1, P_2, \ldots$ and a sequence of *modality names* with prescribed arity, $O_1^{(k_1)}, \ldots, O_n^{(k_n)}, \ldots$ (the arity notation is usually omitted). The formulas of this temporal logic are given by the grammar:

$$\varphi ::= True \mid P \mid \neg\varphi \mid \varphi \wedge \varphi \mid O^{(k)}(\varphi_1, \cdots, \varphi_k)$$

A temporal logic with a finite set of modalities is called a finite (base) temporal logic.

Structures for TL are again linear orders with monadic predicates $M = \langle A, <, P_1, P_2, \ldots, P_n \rangle$, where the predicate P_i are those which are mentioned in the formulas of the logic. Every modality $O^{(k)}$ is interpreted in every structure M as an operator $O_M^{(k)} : [\mathbb{P}(A)]^k \to \mathbb{P}(A)$ which assigns "the set of points where $O^{(k)}[S_1, \ldots, S_k]$ holds" to the k-tuple $\langle S_1, \ldots, S_k \rangle \in \mathbb{P}(A)^k$. (Here $\mathbb{P}$ is the power set notation, and $\mathbb{P}(A)$ denotes the set of all subsets of A.) Once every modality corresponds to an operator the semantics is defined by structural induction:

- for atomic formulas: $\langle M, t \rangle \models_{TL} P$ iff $t \in P$.
- for Boolean combinations the definition is the usual one.
- for $O^{(k)}(\varphi_1, \cdots, \varphi_k)$

$$\langle M, t \rangle \models_{TL} O^{(k)}(\varphi_1, \cdots, \varphi_k) \quad \text{iff} \quad t \in O_M^{(k)}(A_{\varphi_1}, \cdots, A_{\varphi_k})$$

where $A_\varphi = \{ \tau : \langle M, \tau \rangle \models_{TL} \varphi \}$ (we suppressed predicate parameters that may occur in the formulas).

We are interested in a more restricted case; for the modality to be of interest the operator $O^{(k)}$ should reflect some intended connection between the sets A_{φ_i} of points satisfying φ_i and the set of points $O[A_{\varphi_1}, \ldots, A_{\varphi_k}]$. The intended meaning is usually given by a formula in an appropriate predicate logic:

Truth Tables: A formula $\overline{O}(t_0, X_1, \ldots X_k)$ in the predicate logic L is a *Truth Table* for the modality $O^{(k)}$ if for every structure M

$$O_M(A_1, \ldots, A_k) = \{ \tau : M \models_{MLO} \overline{O}[\tau, A_1, \ldots, A_k] \} .$$

The modalities *until* and *since* are most commonly used in temporal logic for computer science. They are defined through the following truth tables:

- The modality X **U** Y, "*X until Y*", is defined by

$$\psi(t_0, X, Y) \equiv \exists t_1 (t_0 < t_1 \wedge Y(t_1) \wedge \forall t(t_0 < t < t_1 \rightarrow X(t))).$$

- The modality X **S** Y, "*X since Y*", is defined by

$$\psi(t_0, X, Y) \equiv \exists t_1 (t_0 > t_1 \wedge Y(t_1) \wedge \forall t(t_1 < t < t_0 \rightarrow X(t))).$$

If the modalities of a temporal logic have truth tables in a predicate logic then the temporal logic is equivalent to a fragment of the predicate logic. Formally:

Proposition 1. *If every modality in the temporal logic TL has a truth table in the logic MLO then to every formula $\varphi(X_1, \ldots, X_n)$ of TL there corresponds effectively (and naturally) a formula $\overline{\varphi}(t_0, X_1, \ldots X_n)$ of MLO such that for every M, $\tau \in M$ and predicates $P_1, \ldots, P_n$*

$$\langle M, \tau, P_1, \ldots, P_n \rangle \models_{TL} \varphi \quad \text{iff} \quad \langle M, \tau, P_1, \ldots, P_n \rangle \models_{MLO} \overline{\varphi} .$$

In particular the temporal logic $TL($ **U** , **S** $)$ with the modalities "until" and "since" corresponds to a fragment of first-order MLO ($FOMLO$).

The two modalities **U** and **S** are also enough to express all the formulas of first-order MLO with one free variable:

Theorem 2. *([12, 6]) The temporal logic $TL($ **U** , **S** $)$ is expressively complete for FOMLO over the two canonical structures: For every formula of FOMLO with at most one free variable, there is a formula of $TL($ **U** , **S** $)$, such that the two formulas are equivalent to each other, over the positive integers (discrete time) and over the positive real line (continuous time).*

2.3   QTL - Quantitative Temporal Logic

The logics MLO and $TL($ **U** , **S** $)$ are not suitable to deal with quantitative statements like "*X will occur within one unit of time*". In [8, 9, 10] we introduced the *Quantitative Temporal Logic*, adding to TL the modalities $\Diamond_1 X$ (X will happen within the next unit of time) and $\overleftarrow{\Diamond}_1 X$ (X happened within the last unit of time):

Definition 3 (Quantitative Temporal Logic). *QTL, quantitative temporal logic is the logic $TL($ **U** , **S** $)$ enhanced by the two modalities: $\Diamond_1 X$ and $\overleftarrow{\Diamond}_1 X$. These modalities are defined by the tables with free variable t_0:*

(3) $\qquad\qquad \Diamond_1 X : \qquad \exists t((t_0 < t < t_0 + 1) \wedge X(t))$

(4) $\qquad\qquad \overleftarrow{\Diamond}_1 X : \qquad \exists t((t < t_0 < t + 1) \wedge X(t)) .$

QTL was the latest in a list of metric logics for continuous time, developed over approximately 15 years. When interpreted carefully all these logics are equivalent. For completeness we list the two main modalities that were suggested before QTL together with their natural truth table:

1. The logic $MITL$ [2] has as modalities $X\ until_{(m,n)}Y$ with natural numbers $m < n$, which holds at t_0 iff

$$\exists t_1[(t_0 + m < t_1 < t_0 + n) \wedge Y(t_1) \wedge \forall t(t_0 < t < t_1 \rightarrow X(t))].$$

Other modalities with closed and half closed intervals as indices, and dual modalities with "since" replacing "until" are defined similarly.

2. Manna and Pnueli [13] base their logic on modalities $[\Gamma(X) > n]$ which holds at t_0 iff

$$\forall t(t_0 - n < t < t_0 \rightarrow X(t)).$$

The dual modality for the future is defined similarly. To these they add modalities $[\Gamma(X) = n]$ saying that X started exactly n units of time ago.

We proved in [9] and [11] that:

1. The logic QTL can express more liberal bounds in time like: "X will happen in the future, within the period that starts in m units of time, and ends in n units of time" $(m < n)$. We may also include or exclude one of both of the endpoints of the period.

2. QTL consumes the different decidable metric temporal logics that we found in the literature, including $MITL$ and the Manna-Pnueli logic described above.

3. There is a natural fragment $QMLO$ (quantitative monadic logic of order), of the classical monadic logic of order with the $+1$ function, that equals in expressive power to QTL.

4. The *validity and satisfiability problem for this logic is decidable*, whether we are interested in systems with *finite variability*, or in all systems evolving in time (a system has finite variability if it changes only at finitely many points, in any finite interval of time).

The advantages of the logic QTL were the subject of [8, 9, 10, 11]. In particular, it is decidable. Here we investigate the limitations of its expressive power.

3   Modalities Which Are Not Expressible in QTL

We start the investigation of the limitations of the temporal logic proving Pnueli's conjecture:

Theorem 4. *The modality $C_2(X)$ is not expressible in QTL.*

Proof. Let M be the real non negative line with the predicate $P(t)$ that is true exactly at the points $n \cdot \frac{2}{3}$ for all natural numbers n. Let us call the following four predicates: $P, \neg P, True, False$ the **trivial predicates**. We show by structural induction that for every statement φ of QTL there is a point t_φ such that from this point on φ is equivalent to one of the trivial predicates.

- This is trivially true for atomic statements.
- The collection of truth sets for the four trivial predicates is closed under Boolean combinations. Therefore the set of formulas satisfying our claim is closed under the Boolean connectors.
- Assume now that $\varphi = (\theta\ Untill\ \psi)$ and t_0 is a point beyond which both θ and ψ are equivalent to one of the trivial predicates. We check the different possibilities for the truth value of φ at a point t beyond t_0. If θ is equivalent to P or to $False$ then φ is false. If θ is equivalent to $\neg P$ or to $True$ then φ is true if ψ is equivalent to either of P, $\neg P$ or $True$, and φ is false if ψ is equivalent to $False$. In every case φ is equivalent either to $True$ or to $False$.
- For $\varphi = (\theta\ Since\ \psi)$ we need only a minor modification: Let t_1 be an even integer beyond t_0 (so that P is true at t_1). Then for points beyond t_1 φ is true if $\theta \equiv True$ and ψ occurred at t_1 or earlier, or if $\theta \equiv \neg P$ and ψ is equivalent to any of the special predicates except $False$ (the choice of t_1 ensures the case that $\psi \equiv P$) in all other cases $\varphi \equiv False$.
- Assume that $\varphi = \Diamond_1\theta$ and from t_0 on θ is equivalent to one of the four trivial predicates. If θ is equivalent to $False$ then φ is equivalent to $False$ from t_0 on. In the other three cases φ is equivalent to $True$ from t_0 on.
- A similar argument works when $\varphi = \overleftarrow{\Diamond}_1\theta$.

On the other hand the statement $C_2(P)$ is false at any point in the interval $(n, n + 1/3)$ if n is even and it is true at any point in the interval $(n, n + 1/3)$ if n is odd. This shows that $C_2(P)$ is not equivalent to any QTL formula.

The method of the proof can be modified to produce a hierarchy of temporal logics, each stronger than the previous.

Definition 5. *The* counting modalities *are the modalities* $C_n(X)$ *for every* n *which state that* X *will be true at least at* n *points within the next unit of time.*

Theorem 6. *The modality* $C_{n+1}(X)$ *is not expressible in* $QTL(C_1 \cdots, C_n)$.

Proof. Let M be the real non negative line with the predicate $P(t)$ that is true exactly at the points $k \cdot \frac{2}{n+1}$ for all natural numbers k. Call again the following four predicates: $P, \neg P, True, False$ the **trivial predicates**, and as before show that every formula of $QTL(C_1 \cdots, C_n)$ is equivalent from some point on to a trivial predicate. On the other hand $C_{n+1}(P)$ is always true on the interval $(k, k + \frac{1}{n+1})$ if k is even, and false on the interval if k is odd.

4   No Finite Temporal Logic Is Fully Expressive

The hierarchy

$$TL < QTL < QTL(C_2) < \cdots < QTL(C_1 \cdots, C_n) < \cdots$$

raises the suspicion that it will be difficult to find a finite temporal logic that includes all these logics. We showed that it is not difficult. It is impossible. To be precise:

Theorem 7. *Let L be the second order monadic logic of order, with an extra predicate $B(t, s)$ that is interpreted on the real line as $s = t + 1$. Let L_1 be a temporal logic with possibly infinitely many modalities, for which there is a natural number m such that all the modalities have truth tables in L, with quantifier depth not larger than m. Then there is some n such that $C_n(X)$ is not equivalent over the real line to any L_1 formula.*

The proof is quite technical, yet close in spirit to the proof of theorem 6: We define an infinite family of very uniform models, with P their only unary predicate. We define for each integer $k > 0$ the model M_k to be the full real line R with $P(t)$ occurring at the points $m\frac{1}{k}$ for every integer m (positive, negative or zero). We show that any pair of models in this class that can be distinguished by some formula in L_1, can also be distinguished by one of finitely many *simple formulas*. It follows that there is an infinite subfamily of models that satisfy the same formulas of L_1. On the other hand for large $n < k$ the model M_n satisfies $C_n(P)$ and the model M_k does not. Hence the formula $C_n(X)$ is not definable in L_1.

Discussion:

1. The theorem says both more and less than what the title of the section says. Less because we confined ourselves to temporal logics with truth tables in the second order monadic logic of order with the addition of the $+1$ function. Allowing more arithmetical operations would produce more modalities. Moreover, modalities need not have truth tables in any predicate logic, and the following is a natural question:

   Is there a finite temporal logic that includes all the modalities
   $P_n(X_1, , \ldots, X_n)$, if we do not require that the modalities are are defined
   by truth tables?

   On the other hand we prove more than is claimed because we prove that no *infinite temporal logic* can express all the counting modalities $C_n(X)$ if the truth tables of the modalities are of bounded quantifier depth.

2. Second order monadic logic of order with the $+1$ function is a much stronger logic than is usually considered when temporal logics are defined. All the temporal logics that we saw in the literature are defined in a fragment of monadic logic, with a very restricted use of the $+1$ function. All the decidable temporal logics in the literature remain decidable when we add the counting modalities $C_n(X)$ [10]. On the other hand second order monadic logic is undecidable over the reals even without the $+1$ function [16]. When the $+1$ function is added even a very restricted fragment of *first order* monadic logic of order is undecidable over the positive reals.

3. A natural way to strengthen the predicate logic is by adding predicates $B_q(t, s)$ for every rational umber q, to express the relation $s = t + q$. We call this logic L_Q. The proof of the theorem will not apply if we replace L by L_Q, and even modalities with truth tables of quantifier depth 2 distinguish

any two models M_k and M_r in our class. On the other hand just as before no *finite* temporal logic defined in this logic can express all the counting modalities. This leaves open the question:

Is the theorem above true when the predicate logic L is replaced by L_Q?

4. It is well known that to say in *predicate logic* "there are at least n elements with a given property" requires quantifier depth that increases with n. We emphasize again that the theorem is much more significant than that. Temporal logics do not have quantifiers, and the expressive power is achieved by deeper nesting of the modalities. Thus to say that P will not occur in the next n units of time requires formulas of predicate logic with quantifier depth that increases with n. On the other hand QTL itself suffices to claim it for any n (with increasing modality nesting), although all the modalities of QTL have very simple truth tables with quantifier depth at most 2. Therefore it is far from obvious that no finite temporal logic expresses all the modalities C_n using unlimited modality nesting.

References

1. R. Alur, T. Feder, T.A. Henzinger. The Benefits of Relaxing Punctuality. Journal of the ACM 43 (1996) 116-146.
2. R. Alur, T.A. Henzinger. Logics and Models of Real Time: a survey. In Real Time: Theory and Practice. Editors de Bakker et al. LNCS 600 (1992) 74-106.
3. Baringer H. Barringer, R. Kuiper, A. Pnueli. A really abstract concurrent model and its temporal logic. Proceedings of the 13th annual symposium on principles of programing languages (1986), 173-183.
4. H.D. Ebbinghaus, J. Flum, Finite Model Theory. Perspectives in mathematical logic, Springer (1991).
5. D.M. Gabbay, I. Hodkinson, M. Reynolds. Temporal Logics volume 1. Clarendon Press, Oxford (1994).
6. D.M. Gabbay, A. Pnueli, S. Shelah, J. Stavi. On the Temporal Analysis of Fairness. 7th ACM Symposium on Principles of Programming Languages. Las Vegas (1980) 163-173.
7. T.A. Henzinger It's about time: real-time logics reviewed. in Concur 98, Lecture Notes in Computer Science 1466, 1998.
8. Y. Hirshfeld and A. Rabinovich, A Framework for Decidable Metrical Logics. In Proc. 26th ICALP Colloquium, LNCS vol.1644, pp. 422-432, Springer Verlag, 1999.
9. Y. Hirshfeld and A. Rabinovich. Quantitative Temporal Logic. In Computer Science Logic 1999, LNCS vol. 1683, pp 172-187, Springer Verlag 1999.
10. Y. Hirshfeld and A. Rabinovich, Logics for Real Time: Decidability and Complexity. Fundam. Inform. 62(1): 1-28 (2004).
11. Y. Hirshfeld and A. Rabinovich, Timer formulas and decidable metric temporal logic. Information and Computation Vol 198(2), pp. 148-178, 2005.
12. H. Kamp. Tense Logic and the Theory of Linear Order. Ph.D. thesis, University of California L.A. (1968).
13. Z. Manna, A. Pnueli. Models for reactivity. Acta informatica 30 (1993) 609-678.

14. R. Koymans. Specifying Real-Time Properties with Metric Temporal Logic. Real-Time Systems 2(4):255-299,1990.
15. A. Rabinovich. Expressive Power of Temporal Logics In Proc. 13th Int. Conf. on Concurrency Theory, vol. 2421 of Lecture Notes in Computer Science, pages 57–75. Springer, 2002.
16. S. Shelah. The monadic theory of order. *Ann. of Math.*, **102**, pp 349-419, 1975.
17. T. Wilke. Specifying Time State Sequences in Powerful Decidable Logics and Time Automata. In Formal Techniques in Real Time and Fault Tolerance Systems. LNCS 863 (1994), 694-715.

Extending Dijkstra's Algorithm to Maximize the Shortest Path by Node-Wise Limited Arc Interdiction

Leonid Khachiyan[1,*], Vladimir Gurvich[2], and Jihui Zhao[1,*]

[1] Department of Computer Science, Rutgers University,
110 Frelinghuysen Road, Piscataway, New Jersey 08854, USA
{leonid, zhaojih}@cs.rutgers.edu
[2] RUTCOR, Rutgers University,
640 Bartholomew Road, Piscataway, New Jersey 08854, USA
gurvich@rutcor.rutgers.edu

Abstract. We consider the problem of computing shortest paths in a directed arc-weighted graph $G = (V, A)$ in the presence of an adversary that can block (interdict), for each vertex $v \in V$, a given number $p(v)$ of the arcs $A_{out}(v)$ leaving v. We show that if all arc-weights are non-negative then the single-destination version of the problem can be solved by a natural extension of Dijkstra's algorithm in time

$$O\big(|A| + |V|\log|V| + \sum_{v \in V \setminus \{t\}} (|A_{out}(v)| - p(v)) \log(p(v) + 1)\big).$$

Our result can be viewed as a polynomial algorithm for a special case of the network interdiction problem where the adversary's budget is node-wise limited. When the adversary can block a given number p of arcs distributed arbitrarily in the graph, the problem (p-most-vital-arcs problem) becomes NP-hard. This result is also closely related to so-called cyclic games. No polynomial algorithm computing the value of a cyclic game is known, though this problem belongs to both NP and coNP.

1 Introduction

1.1 Main Problems

Let $G = (V, A)$ be a directed graph (digraph) with given arc-weights $w(e)$, $e \in A$, and let $s, t \in V$ be two distinguished vertices of G. We consider the problem of maximizing the shortest path from s to t in G by an adversary who can block (interdict), for each vertex $v \in V$, some subsets $X(v)$ of the arcs $A(v) = \{e \in A \mid e = (v, u)\}$ leaving v. We assume that the blocking arc-sets $X(v) \subseteq A(v)$ are selected for all vertices $v \in V$ independently and that for each v, the collection $\mathcal{B}(v)$ of all admissible blocks $X(v)$ forms an *independence system*: if $X(v) \in \mathcal{B}(v)$ is an admissible block at v, then so is any subset of $X(v)$. Hence, we could replace

---

* [On April 29th, 2005, our co-author Leonid Khachiyan passed away with tragic suddenness while we were finalizing this paper].

D. Grigoriev, J. Harrison, and E.A. Hirsch (Eds.): CSR 2006, LNCS 3967, pp. 221–234, 2006.
© Springer-Verlag Berlin Heidelberg 2006

the independence systems $\mathcal{B}(v)$ by the collections of all inclusion-wise maximal blocking arc-sets. In general, we will only assume that the blocking systems $\mathcal{B}(v)$ are given by a *membership oracle* $\mathcal{O}$:

Given a list $X(v)$ of outgoing arcs for a vertex v, the oracle can determine whether or not the arcs in the list belong to $\mathcal{B}(v)$ and hence can be simultaneously blocked.

A similar formalization of blocking sets via membership oracles was introduced by Pisaruk in [27]. We will also consider two special types of blocking systems:

($\mathcal{B}_1$) The blocking system is given by a function $p(v) : V \to \mathcal{Z}_+$, where $p(v) \leq |A(v)| = \text{out-deg}(v)$. For each vertex v, the adversary can block any collection of (at most) $p(v)$ arcs leaving v. The numbers $p(v)$ define *digraphs with prohibitions* considered by Karzanov and Lebedev in [21].

($\mathcal{B}_2$) There are two types of vertices: *control vertices*, where the adversary can choose any outgoing arc $e \in A(v)$ and block all the remaining arcs in $A(v)$, and *regular vertices*, where the adversary can block no arc. This case, considered in [17] and [6], is a special case of $\mathcal{B}_1$: $p(v) = |A(v)| - 1$ for control vertices, and $p(v) = 0$ otherwise.

Let us call a digraph $G' = (V, A')$ *admissible* for $G = (V, A)$ if A' is obtained from A by deleting some sets of outgoing arcs $X(v) \in \mathcal{B}(v)$ for each vertex $v \in V$. Consider the following problem:

Given an arc-weighted digraph $G = (V, A, w)$ and a blocking system $\mathcal{B}$, find an admissible digraph G' that maximizes the distance from a given start vertex s to a given terminal vertex t:

$$d(s, t) \stackrel{\text{def}}{=} \max\{s\text{-}t \text{ distance in } G' \mid G' \text{ is an admissible digraph of } G\}.$$

We call $d(s, t)$ the *blocking distance* from s to t. We will see from what follows that, for any fixed terminal vertex $t \in V$, the adversary can select an optimal admissible digraph that simultaneously maximizes the distances from all start vertices s. In other words, there exists an admissible digraph G^o such that for all vertices $v \in V \setminus \{t\}$, we have [1]

$$d(v, t) \equiv v\text{-}t \text{ distance in } G^o.$$

For this reason, it is convenient to consider the single-destination version of the above problem:

MASPNLAI (Maximizing all shortest paths to a given terminal by node-wise limited arc interdiction): *Given an arc-weighted digraph $G = (V, A, w)$, a terminal vertex $t \in V$, and a blocking system $\mathcal{B}$, find an optimal admissible digraph G^o that maximizes the distances from all vertices $v \in V \setminus \{t\}$ to t.*

1.2   Network Interdiction Problem

MASPNLAI is a special (polynomially solvable) case of the so-called *network interdiction problem*. Interdiction (or inhibition) is an attack on arcs which destroys them, or increases their effective lengths, or decreases capacities. The goal

---

[1] Note, however, that if we fix a start vertex s, then distinct terminal vertices t may require distinct optimal admissible digraphs.

of the interdiction is to expend a fixed budget most efficiently, that is to maximize the shortest path or minimize the maximum flow between two given nodes. The problem was originally motivated by military applications, see McMasters and Mustin [23], Ghare, Montgomery, and Turner [13]. Then models of pollution and drug interdiction were developed by Wood [31], see also [30]. Minimizing the maximum flow was considered by Phillips [26]. Maximizing the shortest path was first studied by Fulkerson and Harding [11] and also by Golden [14]; see Israeli and Wood [18] for a short survey. An important special case of the latter problem is so-called *p-most-vital-arcs problem* [2][3][7][22] when the adversary is allowed to destroy exactly p arcs. For $p = 1$ a polynomial algorithm to maximize the shortest path was given by Corley and Shaw [7], however, in general the problem is NP-hard, as it was shown by Bar-Noy, Khuller, and Schieber [3].

MASPNLAI is the shortest path interdiction problem under the assumption that adversary's budget is node-wise limited. We will show that this special case is polynomially solvable.

To illustrate, suppose that for each arc $e = (u, v)$ we are given a probability $p(e)$ that some undesirable transition (for example, contraband smuggling) from u to v can be carried out undetected. Then, assuming the independence and letting $w(e) = -\log p(e) \geq 0$, we can interpret problem MASPNLAI as the uniform maximization of interception capabilities for a given target t under limited inspection resources distributed over the nodes of G.

1.3   Cyclic Games

Another application of MASPNLAI is related to a class of games on digraphs known as *cyclic* or *mean payoff games* [8][9][17][24][25]. Björklund, Sandberg and Vorobyov [6] observed that this class of games is polynomially reducible to problem MASPNLAI with blocks of type $\mathcal{B}_2$, provided that the arc-weights in G have arbitrary signs. A mean payoff game is a zero-sum game played by two players on a finite arc-weighted digraph G all vertices of which have positive out-degrees. The vertices of the digraph (*positions*) are partitioned into two sets controlled by two players, who move a chip along the arcs of the digraph, starting from a given vertex $s \in V$ (the *initial position*). A *positional strategy* of a player is a mapping which assigns an outgoing arc to each his position. If both players select positional strategies then the sequence of moves (the *play*) settles on a simple directed cycle of G whose average arc-weight is the *payoff* corresponding to the selected strategy.

Ehrenfeucht and Mycielski [8][9] and Moulin [24][25] introduced mean payoff games on bipartite digraphs and proved the existence of the value for such games in positional strategies. Gurvich, Karzanov and Khachiyan [17] extended this result to arbitrary digraphs and suggested a potential-reduction algorithm to compute the value and optimal positional strategies of the players. In many respects this algorithm for mean cycle games is similar to the simplex method for linear programming.

Let us assume that the vertices assigned to the maximizing (respectively, to the minimizing) player are controlled (respectively, regular) vertices for $\mathcal{B}_2$.

Then the determination of an optimal positional strategy for the maximizing player reduces to computing a $\mathcal{B}_2$-admissible digraph $G' = (V, A')$ that maximizes the minimum average arc-cost for the cycles reachable from the initial position s. Beffara and Vorobyov [4] report on computational experiments with the potential-reduction algorithm [17] in which it was used to solve very large instances of mean payoff games. However, for some special instances with exponentially large arc-weights, this algorithm may require exponentially many steps [17][5]. Interestingly, computational experiments [5] seem to indicate that such hard instances become easily solvable if the game is modified into an equivalent one by a random potential transformation.

Karzanov and Lebedev [21] extended the potential-reduction algorithm [17] to so-called mean payoff games with prohibitions, that is to blocking systems of type $\mathcal{B}_1$. Pisaruk [27] further extended these results to blocking systems defined by an arbitrary membership oracle, and showed that in this general setting, the potential-reduction algorithm [17] is pseudo-polynomial. Zwick and Paterson [32] gave another pseudo-polynomial algorithm for blocks of type $\mathcal{B}_2$.

As mentioned above, mean payoff games can be reduced to shortest paths with blocks and arc-weights of arbitrary sign. For instance, if we fix a start vertex s, then determining whether the value of a mean payoff game on $G = (V, A)$ exceeds some threshold ξ is equivalent to the following decision problem:

(ξ) : Is there an admissible digraph G' such that the average arc-weight of each cycle reachable from s in G' is at least ξ?

After the substitution $w(e) \to w(e) - \xi$ we may assume without loss of generality that $\xi = 0$, and then (ξ) becomes equivalent to determining whether or not the blocking distance $d(s, v)$ is equal to $-\infty$ for some vertex $v \in V$.

Björklund, Sandberg and Vorobyov [6] recently showed that mean payoff games can be solved in *expected sub-exponential time*. However, the question as to whether this class of games can be solved in polynomial time remains open, even though the decision problem (ξ) is obviously in NP∩coNP [21][32]. Accordingly, for arc-weights of arbitrary sign and magnitude, no polynomial algorithm is known for MASPNLAI , though a pseudo-polynomial one exists [6].

1.4   Main Results

In this paper, we show that for non-negative arc-weights, MASPNLAI can be solved in strongly polynomial time by a natural extension of Dijkstra's algorithm.

Theorem 1. *Given a digraph $G = (V, A)$, a non-negative weight function $w : A \to \Re_+$, and a terminal vertex $t \in V$,*

(i) *The special case of problem MASPNLAI for blocking systems $\mathcal{B}_1$ can be solved in time*

$$O\left(|A| + |V| \log |V| + \sum_{v \in V \setminus \{t\}} [\text{out-deg}(v) - p(v)] \log(p(v) + 1) \right).$$

In particular, for blocking systems $\mathcal{B}_2$ the problem can be solved in $O(|A| + |V| \log |V|)$ time;

(ii) *For arbitrary blocking systems defined by membership oracles, MASPN-LAI can be solved in $O(|A| \log |V|)$ time and at most $|A|$ monotonically increasing membership tests;*

(iii) *When all of the arcs have unit weight, problem MASPNLAI can be solved in $O(|A| + |V|)$ time and at most $|A|$ monotonically increasing blocking tests. The special cases $\mathcal{B}_1$ and $\mathcal{B}_2$ can be solved in $O(|A| + |V|)$ time.*

We show parts **(ii)** and **(iii)** of the theorem by using an extension of Dijkstra's algorithm and breadth-first search, respectively. As mentioned in the theorem, both of these algorithms employ monotonically increasing membership queries and never de-block a previously blocked arc. This is not the case with the variant of Dijkstra's algorithm used in the proof of part **(i)**. Note also that for blocks of type $\mathcal{B}_1$ and $\mathcal{B}_2$, the above bounds include the blocking tests overhead, and that the bound stated in **(i)** for $\mathcal{B}_2$ is as good as the running time of the fastest currently known strongly-polynomial algorithm by Fredman and Tarjan [10] for the standard shortest path problem, without interdiction.

Let us also mention that by Theorem 1, problem MASPNLAI can be solved in strongly polynomial time for any digraph $G = (V, A)$ that has no negative total arc-weight directed cycles. Indeed, Gallai [12] proved that if G has no negative cycle then all input arc-weights $w(v, u)$ can be made non-negative by a potential transformation $w(v, u) \rightarrow w(v, u) + \varepsilon(v) - \varepsilon(u)$, where $\varepsilon : V \rightarrow \Re$ are some vertex weights (potentials); see [1][28]. Clearly, the weights of all directed cycles remain unchanged and the total weight of a directed path ℓ from s to t is transformed as: $w(\ell(s, t)) \rightarrow w(\ell(s, t)) + \varepsilon(s) - \varepsilon(t)$. Hence, the set of optimal arc blocks for MASPNLAI remains unchanged, too. Karp [20] showed that such a potential transformation can be found in $O(|A||V|)$ time.

1.5 Main Remarks

We proceed with two negative observations.

1) It is well known that the standard shortest path problem is in NC, that is it can be efficiently solved in parallel. In contrast, problem MASPNLAI is P-complete already for blocking systems of type $\mathcal{B}_2$ and acyclic digraphs $G = (V, A)$ of out-degree 2. This is because determining whether the blocking distance between a pair of vertices s, t is finite: $d(s, t) < +\infty$ includes, as a special case, the well-known monotone circuit value problem [15][16].

2) The independence systems $\mathcal{B} \subseteq 2^A$ considered in this paper are Cartesian products of the systems $\mathcal{B}(v) \subseteq 2^{A(v)}$ defined on the sets $A(v)$ of outgoing arcs for each vertex v of $G = (V, A)$, that is $\mathcal{B} = \bigotimes_{v \in V \setminus \{t\}} \mathcal{B}(v)$. When $\mathcal{B} \subseteq 2^A$ is not decomposable as above, maximizing the shortest path becomes NP-hard for very simple blocking systems and unit arc-weights; the problem is NP-complete for both directed or undirected graphs if the adversary can block a given number p of arcs or edges arbitrarily distributed in the input graph (so-called p-most-vital-arcs problem) [3]. However, the following related problem can be solved in polynomial time:

$\mathcal{B}$: *Given a digraph $G = (V, A)$ with two distinguished vertices $s, t \in V$ and positive integers p and q, determine whether there exists a subsets A' of at most p arcs such that any directed path from s to t in G contains at least q arcs of A'.*

Suppose without loss of generality that t is reachable from s in G, and let A' be an arbitrary q-cut, i.e. $|A' \cap P| \geq q$ for any s-t path $P \subseteq A$. Then, denoting by V_i the set of vertices that can be reached from s by using at most i arcs from A', we conclude that A' contains q disjoint s-t cuts $C_i = \text{cut}(V_{i-1}, V_i)$ for $i = 1, \ldots, q$. Conversely, the union of any q arc-disjoint s-t cuts is a q-cut separating t from s. Hence problem $\mathcal{B}$ can be equivalently stated as follows:

$\mathcal{B}$: *Given a digraph $G = (V, A)$, two distinguished vertices $s, t \in V$, and positive integers p and q, determine whether there exist q arc-disjoint s-t-cuts $C_1, \ldots, C_q$ such that $|C_1| + \ldots + |C_q| \leq p$.*

The latter problem is polynomial. Moreover, Wagner [29] showed that its weighted optimization version can be solved in strongly polynomial time.

$\mathcal{B}'_w$: *Given a digraph $G = (V, A)$ with two distinguished vertices $s, t \in V$, a weight function $w : A \to \Re_+$, and a positive integer q, find q arc-disjoint s, t-cuts $C_1, \ldots, C_q$ of minimum total weight $w(C_1) + \ldots + w(C_q)$.*

Finally, let us remark that "the node-wise limited interdiction problems are usually easier than the total ones". For example, given a digraph $G = (V, A)$ and a positive integer p, is it possible to destroy all directed cycles of G by eliminating at most p arcs of A, or in other words, whether G has a feedback of at most p arcs ? This decision problem is NP-hard [19]. However, if instead of p, for each vertex $v \in V$, we are given a number $p(v)$ of outgoing arcs which can be eliminated then it is easy to decide whether all directed cycles can be destroyed. Indeed, they definitely can not be destroyed if $p(v) < \text{out-deg}(v)$ for each $v \in V$. Yet, if $p(v) \geq \text{out-deg}(v)$ for a vertex $v \in V$ then all outgoing arcs in v should be eliminated, since in this case we can eliminate the vertex v itself. Repeating this simple argument we get a linear time algorithm.

2   Proof of Theorem 1

We first describe an extension of Dijkstra's algorithm for MASPNLAI that uses *blocking queues* and may temporarily block and then de-block some arcs. This extension, presented in Section 2.2, is used to show part **(i)** of Theorem 1. Then in Section 2.4 we present another implementation of the extended algorithm to prove part **(ii)** of the theorem. Part **(iii)** is shown in Section 2.5.

2.1   Blocking Queues

Let $\mathcal{B}$ be a blocking (i.e. independence) system on a finite set E, for example on the set $A(v)$ of arcs leaving a given vertex v of G. Given a mapping $k : E \to \Re$, and a set $Y \subseteq E$, let

$$k_{\mathcal{B}}(Y) = \max_{X \in \mathcal{B}} \min_{e \in Y \setminus X} k(e), \tag{1}$$

where, as usual, it is assumed that the minimum over the empty set is $+\infty$. For instance, if $Y = \{e_1, e_2, e_3, e_4\}$ and $(k(e_1), k(e_2), k(e_3), k(e_4)) = (1, 3, 3, 5)$, then

$$k_{\mathcal{B}}(Y) = \begin{cases} 1, & \text{if } \{e_1\} \notin \mathcal{B}; \\ 3, & \text{if } \{e_1\} \in \mathcal{B} \text{ but } \{e_1, e_2, e_3\} \notin \mathcal{B}; \\ 5, & \text{if } \{e_1, e_2, e_3\} \in \mathcal{B} \text{ but } Y \notin \mathcal{B}; \\ +\infty, & \text{if } Y \in \mathcal{B}. \end{cases}$$

Considering the image $\{k(e), e \in Y\}$ as a sets of keys, we define a $\mathcal{B}$-queue as a data structure for maintaining a dynamic set of keys under the following operations:

1. *Make_queue*: Create an empty queue $Y = \varnothing$;
2. *Insert*: Expand Y by adding a new element e with a given key value $k(e)$;
3. *Return $k_{\mathcal{B}}(Y)$*: Compute the right-hand side of (1) for the current key set.

Note that when the independence system is empty, $|\mathcal{B}| = 0$, we obtain the customary definition of a minimum priority queue.

When $\mathcal{B}$ is a blocking system of type $\mathcal{B}_1$, i.e., $X \in \mathcal{B}$ whenever $|X| \leq p$ for some given integer $p \leq |E|$, then

$$k_{\mathcal{B}}(Y) = \begin{cases} +\infty, & \text{if } |Y| \leq p; \\ (p+1)^{st} \text{ smallest key of } Y, & \text{if } |Y| \geq p+1. \end{cases}$$

Hence, by maintaining a regular maximum priority queue of at most $p+1$ elements of E,

> A sequence of $d \geq p$ queue operations for an initially empty $\mathcal{B}_1$-queue can be implemented to run in $O(p + (d-p)\log(p+1))$ time.

For general blocking systems $\mathcal{B}$, each $\mathcal{B}$-queue operation can be performed in $O(\log|Y|)$ time and $O(\log|Y|)$ oracle queries. This can be done by using a balanced binary search tree on the set of keys in Y. Specifically, inserting a new key into Y takes $O(\log|Y|)$ time and no oracle queries, while computing the value of $k_{\mathcal{B}}(Y)$ can be done by searching for the largest key k in the tree for which the oracle can block the set of all keys smaller than k. Note that each query to the blocking oracle can be specified by a *list* of keys if we additionally maintain a sorted list of all keys in Y along with pointers from the search tree to the list.

We close this subsection by defining, for each set $Y \subseteq E$ of keys, a (unique) inclusion-wise minimal blocking set $\hat{X}(Y) \in \mathcal{B}$ such that

$$k_{\mathcal{B}}(Y) = \min_{e \in Y \setminus \hat{X}(Y)} k(e).$$

We will refer to $\hat{X}(Y) \subseteq Y$ as the *lazy block for Y*. For instance, if, as before, $Y = \{e_1, e_2, e_3, e_4\}$ and $(k(e_1), k(e_2), k(e_3), k(e_4)) = (1, 3, 3, 5)$, then

$$
\hat{X}(Y) = \begin{cases} \varnothing, & \text{if } \{e_1\} \notin \mathcal{B}; \\ \{e_1\}, & \text{if } \{e_1\} \in \mathcal{B}, \text{ but } \{e_1, e_2, e_3\} \notin \mathcal{B}; \\ \{e_1, e_2, e_3\}, & \text{if } \{e_1, e_2, e_3\} \in \mathcal{B}, \text{ but } Y \notin \mathcal{B}; \\ Y, & \text{if } Y \in \mathcal{B}. \end{cases}
$$

For an unsorted list of keys $\{k(e),\ e \in Y\}$, the lazy block $\hat{X}(Y)$ can be computed in $O(|Y|)$ time and $O(\log|Y|)$ oracle queries by recursively splitting the keys around the median. For blocking systems $\mathcal{B}_1$ this computation takes $O(|Y|)$ time.

2.2   Extended Dijkstra's Algorithm for MASPNLAI

Given a digraph $G = (V, A)$, a non-negative weight function $w(v) : A \to \Re^+$, a vertex $t \in V$, and a blocking system $\mathcal{B}$, we wish to find an admissible graph G^o that maximizes the distance from each start vertex $v \in V$ to t. In the statement of extended Dijkstra's algorithm below we assume without loss of generality that the out-degree of the terminal vertex t is 0, and the input arc-weights $w(e)$ are all finite. By definition, we let $d(t, t) = 0$.

Similarly to the regular Dijkstra's algorithm, the extended version maintains, for each vertex $v \in V$, an upper bound $\rho(v)$ on the blocking v-t distance:

$$
\rho(v) \geq d(v, t) \stackrel{\text{def}}{=} \max_{G' \ admissible} \{\text{distance from } v \text{ to } t \text{ in } G'\}.
$$

Initially, we let $\rho(t) = 0$ and $\rho(v) = +\infty$ for all vertices $v \in V \setminus \{t\}$. As the regular Dijkstra's algorithm, the extended version runs in at most $|V| - 1$ iterations and (implicitly) partitions V into two subsets S and $T = V \setminus S$ such that $\rho(v) = d(v, t)$ for all $v \in T$. We iteratively grow the initial set $T = \varnothing$ by removing, at each iteration, the vertex u with the smallest value of $\rho(v)$ from S and adding it to T. For this reason, the values of $\rho(v)$, $v \in S$ are stored in a minimum priority queue, e.g., in a Fibonacci heap. Once we remove the minimum-key vertex u from S (and thus implicitly declare that $\rho(u) = d(u, t)$), we update $\rho(v)$ for all those vertices $v \in S$ that are connected to u by an arc in G. Recall that the regular version of Dijkstra's algorithm uses updates of the form $\rho(v) \leftarrow \min\{\rho(v), w(v, u) + \rho(u)\}$. The updates performed by the extended version use blocking queues $Y(v)$ maintained at all vertices $v \in V \setminus \{t\}$. Initially, all these $\mathcal{B}(v)$-queues are empty, and when the value of $\rho(v)$ needs to be updated for some vertex $v \in S$ such that $e = (v, u) \in A$, we first insert arc e with the key value $k(e) = w(v, u) + \rho(u)$ into $Y(v)$, and then let $\rho(v) \leftarrow k_{\mathcal{B}}(Y(v)) \stackrel{\text{def}}{=} \max_{X \in \mathcal{B}(v)} \min_{e \in Y(v) \setminus X} k(e)$. In particular, for the standard shortest path problem, we obtain the regular updates.

Finally, as the regular Dijkstra's algorithm, the extended version terminates as soon as $\rho(u) = \min\{\rho(v),\ v \in S\} = +\infty$ or $|S| = 1$.

EXTENDED DIJKSTRA'S ALGORITHM

Input: A digraph $G = (V, A)$ with arc-weights $\{w(e) \in [0, +\infty), \ e \in A\}$, a destination vertex $t \in V$, and a blocking system $\mathcal{B}$.

Initialization:

1. $\rho(t) \leftarrow 0$;
2. For all vertices $v \in V \setminus \{t\}$ do:
3. $\rho(v) \leftarrow +\infty$; Set up an empty blocking queue $Y(v)$;
4. Build a minimum priority queue (Fibonacci heap) S on the key values $\rho(v)$, $v \in V$.

Iteration loop:

5. While $|S| > 1$ do:
6.   If $\min\{\rho(v), \ v \in S\} = +\infty$, break loop and go to line 12;
7.   Extract the vertex u with the smallest key value $\rho(\cdot)$ from S;
8.   For all arcs $e = (v, u) \in A$ such that $v \in S$, do:
9. $k(e) \leftarrow w(e) + \rho(u)$;
10.   Insert $k(e)$ into $Y(v)$;
11.   Update the value of $\rho(v)$: $\rho(v) \leftarrow k_{\mathcal{B}}(Y(v))$.

Output:

12. For each vertex $v \in V \setminus \{t\}$, return $\rho(v)$ with the lazy block $\hat{X}(Y(v))$.

Bounds on running time for blocks of type $\mathcal{B}_1$. Line 12 and the initialization steps in lines 1-4 take linear time $O(|V| + |A|)$. Let $n \leq |V| - 1$ be the number of iteration performed by the algorithm. Denote by $Y_i(v)$ (the set of key values in) the blocking queue at a fixed vertex $v \in V \setminus \{t\}$ after the execution of iteration $i = 1, \ldots, n$, and let $Y_0(v) = \varnothing$ be the initial queue at v. As $Y_0(v) \subseteq Y_1(v) \subseteq \ldots \subseteq Y_n$, the values of $\rho_i(v) = k_{\mathcal{B}}(Y_i(v))$ are monotonically non-increasing: $+\infty = \rho_0(v) \geq \rho_1(v) \geq \ldots \geq \rho_n(v)$. Since S is a (minimum) Fibonacci heap, the decrease-key operations in line 11 can be executed in constant amortized time per iteration, provided that the values of $k_{\mathcal{B}}(Y_i(v))$ are known. Lines 6 and 7 take $O(1)$ and $O(\log |V|)$ time per iteration, respectively. In view of the bounds on the $\mathcal{B}_1$-queue operations 10-11 stated in Section 2.1, the overall running time of the algorithm is thus within the bound stated in part (**i**) of Theorem 1.

To complete the proof of part (**i**) it remains to show that the extended algorithm is correct.

2.3 Correctness of Extended Dijkstra's Algorithm

Let us show that *upon the termination of the extended Dijkstra algorithm,*

- $\rho(v) = d(v, t) \stackrel{\text{def}}{=} \max_{G' \text{admissible}} \{distance\ from\ v\ to\ t\ in\ G'\}$ *for all vertices* $v \in V$, *and*
- *The digraph* $G^o = \left(V, A \setminus \bigcup_{v \in V \setminus \{t\}} \hat{X}(Y(v))\right)$ *obtained by deleting the lazy blocking sets of arcs* $\hat{X}(Y(v))$ *is an optimal admissible digraph for all vertices:* $d(v, t) \equiv$ *v-t distance in* G^o.

Let S_i and $T_i = V \setminus S_i$ be the vertex partition maintained by the algorithm for $i = 0, 1, \ldots, n \leq |V| - 1$. We have $S_0 = V \supset S_1 = V \setminus \{t\} \supset \ldots \supset S_{n-1} \supseteq S_n$, where $S_{n-1} = S_n$ if and only if the algorithm terminates due to the stopping criterion in line 6. For the given arc weights $w(e)$, $e \in A$, consider the following weight functions $w_i : A \to \Re_+ \cup \{+\infty\}$,

$$w_i(e) = \begin{cases} +\infty, & \text{if both endpoints of } e \text{ are in } S_i, \\ w(e), & \text{otherwise.} \end{cases}$$

Clearly, we have $w_0(e) = +\infty \geq w_1(e) \geq \ldots \geq w_n(e) \geq w(e)$. Let

$$d_i(v,t) \overset{\text{def}}{=} \max_{\{G' \text{ admissible}\}} \{w_i\text{-distance from } v \text{ to } t \text{ in } G'\},$$

then $d_0(v,t) = +\infty \geq d_1(v,t) \geq \ldots \geq d_n(v,t) \geq d(v,t)$ for all $v \in V \setminus \{t\}$. The correctness of the algorithm will follow from the following two invariants: *for all $i = 0, 1, \ldots, n$,*

$\mathcal{I}_i^S$: $\rho_i(v) = d_i(v,t)$ *for all vertices* $v \in S_i$;
$\mathcal{I}_i^T$: *If $v \in T_i = V \setminus S_i$, then $\rho_i(v) = d(v,t)$ and the admissible digraph $G_i^o = \left(V, A \setminus \bigcup_{v \in V \setminus \{t\}} \hat{X}(Y_i(v))\right)$ is an optimal blocking digraph for v. Moreover, $\min\{\rho_i(v), v \in S_i\} \geq \max\{d(v,t), v \in T_i\}$ and for each vertex $v \in T_i$ there exists a shortest v-t path in G_i^o which lies entirely in T_i.*

Note that by $\mathcal{I}_i^T$, the algorithm removes vertices from S and determines their blocking distances in non-decreasing order.

Proof of invariants $\mathcal{I}_i^S$ and $\mathcal{I}_i^T$ is similar to that for the regular Dijkstra's algorithm. Since $T_0 = \varnothing$, invariant $\mathcal{I}_0^T$ holds trivially. $\mathcal{I}_0^S$ follows from the initialization steps of the algorithm: for $S_0 = V$ we have $w_0(e) \equiv +\infty$, and hence $\rho_0(t) = d_0(t,t) = 0$ and $\rho_0(v) = d_0(v,t) = +\infty$ for all vertices $v \in V \setminus \{t\}$.

In order to prove by induction that $\mathcal{I}_{i+1}^S$ and $\mathcal{I}_{i+1}^T$ follow from $\mathcal{I}_i^S$ and $\mathcal{I}_i^T$, let us first suppose that the ith iteration loop breaks due to the stopping criterion in line 6: $\min\{\rho_i(v), v \in S_i\} = +\infty$. Then $i = n - 1$ and $S_{n-1} = S_n$, which means that $d_n(v,t) \equiv d_{n-1}(v,t)$ and $\rho_n(v) \equiv \rho_{n-1}(v)$. Consequently, the statements of $\mathcal{I}_n^S$ and $\mathcal{I}_n^T$ become identical to $\mathcal{I}_{n-1}^S$ and $\mathcal{I}_{n-1}^T$, and we have nothing to prove. Moreover, as all vertices of S_n are disconnected from t in $G^o = G_n^o$, invariant $\mathcal{I}_n^T$ also shows that the algorithm correctly computes the blocking distances and the optimal blocking digraph G^o for all vertices.

We may assume henceforth that $n = |V| - 1$ and $|S_n| = 1$. Consider the vertex u that moves from S_i to T_{i+1} at iteration i:

$$\rho_i(u) = \min\{\rho_i(v), v \in S_i\} < +\infty. \tag{2}$$

To show that $\rho_i(u) = d(u,t)$, observe that by $\mathcal{I}_i^S$, $\rho_i(u) = d_i(u,t) \geq d(u,t)$. In other words, $\rho_i(u)$ is an upper bound on the w-cost of reaching t from u, regardless of any admissible blocks selected by the adversary. So we will have $\rho_i(u) = d(u,t)$ if we can find an admissible digraph G' such that

$$\rho_i(u) = w\text{-distance from } u \text{ to } t \text{ in } G'. \tag{3}$$

Let $G' = G_i^o$ be the admissible digraph defined in $\mathcal{I}_i^T$. Then (3) follows from $\mathcal{I}_i^T$, the non-negativity of the input arc-weights, and the fact that $\rho_i(u) = k(e_*) = w(e_*) + \rho_i(v)$, where $e_* = (u,v) \in A$ is the arc with the smallest key value in the (S_i, T_i)-cut of G'.

After u gets into T_{i+1}, the value of $\rho(u)$ never changes. Hence $\rho_{i+1}(u) = d(u,t)$, as stated in $\mathcal{I}_{i+1}^T$. Note that (2) and invariant $\mathcal{I}_i^T$ also tell us that
$$\min\{\rho_i(v), v \in S_i\} = d(u,t) \geq \max\{d(v,t), \ v \in T_i\}.$$
Let us now show that after the algorithm updates $\rho(v)$ on S_{i+1}, we still have

$$\min\{\rho_{i+1}(v), \ v \in S_{i+1}\} \ \geq \ d(u,t) \ = \max\{d(v,t), \ v \in T_{i+1}\}, \qquad (4)$$

again as stated in $\mathcal{I}_{i+1}^T$. Suppose to the contrary, that $\rho_{i+1}(v) < d(u,t) = \rho_i(u)$ for some vertex $v \in S_{i+1}$. Then from (2) it would follow that $e = (v,u)$ is an arc of $G = (V,A)$ and consequently $Y_{i+1}(v) = Y_i(v) \cup \{e\}$. Moreover, we must have $e \in \hat{X}(Y_{i+1}(v))$, for otherwise the value of $\rho_{i+1}(v) = k_\mathcal{B}(Y_{i+1}(v))$ could not have dropped below the minimum of $\rho_i(v)$ and $k(e) = w(v,u) + d(u,t)$, which is at least $d(u,t)$. But if $e \in \hat{X}(Y_{i+1}(v))$ then again $k_\mathcal{B}(Y_{i+1}(v)) \geq k(e)$, contradiction.

To complete the proof of $\mathcal{I}_{i+1}^T$, it remains to show that G_{i+1}^o is an optimal admissible digraph for each vertex $v \in T_{i+1}$, and that some shortest v-t path in G_{i+1}^o lies in T_{i+1}. This readily follows from (4) and the fact that the sub-graphs of G_i^o and G_{i+1}^o induced by T_{i+1} are identical.

Finally, $\mathcal{I}_{i+1}^S$ follows from the updates $\rho_{i+1}(v) \leftarrow k_\mathcal{B}(Y_{i+1}(v))$ performed by the algorithm in lines 8-11. $\qquad\square$

Since we assume that $n = |V| - 1$ and $|S_n| = 1$, the correctness of the algorithm readily follows from $\mathcal{I}_n^S$ and $\mathcal{I}_n^T$. When S_n is a singleton $s \in V$, then $w_n(e) \equiv w(e)$. Hence $d_n(v,t) \equiv d(v,t)$, and $\mathcal{I}_n^S$ yields $\rho_n(s) = d_n(s,t) = d(s,t)$. By $\mathcal{I}_n^T$, we also have $\rho_n(v) = d(v,t)$ for the remaining vertices $v \in T_n = V \setminus \{s\}$. Invariant $\mathcal{I}_n^T$ also guarantees that $G^o = G_n^o$ is an optimal admissible digraph for all vertices $v \in V$.

2.4   Modified Dijkstra's Algorithm

In this section we prove part **(ii)** of Theorem 1 by modifying the algorithm stated in Section 2.2.

The modified algorithm keep all arcs across the current (S,T)-cut in a minimum priority queue $\mathcal{Q}$, implemented as a binary heap. As in the previous algorithm, each arc $e = (v,v')$ across the cut is assigned the key value $k(e) = w(e) + \rho(v')$, where $\rho(v') = d(v',t)$ for all vertices $v' \in T$. In addition to the arcs in the current cut, $\mathcal{Q}$ may also contain some arcs $e = (v,v')$ for which endpoints v, v' are both in T. In order to compute the vertex u to be moved from S to T, we repeatedly extract the minimum-key arc e from $\mathcal{Q}$, and check whether $e = (v,v')$ belongs to the current cut and can be blocked along with the arcs that have already been blocked at v. The first arc $e = (v,v')$ in the cut that cannot be blocked defines the vertex $u = v$. We then move u to T, insert all arcs $e = (v,u) \in A$ for which $v \in S$ into $\mathcal{Q}$, and iterate.

The outputs of the modified algorithm and the extended Dijkstra's algorithm presented in Section 2.2 are identical. It is also easy to see that the running time and the number of membership tests required by the modified algorithm satisfy the bounds stated in part **(ii)** of Theorem 1.

MODIFIED ALGORITHM

Input: **A digraph $G = (V, A)$ with arc-weights $\{w(e) \in [0, +\infty), \ e \in A\}$, a terminal vertex $t \in V$, and a blocking system $\mathcal{B} \subseteq 2^A$ defined via a membership testing subroutine.**

Initialization:
1. Initialize arrays $T[1 : V] \equiv FALSE$ and $d[1 : V, t] \equiv +\infty$;
2. $T[t] \leftarrow TRUE, \quad d[t, t] \leftarrow 0$;
3. For each vertex $v \in V \setminus \{t\}$ initialize an empty list $\hat{X}(v)$;
4. For each arc $e = (v, t) \in A$, insert e with key $k(e) = w(e)$ into an initially empty binary heap $\mathcal{Q}$.

Iteration loop:
5. While $\mathcal{Q} \neq \varnothing$ do:
6.   Extract the minimum-key arc $e = (u, v)$ from $\mathcal{Q}$;
7.   If $T[u] = FALSE$ and $T[v] = TRUE$ do:
8.   If $\hat{X}(u) \cup \{e\}$ can be blocked at u, insert e into $\hat{X}(u)$
9.   else $\{ T[u] \leftarrow TRUE; \ $ Return $\hat{X}(u)$ and $d[u, t] = k(e)$;
10.   For all arcs $e = (v, u) \in A$ such that $T[v] = FALSE$,
               Insert e with key value $k(e) = w(e) + d[u, t]$ into $\mathcal{Q}\}$.

2.5   Unit Arc-Weights

When $w(e) = 1$ for all $e \in A$, and the blocking systems $\mathcal{B}(v)$ are all empty, the single-destination shortest path problem can be solved in linear time by breadth-first search. The extended Dijkstra's algorithm for problem MASPNLAI can be similarly simplified to prove part **(iii)** of Theorem 1.

BREADTH-FIRST SEARCH FOR MASPNLAI

Input: **A digraph $G = (V, A)$ with a destination vertex $t \in V$, and a blocking system $\mathcal{B}$ defined by a membership subroutine.**
Initialization:
1. Initialize $d(1 : V, t) \equiv +\infty$ and an empty first-in first-out queue $\mathcal{T}$;
2. $d(t, t) \leftarrow 0$; Enqueue t into $\mathcal{T}$;
3. For each vertex $v \in V \setminus \{t\}$ initialize an empty list $\hat{X}(v)$;

Iteration loop:
4. While $\mathcal{T} \neq \varnothing$ do:
5.   Extract the first vertex u from $\mathcal{T}$;
6.   For all arcs $e = (v, u) \in A$, do:
7.   If $d(v, t) = +\infty$ and $\hat{X}(v) \cup \{e\}$ can be blocked, insert e into $\hat{X}(v)$;
8.   else $d(v, t) \leftarrow d(u, t) + 1$, enqueue v into $\mathcal{T}$, and return $d(v, t), \hat{X}(v)$.

    The above algorithm runs in at most $|A|$ iterations. It follows by induction on $d(v, t)$ that it correctly computes the blocking distances and that the admissible digraph $G^o = \left(V, A \setminus \bigcup_{v \in V \setminus \{t\}} \hat{X}(v)\right)$ is optimal.

References

1. R.K. Ahuja, T.L. Magnanti and J.B. Orlin, Network Flows: Theory, Algorithms, and Applications, Prentice Hall, New Jersey, 1993.
2. M.O. Ball, B.L. Golden and R.V. Vohra, Finding the most vital arcs in a network, Operations Research Letters **8** (1989), pp. 73-76.
3. A. Bar-Noy, S. Khuller and B. Schieber, The complexity of finding most vital arcs and nodes, University of Maryland, Institute of Anvanced Computer Studies, College Park, MD, Technical Report CS-TR-3539, 1995.
4. E. Beffara and S. Vorobyov, Adapting Gurvich-Karzanov-Khachiyan's algorithm for parity games: Implementation and experimentation, Technical Report 020, Department of Information Technology, Uppsala University, 2001 (available at https://www.it.uu.se/research /reports/#2001).
5. E. Beffara and S. Vorobyov, Is randomized Gurvich-Karzanov-Khachiyan's algorithm for parity games polynomial? Technical Report 025, Department of Information Technology, Uppsala University, 2001 (available at https://www.it.uu.se/research/reports/#2001).
6. H. Björklund, S. Sandberg and S. Vorobyov, A Combinatorial strongly subexponential strategy improvement algorithm for mean payoff games, DIMACS Technical Report 2004-05 (2004) (available at http://dimacs.rutgers.edu/TechnicalReports/2004.html).
7. H.W. Corely and D.Y. Shaw, Most vital links and nodes in weighted networks, Operations Research Letters **1** (1982), pp. 157-160.
8. A. Eherenfeucht and J. Mycielski, Positional games over a graph, Notices of the American Mathematical Society **20** (1973), A-334.
9. A. Ehrenfeucht and J. Mycielski, Positional strategies for mean payoff games, International Journal of Game Theory **8** (1979), pp. 109-113.
10. M.L. Fredman and R.E. Tarjan, Fibonacci heaps and their uses in improved network optimization algorithms, Journal of the ACM **34(3)** (1987), pp. 596-615.
11. D.R. Fulkerson and G.C. Harding, Maximizing the minimum source-sink path subject to a budget constraint, Mathematical Programming **13** (1977), pp. 116-118.
12. T. Gallai, Maximum-minimum Sätze über Graphen. *Acta Mathematica Academiae Scientiarum Hungaricae* **9** (1958) pp. 395-434.
13. P.M. Ghare, D.C. Montgomery, and T.M. Turner, Optimal interdiction policy for a flow network, Naval Research Logistics Quarterly **18** (1971), pp. 37-45.
14. B.L. Golden, A problem in network interdiction, Naval Research Logistics Quarterly **25** (1978), pp. 711-713.
15. L.M. Goldschlager, The monotone and planar circuit value problem are log space complete for P, SIGACT News **9(2)** (1977), pp. 25-29.
16. R. Greenlaw, H.J. Hoover and W.L. Ruzzo, Limits to Parallel Computation: P-Completeness Theory, Oxford University Press, 1995.
17. V. Gurvich, A. Karzanov and L. Khachiyan, Cyclic games and an algorithm to find minimax cycle means in directed graphs, USSR Computational Mathematics and Mathematical Physics **28** (1988), pp. 85-91.
18. E. Israely and K. Wood, Shortest-path network interdiction, Networks **40(2)** (2002), pp. 97-111.
19. R. Karp, Reducibility among combinatorial problems, in: R.E. Miller and J.W. Thatcher, eds., Complexity of Computer Computations, Plenum Press, New York (1972) pp. 85-103.

20. R. Karp, A Characterization of the Minimum Cycle Mean in a Digraph, Discrete Math. **23** (1978), pp. 309–311.
21. A.V. Karzanov and V.N. Lebedev, Cyclical games with prohibition, Mathematical Programming **60** (1993), pp. 277-293.
22. K. Malik, A.K. Mittal, and S.K. Gupta, The k most vital arcs in the shortest path problem, Operations Research Letters **8** (1989), pp. 223-227.
23. A.W. McMasters and T.M. Mustin, Optimal interdiction of a supply networks, Naval Research Logistics Quarterly 17 (1970), pp. 261-268
24. H. Moulin, Prolongement des jeux à deux joueurs de somme nulle, Bull. Soc. Math. France, Memoire **45**, (1976).
25. H. Moulin, Extension of two person zero sum games, Journal of Mathematical Analysis and Apllication **55 (2)** (1976), pp. 490-507.
26. C.A. Phillips, The network inhibition problem, Proceedings of the 25th Annual ACM Symposium on the Theory of Computing, 1993, pp. 776-785.
27. N.N. Pisaruk, Mean cost cyclical games, Mathematics of Operations Research **24(4)** (1999), pp. 817-828.
28. A. Schrijver, Combinatorial Optimization: Polyhedra and Efficiency, Algorithms and Combinatorics **24**, Springer, 2003.
29. D.K. Wagner, Disjoint (s,t)-cuts in a network, Networks **20** (1990), pp. 361-371.
30. A. Washburn and K. Wood, Two-person zero-sum games for network interdiction, Operations Research **43(2)** (1995), pp. 243-251.
31. R.K. Wood, Deterministic network interdiction, Mathematical and Computer Modelling **17** (1993), pp. 1-18.
32. U. Zwick , M. Paterson, The complexity of mean payoff games on graphs, Theoretical Computer Science **158(1-2)** (1996), pp. 343-359.

Weighted Logics for Traces

Ingmar Meinecke

Institut für Informatik, Universität Leipzig,
D-04109 Leipzig, Germany
meinecke@informatik.uni-leipzig.de

Abstract. We study a quantitative model of traces, *i.e.* trace series which assign to every trace an element from a semiring. We show the coincidence of recognizable trace series with those which are definable by restricted formulas from a weighted logics over traces. We use a translation technique from formulas over words to those over traces, and vice versa. This way, we show also the equivalence of aperiodic and first-order definable trace series.

1 Introduction

Traces as introduced by Mazurkiewicz [19] model concurrency by a global independence relation on a finite alphabet, *i.e.* traces are congruence classes of words modulo the independence relation. A fruitful theory of combinatorics on traces and of trace languages has developed the last twenty years, see [6, 5] for an overview. Droste and Gastin [7] started to explore quantitative aspects of traces a few years ago. They enriched the model with weights from a semiring as it was done for words already in the 1960s by Schützenberger [23]. Droste and Gastin obtained a result in the style of Kleene and Schützenberger, *i.e.* the coincidence of recognizability and a restricted form of rationality. Moreover, they defined and characterized in [8] a weighted concept of aperiodicity for traces. Kuske [16] showed recently the coincidence of recognizable trace series with those recognized by weighted asynchronous cellular automata, both in the non-deterministic and deterministic case. However, a characterization by weighted logics in the lines of Büchi [4] and Elgot [12] was missing even for words. This gap was closed recently by an introduction of weighted logics over words by Droste and Gastin [9]. The semantics of this weighted MSO-logics is a formal power series over words, *i.e.* a function from the free monoid into a semiring. Weighted logics was already extended to trees by Droste and Vogler [10] and to pictures by Mäurer [18].

Naturally, the question arises whether this concept carries over to traces and, therewith, generalizes the results of Droste and Gastin for weighted logics over words [9] on the one hand and the logical characterization of trace languages as done by Ebinger and Muscholl [11] and Thomas [24] on the other hand. Moreover, one could be interested in the execution time of a trace or in the multiplicity of a certain property satisfied by a trace. Such problems can be formulated often better by a logical formula than by a direct construction of a weighted automaton for traces. Therefore, we are interested in weighted logics over traces and in a result that states the coincidence of logically definable and recognizable trace series. Moreover, such a coincidence should be effective in order to open the way to something like quantitative model checking over traces.

D. Grigoriev, J. Harrison, and E.A. Hirsch (Eds.): CSR 2006, LNCS 3967, pp. 235–246, 2006.

Here, we can avoid to repeat the proof of [9] for traces. Instead of this we adapt a technique introduced by Ebinger and Muscholl [11] for their result about the coincidence of definable and recognizable trace languages. There, a formula over traces is translated into an appropriate one over words and vice versa using the lexicographic normal form. This way one is able to transfer the coincidence of definable and recognizable word languages to trace languages. For the weighted case the main problem is to keep the right weighted semantics within the translation of the formulas. Indeed, disjunction and existential quantification result in an addition, whereas conjunction and universal quantification result in multiplication within the underlying semiring. Certainly, these operations are not idempotent in general. Therefore, we are in need of certain "unambiguity" results that will guarantee the right semantics. We obtain such a result for first-order formulas over more general relational structures with a well-order on their elements which is definable by a propositional formula. We apply this result to traces, prove the "translation lemma", and succeed in proving the coincidence of recognizable trace series with trace series defined by restricted monadic second-order formulas. Moreover, for the underlying semiring being either a computable field or being locally finite we will show that decidability results carry over from words to traces. Finally, the coincidence of aperiodic and first-order definable trace series is shown.

For further research the consequences of these results should be explored more in detail. Moreover, application of weighted logics to other models of concurrency like sp-posets [17, 21, 20], MSCs [3], and Σ-DAGs [2] is in work.

2   Basic Concepts

Let Σ be a finite alphabet, Σ^* the free monoid, and $I \subseteq \Sigma^2$ an irreflexive and symmetric relation, called the *independence relation*. Then $D = \Sigma^2 \setminus I$ is reflexive and symmetric and called the *dependence relation*. We define $\sim \subseteq \Sigma^* \times \Sigma^*$ by

$$u \sim v \iff u = w_1 a b w_2 \wedge v = w_1 b a w_2 \text{ for } (a, b) \in I \text{ and } w_1, w_2 \in \Sigma^*.$$

By abuse of notation we denote the reflexive and transitive closure of $\sim$ also by $\sim$. Now $\sim$ is a congruence relation on Σ^* and the resulting quotient is called the *trace monoid* $\mathbb{M} = \mathbb{M}(\Sigma, D)$. Its elements are called *traces*. Let $\varphi : \Sigma^* \to \mathbb{M}$ be the canonical epimorphism with $\varphi(w) = [w]$ where $[w]$ is the congruence class of w. For $t \in \mathbb{M}$ there is a prominent representative among $\varphi^{-1}(t)$, the *lexicographic normal form* LNF(t) of t, *i.e.* the least representative of t with regard to the lexicographic order. The set of all lexicographic normal forms is denoted by LNF. $L \subseteq \mathbb{M}$ is called a *trace language*.

A *semiring* $\mathbb{K} = (K, \oplus, \circ, \mathbb{0}, \mathbb{1})$ is a set K equipped with two binary operations, called *addition* $\oplus$ and *multiplication* $\circ$, such that

1. $(K, \oplus, \mathbb{0})$ is a commutative monoid and $(K, \circ, \mathbb{1})$ a monoid,
2. multiplication distributes over addition: $k \circ (l \oplus m) = (k \circ l) \oplus (k \circ m)$ and $(l \oplus m) \circ k = (l \circ k) \oplus (m \circ k)$ for all $k, l, m \in K$, and
3. $\mathbb{0} \circ k = k \circ \mathbb{0} = \mathbb{0}$ for all $k \in K$.

If the multiplication is commutative we speak of a *commutative semiring*. Examples of semirings are the natural numbers $\mathbb{N} = (\mathbb{N}, +, \cdot, 0, 1)$, the *tropical semiring*

$\mathbb{T} = (\mathbb{N} \cup \{-\infty\}, \max, +, -\infty, 0)$, and the *Boolean semiring* $\mathbb{B} = (\{0, 1\}, \vee, \wedge, 0, 1)$ which equals the two-element Boolean algebra. For an overview about semirings see [13, 14].

A *formal trace series* or just *trace series* over a trace monoid $M(\Sigma, D)$ and a semiring $\mathbb{K}$ is a function $T : M(\Sigma, D) \rightarrow \mathbb{K}$. It is often written as a formal sum

$$T = \sum_{t \in M(\Sigma, D)} (T, t) \, t$$

where $(T, t) = T(t)$. Functions $S : \Sigma^* \rightarrow \mathbb{K}$ are called here *word series*. The collection of formal trace series over M and $\mathbb{K}$ is referred to as $\mathbb{K} \langle\!\langle M \rangle\!\rangle$, and, similarly, $\mathbb{K} \langle\!\langle \Sigma^* \rangle\!\rangle$ is defined. For an overview about formal word series see [22, 15, 1].

For the Boolean semiring $\mathbb{B}$ there is a one-to-one correspondence between trace series $T = \sum_{t \in M(\Sigma, D)} (T, t) \, t$ over $\mathbb{B}$ and their *support* $\mathrm{supp}(T) = \{t \in M(\Sigma, D) \mid (T, t) \neq \mathbb{0}\}$. Vice versa, a trace language $L \subseteq M(\Sigma, D)$ corresponds to its *characteristic series* $\mathbb{1}_L$ where

$$(\mathbb{1}_L, t) = \begin{cases} \mathbb{1} & \text{if } t \in L, \\ \mathbb{0} & \text{otherwise.} \end{cases}$$

Hence, formal power series extend formal language theory.

3 Recognizable Trace Series

Let M be a trace monoid and $\mathbb{K}$ a semiring. Let $\mathbb{K}^{n \times n}$ denote the monoid of $n \times n$-matrices over $\mathbb{K}$ equipped with multiplication. A *recognizable trace series* is a trace series $T \in \mathbb{K} \langle\!\langle M \rangle\!\rangle$ such that there are an $n \in \mathbb{N}$, a monoid homomorphism $\mu : M \rightarrow \mathbb{K}^{n \times n}$, $\lambda \in K^{1 \times n}$, and $\gamma \in K^{n \times 1}$ with $(T, t) = \lambda \mu(t) \gamma$ for all $t \in M$. The triple (λ, μ, γ) is called a *linear representation* of T. For $\varphi : \Sigma^* \rightarrow M$ the canonical epimorphism and $S \in \mathbb{K} \langle\!\langle M \rangle\!\rangle$ we define $\varphi^{-1}(S) \in \mathbb{K} \langle\!\langle \Sigma^* \rangle\!\rangle$ by $(\varphi^{-1}(S), w) = (S, \varphi(w))$. Furthermore, for $S' \in \mathbb{K} \langle\!\langle \Sigma^* \rangle\!\rangle$ we denote by $S'_{|\mathrm{LNF}}$ the *restriction* of S' to LNF, *i.e.*

$$(S'_{|\mathrm{LNF}}, w) = \begin{cases} (S', w) & w \in \mathrm{LNF}, \\ \mathbb{0} & \text{otherwise.} \end{cases}$$

The following theorem is implicit in [7].

Theorem 3.1. *Let* $\mathbb{K}$ *be a commutative semiring. Then* $S \in \mathbb{K} \langle\!\langle M \rangle\!\rangle$ *is recognizable iff* $S' = \varphi^{-1}(S)_{|\mathrm{LNF}} \in \mathbb{K} \langle\!\langle \Sigma^* \rangle\!\rangle$ *is recognizable.*

The next lemma can be shown as for word series, cf. [1, L. III.1.3].

Lemma 3.2. *Let* $\mathbb{K}$ *be a semiring and* $L \subseteq M$ *a recognizable trace language. Then* $\mathbb{1}_L \in \mathbb{K} \langle\!\langle M \rangle\!\rangle$ *is a recognizable trace series.*

Corollary 3.3. *Let* $L_i \subseteq M$ *be recognizable trace languages and* $k_i \in \mathbb{K}$ *for* $i = 1, \ldots, n$. *Then* $S = \sum_{i=1}^n k_i \mathbb{1}_{L_i}$ *is a recognizable trace series.*

The last corollary justifies the name *recognizable step function* for a series of the form $S = \sum_{i=1}^n k_i \mathbb{1}_{L_i}$ with $L_i \subseteq M$ recognizable for all $i = 1, \ldots, n$.

4 Definable Trace Series

We represent every trace $t \in \mathbb{M}(\Sigma, D)$ by its *dependence graph*. A dependence graph is (an isomorphism class of) a node-labeled acyclic graph (V, E, l) where V is an at most countable set of nodes[1], $E \subseteq V \times V$ is the edge relation such that (V, E) is acyclic and the induced partial order is well-founded, $l : V \to \Sigma$ is the node-labeling such that

$$(l(x), l(y)) \in D \iff (x, y) \in E \cup E^{-1} \cup id_V .$$

A concatenation of dependence graphs is defined by the disjoint union provided with additional edges between nodes with dependent labels, *i.e.*

$$(V_1, E_1, l_1) \cdot (V_2, E_2, l_2)$$
$$= (V_1 \,\dot\cup\, V_2, E_1 \,\dot\cup\, E_2 \,\dot\cup\, \{(x, y) \in V_1 \times V_2 \mid (l_1(x), l_2(y)) \in D\}, l_1 \,\dot\cup\, l_2) .$$

The monoid $\mathbb{M}(\Sigma, D)$ of finite traces can be identified with the monoid of finite dependence graphs.

Let $t = (V, E, l) \in \mathbb{M}$ and $w = a_1 \ldots a_n \in \Sigma^*$ with $\varphi(w) = t$. Then we represent w as $(V, <, (R_a)_{a \in \Sigma})$ where $<$ is a strict total order on V (the order of positions) and $R_a = \{v \in V \mid l(v) = a\}$ for all $a \in \Sigma$.

Definition 4.1. *The syntax of formulas of* weighted MSO-logic over traces *from* $\mathbb{M}$ *and over a semiring* $\mathbb{K}$ *is given by*

$$\Phi ::= k \mid P_a(x) \mid \neg P_a(x) \mid E(x, y) \mid \neg E(x, y) \mid x \in X \mid \neg x \in X \mid$$
$$\Phi \vee \Psi \mid \Phi \wedge \Psi \mid \exists x.\Phi \mid \exists X.\Phi \mid \forall x.\Phi \mid \forall X.\Phi$$

with $k \in \mathbb{K}$ *and* $a \in \Sigma$. *This class of formulas is denoted by* $\mathrm{MSO}(\mathbb{K}, \mathbb{M})$.

Remark 4.2. The weighted MSO-logic is a generalization of the usual MSO-logic. Weighted MSO-logic differs in two aspects. Firstly, atomic formulas of type k for $k \in K$ are added. Secondly, negation is applied to "unweighted" atomic formulas only. This is due to the fact that a semantics of something like $\neg k$ cannot be defined properly for arbitrary semirings. Hence, we cannot negate neither k nor general MSO-formulas. Thus negation is pulled through the unweighted atomic formulas and conjunction and universal quantification have to be added.

Note 4.3. A weighted MSO-logic for words, denoted by $\mathrm{MSO}(\mathbb{K}, \Sigma)$ was defined in [9]. It uses k, $x \leq y$, $P_a(x)$, and $x \in X$ as atomic formulas[2]. Here, we do not include the formula $x = y$ in our syntax because for traces this can be written as

$$\bigvee_{a \in \Sigma} (P_a(x) \wedge P_a(y)) \wedge \neg E(x, y) \wedge \neg E(y, x) .$$

---

[1] Here, we deal with finite objects, *i.e.* finite traces, only. But we stick to the more general case, keeping in mind the possibility to consider infinite objects.

[2] Later on, we will use for words $x < y$ instead of $x \leq y$ which is just a slight technical difference.

A variable is *free* in Φ if it is not within the scope of a quantifier. The collection of all free variables of Φ is denoted by free(Φ). Let $\mathcal{V}$ be a finite set of first-order and second-order variables and $t = (V, E, l)$. A $(\mathcal{V}, t)$-assignment σ is a function mapping first-order variables of $\mathcal{V}$ to elements of V and second-order variables of $\mathcal{V}$ to subsets of V. An *update* $\sigma[x \to v]$ for $v \in V$ is defined as $\sigma[x \to v](x) = v$ and $\sigma[x \to v](y) = \sigma(y)$ for all $y \neq x$, and, similarly, for $\sigma[X \to W]$ where $W \subseteq V$. A pair (t, σ) where σ is a $(\mathcal{V}, t)$-assignment will be encoded as a trace over an extended dependence alphabet $\Sigma_\mathcal{V} = \Sigma \times \{0, 1\}^\mathcal{V}$. The new dependence relation $D_\mathcal{V}$ is defined by $(a, \bar{x})D_\mathcal{V}(b, \bar{y})$ iff aDb for $a, b \in \Sigma$ and $\bar{x}, \bar{y} \in \{0, 1\}^\mathcal{V}$. A trace t' over $\Sigma_\mathcal{V}$ will be written as a pair (t, σ) where t is the projection of t' over Σ and σ is the projection over $\{0, 1\}^\mathcal{V}$. Then σ represents a *valid $\mathcal{V}$-assignment* if for any first-order variable $x \in \mathcal{V}$ the x-row of σ contains exactly one 1. Similarly, valid $\mathcal{V}$-assignments are defined for words.

Proposition 4.4. *The trace language* $A_\mathcal{V} = \{(t, \sigma) \mid \sigma$ *is a valid $\mathcal{V}$-assignment*$\}$ *is recognizable.*

For any formula Φ of MSO we simply write $\Sigma_\Phi = \Sigma_{\text{free}(\Phi)}$ and $A_\Phi = A_{\text{free}(\Phi)}$. Now we turn to the semantics of our formulas.

Definition 4.5. *Let* $\Phi \in \text{MSO}(\mathbb{K}, \mathbb{M})$ *and let* $\mathcal{V}$ *be a finite set of variables with* free$(\Phi) \subseteq \mathcal{V}$. *The semantics of* Φ *is a formal trace series* $\llbracket \Phi \rrbracket_\mathcal{V} \in \mathbb{K} \langle\!\langle \mathbb{M}(\Sigma_\mathcal{V}^*, D_\mathcal{V}) \rangle\!\rangle$ *defined as follows: Let* $(t, \sigma) \in \mathbb{M}(\Sigma_\mathcal{V}, D_\mathcal{V})$. *If* σ *is not a valid $\mathcal{V}$-assignment, then* $\llbracket \Phi \rrbracket_\mathcal{V}(t, \sigma) = 0$. *Otherwise, we define* $\llbracket \Phi \rrbracket_\mathcal{V}(t, \sigma)$ *for* $t = (V, E, l)$ *inductively as follows:*

- $\llbracket k \rrbracket_\mathcal{V}(t, \sigma) = k$,

- $\llbracket P_a(x) \rrbracket_\mathcal{V}(t, \sigma) = \begin{cases} 1 & \text{if } l(\sigma(x)) = a, \\ 0 & \text{otherwise,} \end{cases}$

- $\llbracket E(x, y) \rrbracket_\mathcal{V}(t, \sigma) = \begin{cases} 1 & \text{if } (\sigma(x), \sigma(y)) \in E, \\ 0 & \text{otherwise,} \end{cases}$

- $\llbracket x \in X \rrbracket_\mathcal{V}(t, \sigma) = \begin{cases} 1 & \text{if } \sigma(x) \in \sigma(X), \\ 0 & \text{otherwise,} \end{cases}$

- *if* Φ *is of the form* $P_a(x)$, $E(x, y)$, *or* $x \in X$, *then*

$$\llbracket \neg\Phi \rrbracket_\mathcal{V}(t, \sigma) = \begin{cases} 1 & \text{if } \llbracket \Phi \rrbracket_\mathcal{V}(t, \sigma) = 0, \\ 0 & \text{if } \llbracket \Phi \rrbracket_\mathcal{V}(t, \sigma) = 1, \end{cases}$$

- $\llbracket \Phi \vee \Psi \rrbracket_\mathcal{V}(t, \sigma) = \llbracket \Phi \rrbracket_\mathcal{V}(t, \sigma) \oplus \llbracket \Psi \rrbracket_\mathcal{V}(t, \sigma)$,
- $\llbracket \Phi \wedge \Psi \rrbracket_\mathcal{V}(t, \sigma) = \llbracket \Phi \rrbracket_\mathcal{V}(t, \sigma) \circ \llbracket \Psi \rrbracket_\mathcal{V}(t, \sigma)$,
- $\llbracket \exists x.\Phi \rrbracket_\mathcal{V}(t, \sigma) = \bigoplus_{v \in V} \llbracket \Phi \rrbracket_{\mathcal{V} \cup \{x\}}(t, \sigma[x \to v])$,
- $\llbracket \exists X.\Phi \rrbracket_\mathcal{V}(t, \sigma) = \bigoplus_{W \subseteq V} \llbracket \Phi \rrbracket_{\mathcal{V} \cup \{X\}}(t, \sigma[X \to W])$,
- $\llbracket \forall x.\Phi \rrbracket_\mathcal{V}(t, \sigma) = \prod_{v \in V} \llbracket \Phi \rrbracket_{\mathcal{V} \cup \{x\}}(t, \sigma[x \to v])$,
- $\llbracket \forall X.\Phi \rrbracket_\mathcal{V}(t, \sigma) = \prod_{W \subseteq V} \llbracket \Phi \rrbracket_{\mathcal{V} \cup \{X\}}(t, \sigma[X \to W])$.

where we fix some order both on V and on $\mathfrak{P}(V)$ so that the last two products are defined even if $\mathbb{K}$ is not commutative. We simply write $\llbracket \Phi \rrbracket$ *for* $\llbracket \Phi \rrbracket_{\text{free}(\Phi)}$.

If Φ is a sentence, then $[\![\Phi]\!] \in \mathbb{K}\langle\!\langle\mathbb{M}\rangle\!\rangle$. As usual, the semantics of some formula Φ depends on the free variables only. We call $S = \sum_{i=1}^{n} k_i \mathbb{1}_{L_i}$ a *definable step function* if the languages L_i are definable trace languages, or word languages respectively, for all $i = 1, \ldots, n$. For words and traces the notions of recognizable and definable step functions coincide because of the results of Büchi & Elgot [4, 12] and Ebinger & Muscholl [11].

Definition 4.6. *A formula $\Phi \in \mathrm{MSO}(\mathbb{K}, \Sigma)$ or $\Phi \in \mathrm{MSO}(\mathbb{K}, \mathbb{M})$ is called* restricted, *if it contains no universal quantification of second-order $\forall X.\Psi$, and whenever Φ contains a universal first-order quantification $\forall x.\Psi$, then $[\![\Psi]\!]$ is a definable step function.*

Remark 4.7. Droste and Gastin [9] had to use restricted MSO-formulas over words to preserve recognizability of the defined series. For universal FO-quantification $\forall x.\Psi$ they required $[\![\Psi]\!] = \sum_{i=1}^{n} k_i \mathbb{1}_{L_i}$ being a recognizable step function. Since we define a class of formulas, we favor to speak of the logical counterpart, *i.e.* definable step functions.

$\mathrm{RMSO}(\mathbb{K}, \mathbb{M})$ is the class of all restricted formulas from $\mathrm{MSO}(\mathbb{K}, \mathbb{M})$. Moreover, let $\mathrm{REMSO}(\mathbb{K}, \mathbb{M})$ contain all restricted *existential* formulas $\Phi \in \mathrm{RMSO}(\mathbb{K}, \mathbb{M})$, *i.e.* Φ is of the form $\exists X_1. \exists X_2 \ldots \exists X_n.\Psi$ with $\Psi \in \mathrm{RMSO}(\mathbb{K}, \mathbb{M})$ containing no second-order quantification anymore. FO and RFO denote the classes of first-order formulas and restricted first-order formulas, respectively. Similar notations are used for formulas over words.

5   Characteristic Series of FO-Definable Languages

Let $\mathcal{C}$ be a class of finite relational structures. We define formulas of a *weighted* MSO-*logic over* $\mathcal{C}$ in the same manner as for traces, *i.e.* atomic formulas are beside k for $k \in K$, and $x \in X$ the relation symbols of $\mathcal{C}$ and possibly $x = y$, and negation is applied to atomic formulas only. The formulas are provided with the appropriate semantics $S : \mathcal{C} \to \mathbb{K}$ as for traces, *i.e.* atomic formulas are interpreted by the characteristic series of the defined language (a valid $\mathcal{V}$-assignment provided) and the semantics of composed formulas is given as above. Similarly, an unweighted MSO-logic for $\mathcal{C}$ is defined with a semantics of languages $L \subseteq \mathcal{C}$. Moreover, we suppose that there is a propositional formula $\Omega(x, y)$ (*i.e.* one without any quantifier) with free FO-variables x, y such that for any structure $t \in \mathcal{C}$ the binary relation defined by Ω is a linear order on the elements of t. We say that $\mathcal{C}$ has a *simply definable linear order*.

Let $L \subseteq \mathcal{C}$ be a language of $\mathcal{C}$ and $\overline{L} = \mathcal{C} \setminus L$ the complement of L.

Lemma 5.1. *Let $\mathcal{C}$ be a class of finite relational structures with a simply definable linear order. Let $L = L(\Phi)$ be defined by an FO-formula Φ. Then both $\mathbb{1}_L$ and $\mathbb{1}_{\overline{L}}$ are definable in RFO.*

Proof (sketch). Let $L \subseteq \mathcal{C}$ be defined by Φ. We proceed by induction giving for each FO-formula Φ RFO-formulas Φ^+ and Φ^- such that $[\![\Phi^+]\!] = \mathbb{1}_L$ and $[\![\Phi^-]\!] = \mathbb{1}_{\overline{L}}$. The interesting cases are $(\exists x.\Phi)^+$ and $(\forall x.\Phi)^-$. Therefore, we choose the "smallest" element that satisfies Φ^+, and Φ^- respectively, by using $\Omega(x, y)$ defining the linear order $\leq_\Omega$. Since $\Omega(x, y)$ is a propositional formula we can already define $\Omega^+(x, y)$. Now we put

$$\left(\exists x.\Phi(x)\right)^+ = \exists x.\left(\Phi^+(x) \wedge \forall y.\left(\Phi^-(y) \vee \Omega^+(x, y)\right)^+\right).$$

This is an RFO-formula. Indeed, $\Phi^+(x)$ is an RFO-formula by induction hypothesis and so are $\Phi^-(y)$ and $\Omega^+(x, y)$. Moreover, $\left(\Phi^-(y) \vee \Omega^+(x, y)\right)^+$ defines a definable step function by induction hypothesis. Since we choose the "smallest" element x satisfying Φ we get for a valid $\mathcal{V}$-assignment

$$[\![\,(\exists x.\Phi(x))^+\,]\!]_{\mathcal{V}}(t, \sigma) = \begin{cases} 1 & \text{if there is an } v \text{ such that } (t, \sigma[x \to v]) \text{ satisfies } \Phi, \\ 0 & \text{otherwise.} \end{cases}$$

Similarly we proceed for $(\forall x.\Phi)^-$. $\square$

Corollary 5.2. *Let L be an FO-definable trace language. Then $\mathbb{1}_L$ is RFO-definable.*

Proof. For a fixed linear order $\preceq$ on the alphabet Σ put

$$\Omega(x, y) = \bigvee_{(a,b)\in\prec} \left(P_a(x) \wedge P_b(y)\right) \vee \bigvee_{a\in\Sigma} \left(P_a(x) \wedge P_a(y) \wedge \neg E(y, x)\right)$$

and apply Lemma 5.1. $\square$

6 The Coincidence of Recognizable and Definable Trace Series

We will follow the ideas of the proof as given for trace languages, cf. [5, pp. 497–505] and use the result of the previous section.

Lemma 6.1. *Let $\mathbb{K}$ be a commutative semiring, $\varphi : \Sigma^* \to \mathbb{M}$ the canonical epimorphism, and $T \in \mathbb{K}\langle\!\langle\mathbb{M}\rangle\!\rangle$ a trace series. The following are equivalent:*

(i) T is definable in RMSO, and REMSO respectively.
(ii) $\varphi^{-1}(T) \in \mathbb{K}\langle\!\langle\Sigma^*\rangle\!\rangle$ is definable in RMSO, and REMSO respectively.
(iii) $S = \varphi^{-1}(T)|_{\text{LNF}} \in \mathbb{K}\langle\!\langle\Sigma^*\rangle\!\rangle$ is definable in RMSO, and REMSO respectively.

Proof. (i) $\Longrightarrow$ (ii) Let $T \in \mathbb{K}\langle\!\langle\mathbb{M}\rangle\!\rangle$ be defined by some sentence Ψ. Let $t = (V, E, l)$ be any trace and $w \in \Sigma^*$ with $\varphi(w) = t$. We have for $v_1, v_2 \in V$ that $(v_1, v_2) \in E$ iff $v_1 < v_2$ in w and $(l(v_1), l(v_2)) \in D$. Thus, replacing every atomic formula $E(x, y)$ in Ψ by the propositional formula

$$x < y \wedge \bigvee_{(a,b)\in D} (P_a(x) \wedge P_b(y)) \tag{1}$$

yields an new sentence $\tilde{\Psi}$. One shows easily $([\![\,\Psi\,]\!], t) = ([\![\,\tilde{\Psi}\,]\!], w)$ for each $t \in \mathbb{M}, w \in \Sigma^*$ with $\varphi(w) = t$ by structural induction.

It still remains to show that for $\Psi \in \mathrm{RMSO}(\mathbb{K}, \mathbb{M})$ also $\tilde{\Psi} \in \mathrm{RMSO}(\mathbb{K}, \Sigma)$. Clearly, if Ψ contains no universal second-order quantification neither does $\tilde{\Psi}$. Consider $\Psi = \forall x.\Phi$ with $[\![\Phi]\!] = \sum_{i=1}^{n} k_i \mathbb{1}_{L_i}$ for definable and, hence, recognizable trace languages $L_i \subseteq \mathbb{M}$ for $i = 1, \ldots, n$. As we have shown, $[\![\tilde{\Psi}]\!](w, \sigma) = [\![\Psi]\!](t, \sigma)$ where $\varphi(w) = t$. Consider the word series $S = \sum_{i=1}^{n} k_i \mathbb{1}_{\varphi^{-1}(L_i)}$. It is a recognizable and, hence, definable step function over words since $\varphi^{-1}(L_i)$ is recognizable for $i = 1, \ldots, n$. Moreover, $(w, \sigma) \in \varphi^{-1}(L_i)$ for some i implies that σ is a valid $\mathcal{V}$-assignment. For (w, σ) with σ a valid $\mathcal{V}$-assignment we have

$$S(w, \sigma) = \bigoplus_{i=1}^{n} k_i \mathbb{1}_{\varphi^{-1}(L_i)}(w, \sigma) = \bigoplus_{\{i \mid w \in \varphi^{-1}(L_i)\}} k_i = \bigoplus_{\{i \mid t = \varphi(w) \in L_i\}} k_i = [\![\Phi]\!](t, \sigma).$$

Hence, $S = \tilde{\Phi}$ is a definable step function. Thus, if Ψ is reduced so is $\tilde{\Psi}$. Moreover, for $\Psi \in \mathrm{REMSO}(\mathbb{K}, \mathbb{M})$ also $\tilde{\Psi} \in \mathrm{REMSO}(\mathbb{K}, \Sigma)$ because (1) is a propositional formula.

(ii) $\implies$ *(iii)* Let $\varphi^{-1}(T) \in \mathbb{K} \langle\!\langle \Sigma^* \rangle\!\rangle$ be defined by an RMSO-formula Φ, let $S = \varphi^{-1}(T)_{|\mathrm{LNF}}$, and let $\preceq$ be the fixed order on Σ. The language LNF of all lexicographic normal forms is defined by the FO-sentence

$$\forall i \forall k.\Big[(i \le k) \longrightarrow \Big(l(i) \preceq l(k) \lor \exists j.(i \le j < k \land (l(j), l(k)) \in D)\Big)\Big]$$

where implication $\longrightarrow$, $l(i) \preceq l(k)$, and $(l(j), l(k)) \in D$ are obvious abbreviations. By Corollary 5.2, there is an RFO-formula Λ with $[\![\Lambda]\!] = \mathbb{1}_{\mathrm{LNF}}$. Hence, $S = [\![\Phi \land \Lambda]\!]$. If $\Phi = \exists X_1 \ldots \exists X_n.\Psi$ is an REMSO-formula, then S is defined by the REMSO-formula $\exists X_1 \ldots \exists X_n.(\Psi \land \Lambda)$ because Λ is from RFO.

(iii) $\implies$ *(i)* Let $S = \varphi^{-1}(T)_{|\mathrm{LNF}}$ be defined by $\Phi \in \mathrm{RMSO}$. Then $\mathrm{supp}(S) \subseteq \mathrm{LNF}$. We replace every atomic formula $x < y$ in Φ by a new formula $\mathrm{lex}(x, y)$ that models the order in the lexicographic normal form, *i.e.* for every t and a valid $\mathcal{V}$-assignment σ we get $(t, \sigma) \models \mathrm{lex}(x, y)$ iff $\sigma(x) < \sigma(y)$ in $\mathrm{LNF}(t)$. The formula $\mathrm{lex}(x, y)$ can be found in the literature (cf.[5, p. 502]) and is an FO-formula because transitive closure of E can be expressed for traces in FO. Hence, we apply Corollary 5.2 and obtain an RFO-formula $\mathrm{lex}^+(x, y)$ defining $\mathbb{1}_{L(\mathrm{lex}(x,y))}$, and similarly $\mathbb{1}_{L(\neg\,\mathrm{lex}(x,y))}$ can be defined by an RFO-formula $\mathrm{lex}^-(x, y)$. Let Ψ be any formula over words and $\tilde{\Psi}$ the formula over traces obtained from Ψ by replacing every occurence of the atomic formula $x < y$ by $\mathrm{lex}^+(x, y)$, and any occurence of $\neg(x < y)$ by $\mathrm{lex}^-(x, y)$. Then we get for every trace t and every valid $\mathcal{V}$-assignment σ

$$[\![\tilde{\Psi}]\!]_{\mathcal{V}}(t, \sigma) = [\![\Psi]\!]_{\mathcal{V}}(\mathrm{LNF}(t), \sigma). \tag{2}$$

We still have to show that $\tilde{\Psi}$ is restricted. For an RMSO-formula $\forall x.\Psi$ over words $[\![\Psi]\!]_{\mathcal{V}}$ is a definable and recognizable step function, *i.e.* $[\![\Psi]\!]_{\mathcal{V}}(t, \sigma) = \sum_{i=1}^{n} k_i \mathbb{1}_{L_i}$ with recognizable word languages L_i $(i = 1, \ldots, n)$. Now we have by Equation (2) $[\![\tilde{\Psi}]\!]_{\mathcal{V}} = \sum_{i=1}^{n} k_i \mathbb{1}_{\varphi(L_i \cap \mathrm{LNF})}$. By [6, Thm. 6.3.12] the trace languages $\varphi(L_i \cap \mathrm{LNF})$ are recognizable languages. Thus $[\![\tilde{\Psi}]\!]_{\mathcal{V}}$ is a recognizable and, hence, definable step function. Hence, if $S \in \mathbb{K} \langle\!\langle \Sigma^* \rangle\!\rangle$ with $\mathrm{supp}(S) \subseteq \mathrm{LNF}$ is defined by some sentence Φ from RMSO then $T = \varphi(S) \in \mathbb{K} \langle\!\langle \mathbb{M} \rangle\!\rangle$ is defined by the RMSO-sentence $\tilde{\Phi}$. Certainly, if Φ is in REMSO so is $\tilde{\Phi}$. $\qquad\square$

Remark 6.2. The translations of formulas over traces to those over words and vice versa as given in the proof of Lemma 6.1 are effective.

By Lemma 6.1 and the result for word series [9, Thm. 3.7] we get:

Theorem 6.3. *Let* $\mathbb{K}$ *be a commutative semiring and* $T \in \mathbb{K} \langle\!\langle \mathbb{M} \rangle\!\rangle$. *The following are equivalent:*

(i) T is recognizable,
(ii) T is definable by some sentence of RMSO,
(iii) T is definable by some sentence of REMSO.

Proof. Let T be recognizable. Then $S = \varphi^{-1}(T)_{|\,\mathrm{LNF}}$ is a recognizable word series by Theorem 3.1. Hence, S is definable in RMSO and REMSO by the main result of [9]. By Lemma 6.1, T is definable in RMSO and REMSO, respectively.

Conversely, let T be definable in RMSO and REMSO, respectively. Then the series $\varphi^{-1}(T)_{|\,\mathrm{LNF}}$ is definable by Lemma 6.1, hence recognizable by [9]. Now, Theorem 3.1 shows the recognizability of T. □

Example 6.4. Let $\mathbb{K} = (\mathbb{N} \cup \{-\infty\}, \max, +, -\infty, 0)$. We show that $H \in \mathbb{K} \langle\!\langle \mathbb{M} \rangle\!\rangle$ mapping every $t \in \mathbb{M}$ to height(t), *i.e.* the length of the longest chain in t, is recognizable. Let

$$\mathrm{chain}(X) = \forall x, y \in X. \big(x = y \vee (x, y) \in E^+ \vee (y, x) \in E^+ \big)$$

be an unweighted formula stating that X is a chain. Since transitive closure of E can be expressed for traces by an FO-formula (cf. [5, p. 501]), chain(X) is an FO-formula. By Corollary 5.2, there is an RFO-formula chain$(X)^+$ defining $\mathbb{1}_{L(\mathrm{chain}(X))}$. Moreover, the formula

$$\mathrm{card}(X) = \forall x. \big((x \in X \longrightarrow 1) \wedge (\neg x \in X \longrightarrow 0) \big)$$

has the semantics $|X|$ over $\mathbb{K}$. Hence, $H = \sum_{t \in \mathbb{M}} \mathrm{height}(t)\, t$ is defined by

$$\Phi = \exists X.\, \mathrm{chain}(X)^+ \wedge \mathrm{card}(X).$$

By Theorem 6.3, $H \in \mathbb{K} \langle\!\langle \mathbb{M} \rangle\!\rangle$ is recognizable.

7   Some Notes About Decidability

Given a weighted MSO-formula Φ over traces, there are two immediate questions:

– It is decidable whether Φ is an RMSO-formula?
– If Φ is in RMSO, can we effectively compute the semantics of Φ, *i.e.* compute $(\llbracket \Phi \rrbracket, t)$ for every trace $t \in \mathbb{M}$?

Droste and Gastin [9] answer these questions for weighted logics over words where the underlying semiring is either a computable field or a locally finite semiring. We cannot expect to do any better. By the effective translation of formulas the results carry over.

Proposition 7.1. *Let* $\mathbb{K}$ *be a computable field, and let* $\Phi \in \mathrm{MSO}(\mathbb{K}, \mathbb{M})$. *It is decidable whether* Φ *is reduced. In this case we can compute effectively for every trace* $t \in \mathbb{M}$ *the coefficient* $(\llbracket \Phi \rrbracket, t)$ *in a uniform way.*

Corollary 7.2. *Let* $\mathbb{K}$ *be a computable field, and let* $\Phi, \Psi \in \mathrm{RMSO}(\mathbb{K}, \mathbb{M})$. *It is decidable whether* $\llbracket \Phi \rrbracket$ *has empty support, whether* $\llbracket \Phi \rrbracket = \llbracket \Psi \rrbracket$, *and whether* $\llbracket \Phi \rrbracket$ *and* $\llbracket \Psi \rrbracket$ *differ for finitely many traces only.*

Recall that a semiring $\mathbb{K}$ is *locally finite*, if each finitely generated subsemiring of $\mathbb{K}$ is finite. A monoid M is locally finite, if each finitely generated submonoid of M is finite. Clearly, a semiring $(K, \oplus, \circ, \mathbb{0}, \mathbb{1})$ is locally finite iff both monoids $(K, \oplus, 0)$ and $(K, \circ, \mathbb{1})$ are locally finite. Now even every MSO-definable trace series is recognizable as it is true for word series [9, Thm. 6.4].

Theorem 7.3. *Let* $\mathbb{K}$ *be a locally finite commutative semiring and* $T \in \mathbb{K} \langle\langle \mathbb{M} \rangle\rangle$. *Then the following are equivalent:*

(i) T *is definable in* MSO.
(ii) T *is recognizable.*

Proposition 7.4. *Let* $\mathbb{K}$ *be a locally finite commutative semiring and* $\Phi \in \mathrm{MSO}(\mathbb{K}, \mathbb{M})$. *Then the coefficient* $(\llbracket \Phi \rrbracket, t)$ *can be computed effectively for every* $t \in \mathbb{M}$ *in a uniform way. Moreover, it is decidable*

(a) *whether two* $\mathrm{MSO}(\mathbb{K}, \mathbb{M})$-*formulas* Φ *and* Ψ *satisfy* $\llbracket \Phi \rrbracket = \llbracket \Psi \rrbracket$, *and*
(b) *whether an* $\mathrm{MSO}(\mathbb{K}, \mathbb{M})$-*formula* Φ *satisfies* $\mathrm{supp}(\llbracket \Phi \rrbracket) = \mathbb{M}$.

8   FO-Definable Trace Series

By considering Lemma 6.1 and its proof we get:

Lemma 8.1. *Let* $\mathbb{K}$ *be a commutative semiring and* $\varphi : \Sigma^* \to \mathbb{M}$ *the canonical epimorphism. The following are equivalent:*

(i) $T \in \mathbb{K} \langle\langle \mathbb{M} \rangle\rangle$ *is definable in* RFO *(in* FO, *respectively).*
(ii) $\varphi^{-1}(T) \in \mathbb{K} \langle\langle \Sigma^* \rangle\rangle$ *is definable in* RFO *(in* FO, *respectively).*
(iii) $\varphi^{-1}(T)_{| \mathrm{LNF}} \in \mathbb{K} \langle\langle \Sigma^* \rangle\rangle$ *is definable in* RFO *(in* FO, *respectively).*

Droste and Gastin showed that the classes of aperiodic word series, RFO-definable and FO-definable word series coincide for commutative weakly bi-aperiodic semirings [9, Thm. 7.8]. A monoid M is *weakly aperiodic*, if for each $m \in M$ there is an $n \geq 0$ such that $m^n = m^{n+1}$. M is *aperiodic* if there is an $n \geq 0$ such that $m^n = m^{n+1}$ for all $m \in M$. A semiring $\mathbb{K}$ is *weakly bi-aperiodic*, if both $(K, \oplus)$ and $(K, \circ)$ are weakly aperiodic monoids. Note that every commutative weakly aperiodic semiring K is locally finite. Let $S \in \mathbb{K} \langle\langle M \rangle\rangle$ be a recognizable series over an arbitrary monoid M. Then S is called *aperiodic* if there exists a representation $S = (\lambda, \mu, \gamma)$ with $\mu(M)$ aperiodic, *i.e.* there is some integer $n \geq 0$ such that $\mu(u^n) = \mu(u^{n+1})$ for all $u \in M$. A recognizable series S is *weakly aperiodic* if there exists some integer $n \geq 0$ such that $(S, uv^n w) = (S, uv^{n+1}w)$ for all $u, v, w \in M$. Clearly, every aperiodic series is also weakly aperiodic. The converse is true for locally finite semirings as already Droste and Gastin noted [8, Sect. 3].

Lemma 8.2. *Let $\mathbb{K}$ be a locally finite semiring, M a finitely generated monoid, and $S \in \mathbb{K}\langle\!\langle M \rangle\!\rangle$ a recognizable series. Then S is aperiodic iff S is weakly aperiodic.*

Using Lemma 8.2, we can clarify the relation between aperiodic trace and aperiodic word series.

Proposition 8.3. *Let $\mathbb{K}$ be a locally finite semiring. Then $T \in \mathbb{K}\langle\!\langle \mathbb{M} \rangle\!\rangle$ is aperiodic iff $\varphi^{-1}(T) \in \mathbb{K}\langle\!\langle \Sigma^* \rangle\!\rangle$ is aperiodic.*

Proof. If T has the aperiodic representation $T = (\lambda, \mu, \gamma)$ such that there is an $r \in \mathbb{N}$ with $\mu(t^r) = \mu(t^{r+1})$ for all $t \in \mathbb{M}$. Then $\varphi^{-1}(T)$ has the aperiodic representation $(\lambda, \mu \circ \varphi, \gamma)$. Vice versa, let $\varphi^{-1}(T)$ be aperiodic and, hence, also weakly aperiodic, *i.e.* there is some $r \in \mathbb{N}$ with $(\varphi^{-1}(T), uv^r w) = (\varphi^{-1}(T), uv^{r+1}w)$ for all $u, v, w \in \Sigma^*$. By t' we denote some representative for the trace t. Now we have

$$(T, xt^r y) = (\varphi^{-1}(T), (xt^r y)') = (\varphi^{-1}(T), x't'^r y') = (\varphi^{-1}(T), x't'^{r+1}y')$$
$$= (\varphi^{-1}(T), (xt^{r+1}y)') = (T, xt^{r+1}y)$$

and, thus, T is weakly aperiodic. By Lemma 8.2, T is aperiodic. □

Theorem 8.4. *Let $\mathbb{K}$ be a commutative, weakly bi-aperiodic semiring and $T \in \mathbb{K}\langle\!\langle \mathbb{M} \rangle\!\rangle$. Then the following are equivalent:*

(i) T is aperiodic.
(ii) T is weakly aperiodic.
(iii) T is RFO-definable.
(iv) T is FO-definable.

Proof. Recall that a commutative, weakly bi-aperiodic semiring $\mathbb{K}$ is locally finite. Then the equivalence of (i) and (ii) is clear by Lemma 8.2. Now, let T be aperiodic. Then $\varphi^{-1}(T) \in \mathbb{K}\langle\!\langle \Sigma^* \rangle\!\rangle$ is aperiodic by Proposition 8.3. Now, [9, Thm. 7.8] implies RFO- and FO-definability of $\varphi^{-1}(T)$. By Lemma 8.1, T is RFO- and FO-definable, respectively. The converse direction follows similarly. □

Acknowledgements. The author would like to thank Dietrich Kuske for a lot of fruitful discussions.

References

1. J. Berstel and C. Reutenauer. *Rational Series and Their Languages*, volume 12 of *EATCS Monographs on Theoret. Comp. Sc.* Springer, 1988.
2. B. Bollig. On the expressiveness of asynchronous cellular automata. In *Proceedings of the 15th International Symposium on Fundamentals of Computation Theory (FCT'05), Lübeck, Germany, August 2005*, volume 3623 of *Lecture Notes in Comp. Sc.*, pages 528–539. Springer, 2005.
3. B. Bollig and M. Leucker. Message-passing automata are expressively equivalent to EMSO logic. *Theoret. Comp. Sc.*, 2006. to appear.

4. J.R. Büchi. Weak second-order arithmetic and finite automata. *Z. Math. Logik Grundlagen Math.*, (6):66–92, 1960.
5. V. Diekert and Y. Métivier. Partial commutation and traces. In G. Rozenberg and A. Salomaa, editors, *Handbook of Formal Languages, Beyond Words*, volume 3, chapter 8, pages 457–534. Springer, 1997.
6. V. Diekert and G. Rozenberg, editors. *The Book of Traces*. World Scientific, 1995.
7. M. Droste and P. Gastin. The Kleene-Schützenberger theorem for formal power series in partially commuting variables. *Information and Computation*, 153:47–80, 1999.
8. M. Droste and P. Gastin. On aperiodic and star-free formal power series in partially commuting variables. In *Formal Power Series and Algebraic Combinatorics (Moscow 2000)*, pages 158–169. Springer Berlin, 2000.
9. M. Droste and P. Gastin. Weighted automata and weighted logics. In *Automata, Languages and Programming (32nd ICALP, Lissabon)*, volume 3580 of *Lecture Notes in Comp. Sc.*, pages 513–525. Springer, 2005.
10. M. Droste and H. Vogler. Weighted tree automata and weighted logics. submitted, 2005.
11. W. Ebinger and A. Muscholl. Logical definability on infinite traces. *Theoret. Comp. Sc.*, (154):67–84, 1996.
12. C.C. Elgot. Decision problems of finite automata design and related arithmetics. *Trans. Amer. Math. Soc.*, (98):21–52, 1961.
13. J.S. Golan. *Semirings and their Applications*. Kluwer Academic Publishers, 1999.
14. U. Hebisch and H.J. Weinert. *Semirings: Algebraic Theory and Application*. World Scientific, 1999.
15. W. Kuich and A. Salomaa. *Semirings, Automata, Languages*, volume 5 of *EATCS Monographs on Theoret. Comp. Sc.* Springer, 1986.
16. D. Kuske. Weighted asynchronous cellular automata. In *STACS 2006*, volume 3884 of *Lecture Notes in Comp. Sc.*, pages 685–696. Springer, 2006.
17. D. Kuske and I. Meinecke. Branching automata with costs – a way of reflecting parallelism in costs. *Theoret. Comp. Sc.*, 328:53–75, 2004.
18. I. Mäurer. Weighted picture automata and weighted logics. In *STACS 2006*, Lecture Notes in Comp. Sc. Springer, 2006. to appear.
19. A. Mazurkiewicz. Trace theory. In *Advances in Petri Nets 1986, Part II on Petri Nets: Applications and Relationships to Other Models of Concurrency*, volume 255 of *Lecture Notes in Comp. Sc.*, pages 279–324. Springer, 1987.
20. I. Meinecke. *Weighted Branching Automata – Combining Concurrency and Weights*. Dissertation, Technische Universität Dresden, Germany, December 2004.
21. I. Meinecke. The Hadamard product of sequential-parallel series. *J. of Automata, Languages and Combinatorics*, 10(2), 2005. To appear.
22. A. Salomaa and M. Soittola. *Automata-Theoretic Aspects of Formal Power Series*. Texts and Monographs in Computer Science. Springer, 1978.
23. M.P. Schützenberger. On the definition of a family of automata. *Information and Control*, 4:245–270, 1961.
24. W. Thomas. On logical definability of trace languages. In V. Diekert, editor, *Proceedings of a workshop of the ESPRIT BRA No 3166: Algebraic and Syntactic Methods in Computer Science (ASMICS) 1989*, pages 172–182. Technical University of Munich, 1990.

On Nonforgetting Restarting Automata That Are Deterministic and/or Monotone

Hartmut Messerschmidt and Friedrich Otto

Fachbereich Mathematik/Informatik, Universität Kassel, D-34109 Kassel
{hardy, otto}@theory.informatik.uni-kassel.de

Abstract. The nonforgetting restarting automaton is a restarting automaton that is not forced to reset its internal state to the initial state when executing a restart operation. We analyse the expressive power of the various deterministic and/or monotone variants of this model.

1 Introduction

The restarting automaton was introduced by Jančar et. al. [4] to model the *analysis by reduction*, which is a technique used in linguistics to analyse sentences of natural languages. According to this technique a sentence is processed by simplifying it stepwise, in each step preserving the correctness or incorrectness of the sentence. In Czech and German linguistics already several programs use the idea of restarting automata [9, 13].

A (two-way) restarting automaton M consists of a *finite-state control* Q, a finite *tape alphabet* Γ containing an *input alphabet* Σ, a *flexible tape* with a *left border marker* ¢ and a *right border marker* $, and a *read/write window* of a fixed size $k \geq 1$. Its actions are governed by a *transition relation* δ that assigns to each pair (q, u) consisting of a state $q \in Q$ and a possible contents u of the window a finite set of transition steps, of which there are five types: *move-right* and *move-left steps*, which change the internal state of M and shift the window one position to the right or to the left, respectively, *rewrite steps* that change the internal state and replace the contents of the window by some shorter string, thereby shortening the tape, *restart steps* that place the window over the left end of the tape and reset the internal state to the initial state q_0, and *accept steps* that cause M to halt and accept. Thus, each computation of M proceeds in *cycles*: starting from a (restarting) configuration of the form $q_0 ¢ x \$$, M performs move-right, move-left and rewrite steps until a restart operation takes it back to a restarting configuration of the form $q_0 ¢ y \$$. As part of the definition it is required that in each such cycle M executes *exactly one* rewrite operation. The part of a computation that follows after the last application of a restart step will be denoted as the *tail* of the computation. By now many restricted variants of the restarting automaton have been studied, and it has been shown that many well studied language classes can be characterised by variants of the restarting automaton (see, e.g., [10, 11]). Actually the main variants of the restarting automaton are obtained by combining two types of restrictions:

D. Grigoriev, J. Harrison, and E.A. Hirsch (Eds.): CSR 2006, LNCS 3967, pp. 247–258, 2006.

(a) Restrictions on the movement of the read/write window (expressed by the first part of the class name):
- RL- means no restriction,
- RR- means that no move-left operations are available,
- R- means that no move-left operations are available and that each rewrite step is followed immediately by a restart step.

(b) Restrictions on the rewrite instructions (expressed by the second part of the class name, where λ denotes the empty string):
- -WW means no restriction,
- -W means that no auxiliary symbols are available (that is, $\Gamma = \Sigma$),
- -λ means that each rewrite step simply deletes some symbols, that is, if $(q', v) \in \delta(q, u)$, then v is a scattered proper subword of u.

Obviously, each restarting automaton M can be simulated by a linear-bounded automaton, and so the language $L(M)$ of strings accepted by M is included in the class CSL of context-sensitive languages. In fact, the restarting automaton is more restricted than a linear-bounded automaton in two ways:

- each rewrite operation of a restarting automaton strictly reduces the length of the tape (inscription), and
- between any two rewrite operations a restarting automaton must execute a restart step, which has two effects:

  (a) the automaton does not remember the place of the last rewrite operation, as the window is moved back to the left end of the tape, and
  (b) the automaton does not remember that it has at all done some rewrite steps, as its internal state is reset to the initial state.

Thus, a restarting automaton must encode this information in the tape inscription, but it can do so only in a very restricted way, as it can only perform a single rewrite operation between any two restart steps, and, in addition, this rewrite operation is length-reducing.

Here we investigate the influence of the first effect of the restart operation listed above on the expressive power of the restarting automaton. For this we introduce a more general model, the so-called *nonforgetting restarting automaton*. As the standard model it is required to perform exactly one rewrite operation between any two restart steps, but the restart operation is more powerful than in the standard model as it involves a change of the internal state just like any of the other operations. Hence, when executing a restart step, such an automaton does not forget the information it has collected internally about the state of its actual computation, but only the place of the latest rewrite operation is forgotten. How much does this influence the expressive power of the various types of restarting automata?

We will restrict our attention to nonforgetting restarting automata that are deterministic or monotone. Deterministic RWW- and RRWW-automata accept exactly the *Church-Rosser languages* (see, e.g., [10]), while monotone deterministic R(R)(W)(W)-automata all accept the determinisitic context-free languages [5]. Here we will see that deterministic nonforgetting RWW-automata are strictly more expressive than the class CRL of Church-Rosser languages, while deterministic monotone nonforgetting R-, RW-, and RWW-automata still characterize the class DCFL of deterministic context-free languages. On the other hand, the deterministic monotone nonforgetting RR-, RRW-, and RRWW-automata form a strict hierarchy above DCFL. In particular, this means that the deterministic monotone nonforgetting RWW-automaton is stricly weaker in expressive power than the corresponding RRWW-automaton, which represents the first separation result in the literature between an RWW- and an RRWW-automaton of the same restricted form. Further, we will see that the nondeterministic monotone nonforgetting RWW- and RRWW-automata are just as powerful as the corresponding standard models in that they still characterize the class CFL of context-free languages.

2   Basic Notions and Examples

In [8] the *shrinking restarting automaton* was introduced as another generalization of the restarting automaton. For such an automaton it is only required that each rewrite operation reduces the weight of the tape content with respect to some weight function. This turned out to be a very robust model in that the class of languages accepted by shrinking restarting automata is not even effected by allowing the automata to perform several (but constantly many) rewrite operations between any two restart steps, or by making them nonforgetting [7]. As these automata are equivalent in expressive power to the so-called *finite-change automata* of [1], it follows that nonforgetting restarting automata only accept deterministic context-sensitive languages.

In the following we will use the prefix nf- to denote classes of nonforgetting restarting automata, and the prefix det- to denote classes of deterministic (nonforgetting) restarting automata.

Each cycle of each computation of an nf-RLWW-automaton M consists of four phases: 1. scan the tape, 2. perform a rewrite operation, 3. scan the tape again, 4. execute a restart step. During the first and the third of these phases M behaves like a nondeterministic two-way finite-state acceptor. Hence, in analogy to the proof that the language accepted by a two-way finite-state acceptor is regular (see, e.g, [3]), it can be shown that, for each nf-RLWW-automaton M, there exists an nf-RRWW-automaton M' that executes exactly the same cycles in each computation as M. Thus, in particular, the languages $L(M)$ and $L(M')$ coincide. This is a slight generalization of a result of [12]. It should be pointed out that a corresponding result does not hold for deterministic restarting automata.

For an nf-RRWW-automaton M, the first and the third phase of each cycle consist only of MVR-steps, that is, during these phases M behaves just like a (one-way) finite-state acceptor. Thus, the transition relation of M can be described more compactly by so-called *meta-instructions*.

A *meta-instruction* for an nf-RRWW-automaton M is either of the form $(p_1, E_1, u \rightarrow v, E_2, p_2)$ or $(p_1, E_1, \text{Accept})$, where p_1 and p_2 are internal states, E_1 and E_2 are regular expressions, and $u, v \in \Gamma^*$ are words satisfying $|u| > |v|$, which stand for a rewrite step of M. To execute a cycle M chooses a meta-instruction of the form $(p_1, E_1, u \rightarrow v, E_2, p_2)$. On trying to execute this meta-instruction M will get stuck (and so reject) starting from a configuration $p_1 \mathcal{c} w \$$, if w does not admit a factorization of the form $w = w_1 u w_2$ such that $\mathcal{c} w_1 \in E_1$ and $w_2 \$ \in E_2$. On the other hand, if w does have factorizations of this form, then one such factorization is chosen nondeterministically, and $p_1 \mathcal{c} w \$$ is transformed into $p_2 \mathcal{c} w_1 v w_2 \$$. In order to describe the tails of accepting computations we use meta-instructions of the form $(p_1, E_1, \text{Accept})$, where the strings from the regular language E_1 are accepted by M in tail computations, that is, without a restart. As an nf-RWW-automaton restarts immediately after executing a rewrite operation, the meta-instructions describing cycles of such an automaton are of the simplified form $(p_1, E_1, u \rightarrow v, p_2)$.

Example 1 (Exponential languages). The language $L_{\text{expo}} := \{ a^{2^n} \mid n \in \mathbb{N} \}$ is accepted by the det-nf-RWW-automaton with one auxiliary symbol B that is given through the following meta-instructions:

(i) $(q_0, \mathcal{c} a^*, aaaa\$ \rightarrow Baa\$, q_1)$,   (iv) $(q_1, \mathcal{c} a^*, aaB \rightarrow Ba, q_1)$,
(ii) $(q_1, \mathcal{c}, B \rightarrow \lambda, q_0)$,   (v) $(q_0, \mathcal{c} aa\$, \text{Accept})$.
(iii) $(q_0, \mathcal{c} a\$, \text{Accept})$,

The language $L'_{\text{expo}} := \{ a^{2^n} b \mid n \in \mathbb{N} \}$ is accepted by the det-nf-RW-automaton given through the following meta-instructions:

(i) $(q_0, \mathcal{c} a^+, aab\$ \rightarrow ba\$, q_1)$,   (v) $(q_1, \mathcal{c} a^*, aab \rightarrow ba, q_1)$,
(ii) $(q_1, \mathcal{c}, b \rightarrow \lambda, q_1)$,   (vi) $(q_1, \mathcal{c} a^*, aaaa\$ \rightarrow baa\$, q_1)$,
(iii) $(q_1, \mathcal{c} aa\$, \text{Accept})$,   (vii) $(q_0, \mathcal{c} ab\$, \text{Accept})$.
(iv) $(q_0, \mathcal{c} aab\$, \text{Accept})$,

3   Deterministic Restarting Automata

Deterministic RWW- and RRWW-automata accept the *Church-Rosser languages* (CRL) (see, e.g., [10]), which are the deterministic variants of the *growing context-sensitive languages* (GCSL). While the language $L_{\text{copy}} := \{ w \# w \mid w \in \{a, b\}^* \}$ is not even growing context-sensitive [2], it is accepted by the det-nf-R-automaton that is given through the following meta-instructions:

(i) $(q_0, \mathcal{c} a \cdot \{a, b\}^* \cdot \#, a \rightarrow \lambda, q_1)$,   (iv) $(q_1, \mathcal{c}, a \rightarrow \lambda, q_0)$,
(ii) $(q_0, \mathcal{c} b \cdot \{a, b\}^* \cdot \#, b \rightarrow \lambda, q_1)$,   (v) $(q_1, \mathcal{c}, b \rightarrow \lambda, q_0)$.
(iii) $(q_0, \mathcal{c} \# \$, \text{Accept})$,

As it is easily seen that the language L_{expo} (L'_{expo}) is not accepted by any nf-RW-automaton (nf-R-automaton), we have the situation depicted in Figure 1, where $\mathcal{L}_1 \rightarrow \mathcal{L}_2$ expresses the fact that the language class $\mathcal{L}_1$ is properly contained in the class $\mathcal{L}_2$, while $\mathcal{L}_1 \underset{?}{\rightarrow} \mathcal{L}_2$ indicates an inclusion for which it is still open whether or not it is proper.

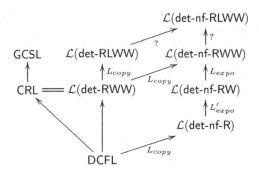

Fig. 1. The taxonomy of deterministic nonforgetting restarting automata

In addition, it is also open whether, for any choice of $X \in \{WW, W, \lambda\}$, any of the inclusions $\mathcal{L}(\text{det-nf-RX}) \subseteq \mathcal{L}(\text{det-nf-RRX}) \subseteq \mathcal{L}(\text{det-nf-RLX})$ is proper.

To further illustrate the expressive power of det-nf-RW-automata, we now consider the language $L_{copy^*} := \{\ (\omega\#)^+ \mid \omega \in \{a, b\}^*\ \}$.

Lemma 1. $L_{copy^*} \in \mathcal{L}(\text{det-nf-RW})$.

Proof. A det-nf-RW-automaton M for the language L_{copy^*} proceeds as follows. Given an input of the form $\omega_1 \# \omega_2 \# \ldots \# \omega_m \#$, the last two symbols of ω_1 are replaced by a new copy of the symbol $\#$, thus creating an occurrence of a factor $\#\#$, and stored in the state of M. In the next cycle M compares the stored symbols to the last two symbols of ω_2. If they do not agree, then M halts and rejects, otherwise the last two symbols of ω_2 are also replaced by the symbol $\#$. This continues until all syllables $\omega_2, \ldots, \omega_m$ have been processed. Then M enters a new restart state, which causes M to replace each factor of the form $\#\#$ by the symbol $\#$, one at a time, proceeding from left to right. Once this task is completed, every syllable ω_i has been shortened by two symbols, which have been verified to agree for all syllables. Now M reenters the initial state, proceeding to process the now shortened word. This process continues until either a disagreement between some factors is found, or until all factors have been shortened to length at most 2, and their agreement can be checked simply by scanning the tape from left to right.   □

Essentially the same method can be used to design a det-nf-RW-automaton for the language $\text{VALC}(T)$ of all valid computations of a single-tape Turing machine T. This language consists of all words of the form $\omega_0 \# \omega_1 \# \ldots \# \omega_n \#$, where ω_0 is an initial configuration of T, ω_n is an accepting configuration of T, and ω_{i+1} is an immediate successor configuration of the configuration ω_i for all $0 \leq i \leq n - 1$. Each configuration ω_i is of the form $t_0 t_1 \ldots t_{j-1} q t_j t_{j+1} \ldots t_m$, where $t_0 t_1 \ldots t_m$ is the support of the tape inscription and q is the current state of T, scanning t_j.

Lemma 2. $\text{VALC}(T) \in \mathcal{L}(\text{det-nf-RW})$.

Proof. To accept VALC(T) a det-nf-RW-automaton M first checks that all syllables ω_i are configurations of T. This is done in the first cycle. Also M checks in this cycle whether, for all $0 \le i \le n-1$, ω_{i+1} is a possible successor configuration of ω_i. This means that, if ω_i contains the factor $t_{j-1}qt_jt_{j+1}$, where q is a state of T, and ω_{i+1} contains the factor $t_{l-1}pt_lt_{l+1}$, where p is a state of T, then either

- $\delta_T(q, t_j) = (p, t_{l-1}, R)$ and $t_{j+1} = t_l$, or
- $\delta_T(q, t_j) = (p, t_l, N)$, $t_{j-1} = t_{l-1}$, and $t_{j+1} = t_{l+1}$, or
- $\delta_T(q, t_j) = (p, t_{l+1}, L)$ and $t_{j-1} = t_l$

holds, where δ_T denotes the transition function of T. When M reaches the end of the tape, the last occurrence of the symbol $\#$ is removed, and M restarts in a non-initial state, if all these tests were positive. In this first cycle M cannot possibly check whether the various configurations are consistent with each other. This is done in the next phase, where a variant of the method to accept the language L_{copy^*} is used to shorten the tape content and to verify letter by letter that the states occur at the correct places within the various configurations, and that the tape content of each configuration is consistent with the tape content of the next configuration.   □

Actually, using a properly chosen encoding (see, e.g., [11]) each det-nf-RW-automaton can be simulated by a det-nf-R-automaton. These observations yield the following undecidablility results.

Theorem 1. *The following problems are in general undecidable:*

INSTANCE   : *A* det-nf-R-*automaton* M.
QUESTION 1 : *Is the language* $L(M)$ *non-empty?*
QUESTION 2 : *Is the language* $L(M)$ *finite?*
QUESTION 3 : *Is the language* $L(M)$ *regular?*
QUESTION 4 : *Is the language* $L(M)$ *context-free?*

Observe that for a det-R-automaton, emptiness of the language accepted is easily decidable.

4   Monotone Restarting Automata

Each cycle C of each computation of a restarting automaton M contains a unique configuration of the form $\text{¢}x_1qx_2\$$ in which a rewrite step is applied. Hence, we can associate with this cycle the number $D_r(C) := |x_2| + 1$, called the *right distance* of C. A sequence $C_1, C_2, \ldots, C_m$ of cycles of M is *monotone* if $D_r(C_1) \ge D_r(C_2) \ge \ldots \ge D_r(C_m)$ holds. A computation of M is *monotone* if the corresponding sequence of cycles is monotone, and M itself is *monotone* if each of its computations that starts from an initial configuration is monotone. It is known that monotone RWW- and RRWW-automata accept exactly the context-free languages [5]. Extending the proof of this result (cf. the proof of Theorem 3 below) the following generalization can be obtained.

Theorem 2. CFL $= \mathcal{L}(\text{mon-nf-RWW}) = \mathcal{L}(\text{mon-nf-RRWW})$.

Monotonicity is a decidable property of RRWW-automata (see, e.g., [11]), and this result extends to nonforgetting RRWW-automata.

For deriving some proper inclusion results we now consider the following example languages:

$L_{\text{pal}} := \{ \omega\omega^R \mid \omega \in \{a,b\}^* \}$,
$L_{\text{pal}'} := \{ \omega\omega^R \# \mid \omega \in \{a,b\}^* \}$, and
$L_{\text{pal}''} := \{ \omega c\phi(\omega)^R \mid \omega \in \{a,b,c\}^* \}$, where $\phi(a) := b$, and $\phi(b) := \phi(c) := a$.

The language L_{pal} is context-free, but based on the Error-Preserving Property the following negative result can be shown.

Lemma 3. *The language L_{pal} is not accepted by any monotone nf-RRW-automaton.*

Proof. Let M be a monotone nf-RRW-automaton. The Error Preserving Property for M tells us that, starting from an initial configuration that corresponds to an input word $w \notin L(M)$, M cannot reach any configuration that also occurs in an accepting computation of M. Assume that M accepts the language L_{pal}. We will now argue that the Error Preserving Property is necessarily violated for M.

Let p be an integer that is larger than the number of states of M. Observe that there exist strings $\omega \in L_{\text{pal}}$ such that each string $u \in \{a,b\}^k$ occurs as a factor in ω at more than p positions, where k is the size of the read/write window of M. Thus, if such a string $\omega = xx^R$ is given as input, then there is no way that M can mark the middle of its tape, as M has no auxiliary symbols. On the other hand, as M is monotone, it cannot apply rewrite steps alternatingly to the prefix of ω and to the suffix of ω.

Assume that in an accepting computation on input ω, M performs a large number of rewrite steps within the prefix x, thus deriving the configuration $q¢yx^R\$$. Then there exists a pumping factor within x such that by deleting this factor we obtain a word $\omega' = x'x^R \notin L_{\text{pal}}$ such that, when starting from input ω', M will reach the same configuration $q¢yx^R\$$. This, however, contradicts the Error Preserving Property.

Finally assume that in an accepting computation on input ω, M first executes a limited number of rewrite steps within the prefix x, which yields a configuration of the form $C_1 := q¢x_1ux_2vx_3\$$. Here the factor u displayed lies in the middle part of the tape. Further, assume that in the next cycle M rewrites the factor u displayed into v, which leads to the restarting configuration $C_2 := q'¢x_1vx_2vx_3\$$. As there are many occurrences of the factors u and v within x_1, x_2, and x_3, there exists a string $\omega' \notin L_{\text{pal}}$ such that, when starting with input ω', M will perform the same rewrite steps within the prefix of the tape content and reach the configuration $C_3 := q¢x_1vx_2ux_3\$$. As M cannot possibly distinguish between the configurations C_1 and C_3, it can also rewrite the factor u displayed in C_3 into v and in this way reach the restarting configuration C_2. However, this contradicts the Error Preserving Property again. This completes the proof of Lemma 3.   □

For processing $L_{\text{pal}'}$ a mon-nf-RW-automaton can use the #-symbol to mark its guess for the middle of the input. However, a mon-nf-RR-automaton is again too weak for accepting this language. Finally, the language $L_{\text{pal}''}$ is accepted by the mon-RR-automaton that is given through the following sequence of meta-instructions. Essentially it guesses the rightmost occurrence of the symbol c in its input and then checks whether its guess was correct:

(i) $(\mathrm{¢} \cdot \{a,b,c\}^*, acb \to c, \{a,b\}^* \cdot \$)$, (iii) $(\mathrm{¢} \cdot \{a,b,c\}^*, bca \to c, \{a,b\}^* \cdot \$)$,
(ii) $(\mathrm{¢} \cdot \{a,b,c\}^*, cca \to c, \{a,b\}^* \cdot \$)$, (iv) $(\mathrm{¢}c\$, \mathsf{Accept})$.

On the other hand, the language $L_{\text{pal}''}$ is not accepted by any mon-nf-RW-automaton, as such an automaton cannot verify whether it has found the rightmost occurrence of the symbol c correctly. Even the option of performing rewrites does not help, as it has no auxiliary symbol that it could use to mark this position. Of course, a det-mon-RL-automaton can easily find the rightmost occurrence of the symbol c, and so it can accept this language. These results yield the taxonomy for monotone nonforgetting restarting automata displayed in Figure 2.

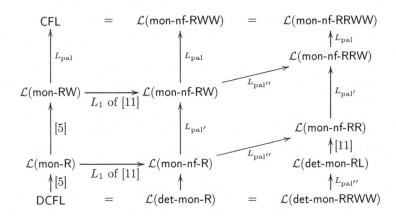

Fig. 2. The taxonomy of monotone nonforgetting restarting automata

Actually, $\mathcal{L}(\text{mon-nf-R}) \not\subseteq \mathcal{L}(\text{mon-RRW})$ as seen from the example language L_2 considered below, but it is still open where exactly the classes $\mathcal{L}(\text{mon-RRW})$ and $\mathcal{L}(\text{mon-RW})$ are located in relation to the various monotone nonforgetting restarting automata.

5 Deterministic Monotone Restarting Automata

All types of deterministic monotone R- and RR-automata accept exactly the deterministic context-free languages [5]. For nonforgetting restarting automata this result only carries over to R-automata.

Theorem 3. *For each* $\mathsf{X} \in \{\lambda, \mathsf{W}, \mathsf{WW}\}$, DCFL $= \mathcal{L}(\text{det-mon-nf-RX})$.

Proof. As DCFL = $\mathcal{L}$(det-mon-R), the inclusions from left to right are obvious. It remains to show that $\mathcal{L}$(det-mon-nf-RWW) $\subseteq$ DCFL holds.

So let M be a det-mon-nf-RWW-automaton with window of size k and the set Q of internal states. We simulate M by a deterministic pushdown automaton P, which uses a buffer of size k in its finite control. On starting the simulation P reads the first k input symbols into this buffer. When simulating a MVR-step of M, P reads the next input symbol into its buffer, pushing the leftmost symbol from the buffer onto its pushdown store. Together with this symbol P also writes, for each state $q \in Q$, that state on the pushdown in which M would be at this point had it started the current cycle in state q, using $|Q|$ many extra tracks on its pushdown store. This continues until M executes a rewrite/restart step $(p, u) \rightarrow (q, v)$, which P simulates within its buffer, refilling the buffer from the pushdown store by popping the topmost $|u| - |v|$ symbols. Afterwards P continues with the simulation of M by using the state from the track of the pushdown store that corresponds to state q. As M is monotone, the next rewrite step cannot occur to the left of the previous rewrite step. It follows that P accepts the language $L(M)$.   □

Next we will see, however, that deterministic monotone nf-RR-automata are strictly more powerful than the corresponding standard variant. For this we consider the example language $L_{\overline{pal}} := \{ \, wc\phi(w)^R \mid w \in \{a, b, c\}^*, |w|_c \geq 1 \, \}$, where $|w|_c$ denotes the number of c-symbols occurring in w, and ϕ is the morphism defined by $\phi(a) := a$, $\phi(b) := b$, and $\phi(c) := \lambda$.

Lemma 4. $L_{\overline{pal}} \in \mathcal{L}$(det-mon-nf-RR) $\smallsetminus$ DCFL.

Proof. The language $L_{\overline{pal}}$ is accepted by the det-mon-nf-RR-automaton that is given through the following meta-instructions:

(i) $(q_0, \mathbb{c} \cdot \{a, b\}^*, c \rightarrow \lambda, (\{a, b\}^* \cdot c)^2 \cdot \{a, b, c\}^* \cdot \$, q_0)$,
(ii) $(q_0, \mathbb{c} \cdot \{a, b\}^*, c \rightarrow \lambda, \{a, b\}^* \cdot c \cdot \{a, b\}^* \cdot \$, q_1)$,
(iii) $(q_1, \mathbb{c} \cdot \{a, b\}^*, aca \rightarrow c, \{a, b\}^* \cdot \$, q_1)$,
(iv) $(q_1, \mathbb{c} \cdot \{a, b\}^*, bcb \rightarrow c, \{a, b\}^* \cdot \$, q_1)$,
(v) $(q_1, \mathbb{c}c\$, \mathsf{Accept})$.

It remains to show that the language $L_{\overline{pal}}$ is not deterministic context-free. As DCFL = $\mathcal{L}$(det-mon-R), it suffices to prove that this language is not accepted by any deterministic monotone R-automaton.

So assume to the contrary that M is a deterministic monotone R-automaton that accepts the language $L_{\overline{pal}}$. For a sufficiently large integer m, the word $\omega := ca^mca^m \in L_{\overline{pal}}$ cannot be accepted by M in a tail computation. Hence, starting from the initial configuration $q_0\mathbb{c}ca^mca^m\$$, M will execute a cycle $q_0\mathbb{c}\omega\$ \vdash_M^c q_0\mathbb{c}\omega'\$$. As M is deterministic, $\omega' \in L_{\overline{pal}}$, implying that $\omega' = ca^{m-i}ca^{m-i}$ for some integer $i > 0$. As M restarts immediately after executing a rewrite step, M will apply the same rewrite operation when starting from the initial configuration $q_0\mathbb{c}ca^mca^mca^{m-i}a^{m-i}\$$, that is, it will execute the cycle

$$q_0ca^mca^mca^{m-i}a^{m-i}\$ \vdash_M^c q_0ca^{m-i}ca^{m-i}ca^{m-i}a^{m-i}\$.$$

This contradicts the Error Preserving Property, as $ca^m ca^m ca^{m-i} a^{m-i} \notin L_{\overline{pal}}$, while $ca^{m-i} ca^{m-i} ca^{m-i} a^{m-i} \in L_{\overline{pal}}$.   □

For establishing further separation results, we consider the following variants of the above language $L_{\overline{pal}}$, where $\psi : \{a, b, c\}^* \to \{a, b\}^*$ is the transduction that is defined through the finite transducer with a single state and the transition relation $a \mapsto a$, $b \mapsto b$, $ca \mapsto b$, $cb \mapsto a$, and $cc \mapsto \lambda$, and $\phi : \{a, b, c\}^* \to \{a, b\}^*$ is the morphism defined by $a \mapsto a$, $b \mapsto b$, and $c \mapsto a$:

$$L'_{\overline{pal}} := \{\, wc\psi(w)^R \mid w \in \{a, b, c\}^*, \ |w|_c \geq 1, \ w \text{ does not end with a } c \,\},$$
$$L''_{\overline{pal}} := \{\, wc\phi(w)^R \mid w \in \{a, b, c\}^*, \ |w|_c \geq 1 \,\}.$$

Lemma 5. (a) $L'_{\overline{pal}} \in \mathcal{L}(\text{det-mon-nf-RRW}) \smallsetminus \mathcal{L}(\text{det-mon-nf-RR})$.
    (b) $L''_{\overline{pal}} \in \mathcal{L}(\text{det-mon-nf-RRWW}) \smallsetminus \mathcal{L}(\text{det-mon-nf-RRW})$.

Thus, we obtain the following chain of proper inclusions, which represents the first separation between the RWW- and the RRWW-variant for any type of restarting automaton.

Theorem 4. $\text{DCFL} = \mathcal{L}(\text{det-mon-nf-RWW}) \subset \mathcal{L}(\text{det-mon-nf-RR})$
    $\subset \mathcal{L}(\text{det-mon-nf-RRW}) \subset \mathcal{L}(\text{det-mon-nf-RRWW}).$

All the above languages belong to the class $\mathcal{L}(\text{det-mon-nf-RL})$. Surprisingly, the det-mon-nf-RLWW-automaton is not more powerful than the standard det-mon-RL-automaton.

Theorem 5. $\mathcal{L}(\text{det-mon-nf-RLWW}) = \mathcal{L}(\text{det-mon-RL})$.

Proof. It is shown in [6] that $\mathcal{L}(\text{det-mon-RL}) = \mathcal{L}(\text{det-mon-RLWW})$ holds. In fact, this equality even extends to *shrinking* restarting automata (cf. Section 2). This extension is obtained as follows.

A language L is accepted by a deterministic monotone shrinking RLWW-automaton (sRLWW) if and only if L^R is accepted by a deterministic left-monotone sRLWW-automaton (cf. Lemma 1 of [6]). From the main results of [8] we see that $\mathcal{L}(\text{det-left-mon-sRLWW}) = \mathcal{L}(\text{det-left-mon-RLWW})$, while Theorem 3 of [6] yields $\mathcal{L}(\text{det-left-mon-RLWW}) = \mathcal{L}(\text{det-left-mon-RL})$. Hence, we see that $\mathcal{L}(\text{det-left-mon-sRLWW}) = \mathcal{L}(\text{det-left-mon-RL})$. By using Lemma 1 of [6] again we obtain $\mathcal{L}(\text{det-mon-RL}) = \mathcal{L}(\text{det-mon-sRLWW})$.

Thus, it remains to show that every deterministic monotone nonforgetting sRLWW-automaton can be simulated by a deterministic monotone sRLWW-automaton. Let M by a det-mon-nf-sRLWW-automaton. It is simulated by a deterministic monotone sRLWW-automaton M' that uses two cycles for simulating a cycle of M.

First M' scans the tape from right to left searching for symbols that encode a state and a tape symbol of M. In its internal state M' remembers the rightmost state found in this way, and if no such symbol is found, then M' remembers the initial state of M. Next M' starts with the simulation of M, either at the left end of the tape or at the position where the state symbol of M was found, if this state is entered by M after executing a rewrite operation (see below). In the first case M' continues with the simulation until it reaches the configuration in which M would perform a rewrite operation. M' executes this rewrite operation, encoding the state that M reaches afterwards as part of the new tape inscription together with an indicator that it is currently in phase 1 of the actual simulation. In the second case M' continues with the simulation until it reaches a configuration in which M would now execute a restart operation. This M' does not execute yet, but it remembers the new state that M would now enter, returns to the place where the last rewrite operation was performed, encodes the new state of M into the rightmost symbol produced by that rewrite operation, and restarts. As M is monotone, the newly encoded state is the rightmost state encoded in the tape inscription, and therewith it will be chosen correctly in the next cycle. It is easily seen that M' simulates M cycle by cycle and that it accepts the same language as M. □

It is known that $\mathcal{L}(\text{det-mon-RL})$ is contained in $\mathsf{CRL} \cap \mathsf{CFL}$ [11]. On the other hand, the context-free language $L_2 := \{\, a^n b^n \mid n \geq 0 \,\} \cup \{\, a^n b^m \mid m > 2n \geq 0 \,\}$ is Church-Rosser, but it is not accepted by any RLW-automaton [5], implying that it does not belong to $\mathcal{L}(\text{det-mon-RL})$. This yields the following proper inclusion.

Corollary 1. $\mathcal{L}(\text{det-mon-nf-RLWW}) \subset \mathsf{CRL} \cap \mathsf{CFL}$.

Actually as seen in the proof of Theorem 5 the above results even extend to shrinking restarting automata. Note, however, that it is still open whether the inclusion $\mathcal{L}(\text{det-mon-nf-RRWW}) \subseteq \mathcal{L}(\text{det-mon-nf-RLWW})$ is proper or not.

6 Concluding Remarks

As we have seen, deterministic nonforgetting restarting automata are in general more powerful than their standard counterparts. This makes them an interesting variant, although, as seen in Section 3, already for det-nf-R-automata many algorithmic problems are undecidable.

References

1. B. von Braunmühl and R. Verbeek. Finite-change automata. In: K. Weihrauch (ed.), *4th GI Conference, Proc., Lecture Notes in Computer Science* 67, Springer, Berlin, 1979, 91–100.
2. G. Buntrock and F. Otto. Growing context-sensitive languages and Church-Rosser languages. *Information and Computation* 141, 1–36, 1998.
3. J. Hopcroft and J. Ullman. *Introduction to Automata Theory, Languages, and Computation*, Addison-Wesley, Reading, MA, 1979.
4. P. Jančar, F. Mráz, M. Plátek, and J. Vogel. Restarting automata. In: H. Reichel (ed.), *FCT'95, Proc., Lecture Notes in Computer Science 965*, Springer, Berlin, 1995, 283–292.

5. P. Jančar, F. Mráz, M. Plátek, and J. Vogel. On monotonic automata with a restart operation. *J. Automata, Languages and Combinatorics* 4, 287–311, 1999.

6. T. Jurdziński, F. Mráz, F. Otto, and M. Plátek. Monotone deterministic RL-automata don't need auxiliary symbols. In: C. De Felice and A. Restivo (eds.), *DLT 2005, Proc.*, *Lecture Notes in Computer Science 3572*, Springer, Berlin, 2005, 284–295.

7. T. Jurdziński and F. Otto. Shrinking restarting automata. In: J. Jędrzejowicz and A. Szepietowski (eds.), *MFCS 2005, Proc.*, *Lecture Notes in Computer Science* 3618, Springer, Berlin, 2005, 532–543.

8. T. Jurdziński, F. Otto, F. Mráz, and M. Plátek. On left-monotone deterministic restarting automata. In: C.S. Calude, E. Calude, and M.J. Dinneen (eds.), *DLT 2004, Proc.*, *Lecture Notes in Computer Science 3340*, Springer, Berlin, 2004, 249–260.

9. K. Oliva, P. Květoň, and R. Ondruška. The computational complexity of rule-based part-of-speech tagging. In: V. Matoušek and P. Mautner (eds.), *TSD 2003, Proc.*, *Lecture Notes in Computer Science 2807*, Springer, Berlin, 2003, 82–89.

10. F. Otto. Restarting automata and their relations to the Chomsky hierarchy. In: Z. Ésik and Z. Fülöp (eds.), *DLT 2003, Proc.*, *Lecture Notes in Computer Science* 2710, Springer, Berlin, 2003, 55–74.

11. F. Otto. Restarting Automata - Notes for a Course at the 3rd International PhD School in Formal Languages and Applications. *Mathematische Schriften Kassel* 6/04, Universität Kassel, 2004.

12. M. Plátek. Two-way restarting automata and j-monotonicity. In: L. Pacholski and P. Ružička (eds.), *SOFSEM'01, Proc.*, *Lecture Notes in Computer Science 2234*, Springer, Berlin, 2001, 316–325.

13. M. Plátek, M. Lopatková, and K. Oliva. Restarting automata: motivations and applications. In: M. Holzer (ed.), *Workshop 'Petrinetze' and 13. Theorietag 'Formale Sprachen und Automaten'*, *Proc.*, Institut für Informatik, Technische Universität München, 2003, 90-96.

Unwinding a Non-effective Cut Elimination Proof

Grigori Mints

Stanford University, Stanford Ca 94305
mints@csli.stanford.edu

Abstract. Non-effective cut elimination proof uses Koenig's lemma to obtain a non-closed branch of a proof-search tree $\mathcal{T}$ (without cut) for a first order formula A, if A is not cut free provable. A partial model (semi-valuation) corresponding to this branch and verifying $\neg A$ is extended to a total model for $\neg A$ using arithmetical comprehension. This contradicts soundness, if A is derivable with cut. Hence $\mathcal{T}$ is a cut free proof of A. The same argument works for Herbrand Theorem. We discuss algorithms of obtaining cut free proofs corresponding to this schema and quite different from complete search through all possible proofs.

1 Introduction

Unwinding means here obtaining explicit bounds from non-constructive proofs of existential statements. We apply existing proof theoretic techniques to a non-effective proof of the basic proof-theoretic result, cut elimination theorem, to obtain an algorithm transforming any first order derivation of a first order formula A into a cut free derivation of A. The first proof of this result by Gentzen was constructive, but seemed rather complicated combinatorially. Gentzen's Extended Main Theorem was a version of Herbrand's theorem. Gödel's proof of his completeness theorem and related work by Skolem implicitly contained a version of Herbrand's theorem proved in a non-constructive way. These proofs led via work by Schütte to non-constructive cut elimination proofs for higher order systems. by Prawitz and Takahashi. The problem of constructivization of non-constructive proofs has much longer history. Modern period began with negative translation and Brouwer-Heyting-Kolmogorov interpretation of sentences followed by functional interpretations including Kleene's realizability and Godel's Dialectica interpretation.

Intuitionism and constructivism introduced a program of constructivizing classical results preserving constructive proofs but sometimes drastically changing non-constructive arguments. G. Kreisel promoted (in [6, 7, 8] and other papers) a program of unwinding non-effective existence proofs using techniques from proof theory including cut elimination and Dialectica interpretation which transform a classical proof of an arithmetical Σ_1^0–formula $\exists x R(x)$ into a number n satisfying $R(n)$. A new tool was provided by H.Friedman [2] and A. Dragalin [1].

D. Grigoriev, J. Harrison, and E.A. Hirsch (Eds.): CSR 2006, LNCS 3967, pp. 259–269, 2006.
© Springer-Verlag Berlin Heidelberg 2006

The potential of application of all methods mentioned above was drastically increased by emergence of proof construction and manipulation systems such as Coq, Isabel,MIZAR, Minlog etc.

A General Schema of Unwinding and Necessary Modifications

A straightforward schema looks as follows.

1. Starting with a non-effective proof $d : A$ of a sentence A in some formal system S one gets first an effective proof $d^- : A^{neg}$ of a negative translation of A in an intuitionistic version S^{int} of S.
2. A proof $d^+ : A'$ of an effective modification A' of A^{neg} is obtained for example by methods of [2, 1].
3. A functional interpretation of the system S^{int} is applied to d^+ to extract an effective content of the original proof d.

For $A = \forall x \exists y R(x, y)$ with a primitive recursive R one has

$$A^{neg} = \forall x \neg \neg \exists y R(x, y), \qquad A' = \forall x \exists y R(x, y) \text{ (up to equivalence)},$$

hence the Dialectica interpretation provides a function Y such that $R(x, Y(x))$. The three steps are often combined into a functional interpretation of the original classical system S. In this paper we use Shoenfield's interpretation [13].

In practice for every non-trivial A each stage of this process presents significant technical difficulties even when there are no difficulties in principle, that is negative translation, the method of Friedman-Dragalin and a suitable functional interpretation is applicable to the original proof $d : A$. Before the first stage d is usually incomplete or the complete formalization in one of proof manipulation systems is not human readable. The stage 1 itself turns d into much less readable (and often less complete) proof d^-, so that the functional interpretation is very difficult to apply, and even if a result is obtained, it is difficult to interpret. This is sometimes complicated by difficulties in principle: standard proof theoretic transformations have to be modified to treat a given system S.

These difficulties are dealt with by using shortcuts at each stage. According to an observation by G. Kreisel they are often mathematically insignificant. This refers both to "equivalent" reformulation of A and changes in the proofs d, d^{neg}, d^+ that seem inessential to researchers establishing results like A by proofs like d. From the logical point of view the effect of such shortcuts can be quite dramatic. In our case the expected complexity ACA (Arithmetic Comprehension Axiom) is reduced to PRA (Primitive Recursive Arithmetic).

In the present paper proof theory is applied to unwind a non-effective proof of cut elimination theorem. To simplify notation we work with existential (Skolemized) formulas. This proof uses Koenig's lemma to obtain a non-closed branch of a *proof-search tree* $\mathcal{T}$ (without cut) for a first order formula A, if A is not cut free provable. A partial model (semi-valuation) $\mathbf{B}$ corresponding to this branch and verifying $\neg A$ is extended to a total model for $\neg A$ using arithmetical comprehension. This contradicts soundness, if A is derivable with cut. Hence the proof search tree should close (with axioms in all branches) after finite number of steps. This provides a cut free proof for A. The same argument works for Herbrand Theorem.

It may seem to correspond to complete search through all possible cut free proofs. We show it give rise to a more sophisticated algorithm transforming a given derivation d of the formula A with cut into a cut free derivation . When d itself is cut free, a bound for the size of the proof search tree T is obtained using considerations in [11]: every branch of d is contained in some branch of T which is obtained primitive recursively. When d contains cuts, computations of the values for functional interpretation (of the metamathematical argument above) roughly corresponds to standard cut elimination steps applied to d, similarly to [10].

Combining all this leads to the following *primitive recursive algorithm*: apply the standard cut elimination to the given proof d with cut, then compute a primitive recursive bound for T by the method of [11] providing a cut free proof. This looks very different from the method given by the literal application of the functional interpretation. The proof along the lines of [10] that the resulting bound for T (not the time of computation) is approximately the same is a topic for future work. We gave explicit construction for some stages of the Shoenfield's interpretations to make it plausible. On the other hand, the rough bounds obtained by straightforward arguments have much higher computation complexity than our primitive recursive algorithm. Indeed, the lemma on extension of an *arbitrary* semivaluation V_0 to a total model V' is non-arithmetical. If for example V_0 is a standard model of arithmetic which is primitive recursive on atomic formulas, then its total extension is a truth definition for arithmetic, hence non-arithmetical by Tarski's theorem. This shows why the conservative extension theorem for WKL_0 over PRA is not sufficient in our context and may explain why non-effective cut elimination proof did not yield up to now to U. Kohlenbach's methods.

In section 2 we describe a familiar construction of a proof serch tree, section 3 presents Shoenfield's interpretation and its important instance. Section 4 computes Shoenfield's interpretation of the Weak Köning's Lemma, a basic ingredient of many projects in proof construction and proof mining. In section 5 the basic part of a non-effective proof of cut elimination is presented in sufficient detail to make possible computing Shoenfield interpretation and justify our algorithm.

Most of the results in this work were obtained when the author was on sabbatical leave from Stanford University in Ludwig-Maximillian University, Munich, Germany. Discussions with G. Kreisel, H. Schwichtenberg, W. Buchholz, P. Schuster, U. Berger were especially useful. Remarks by the anonymous referee helped to focus the presentation.

2 Tait Calculus

We consider first order formulas in positive normal form (negations only at atomic formulas). Negation $\bar{A}$ of a formula A is defined in a standard way by de-Morgan rules. Derivable objects are *sequents*, that is multisets of formulas.

Axioms: $A, \neg A, \Gamma$ for atomic A **Inference Rules:**

$$\frac{A, \Gamma \quad B, \Gamma}{A \& B, \Gamma} \ \& \qquad \frac{A, B, \Gamma}{A \vee B, \Gamma} \ \vee \qquad \frac{M(t), \exists x B(x), \Gamma}{\exists x B(x), \Gamma} \ \exists$$

$$\frac{B(a), \Gamma}{\forall x B(x), \Gamma} \ \forall \qquad \frac{C, \Gamma \quad \bar{C}, \Gamma}{\Gamma} \ cut$$

The variable a in $\forall$ inference should be fresh. The term t in the rule $\exists$ is called *the term* of that rule. We are primarily interested in Skolemized Σ–formulas with a quantifier free matrix B.

$$A = \exists x_1 \ldots \exists x_q B := \exists \mathbf{x} B \tag{2.1}$$

Definition 2.1. *The* Herbrand Universe **HU**(A) *of a formula A consists of all terms generated from constants and free variables occurring in A by function symbols occurring in A. If the initial supply is empty, add a free variable.*

For a given formula A and given q (cf. (2.1)) list all q-tuples of terms in **HU**(A) in a sequence

$$\mathbf{t}_1, \ldots, \mathbf{t}_i, \ldots \tag{2.2}$$

2.1 Canonical Tree $\mathcal{T}$

The *canonical tree* $\mathcal{T}$ for a Σ–formula $A = \exists \mathbf{x} B(\mathbf{x})$ is constructed by bottom-up application of the rules $\&, \vee$ (first) and $\exists$ when $\&, \vee$ are not applicable.

$\mathcal{T}$ assigns sequents to nodes a of the tree of finite binary sequences (cf. beginning of section 4). To express a $\notin \mathcal{T}$ (when a is situated over an axiom or the second premise of a one-premise inference rule) we write $\mathcal{T}(a) = 0$.

$\mathcal{T}(<>)$ contains sequent A. If $\mathcal{T}(a)$ is already constructed and is not an axiom ($:=closed$ node or branch), then it is extended preserving all existing formulas. Principal formulas of the propositional rules are preserved (for bookkeeping). If all branches of $\mathcal{T}$ are closed, then the whole *tree is closed*.

The following *fairness conditions* are assumed. There exists a Kalmar elementary function f such that for each a $\in TT$ and every b $\supseteq$ a, b $\in TT$ with $lth(b) \geq lth(a) + F(a)$ (we use notation at the beginning of section 4):

1. If $C\&D \in \mathcal{T}(a)$ then $C \in \mathcal{T}(b)$ or $D \in \mathcal{T}(b)$,
2. If $C \vee D \in \mathcal{T}(a)$ then $C \in \mathcal{T}(b)$ and $D \in \mathcal{T}(b)$,
3. $B(\mathbf{t}_i) \in \mathcal{T}(b)$ for every $i \leq lth(a)$.

3   Shoenfield's Variant of the Dialectica Interpretation

We use here the notation and results from [13] Section 8.3. where a version of Dialectica interpretation is defined for classical arithmetic. Primitive recursive *terms of higher types* are constructed from constants and variables of all finite types by substitution, primitive recursion and explicit definitions. *Arithmetical formulas* are formulas of the first order arithmetic (with constants for the primitive recursive functions of natural numbers). *Generalized formulas* are expressions $\forall x_1 \ldots \forall x_n \exists y_1 \ldots \exists y_m A$, or shorter $\forall \mathbf{x} \exists \mathbf{y} A$, where A is a quantifier free formula of finite type. Note that this form is different from the "Skolem" form $\exists \mathbf{x} \forall \mathbf{y} A$ used in Gödel's Dialectica interpretation for first order arithmetic.

Definition 3.1. *To every formula* A *of the arithmetic of finite type with connectives* $\{\forall, \vee, \&, \neg\}$ *a generalized formula* A^* *is inductively assigned.*

1. $A^* := A$ *for atomic* A.
2. *If* $A = \neg B$, $B^* = \forall \mathbf{x} \exists \mathbf{y} B'[\mathbf{x}, \mathbf{y}, \mathbf{z}]$, *then* $A^* := \forall \mathbf{y}' \exists \mathbf{x} \neg B'[\mathbf{x}, \mathbf{y}'(x), \mathbf{z}]$ *where* $\mathbf{y}'$ *is a sequence of fresh variables of appropriate types.*
3. *If* $A = B \vee C$,

$$B^* = \forall \mathbf{x} \exists \mathbf{y} B'[\mathbf{x}, \mathbf{y}, \mathbf{z}], \quad C^* = \forall \mathbf{x}' \exists \mathbf{y}' C'[\mathbf{x}', \mathbf{y}', \mathbf{z}], \tag{3.1}$$

   then $A^* := \forall \mathbf{x}\mathbf{x}' \exists \mathbf{y}\mathbf{y}'(B'[\mathbf{x}, \mathbf{y}, \mathbf{z}] \vee C'[\mathbf{x}', \mathbf{y}', \mathbf{z}])$
4. *If* $A = B \& C$ *then* $A^* := \forall \mathbf{x}\mathbf{x}' \exists \mathbf{y}\mathbf{y}'(B'[\mathbf{x}, \mathbf{y}, \mathbf{z}]\& C'[\mathbf{x}', \mathbf{y}', \mathbf{z}])$.
5. *If* $A = \forall \mathbf{w} B$ *then* $A^* := \forall \mathbf{w}\mathbf{x} \exists \mathbf{y} B'[\mathbf{x}, \mathbf{y}, \mathbf{z}]$.

The &-clause is added here compared to [13], but the next theorem is still true after obvious addition to its proof in [13].

Theorem 3.2. *If* A *is derivable in the classical arithmetic, and*

$$A^* = \forall \mathbf{x} \exists \mathbf{y} A'[\mathbf{x}, \mathbf{y}, \mathbf{z}],$$

then there is a sequence **a** *of primitive recursive terms of higher types such that*

$$A'[\mathbf{x}, \mathbf{a}, \mathbf{z}] \tag{3.2}$$

is derivable in quantifier-free arithmetic of finite types (Gödel's system T). ⊣

Definition 3.3. *Terms* **a** *in (3.2) are called* realization *of the formula* A.

Comment. Often **a** is written as $Y(\mathbf{x})$ or $Y(\mathbf{x}, \mathbf{z})$.

Let's compute Shoenfield translation of classical existence formulas. Recall that classically $\exists w B \iff \neg \forall w \neg B$, hence if $B^* = \forall \mathbf{x} \exists \mathbf{y} B'[\mathbf{x}, \mathbf{y}, \mathbf{z}, w]$ then

$$(\exists w B)^* = \forall \mathbf{x}' \exists w \mathbf{y}' B'[\mathbf{x}'(w, \mathbf{y}'), \mathbf{y}'(\mathbf{x}'(w, \mathbf{y}')), \mathbf{z}, w]$$

since $(\neg B)^* = \forall \mathbf{y}' \exists \mathbf{x} \neg B'[\mathbf{x}, \mathbf{y}'(\mathbf{x}), \mathbf{z}, w]$,

$$(\neg \forall w \forall \mathbf{y}' \exists \mathbf{x} \neg B'[\mathbf{x}, \mathbf{y}'(\mathbf{x}), \mathbf{z}, w])^* = \forall \mathbf{x}' \exists w \mathbf{y}' B'[\mathbf{x}'(w, \mathbf{y}'), \mathbf{y}'(\mathbf{x}'(w, \mathbf{y}')), \mathbf{z}, w],$$

Let's also recall that the implication is *defined* in terms of $\vee, \neg$ by $B \rightarrow C := \neg A \vee B$, so (3.1) implies $(B \rightarrow C)^* := \forall \mathbf{Y}\mathbf{x}' \exists \mathbf{x}\mathbf{y}'(\neg B'[\mathbf{x}, \mathbf{Y}(\mathbf{x}), \mathbf{z}] \vee C'[\mathbf{x}', \mathbf{y}', \mathbf{z}])$.

3.1 Kreisel's Trick

An especially important case of a functional interpretation is case distinction for Σ_1^0–formulas. It is similar to Hilbert's Ansatz (approach) to epsilon substitution method [4] and was made explicit by G. Kreisel who empoyed it to analyze a proof by Littlewood that the difference $\pi(x) - li(x)$ changes sign. The statement to be proved can be expressed in Σ_1^0-form $\exists y P(y)$. Littlewood's proof consisted of two parts: If Riemann Hypothesis RH is true, then $\exists y P(y)$; If Riemann Hypothesis RH is false, then $\exists y P(y)$.

Kreisel's analysis used the fact that RH can be expressed in Π_1^0-form $\forall x R(x)$ to deal with $\exists x \bar{R}(x) \rightarrow \exists y P(y)$ and $\forall x R(x) \rightarrow \exists y P(y)$. The first of these relations provided a computable function f such that

$$\forall x (\bar{R}(x) \rightarrow P(f(x))) \tag{3.3}$$

The second relation provided natural numbers x_0, y_0 such that

$$\bar{R}(x_0) \vee P(y_0) \tag{3.4}$$

Combination of the latter two relations yields $P(z_0)$ where

$$z_0 = \text{ the least } z \in \{f(x_0), y_0\} \text{ such that } P(z) \tag{3.5}$$

4 König's Lemma

Here we develop Shoenfield's interpretation for a version of König's Lemma for binary trees. Let's recall some notation concerning finite sequences of natural numbers. We use a, b, c as variables for *binary finite sequences*

a $=< a_0, \ldots, a_n >$ where $a_i \in \{0,1\}$, $lh(\mathrm{a}) := n+1$, $(\mathrm{a})_i := a_i$.

Concatenation $*$:

$< a_0, \ldots, a_n > * < b_0, \ldots, b_m > := < a_0, \ldots, a_n, b_0, \ldots, b_m >$.

$<>$ is the empty sequence with $lh <> = 0$. $\mathrm{a} \subseteq \mathrm{b} : \Longleftrightarrow \exists \mathrm{c}\, \mathrm{b} = \mathrm{a} * \mathrm{c}$;

a $<$ b iff a if *lexicographically strictly precedes* b, that is situated strictly to the left in the tree of all finite sequences:

a $\subset$ b or for some $j < lh\mathrm{a}$, $(a)_i = (b)_i$ for all $i < j$, and $(a)_j < (b)_j$.

In this section $\mathcal{T}$ (with subscripts etc.) denotes primitive recursive trees of *binary sequences* with the root $<>$:

$$\mathrm{b} \in \mathcal{T} \& \mathrm{a} \subseteq \mathrm{b} \rightarrow \mathrm{a} \in \mathcal{T}; \qquad \mathrm{a} \in \mathcal{T} \rightarrow (\forall i < lh\mathrm{a})(\mathrm{a})_i \leq 1$$

$\mathcal{T}_\mathrm{a}$ is the subtree of $\mathcal{T}$ with the root a: $\{\mathrm{b} \in \mathcal{T} : \mathrm{a} \subseteq \mathrm{b}\}$.

In fact we use labeled trees. $\mathcal{T}(\mathrm{a}) = 0$ means $\mathrm{a} \notin \mathcal{T}$, while $\mathcal{T}(\mathrm{a}) \neq 0$ means that $\mathrm{a} \in \mathcal{T}$ and contains some additional information. A node $\mathrm{a} \in \mathcal{T}$ is a *leaf* if $\mathrm{b} \supset \mathrm{a}$ implies $\mathrm{b} \notin \mathcal{T}$. In this case all btanches of $\mathcal{T}$ through a are *closed*.

$$\mathcal{T} < l := (\forall \mathrm{a} : lh\mathrm{a} = l)(\mathrm{a} \notin \mathcal{T}); \qquad \mathcal{T} > l := (\exists \mathrm{a} : lh\mathrm{a} = l+1)(\mathrm{a} \in \mathcal{T})$$

and similar bounded formulas with replacement of $<, >$ by $\leq, \geq$.

We use an adaptation of König's Lemma in the form:
For every infinite binary tree its leftmost non-closed branch **B** *is infinite.*
Let

$$\text{Left}(a, m) := (\forall a' < a)\mathcal{T}_{a'} < m; \quad (a, m) \in \mathbf{B} := \text{Left}(a, m) \& \forall l(\mathcal{T}_a > l). \quad (4.1)$$

Now the statement above can be expressed as follows:

$$\forall l(\mathcal{T} > l) \rightarrow \forall k \exists m(\exists a : lha = k)((a, m) \in \mathbf{B}) . \quad (4.2)$$

Let's state some properties of this construction and compute their Shoenfield's interpretations and realizations.

Lemma 4.1.   *1.* **B** *is a branch:*

$$(a, m) \in \mathbf{B} \& (b, n) \in \mathbf{B} \rightarrow a \subseteq b \vee b \subseteq a \quad (4.3)$$

2. If $\mathcal{T}$ *is infinite then* **B** *is infinite:*

$$\forall l_0(\mathcal{T} > l_0) \rightarrow \forall k \exists m(\exists b : lhb = k)((b, m) \in \mathbf{B}) \quad (4.4)$$

3. Realizations of (4.3,4.4) are primitive recursive

Proof.

1. Assume $(a, m) \in \mathbf{B}, (b, n) \in \mathbf{B}$, $a < b$, that is a is to the left of b. Then $(b, n) \in \mathbf{B}$ implies $\mathcal{T}_a < n$ contradicting $\forall l(\mathcal{T} > l)$ with $l := n$. Similarly $(a, m) \in \mathbf{B}$ contradicts $\forall l_1(\mathcal{T}_b > l_1)$ with $l_1 := m$. This proves (4.3) and provides realizations for the quantifiers $\forall l, \forall l_1$.
2. We use induction on k. For $k = 0$ put $b = <>, m := 0$. For the induction step assume the values $b(k), m(k)$ for k are given and note that

$$\forall l(\mathcal{T}_b > l) \quad (4.5)$$

   If $b := b(k)$ has one son in $\mathcal{T}$, say $b * \{0\}$, put

$$b(k + 1) := b * \{0\}; \qquad m(k + 1) := m(k) \quad (4.6)$$

   Assume there are two sons. If $\forall l(\mathcal{T}_{b*\{0\}} > l)$, use (4.6) again. Otherwise there is an l_1 such that $\mathcal{T}_{b*\{0\}} < l_1$, and (4.5) implies $\forall l \mathcal{T}_{b*\{1\}} > l$. Put

$$b(l + 1) := b * \{1\}; \qquad m(l + 1) := \max(m(l), l_1). \quad (4.7)$$

Extracting realization from this non-effective proof, amounts to Kreisel's trick (see Section 3.1). Shoenfield's interpretation of (4.4) has a form:

$$\forall k \forall L \exists l_0 \exists m[\neg(\mathcal{T} < l_0) \vee (\exists b : lhb = k)(\text{Left}(b, m) \& \mathcal{T}_b > L(m))] \quad (4.8)$$

Define realizations $M(k, L), L_0(k, L)$ for m, l_0 and a function $b(K, L)$ by recursion. Put

$$M(0, L) := 0, b(0, L) :=<>, \ L_0(0, L) = 1,$$

which covers also the case when $\mathcal{T}$ stops at $<>$.

Assume the values for k, L are defined and satisfy the condition in (4.8). Consider possible cases.

(a) $\mathcal{T} < L_0(0, L)$. We are done. Make all realizations constant (the same as for k) for all $k' > k$.

(b) b := b(k, L) has one son in $\mathcal{T}$, say b $* \{0\}$.

i. $\mathcal{T}_{\text{b}*\{0\}} > L(M(k, L))$. Put b$(k+1, 0) :=$ b $* \{0\}$ and preserve the rest: $M(k+1, L) := M(k, L), L_0(k+1, L) = L_0(k, L)$.

ii. $\mathcal{T}_{\text{b}*\{0\}} \leq L(M(k, L))$.

If $\mathcal{T} \leq \max(k + 1, M(k, L), L(M(k, L))) := M_0$ then we are done. Put $L_0(k+1, L) := M' + 1$ as in the case 2a. Otherwise list all a $\in \mathcal{T}$ with b $* \{0\} \leq$ a, $lha = k + 1$ in the list

$$\text{b} * \{0\} := \text{a}_0 < \text{a}_1 < \ldots < \text{a}_k$$

and for $i < k$ put: $M_{i+1} := \max(M_i, L(M_i))$. If $\mathcal{T} < M_i$ for some i, we finish as before by defining $L_0(k+1, L)$. Otherwise take the first i such that $\mathcal{T}_{\text{a}_i} > M_i$ and define

$$\text{b}(k + 1, L) := \text{a}_i, \; M(k + 1, L) := M_{i-1}, L_0(k + 1, L) = L_0(k, L)$$

(c) b := b(k, L) has two sons in $\mathcal{T}$, namely b $* \{0\}$, b $* \{1\}$. Similar to the previous case with the case distinctions according to which of b $* \{0\}$, b $* \{1\}$ or a node to the right of b $* \{1\}$ fits as b$(k+1, L)$.   ⊣

5   Bounds for the Proof Search Tree

In this section we assume a Σ–formula F of first order logic to be fixed and $\mathcal{T}$ to be the canonical *proof search tree* for F, see section 2.1. Now every node a $\in \mathcal{T}$ contains a sequent $A_1, \ldots, A_n$ and we write $A_i \in \mathcal{T}(\text{a})$, $i \leq n$. The main property of the proof search tree is its *fairness*: there is a primitive recursive function *split* insuring that all necessary splittings of formulas are eventually made in the process of proof search, see (5.3), (5.4) below.

We define the leftmost non-closed branch **B** by (4.1) above. Satisfaction for literals in a corresponding semivaluation is defined in a standard way: Pt is satisfied iff $(\neg Pt) \in \mathcal{T}(\text{a})$ for some m with $(\text{a}, m) \in \textbf{B}$. Construction of realizations is slightly simplified if satisfaction is defined separately for the two kinds of literals:

$$\textbf{B} \models Pt : \iff \exists \text{a} \exists m((\neg Pt) \in \mathcal{T}(\text{a}) \& (\text{a}, m) \in \textbf{B}) \tag{5.1}$$

$$\textbf{B} \models \neg Pt : \iff \forall \text{a} \forall m[\neg((\neg Pt) \in \mathcal{T}(\text{a})) \& (\text{a}, m) \in \textbf{B})] \tag{5.2}$$

Let's write down interpretations of these satisfaction relations. If

$$\text{Confirm}(A, \text{a}, m, l) := A \in \mathcal{T}(\text{a}) \& \text{Left}(\text{a}, m) \& \mathcal{T}_\text{a} > l,$$

then

$$(\textbf{B} \models Pt)^* = \forall L \exists \text{a} \exists m \text{Confirm}(\neg Pt, \text{a}, m, L(\text{a}, m))$$

$$(\textbf{B} \models \neg Pt)^* = \forall \text{a} \forall m \exists l \neg \text{Confirm}(\neg Pt, \text{a}, m, l)$$

The satisfaction relation for composite formulas is defined in ordinary way:

$$\mathbf{B} \models (A \vee B) := \mathbf{B} \models A \vee \mathbf{B} \models B; \qquad \mathbf{B} \models (A \& B) := \mathbf{B} \models A \& \mathbf{B} \models B;$$

$$\mathbf{B} \models \neg A := \mathbf{B} \not\models A;$$

$$\mathbf{B} \models \forall x A(x) := \forall t (\mathbf{B} \models A(t)); \qquad \mathbf{B} \models \exists x A(x) := \exists t (\mathbf{B} \models A(t))$$

with a quantifier over all terms t in the Herbrand universe of the formula F for which the search tree $\mathcal{T}$ is constructed.

Lemma 5.1. *For every formula A, $(\mathrm{a}, m) \in \mathbf{B} \& A \in \mathcal{T}(\mathrm{a}) \rightarrow \mathbf{B} \not\models A$ with a primitive recursive realization.*

Proof. Induction on A. For computing realizations it is convenient to restate the Lemma in the form

$$\mathrm{Left}(\mathrm{a}_0, m_0), A \in \mathcal{T}(\mathrm{a}_0), \mathbf{B} \models A \rightarrow \exists l \mathcal{T}_{\mathrm{a}_0} < l$$

and construct realization for l in terms of hypothetical realizations of $\mathbf{B} \models A$.

1. $A = P\mathbf{t}$. Take $(\mathrm{a}', m') \in \bar{\mathbf{B}}$ with $\neg P\mathbf{t} \in \mathcal{T}(\mathrm{a}')$. By Lemma 4.1.1 $\mathrm{a}_0, \mathrm{a}'$ are in the same branch. Hence at the level $\max(lh\mathrm{a}_0, lh\mathrm{a}')$ there is an axiom $P\mathbf{t}, \neg P\mathbf{t}, \Gamma \in \mathbf{B}$. This contradicts Lemma 4.1.2.
   Realization of $(\mathbf{B} \models P\mathbf{t})^*$ provides $\mathcal{A}, \mathcal{M}$ such that

   $$\forall L \mathrm{Confirm}(\neg P\mathbf{t}, \mathcal{A}(L), \mathcal{M}(L), L(\mathcal{A}(L), \mathcal{M}(L))).$$

   To realize $\exists l \mathcal{T} < l$ put: $L(\mathrm{a}, m) := \max(lh\mathrm{a}, lh\mathrm{a}_0, m, m_0)$, $l := L(\mathcal{A}(L)) + 1$. To follow realization algorithm more literally one should use realizations from the Lemma 4.1.
2. $A = \neg P\mathbf{t}$. Take $\mathrm{a} := \mathrm{a}_0, m := m_0$ in $\mathbf{B} \models \neg P\mathbf{t}$. With $A \in \mathcal{T}(\mathrm{a}_0)$ this implies $\mathrm{a}_0 \notin \mathbf{B}$, a contradiction.
   Realization of $\mathbf{B} \models \neg P\mathbf{t}$ provides a functional $L(\mathrm{a}, m)$ such that for all a, m, $\neg \mathrm{Confirm}(\neg P\mathbf{t}, \mathrm{a}, m, L(\mathrm{a}, m))$. Put $l := L(\mathrm{a}_0, m_0)$.
3. $A = B \vee C$. From $(\mathrm{a}_0, m) \in \mathbf{B}$, $A \in \mathcal{T}(\mathrm{a}_0)$ we have

   $$\forall \mathrm{a}'(\mathrm{a}' \supseteq \mathrm{a}_0 \& lh(\mathrm{a}') \geq split(A, 0) \& \mathcal{T}(\mathrm{a}') \neq 0 \rightarrow B \in \mathcal{T}(\mathrm{a}') \& C \in \mathcal{T}(\mathrm{a}'))$$
   $$(5.3)$$

   In particular, this is true for $\mathrm{a}' \in \mathbf{B}$ with $lh(\mathrm{a}') = split(A, 0)$ which exists by Lemma 4.1. By IH applied to one of the formulas B, C depending of $\mathbf{B} \models B$ or $\mathbf{B} \models C$ we get the value of l. Realization is constructed in a similar way.
4. $A = B \& C$. This time IH is applied to both B and C to produce two bounds l_B, l_C for l, then put $l := \max(l_B, l_C)$.
5. $A = \exists z B(z)$. Similar to the $\vee$–case. $A \in \mathcal{T}_{\mathrm{a}_0}$ implies

   $$\forall t \forall \mathrm{a}'(\mathrm{a}' \supseteq \mathrm{a} \& lh(\mathrm{a}') \geq split(A, t) \& \mathcal{T}(\mathrm{a}') \neq 0 \rightarrow B(t) \in \mathcal{T}(\mathrm{a}'))   (5.4)$$

   Apply IH to t such that $\mathbf{B} \models B(t)$. *dashv*

The next statement of soundness of predicate logic does not assume that the canonical tree is unbounded. Instead of computing a realization we use a shortcut below.

Lemma 5.2. $d : A_1, \ldots, A_n \to \mathbf{B} \models A_1, \ldots, A_n$

Proof. Standard induction on d. Every formal inference in d is simulated by the inference using the same rule to obtain $\mathbf{B} \models A_1, \ldots, A_n$.   ⊣

Corollary 5.3. *1.* $(a, m) \in \mathbf{B} \& A_1, \ldots, A_n \in \mathcal{T}(a) \to \mathbf{B} \not\models A_1, \ldots, A_n$
2. $d : A \to \exists l \mathcal{T} < l.$

Proof.

1. Use Lemma 5.1 and $\mathbf{B} \models A_1, \ldots, A_n \iff \mathbf{B} \models A_1 \vee \ldots \vee \mathbf{B} \models A_n.$
2. Use Lemma 5.2 and the part 1 for $a = <>, m = 0$ to obtain a contradiction from $(<>, 0) \in \mathbf{B}$, that is from $\forall l \mathcal{T} < l.$ *dashv*

The next series of arguments was obtained essentially by permuting the induction used in the proof of Lemma 5.2, the cut over the formula $\mathbf{B} \models A_1, \ldots, A_n$ used in the argument for the Corollary 5.3.2 and computing realization. Only the final result of this process is presented here.

Definition 5.4. *We say that a sequent Γ is saturated with respect to a derivation d, written $Sat(\Gamma, d)$ if Γ contains the endsequent of d and all side formulas of $\exists$-inferences in d and is closed under propositional inferences:*
   $(A \& B) \in \Gamma$ *implies that $A \in \Gamma$ or $B \in \Gamma$;*
   $(A \vee B) \in \Gamma$ *implies that $A \in \Gamma$ and $B \in \Gamma$.*

The next Lemma relies on (but does not use explicitly) the following intuitions derived from [9],[11]. Infinite branches in the proof search tree $\mathcal{T}$ for a formula A correspond to semivaluations refuting A. Every branch of a cut free derivation $d : A$ is contained in each sufficiently long non-closed branch of $\mathcal{T}$.

Lemma 5.5. $d : \Gamma \& \text{Left}(a, m) \& \Gamma \subseteq \mathcal{T}(a) \& Sat(\mathcal{T}(a), d)$ *implies that $\mathcal{T}(a)$ is an axiom.*

Proof. Induction on d. If d is an axiom, use the fact that $\mathcal{T}(a)$ contains the last sequent of d. Otherwise use IH which is applicable in view of $Sat(\mathcal{T}(a), d)$.   ⊣

Theorem 5.6. *There is a primitive recursive function CE such that $\mathcal{T} < \text{CE}(d)$. for any derivation $d : A$ of a Σ-formula A and the proof search tree $\mathcal{T}$ for A.*

Proof. Assume first d is cut free. Substituting an element of $\mathbf{HU}(A)$ for terms not in $\mathbf{HU}(A)$ we can assume all $\exists$ inferences in d have terms in $\mathbf{HU}(A)$. There is only a finite number of such terms in d. Using the function *split* find a number l such that every sequent $\mathcal{T}(a)$ at the level l in $\mathcal{T}$ is saturated with respect to d. By Lemma 5.5 every such sequent is an axiom, hence $\mathcal{T} < l + 1$. This construction is primitive recursive.

If d contains cuts, eliminate them by a standard reduction procedure which is primitive recursive.   ⊣

References

1. A. Dragalin, New forms of realizability and Markov's rule, Soviet Math. Dokl., 1980, v. 21, no 2, p. 461-463
2. H. Friedman, Classically and intuitionistically provably recursive functions, In: Higher Set Theory, SLNCS v. 699, 1978, p.21-28
3. J.-Y. Girard, Proof Theory and Logical Complexity , vol.1, Bibliopolis, Naples,1987
4. D.Hilbert, P.Bernays, Grundlagen der Mathematik, Bd.2, Springer, 1970
5. U. Kohlenbach, Some logical metatheorems with applications to functional analysis, Trans. Amer. Math. Soc., v.357,no.1, 2005, p. 89-128
6. G. Kreisel, On the Interpretation of Non-finitist Proofs I, J. Symbolic Logic 16, 1951, p.241-267
7. G. Kreisel, On the Interpretation of Non-finitist Proofs II, J. Symbolic Logic 17, 1952, p.43-58
8. G. Kreisel, Logical Aspects of Computations: Contributions and Distractions, in: P. Odifreddi, Ed., Logic and Computer Science, Academic Press, London, 1990, p. 205-278
9. G. Kreisel,G. Mints, S. Simpson, The Use of Abstract Language in Elementary Metamathematics. Lecture Notes in Mathematics, 253, 1975, p.38-131
10. G. Mints, E-theorems, J. Soviet Math. v. 8,no. 3, 1977, p. 323-329
11. G. Mints, The Universality of the Canonical Tree.Soviet Math. Dokl. 1976, 14, p.527-532
12. H. Schwichtenberg, W. Buchholz, Refined program extraction from classical proofs, Annals of Pure and Applied Logic, 2002, v. 114
13. J. Shoenfield, Mathematical Logic, Addison-Vesley, 1967

Enumerate and Expand: Improved Algorithms for Connected Vertex Cover and Tree Cover*

Daniel Mölle, Stefan Richter, and Peter Rossmanith

Dept. of Computer Science, RWTH Aachen University, Germany
{moelle, richter, rossmani}@cs.rwth-aachen.de

Abstract. We present a new method of solving graph problems related to VERTEX COVER by enumerating and expanding appropriate sets of nodes. As an application, we obtain dramatically improved runtime bounds for two variants of the VERTEX COVER problem: In the case of CONNECTED VERTEX COVER, we take the upper bound from $O^*(6^k)$ to $O^*(3.2361^k)$ without large hidden factors. For TREE COVER, we show exactly the same complexity, improving vastly over the previous bound of $O^*((2k)^k)$. In the process, faster algorithms for solving subclasses of the Steiner tree problem on graphs are investigated.

1 Vertex Cover Variants and Enumeration

VERTEX COVER arguably constitutes the most intensely studied problem in parameterized complexity [5], which is most perspicuously reflected by a long history of improved runtime bounds [1, 2, 3, 10, 11] culminating in the algorithm by Chen, Kanj, and Xia with running time $O(1.2738^k + kn)$ [4]. This naturally led to the investigation of generalizations such as CONNECTED VERTEX COVER, CAPACITATED VERTEX COVER, and MAXIMUM PARTIAL VERTEX COVER. Guo, Niedermeier, and Wernicke recently proved upper bounds of $O(6^k n + 4^k n^2 + 2^k n^2 \log n + 2^k nm)$, $O((2k)^k kn^2)$ and $O(1.2^{k^2} + n^2)$ for CONNECTED VERTEX COVER, TREE COVER and CAPACITATED VERTEX COVER, respectively, as well as W[1]-hardness for MAXIMUM PARTIAL VERTEX COVER [8].

Let us define those three variants that are to play a rôle in what follows: Given a graph $G = (V, E)$ and a natural number k, the problem VERTEX COVER asks for a set $C \subseteq V$, $|C| \leq k$, which intersects every edge. CONNECTED VERTEX COVER introduces the additional constraint that the subgraph $G[C]$ induced by C be connected. In order for C to also be a solution to the TREE COVER problem with weight bound W, $G[C]$ needs to be spanned by a tree of weight at most W.

Fernau has investigated the complexity of enumerating vertex covers [7]: Given a graph $G = (V, E)$ and a number k, all minimal vertex covers of size at most k can be enumerated in time $O(2^k k^2 + kn)$. A lower bound of 2^k can be shown using the fact that the graph consisting of k disjoint P_2's simply *has* 2^k different minimal vertex covers of size k.

---

* Supported by the DFG under grant RO 927/6-1 (TAPI).

D. Grigoriev, J. Harrison, and E.A. Hirsch (Eds.): CSR 2006, LNCS 3967, pp. 270–280, 2006.

Now this lower bound is derived from graphs that are not even connected, while algorithms employed to solve graph problems are usually designed to consider each component by itself. In the case of connected graphs, the worst example which comes to mind is that of a path P_n, which has $\binom{k+1}{n-k}$ minimal vertex covers of size exactly $k+1$. These are $2^k n^{O(1)}$ many for $n = \frac{3}{2}k$. Hence, Fernau's bound also holds for connected graphs up to a polynomial factor, suggesting that the complexity of enumerating all minimal vertex covers remains close to 2^k unless restricted to very special graphs.

In order to accelerate algorithms that rely on the enumeration of vertex covers, we thus take a different approach: Instead of going through all the minimal vertex covers, we just enumerate subsets of vertex covers that can easily be expanded to complete vertex covers. We call this method *Enumerate and Expand*.

Definition 1. *Let $\mathcal{C}$ be a graph class. A $\mathcal{C}$-cover for a graph $G = (V, E)$ is a subset $C \subseteq V$ such that $G[V \setminus C] \in \mathcal{C}$.*

That is, a $\mathcal{C}$-cover does not have to cover all edges, but the uncovered part of the graph must be in $\mathcal{C}$. For instance, C is a vertex cover if and only if C is an $\mathcal{I}$-cover, where $\mathcal{I}$ denotes the class of all graphs without edges.

Definition 2. *Let $G = (V, E)$ a graph, $k \in \mathbf{N}$. A family of subsets of V is k-representative if it contains a subset of every vertex cover C of G with $|C| \le k$.*

Theorem 1. *Let $G = (V, E)$ be a graph, $k \in \mathbf{N}$, and $\mathcal{C}$ a graph class that contains all graphs with degree at most d and has a linear-time membership test. A k-representative family of $\mathcal{C}$-covers for G can be enumerated in $O(\zeta^k k^2 + kn)$ time, where ζ is the unique positive root of the polynomial $z^{d+1} - z^d - 1$.*

Proof. We first employ some preprocessing which resembles Buss's kernelization for vertex cover [5] and takes $O(kn)$ time: Repeatedly remove vertices of degree more than k from G until none remain. Call the set of removed vertices H. This kernelization leaves a graph with at most $O(k^2)$ vertices, because if $G[V \setminus H]$ contains more than $k(k+1)$ vertices, then G obviously has no vertex cover of size at most k, and $\emptyset$ is a k-representative set of $\mathcal{C}$-covers. As H is contained in *every* vertex cover of size at most k, it is sufficient to first enumerate a $(k - |H|)$-representative family of $\mathcal{C}$-covers for $G[V \setminus H]$ and then add H to each of its members.

On the remaining graph, apply the algorithm from Table 1. Let us prove that the algorithm enumerates k-representative $\mathcal{C}$-covers for a given graph G and a given number k. It is clear that the algorithm outputs only $\mathcal{C}$-covers. To show that it outputs a $\mathcal{C}$-cover that is a subset of any given vertex cover I with $|I| \le k$, we follow an appropriate branch in the recursion tree: Whenever the algorithm chooses v, follow the $G \setminus \{v\}$ branch if $v \in I$ and the $G \setminus N[v]$ branch otherwise, which implies $N(v) \subseteq I$. In the branch selected in this fashion, $C \subseteq I$ holds at each stage. Since I is a $\mathcal{C}$-cover and C grows in each recursion step, the node set C eventually becomes a $\mathcal{C}$-cover itself and will be output.

The runtime bound is easily shown: Obviously, we have $|N[v]| > d$ whenever the last line is executed, because otherwise $G \in \mathcal{C}$ (and the algorithm would have

Table 1. *Enumerate*$(G, k, \emptyset)$ *enumerates a* k-*representative family of* $\mathcal{C}$-*covers for* G

Input: a graph $G = (V, E)$, $k \in \mathbf{N}$, a set $C \subseteq V$

if $G \in \mathcal{C}$ then output C; return fi;
if $k \leq 0$ then return fi;
choose $v \in V$ with maximum degree;
Enumerate$(G \setminus \{v\}, k - 1, C \cup \{v\})$;
Enumerate$(G \setminus N[v], k - |N(v)|, C \cup N(v))$

terminated in the first line). Hence, we can bound the number of leaves in the recursion tree by $r_k \leq r_{k-1} + r_{k-d-1}$. Standard techniques show that $r_k = O(\zeta^k)$. Choosing v, checking $G \in \mathcal{C}$, and computing $G \setminus \{v\}$, $G \setminus N[v]$, etc. can be done in linear time, resulting in an overall running time of $O(\zeta^k k^2 + kn)$. $\square$

Let $\mathcal{M}$ be the class of all graphs with maximum degree one. Clearly, a graph is in $\mathcal{M}$ if and only if its edges constitute a matching. We will see that the enumeration of any k-representative family of $\mathcal{M}$-covers is sufficient to compute optimal connected vertex covers as well as tree covers. From now on, let $\phi = (1 + \sqrt{5})/2$ denote the golden ratio.

Corollary 1. *Given a graph* G, *a* k-*representative family of* $\mathcal{M}$-*covers of size at most* k *can be enumerated in* $O(\phi^k k^2 + kn)$.

Even though we focus on $\mathcal{M}$-covers in this paper, the above ideas obviously lend themselves to generalization. In general, it takes two steps to apply our technique. Firstly, choose an appropriate graph class $\mathcal{C}$ and prove that a representative family of $\mathcal{C}$-covers can be enumerated efficiently. Secondly, find a way to decide efficiently whether a $\mathcal{C}$-cover can be expanded to a solution to the problem at hand. The following meta theorem shows how to put the pieces together.

Theorem 2. *Let (1)* $\mathcal{C}$ *be a graph class and* P *a property of node sets,*[1]
(2) $T_1(n, k)$ *denote the running time it takes to enumerate a* k-*representative family of* $\mathcal{C}$-*covers of size at most* k *in an* n-*node graph, and*
(3) $T_2(n, k')$ *bound the running time it takes to determine whether we can add at most* k' *nodes to a given* $\mathcal{C}$-*cover for an* n-*node graph in order to obtain a vertex cover of size at most* k *that has property* P.
Given a graph $G = (V, E)$ *with* $n = |V|$, *it is then possible to check whether* G *has a size-*k *vertex cover with property* P *in time*

$$O\Big(\sum_{i=0}^{k} T_1(n, i) \cdot T_2(n, k - i)\Big).$$

Proof. Since any size-k vertex cover with property P is a superset of some $\mathcal{C}$-cover and thus a superset of some minimal $\mathcal{C}$-cover of size at most k, it suffices

---

[1] More precisely, a property defined on node sets and graphs, such as independence, being a dominating set, or connectedness.

to enumerate a k-representative set of C-covers and check whether they can be expanded into a size-k vertex cover with property P. In order to obtain a bound tighter than the trivial $O(T_1(n,k) \cdot T_2(n,k))$ one, we instruct the meta algorithm to look at all possible sizes i of minimal C-covers independently. The above runtime bound follows from the fact that in our context, a C-cover of size i may only be expanded by $k - i$ nodes. □

Please note that, if $T_1(n,k) = f_1(k)poly(n)$ and $T_2(n,k') = f_2(k)poly(n)$ (meaning that both parts are fixed-parameter tractable), then checking the existence of a size-k vertex cover with property P is also fixed-parameter tractable. Moreover, if the second part is even fixed-parameter tractable in k'—the number of nodes *to add*— and the bounds are, say, $T_1(n,k) = O^*(c_1^k)$ and $T_2(n,k') = O^*(c_2^{k'})$, then the overall runtime becomes $O^*((\max\{c_1, c_2\})^k)$. When applying the theorem, one should thus check whether adding k' nodes to achieve property P is fixed-parameter tractable with respect to k'. Unfortunately, this turns out not to be the case in our application: It is W[2]-hard for both CONNECTED VERTEX COVER and TREE COVER, as detailed in Section 5.

2   Restricted Steiner Tree Problems

In order to check whether a given vertex cover C is a subset of a *connected* vertex cover of size at most k, Guo, Niedermeier, and Wernicke compute an optimum Steiner tree for the terminal set C [8]. The answer to this question is "Yes" if and only if the resulting tree has k or less nodes. Enumerating the up to 2^k minimal vertex covers takes $O(2^k k^2 + kn)$ time, and computing the optimum Steiner tree takes $O(3^k n + 2^k n^2 + n^3)$ time for each of these when employing the Dreyfus-Wagner algorithm [6]. The overall running time is thus $O^*(6^k)$ and, using an asymptotically faster Steiner tree algorithm [9], can be improved to $O^*((4+\varepsilon)^k)$ for arbitrarily small $\varepsilon > 0$ at the price of an exorbitant polynomial factor.

Their approach, however, does not exploit the fact that only a restricted version of the Steiner tree problem needs to be solved: The terminals form a vertex cover. In this section, we present an improved algorithm for this variant that takes $O^*(2^k)$ time. In order to allow for the propagation of future improvement on this bound and to reflect the interplay of the Steiner tree problem and variants of VERTEX COVER, we agree upon the following definitions:

Definition 3. *Let $S(n,k)$ denote the time it takes to compute an optimum Steiner tree for any set C of k terminals that also forms a vertex cover in any n-node network (G, ℓ) with $G = (V, E)$ and weight function $\ell: E \to \mathbf{Q}^+$. Likewise, $S(n,k,n')$ denotes the runtime bound if we further restrict the class of eligible Steiner trees to contain at most n' nodes.*

Theorem 3. $S(n,k) \in O(2^k poly(n))$.

Proof. Consider the algorithm in Table 2. It is easy to see that its running time is $O(2^k k^3 n)$ steps on a standard RAM.

Table 2. Computing an optimum Steiner tree for a terminal set C that is a vertex cover

```
for all v ∈ V, y, y' ∈ C do
    T({y, v, y'}) ← an optimum Steiner tree for {y, v, y'};
    T({y, y'}, v) ← an optimum Steiner tree for {y, v, y'}
od;
for i = 3, ..., |C| do
    for all Y ⊆ (C/i) do
        T(Y, v) ← E;
        for all v ∈ V, y ≠ y' ∈ Y do
            T₁ ← T(Y \ {y'}) ∪ T({y, v, y'});
            T₂ ← T(Y \ {y'}, v) ∪ T({v, y'});
            if ‖T₁‖ ≤ ‖T(Y, v)‖ then T(Y, v) ← T₁ fi;
            if ‖T₂‖ ≤ ‖T(Y, v)‖ then T(Y, v) ← T₂ fi
        od;
        T(Y) ← E;
        for all v ∈ V do
            if ‖T(Y, v)‖ ≤ ‖T(Y)‖ then T(Y) ← T(Y, v) fi
        od
    od
od
return T(C)
```

To see the correctness, let us call a set of nodes Y *regional* if there is an optimal Steiner tree for Y that is a subtree of a globally optimal Steiner tree for C. By induction on i, the number of terminals processed in the main part of the algorithm, we want to show that $T(Y)$ contains an optimum Steiner tree for the terminal set Y whenever Y is regional.

The algorithm tries to compute an optimum Steiner tree for Y' by adding new terminals to previously computed trees $T(Y)$ with $Y \subseteq Y'$ and $|Y| = |Y'| - 1$. In any case, it computes a Steiner graph for Y'. Unfortunately, we cannot be sure that there is a regional Y' of every size along the way. However, since C is a vertex cover, the distance between Y and a newly added terminal y' cannot exceed two. That is, for any size i, there is a set $Y \subset C$, such that either Y or $Y \cup v$ for a suitable $v \in V$ is regional.

After strengthening the induction hypothesis to moreover have $T(Y, v)$ contain a minimum Steiner tree for $Y \cup v$ whenever this set is regional and v is a neighbor of a terminal from Y, it suffices to distinguish the three cases depicted in Figure 1. In whichever applies, we are assured that two smaller regional terminal sets are united to form a larger regional set. By the induction hypothesis, the tables hold optimal Steiner trees for the subsets. Because subtrees for regional sets can be replaced by any locally optimal Steiner tree in a globally minimum Steiner tree without changing the costs, their union is also optimal for the superset.   □

Corollary 2. $S(n, k, n') \in O(n' \cdot 2^k poly(n))$.

Proof. Modify the above algorithm as follows: Instead of using one table of optimum Steiner trees, keep n' different tables for optimum Steiner trees consisting

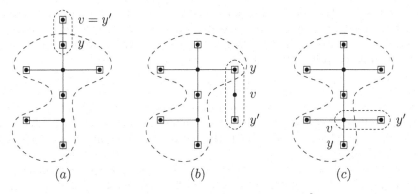

Fig. 1. Three ways to add a terminal y' to $T(Y)$ or $T(Y, v)$

of $1, 2, \ldots, n'$ nodes. It is easy to see that upon termination of the modified algorithm, the n' table cells for the entire terminal set C contain only valid Steiner trees, at least one of which constitutes an optimum solution.   □

3 Connected Vertex Cover

In this section, we present an improved algorithm for CONNECTED VERTEX COVER, where the improvement originates from two different modifications to the algorithm by Guo, Niedermeier, and Wernicke. The first modification is that we apply the Enumerate and Expand method. As suggested in Section 1, we thus do not enumerate all minimal vertex covers, but only minimal $\mathcal{M}$-covers, of size at most k.

On the positive side, this reduces the number of enumerated entities drastically. On the negative side, finding out whether an $\mathcal{M}$-cover—rather than a regular vertex cover—can be expanded to a connected vertex cover is obviously harder, since simply computing an optimum Steiner tree for the cover does not suffice any longer. We solve this problem by transforming the graph and the $\mathcal{M}$-cover into a different graph and a terminal set, such that an optimum Steiner tree for the new instance reflects a CONNECTED VERTEX COVER for the original graph.

The second modification lies in replacing a general algorithm for the Steiner tree problem by the specialized but more efficient one from Section 2. Recall that this algorithm has a running time bounded by $O^*(2^k)$, but requires the terminal set to form a vertex cover. Whereas this condition is obviously met in the approach taken by Guo, Niedermeier, and Wernicke, we need to make sure that it is not violated by the aforementioned transformation.

Lemma 1. *The following problem can be solved in* $S(n, k)$ *time:*

Input:	A graph $G = (V, E)$, an $\mathcal{M}$-cover C, a number $k \in \mathbf{N}$
Parameter: k	
Question:	Is there a connected vertex cover $\hat{C} \supseteq C$
	of size at most k?

Proof. Let $M = \{\{s_1, t_1\}, \ldots, \{s_r, t_r\}\}$ denote the matching obtained by removing C from G. Since any connected vertex cover $\hat{C} \supseteq C$ obviously contains at least $|C| + r$ nodes, the answer is "No" if $|C| + r > k$. Without loss of generality, we can assume C to be nonempty. Now let $G' = (V', E')$ with

$$V' = V \cup X \text{ and } E' = (E \setminus M) \cup \{\{s_1, x_1\}, \{x_1, t_1\}, \ldots, \{s_r, x_r\}, \{x_r, t_r\}\},$$

where $X = \{x_1, \ldots, x_r\}$ are new nodes meant to subdivide the matching edges. The claim follows if we can show that the number of nodes in an optimum Steiner tree for $C \cup \{x_1, \ldots, x_r\}$ in G' exceeds the size of any minimum connected vertex cover $\hat{C} \supseteq C$ for G by r.

Let $\hat{C} \supseteq C$ a connected vertex cover for G. Since both $\hat{C}$ and the subgraph of G induced by $\hat{C}$ are connected, G contains a $(|\hat{C}| - 1)$-edge spanning tree T for $\hat{C}$. Construct a corresponding tree T' for G' as illustrated in Figure 2: First, copy every edge from $E[T] \setminus M$. Second, insert new edges for $i \in \{1, \ldots, r\}$:(1) $\{s_i, x_i\}$ if $\{s_i, t_i\} \notin T$ and $s_i \in \hat{C}$, (2) $\{x_i, t_i\}$ if $\{s_i, t_i\} \notin T$ and $t_i \in \hat{C}$, and (3) $\{s_i, x_i\}$ and $\{x_i, t_i\}$ if $\{s_i, t_i\} \in T$. Obviously, T' constitutes a $(|\hat{C}| + r)$-node Steiner graph for $C \cup X$.

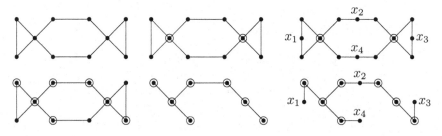

Fig. 2. The first row shows a graph G, an $\mathcal{M}$-cover C for G, and the respective graph G'. The second row shows a connected vertex cover $\hat{C} \supseteq C$ for G, a $(|\hat{C}| - 1)$-edge spanning tree for $\hat{C}$ in G, and the respective tree T' in G'.

On the other hand, let T' a Steiner tree for $C \cup X$ in G'. It remains to check that $V[T'] \setminus X$ is a connected vertex cover for G. It is a vertex cover: C alone covers all edges except the ones in M. Moreover, for each i, at least one of s_i and t_i is contained in T', since x_i can only be reached in T' via these. It is also connected: Consider any path from u to v in T'. Replacing any $s_i - x_i - t_i$ bridge in this path by the simple edge $s_i - t_i$ yields another path in G.   □

Theorem 1 and Lemma 1 comprise the building blocks for an application of the meta theorem (Theorem 2). We immediately obtain the following result:

Corollary 3. *The decision problem* CONNECTED VERTEX COVER *can be solved in* $O^*(\phi^k S(n, k)) = O^*((2\phi)^k)$ *steps.*

4   Tree Cover

Whereas we improved upon the running time for CONNECTED VERTEX COVER essentially in only the base of the exponential part, we are now able to present a much more dramatical speed-up for TREE COVER. In fact, we establish the first parameterized algorithm with exponential running time, as opposed to $O((2k)^k kn^2)$ [8].

Intuitively, there are several aspects that make tackling TREE COVER harder than CONNECTED VERTEX COVER. If we already have a vertex cover C that is a subset of an optimal CONNECTED VERTEX COVER or TREE COVER, then an optimum Steiner tree for C yields a connected vertex cover—optimal in the number of nodes *and* edges, because these numbers differ by exactly one in any tree. In the case of TREE COVER, however, the two optimization goals of having at most k vertices while minimizing the edge weights can be conflicting. This difficulty, which was overcome at the price of a heavily increased running time in earlier scholarship, is surprisingly easy to handle using Corollary 2. However, the conflict springs up again when trying to find a one-to-one correspondence between tree covers that expand $\mathcal{M}$-covers and Steiner trees on modified graphs. The solution lies in a case distinction, i.e., deciding beforehand which matching edges would wind up in the tree cover.

Lemma 2. *It takes no more than $S(n, |C| + |M|, k + |M_2|)$ steps to solve the following problem:*

Input:	A network (G, ℓ), an $\mathcal{M}$-cover C with matching M, a bipartition $M = M_1 \cup M_2$, a number k, and a number W
Question:	Is there a tree cover for G with at most k nodes and weight at most W that contains all nodes in C, all edges from M_1, and no edges from M_2?

Proof. If $|C| + 2|M_1| + |M_2| > k$, the answer is "No."

Let $M_1 = \{\{q_1, q_1'\}, \ldots, \{q_s, q_s'\}\}$. Let $M_2 = \{\{s_1, t_1\}, \ldots, \{s_r, t_r\}\}$. Without loss of generality, we can assume C to be nonempty. Now let $G' = (V', E')$ with $V' = (V \setminus \{q_1', \ldots, q_s'\}) \cup X$ and $E' = ((E \setminus M) \setminus Q') \cup Q \cup E_X$, where $X = \{x_1, \ldots, x_r\}$ are new nodes meant to subdivide the edges in M_2, $Q' = \{\{q_i', v\} \in E \mid \{q_i, q_i'\} \in M_1\}$ is replaced by $Q = \{\{q_i, v\} \mid \{q_i', v\} \in Q'\}$, and $E_X = \{\{s_1, x_1\}, \{x_1, t_1\}, \ldots, \{s_r, x_r\}, \{x_r, t_r\}\}$. We set

$$\ell'(e) = \begin{cases} W & \text{if } e \in E_X, \\ \ell(e) & \text{otherwise.} \end{cases}$$

We claim that the answer is "Yes," if and only if there is a Steiner tree for $C \cup \{q_1, \ldots, q_s\} \cup X$ in G' with at most $k + r$ nodes and weight at most $W - \sum_{e \in M_1} \ell(e) + rW$.

Let T be a tree cover for G that contains C, M_1, and no edges from M_2, has at most k nodes, and weight at most W. Construct T' in G' as follows: First, adapt the edges from T in $((E[T] \setminus M) \setminus Q') \cup Q$. Second, insert new edges for $i \in \{1, \ldots, r\}$: (1) $\{s_i, x_i\}$ if $s_i \in T$, and (2) $\{x_i, t_i\}$ if $s_i \notin T$ and $t_i \in T$. Obviously, T' constitutes a $k + r$ node Steiner graph for $C \cup V[M_1] \cup X$ with weight $W - \sum_{e \in M_1} \ell(e) + rW$.

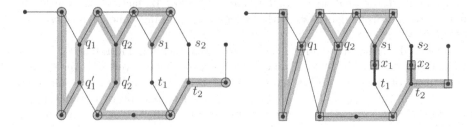

Fig. 3. A graph G with $\mathcal{M}$-cover C (marked) and $|M_1| = |M_2| = 2$, and the corresponding graph G'. An optimal tree cover and the corresponding Steiner tree are highlighted.

On the other hand, let T' a Steiner tree of the above kind in G'. It remains to check that $T = (V_T, E_T)$ with $V_T = (V[T'] \setminus X) \cup V[M_1]$ and $E_T = (E[T'] \setminus E_X) \cup M_1$ is a tree cover for G. It is a vertex cover: C alone covers all edges except the ones in M. Moreover, for each i, at least one of s_i and t_i is contained in T', since x_i can only be reached in T' via these. It is also obviously a tree, and the node and weight bound are easily verified.   □

Fortunately, it turns out that the additional case distinction does not increase the asympotic runtime, when analyzed properly.

Theorem 4. TREE COVER *can be solved in* $O^*((2\phi)^k)$ *steps.*

Proof. Enumerate a k-representative family $\mathcal{C}$ of $\mathcal{M}$-covers. Each $C \in \mathcal{C}$ has a corresponding matching M. For each pair (C, M), cycle through all partitions $M_1 \cup M_2 = M$. Check whether there exists a tree cover of size at most k and weight at most W that contains all nodes from C, all edges from M_1, and no edge from M_2. If such a tree cover is found, output "Yes." Otherwise, if no such cover is found for any pair (C, M), output "No."

To see the correctness, observe that any feasible tree cover T also constitutes a vertex cover, implying that at some point, the above algorithm looks at an $\mathcal{M}$-cover $C \subseteq V[T]$ and a corresponding matching M. Since there is a partition $M_1 \cup M_2 = M$ such that $M_1 \subseteq E[T]$ and $M_2 \cap E[T] = \emptyset$, the algorithm finds T or another feasible tree cover.

According to Lemma 2, the running time is thus bounded by

$$\sum_{(C,M)} \sum_{M_1 \dot\cup M_2 = M} S(n, |C| + |M|, k + |M_2|) =$$

$$\sum_{(C,M)} \left(\sum_{\substack{M_1 \dot\cup M_2 = M \\ |M_1| \le \frac{2}{11}|M|}} S(n, |C| + |M|, k + |M_2|) + \sum_{\substack{M_1 \dot\cup M_2 = M \\ |M_1| > \frac{2}{11}|M|}} S(n, |C| + |M|, k + |M_2|) \right).$$

If $|M_1| \le \frac{2}{11}|M|$, then there are at most $\binom{|M|}{2|M|/11} = O^*(1.6067^{|M|})$ ways to choose M_1. By Corollary 1, there are no more than $\phi^{|C|}$ ways to choose C. Thus, the first sum is bounded by $O^*(\phi^{|C|+|M|}S(n, |C|+|M|, k+|M_2|)) = O^*(\phi^k 2^k)$ according to Corollary 2.

In the second sum, we have $|M_1| > \frac{2}{11}|M|$, and we can assume that $|M| < \frac{11}{13}(k - |C|)$ because otherwise, $|C| + |M| + |M_1| > k$—which already implies that there is no feasible solution. There are no more than $2^{|M|}$ ways to choose M_1, and again at most $\phi^{|C|}$ ways to choose C. We also have that $S(n, |C|+|M|, k+|M_2|) \leq 2^{|C|+|M|}$. Hence, the second sum is bounded by $O^*((2\phi)^{|C|}4^{|M|})$. This is easily seen to be $O^*((2\phi)^{|C|}4^{\frac{11}{13}(k-|C|)}) = O^*((2\phi)^k)$. $\square$

5   Results on Hardness

In terms of our meta theorem (Theorem 2), it would be nice if the restriction of the Steiner tree problem encountered in the preceding sections was fixed-parameter tractable even in the number k' of *non-terminals* in the Steiner tree. More precisely, if this particular Steiner tree problem—where the terminal set forms a vertex cover—could be solved in $O^*(2^{k'})$ time, we would get a new runtime bound of $O^*(\phi^{|C|}2^{k-|C|}) = O^*(2^k)$ for CONNECTED VERTEX COVER immediately.

Unfortunately, this seems very unlikely, since this variant of the Steiner tree problem turns out to be W[2]-hard. Moreover, because TREE COVER can be seen as a generalization of CONNECTED VERTEX COVER, we thus cannot expect to improve the runtime bound for TREE COVER this way, either.

Theorem 5. *The following problem is* W[2]*-hard:*

Input:	A graph G, a vertex cover C, a number $k' \in \mathbf{N}$
Parameter:	k'
Question:	Is there a Steiner tree of size at most $|C| + k'$ for the terminal set C?

Proof. Given a finite family of sets $S = S_1, \ldots, S_n$ comprised of elements from a universe $U = \{u_1, \ldots, u_m\}$ and a number k, the problem HITTING SET is to decide whether there exists a $T \subseteq U$, $|T| \leq k$, such that T and S_i share at least one element for every $i \in \{1, \ldots, n\}$. Since HITTING SET is W[2]-complete, we can prove the above statement via reduction from HITTING SET.

Assume $U = S_1 \cup \cdots \cup S_n$ and $S_i \neq \emptyset$ for all $i \in \{1, \ldots, n\}$ without loss of generality. Construct the bipartite graph $G = (V_1, V_2, E)$ with $V_1 = \{v_0, v_1, \ldots, v_n\}$, $V_2 = U = \{u_1, \ldots, u_m\}$, and $\{v_i, u_j\} \in E$ whenever $u_j \in S_i$ or $i = 0$. Clearly, V_1 is a vertex cover for G, and a set $T \subseteq U = V_2$ constitutes a hitting set for S if and only if its neighborhood in G contains all of V_1. Observe that any Steiner tree for the terminal set V_1 consists of all the nodes of V_1 as well as a subset of V_2 whose neighborhood contains all of V_1, and vice versa. Hence, a k-hitting set T for S exists if and only if there is a Steiner tree for V_1 comprised of $|V_1| + k$ nodes. $\square$

In the case of CONNECTED VERTEX COVER, we can go even further. Theorem 5 suggests that, in order to obtain even better runtime bounds, we would have to let go of reducing to the first variant of the Steiner tree problem presented in Section 2. After all, CONNECTED VERTEX COVER could very well be fixed-parameter tractable for the parameter $k - |C|$. This, however, can be proven not to be the case, unless FPT = W[2]:

Theorem 6. *The following problem is* W[2]*-hard:*

Input: A graph $G = (V, E)$, an $\mathcal{M}$-cover C, a number $k' \in \mathbf{N}$
Parameter: k'
Question: Is there a connected vertex cover $\hat{C} \supseteq C$
of size at most $|C| + k'$?

Proof. This is a generalization of the problem addressed in Theorem 5: If C happens to be a vertex cover, then $\hat{C} \supseteq C$ is a connected vertex cover iff there is a Steiner tree for C with node set $\hat{C}$. □

References

1. R. Balasubramanian, M. R. Fellows, and V. Raman. An improved fixed parameter algorithm for vertex cover. *Inf. Process. Lett.*, 65(3):163–168, 1998.
2. L. S. Chandran and F. Grandoni. Refined memorization for vertex cover. *Inf. Process. Lett.*, 93:125–131, 2005.
3. J. Chen, I. A. Kanj, and W. Jia. Vertex cover: Further observations and further improvements. *Journal of Algorithms*, 41:280–301, 2001.
4. J. Chen, I. A. Kanj, and G. Xia. Simplicity is beauty: Improved upper bounds for vertex cover. Technical Report TR05-008, School of CTI, DePaul University, 2005.
5. R. G. Downey and M. R. Fellows. *Parameterized Complexity.* Springer-Verlag, 1999.
6. S. E. Dreyfus and R. A. Wagner. The Steiner problem in graphs. *Networks*, 1:195–207, 1972.
7. H. Fernau. On parameterized enumeration. In *Proc. of 8th COCOON*, number 2387 in LNCS, pages 564–573. Springer, 2002.
8. J. Guo, R. Niedermeier, and S. Wernicke. Parameterized complexity of generalized vertex cover problems. In *Proc. of 9th WADS*, number 3608 in LNCS, pages 36–48, Waterloo, Canada, 2005. Springer.
9. D. Mölle, S. Richter, and P. Rossmanith. A faster algorithm for the Steiner tree problem. In *Proc. of 23rd STACS*, number 3884 in LNCS, pages 561–570. Springer, 2006.
10. R. Niedermeier and P. Rossmanith. Upper bounds for Vertex Cover further improved. In *Proc. of 16th STACS*, number 1563 in LNCS, pages 561–570. Springer, 1999.
11. R. Niedermeier and P. Rossmanith. On efficient fixed parameter algorithms for Weighted Vertex Cover. *Journal of Algorithms*, 47:63–77, 2003.

Shannon Entropy vs. Kolmogorov Complexity

An. Muchnik[1] and N. Vereshchagin[2,*]

[1] Institute of New Technologies, 10 Nizhnyaya Radischewskaya,
Moscow, Russia 109004
[2] Department of Mathematical Logic and Theory of Algorithms,
Faculty of Mechanics and Mathematics, Moscow State University,
Leninskie Gory, Moscow, Russia 119992
ver@mccme.ru

Abstract. Most assertions involving Shannon entropy have their Kolmogorov complexity counterparts. A general theorem of Romashchenko [4] states that every information inequality that is valid in Shannon's theory is also valid in Kolmogorov's theory, and vice verse. In this paper we prove that this is no longer true for ∀∃-assertions, exhibiting the first example where the formal analogy between Shannon entropy and Kolmogorov complexity fails.

1 Introduction

Since the very beginning the notion of complexity of finite objects was considered as an algorithmic counterpart to the notion of Shannon entropy [9]. Kolmogorov's paper [6] was called "Three approaches to the quantitative definition of information"; Shannon entropy and algorithmic complexity were among these approaches. Let us recall the main definitions.

Let α be a random variable with a finite range $a_1, \ldots, a_N$. Let p_i be the probability of the event $\alpha = a_i$. Then the Shannon entropy of α is defined as

$$H(\alpha) = -\sum_i p_i \log p_i$$

(All logarithms in the paper are base 2.) Using the concavity of the function $p \mapsto -p \log p$, one can prove that the Shannon entropy of every random variable does not exceed its *max-entropy*, $H_0(\alpha)$, defined as the logarithm of the cardinality of the range of α (and is equal to $H_0(\alpha)$ only for uniformly distributed variables).

Let β be another variable with a finite range $b_1, \ldots, b_M$ defined on the same probabilistic space as α is. We define $H(\alpha|\beta = b_j)$ in the same way as $H(\alpha)$; the only difference is that p_i is replaced by the conditional probability $\Pr[\alpha = a_i|\beta = b_j]$. Then we define the conditional entropy as

$$H(\alpha|\beta) = \sum_j \Pr[\beta = b_j] \cdot H(\alpha|\beta = b_j).$$

* Supported in part by Grants 03-01-00475, 06-01-00122, NSh-358.2003.1 from Russian Federation Basic Research Fund.

D. Grigoriev, J. Harrison, and E.A. Hirsch (Eds.): CSR 2006, LNCS 3967, pp. 281–291, 2006.

It is easy to check that

$$H(\langle \alpha, \beta \rangle) = H(\beta) + H(\alpha|\beta). \tag{1}$$

Using the concavity of logarithm function, one can prove that

$$H(\alpha|\beta) \leq H(\alpha), \tag{2}$$

and that $H(\alpha|\beta) = H(\alpha)$ if and only if α and β are independent. This inequality may be rewritten as

$$H(\langle \alpha, \beta \rangle) \leq H(\alpha) + H(\beta). \tag{3}$$

All these notions have their counterparts in Kolmogorov complexity theory.

Roughly speaking, the Kolmogorov complexity of a binary string a is defined as the minimal length of a program that generates a; the conditional complexity $K(a|b)$ of a conditional to b is the minimal length of a program that produces a having b as input. There are different refinements of this idea (called *simple* Kolmogorov complexity, *monotone* complexity, *prefix* complexity, *decision* complexity, see [5], [11]). However, for our purposes the difference is not important, since all these complexity measures differ only by $O(\log n)$ where n is the length of a.

Now we define these notions rigorously. A *conditional description method* is a partial computable function F mapping pairs of binary strings to binary strings. A string p is called a *description of a conditional to b* with respect to F if $F(p, b) = a$. The complexity of a conditional to b with respect to F is defined as the minimal length of a description of a conditional to b with respect to F:

$$K_F(a|b) = \min\{l(p) \mid F(p, b) = a\}.$$

A conditional description method F is called *optimal* if for all other conditional description methods G there is a constant C such that

$$K_F(a|b) \leq K_G(a|b) + C$$

for all a, b. The Solomonoff–Kolmogorov theorem [6, 10] (see also the textbook [5]) states that optimal methods exist. We fix an optimal F and define conditional Kolmogorov complexity $K(a|b)$ as $K_F(a|b)$. The (unconditional) Kolmogorov complexity $K(a)$ is defined as Kolmogorov complexity of a conditional to the empty string. Comparing the optimal function F with the function $G(p, b) = p$ we see that Kolmogorov complexity does not exceed the length:

$$K(a) \leq l(a) + O(1).$$

Fix a computable injective function $a, b \mapsto [a, b]$ encoding pairs of binary strings by binary strings (different computable encodings lead to complexities of

$K([a, b])$ that differ only by $O(1)$). The inequalities (1), (2), and (3) translate to Kolmogorov complexity as follows

$$K([a, b]) = K(b) + K(a|b) + O(\log n), \tag{4}$$
$$K(a|b) \leq K(a) + O(1), \tag{5}$$
$$K([a, b]) \leq K(a) + K(b) + O(\log n). \tag{6}$$

Here $n = l(x) + l(y)$. The inequalities (5) and (6) are easy. The inequality (4) is easy in one direction:

$$K([a, b]) \leq K(b) + K(a|b) + O(\log n).$$

The inverse inequality is the famous theorem of Kolmogorov and Levin, see [5].

Following this analogy between Shannon entropy and Kolmogorov complexity, Romashchenko proved in [4] that the class of linear inequalities for Shannon entropy coincides with the class of inequalities for Kolmogorov complexity. To state this result rigorously, we introduce the following notation. Let $\alpha_1, \alpha_2, \ldots, \alpha_m$ be random variables having a joint distribution. For a set $A \subset \{1, 2, \ldots, m\}$ let α_A denote the tuple $\langle \alpha_i \mid i \in A \rangle$. For instance, $\alpha_{\{1,2,4\}} = \langle \alpha_1, \alpha_2, \alpha_4 \rangle$. Similarly, for a sequence $x_1, \ldots, x_n$ of binary strings x_A denotes $[x_i \mid i \in A]$, for example, $x_{\{1,2,4\}} = [[x_1, x_2], x_4]$.

Theorem 1 (Romashchenko). *If an inequality of the form*

$$\sum_{A,B} \lambda_{A,B} H(\alpha_A | \alpha_B) \leq 0 \tag{7}$$

is true for all random variables $\alpha_1, \ldots, \alpha_m$ then for some function $f(n) = O(\log n)$ the inequality

$$\sum_{A,B} \lambda_{A,B} K(x_A | x_B) \leq f(n) \tag{8}$$

holds for all binary strings $x_1, \ldots, x_m$. Here n stands for the sum of lengths of x_i, the summation is over all subsets A, B of $\{1, 2, \ldots, m\}$, and $\lambda_{A,B}$ denote arbitrary real numbers. Conversely, if for some function $f(n) = o(n)$ the inequality (8) is true for all $x_1, \ldots, x_m$ then (7) holds for all $\alpha_1, \ldots, \alpha_m$.

This theorem shows that all "information inequalities" for Shannon entropy of random variables are true for Kolmogorov complexity of binary strings with logarithmic accuracy, and vice versa. Information inequalities can be considered as universal formulas in a language having $\leq$ as the only predicate symbol and terms of the form $\sum_{A,B} \lambda_{A,B} H(\alpha_A | \alpha_B)$. In this paper we compare Shannon's and Kolmogorov's information theories using $\forall \exists$-formulas in this language. We show that there is $\forall \exists$-formula that is valid in Kolmogorov's theory but is false in Shannon's theory. Then we exhibit another $\forall \exists$-formula that is true in Shannon's theory (assuming that all universal quantifiers range over sequences of independent identically distributed random variables) but is false in Kolmogorov's theory.

2 Relating Shannon Entropy and Kolmogorov Complexity Using $\forall\exists$-Formulas

Consider $\forall\exists$-formulas with atomic formulas being OR of ANDs of information inequalities:

$$\forall\alpha_1\ldots\forall\alpha_k\exists\alpha_{k+1}\ldots\exists\alpha_{k+l} \bigvee_i \bigwedge_j \sum_{A,B} \lambda^{ij}_{A,B} H(\alpha_A|\alpha_B) \leq 0. \tag{9}$$

Here the summation is over all subsets A, B of $\{1, 2, \ldots, k+l\}$, and $\lambda_{A,B}$ denote arbitrary real numbers. This formula expresses in a succinct form the following statement: For all finite sets $A_1, \ldots, A_k$ and jointly distributed random variables $\tilde{\alpha}_1, \ldots, \tilde{\alpha}_k$ in $A_1, \ldots, A_k$ there are finite sets $A_{k+1}, \ldots, A_{k+l}$ and jointly distributed random variables $\alpha_1, \alpha_2, \ldots, \alpha_{k+l}$ in $A_1, A_2, \ldots A_{k+l}$ such that the marginal distribution of $\langle\alpha_1, \ldots, \alpha_k\rangle$ is the same as that of $\langle\tilde{\alpha}_1, \ldots, \tilde{\alpha}_k\rangle$ and $\bigvee_i \bigwedge_j \sum_{A,B} \lambda^{ij}_{A,B} H(\alpha_A|\alpha_B) \leq 0$. For every such formula consider the corresponding formula for Kolmogorov complexity:

$$\forall x_1\ldots\forall x_k\exists x_{k+1}\ldots\exists x_{k+l} \bigvee_i \bigwedge_j \sum_{A,B} \lambda^{ij}_{A,B} K(x_A|x_B) \leq o(n). \tag{10}$$

Here n denotes $l(x_1) + \cdots + l(x_k)$. Note that we include in the sum only the length of strings under universal quantifiers. Otherwise, if we included also the length of other strings, the assertion could become much weaker. We could choose $x_{j+1}, \ldots, x_{j+m}$ of length much greater than that of $x_1, \ldots, x_j$, and the accuracy $o(n)$ might become larger than $K(x_i)$ for $i \leq k$.

Is it true that for all m and $\lambda^{ij}_{A,B}$ Equation (9) holds if and only if Equation (9) holds? The following trivial counter-example shows that this is not the case. Consider the formula:

$$\forall\alpha\exists\beta \ H(\beta) = H(\alpha)/2, \ H(\beta|\alpha) = 0.$$

This statement is false: let α be the random variable with 2 equiprobable outcomes, thus $H(\alpha) = 1$. If $H(\beta|\alpha) = 0$ then β is a function of α and thus $H(\beta)$ can take only values $0, 1$. On the other hand, the similar assertion for Kolmogorov complexity is true:

$$\forall x\exists y \ K(y) = K(x)/2 + O(\log n), \ K(y|x) = O(\log n),$$

where $n = l(x)$. Indeed, as y we can take the first half of the shortest description of x. However, we think that this counter-example is not honest. Indeed, the statement holds for Shannon entropy with accuracy $O(1)$:

$$\forall\alpha\exists\beta \ H(\beta) = H(\alpha)/2 + O(1), \ H(\beta|\alpha) = 0.$$

To prove this, define a sequence $\beta_0, \beta_1, \ldots, \beta_N$ of random variables, where N is the number of outcomes of α, as follows. Let $\beta_0 = \alpha$ and β_{i+1} is obtained from β_i

by gluing any two outcomes of β_i. Each β_i is a function of α, hence $H(\beta_i|\alpha) = 0$. It is easy to verify that gluing any two outcomes can decrease the entropy at most by 1. As $H(\beta_0) = H(\alpha)$ and $H(\beta_n) = 0$ there is i with $H(\beta_i) = H(\alpha)/2 \pm 1$.

We think that it is natural, in the comparison of Shannon and Kolmogorov theories, to consider the information inequalities for Shannon entropy also with accuracy $o(n)$ where n is the sum of "lengths" of the involved random variables. As a "length" of a random variable α it is natural to consider its max-entropy $H_0(\alpha)$. Thus, instead of Equation (9) we will consider the following formula:

$$\forall \alpha_1 \ldots \forall \alpha_k \exists \alpha_{k+1} \ldots \exists \alpha_{k+l} \bigvee_i \bigwedge_j \sum_{A,B} \lambda_{A,B}^{ij} H(\alpha_A|\alpha_B) \leq o(n) \qquad (11)$$

where $n = H_0(\alpha_1) + \cdots + H_0(\alpha_k)$. This formula is a succinct representation of the following assertion: there is a function $f(n) = o(n)$ such that the formula

$$\forall \alpha_1 \ldots \forall \alpha_k \exists \alpha_{k+1} \ldots \exists \alpha_{k+l} \bigvee_i \bigwedge_j \sum_{A,B} \lambda_{A,B}^{ij} H(\alpha_A|\alpha_B) \leq f(n)$$

holds in the same sense, as (9) does. Is it true that Equation (11) holds if and only if Equation (10) holds? This is true for formulas without existential quantifiers ($l = 0$), as Romashchenko's theorem holds (with the same proof) if we replace 0 by $o(n)$ in the right hand side of (7).

3   Separating Shannon's and Kolmogorov's Information Using Max-Entropy in Formulas

It is easy to find a counter-example if we allow to use the max-entropy in formulas (and the length of strings in the corresponding formulas for Kolmogorov complexity). Namely, in Kolmogorov theory, for every string x it is possible to extract about $K(x)$ bits of randomness from x: For every string x there is a string y with

$$K(y|x) = O(\log l(x)), \ K(y) = l(y) + O(1) = K(x) + O(1)$$

(let y to be the shortest description of x). This property of Kolmogorov complexity translates to Shannon theory as follows. For every random variable α there is a random variable β with

$$H(\beta|\alpha) = o(n), \ H(\beta) = H_0(\beta) + o(n) = H(\alpha) + o(n), \qquad (12)$$

where $n = H_0(\alpha)$. This statement is false. This is implied by the following Theorem 2. Indeed, the inequalities (12) and the equality (1) imply that $H(\alpha|\beta) = o(n)$. Thus the left hand side of the inequality (13) is equal to $H_0(\beta) + o(n) = H(\alpha) + o(n)$, which is much less than its right hand side.

Theorem 2. *For every n there is a random variable α with $2^n + 1$ outcomes such that for all random variables β it holds*

$$H_0(\beta) + 64H(\alpha|\beta) > H(\alpha) + n/2 - 2. \qquad (13)$$

Proof. Let the outcomes of α be $a_0, a_1, \ldots, a_{2^n}$ and have probabilities $p_0 = 1/2$ and $p_i = 2^{-n-1}$ for $i = 1, \ldots, 2^n$. Obviously, $H(\alpha) = n/2 + 1$. Thus, given a random variable β in a set B of cardinality 2^d, we have to prove that $d + 64H(\alpha|\beta) > n - 1$.

Let A denote the set $\{a_1, \ldots, a_{2^n}\}$ (the element a_0 is not included). Divide all the pairs $\langle a, b \rangle$ in the set $A \times B$ into three groups:

(1) the pairs $\langle a, b \rangle$ with $H(\alpha|\beta = b) \geq 8H(\alpha|\beta)$;
(2) the pairs $\langle a, b \rangle$ outside (1) with with $\Pr[\alpha = a|\beta = b] \leq 2^{-64H(\alpha|\beta)}$;
(3) the pairs $\langle a, b \rangle$ with $\Pr[\alpha = a|\beta = b] > 2^{-64H(\alpha|\beta)}$.

The sum of probabilities of pairs in (1) is at most $1/8$, as the probability that $H(\alpha|\beta = b)$ exceeds its expectation 8-fold is at most $1/8$. The same argument applies for pairs $\langle a, b \rangle$ in (2): for every fixed b the value $-\log \Pr[\alpha = a|\beta = b]$ is more than $64H(\alpha|\beta)$ hence exceeds its expectation $H(\alpha|\beta = b)$ more than 8-fold. And the total probability of pairs in (3) is at most $2^{d+64H(\alpha|\beta)-n-1}$. Indeed, for every $b \in B$ there are less than $2^{64H(\alpha|\beta)}$ pairs $\langle a, b \rangle$ in (3). Thus, the total number of pairs in (3) is less than $2^{d+64H(\alpha|\beta)}$. The probability of each of them is at most 2^{-n-1}. Summing all probabilities we should obtain at least $1/2$, the probability of $A \times B$. Thus we have

$$1/8 + 1/8 + 2^{d+64H(\alpha|\beta)-n-1} > 1/2 \; \Rightarrow \; d + 64H(\alpha|\beta) > n - 1. \qquad \square$$

It is worth to mention here that a result on implicit extractors from [1] implies that in Shannon's theory it is possible to extract about $H_\infty(\alpha)$ random bits from every random variable α. Here $H_\infty(\alpha)$ denotes the min-entropy of α defined as $\min\{-\log p_1, \ldots, -\log p_N)$ where $p_1, \ldots, p_N$ are probabilities of outcomes of α. More specifically, the following is true.

Theorem 3. *For every random variable α there is a random variable β with*

$$H(\beta|\alpha) = O(\log n), \; H(\beta) = H_0(\beta) + O(\log n) = H_\infty(\alpha) + O(\log n),$$

where $n = H_0(\alpha)$.

Proof. We use the following theorem on extractors from [1]. For all integer $n \geq m$ and positive ε there is a set C of cardinality $O(n/\varepsilon^2)$ and a function $f : \{0,1\}^n \times C \to \{0,1\}^m$ with the following property. Let α be a random variable in $\{0,1\}^n$ with min-entropy at least m and let u be uniformly distributed in C and independent of α. Then the distribution of $f(\alpha, u)$ is at most ε apart from the uniform distribution over $\{0,1\}^m$. This means that for every subset B of $\{0,1\}^m$ the probability that $f(\alpha, u)$ gets into B is $|B|2^{-m} \pm \varepsilon$.

Apply this theorem to $n = \lceil H_0(\alpha) \rceil$, $m = \lfloor H_\infty(\alpha) \rfloor$ and $\varepsilon = \log m/m$. Let $\beta = f(\alpha, u)$ (we may assume that α takes values in $\{0,1\}^n$). Then we have $H(\beta|\alpha) \leq \log |C| \leq \log n - 2\log \varepsilon + O(1) = O(\log n)$.

To estimate $H(\beta)$ we need the following

Lemma 1. *If β is at most ε apart from the uniform distribution over $\{0,1\}^m$ then $H(\beta) \geq m(1 - \varepsilon) - 1$.*

Proof. Let μ stand for the probability distribution of β, that is, $\mu(b) = \Pr[\beta = b]$. Let B_i denote the set of all $b \in \{0,1\}^m$ with $\mu(b) > 2^{i-m}$. As β is at most ε apart from the uniform distribution, we can conclude that $\mu(B_i) \leq |B_i|2^{-m} + \varepsilon < 2^{-i} + \varepsilon$. For all b outside B_i we have $-\log\mu(b) \geq m - i$. Thus the entropy of β can be lower bounded as

$$H(\beta) \geq m - \sum_{i=1}^{m} i \cdot \mu(B_{i-1} - B_i) = m - \sum_{i=1}^{m} i \cdot (\mu B_{i-1} - \mu B_i)$$

$$= m - \sum_{i=1}^{m-1} \mu B_i > m - \sum_{i=1}^{m-1}(2^{-i} + \varepsilon) \geq m - 1 - \varepsilon m.$$

The lemma implies that

$$H(\beta) \geq m - \log m - 1 \geq H_0(\beta) - O(\log n).$$

As $H_\infty(\alpha) = m + O(1)$ and $H(\beta) \leq H_0(\beta)$, this inequality implies that the difference between $H(\beta), H_0(\beta), H_\infty(\alpha)$ is $O(\log n)$.

4 Separating Shannon's and Kolmogorov's Information Theories

Looking for an assertion of the form (11) that distinguishes Shannon entropy and Kolmogorov complexity, it is natural to exploit the following difference between Shannon and Kolmogorov definitions of conditional entropy. In Kolmogorov's approach conditional complexity $K(a|b)$ is defined as the length of a string, namely the shortest description of a conditional to β. In Shannon's approach $H(\alpha|\beta)$ is not defined as the max-entropy or Shannon entropy of any random variable. Thus the following easy statement could distinguish Kolmogorov's and Shannon's theories:

$$\forall x \forall y \exists z \ K(z) \leq K(x|y) + O(1), \ K(x|[y,z]) = O(1)$$

(let z be the shortest description of x conditional to y). However it happens that its analog holds also in Shannon's theory:

Theorem 4. *For all random variables α, β there is random variable γ such that $H(\gamma) \leq H(\beta|\alpha) + O(\log n)$ and $H(\beta|\langle\alpha,\gamma\rangle) = 0$, where $n = H(\beta|\alpha) \leq H_0(\alpha) + H_0(\beta)$.*

Proof. Let A, B denote the set of outcomes of α, β, respectively. Fix $a \in A$ and let β_a denote the random variable in B which takes every value $b \in B$ with probability $\Pr[\beta = b|\alpha = a]$. Using Shannon or Fano code we can construct, for each a, an injective mapping f_a from B to the set of binary strings such that the expected length of $f_a(\beta_a)$ is at most $H(\beta_a) + O(1)$. Let $\gamma = f_a(\beta)$. By construction the outcomes of α and γ together determine the outcome of β uniquely. This shows that $H(\beta|\langle\alpha,\gamma\rangle) = 0$.

It remains to show that $H(\gamma) \leq H(\beta|\alpha) + O(\log n)$. Let us first upper bound the expectation of $l(\gamma)$. It is less than the expectation of $H(\beta_\alpha) + O(1)$, which is equal to $H(\beta|\alpha) + O(1)$. Thus it suffices to show that Shannon entropy of every random variable γ in the set of binary strings with expected length n is at most $n + O(\log n)$. To this end consider the following "self-delimiting" encoding $\bar{x}$ of a binary string x. Double each bit of binary representation of the length of x then append the string 10 to it, and then append x. Obviously $l(\bar{x}) \leq 2\log l(x) + 2 + l(x)$. The set of all strings of the form $\bar{x}$ is a prefix code. Thus the set of all strings $\bar{c}$ where c is an outcome of γ is a prefix code, too. By the Shannon noiseless coding theorem [9] its expected length is at least $H(\gamma)$. Therefore $H(\gamma)$ is less than the expectation of $l(\gamma) + 2\log l(\gamma) + 2$. The expectation of the first term here is equal to n. The expectation of the second term is at most $2\log n$ by concavity of the logarithm function.

The previous discussion shows that it is not so easy to find a counter-example. Looking for a candidate in the literature we find the following:

Theorem 5 ([2]). *For all strings x, y there is a string z such that $K(z) \leq \max\{K(x|y), K(y|x)\} + O(\log n)$ and $K(y|z, x) = O(\log n)$, $K(x|z, y) = O(\log n)$ where $n = K(x|y) + K(y|x)$.*

This theorem implies the following statement

$$\forall x \forall y \exists z \; K(z) + K(y|z, x) + K(x|z, y) \leq \max\{K(x|y), K(y|x)\} + O(\log n)$$

where $n = l(x) + l(y)$. The inner quantifier free formula here can be expressed as an OR of two inequalities. Thus this formula has the form (10). And the analogous statement for Shannon entropy is false:

Theorem 6. *For every n there are random variables α, β with $2^n + 1$ outcomes each such that for every random variable γ we have*

$$H(\gamma) + H(\alpha|\beta, \gamma) + H(\beta|\alpha, \gamma) \geq \max\{H(\alpha|\beta), H(\beta|\alpha)\} + n/2. \quad (14)$$

Proof. Let δ be a random variable having two equiprobable outcomes 0,1. The random variables α and β have the range $\{a_0, a_1, \ldots, a_{2^n}\}$ and are defined as follows. If $\delta = 0$ then α is equal to a_0 and β is uniformly distributed in $\{a_1, \ldots, a_{2^n}\}$. If $\delta = 1$ then β is equal to a_0 and α is uniformly distributed in $\{a_1, \ldots, a_{2^n}\}$. Note that $H(\alpha|\beta) = H(\alpha|\beta) = n/2$, thus the right hand side of Equation (14) is equal to n.

Let γ be a random variable. If $\delta = 0$ then α is constant and β is uniformly distributed in a set of cardinality 2^n, therefore

$$n = H(\beta|\alpha, \delta = 0) \leq H(\beta|\alpha, \gamma, \delta = 0) + H(\gamma|\delta = 0) \leq 2H(\beta|\alpha, \gamma) + H(\gamma|\delta = 0).$$

In a similar way we have

$$n \leq 2H(\alpha|\beta, \gamma) + H(\gamma|\delta = 1).$$

Taking the arithmetical mean of these inequalities we get

$$n \leq H(\beta|\alpha, \gamma) + H(\alpha|\beta, \gamma) + H(\gamma|\delta) \leq H(\beta|\alpha, \gamma) + H(\alpha|\beta, \gamma) + H(\gamma). \quad \square$$

5 Sequences of Identically Distributed Independent Random Variables

A large part of the classical information theory is devoted to the study of sequences of independent identically distributed random variables. Following this line, assume that the universal quantifiers in (11) range over sequences of i.i.d. variables. More specifically let ξ_s^n, $s = 1, \ldots, k$, $n = 1, 2, \ldots$ be random variables such that the k-tuples $\langle \xi_1^n, \ldots, \xi_k^n \rangle$ for $n = 1, 2, \ldots$, have the same distribution and are independent. Let α_s^n denote the sequence of n first outcomes of ξ_s:

$$\alpha_s^n = \xi_s^{(n)} = \xi_s^1, \ldots, \xi_s^n.$$

Consider the following formula:

$$(\forall \text{ i.i.d. } \langle \xi_1^n, \ldots, \xi_k^n \rangle) \, \exists \alpha_{k+1}^n \ldots \exists \alpha_{k+l}^n \bigvee_i \bigwedge_j \sum_{A,B} \lambda_{A,B}^{ij} H(\alpha_A^n | \alpha_B^n) \leq o(n). \quad (15)$$

This formula represents the following statement: For all random variables ξ_s^n, $s = 1, \ldots, k$, $n = 1, 2, \ldots$ such that the k-tuples $\langle \xi_1^n, \ldots, \xi_k^n \rangle$ for $n = 1, 2, \ldots$, have the same distribution and are independent there are sequences of random variables $\alpha_1^n, \ldots, \alpha_{k+l}^n$, $n = 1, 2, \ldots$, and a function $f(n) = o(n)$ with $\bigvee_i \bigwedge_j \sum_{A,B} \lambda_{A,B}^{ij} H(\alpha_A^n | \alpha_B^n) \leq f(n)$, and $\langle \alpha_1^n, \ldots, \alpha_k^n \rangle$ having the same distribution as $\langle \xi_1^{(n)}, \ldots, \xi_k^{(n)} \rangle$ has (for all n).

An example of (15) is the Slepian—Wolf theorem [7]: for every sequence of i.i.d. pairs $\langle \xi^n, \eta^n \rangle$, $n = 1, 2, \ldots$ of random variables there is a sequence of random variables $\{\beta^n\}$ such that

$$H(\beta^n) = H(\xi^{(n)} | \eta^{(n)}) + o(n), \; H(\beta^n | \xi^{(n)}) = o(n), \; H(\xi^{(n)} | \langle \eta^{(n)}, \beta^n \rangle) = o(n).$$

(To fit in our framework, we give here a formulation of Slepian—Wolf theorem that differs slightly from that in [7].)

Is it true that for every theorem of the form (15) the analogous statement (10) for Kolmogorov complexity is also true, and vice versa? We will show that this is not the case. As a counter-example, it is natural to try the Slepian-Wolf theorem, since its proof is very Shannon-theory-specific. Surprisingly, it turns out that the analogous theorem holds for Kolmogorov complexity, too:

Theorem 7 ([8]). *Let x and y be arbitrary strings of length less than n. Then there exists a string z of complexity $K(x|y) + O(\log n)$ such that $K(z|x) = O(\log n)$ and $K(x|z, y) = O(\log n)$. (The constants in $O(\log n)$-notation do not depend on n, x, y.)*

The following easy fact about i.i.d. sequences of random variables gives an example of a true statement of the form (15) whose analog is false for Kolmogorov complexity.

Theorem 8. *For every sequence of i.i.d. pairs $\langle \xi^n, \eta^n \rangle$, $n = 1, 2, \ldots$ of random variables there is a sequence $\{\beta^n\}$ of random variables such that*

$$H(\beta^n) \leq \frac{H(\xi^{(n)}) + H(\eta^{(n)})}{2} + O(1),$$

$$H(\xi^{(n)}|\beta^n) \leq \frac{H(\xi^{(n)}|\eta^{(n)})}{2} + O(1), \quad H(\eta^{(n)}|\beta^n) \leq \frac{H(\eta^{(n)}|\xi^{(n)})}{2} + O(1).$$

Proof. Let $\beta^n = \xi^1, \xi^2, \ldots, \xi^{n/2}, \eta^{n/2+1}, \eta^{n/2+2}, \ldots, \eta^n$.

On the other hand, the similar statement for Kolmogorov complexity is false:

Theorem 9. *There are sequences of strings $\{x_n\}, \{y_n\}$ of length $O(n)$ such that there is no sequence $\{z_n\}$ with*

$$K(z_n) \leq \frac{K(x_n) + K(y_n)}{2} + o(n),$$

$$K(x_n|z_n) \leq \frac{K(x_n|y_n)}{2} + o(n), \quad K(y_n|z_n) \leq \frac{K(y_n|x_n)}{2} + o(n). \tag{16}$$

Proof. The proof easily follows from a theorem from [3]:

Theorem 10 ([3]). *There are sequences of strings $\{x_n\}, \{y_n\}$ such that $l(x_n) = l(y_n) = 2n + O(\log n)$, $K(x_n|y_n) = K(y_n|x_n) = n + O(\log n)$ and for all but finitely many n there is no z_n satisfying the inequalities*

$$K(z_n) + K(x_n|z_n) + K(y_n|z_n) < 4n,$$

$$K(z_n) + K(x_n|z_n) < 3n, \quad K(z_n) + K(y_n|z_n) < 3n.$$

Let x_n, y_n be the sequences from Theorem 10. Assume that there is z_n satisfying (16). Then

$$K(z_n) \leq (K(x_n) + K(y_n))/2 + o(n) \leq 2n + o(n),$$
$$K(x_n|z_n) \leq K(x_n|y_n)/2 + o(n) \leq n/2 + o(n),$$
$$K(y_n|z_n) \leq K(y_n|x_n)/2 + o(n) \leq n/2 + o(n).$$

and

$$K(z_n) + K(x_n|z_n) + K(y_n|z_n) \leq 3n + o(n) \ll 4n$$
$$K(z_n) + K(x_n|z_n) \leq 5n/2 + o(n) \ll 3n$$
$$K(z_n) + K(y_n|z_n) \leq 5n/2 + o(n) \ll 3n.$$

□

6   Conclusion and Open Problems

We have shown that Equation (15) does not imply Equation (10) and that Equation (10) does not imply Equation (11). Are the inverse implications always true? The implication (11) $\Rightarrow$ (15) is straightforward. Can it be split into two implications: (11) $\Rightarrow$ (10) $\Rightarrow$ (15)?

References

1. M.Sipser, "Expanders, randomness, or time versus space", *J. Comput. and System Sci.*, 36 (1988) 379–383.
2. C.H. Bennett, P. Gács, M. Li, P. Vitányi and W. Zurek, "Information Distance", *IEEE Transactions on Information Theory* 44:4 (1998) 1407–1423.
3. A. Chernov, An. Muchnik, A. Romashchenko, A. Shen, and N. Vereshchagin, "Upper semi-lattice of binary strings with the relation 'x is simple conditional to y", *Theoretical Computer Science* 271 (2002) 69–95. Preliminary version in: *14th Annual IEEE Conference on Computational Complexity*, Atlanta, May 4-6, 1999, 114–122.
4. D. Hammer, A. Romashchenko, A. Shen, and N. Vereshchagin, "Inequalities for Shannon entropy and Kolmogorov complexity", *Journal of Computer and Systems Sciences* 60 (2000) 442–464.
5. M. Li and P.M.B. Vitányi, *An Introduction to Kolmogorov Complexity and its Applications*, Springer-Verlag, New York, 2nd Edition, 1997.
6. A.N. Kolmogorov, "Three approaches to the quantitative definition of information", *Problems Inform. Transmission* 1:1 (1965) 1–7.
7. D. Slepian and J.K. Wolf, "Noiseless Coding of Correlated Information Sources", *IEEE Trans. Inform. Theory* IT-19 (1973) 471–480.
8. A.A. Muchnik, "Conditional complexity and codes", *Theoretical Computer Science* 271 (2002) 97–109.
9. C. E. Shannon, "A mathematical theory of communication", *Bell Sys. Tech. J.* 27 (1948) 379–423 and 623–656.
10. R.J. Solomonoff, "A formal theory of inductive inference", Part 1 and Part 2, *Information and Control* 7 (1964) 1–22 and 224–254.
11. V. A. Uspensky, A. Shen. "Relations Between Varieties of Kolmogorov Complexities", *Mathematical Systems Theory* 29(3) (1996) 271–292.

Language Equations with Symmetric Difference*

Alexander Okhotin

Department of Mathematics, University of Turku, FIN-20014 Turku, Finland
alexander.okhotin@utu.fi

Abstract. Systems of language equations used by Ginsburg and Rice
("Two families of languages related to ALGOL", *JACM*, 1962) to represent context-free grammars are modified to use the symmetric difference
operation instead of union. Contrary to a natural expectation that these
two types of equations should have incomparable expressive power, it is
shown that equations with symmetric difference can express every recursive set by their unique solutions, every recursively enumerable set
by their least solutions and every co-recursively-enumerable set by their
greatest solutions. The solution existence problem is Π_1-complete, the
existence of a unique, a least or a greatest solution is Π_2-complete, while
the existence of finitely many solutions is Σ_3-complete.

1 Introduction

The study of language equations began in early 1960s with a paper by Ginsburg
and Rice [8], who represented context-free grammars as systems of equations of
the form

$$X_i = \alpha_{i1} \cup \ldots \cup \alpha_{im_i} \quad (1 \leqslant i \leqslant n, \ m_i \geqslant 1), \tag{*}$$

where α_{ij} are concatenations of variables $X_1, \ldots, X_n$ and terminal symbols. Such
equations actually give a more natural semantics of context-free grammars than
the Chomskian derivation: (*) explicitly states that a word w has property X_i
if and only if it has the property α_1 or α_2 or $\ldots$ or α_n. In particular, this
definition exposes logical disjunction inherent to context-free grammars, which
suggests one to consider context-free grammars with other propositional connectives. Such extensions were successfully defined: *conjunctive grammars* [12]
with disjunction and conjunction, *Boolean grammars* [15] further augmented
with a restricted form of negation, and *dual concatenation grammars* [16] with
disjunction, conjunction, concatenation and the logical dual of concatenation.

What if the logical OR operation used in context-free grammars is replaced
with a related operation, the *exclusive* OR, i.e., sum modulo two? That is, equations of the form (*) are replaced with

$$X_i = \alpha_{i1} \bigtriangleup \ldots \bigtriangleup \alpha_{im_i} \quad (1 \leqslant i \leqslant n, \ m_i \geqslant 1), \tag{**}$$

where, by definition, $K \bigtriangleup L = (K \setminus L) \cup (L \setminus K)$. As an example, consider the
following two systems of language equations, one representing a context-free

---

* Supported by the Academy of Finland under grant 206039.

D. Grigoriev, J. Harrison, and E.A. Hirsch (Eds.): CSR 2006, LNCS 3967, pp. 292–303, 2006.

grammar, and the other obtained from it by replacing union with symmetric difference:

$$
\begin{cases}
S = AB \cup DC \\
A = aA \cup \varepsilon \\
B = bBc \cup \varepsilon \\
C = cC \cup \varepsilon \\
D = aDb \cup \varepsilon
\end{cases}
\qquad
\begin{cases}
S = AB \triangle DC \\
A = aA \triangle \varepsilon \\
B = bBc \triangle \varepsilon \\
C = cC \triangle \varepsilon \\
D = aDb \triangle \varepsilon
\end{cases}
$$

Each system has a unique solution, and the last four components of their solutions are the same: $A = a^*$, $B = \{b^n c^n \mid n \geqslant 0\}$, $C = c^*$ and $D = \{a^m b^m \mid m \geqslant 0\}$. However, the first component is different: it is $S_\cup = \{a^i b^j c^k \mid i = j \text{ or } j = k\}$ for the system with union and $S_\triangle = \{a^i b^j c^k \mid \text{either } i = j \text{ or } j = k, \text{ but not both}\}$ for the system with symmetric difference. The language $S_\triangle$ is clearly not a context-free language, so equations (**) can specify something outside of the scope of (*). On the other hand, $S_\cup$ is an inherently ambiguous context-free language, and it is not clear how to specify it using the symmetric difference only. It is natural to ask whether there exist any context-free languages that cannot occur in unique solutions of systems (**).

From the theory of functional systems of propositional logic developed by Post [20] and brought to perfection in the Russian mathematical school [21], it is known that disjunction and sum modulo two cannot be expressed through each other. So a natural expectation would be that equations of the form (*) and (**) define incomparable families of languages. However, we shall see that this is not so, and language equations with symmetric difference only are strictly more powerful than those with union only, in fact as powerful as the most general known language equations [13].

This will lead us from formal grammars to the theory of language equations of the general form, which is closely connected to computability and complexity, and has recently received a considerable attention. The complexity of some decidable problems for language equations has been studied by Baader and Küsters [3], Bala [5], Meyer and Rabinovich [11] and Okhotin and Yakimova [19]. The first undecidability result for language equations was obtained by Charatonik [6] in connection with the related research on set constraints, see Aiken et al. [1]. A systematic study of the hardness of decision problems for language equations of the general form is due to the author [13, 16, 17, 18], along with the results on the computational universality in several classes of language equations. A very interesting particular case of universality was found by Kunc [10] as an unexpected solution to a problem raised by Conway [7].

Symmetric difference in language equations has not yet been studied. Though van Zijl [22] used language equations with symmetric difference and one-sided concatenation, the object of her study were finite automata and their descriptional complexity, and language equations per se were not considered. Equations of the form (**) constitute one of the basic uninvestigated cases of language equations [14], and their fundamental properties are determined in this paper, filling quite an important gap in the emerging theory of such equations.

The technical foundation of this study is an method of simulating intersection via symmetric difference and concatenation, which is given in Section 3. In particular, this allows us to specify a variant of the language of computation histories of Turing machines [4, 9]. This language is then used in Section 4 to obtain hardness results for the main decision problems for these equations. All problems considered are as hard as for equations with all Boolean operations [13, 18], and the new proofs are peculiar elaborations of earlier ideas.

Next, in Section 5 the expressive power of these equations is determined. It turns out that they do not define any new class of languages, and their expressive power is the same as in the case of all Boolean operations [13]. This completes the classification of systems of the form $X_i = \varphi(X_1, \ldots, X_n)$ $(1 \leqslant i \leqslant n)$ with singleton constants, concatenation and different families of Boolean operations.

2   Language Equations and Boolean Operations

Let Σ be an alphabet, let $(X_1, \ldots, X_n)$ be a vector of variables that assume values of languages over Σ. We shall consider systems of language equations of the form

$$\begin{cases} X_1 = \varphi_1(X_1, \ldots, X_n) \\ \quad \vdots \\ X_n = \varphi_n(X_1, \ldots, X_n) \end{cases} \tag{1}$$

where φ_i are expressions that may contain arbitrarily nested concatenation, symmetric difference and regular constant languages. A solution of a system is a vector of languages $(L_1, \ldots, L_n)$, such that the substitution $X_j = L_j$ for all j turns every equation in (1) into an equality.

Proposition 1 (Basic properties of symmetric difference).

  i. *For all $K, L, M \subseteq \Sigma^*$, $K \triangle (L \triangle M) = (K \triangle L) \triangle M$ and $K \triangle L = L \triangle K$, i.e., the symmetric difference is associative and commutative;*
 ii. *For all $L \subseteq \Sigma^*$, $L \triangle L = \varnothing$;*
iii. *For all $L \subseteq \Sigma^*$, $L \triangle \varnothing = L$ and $L \triangle \Sigma^* = \overline{L}$;*
 iv. *For all $K, L \subseteq \Sigma^*$, such that $K \cap L = \varnothing$, $K \cup L = K \triangle L$, i.e., disjoint union can be simulated by symmetric difference;*
  v. *For all $K, L, M \subseteq \Sigma^*$, $K \triangle L = M$ if and only if $K = L \triangle M$, i.e., terms can be freely moved between the sides of an equation.*
 vi. *$(K \triangle L) \cdot M \neq (K \triangle M) \cdot (L \triangle M)$ for some $K, L, M \subseteq \Sigma^*$, i.e., concatenation is not distributive over symmetric difference.*

It is interesting to see that the resolved form of equations (1), which is of a great importance in the case of union and concatenation [8] and intersection [12], means nothing when the symmetric difference is allowed, since terms can be moved around by Proposition 1(v). One can thus transform an equation $X_i = \varphi_i(X_1, \ldots, X_n)$ to the form $\varphi_i(X_1, \ldots, X_n) \triangle X_i = \varnothing$. Furthermore, if $|\Sigma| \geqslant 2$, then multiple equations of the form $\psi(X_1, \ldots, X_n) = \varnothing$ can be joined into one: a system $\{\psi_1(X_1,\ldots,X_n) = \varnothing, \ldots, \psi_m(X_1,\ldots,X_n) = \varnothing\}$ is equivalent to a single equation $x_1\varphi(X_1,\ldots,X_n) \triangle \ldots \triangle x_n\psi(X_1,\ldots,X_n) = \varnothing$, for any pairwise distinct $x_1,\ldots,x_n \in \Sigma^*$ of equal length. Thus every system (1) can be reformulated as

$$\xi(X_1,\ldots,X_n) = \varnothing \qquad (2)$$

The converse transformation is also possible: an equation $\xi(X_1,\ldots,X_n) = \varnothing$ can be trivially "resolved" as $\{X_1 = X_1 \triangle \xi(X_1,\ldots,X_n), X_2 = X_2, \ldots, X_n = X_n\}$.

We shall, in particular, consider a restricted type of equations with *linear concatenation*, that is, where one of the arguments of every concatenation used in the right-hand sides of (1) must be a constant language. We shall also use the following logical dual of concatenation [16], which can be expressed via concatenation and symmetric difference:

Definition 1. *For any languages $K, L \subseteq \Sigma^*$, their dual concatenation is defined as*

$$K \odot L = \overline{\overline{K} \cdot \overline{L}} = \{w \mid \forall u, v : w = uv \Rightarrow u \in K \text{ or } v \in L\}$$

The equivalence of these two definitions is known [16, Theorem 1].

Let us now construct some language equations with symmetric difference that have interesting solutions.

3 Basic Expressive Power

The symmetric difference cannot express union, but we know that disjoint union can be expressed. This allows us to simulate those context-free grammars where union is assumed to be disjoint: the unambiguous grammars.

Lemma 1. *For every unambiguous linear context-free grammar G there exists and can be effectively constructed a system of language equations with symmetric difference and linear concatenation, such that $L(G)$ is the first component of its unique solution.*

Proof (a sketch). Assume, without loss of generality, that the grammar contains no chain rules of the form $A \to B$, i.e., each rule in the grammar is of the form $A \to uBv$, where $uv \in \Sigma^+$, or of the form $A \to \varepsilon$. Then the system of language equations representing this grammar is of the form

$$X_i = (\{\varepsilon\} \text{ or } \varnothing) \cup \bigcup_{j=1}^{\ell_i} u_{ij} L_{k_{ij}} v_{ij} \qquad (3)$$

It is a so-called *strict system* [2, 12], which is known to have a unique solution. Let $(L_1,\ldots,L_n)$ be this unique solution.

Replace union with symmetric difference everywhere, obtaining a new system

$$X_i = (\{\varepsilon\} \text{ or } \varnothing) \triangle \bigtriangleup_{j=1}^{\ell_i} u_{ij} L_{k_{ij}} v_{ij} \qquad (3')$$

This system is also strict, so it has a unique solution as well. To show that the unique solutions of (3) and (3') coincide, it is sufficient to prove that $(L_1, \ldots, L_n)$ satisfies (3').

Consider every i-th equation. If the right-hand side of (3) is instantiated with $(L_1, \ldots, L_n)$, the union is disjoint, because the original grammar is unambiguous. Therefore, by Proposition 1(iv), the right-hand side of (3') instantiated with $(L_1, \ldots, L_n)$ has the same value as (3), i.e., L_i. This proves that $(L_1, \ldots, L_n)$ is a solution of (3').   □

According to the theory of Boolean functions [20, 21], neither union nor intersection can be expressed via the symmetric difference. However, following is a restricted case in which intersection can be simulated using symmetric difference and dual concatenation. Denote $\Sigma^{\geqslant k} = \{ w \mid w \in \Sigma^*, |w| \geqslant k \}$.

Lemma 2. *Let $i \in \{0, 1, 2\}$ and let $K, L \subseteq \Sigma^i(\Sigma^3)^*$, i.e., K and L are languages over Σ and all words in K and in L are of length i modulo 3. Assume $\varepsilon \notin K$. Then, for every $a \in \Sigma$,*

$$a(K \cap L) = \Sigma^{\geqslant 2} \odot (K \cup aL)$$

Proof. Suppose $aw \in a(K \cap L)$, i.e., $w \in K$ and $w \in L$, and let us show that for every factorization $w = xy$, such that $x \notin \Sigma^{\geqslant 2}$, it holds that $x \in K \cup aL$. Since $x \notin \Sigma^{\geqslant 2}$ means $|x| \leqslant 1$, there are only two factorizations to consider: for the factorization $aw = \varepsilon \cdot aw$ we have $aw \in aL$, while for $aw = a \cdot w$ we have $w \in K$. Then $w \in \Sigma^{\geqslant 2} \odot (K \cup aL)$ by the definition of dual concatenation, which proves one direction of inclusion.

Conversely, let $w \in \Sigma^{\geqslant 2} \odot (K \cup aL)$. Clearly, $w \notin \varepsilon$, since $\varepsilon \notin \Sigma^{\geqslant 2}$ and $\varepsilon \notin K \cup aL$. Then $w = bu$ ($b \in \Sigma$) and $bu, u \in K \cup aL$. Consider three cases:

- if $|u| = i$ (mod 3), then $u \in K$ and $bu \in aL$, i.e., $b = a$ and $u \in K \cap L$;
- if $|u| = i + 1$ (mod 3), then $|bu| = i + 2$ (mod 3), and therefore $bu \notin K \cup aL$, which forms a contradiction, so this case is impossible;
- if $|u| = i + 2$ (mod 3), then $u \notin K \cup aL$, and this case is also impossible.

Therefore, $w = au$ for $u \in K \cap L$, i.e., $w \in a(K \cap L)$.   □

Corollary 1. *Under the conditions of Lemma 2,*

$$a(K \cap L) = \left(\Sigma^{\leqslant 1} \cdot (K \bigtriangleup aL \bigtriangleup \Sigma^*) \right) \bigtriangleup \Sigma^*$$

This special case of intersection is sufficient to represent a very important class of languages. These are the languages of computation histories of Turing machines, first used by Hartmanis [9] to obtain a series of fundamental undecidability and succinctness results for context-free grammars. These languages have already played a crucial role in the study of language equations [6, 13, 17, 18].

Following is a strengthened statement on the representation of these languages, derived from Hartmanis [9] and Baker and Book [4]:

Lemma 3. *For every TM T over an input alphabet Σ there exists an alphabet Γ and an encoding of computations $C_T : \Sigma^* \to \Gamma^*$, such that the language*

$$\mathrm{VALC}(T) = \{w \natural C_T(w) \mid w \in \Sigma^* \text{ and } C_T(w) \text{ is an accepting computation}\}$$

over the alphabet $\Omega = \Sigma \cup \Gamma \cup \{\natural\}$ is an intersection of two LL(1) linear context-free languages $L_1, L_2 \subseteq \Sigma^* \natural \Gamma^*$. Given T, the corresponding LL(1) linear context-free grammars can be effectively constructed.

The proof is based upon the idea of Baker and Book [4, Th.1], though the result is somewhat stronger, since Baker and Book did not require L_1, L_2 to be deterministic. Because this idea is well-known, the proof is omitted.

The plan is to use Lemma 2 to specify the intersection $L_1 \cap L_2$. In order to meet the requirements of the lemma, one has to observe a stricter form of these computations, and to use the following language instead:

Lemma 4. *For every TM T over an input alphabet Σ there exists an alphabet Γ and an encoding of computations $C_T : \Sigma^* \to (\Gamma^3)^*$, such that for every $u \in \Sigma^*$ the language*

$$\mathrm{VALC}_u(T) = \{v \natural C_T(uv) \mid v \in (\Sigma^3)^* \text{ and } C_T(uv) \text{ is an accepting computation}\}$$

is an intersection of two LL(1) linear context-free languages $L_1, L_2 \subseteq (\Sigma^3)^* \natural (\Gamma^3)^*$. Given T and u, the corresponding LL(1) linear context-free grammars can be effectively constructed.

This lemma is proved by a variant of the same construction as in Lemma 3.

Let us now put together all technical results of this section to specify the language $\mathrm{VALC}_u(T)$ by our equations.

Theorem 1. *For every TM T over Σ and for every $u \in \Sigma^*$ there exists and can be effectively constructed a language equation with symmetric difference and linear concatenation, which has a unique solution that contains the language $\mathrm{VALC}_u(T)$ as one of its components.*

Proof. By Lemma 4, $\mathrm{VALC}_u(T) = L(G_1) \cap L(G_2)$ for some LL(1) linear context-free grammars G_1, G_2. Since G_1 and G_2 are LL(1), they are unambiguous, and hence, by Lemma 1, can be specified by language equations with symmetric difference and linear concatenation. Let us combine these equations in a single equation $\varphi(X_1, X_2, Y_1, \ldots, Y_n) = \varnothing$ with a unique solution $(L_1, L_2, K_1, \ldots, K_n)$. Introduce a new variable Z and the following equation:

$$aZ = \left(\Omega^{\leqslant 1} \cdot (X_1 \bigtriangleup aX_2 \bigtriangleup \Omega^*)\right) \bigtriangleup \Omega^* \quad \text{(for some } a \in \Sigma)$$

By Corollary 1, this is equivalent to $aZ = a(X_1 \cap X_2)$, i.e., $Z = X_1 \cap X_2$, and therefore $(L_1, L_2, K_1, \ldots, K_n, L_1 \cap L_2)$, is the unique solution of the constructed system. The last component is the required language $\mathrm{VALC}_u(T)$.   □

Since the language of computation histories is "sliced" according to the length of the input and a certain prefix u, let us introduce a corresponding notation for slices of the language of words accepted by a Turing machine.

Definition 2. *Let T be a Turing machine over an input alphabet Σ. For every $u \in \Sigma^*$, define $L_u(T) = \{v \mid v \in (\Sigma^3)^*, \ uv \in L(T)\}$.*

It is easy to see that the main decision problems for the slices $L_u(T)$, for a given T, are as hard as the corresponding questions for $L(T)$. This fact will significantly simplify the proofs in the next section.

Lemma 5. *Let Σ be an alphabet, let $u \in \Sigma^*$. The problem of testing whether $L_u(T) = \varnothing$ for a given T is Π_1-complete. The problem of whether $L_u(T) = (\Sigma^3)^*$ is Π_2-complete. The problem of whether $(\Sigma^3)^* \setminus L_u(T)$ is finite is Σ_3-complete.*

The hardness part is proved by reduction from the corresponding problems for $L(T)$, while the containment is straightforward.

4   Decision Problems

The main decision problems for language equations of a general form with all Boolean operations and unrestricted concatenation have recently been systematically studied [13, 18]. It was demonstrated that the problem whether a given system has a solution is Π_1-complete [13], the problems of whether a system has a unique solution, a least solution (under a partial order of componentwise inclusion) or a greatest solution are all Π_2-complete [13], while the problem of whether such a system has finitely many solutions is Σ_3-complete [18].

On the other hand, for the language equations of Ginsburg and Rice [8] with union and concatenation only, the problems of existence of least and greatest solutions are trivial (there always are such solutions), while the solution uniqueness problem is undecidable, though only Π_1-complete [17]. For unresolved equations with the same operations all these problems are undecidable, but their exact hardness depends upon restrictions on concatenation [17].

It will now be demonstrated that for our equations with symmetric difference all these problems are as hard as in the case of all Boolean operations [13], even when the concatenation is restricted to linear.

Theorem 2. *The problem of testing whether a system of language equations with symmetric difference and concatenation, whether unrestricted or linear, has a solution is Π_1-complete.*

Proof. The membership in Π_1 is known [13], while the Π_1-hardness is proved by reduction from the following problem: "Given a Turing machine T over an alphabet Σ, determine whether $L_\varepsilon(T) = \varnothing$", which is Π_1-complete by Lemma 5.

Given T, consider the language $\mathrm{VALC}_\varepsilon(L)$ that contains valid accepting computations of T on words of length 0 modulo 3. By Theorem 1, construct the language equation $\varphi(X, Y_1, \ldots, Y_n)$, such that the first component of its unique solution is $\mathrm{VALC}_\varepsilon(L)$. Then add the equation $X = \varnothing$. The resulting system has a solution if and only if $\mathrm{VALC}_\varepsilon(L) = \varnothing$, which is equivalent to $L_\varepsilon(T) = \varnothing$. This proves the correctness of the reduction.   □

The rest of the proofs in this paper require a much more elaborate construction. The following lemma contains the key technical element of that construction, to be used in all subsequent theorems.

Lemma 6. *For all T, u, there exists and can be effectively constructed a language equation $\varphi(X, Z_1, \ldots, Z_n) = \varnothing$ with linear concatenation and symmetric difference that has the set of solutions*

$$\{(X, f_1(X), \ldots, f_n(X)) \mid L_u(T) \subseteq X \subseteq (\Sigma^3)^*\} \tag{4}$$

for some monotone functions $f_i : 2^{\Sigma^*} \to 2^{\Sigma^*}$.

Proof. Consider the language $\mathrm{VALC}_u(T) \subseteq (\Sigma^3)^* \natural (\Gamma^3)^*$, let $a \in \Sigma$ and let $\Omega = \Sigma \cup \Gamma \cup \{\natural\}$. The proof of the lemma is based upon the following claim: for every $X, Y \subseteq \Omega^*$,

$$Y = a^3 X \cup \varepsilon \tag{5a}$$

$$Y(\Sigma^3)^* = (\Sigma^3)^* \tag{5b}$$

$$\Omega^{\geqslant 2} \odot \left(a(Y \triangle (\Sigma^3)^*)\natural(\Gamma^3)^* \triangle a^3\mathrm{VALC}_u(T)\right) = \varnothing \tag{5c}$$

holds if and only if

$$Y = a^3 X \cup \varepsilon \tag{6a}$$

$$X \subseteq (\Sigma^3)^* \tag{6b}$$

$$L_u(T) \subseteq X \tag{6c}$$

First, (5a,5b) is equivalent to (6a,6b). This allows us to use Lemma 2 with $i = 1$, $K = a^3\mathrm{VALC}_u(T)$ and $L = (Y \triangle (\Sigma^3)^*)\natural(\Gamma^3)^*$ to show that (5c) is equivalent to

$$(Y \triangle (\Sigma^3)^*)\natural(\Gamma^3)^* \cap a^3\mathrm{VALC}_u(T) = \varnothing, \tag{7}$$

which essentially means that for every word w in $L_u(T)$, a^3w must be in Y (otherwise the intersection (7) contains $a^3w\natural C_T(uw)$), that is, w must be in X. The latter condition is exactly (6c), which completes the proof of equivalence of (5) and (6).

Now (5) can serve as a model for a system of equations with symmetric difference and linear concatenation. (5a) is replaced by $Y = a^3X \triangle \varepsilon$. (5b) is already in the required form. In order to represent (5c), let us first specify $\mathrm{VALC}_u(T)$: by Theorem 1, there exists an equation $\xi(V_1, \ldots, V_n) = \varnothing$, which has a unique solution with $\mathrm{VALC}_u(T)$ as its first component. Then (5c) can be written as

$$\Omega^{\leqslant 1} \cdot \left(a(Y \triangle (\Sigma^3)^*)\natural(\Gamma^3)^* \triangle a^3V_1 \triangle \Omega^*\right) = \Omega^*$$

The resulting system uses variables $X, Y, V_1, \ldots, V_n$ and the conditions on X and Y are equivalent to (6). In particular, the conditions on X, (6b,6c), are as in (4), Y is defined as (6a), which is a monotone function of X, and V_i are determined uniquely. Hence, the set of solutions is exactly as in (4). □

Theorem 3. *The problem of testing whether a system of language equations with symmetric difference and concatenation, whether unrestricted or linear, has a unique solution (a least solution, a greatest solution) is Π_2-complete.*

Proof. All these problems are in Π_2 for more general language equations [13]. To prove their Π_2-hardness, let us use a reduction from the following problem: "Given a TM T over Σ, determine whether $L(T) \cap (\Sigma^3)^* = (\Sigma^3)^*$".

Use Lemma 6 to construct an equation with the set of solutions (4). It is easy to see that the bounds on X are tight if and only if $L_u(T) = (\Sigma^3)^*$, in which case the unique solution is $((\Sigma^3)^*, f_1((\Sigma^3)^*), \ldots, f_n((\Sigma^3)^*))$, and if $L_u(T) \subset (\Sigma^3)^*$, then $X = L_u(T)$ and $X = (\Sigma^3)^*$ give rise to different solutions.

If a new variable Y with an equation $Y = X \triangle (\Sigma^3)^*$ is added, then these different solutions become pairwise incomparable, which extends the hardness argument to the case of least and greatest solutions. $\qquad\Box$

Theorem 4. *The problem of testing whether a system of language equations with symmetric difference and concatenation, whether unrestricted or linear, has finitely many solutions is Σ_3-complete.*

Proof. The problem is known to be in Σ_3 for a much more general class of equations [18]. The reduction from the problem "Given T over Σ, determine whether $(\Sigma^3)^* \setminus L(T)$ is finite" is exactly as in the proof of Theorem 3: the equation given by Lemma 6 has finitely many solutions (4) if and only if $(\Sigma^3)^* \setminus L(T)$ is finite. $\qquad\Box$

5   Characterization of Expressive Power

We have seen that the main decision problems for language equations with linear concatenation and symmetric difference are as hard as in the case of all Boolean operations and unrestricted concatenation. The proof was based upon the fact that the questions about "slices" of a language accepted by a Turing machine are as hard as the questions about the entire language.

Now this method will be applied to showing that the classes of languages represented by solutions are the same for the general and the restricted class of equations. The goal is now to represent exact languages rather than their slices, and each language will be assembled from finitely many slices by the means of disjoint union.

Theorem 5. *The class of languages representable by least (greatest) solutions of language equations with symmetric difference and concatenation, whether linear or unrestricted, is the class of recursively enumerable (co-r.e., respectively) sets.*

Proof. It is known that least solutions of more general classes of language equations are recursively enumerable [13], so it is left to represent every r.e. language $L \subseteq \Sigma^*$ using equations with symmetric difference and linear concatenation.

Consider any Turing machine T. For each $u \in \Sigma^{\leqslant 2}$, by Lemma 6, there exists a language equation $\varphi_u(X_u, Z_1, \ldots, Z_n) = \varnothing$, such that its set of solutions is $\{(X_u, f_1(X_u), \ldots, f_n(X_u)) \mid L_u(T) \subseteq X_u \subseteq (\Sigma^3)^*\}$. Let us assemble a system of these equations for all u, and add one more variable X with the following equation:

$$X = \bigwedge_{|u| \leqslant 2} u X_u. \tag{8}$$

Since the union $\bigcup_{|u| \leqslant 2} u X_u$ is disjoint regardless of the values of $X_u \subseteq (\Sigma^3)^*$, X is always equal to $\bigcup X_u$, and thus the entire system has the set of solutions

$$\{(X, X_u, f_i(X_u))_{|u| \leqslant 2} \mid L_u(T) \subseteq X_u \subseteq (\Sigma^3)^* \text{ for all } u; X = \bigcup X_u\},$$

the least of which has $X_u = L_u(T)$ and $X = L(T)$.

The case of greatest solutions, which are known to be co-r.e. in a broader setting [13], is handled by the same construction. For each u, Lemma 6 gives an equation $\psi_u(X_u, Z_1, \ldots, Z_n) = \varnothing$. Let us modify it as follows: $\psi_u(X_u \triangle (\Sigma^3)^*, Z_1, \ldots, Z_n) = \varnothing$; then the set of solutions becomes $\{(X_u, f_1(X_u), \ldots, f_n(X_u)) \mid X_u \subseteq (\Sigma^3)^* \setminus L_u(T)\}$. Once these equations are assembled into a system and (8) is added, the greatest solution of the system has $X_u = (\Sigma^3)^* \setminus L_u(T)$ and $X = \Sigma^* \setminus L(T)$. □

Theorem 6. *The class of languages representable by unique solutions of language equations with symmetric difference and concatenation, whether linear or unrestricted, is the class of recursive sets.*

Proof. Unique solutions are always recursive [13, 18] for much more general classes of language equations. Every recursive language $L \subseteq \Sigma^*$ can be represented using our equations as follows.

Consider two Turing machines, T and $\widehat{T}$, such that $L(T) = L$ and $L(\widehat{T}) = \Sigma^* \setminus L$. For each $u \in \Sigma^{\leqslant 2}$, use Lemma 6 for T to construct a language equation $\varphi_u(X_u, Z_1, \ldots, Z_m) = \varnothing$, and then for $\widehat{T}$ to obtain an equation $\psi_u(X_u \triangle (\Sigma^3)^*, Z_{m+1}, \ldots, Z_n) = \varnothing$. These equations share the same variable X_u, and the system $\{\varphi_u = \varnothing, \psi_u = \varnothing\}$ has the set of solutions $\{(X_u, f_1(X_u), \ldots, f_n(X_u)) \mid L_u(T) \subseteq X_u \subseteq (\Sigma^3)^* \setminus L_u(\widehat{T})\}$. Since $L(\widehat{T}) = \Sigma^* \setminus L$, the bounds on X_u are tight, and the unique solution is $L_u(T)$.

It is left to assemble equations for all $u \in \Sigma^{\leqslant 2}$ to obtain $L(T)$. □

Corollary 2. *The class of languages represented by unique solutions of systems of language equations with symmetric difference and concatenation (whether linear or unrestricted) is closed under concatenation and all Boolean operations.*

Moreover, this closure is even effective: given two systems with unique solutions, one can construct Turing machines that recognize their solutions [13], then apply union, intersection or concatenation, obtaining a new Turing machine, and finally convert it back to a language equation using the construction of Theorem 6. It is interesting to compare this with the fact that, outside the context of language equations, concatenation cannot be expressed via linear concatenation and neither union nor intersection can be expressed via symmetric difference.

6   Conclusions and Open Problems

A small variation of the algebraic definition of the context-free grammars [8] turned out to have completely different properties: the full expressive power of language equations with all Boolean operations is attained.

This fills an important gap in our knowledge on language equations. First of all, it provides a definite answer to the question of what Boolean operations are needed to attain computational universality in resolved language equations [14]. Looking at Post's lattice [20, 21] one can see that the only nonmonotone clone below $\{\oplus\}$ and its variations is $\{\neg\}$, and recently it was proved that language equations with complementation and concatenation are not computationally universal [19]. The rest of nonmonotone classes were shown to be universal [14] before, and the following criterion of universality can now be formulated:

Proposition 2. *A class of systems of language equations of the form $X_i = \varphi_i(X_1, \ldots, X_n)$ with concatenation and a certain set of Boolean operations $\mathcal{F}$ is computationally universal if and only if $x \oplus y \oplus z$, $x \vee (y \& \neg z)$ or $x \& (y \vee \neg z)$ can be expressed as a superposition of functions in $\mathcal{F}$.*

This result, which is quite in the spirit of the Russian school of discrete mathematics [21], brings us much closer to a complete classification of language equations with Boolean operations. For unrestricted concatenation and singleton constants, it can now be stated that for different sets of Boolean operations these equations define exactly *seven* distinct families of languages: the context-free languages (union only [8]), the conjunctive languages (union and intersection [12]), the recursive languages (all Boolean operations [13] or symmetric difference only), a strange non-universal class with complement only [19], and three simple subregular classes [14]. The results obtained in this paper shall be useful in extending this preliminary classification to other natural cases, which will help to establish a general theory of language equations.

The new results suggest several questions to study. First, does there exist any direct translation between language equations with all Boolean operations and those with symmetric difference only? Could there be a stronger equivalence between these families of equations: is it true that for every language equation $\varphi(X_1, \ldots, X_n) = \varnothing$ with all Boolean operations there exists a language equation $\psi(X_1, \ldots, X_n) = \varnothing$ with symmetric difference only that has the same set of solutions?

Another question concerns *strongly unique* solutions of equations, i.e., those that are unique modulo every finite language [15]. If all Boolean operations are allowed, these solutions provide semantics for *Boolean grammars* [15], which are an extension of context-free grammars that still admits cubic-time parsing. Will the expressive power remain the same if only symmetric difference is allowed?

One more interesting question is whether the characterizations of the expressive power given in Theorems 5 and 6 could be established without adding any auxiliary symbols to the alphabet. For systems with all Boolean operations this is known to be true for $|\Sigma| \geqslant 2$ [13], and it is open for a unary alphabet.

References

1. A. Aiken, D. Kozen, M. Y. Vardi, E. L. Wimmers, "The complexity of set constraints", *Computer Science Logic* (CSL 1993, Swansea, UK, September 13–17, 1993), LNCS 832, 1–17.
2. J. Autebert, J. Berstel, L. Boasson, "Context-free languages and pushdown automata", in: Rozenberg, Salomaa (Eds.), *Handbook of Formal Languages*, Vol. 1, Springer-Verlag, Berlin, 1997, 111–174.
3. F. Baader, R. Küsters, "Unification in a description logic with transitive closure of roles", *LPAR 2001* (Havana, Cuba, December 3–7, 2001), LNCS 2250, 217–232.
4. B. S. Baker, R. V. Book, "Reversal-bounded multipushdown machines", *Journal of Computer and System Sciences*, 8 (1974), 315–332.
5. S. Bala, "Regular language matching and other decidable cases of the satisfiability problem for constraints between regular open terms", *STACS 2004* (Montpellier, France, March 25–27, 2004), LNCS 2996, 596–607.
6. W. Charatonik, "Set constraints in some equational theories", *Information and Computation*, 142 (1998), 40–75.
7. J. H. Conway, *Regular Algebra and Finite Machines*, Chapman and Hall, 1971.
8. S. Ginsburg, H. G. Rice, "Two families of languages related to ALGOL", *Journal of the ACM*, 9 (1962), 350–371.
9. J. Hartmanis, "Context-free languages and Turing machine computations", *Proceedings of Symposia in Applied Mathematics*, Vol. 19, AMS, 1967, 42–51.
10. M. Kunc, "The power of commuting with finite sets of words", *STACS 2005* (Stuttgart, Germany, February 24–26, 2005), LNCS 3404, 569–580.
11. A. R. Meyer, A. M. Rabinovich, "Valid identity problem for shuffle regular expressions", *Journal of Automata, Languages and Combinatorics* 7:1 (2002), 109–125.
12. A. Okhotin, "Conjunctive grammars and systems of language equations", *Programming and Computer Software*, 28:5 (2002), 243–249.
13. A. Okhotin, "Decision problems for language equations with Boolean operations", *Automata, Languages and Programming* (ICALP 2003, Eindhoven, The Netherlands, June 30–July 4, 2003), LNCS 2719, 239–251; full journal version submitted.
14. A. Okhotin, "Sistemy yazykovykh uravnenii i zamknutye klassy funktsii algebry logiki" (Systems of language equations and closed classes of logic algebra functions), in Russian, Proceedings of the Fifth International conference "Discrete models in the theory of control systems", 2003, 56–64.
15. A. Okhotin, "Boolean grammars", *Information and Computation*, 194:1 (2004), 19–48.
16. A. Okhotin, "The dual of concatenation", *Theoretical Computer Science*, 345:2–3 (2005), 425–447.
17. A. Okhotin, "Unresolved systems of language equations: expressive power and decision problems", *Theoretical Computer Science*, 349:3 (2005), 283–308.
18. A. Okhotin, "Strict language inequalities and their decision problems", *Mathematical Foundations of Computer Science* (MFCS 2005, Gdansk, Poland, August 29–September 2, 2005), LNCS 3618, 708–719.
19. A. Okhotin, O. Yakimova, "On language equations with complementation", TUCS Technical Report No 735, Turku, Finland, December 2005.
20. E. L. Post, *The two-valued iterative systems of mathematical logic*, 1941.
21. S. V. Yablonski, G. P. Gavrilov, V. B. Kudryavtsev, *Funktsii algebry logiki i klassy Posta* (Functions of logic algebra and the classes of Post), 1966, in Russian.
22. L. van Zijl, "On binary ⊕-NFAs and succinct descriptions of regular languages", *Theoretical Computer Science*, 313:1 (2004), 159–172.

On Primitive Recursive Realizabilities

Valery Plisko*

Faculty of Mechanics and Mathematics, Moscow State University,
Moscow 119992, Russia

Abstract. An example of arithmetic sentence which is deducible in intuitionistic predicate calculus with identity but is not strictly primitive
recursively realizable by Z. Damnjanovic is proposed. It is shown also
that the notions of primitive recursive realizability by Z. Damnjanovic
and by S. Salehi are essentially different.

1   Introduction

Constructive semantics of formal languages are widely used in intuitionistic
proof threory. Now the interest in such semantics is growing because of their
applications in theoretical computer science, especially in extracting algorithms
from constructive proofs. Historically, the first precise constructive semantics
of the language of formal arithmetic was recursive realizability introduced by
S. C. Kleene [4] in 1945. The main idea of recursive realizability is coding of information on intuitionistic truth of an arithmetic statement by a natural number
called its *realization* and using recursive functions instead of rather vague intuitionistic concept of effective operation. In this case a realization of the statement
$\Phi \to \Psi$ is the Gödel number of a partial recursive function which maps every
realization of Φ to a realization of Ψ and this corresponds to the intuitionistic
treatment of the truth of an implication $\Phi \to \Psi$ as existence of an effective operation which allows to get a justification of the conclusion from the justification of
the premise. A realization of a universal statement $\forall x \Phi(x)$ is the Gödel number
of a recursive function mapping every natural number n to a realization of $\Phi(n)$
and this corresponds to the intuitionistic treatment of the truth of a universal
statement $\forall x \Phi(x)$ as existence of an effective operation which allows to get a
justification of the statement $\Phi(n)$ for every n. Therefore an intuitionistic concept of effectiveness is made more precise in Kleene's definition of realizability
by means of recursive functions. In mathematics other more restricted classes of
computable functions are considered, for example, the primitive recursive functions. One can to try to define a notion of primitive recursive realizability by
analogy with Kleene's realizability.

In attempting to define a notion of realizability based on the primitive recursive functions an obstacle appears because there is no universal primitive

---
* Partially supported by RFBR grant 05-01-00624 and NWO/RFBR grant
047.017.014.

D. Grigoriev, J. Harrison, and E.A. Hirsch (Eds.): CSR 2006, LNCS 3967, pp. 304–312, 2006.

recursive function and, for example, we can not perform an application of a realization of the implication to a realization of its premise in a primitive recursive way.

Z. Damnjanovic [2] introduced the notion of strictly primitive recursive realizability. He overcomes the obstacle mentioned above by a combination of ideas of realizability and Kripke models. Namely the Grzegorczyk hierarchy of the primitive recursive functions is considered. For every Grzegorczyk's class, there exists a universal function in the next class. This fact allows to define a primitive recursive realizability without using any concept of computability except primitive recursiveness. It was proved in [2] that the primitive recursive intuitionistic arithmetic is correct relative to the strictly primitive recursive realizability. Unfortunately there are gaps in that proof. In this paper we propose an example of arithmetic sentence which is provable by means of the intuitionistic predicate calculus with identity but is not strictly primitive recursively realizable. Hence the Damnjanovic result needs a correction. It is possible that intuitionistic logic must be replaced by a weaker logical system.

S. Salehi [7] introduced another notion of primitive recursive realizability for the language of Basic Arithmetic. He used an indexing of the primitive recursive functions relative to a recursive universal function. The correctness of the basic predicate logic relative to Salehi's realizability was proved. We show in this paper that the notions of primitive recursive realizability by Damnjanovic and by Salehi are essentialy different: there are arithmetic sentences realizable in one sense but not in another. We begin with a description of an indexing of the primitive recursive functions equally acceptable for both notions of primitive recursive realizability.

2   Indexing of the Primitive Recursive Functions

Primitive recursive functions are the functions obtained by substitution and recursion from the following *basic functions*: the constant $O(x) = 0$, the successor operation $S(x) = x + 1$, and the family of the projection functions $I_m^i(x_1, \ldots, x_m) = x_i$ ($m = 1, 2, \ldots$; $1 \leq i \leq m$). The class of *elementary* (by Kalmar) functions is defined as the least class containing the constant $f(x) = 1$, the projection functions I_m^i, and the functions $f(x, y) = x + y$, $f(x, y) = x \div y$, where $x \div y = \begin{cases} 0 \text{ if } x < y, \\ x - y \text{ if } x \geq y, \end{cases}$ and closed under substitution, summation $\varphi(\boldsymbol{x}, y) = \sum_{i=0}^{y} \psi(\boldsymbol{x}, i)$, and multiplication $\varphi(\boldsymbol{x}, y) = \prod_{i=0}^{y} \psi(\boldsymbol{x}, i)$, where $\boldsymbol{x}$ is the list $x_1, \ldots, x_m$. If $a_0, \ldots, a_n$ are natural numbers, then $\langle a_0, \ldots, a_n \rangle$ denotes the number $p_0^{a_0} \cdot \ldots \cdot p_n^{a_n}$, where $p_0, \ldots, p_n$ are sequential prime numbers ($p_0 = 2, p_1 = 3, p_2 = 5, \ldots$). Note that the functions $\pi(i) = p_i$ and $f(x, y) = \langle x, y \rangle$ are elementary. In what follows, for $a \geq 1$ and $i \geq 0$ let $[a]_i$ denote the exponent of p_i under the decomposition of a into prime factors. Therefore $[a]_i = a_i$ if $a = \langle a_0, \ldots, a_n \rangle$. For the definiteness, let $[0]_i = 0$ for every i. Note that the function $\exp(x, i) = [x]_i$ is elementary.

An $(m+1)$-ary function f is obtained by *bounded recursion* from an m-ary function g, an $(m+2)$-ary function h, and an $(m+1)$-ary function j if the following conditions are fulfilled:

$$f(0, x_1, \ldots, x_m) = g(x_1, \ldots, x_m);$$

$$f(y+1, x_1, \ldots, x_m) = h(y, f(y, x_1, \ldots, x_m), x_1, \ldots, x_m);$$

$$f(y, x_1, \ldots, x_m) \leq j(y, x_1, \ldots, x_m)$$

for every $x_1, \ldots, x_m, y$. For given functions $\theta_1, \ldots, \theta_k$, let $\mathbf{E}[\theta_1, \ldots, \theta_k]$ denote the least class containing $\theta_1, \ldots, \theta_k$, the function $S(x)$, all the constant functions and the projection functions and closed under substitution and bounded recursion. Consider a sequence of functions

$$f_0(x, y) = y + 1; \quad f_1(x, y) = x + y; \quad f_2(x, y) = (x+1) \cdot (y+1);$$

$$f_{n+1}(y, 0) = f_n(y+1, y+1); \quad f_{n+1}(y, x+1) = f_{n+1}(f_{n+1}(x, y), x)$$

for $n \geq 2$. A. Grzegorczyk [3] introduced a hierarchy of the classes of functions $\mathcal{E}^n$, where $\mathcal{E}^n = \mathbf{E}[f_n]$. The class $\mathcal{E}^3$ contains all the elementary functions. It was shown by Grzegorczyk [3] that the union of the classes $\mathcal{E}^n$ is exactly the class of primitive recursive functions.

P. Axt [1] improved the description of the Grzegorczyk hierarchy in two directions. First he showed that, for $n \geq 4$, the usual bounded recursion in the definition of the classes $\mathcal{E}^n$ can be replaced by the following scheme applicable to every triple of suitable functions g, h, j:

$$f(0, \boldsymbol{x}) = g(\boldsymbol{x});$$

$$f(y+1, \boldsymbol{x}) = h(y, f(y, \boldsymbol{x}), \boldsymbol{x}) \cdot \mathrm{sg}(j(y, \boldsymbol{x}) \div f(y, \boldsymbol{x})) \cdot \mathrm{sg}(f(y, \boldsymbol{x})),$$

where $\mathrm{sg}(x) = \begin{cases} 0 \text{ if } x = 0; \\ 1 \text{ if } x > 0. \end{cases}$

The second improvement proposed by Axt was the construction of the Grzegorczyk classes by a general scheme of constructing hierarchies of the classes of functions described by Kleene [5]. Namely for every collection of functions $\Theta = \{\theta_1, \ldots, \theta_m\}$, we denote by $\mathbf{E}^4[\Theta]$ the least class including Θ, containing $S(x)$, all the constant functions and the projection functions, the functions sg, $\div$, f_4 and closed under substitution and Axt's bounded recursion. In [5] it is proposed a way of indexing the functions which are primitive recursive relative to $\theta_1, \ldots, \theta_m$. This way can be adopted to an indexing of the class $\mathbf{E}^4[\Theta]$. The functions in $\mathbf{E}^4[\Theta]$ obtain indexes according to their definition from the basic functions. We list below the possible defining schemes for such functions and indicate on the right the indexes of the defined functions.

( )	$\varphi(x_1, \ldots, x_{k_i}) = \theta_i(x_1, \ldots, x_{k_i})$	$\langle 0, k_i, i \rangle$
(I)	$\varphi(x) = x + 1$	$\langle 1, 1 \rangle$
(II)	$\varphi(x_1, \ldots, x_n) = q$	$\langle 2, n, q \rangle$

(III) $\varphi(x_1,\ldots,x_n) = x_i$ (where $1 \le i \le n$) $\langle 3, n, i \rangle$

$$\text{(IV)} \qquad \varphi(x) = \text{sg}(x) \qquad\qquad \langle 4, 1 \rangle$$

$$\text{(V)} \qquad \varphi(x,y) = x \div y \qquad\qquad \langle 5, 2 \rangle$$

$$\text{(VI)} \qquad \varphi(x,y) = f_4(x,y) \qquad\qquad \langle 6, 2 \rangle$$

$$\text{(VII)} \qquad \varphi(\boldsymbol{x}) = \psi(\chi_1(\boldsymbol{x}),\ldots,\chi_k(\boldsymbol{x})) \qquad \langle 7, m, g, h_1, \ldots, h_k \rangle$$

(VIII) $\begin{cases} \varphi(0,\boldsymbol{x}) = \psi(\boldsymbol{x}) \\ \varphi(y+1,\boldsymbol{x}) = \chi(y, f(y,\boldsymbol{x}), \boldsymbol{x}) \cdot \\ \;\cdot \text{sg}(\xi(y,\boldsymbol{x}) \div \varphi(y,\boldsymbol{x})) \cdot \text{sg}(\varphi(y,\boldsymbol{x})) \end{cases} \qquad \langle 8, m+1, g, h, j \rangle$

Here $g, h_1, \ldots, h_k, h, j$ are respectively indexes of $\psi, \chi_1, \ldots, \chi_k, \chi, \xi$.

Let $In^\Theta(b)$ mean that b is an index of some function in the described indexing of the class $\mathbf{E}^4[\Theta]$. It is shown in [1] that $In^\Theta(b)$ is an elementary predicate. If $In^\Theta(b)$, then ef_b^Θ denotes the $[b]_1$-ary function in $\mathbf{E}^4[\Theta]$ indexed by b. Following [1], we set

$$\text{ef}^\Theta(b,a) = \begin{cases} \text{ef}_b^\Theta([a]_0, \ldots, [a]_{[b]_1 \div 1}), & \text{if } In^\Theta(b); \\ 0 & \text{else.} \end{cases}$$

Therefore the function ef^Θ is universal for the class $\mathbf{E}^4[\Theta]$ and it is not in this class. Every function in $\mathbf{E}^4[\Theta]$ has obviously infinetely many indexes relative to the universal function ef^Θ. Now, following Axt, we define for every n a function of two variables e_n by letting $e_0(b,a) = 0$, $e_{n+1}(b,a) = \text{ef}^{e_0,\ldots,e_n}(b,a)$. Finally the class $\mathbf{E}_n$ is defined as $\mathbf{E}^4[e_0,\ldots,e_n]$. Axt proved (see [1, p. 58]) that $\mathbf{E}_n = \mathcal{E}^{n+4}$ for every $n \ge 0$ and this is an improvement of the definition of the Grzegorczyk classes mentioned above.

Let $In(n,b)$ mean that b is an index of a function in $\mathbf{E}_n$ (or is an n-index). It is shown in [1] that the predicate $In(n,b)$ is elementary. It folllows immediately from the definition of the indexing of the classes $\mathbf{E}^4[\Theta]$ that every n-index is also an m-index of the same function for every $m > n$.

We see that the function e_{n+1} is not in the class $\mathbf{E}_n$ but, for every b, the unary function $\psi_b^n(x) = e_{n+1}(b,x)$ is in $\mathbf{E}_n$. Namely, $e_{n+1}(b,x)$ is the constant function $O(x) = 0$ if b is not an index of any function in $\mathbf{E}_n$ and $e_{n+1}(b,x)$ is the function $\varphi([x]_0, \ldots, [x]_{m-1})$ if b is an index of an m-ary function $\varphi(x_1,\ldots,x_m)$ in $\mathbf{E}_n$. Note that an n-index of the function ψ_b^n can be found primitive recursively from b. Namely let d_0 be a 0-index of the function exp and $k_i = \langle 7, 1, d_0, \langle 3, 1, 1 \rangle, \langle 2, 1, i \rangle \rangle$. Obviously, k_i is a 0-index of the function $[x]_i$, then $\langle 7, 1, b, k_1, \ldots, k_m \rangle$ is an index of $\varphi([x]_0, \ldots, [x]_{m-1})$ if $In(n,b)$. Unfortunately, m depends on b. We avoid this obstacle by defining a function $\xi(b,i)$ as

$$\xi(b,0) = \langle 7, 1, b, k_0 \rangle, \quad \xi(b, i+1) = \xi(b,i) \cdot p_{i+3}^{k_i}$$

and letting $\varepsilon(b) = \xi(b, [b]_1 \div 1)$. Now let $\text{in}(n,b)$ be an elementary function such that $\text{in}(n,b) = 1$ if $In(n,b)$ and $\text{in}(n,b) = 0$ else. We define a binary function α in the following way:

$$\alpha(n,b) = (1 \div \text{in}(n,b)) \cdot \langle 2, 1, 0 \rangle + \text{in}(n,b) \cdot \varepsilon(b).$$

Clearly, $\alpha(n, b)$ is an n-index of ψ_b^n and α is a primitive recursive function. Let $\alpha_n(b) = \alpha(n, b)$.

3   Strictly Primitive Recursive Realizability

Damnjanovic [2] defined a relation $t \vdash_n A$, where t, n are natural numbers, A is a closed formula of the first-order language of arithmetic containing symbols for all the primitive recursive functions and using logical connectives $\&, \vee, \rightarrow, \forall, \exists$, the formula $\neg A$ being considered as an abbreviation for $A \rightarrow 0 = 1$. The relation $t \vdash_n A$ is defined by induction on the number of logical symbols in A.

If A is atomic, then $t \vdash_n A \rightleftharpoons t = 0$ and A is true.

$t \vdash_n (B \,\&\, C) \rightleftharpoons [t]_0 \vdash_n B$ and $[t]_1 \vdash_n C$.

$t \vdash_n (B \vee C) \rightleftharpoons [t]_0 = 0$ and $[t]_1 \vdash_n B$ or $[t]_0 = 1$ and $[t]_1 \vdash_n C$.

$t \vdash_n (B \rightarrow C) \rightleftharpoons In(n, t)$ and $(\forall j \geq n) In(j, e_{j+1}(t, \langle j \rangle))$ and $(\forall j \geq n) \forall b [b \vdash_j B \Rightarrow e_{j+1}(e_{j+1}(t, \langle j \rangle), \langle b \rangle) \vdash_j C]$.

$t \vdash_n \exists x B(x) \rightleftharpoons \exists m([t]_1 \vdash_n B(m)$ and $[t]_0 = m)$.

$t \vdash_n \forall x B(x) \rightleftharpoons In(n, t)$ and $\forall m\, e_{n+1}(t, \langle m \rangle) \vdash_n B(m)$.

A closed arithmetic formula A is called *strictly primitive recursively realizable* if $t \vdash_n A$ for some t and n. Note that it follows from $t \vdash_n A$ that $t \vdash_m A$ for every $m > n$. The following facts are implied by the definition:

  1) for every closed arithmetic formula A, the formula $\neg A$ is strictly primitive recursively realizable if and only if A is not strictly primitive recursively realizable;

  2) a closed arithmetic formula A is strictly primitive recursively realizable if and only if $\neg\neg A$ is strictly primitive recursively realizable;

  3) if a closed arithmetic formula A is not strictly primitive recursively realizable, then $a_0 \vdash_0 \neg A$, where $a_0 = \langle 2, 1, \langle 2, 1, 0 \rangle \rangle$.

  This facts imply the following

Proposition 1. *An arithmetic formula A is strictly primitive recursively realizable if and only if $a_0 \vdash_0 \neg\neg A$.*

It is known that in the language of arithmetic containing symbols for the primitive recursive functions every recursively enumerable predicate $P(\boldsymbol{x})$ is expressed by a Σ_1-formula of the form $\exists y A(\boldsymbol{x}, y)$, where $A(\boldsymbol{x}, y)$ is an atomic formula. This means that for every finite sequence of natural numbers $\boldsymbol{m}$, $P(\boldsymbol{m})$ holds if and only if the formula $\exists y A(\boldsymbol{m}, y)$ is true.

Proposition 2. *Let a recursively enumerable predicate $P(\boldsymbol{x})$ be expressed by a Σ_1-formula $\exists y A(\boldsymbol{x}, y)$, where $A(\boldsymbol{x}, y)$ is an atomic formula. Then for every $\boldsymbol{m}$, $P(\boldsymbol{m})$ holds if and only if the formula $\neg\neg\exists y A(\boldsymbol{x}, y)$ is strictly primitive recursively realizable.*

Proof. Let $P(\boldsymbol{m})$. Then $\exists y A(\boldsymbol{m}, y)$ is true. This means that atomic formula $A(\boldsymbol{m}, n)$ is true for some n. Therefore $0 \vdash_0 A(\boldsymbol{m}, n)$ and $\langle n, 0 \rangle \vdash_0 \exists y A(\boldsymbol{m}, y)$. Thus the formula $\exists y A(\boldsymbol{m}, y)$ is strictly primitive recursively realizable. By Proposition 1, $a_0 \vdash_0 \neg\neg\exists y A(\boldsymbol{m}, y)$, i. e. $\neg\neg\exists y A(\boldsymbol{x}, y)$ is strictly primitive recursively realizable. Conversely, if $\neg\neg\exists y A(\boldsymbol{m}, y)$ is strictly primitive recursively realizable, then $\exists y A(\boldsymbol{m}, y)$ is strictly primitive recursively realizable too, consequently $t \vdash_n \exists y A(\boldsymbol{m}, y)$ for some t, n. Then $[t]_1 \vdash_n A(\boldsymbol{m}, [t]_0)$. As A is atomic, the formula $A(\boldsymbol{m}, [t]_0)$ is true. Therefore $\exists y A(\boldsymbol{m}, y)$ is also true and $P(\boldsymbol{m})$ holds. □

Proposition 3. *Let a recursively enumerable predicate $P(\boldsymbol{x})$ be expressed by a Σ_1-formula $\exists y A(\boldsymbol{x}, y)$, where $A(\boldsymbol{x}, y)$ is an atomic formula. Then for every $\boldsymbol{m}$, the following conditions are equivalent:*
  1) $P(\boldsymbol{m})$ holds;
  2) the formula $\neg\neg\exists y A(\boldsymbol{m}, y)$ is strictly primitive recursively realizable;
  3) $a_0 \vdash_0 \neg\neg\exists y A(\boldsymbol{m}, y)$.

This is an immediate consequence of Propositions 1 and 2 and the facts listed above.

4 Strictly Primitive Recursive Realizability and Intuitionistic Logic

Now we prove the main result of this paper.

Theorem 1. *There exists a closed arithmetic formula which is deducible in the intuitionistic predicate calculus with identity but is not strictly primitive recursively realizable.*

Proof. Consider a ternary predicate $e_{x+1}([y]_0, [y]_1) = z$. This predicate is obviously decidable. It is expressed by an arithmetic Σ_1-formula $\exists u A(x, y, z, u)$, the formula $A(x, y, z, u)$ being atomic. Let $B(x, y, z)$ be $\neg\neg\exists u A(x, y, z, u)$. Now we define a formula Φ as $\forall x(\forall y(0 = 0 \to \exists z B(x, y, z)) \to \forall y \exists z B(x, y, z))$. Evidently, Φ is deducible in the intuitionistic predicate calculus with identity. We shall prove that Φ is not strictly primitive recursively realizable.

We have defined in the section 2 a primitive recursive function α. Let n_0 be a natural number such that $\alpha \in \mathbf{E}_{n_0}$. Then, obviously, $\alpha_n \in \mathbf{E}_{n_0}$ for every n.

Lemma 1. *If $n \geq n_0$, then there exists a natural t such that*

$$t \vdash_n \forall y(0 = 0 \to \exists z B(n, y, z)).$$

Proof. Let $n \geq n_0$ be fixed. Let $\delta(y) = \langle 7, 1, \alpha_n([y]_0), \langle 2, 1, [y]_1 \rangle \rangle$. Obviously, for every b, $\delta(b)$ is an n-index of the constant function $\varphi(x) = \psi_{[b]_0}^n([b]_1)$. Further, let c_0 be a 0-index of the function $\zeta(x) = 2^x \cdot 3^{a_0}$ and $\beta(y) = \langle 7, 1, c_0, \delta(y) \rangle$. For every b, the value $\beta(b)$ is an n-index of the constant function $\xi(x) = 2^{\psi_{[b]_0}^n([b]_1)} \cdot 3^{a_0}$. Finally, let $\gamma(y) = \langle 2, 1, \beta(y) \rangle$. Clearly, for every b, the value $\gamma(b)$ is an n-index of the constant function $\phi(x) = \beta(b)$. Therefore

$$\gamma(y) = \langle 2, 1, \langle 7, 1, c_0, \langle 7, 1, \alpha_n([y]_0), \langle 2, 1, [y]_1 \rangle \rangle \rangle \rangle.$$

We see that γ is in $\mathbf{E}_{n_0}$, hence it is in $\mathbf{E}_n$. Let t be an n-index of γ. Let us prove that

$$t \vdash_n \forall y (0 = 0 \to \exists z B(n, y, z)). \tag{1}$$

By the definition of the strictly primitive recursive realizability, (1) means that $In(n, t)$ holds (this condition is obviously fulfilled) and

$$e_{n+1}(t, < b >) \vdash_n 0 = 0 \to \exists z B(n, b, z)$$

for every b. However $e_{n+1}(t, < b >) = \gamma(b)$. Thus we have to prove, for every b, that

$$\gamma(b) \vdash_n 0 = 0 \to \exists z B(n, b, z). \tag{2}$$

The condition (2) means that
1) $\gamma(b)$ is an n-index of a function $\phi(j)$ in $\mathbf{E}_n$ such that
2) for any $j \geq n$, $\phi(j)$ is a j-index of a unary function ξ in $\mathbf{E}_j$ such that for every a, if $a \vdash_j 0 = 0$, then $\xi(a) \vdash_j \exists z B(n, b, z)$.

As it was remarked, $\gamma(b)$ is an n-index of the constant function $\phi(x) = \beta(b)$ in $\mathbf{E}_0$. Thus 1) is fulfilled. Moreover, for every j, the value $\phi(j)$ is $\beta(b)$ being an n-index of the constant function $\xi(x) = 2^{\psi_{[b]_0}^n([b]_1)} \cdot 3^{a_0}$. Hence ξ is in $\mathbf{E}_0$. Let $a \vdash_j 0 = 0$. Then $\xi(a) = 2^{\psi_{[b]_0}^n([b]_1)} \cdot 3^{a_0}$. Let $d = \psi_{[b]_0}^n([b]_1)$. Thus we have to prove that $2^d \cdot 3^{a_0} \vdash_j \exists z B(n, b, z)$, i. e. $a_0 \vdash_j B(n, b, d)$. Note that $B(n, b, d)$ is the formula $\neg\neg \exists u A(x, y, z, u)$. By Proposition 3, it is sufficient to prove that $e_{n+1}([b]_0, [b]_1) = d$, but it is evident because $e_{n+1}([b]_0, [b]_1) = \psi_{[b]_0}^n([b]_1)$. Thus 2) is also fulfilled.   □

Lemma 2. *For every n, there exists no natural t such that $t \vdash_n \forall y \exists z B(n, y, z)$.*

Proof. Let $t \vdash_n \forall y \exists z B(n, y, z)$. This means that t is an n-index of a function f in $\mathbf{E}_n$ such that $f(b) \vdash_n \exists z B(n, b, z)$, i. e. $[f(b)]_1 \vdash_n B(n, b, [f(b)]_0)$. Thus for every b, the formula $B(n, b, [f(b)]_0)$ is strictly primitive recursive realizable. By Proposition 3, in this case $e_{n+1}([b]_0, [b]_1) = [f(b)]_0$ for every b. In particular,

$$e_{n+1}(x, y) = [f(\langle x, y \rangle)]_0 \tag{3}$$

for every x, y. The expression on the right in (3) defines a function in the class $\mathbf{E}_n$, consequently $e_{n+1} \in \mathbf{E}_n$. This contradiction completes the proof of Lemma 2.   □

To complete the proof of Theorem 2, let us assume that $t \vdash_n \Phi$ for some t, n. We may suppose that $n \geq n_0$. Then t is an n-index of a function $g \in \mathbf{E}_n$ such that

$$g(m) \vdash_n \forall y(0 = 0 \rightarrow \exists z B(m, y, z)) \rightarrow \forall y \exists z B(m, y, z)$$

for every m. In paricular, $g(n) \vdash_n \forall y(0 = 0 \rightarrow \exists z B(n, y, z)) \rightarrow \forall y \exists z B(n, y, z)$. This implies that $e_{n+1}(g(n), \langle n \rangle)$ is an n-index of a function h such that

$$h(a) \vdash_n \forall y \exists z B(n, y, z) \tag{4}$$

if $a \vdash \forall y(0 = 0 \rightarrow \exists z B(n, y, z))$. By Lemma 1, there exists such a. It yields (4) in a contradiction with Lemma 2. $\square$

5   Primitive Recursive Realizability by Salehi

Another notion of primitive recursive realizability was introduced by S. Salehi [7] for the formulas of Basic Arithmetic. Basic Arithmetic is a formal system of arithmetic based on basic logic which is weaker than intuitionistic logic. The language of Basic Arithmetic contains symbols for the primitive recursive functions and differs from the usual language of arithmetic by the mode of using universal quantifier. Namely, the quantifier $\forall$ is used only in the formulas of the form $\forall \mathbf{x}(A \rightarrow B)$, where $\mathbf{x}$ is a finite (possibly empty) list of variables, A and B being formulas. If the list $\mathbf{x}$ is empty, then $\forall \mathbf{x}(A \rightarrow B)$ is written merely as $(A \rightarrow B)$. Obviously, every formula of Basic Arithmetic using universal quantifiers only with empty or one-element lists of variables is also a formula of the usual arithmetic language.

Salehi [7] defined a relation $t \, \mathbf{r}^{PR} A$, where t is a natural number, A is a closed formula of the language of Basic Arithmetic. The definition is by induction on the number of logical symbols in A. Let $PR(b)$ mean $\exists n In(n, b)$ and $[b]_1 = 1$; if $PR(b)$, let ψ_b be the function ψ_b^n, where n is such that $In(n, b)$.

If A is atomic, then $t \, \mathbf{r}^{PR} A \rightleftharpoons A$ is true.

$t \, \mathbf{r}^{PR} (B \, \& \, C) \rightleftharpoons [t]_0 \, \mathbf{r}^{PR} B$ and $[t]_1 \, \mathbf{r}^{PR} C$.

$t \, \mathbf{r}^{PR} (B \vee C) \rightleftharpoons [t]_0 = 0$ and $[t]_1 \, \mathbf{r}^{PR} B$ or $[t]_0 \neq 0$ and $[t]_1 \, \mathbf{r}^{PR} C$.

$t \, \mathbf{r}^{PR} \exists x B(x) \rightleftharpoons [t]_1 \, \mathbf{r}^{PR} B([t]_0)$.

$t \, \mathbf{r}^{PR} \forall \mathbf{x}(B(\mathbf{x}) \rightarrow C(\mathbf{x})) \rightleftharpoons PR(t)$ and $\forall b, \mathbf{x}(y \, \mathbf{r}^{PR} B(\mathbf{x}) \Rightarrow \psi_t(\langle y, \mathbf{x} \rangle) \, \mathbf{r}^{PR} C(\mathbf{x}))$.

A closed arithmetic formula A is called *primitive recursively realizable* if $t \, \mathbf{r}^{PR} A$ for some t. Salehi proved that every formula deducible in Basic Arithmetic is primitive recursively realizable.

Theorem 2. *There exists a closed arithmetic formula which is strictly primitive recursively realizable but is not primitive recursively realizable by Salehi.*

Proof. Consider a ternary predicate $e_{x+1}([y]_0, [y]_1) = z$. It is recursive, therefore it is expressed by an arithmetic Σ_1-formula $\exists u A(x, y, z, u)$ with atomic $A(x, y, z, u)$. Let $B(x, y, z)$ be $\neg\neg\exists u A(x, y, z, u)$.

Consider a binary predicate $e_x(x, x) = y$. It is expressed by an arithmetic Σ_1-formula $\exists v C(x, y, v)$ with atomic $C(x, y, v)$. Let $D(x, y)$ be $\neg\neg\exists v C(x, y, v)$.

Consider the formula

$$\forall x (\forall y (0 = 0 \rightarrow \exists z B(x, y, z)) \rightarrow \exists y D(x, y)). \tag{5}$$

Clearly, this formula is also a formula of Basic Arithmetic, thus both concepts of primitive recursive realizability are defined for it. The formula (5) is strictly primitive recursively realizable but is not primitive recursively realizable by Salehi. A detailed proof of this fact is stated in [6]. Note that the negation of (5) is primitive recursively realizable by Salehi but is not strictly primitive recursively realizable. □

References

1. Axt, P.: Enumeration and the Grzegorczyk hierarchy. Z. math. Logik und Grundl. Math. **9** (1963) 53–65
2. Damnjanovic, Z.: Strictly primitive recursive realizability. I. J. Symbol. Log. **59** (1994) 1210–1227
3. Grzegorczyk, A.: Some classes of recursive functions. Rozprawy matematyczne **4** (1953) 1–46
4. Kleene, S. C.: On the interpretation of intuitionistic number theory. J.Symbol. Log. **10** (1945) 109–124
5. Kleene, S. C.: Extension of an effectively generated class of functions by enumeration. Colloquium mathematicum **6** (1958) 67–78
6. Plisko, V. E.: On the relation between two notions of primitive recursive realizability. Vestnik Moskovskogo Universiteta. Ser. 1. Matematika. Mekhanika. N 1 (2006) 6–11
7. Salehi, S.: A generalized realizability for constructive arithmetic. Bull. Symbol. Log. **7** (2001) 147–148

Evidence Reconstruction of Epistemic Modal Logic S5

Natalia Rubtsova

Department of Mathematical Logic, Faculty of Mechanics and Mathematics,
Lomonosov Moscow State University, Leninskie Gory, Moscow, 119992, Russia
Natalya_Rubtsova@mail.ru

Abstract. We introduce the logic of proofs whose modal counterpart is
the modal logic S5. The language of Logic of Proofs LP is extended by
a new unary operation of *negative checker* "?". We define Kripke-style
models for the resulting logic in the style of Fitting models and prove the
corresponding Completeness theorem. The main result is the Realization
theorem for the modal logic S5.

1 Introduction

The Logic of Proofs LP was defined by S. Artemov in [1, 2]. It is formulated in the
propositional language enriched by formulas of the form $[\![t]\!]F$ with the intended
meaning *"t is a proof of F"*. Here t is a *proof term* which represents arithmetical
proofs. Proof terms are constructed from *proof variables* and *proof constants*
by means of the three functional symbols representing elementary computable
operations on proofs: binary "·", "+", and unary "!". The Logic of Proofs LP
is axiomatized over propositional calculus by the weak reflexivity principle and
axioms for operations "·", "+" and "!"

The intended semantics for LP is formalized in Peano Arithnmetic **PA**; $[\![t]\!]F$
is interpreted by an arithmetical proof predicate which numerates theorems of
PA. It is proved in [2] that LP is complete with respect to arithmetical interpretations based on multi-conclusion proof predicates. It is also shown in [2] that the
modal counterpart of LP is Gödel's provability logic S4. For every LP-formula
φ holds LP $\vdash \varphi \Leftrightarrow$ S4 $\vdash \varphi^\circ$. Here φ° is the *forgetful projection*, i. e. the translation that replace all subformulas of the form $[\![t]\!]F$ by $\Box F$. This fact provides
S4 and, therefore, intuitionistic logic with the exact provability semantics. The
implication from right to left is proven using cut-elimination theorem for S4: the
algorithm assigns terms to all $\Box$'s in a cut-free S4-derivation in such a way that
all formulas sequens become derivable in LP.

A symbolic semantics for LP was proposed by A. Mkrtychevin in [6]. A model
consists of an evidence function which describes possible evidences for a sentence
and a truth evaluation of sentence letters. It is proven that LP is complete with
respect to these models. The semantical approach was further developed by
M. Fitting in [5] who proposed Kripke-style models for LP. It was shown in [5]
that LP is complete with respect to these models. This approach made it possible

D. Grigoriev, J. Harrison, and E.A. Hirsch (Eds.): CSR 2006, LNCS 3967, pp. 313–321, 2006.
© Springer-Verlag Berlin Heidelberg 2006

to re-established the Realization theorem for S4 by semantical means and does not involve cut-elimination.

Our work is a part of a research effort to introduce justification into formal epistemology. Hintikka's style modal logic approach to knowledge captures only two parts of Plato's tripartate definition of knowledge as "justified true belief". Namely, the "justified" part was missing in formal epistemology despite a long expressed desire to have it there too. The Logic of Proofs LP which was originally designed to express in logic the notion of a proof became a basis for the theory of knowledge with justification now extended to multi-agent systems; this approach also provided a fresh look at the common knowledge phenomenon. S. Artemov in [3] proposed to consider LP as the logic of so-called "evidence based knowledge" (EBK), where the meaning of the formula $[\![t]\!]F$ is "*t is an evidence for F*". An (EBK)-system is obtained by augmenting a multi-agent logic of knowledge with an evidence component described by LP with the following connection: if a formula has an evidence then all agents know it. Note, that in all these systems, the logic for the evidence part corresponds to S4, whereas the base knowledge logics could be weaker (T), equal to (S4), or stronger (S5) then LP. The "forgetful counterparts" of the above EBK-systems were also considered in [3]. They are obtained by replacing all evidence terms with a new modality J. The epistemic semantics of JF is "there is a justification for F." The forgetful EBK-systems are normal modal logics that respect the standard Kripke-style semantics. In [3] it is shown that given a theorem φ in the forgetful language, one could recover an evidence-carrying formula ψ such that $\psi = \varphi^\circ$ and ψ is derivable in the corresponding EBK-system.

Our purpose is to describe a logic of proofs whose modal counterpart is the modal logic S5. The first logic of explicit proof terms for S5 was suggested by S. Artemov, E. Kazakov, and D. Shapiro in [4]. In the calculus LPS5, introduced in that paper, the negative checker operation was axiomatized by the principle $[\![t]\!](F \to \neg[\![s]\!]G) \to (F \to [\![?t]\!]\neg[\![s]\!]G)$ which corresponds to the modal formula $\Box(F \to \neg\Box G) \to (F \to \Box\neg\Box G)$ that can be taken as an alternative to the traditional axiom $\neg\Box F \to \Box\neg\Box F$ for S5. Their realization theorem for S5 used heavy proof-theoretical methods (e.g., Mints' cut-free treatment of S5); it is not clear whether these methods are extendable to multi-agent situations without extra proof-theoretical work.

An alternative light-weight approach to finding the logic of justification terms for S5 was suggested in 2005 independently by Eric Pacuit and Natalia Pubtsova. The language of LP was extended by a new unary operation of *negative checker* "?" described by the axiom

$$\neg[\![t]\!]F \to [\![?t]\!]\neg[\![t]\!]F.$$

The resulting logic is denoted by LP(S5). In this paper we define Kripke-style models for LP(S5) in the style of [5] and prove the corresponding Completeness theorem. The main result of this paper is the Realization theorem for the modal logic S5, i. e. we prove that LP(S5) is sufficient to restore evidences in every theorem of S5.

2   Logic LP(S5)

In this section we define logic of proofs with negative checker LP(S5) and establish Internalization property for it.

Definition 1. *The language of the logic* LP(S5) *contains objects of two sorts: propositions and evidences. It consists of propositional variables* $P_1, P_2, \ldots$, *propositional constant* $\bot$, *logical connective* $\to$, *evidence variables* $x_1, x_2, \ldots$, *evidence constants* $c_1, c_2, \ldots$, *function symbols* $+, \cdot$ *(binary) and* $!, ?$ *(monadic) and operator symbol of the type* $[\![\langle term \rangle]\!] \langle formula \rangle$.

Evidence terms are built up from evidence constants and evidence variables using the function symbols. Ground terms are those without variables. Formulas are constructed from propositional variables and propositional constant using $\to$ *and according to the rule: if* t *is a term and* F *is a formula then* $[\![t]\!]F$ *is a formula too. Formulas of the form* $[\![t]\!]F$ *are called q-atomic.*

Definition 2. *The logic* LP(S5) *is axiomatized by the following schemas* $(X, Y$ *are formulas,* t, s *are evidence terms):*

A0 **Classical**	*classical propositional axioms*
A1 **Reflexivity**	$[\![t]\!]X \to X$
A2 **Application**	$[\![t]\!](X \to Y) \to ([\![s]\!]X \to [\![t \cdot s]\!]Y)$
A3 **Sum**	$[\![t]\!]X \to [\![t + s]\!]X$
	$[\![s]\!]X \to [\![t + s]\!]X$
A4 **Positive Checker**	$[\![t]\!]X \to [\![!t]\!][\![t]\!]X$
A5 **Negative Checker**	$\neg[\![t]\!]X \to [\![?t]\!]\neg[\![t]\!]X$

There are two rules of inference

R1 **Modus Ponens** $\vdash X, X \to Y \Rightarrow \vdash Y$
R2 **Axiom Necessitation** $\vdash [\![c]\!]A$ *where* A *is an axiom* $A0 - A5$
                             and c *is an evidence constant*

Proposition 1 (Internalization). *Given* LP(S5) $\vdash X$ *there is an evidence term* t *such that* LP(S5) $\vdash [\![t]\!]X$.

Proof. Induction on a derivation of a formula X.

3   Symbolic Semantics for LP(S5)

In this section we introduce models for LP(S5) which are Fitting models (see [5]). The main idea of this semantic is to add an evidence function to Kripke-style models. The evidence function assigns "admissible" evidence terms to a formula. In general case a truth value of a formula does not play any role, but in our case only true statements have evidences. A formula $[\![t]\!]X$ holds in a given world Γ iff both of the following conditions are met: 1) t is an admissible evidence for X in Γ; 2) X holds in all worlds accessible from Γ. We add one more requirement that is specific for LP(S5): if a sentence X has an evidence in a world Γ then X must be true in all worlds accessible from Γ.

A *frame* is a structure $\langle \mathcal{G}, \mathcal{R} \rangle$, where $\mathcal{G}$ is a non-empty set of *possible worlds* and $\mathcal{R}$ is a binary accessibility relation on $\mathcal{G}$. The relation $\mathcal{R}$ is reflexive, transitive and symmetric. Given a frame $\langle \mathcal{G}, \mathcal{R} \rangle$, a *possible evidence* function $\mathcal{E}$ is a mapping from possible worlds and terms to sets of formulas. We can read $X \in \mathcal{E}(\Gamma, t)$ as "X is one of the formulas for which t serves as a possible evidence in world Γ." An evidence function must obey the conditions that respect the intended meaning of the operations on evidence terms.

Definition 3. $\mathcal{E}$ *is an evidence function on* $\langle \mathcal{G}, \mathcal{R} \rangle$ *if, for all terms s and t, for all formulas X and Y, and for all* $\Gamma, \Delta \in \mathcal{G}$:

1. **Monotonicity** $\Gamma \mathcal{R} \Delta$ *implies* $\mathcal{E}(\Gamma, t) \subseteq \mathcal{E}(\Delta, t)$
2. **Application** $(X \to Y) \in \mathcal{E}(\Gamma, t)$ *and* $X \in \mathcal{E}(\Gamma, s)$ *implies* $Y \in \mathcal{E}(\Gamma, t \cdot s)$
3. **Sum** $\mathcal{E}(\Gamma, t) \cup \mathcal{E}(\Gamma, s) \subseteq \mathcal{E}(\Gamma, t + s)$
4. **Positive Checker** $X \in \mathcal{E}(\Gamma, t)$ *implies* $[\![t]\!]X \in \mathcal{E}(\Gamma, !t)$
5. **Negative Checker** $X \notin \mathcal{E}(\Gamma, t)$ *implies* $\neg[\![t]\!]X \in \mathcal{E}(\Gamma, ?t)$

A structure $\mathcal{M} = \langle \mathcal{G}, \mathcal{R}, \mathcal{E}, \mathcal{V} \rangle$ is called a *pre-model*, if $\langle \mathcal{G}, \mathcal{R} \rangle$ is a frame, $\mathcal{E}$ is an evidence function on $\langle \mathcal{G}, \mathcal{R} \rangle$ and $\mathcal{V}$ is a mapping from propositional variables to subsets of $\mathcal{G}$.

Definition 4. *Given a pre-model* $\mathcal{M} = \langle \mathcal{G}, \mathcal{R}, \mathcal{E}, \mathcal{V} \rangle$, *a forcing relation is defined by the following rules. For each* $\Gamma \in \mathcal{G}$:

1. $\mathcal{M}, \Gamma \Vdash P$ *for a propositional variable* P *if* $\Gamma \in \mathcal{V}(P)$.
2. $\mathcal{M}, \Gamma \nVdash \bot$.
3. $\mathcal{M}, \Gamma \Vdash (X \to Y)$ *iff* $\mathcal{M}, \Gamma \nVdash X$ *or* $\mathcal{M}, \Gamma \Vdash Y$.
4. $\mathcal{M}, \Gamma \Vdash [\![t]\!]X$ *iff* $X \in \mathcal{E}(\Gamma, t)$ *and for every* $\Delta \in \mathcal{G}$ *with* $\Gamma \mathcal{R} \Delta$, $\mathcal{M}, \Delta \Vdash X$.

So, $[\![t]\!]X$ is true at a given world Γ iff t is an acceptable evidence for X in Γ and X is true at all worlds Δ accessible from Γ. We say that X is true at a world $\Gamma \in \mathcal{G}$ if $\Gamma \Vdash X$; otherwise, X is false at Γ.

Definition 5. *A pre-model* $\mathcal{M}$ *is called a model if* $X \in \mathcal{E}(\Gamma, t)$ *implies* $\Delta \Vdash X$ *for all* $\Delta \in W$ *with* $\Gamma \mathcal{R} \Delta$.

As in [1, 2], proof constants are intended to represent evidences for elementary truths. Those truths we know for reasons we do not further analyse. It is allowed that a proof constant serves as an evidence for more than one formula, or for nothing at all.

Definition 6. *A constant specification* $\mathcal{CS}$ *is a mapping from evidence constants to (possibly empty) sets of axioms of* LP(S5). *Given a constant specification* $\mathcal{CS}$, *a model* $\mathcal{M}$ *meets* $\mathcal{CS}$ *if* $\mathcal{M} \Vdash [\![c]\!]A$ *whenever* $A \in \mathcal{CS}(c)$. *A derivation meets* $\mathcal{CS}$ *if whenever rule R2 is used to produce* $[\![c]\!]A$, *then* $A \in \mathcal{CS}(c)$.

Definition 7. *A constant specification* $\mathcal{CS}$ *is full (cf [3]), if it entails internalization, namely, for every formula X if* $lp \vdash X$ *then there exists an evidence term t such that* LP(S5) $\vdash [\![t]\!]X$ *meeting* $\mathcal{CS}$.

Remark 1. Note, that a constant specification is full iff it assigns a constant to each axiom.

A set S of formulas is *$\mathcal{CS}$-satisfiable* if there is an LP(S5) model $\mathcal{M}$, meeting $\mathcal{CS}$, and a world Γ in it such that $\mathcal{M}, \Gamma \Vdash X$ for all $X \in S$. A formula X is valid in a model $\mathcal{M}$ if $\mathcal{M}, \Gamma \Vdash X$ for every possible world Γ of $\mathcal{M}$. A formula X is *$\mathcal{CS}$-valid* if X is valid in every model that meets $\mathcal{CS}$.

Theorem 1 (Completeness theorem). *Let $\mathcal{CS}$ be a constant specification. A formula X is provable in* LP(S5) *meeting $\mathcal{CS}$ iff X holds in all* LP(S5) *models meeting $\mathcal{CS}$.*

Proof. **Soundness** is proved by standard induction on derivation of X.

Completeness. A set of formulas S is called *$\mathcal{CS}$-inconsistent* if there is a finite subset $\{X_1, \ldots, X_n\} \subseteq S$ such that $(X_1 \wedge \ldots \wedge X_n) \rightarrow \bot$ is provable in LP(S5) meeting $\mathcal{CS}$ (with $\wedge$ defined from $\rightarrow$ and $\bot$ in the usual way). A set S is called *$\mathcal{CS}$-consistent* if it is not $\mathcal{CS}$-inconsistent. A $\mathcal{CS}$-consistent set S is *maximal $\mathcal{CS}$-consistent* if for any formula X, either $X \in S$ or $\neg X \in S$. $\mathcal{CS}$-consistent sets can be extended to maximal $\mathcal{CS}$-consistent sets, via the standard Lindenbaum construction. Note that any maximal $\mathcal{CS}$-consistent set contains all axioms of the logic LP(S5) and is closed under modus ponens.

Now we define the *canonical model* $\mathcal{M} = \langle \mathcal{G}, \mathcal{R}, \mathcal{E}, \mathcal{V} \rangle$ for LP(S5) with a given constant specification $\mathcal{CS}$. Let $\mathcal{G}$ be the set of all maximal consistent sets of LP(S5)-formulas. If $\Gamma \in \mathcal{G}$, let $\Gamma^\# = \{X \mid [\![t]\!]X \in \Gamma, \text{ for some } t\}$. Define

$$\Gamma \mathcal{R} \Delta \text{ iff } \Gamma^\# \subseteq \Delta.$$

Let us check that the relation $\mathcal{R}$ is reflexive, transitive and symmetric.

The relation $\mathcal{R}$ is reflexive, since $[\![t]\!]X \rightarrow X$ is an axiom of LP(S5) and members of $\mathcal{G}$ are closed under modus ponens. Using positive checker axiom $[\![t]\!]X \rightarrow [\![!t]\!][\![t]\!]X$ we obtain that $\mathcal{R}$ is also transitive. Let us show that $\mathcal{R}$ is symmetric. Let $\Gamma \mathcal{R} \Delta$ and $[\![s]\!]Y \in \Delta$, but $[\![s]\!]Y \notin \Gamma$. Since Γ is a maximal consistent set we have $\neg[\![s]\!]Y \in \Gamma$. Using negative checker axiom $\neg[\![s]\!]Y \rightarrow [\![?s]\!]\neg[\![s]\!]Y$ we obtain $[\![?s]\!]\neg[\![s]\!]Y \in \Gamma$ and $\neg[\![s]\!]Y \in \Gamma^\#$. Thus $\neg[\![s]\!]Y \in \Delta$, contradiction. So we have $\Delta^\# \subseteq \Gamma$, hence $\Delta \mathcal{R} \Gamma$.

This gives us a frame $\langle \mathcal{G}, \mathcal{R} \rangle$. For an evidence function $\mathcal{E}$, simply set

$$\mathcal{E}(\Gamma, t) = \{X \mid [\![t]\!]X \in \Gamma\}.$$

The claim is that $\mathcal{E}$ satisfies the conditions for evidence functions from Definition 3.

Here we verify only condition 5, the others are similar. Suppose $X \notin \mathcal{E}(\Gamma, t)$. By definition of $\mathcal{E}$, we have $[\![t]\!]X \notin \Gamma$. Since Γ is maximal, we obtain $\neg[\![t]\!]X \in \Gamma$. Using the axiom $\neg[\![t]\!]X \rightarrow [\![?t]\!]\neg[\![t]\!]X$ we conclude that $[\![?t]\!]\neg[\![t]\!]X \in \Gamma$, and hence $\neg[\![t]\!]X \in \mathcal{E}(\Gamma, ?t)$.

Finally, define a mapping $\mathcal{V}$ in the following way: $\Gamma \in \mathcal{V}(P)$ if $P \in \Gamma$.

Lemma 1. *For each formula X and each $\Gamma \in \mathcal{G}$*

$$X \in \Gamma \iff \mathcal{M}, \Gamma \Vdash X$$

Proof. Induction on the construction of X. The base and the Boolean case are standard. Let us consider the case $X = [\![t]\!]Y$.

Suppose first that $[\![t]\!]Y \in \Gamma$. Then $Y \in \Gamma^{\#}$, so if Δ is an arbitrary member of $\mathcal{G}$ with $\Gamma\mathcal{R}\Delta$ we have $\Gamma^{\#} \subseteq \Delta$ and hence $Y \in \Delta$. By the induction hypothesis, $\mathcal{M}, \Delta \Vdash Y$. Also since $[\![t]\!]Y \in \Gamma$, we have $Y \in \mathcal{E}(\Gamma, t)$. So, $\mathcal{M}, \Gamma \Vdash [\![t]\!]Y$.

Next, suppose $\mathcal{M}, \Gamma \Vdash [\![t]\!]Y$. This case is trivial. By the definition of $\Vdash$ we have $Y \in \mathcal{E}(\Gamma, t)$, so $[\![t]\!]Y \in \Gamma$ by the definition of $\mathcal{E}$.

Let us verify that the pre-model $\mathcal{M}$ is a model. Suppose $X \in \mathcal{E}(\Gamma, t)$. By the definition of $\mathcal{E}$ we have $[\![t]\!]X \in \Gamma$, so $X \in \Gamma^{\#}$. Let $\Delta \in \mathcal{G}$ be an arbitrary world with $\Gamma\mathcal{R}\Delta$. Then $X \in \Delta$ and by Lemma 1 it follows that $\Delta \Vdash X$.

It is easy to see now that the model $\mathcal{M}$ meets the constant specification $\mathcal{CS}$. Indeed, by the definition of a consistent set, $\mathcal{CS} \subseteq \Gamma$, for each $\Gamma \in \mathcal{G}$. Then $\Gamma \Vdash \mathcal{CS}$ by Lemma 1.

The standard argument shows that if $\mathsf{LP(S5)} \nvdash X$ meeting $\mathcal{CS}$, then $\{X \to \bot\}$ is a consistent set. Extend it to a maximal consistent set Γ. Then $\Gamma \in \mathcal{G}$ and $\mathcal{M}, \Gamma \nVdash X$ by Lemma 1.

Corollary 1 (Compactness). *For a given constant specification $\mathcal{CS}$, a set of formulas S is $\mathcal{CS}$-satisfiable iff any finite subset of S is $\mathcal{CS}$-satisfiable.*

Definition 8. *A model is Fully Explanatory provided that, whenever $\Delta \Vdash X$ for every Δ such that $\Gamma\mathcal{R}\Delta$, then $\Gamma \Vdash [\![t]\!]X$ for some evidence term t.*

Corollary 2 (Fully Explanatory property). *For any full constant specification $\mathcal{CS}$, the canonical $\mathcal{CS}$-model is Fully Explanatory.*

4   Realization Theorem

Definition 9. *A forgetful projection $()^{\circ}$ of the language $\mathsf{LP(S5)}$ into the modal language maps $[\![t]\!]X$ to $\Box X$ and commutes with $\to$.*

Theorem 2 (Realization theorem). $\mathsf{LP(S5)}^{\circ} = \mathsf{S5}$ *i.e., $\mathsf{S5}$ is forgetful projection of $\mathsf{LP(S5)}$.*

Proof. The proof of $\mathsf{LP(S5)}^{\circ} \subseteq \mathsf{S5}$ is given by a straightforward induction on derivations in $\mathsf{LP(S5)}$. The existence of an $\mathsf{LP(S5)}$-realization of any theorems of $\mathsf{S5}$ can be established semantically by methods developed in [5]. The main ingredients of semantical realizability proof from [5] are the Fully Explanatory property of $\mathsf{LP(S5)}$-models with full constant specifications (Corollary 2) and the Compactness property (Corollary 1).

Definition 10. *By $\mathsf{LP(S5)}^{-}$ we mean a variant of $\mathsf{LP(S5)}$ for restricted language without $+$ (so, the axioms A3 are also omitted). Models of $\mathsf{LP(S5)}^{-}$ are the same as for $\mathsf{LP(S5)}$ except that the evidence function is not required to satisfy the Sum condition.*

Note that such features as Internalization and the Fully Explanatory property of the canonical model hold for LP(S5)$^-$ and LP(S5)$^-$-models as well.

Let φ be a modal formula in the standard language of modal logic, with $\Box$ as the only modal operator and no evidence terms. Assume that φ is fixed for the rest of proof of Theorem 2. We work with φ and its subformulas, but by subformulas we mean subformula occurrences. In what follows, A is any assignment of evidence variables to each subformula of φ of the form $\Box X$ that is in a negative position. It is assumed that A assigns different variables to different subformulas (the specific A will be chosen in the proof of Lemma 3). Relative to A, we define two mappings w_A and v_A.

Definition 11. *Both w_A and v_A assign a set of* LP(S5) *formulas to each subformula of φ according to the following rules.*

1. *If P is an atomic subformula of φ then $w_A(P) = v_A(P) = \{P\}$ (this includes the case when P is $\bot$).*
2. *If $X \rightarrow Y$ is a subformula of φ, put*
   $$w_A(X \rightarrow Y) = \{X' \rightarrow Y' \mid X' \in w_A(X), \ Y' \in w_A(Y)\}$$
   $$v_A(X \rightarrow Y) = \{X' \rightarrow Y' \mid X' \in v_A(X), \ Y' \in v_A(Y)\}.$$
3. *If $\Box X$ is a negative subformula of φ, put*
   $$w_A(\Box X) = \{[\![x]\!]X' \mid A(\Box X) = x \ and \ X' \in w_A(X)\}$$
   $$v_A(\Box X) = \{[\![x]\!]X' \mid A(\Box X) = x \ and \ X' \in v_A(X)\}.$$
4. *If $\Box X$ is a positive subformula of φ, put*
   $$w_A(\Box X) = \{[\![t]\!]X' \mid X' \in w_A(X) \ and \ t \ is \ any \ evidence \ term\}$$
   $$v_A(\Box X) = \{[\![t]\!](X_1 \vee \ldots \vee X_n) \mid X_1, \ldots, X_n \in v_A(X) \ and \ t \ is \ any \ evidence \ term\}.$$

Fix a constant specification $\mathcal{CS}$. Let $\mathcal{M} = \langle \mathcal{G}, \mathcal{R}, \mathcal{E}, \mathcal{V} \rangle$ be the canonical LP(S5)$^-$-model built from the given $\mathcal{CS}$ as in the proof of Theorem 1. Since $\mathcal{R}$ is reflexive, transitive and symmetric, we can also think of it as an S5-model ($\mathcal{E}$ plays no role in this).

If X is a subformula of φ we will write $\neg v_A(X)$ for $\{\neg X' \mid X' \in v_A(X)\}$. Also, if S is a set of LP(S5)$^-$-formulas, we will write $\mathcal{M}, \Gamma \Vdash S$ if $\mathcal{M}, \Gamma \Vdash Z$ for every formula $Z \in S$. Since $\mathcal{M}$ is canonical, $\mathcal{M}, \Gamma \Vdash S$ is equivalent to $S \subseteq \Gamma$, by Lemma 1.

Lemma 2. *Let $\mathcal{CS}$ be a full constant specification of* LP(S5)$^-$ *and $\mathcal{M}$ be a canonical model for* LP(S5)$^-$ *that meets $\mathcal{CS}$. Then for each $\Gamma \in \mathcal{G}$:*

1. *If ψ is a positive subformula of φ and $\mathcal{M}, \Gamma \Vdash \neg v_A(\psi)$ then $\mathcal{M}, \Gamma \Vdash \neg \psi$.*
2. *If ψ is a negative subformula of φ and $\mathcal{M}, \Gamma \Vdash v_A(\psi)$ then $\mathcal{M}, \Gamma \Vdash \psi$.*

Proof. The proof is by induction on the complexity of ψ. The atomic case is trivial. Implication is treated similarly to [5].

Positive Necessity. Suppose ψ is $\Box X$, ψ is a positive subformula of φ, $\mathcal{M}, \Gamma \Vdash \neg v_A(\Box X)$, and the result is known for X (which also occurs positively in φ).

The key item to show is that $\Gamma^\# \cup \neg v_A(X)$ is consistent. Then we can extend it to a maximal consistent set Δ, and since $\mathcal{M}$ is the canonical model we will have

$\Delta \in \mathcal{G}$, $\Gamma \mathcal{R} \Delta$ and $\mathcal{M}, \Delta \Vdash \neg v_A(X)$. By the induction hypothesis, $\mathcal{M}, \Delta \nVdash X$, hence $\mathcal{M}, \Gamma \nVdash \Box X$. So now we concentrate on showing this key item.

Suppose $\Gamma^\# \cup \neg v_A(X)$ is inconsistent. Then for some $Y_1, \ldots Y_k \in \Gamma^\#$ and $X_1, \ldots X_n \in v_A(X)$, $\mathsf{LP(S5)}^- \vdash (Y_1 \wedge \ldots \wedge Y_k \wedge \neg X_1 \wedge \ldots \neg X_n) \to \bot$. Then $\mathsf{LP(S5)}^- \vdash Y_1 \to (Y_2 \to \ldots \to (Y_k \to X_1 \vee \ldots \vee X_n) \ldots)$. By Proposition 1, there is an evidence term s such that $\mathsf{LP(S5)}^- \vdash [\![s]\!][Y_1 \to (Y_2 \to \ldots \to (Y_k \to X_1 \vee \ldots \vee X_n) \ldots)]$. Consider terms $t_1, t_2, \ldots, t_k$ such that $[\![t_1]\!]Y_1, [\![t_2]\!]Y_2, \ldots, [\![t_k]\!]Y_k \in \Gamma$. By Axiom A2 and propositional reasoning, $\mathsf{LP(S5)}^- \vdash [\![t_1]\!]Y_1 \wedge [\![t_2]\!]Y_2 \wedge \ldots \wedge [\![t_k]\!]Y_k \to [\![s \cdot t_1 \cdot t_2 \cdot \ldots \cdot t_k]\!](X_1 \vee \ldots \vee X_n)$. Therefore $\Gamma \Vdash [\![s \cdot t_1 \cdot t_2 \cdot \ldots \cdot t_k]\!](X_1 \vee \ldots \vee X_n)$ which is impossible since $[\![s \cdot t_1 \cdot t_2 \cdot \ldots \cdot t_k]\!](X_1 \vee \ldots \vee X_n) \in v_A(\Box X)$. Thus $\Gamma^\# \cup \neg v_A(X)$ is consistent and the case is done.

Negative Necessity. Suppose ψ is $\Box X$, ψ is a negative subformula of φ, $\mathcal{M}, \Gamma \Vdash v_A(\Box X)$ and the result is known for X (which also occurs negatively in φ).

Let X' be an arbitrary member of $v_A(X)$. Since $\Box X$ is a negative subformula of φ, $[\![x]\!]X' \in v_A(\Box X)$, where $x = A(\Box X)$, and so $\mathcal{M}, \Gamma \Vdash [\![x]\!]X'$. Now if Δ is an arbitrary member of $\mathcal{G}$ with $\Gamma \mathcal{R} \Delta$, we must have $\mathcal{M}, \Delta \Vdash X'$. Thus $\mathcal{M}, \Delta \Vdash v_A(X)$, so by the induction hypothesis, $\mathcal{M}, \Delta \Vdash X$. Since Δ was arbitrary, $\mathcal{M}, \Gamma \Vdash \Box X$.

Corollary 3. *Let $\mathcal{CS}$ be a full constant specification. If $\mathsf{S5} \vdash \varphi$ then there are $\varphi_1, \ldots, \varphi_n \in v_A(\varphi)$ such that $\mathsf{LP(S5)}^- \vdash \varphi_1 \vee \ldots \vee \varphi_m$ meeting $\mathcal{CS}$.*

Proof. Suppose $\mathsf{S5} \vdash \varphi$ but $\mathsf{LP(S5)}^- \nvdash (\varphi_1 \vee \ldots \vee \varphi_n)$ for every $\varphi_1, \ldots, \varphi_n \in v_A(\varphi)$ with a given full constant specification $\mathcal{CS}$. Then every finite subset of $\neg v_A(\varphi)$ is satisfiable. By Corollary 1 adapted to $\mathsf{LP(S5)}^-$, there is a world Γ in the canonical model $\mathcal{M}$ for $\mathsf{LP(S5)}^-$ with $\mathcal{CS}$ such that $\mathcal{M}, \Gamma \Vdash \neg v_A(\varphi)$. By Lemma 2, $\mathcal{M}, \Gamma \nVdash \varphi$. Therefore, since φ is the theorem of $\mathsf{S5}$, there are $\varphi_1, \ldots, \varphi_n \in v_A(\varphi)$ such that $\mathsf{LP(S5)}^- \vdash (\varphi_1 \vee \ldots \vee \varphi_n)$.

Lemma 3. *For every subformula ψ of φ and each $\psi_1, \ldots, \psi_n \in v_A(\psi)$, there is a substitution σ of evidence terms for proof variables and a formula $\psi' \in w_A(\psi)$ such that:*

   1. If ψ is a positive subformula of φ then $\mathsf{LP(S5)} \vdash (\psi_1 \vee \ldots \vee \psi_n)\sigma \to \psi'$.
   2. If ψ is a negative subformula of φ then $\mathsf{LP(S5)} \vdash \psi' \to (\psi_1 \wedge \ldots \wedge \psi_n)\sigma$.

Proof. We use the fact that proof variables assigned to different (occurrences of) subformulas ψ in φ are all different.

By induction on the complexity of ψ. If ψ is atomic then the result is immediate, since $v_A(\psi)$ and $w_A(\psi)$ are both $\{\psi\}$, so one can take ψ' to be ψ, and use the empty substitution. Implication is treated similarly to [5].

Positive Necessity. Suppose ψ is $\Box X$, ψ is a positive subformula of φ, and the result is known for X (which also occurs positively in φ).

In this case $\psi_1, \ldots, \psi_n$ are of the form $[\![t_1]\!]D_1, \ldots, [\![t_n]\!]D_n$, where each of D_i is a disjunction of members of $v_A(X)$. Let $D = D_1 \vee \ldots \vee D_n$. Then D is a disjunction of formulas from $v_A(X)$, so by the induction hypothesis there is some substitution σ and $X' \in w_A(X)$ such that $\mathsf{LP(S5)} \vdash D\sigma \to X'$. Consequently for each i, $\mathsf{LP(S5)} \vdash D_i\sigma \to X'$ and so, by Proposition 1, there is a proof polynomial u_i such that $\mathsf{LP(S5)} \vdash [\![u_i]\!](D_i\sigma \to X')$. But then $\mathsf{LP(S5)} \vdash ([\![t_i]\!]D_i)\sigma \to [\![u_i \cdot t_i\sigma]\!]X'$. Let s be the evidence term $(u_1 \cdot t_1\sigma) + \ldots + (u_n \cdot t_n\sigma)$. For each i we have $\mathsf{LP(S5)} \vdash ([\![t_i]\!]D_i)\sigma \to [\![s]\!]X'$, and hence $\mathsf{LP(S5)} \vdash ([\![t_1]\!]D_1 \vee \ldots \vee [\![t_n]\!]D_n)\sigma \to [\![s]\!]X'$. Since $[\![s]\!]X' \in w_A(\Box X)$, this concludes the positive necessity case.

Negative Necessity. Suppose ψ is $\Box X$, ψ is a negative subformula of φ, and the result is known for X (which also occurs negatively in φ).

In this case $\psi_1, \ldots, \psi_n$ are of the form $[\![x]\!]X_1, \ldots, [\![x]\!]X_n$, where $X_i \in v_A(X)$. By the induction hypothesis there is some substitution σ and $X' \in w_A(X)$ such that $\mathsf{LP(S5)} \vdash X' \to (X_1 \wedge \ldots \wedge X_n)\sigma$. Put $A(\Box X) = x$, where the evidence variable x is not in the domain of σ. Now, for each $i = 1, \ldots, n$, $\mathsf{LP(S5)} \vdash X' \to X_i\sigma$, and so by Proposition 1 there is a proof polynomial t_i such that $\mathsf{LP(S5)} \vdash [\![t_i]\!](X' \to X_i\sigma)$. Let s be the proof polynomial $t_1 + \ldots + t_n$; then $\mathsf{LP(S5)} \vdash [\![s]\!](X' \to X_i\sigma)$, for each i. It follows that $\mathsf{LP(S5)} \vdash [\![x]\!]X' \to [\![s \cdot x]\!](X_i\sigma)$, for each i. Consider a new substitution $\sigma' = \sigma \cup \{x/(s \cdot x)\}$, then we have $\mathsf{LP(S5)} \vdash [\![x]\!]X' \to ([\![x]\!]X_i)\sigma'$ for each i, and hence $\mathsf{LP(S5)} \vdash [\![x]\!]X' \to ([\![x]\!]X_1 \wedge \ldots \wedge [\![x]\!]X_n)\sigma'$, which establishes the result in this case.

To conclude the proof of Theorem 2, assume that $\mathsf{S5} \vdash \varphi$. Then by Corollary 3 there are $\varphi_1, \ldots, \varphi_n \in v_A(\varphi)$ such that $\mathsf{LP(S5)} \vdash \varphi_1 \vee \ldots \vee \varphi_n$. By Lemma 3, there is a substitution σ and $\varphi' \in w_A(\varphi)$ such that $\mathsf{LP(S5)} \vdash (\varphi_1 \vee \ldots \vee \varphi_n)\sigma \to \varphi'$. Since $\mathsf{LP(S5)}$ is closed under substitution, $\mathsf{LP(S5)} \vdash \varphi'$.

Acknowledgments

I would like to thank Vladimir Krupsky and Tatyana Yavorskaya-Sidon for helpful discussions. I am thankful to Tatyana Yavorskaya-Sidon for reading the draft of this paper. I am also grateful to the Referee for valuable remarks concerning the introductional part of the paper.

The research was partially supported by Russian Foundation for Basic Research.

References

1. Artemov, S.: Uniform provability realization of intuitionistic logic, modality and λ-terms. Electronic Notes in Theoretical Computer Science. **23** (1999)
2. Artemov S.: Explicit provability and constructive semantics. Bulletine of Symbolic Logic. **7** (2001), 1–36
3. Artemov S.: Evidence-based common knowledge. Technical Report TR-2004018 CUNY Ph.D. Program in Computer Science (2005)
4. Artemov S., Kazakov E., Shapiro D.: On logic of knowledge with justifications. Technical Report CFIS 99-12, Cornell University (1999)
5. Fitting M.: The logic of proofs, semantically. Annals of Pure and Applied Logic, **132(1)** (2005), 1–25
6. Mkrtychev A.: Models for the Logic of Proofs. Lecture Notes in Computer Science, **1234** (1997). Logical Foundations of Computer Science, Yaroslavl, (1997) 266–275

Linear Temporal Logic with Until and Before on Integer Numbers, Deciding Algorithms

V. Rybakov

Department of Computing and Mathematics,
Manchester Metropolitan University,
John Dalton Building, Chester Street, Manchester M1 5GD, UK
V.Rybakov@mmu.ac.uk

Abstract. As specifications and verifications of concurrent systems employ Linear Temporal Logic (LTL), it is increasingly likely that logical consequence in LTL will be used in description of computations and parallel reasoning. We consider the linear temporal logic $\mathcal{LTL}_{N,\,N-1}^{U,B}(\mathcal{Z})$ extending the standard LTL by operations $\mathbf{B}$ (before) and $\mathbf{N}^{-1}$ (previous). Two sorts of problems are studied: (i) satisfiability and (ii) description of logical consequence in $\mathcal{LTL}_{N,\,N-1}^{U,B}(\mathcal{Z})$ via admissible logical consecutions (inference rules). The model checking for LTL is a traditional way of studying such logics. Most popular technique based on automata was developed by M.Vardi (cf. [39, 6]). Our paper uses a reduction of logical consecutions and formulas of LTL to consecutions of a uniform form consisting of formulas of temporal degree 1. Based on technique of Kripke structures, we find necessary and sufficient conditions for a consecution to be not admissible in $\mathcal{LTL}_{N,\,N-1}^{U,B}(\mathcal{Z})$. This provides an algorithm recognizing consecutions (rules) admissible in $\mathcal{LTL}_{N,\,N-1}^{U,B}(\mathcal{Z})$ by Kripke structures of size linear in the reduced normal forms of the initial consecutions. As an application, this algorithm solves also the satisfiability problem for $\mathcal{LTL}_{N,\,N-1}^{U,B}(\mathcal{Z})$.

Keywords: logic in computer science, algorithms, linear temporal logic, logical consequence, inference rules, consecutions, admissible rules.

1 Introduction

Temporal logics were applied in study of many problems concerning computing and reasoning (cf. Manna and Pnueli [21, 22], Pnueli [26], Clark E. et al., [4], Goldblatt [11]). Linear temporal logic (LTL) has been quite successful in dealing with applications to systems specifications and verification (cf. [26, 19]), with model checking (cf. [2, 4]). Also temporal logic is a natural logic for hardware verification (cf. Cyrluk, Natendran [5]); temporal logic has numerous applications to safety, liveness and fairness (cf. Emerson [7]), to various problems arising in computing (cf. Eds. Barringer, Fisher, Gabbay and Gough, [1, 3]).

An effective automata-theoretic approach to LTL based at automata on infinite worlds, was found by M.Vardi [39, 6]. This allows to show decidability of LTL via algorithms for satisfiability problem. Techniques for the model checking

D. Grigoriev, J. Harrison, and E.A. Hirsch (Eds.): CSR 2006, LNCS 3967, pp. 322–333, 2006.
© Springer-Verlag Berlin Heidelberg 2006

of knowledge and time was presented, for instance, in (Hoek, Wooldridge [13]). It would be not an exaggeration to say that one of prime questions concerning temporal logics is the question about decidability (cf. [18]). In our paper we study the question about decidability of linear temporal logics, but in more general form, - decidability w.r.t. admissible logical consecutions.

We consider the linear temporal logic $\mathcal{LTL}_{N, N^{-1}}^{U,B}(\mathcal{Z})$ which is an expansion of the standard LTL by new operations: $\mathbf{B}$ (before) and $\mathbf{N}^{-1}$ (previous). So, $\mathcal{LTL}_{N, N^{-1}}^{U,B}(\mathcal{Z})$ has temporal operations $\mathbf{N}$ (next), $\mathbf{U}$ (until), $\mathbf{N}^{-1}$ (previous) and $\mathbf{B}$ (before). Main problem we are focused on is description of logical consequence in $\mathcal{LTL}_{N, N^{-1}}^{U,B}(\mathcal{Z})$ via admissible logical consecutions (inference rules). This approach puts in the base the notion of logical consecution, inference rule, correct sequent. Usage of inference rules gives an opportunity to describe subtle properties of models which are problematic to be expressed by formulas. A good example is the Gabbay's *irreflexivity rule* (cf. [9]) $(ir) := \neg(p \to \Diamond p) \to \varphi / \varphi$ (where p does not occur in the formula φ). This rule is actually saying that any element of a model, where φ is not valid, should be irreflexive; it was implemented in [9] for the proof of the completeness theorem. Admissible rules form the greatest class of rules which are compatible with the set of the theorems (valid formulas) of a given logic, therefore we are interested to recognize which rules (consecutions) are admissible.

Admissible consecutions have been investigated reasonably deeply for numerous modal and superintuitionistic logics. The history could be dated since Harvey Friedman's question (1975,[8]) about existence of algorithms which could distinguish rules admissible in the intuitionistic propositional logic IPC, and since Harrop's examples [14] of rules admissible but not derivable in standard Hilbert-style axiomatic systems for IPC. In the middle of 70-th, G. Mints [25] found strong sufficient conditions for derivability in IPC admissible rules in special form. H. Friedman's question was answered affirmatively by V. Rybakov (1984, [28]) and later S. Ghilardi [10] found another solution. Since then, the questions concerning admissible rules (decidability, description of bases, inheritance, structural completeness, etc.) were investigated profoundly, but primary only for transitive modal and superintuitionistic logics (cf. [10, 15, 16, 29, 30, 31, 32, 33, 34, 35, 36, 37]). In our paper we make an attempt to find an approach to the linear temporal logics $\mathcal{LTL}_{N, N^{-1}}^{U,B}(\mathcal{Z})$. The obtained results are briefly described in the abstract above.

2   Notation, Preliminaries

Linear temporal logic (LTL in the sequel) has the language including Boolean logical operations and the temporal operations Next and Until. Formulas of LTL are built from a set *Prop* of atomic propositions and are closed under the application of Boolean operations, the unary operation $\mathbf{N}$ (next) and the binary operation $\mathbf{U}$ (until). We also will consider the following counterpart operations to $\mathbf{N}$ and $\mathbf{U}$: $\mathbf{N}^{-1}$ (previous) and $\mathbf{B}$ (before), so for any wffs φ and ψ, $\mathbf{N}^{-1}\varphi$ and $\varphi\mathbf{B}\psi$ are also wffs. The semantics for formulas of this language, which we suggest, is based on the set of all integer numbers with the standard linear order. The basic Kripke-Hinttikka structure (model) for our logic is the following tuple $\mathcal{Z} := \langle \mathcal{Z}, \leq, Next, Prev, V \rangle$, where $\mathcal{Z}$ is the set of integer numbers with the standard linear order $\leq$. The relations $Next$ and $Prev$ are also standard: $a \; Next \; b \Leftrightarrow b = n + 1$, $a \; Prev \; b \Leftrightarrow b = a - 1$, and V is a valuation of a subset S of $Prop$, which assigns truth values to elements of S. The computational interpretation of $\mathcal{Z}$ is as follows. The elements of $\mathcal{Z}$ are *states*, $\leq$ is the *transition* relation (which is linear in our case), and V can be interpreted as *labeling* of the states with atomic propositions. If we refer to $\mathcal{Z}$ as the Kripke frame, we mean the frame $\mathcal{Z}$ without valuations. For $\mathcal{Z}$, the truth values can be extended from propositions of S to arbitrary formulas constructed from these propositions as follows:

$$\forall p \in Prop, \; \forall a \in \mathcal{Z}, \; (\mathcal{Z}, a) \Vdash p \Leftrightarrow a \in V(p);$$

$$(\mathcal{Z}, a) \Vdash \varphi \wedge \psi \Leftrightarrow (\mathcal{Z}, a) \Vdash \varphi \text{ and } (\mathcal{Z}, a) \Vdash \psi;$$

$$(\mathcal{Z}, a) \Vdash \neg \varphi \Leftrightarrow not[(\mathcal{Z}, a) \Vdash \varphi];$$

$$(\mathcal{Z}, a) \Vdash \mathbf{N}\varphi \Leftrightarrow [(a \; Next \; b) \Rightarrow (\mathcal{Z}, b) \Vdash \varphi];$$

$$(\mathcal{Z}, a) \Vdash \mathbf{N}^{-1}\varphi \Leftrightarrow [(a \; Prev \; b) \Rightarrow (\mathcal{Z}, b) \Vdash \varphi];$$

$$(\mathcal{Z}, a) \Vdash \varphi \mathbf{U} \psi \Leftrightarrow \exists b[(a \leq b) \wedge (\mathcal{Z}, b) \Vdash \psi)] \wedge \forall c[(a \leq c < b) \Rightarrow (\mathcal{Z}, c) \Vdash \varphi];$$

$$(\mathcal{Z}, a) \Vdash \varphi \mathbf{B} \psi \Leftrightarrow \exists b[(b \leq a) \wedge (\mathcal{Z}, b) \Vdash \psi)] \wedge \forall c[(b \leq c < a) \Rightarrow (\mathcal{Z}, c) \Vdash \varphi];$$

We can define all standard temporal and modal operations using $\mathbf{U}$ and $\mathbf{B}$. For instance, $\mathbf{F}\varphi$, $- \varphi$ *holds eventually* (in terms of modal logic, $- \varphi$ is possible $- \diamondsuit^{+}\varphi$), can be described as $true\mathbf{U}\varphi$. $\mathbf{G}\varphi$, $- \varphi$ *holds henceforth* $-$, can be defined as $\neg\mathbf{F}\neg\varphi$. $true\mathbf{B}\varphi$ says that φ was true at least once in past, so we can use $true\mathbf{B}\varphi$ to express temporal operation $\mathbf{P}\varphi$. And $\mathbf{H}\varphi - \varphi$ has always been, $-$ may be described as $\neg\mathbf{P}\neg\varphi$. We can describe within this language various properties of transition systems and Kripke structures. For instance, the formula $\mathbf{G}(\neg request \vee (request \; \mathbf{B} \; grant))$ says that whenever a request is made it holds continuously until it is eventually granted.

Definition 1. For a Kripke-Hinttikka structure $\mathcal{Z} := \langle \mathcal{Z}, \leq, Next, Prev, V \rangle$ and a formula φ in the language of $\mathcal{LTL}_{N, \, N-1}^{U, B}(\mathcal{Z})$, we say that φ is **satisfiable in** $\mathcal{Z}$ (denotation $- \mathcal{Z} \Vdash_{Sat}\varphi$) if there is a state b of $\mathcal{Z}$ ($b \in \mathcal{Z}$) where φ is true: $(\mathcal{Z}, b) \Vdash_{V} \varphi$. A formula φ is **valid in** $\mathcal{Z}$ (denotation $- \mathcal{Z} \Vdash \varphi$) if, for any element b of $\mathcal{Z}$ ($b \in \mathcal{Z}$), φ is true at b. A formula of $\mathcal{LTL}_{N, \, N-1}^{U, B}(\mathcal{Z})$ is *satisfiable* iff there is a Kripke structure based on $\mathcal{Z}$ where φ is satisfiable.

Definition 2. The linear temporal logic $\mathcal{LTL}_{N, \, N-1}^{U, B}(\mathcal{Z})$ is the set of all formulas in the described language which are valid in any Kripke structure based on the frame $\mathcal{Z}$.

In the sequel we will also use the standard Kripke semantics for multi-modal logics which consists of arbitrary Kripke structures and Kripke frames. For instance, if we consider a Kripke structure $\mathcal{M} := \langle M, \leq, Next, Prev, V \rangle$, M could be any set, and $\leq$, $Next$ and $Prev$ are some binary relations on M. The truth values of formulas (in the language of LTL) in such Kripke structures can be defined similar to above. We will explain later for which Kripke structures this approach works. The aims of our research are:

(i) To develop a new technique for satisfiability problem in $\mathcal{LTL}^{\mathrm{U,B}}_{\mathrm{N,\,N-1}}(\mathcal{Z})$, to get efficient algorithms solving satisfiability. These algorithms will be based on verification of validity for consecutions in special finite models of size linear in the size of consecutions.

(ii) To study logical consequence in $\mathcal{LTL}^{\mathrm{U,B}}_{\mathrm{N,\,N-1}}(\mathcal{Z})$ in terms of logical consecutions, admissible inference rules. To construct algorithms distinguishing admissible in $\mathcal{LTL}^{\mathrm{U,B}}_{\mathrm{N,\,N-1}}(\mathcal{Z})$ consecutions.

Firstly, we describe what we do towards problem (i) and then we will turn to (ii). How we could get an algorithm checking satisfiability in $\mathcal{LTL}^{\mathrm{U,B}}_{\mathrm{N,\,N-1}}(\mathcal{Z})$? The structures based on $\mathcal{Z}$ are infinite, the language of $\mathcal{LTL}^{\mathrm{U,B}}_{\mathrm{N,\,N-1}}(\mathcal{Z})$ with $\mathbf{U}$ (until) and $\mathbf{B}$ (before) is enough expressible, so we cannot answer this question immediately. Earlier (in [38]), we studied the satisfiability problem for the temporal Tomorrow/Yesterday logic $\mathbf{TYL}$, which is based on all finite intervals of integer numbers, but has no operations $\mathbf{U}$ and $\mathbf{B}$. In was established (Theorem 2 [38]) that for a satisfiable formula φ there is a model for φ of a size linear from φ. In the current paper, we cannot use directly the approach from [38] because our Kripke structures are infinite and because of presence both operations $\mathbf{U}$ (until) and $\mathbf{B}$ (before). The obtained result from the our current paper is

Theorem 1. Small Models Theorem. A formula φ is satisfiable in $\mathcal{LTL}^{\mathrm{U,B}}_{\mathrm{N,\,N-1}}(\mathcal{Z})$ iff there is a special finite Kripke structure $\mathcal{M}$, where φ is satisfiable, and the size of $\mathcal{M}$ is double exponential in the length of φ.

This result follows from technique of the next sections where the structure of $\mathcal{M}$ will be specified. So, comparing to $\mathbf{TYL}$, the situation is worse, the bound is double exponential but not linear. Anyway, we reduce the satisfiability to checking truth of consecutions on models of double exponential size in testing formula, so we get

Corollary 1. The linear temporal logic $\mathcal{LTL}^{\mathrm{U,B}}_{\mathrm{N,\,N-1}}(\mathcal{Z})$ is decidable.

3   Logical Consecutions in Temporal Logic

The basic problem we are dealing here is how to characterize that a formula (a statement) is a logical consequence of a given collection of formulas (statements). A **consecution**, (or, synonymously, – a **rule**) $\mathbf{c}$ is an expression

$$\mathbf{c} := \frac{\varphi_1(x_1, ..., x_n), ..., \varphi_m(x_1, ..., x_n)}{\psi(x_1, ..., x_n)},$$

where $\varphi_1(x_1, ..., x_n), \ldots, \varphi_m(x_1, ..., x_n)$ and $\psi(x_1, ..., x_n)$ are some formulas.

The formula $\psi(x_1, ..., x_n)$ is the conclusion of **c**, formulas $\varphi_j(x_1, ..., x_n)$ are the premises of **c**. Consecutions are supposed to describe the logical consequences, an informal meaning of a consecution is: *the conclusion* logically follows *from the premises*. The questions what *logically follows* means is crucial and has no evident and unique answer. We consider and compare below some approaches. Let $\mathcal{F}$ be a Kripke frame (e.g. our linear frame $\mathcal{Z}$), with a valuation V of all variables from a consecution $\mathbf{c} := \varphi_1, ... \varphi_n/\psi$.

Definition 3. The consecution **c** is said to be **valid** in the Kripke structure $\langle \mathcal{F}, V \rangle$ (we will use notation $\langle \mathcal{F}, V \rangle \Vdash \mathbf{c}$, or $\mathcal{F} \Vdash_V \mathbf{c}$) if $(\mathcal{F} \Vdash_V \bigwedge_{1 \le i \le m} \varphi_i) \Rightarrow (\mathcal{F} \Vdash_V \psi)$. Otherwise we say **c** is **refuted** in $\mathcal{F}$, or **refuted in** $\mathcal{F}$ **by** V, and write $\mathcal{F} \not\Vdash_V \mathbf{c}$.

A consecution **c** is **valid** in a frame $\mathcal{F}$ (notation $\mathcal{F} \Vdash \mathbf{c}$) if, for any valuation V, $\mathcal{F} \Vdash_V \mathbf{c}$. A consecution $\mathbf{c} := \varphi_1, ... \varphi_n/\psi$ is *valid* in a logic $\mathcal{L}(\mathcal{K})$ generated by a class of frames $\mathcal{K}$ if $\forall \mathcal{F} \in \mathcal{K}(\mathcal{F} \Vdash \mathbf{c})$. For many logics $\mathcal{L}$, if **c** is valid for $\mathcal{L}$ w.r.t. a class $\mathcal{K}$ generating $\mathcal{L}$, **c** will be also valid in all $\mathcal{L}$-Kripke frames. Note that this definition of valid consecutions equivalent to the notion of valid modal sequent from [17], where a theory of sequent-axiomatic classes is developed. Also the notion of valid consecutions can be reduced to validity of formulas in the extension of the language with universal modality (cf. Goranko and Passy, [12]). Based on these results, some relevant results of validity of consecutions can be derived. This is easy to accept that valid consecutions correctly describe logical consequence (since, in particular, $\mathcal{L}(\mathcal{K})$ is closed w.r.t. valid consecutions). But a reasonable question is whether we could restrict ourselves by only such consecutions studying a given logic $\mathcal{L}(\mathcal{K})$. Lorenzen (1955) [20] proposed to consider so called admissible consecutions, the definition is as follows. Given a logic $\mathcal{L}$, $Form_{\mathcal{L}}$ is the set of all formulas in the language of $\mathcal{L}$.

Definition 4. A consecution

$$\mathbf{c} := \frac{\varphi(x_1, ..., x_n), ..., \varphi_m(x_1, ..., x_n)}{\psi(x_1, ..., x_n)},$$

is said to be **admissible** for a logic $\mathcal{L}$ if, $\forall \alpha_1 \in Form_{\mathcal{L}}, ..., \forall \alpha_n \in Form_{\mathcal{L}}$,

$$\bigwedge_{1 \le i \le m}[\varphi_i(\alpha_1, ..., \alpha_n) \in \mathcal{L}] \Longrightarrow [\psi(\alpha_1, ..., \alpha_n) \in \mathcal{L}].$$

Thus, for any admissible consecution, any instance into the premises making all of them theorems of $\mathcal{L}$ also makes the conclusion to be a theorem. It is *most strong type of structural logical consecutions:* a consecution **c** is admissible in $\mathcal{L}$ iff $\mathcal{L}$, as the set of its own theorems, is closed with respect to **c**. It is evident that any valid consecution is admissible. The converse is not always true. Before to discuss it, we would like to describe another sort of consecutions: derivable consecutions.

For a logic $\mathcal{L}$ with a fixed axiomatic system $Ax_{\mathcal{L}}$ and a given consecution $\mathbf{cs} := \varphi_1, ..., \varphi_n/\psi$, $\mathbf{cs}$ is said to be *derivable* if $\varphi_1, ..., \varphi_n \vdash_{Ax_{\mathcal{L}}} \psi$ (i.e. if we can derive ψ from $\varphi_1, \ldots, \varphi_n$ in the given axiomatic system $Ax_{\mathcal{L}}$). The derivable consecutions are safely correct. But it could happen, that, for a logic $\mathcal{L}$, with a given axiomatic system, a formula ψ is not derivable from the premises $\varphi_1, \ldots, \varphi_m$, but still the rule $\mathbf{cs} := \varphi_1, \ldots, \varphi_m / \psi$ is admissible: $\mathbf{cs}$ derives $\mathcal{L}$-provable conclusions from $\mathcal{L}$-provable premises. Derivable consecution must be valid, but again not obligatory admissible. The earliest example of a consecution which is admissible in the intuitionistic propositional logic (IPC, in sequel) but not derivable in the Heyting axiomatic system for IPC is the Harrop's rule (1960, [14]):

$$ r := \frac{\neg x \rightarrow y \vee z}{(\neg x \rightarrow y) \vee (\neg x \rightarrow z)}. $$

That is, $\neg x \rightarrow y \vee z \nvdash_{IPC} (\neg x \rightarrow y) \vee (\neg x \rightarrow z)$, were $\vdash_{IPC}$ is the notation for derivability in the Heyting axiomatic system for IPC. But, for any α, β and γ, if $\vdash_{IPC} \neg \alpha \rightarrow \beta \vee \gamma$, then $\vdash_{IPC} (\neg \alpha \rightarrow \beta) \vee (\neg \alpha \rightarrow \gamma)$. G.Mints (1976, [25]) found another nice consecution

$$ \frac{(x \rightarrow y) \rightarrow x \vee y}{((x \rightarrow y) \rightarrow x) \vee ((x \rightarrow y) \rightarrow z)}, $$

which is not derivable but admissible in IPC. The Lemmon-Scott rule (cf. [33])

$$ \frac{\Box(\Box(\Box \Diamond \Box p \rightarrow \Box p) \rightarrow (\Box p \vee \Box \neg \Box p))}{\Box \Diamond \Box p \vee \Box \neg \Box p} $$

is admissible but not derivable in the standard axiomatizations for modal logics $S4$, $S4.1$, Grz.

Notice that, for the case of the temporal linear logic $\mathcal{LTL}^{U,B}_{N,\,N-1}(\mathcal{Z})$, there are consecutions which are invalid (in particular, they are not derivable rules for any possible axiomatic system for $\mathcal{LTL}^{U,B}_{N,\,N-1}(\mathcal{Z})$ where postulated inference rules preserve the truth values of formulas in the Kripke frame $\mathcal{Z}$), which, nevertheless, are admissible.

For instance, the consecution

$$ c_1 := \frac{\Box \Diamond (true \mathbf{U} x) \wedge \Box \Diamond (true \mathbf{U} \neg x)}{y} $$

is admissible but invalid in $\mathcal{LTL}^{U,B}_{N,\,N-1}(\mathcal{Z})$. Therefore, consecutions admissible in $\mathcal{LTL}^{U,B}_{N,\,N-1}(\mathcal{Z})$ are stronger than just valid ones. Also the connection with satisfiability is evident: φ is satisfiable in a logic $\mathcal{L}$ iff $p \rightarrow p / \neg \varphi$ is not admissible for $\mathcal{L}$.

4   Algorithm Recognizing Admissibility in $\mathcal{LTL}^{U,B}_{N,\,N-1}(\mathcal{Z})$

We describe below all necessary mathematical constructions involved in our technique and circumscribe the sequence of statements which allow us to construct the deciding algorithm for the logic $\mathcal{LTL}^{U,B}_{N,\,N-1}(\mathcal{Z})$. A Kripke structure $\mathcal{M}$ is said to be **definable** if any state $c \in \mathcal{M}$ is definable in $\mathcal{M}$, i.e. there is a formula ϕ_a which in true in $\mathcal{M}$ only at the element a. Given a Kripke structure $\mathcal{M} := \langle \mathcal{F}, V \rangle$ based upon a Kripke frame $\mathcal{F}$ and a new valuation V_1 in $\mathcal{F}$ of a set of propositional letters q_i. The valuation V_1 is **definable** in $\mathcal{M}$ if, for any q_i, $V_1(q_i) = V(\phi_i)$ for some formula ϕ_i

Definition 5. Given a logic $\mathcal{L}$ and a Kripke structure $\mathcal{M}$ with a valuation defined for a set of letters $p_1, \ldots, p_k$. $\mathcal{M}$ is said to be k-**characterizing** for $\mathcal{L}$ if the following holds. For any formula $\varphi(p_1, \ldots, p_k)$ built using letters $p_1, \ldots, p_k$,
$$\varphi(p_1, \ldots, p_k) \in \mathcal{L} \text{ iff } \mathcal{M} \Vdash \varphi(p_1, \ldots, p_k).$$

Lemma 1. (cf., for instance, [33]) A consecution **cs** is not admissible in a logic $\mathcal{L}$ iff, for any sequence of k-characterizing models, there are a number n and an n-*characterizing model* $Ch_{\mathcal{L}}(n)$ from this sequence such that the frame of $Ch_{\mathcal{L}}(n)$ refutes **cs** by a certain definable in $Ch_{\mathcal{L}}(n)$ valuation.

Being based at this lemma, in order to describe consecutions **cs** admissible in $\mathcal{LTL}_{N,\,N-1}^{U,B}(\mathcal{Z})$ we need a sequence of k-characterizing for $\mathcal{LTL}_{N,\,N-1}^{U,B}(\mathcal{Z})$ models with *some good* properties. Take the all Kripke structures $\mathcal{Z}_i, i \in I$, all of which are based on the frame $\mathcal{Z}$, with all possible valuations V of letters $p_1, \ldots, p_k$. Take the disjoint union $\bigsqcup_{i \in I} \mathcal{Z}_i$ of all such non-isomorphic Kripke structures. We denote this Kripke structure by $Ch_k(\mathcal{LTL}_{N,\,N-1}^{U,B}(\mathcal{Z}))$.

Lemma 2. The structure $Ch_k(\mathcal{LTL}_{N,\,N-1}^{U,B}(\mathcal{Z}))$ is k-characterizing for the logic $\mathcal{LTL}_{N,\,N-1}^{U,B}(\mathcal{Z})$.

Lemma 3. The Kripke structure $Ch_k(\mathcal{LTL}_{N,\,N-1}^{U,B}(\mathcal{Z}))$ is not definable.

Therefore we cannot directly implement technique from [38] to describe admissible in the logic $\mathcal{LTL}_{N,\,N-1}^{U,B}(\mathcal{Z})$ consecutions. Next instrument we need is a reduction of consecutions to equivalent reduced normal forms. A consecution **c** is said to have the *reduced normal form* if $\mathbf{c} = \varepsilon_c / x_1$ where

$$\varepsilon_c := \bigvee_{1 \leq j \leq m} \left(\bigwedge_{1 \leq i, k \leq n, i \neq k} [x_i^{k(j,i,0)} \wedge (\mathbf{N} x_i)^{k(j,i,1)} \wedge (\mathbf{N}^{-1} x_i)^{k(j,i,2)} \wedge \right.$$

$$\left. (x_i \mathbf{U} x_k)^{k(j,i,k,3)} \wedge (x_i \mathbf{B} x_k)^{k(j,i,k,4)}] \right),$$

and x_s are certain variables, $k(j,i,z), k(j,i,k,z) \in \{0,1\}$ and, for any formula α above, $\alpha^0 := \alpha$, $\alpha^1 := \neg\alpha$.

Definition 6. *Given a consecution* $\mathbf{c}_{nf}$ *in the reduced normal form,* $\mathbf{c}_{nf}$ *is said to be a* **normal reduced form for a consecution** $\mathbf{c}$ *iff, for any temporal logic* $\mathcal{L}$, $\mathbf{c}$ *is admissible in* $\mathcal{L}$ *iff* $\mathbf{c}_{nf}$ *is so.*

Using the ideas of proofs for Lemma 3.1.3 and Theorem 3.1.11 from [33] we can derive

Theorem 2. There exists a double exponential algorithm which, for any given consecution **c**, constructs its normal reduced form $\mathbf{c_{nf}}$.

To describe further the construction of our algorithm distinguishing consecutions admissible in $\mathcal{LTL}^{U,B}_{N,\,N-1}(\mathcal{Z})$, we need the following notations and results.

Definition 7. *Let two Kripke structures* $\mathcal{K}_1 := \langle K_1, R_1, Next_1, Prev_1, V_1 \rangle$ *and* $\mathcal{K}_2 := \langle K_2, R_2, Next_2, Prev_2, V_2 \rangle$ *with designated elements trm (terminal) of* K_1 *and ent (entry) of* K_2 *be given. The* **sequential concatenation** $\mathcal{K}_1$ *and* $\mathcal{K}_2$ *by* (trm, ent) *is the structure*
$$\mathcal{K} := \langle K, R, Next, Prev, V \rangle, \text{ where } K := K_1 \cup K_2,$$

$$V(p) := V_1(p) \cup V_2(p) \text{ for all } p; Next := Next_1 \cup Next_2 \cup \{(trm, ent)\},$$

$$Prev := Prev_1 \cup Prev_2 \cup \{(ent, term)\},$$

$$R := R_1 \cup R_2 \cup \{(a,b) \mid a \in K_1, b \in K_2\}.$$

We will denote $\mathcal{K}$ *by* $\mathcal{K}_1 \oplus_{trm,ent} \mathcal{K}_2$.

Similarly we define the sequential concatenation of frames. For $n, m \in \mathcal{Z}$, $[n, m]$ is the Kripke frame based on all natural numbers situated between n and m with standard $\leq$ and $Next$ and $Prev$. If there us a valuation of a set of letters in $[n, m]$, we refer to $[n, m]$ as a Kripke structure. For any $n, m \in \mathcal{Z}$ with $n < m$, $C[n, m]$ is the 3-modal Kripke frame, where the base set of $C[n, m]$ is $[n, m]$, and the relations $Next_{C[n,m]}$, $Prev_{C[n,m]}$, and $R_{C[n,m]}$ are as follows.

$$mNext_{C[n,m]}n, \ \forall k, n \leq k < m \Rightarrow kNext_{C[n,m]}k + 1;$$

$$nPrev_{C[n,m]}m, \ \forall k, n < k \leq m \Rightarrow kPrev_{C[n,m]}k - 1;$$

$$\forall a, b \in [n, m], \ aR_{C[n,m]}b \Leftrightarrow \exists k(aNext^k_{C[n,m]}b).$$

If there is a valuation of a set of letters on $C[n, m]$, we refer to $C[n, m]$ as a Kripke structure. This definition is a bit confusing for readers which only experienced with standard multi-modal Kripke structure. This is because $C[n, m]$ resembles the time cluster, but actually it is not - to care about **U** and **B** we have to fix the direction in this quasi-cluster, so we choose clockwise. But as soon as the direction is fixed and $R_{C[n,m]}$ works in accordance with $Next_{C[n,m]}$ and $Prev_{C[n,m]}$, we can define the truth values of formulas (in the language of $\mathcal{LTL}^{U,B}_{N,\,N-1}(\mathcal{Z})$) in $C[n, m]$ in the standard manner.

We also need the following notations and results. For any consecution $\mathbf{c_{nf}}$ in normal reduced form, $Pr(c_{nf}) = \{\varphi_i \mid i \in I\}$ is the set of all disjunctive members of the premise of $\mathbf{c_{nf}}$. $Sub(c_{nf})$ is the set of all subformulas of $\mathbf{c_{nf}}$. For any Kripke frame $\mathcal{F}$ and any valuation V of the set of propositional letters of a formula φ, the expression $(\mathcal{F} \Vdash_V \varphi)$ is the abbreviation for $\forall a \in \mathcal{F}((\mathcal{F}, a) \Vdash_V \varphi)$.

Lemma 4. *For any Kripke frame* $\mathcal{F}$ *with a valuation* V, *where*

$$\mathcal{F} \Vdash_V \bigvee Pr(cs_{nf}),$$

for any $a \in \mathcal{M}$, *there is a unique disjunct* D *from* $Pr(c_{nf})$ *such that* $(\mathcal{F}, a) \Vdash_V D$.

In the sequel we will denote this unique disjunct by $D_{\mathcal{F}}^{cs_{nf},V}(a)$.

Lemma 5. If a consecution $\mathbf{c_{nf}}$ in the normal reduced form is not admissible in $\mathcal{LTL}_{N, N-1}^{U,B}(\mathcal{Z})$ then, there are integer numbers $k, n, m \in \mathcal{Z}$, where $-k < -1, 3 < n, n+1 < m$, and there exists a finite Kripke structure $\mathcal{K}_{\mathbf{c_{nf}}} := \langle \mathcal{F}_{\mathbf{c_{nf}}}, V_1 \rangle$ which refutes $\mathbf{c_{nf}}$ by the valuation V_1, where

(a) $\mathcal{F}_{\mathbf{c_{nf}}} := C[-k, 0] \oplus_{0,1} [1, n] \oplus_{n,n+1} C[n+1, m]$,
(b) $D_{\mathcal{F}_{\mathbf{c_{nf}}}}^{c_{nf},V_1}(-k) = D_{\mathcal{F}_{\mathbf{c_{nf}}}}^{c_{nf},V_1}(1); \quad D_{\mathcal{F}_{\mathbf{c_{nf}}}}^{c_{nf},V_1}(n) = D_{\mathcal{F}_{\mathbf{c_{nf}}}}^{c_{nf},V_1}(m); ,$
(c) k, n and m are linearly computable from the size of $\mathbf{c_{nf}}$.

The 3-modal frame **1** is the frame based at the single element set $\{a_1\}$ where $a_1 Next\ a_1$, $a_1 Prev\ a_1$ and $a_1 \leq a_1$. Again we can define truth values of formulas in the language of $\mathcal{LTL}_{N, N-1}^{U,B}(\mathcal{Z})$ in **1** in the standard manner.

Lemma 6. If a consecution $\mathbf{c_{nf}}$ in the normal reduced form is not admissible in $\mathcal{LTL}_{N, N-1}^{U,B}(\mathcal{Z})$ then there exists a valuation V_0 of letters of $\mathbf{c_{nf}}$ in the frame **1** where

$$\mathbf{1} \Vdash_{V_0} \bigvee \{\varphi_i \mid \varphi_i \in Pr(c_{nf})\}$$

Lemma 7. If a consecution $\mathbf{c_{nf}}$ in normal reduced form satisfies the conclusions of Lemma 5 and Lemma 6 then $\mathbf{c_{nf}}$ is not admissible in $\mathcal{LTL}_{N, N-1}^{U,B}(\mathcal{Z})$.

Based on Theorem 2, Lemmas 5, 6 and 7 we derive

Theorem 3. *There is an algorithm recognizing consecutions admissible in the linear temporal logic $\mathcal{LTL}_{N, N-1}^{U,B}(\mathcal{Z})$. In particular, this algorithm solves the satisfiability problem for $\mathcal{LTL}_{N, N-1}^{U,B}(\mathcal{Z})$.*

To comment the complexity of the deciding algorithm, for any consecution $\mathbf{c}$ we first transform $\mathbf{c}$ into the reduced normal form $\mathbf{c_{nf}}$ (complexity is double exponential, cf. Theorem 2). Then we verify conditions of Lemmas 5 and 6 in the models $\mathcal{K}_{\mathbf{c_{nf}}} := \langle \mathcal{F}_{\mathbf{c_{nf}}}, V_1 \rangle$ which size is linear in the size of $\mathbf{c_{nf}}$. So, we have to perform the algorithm of verification the validity for consecutions on models in size linear from $\mathbf{c_{nf}}$.

Conclusion, Future Work: We investigated the linear temporal logic $\mathcal{LTL}_{N, N-1}^{U,B}(\mathcal{Z})$ which extends the standard logic LTL by operations **B** (before) and $\mathbf{N^{-1}}$ (previous). The prime questions which we been focused were (i) problem of satisfiability in $\mathcal{LTL}_{N, N-1}^{U,B}(\mathcal{Z})$ and (ii) problem of description of logical consequence in $\mathcal{LTL}_{N, N-1}^{U,B}(\mathcal{Z})$ via admissible logical consecutions (inference rules). A reduction of consecutions and formulas of $\mathcal{LTL}_{N, N-1}^{U,B}(\mathcal{Z})$ to simple consecutions in formulas of temporal degree 1 is suggested. Based on technique of Kripke structures we find necessary and sufficient conditions for consecutions to be not admissible in $\mathcal{LTL}_{N, N-1}^{U,B}(\mathcal{Z})$. These conditions lead to an algorithm which recognizes admissible consecutions through verification of validity of consecutions in Kripke structures of size linear in the reduced normal forms of the initial consecutions. The obtained results extend previous research of the author concerning transitive modal logics (cf. [29, 30, 31, 32, 33, 34, 35, 36, 37] where technique for construction of algorithms recognizing admissible rules via models with special structure was worked out. The paper Ghilardi [10] and others his papers provide another technique to prove decidability of the modal logic $S4$ and intuitionistic logic IPC w.r.t. admissibility via properties of projective formulas. This technique also led to work out some approach for construction of explicit bases for admissible rules in transitive modal logics, as it is done, e.g. in Jerabek [16]. But our technique based on finite models with special structure, semantic one, woks nicely for construction of explicit bases as well, as it is demonstrated in Rybakov [37], where an explicit bases for rules of $S4$ is found. Our research in the current paper extends the area of admissible rules to linear temporal logics, but we are working here only with algorithms recognizing admissible rules.

The technique developed in this paper can be applied for other similar linear temporal logics. We studied only one natural, maybe most intuitive, linear temporal logic. However there are other linear temporal logics similar to $\mathcal{LTL}_{N,\ N-1}^{U,B}(\mathcal{Z})$, but with other logical operations or based at frames different from $\mathcal{Z}$, which do not yet obey the obtained technique. We plan to study these questions with an attempt to approach the problems and to find deciding algorithms, to construct explicit bases for admissible consecutions.

References

1. Barringer H, Fisher M, Gabbay D., Gough G. *Advances in Temporal Logic*, Vol. 16 of Applied logic series, Kluwer Academic Publishers, Dordrecht, December 1999. (ISBN 0-7923-6149-0).

2. Bloem R., Ravi K, Somenzi F, *Efficient Decision Procedures for Model Checking of Linear Time Logic Properties.-* In: Conference on Computer Aided Verification (CAV), LNCS 1633, Trento, Italy, 1999, Springer-Verlag.

3. Bruns G. and Godefroid P. *Temporal Logic Query-Checking.-* In Proceedings of 16th Annual IEEE Symposium on Logic in Computer Science (LICS'01), pages 409–417, Boston, MA, USA, June 2001. IEEE Computer Society.

4. Clarke E., Grumberg O., Hamaguchi K. P. *Another look at LTL Model Checking.* - In: Conference on Computer Aided Verification (CAV), LNCS 818, Stanford, California, 1994, Springer-Verlag.

5. Cyrluk David, Narendran Paliath. *Ground Temporal Logic: A Logic for Hardware Verification-* Lecture Notes in computer Science, V. 818. From Computer-aided Verification (CaV'94), Ed. David Dill, Springer-Verlag, Stanford, CA, 1994, p. 247 - 259.

6. Daniele M, Giunchiglia F, Vardi M. *Improved Automata Gneration for Linear Temporal Logic.* - In book: (CAV'99), International Conference on Computer-Aided Verification, Trento, Italy, July 7-10, 1999.

7. Emerson E.A. *Temporal and Modal Logics.-* in: Handbook of Theoretical Computer Science. J. van Leenwen, Ed., Elsevier Science Publishers, the Netherlands, 1990, 996 - 1072.

8. Friedman H., *One Hundred and Two Problems in Mathematical Logic.-* Journal of Symbolic Logic, Vol. 40, 1975, No. 3, 113 - 130.

332   V. Rybakov

9. Gabbay D. *An Irreflevivity Lemma with Applications to Axiomatizations of Conditions of Linear Frames.*-Aspects of Phil. Logic (Ed. V.Monnich), Reidel, Dordrecht, 1981, 67 - 89.

10. Ghilardi S. *Unification in Intuitionistic logic.*- Journal of Symbolic Logic, Vo. 64, No. 2 (1999), pp. 859-880.

11. Goldblatt R. *Logics of Time and Computation.*- CSLI Lecture Notes, No.7, 1992.

12. Goranko V., Passy S. *Using the Universal Modality: Gains and Questions.*- J. Log. Comput. 2(1): 5-30, (1992).

13. van der Hoek W., Wooldridge M. *Model Checking Knowledge and Time.* - In SPIN 2002, Proc. of the Ninth International SPIN Workshop on Model Checking of Software, Grenoble, France, 2002.

14. Harrop R. *Concerning Formulas of the Types $A \rightarrow B \lor C$, $A \rightarrow \exists x B(x)$ in Intuitionistic Formal System.*- J. of Symbolic Logic, Vol. 25, 1960, 27-32.

15. Iemhoff R. *On the admissible rules of Intuitionistic Propositional Logic.* - J.of Symbolic Logic Vol. 66, 2001, pp. 281-294.

16. Jerabek E. *Admissible Rules of Modal Logics.* -J. of Logic and Computation, 2005, Vol. 15. pp. 411-431.

17. Kapron B.M. *Modal Sequents and Definability*, J.of Symbolic Logic, 52(3), 756 - 765, (1987).

18. Lichtenstein O., Pnueli A. *Propositional temporal logics: Decidability and completeness.* - Logic Journal of the IGPL, 8(1):55-85, 2000.

19. Francois Laroussinie, Nicolas Markey, Philippe Schnoebelen. *Temporal Logic with Forgettable Past .*- IEEE Symp. Logic in Computer Science (LICS'2002).

20. Lorenzen P. *Einführung in die operative Logik und Mathematik.* - Berlin-Göttingen, Heidelberg, Springer-Verlag, 1955.

21. Manna Z, Pnueli A. *Temporal Verification of Reactive Systems: Safety,* - Springer-Verlag, 1995.

22. Manna Z., Pnueli A. *The Temporal Logic of Reactive and Concurrent Systems: Specification.* - Springer-Verlag, 1992.

23. Nikolaj Bjorner, Anca Browne, Michael Colon, Bernd Finkbeiner, Zohar Manna, Henny Sipma, Tomas Uribe. *Verifying Temporal Properties of Reactive Systems: A Step Tutorial.* - In Formal Methods in System Design, vol 16, pp 227-270. 2000

24. Manna Z., Sipma H. *Alternating the Temporal Picture for Safety.* - In Proc. 27th Intl. Colloq. Aut. Lang. Prog.(ICALP 2000). LNCS 1853, Springer-Verlag, pp. 429-450.

25. Mints G.E. *Derivability of Admissible Rules.*- J. of Soviet Mathematics, V. 6, 1976, No. 4, 417 - 421.

26. Pnueli A. *The Temporal Logic of Programs.*- In Proc. of the 18th Annual Symp. on Foundations of Computer Science, 46 - 57, IEEE, 1977.

27. Pnueli A., Kesten Y. *A deductive proof system for CTL^*.*- In Proc. 13th Conference on Concurrency Theory, volume 2421 of Lecture Notes in Computer Science, pages 24-40, Brno, Czech Republic, August 2002.

28. Rybakov V.V. *A Criterion for Admissibility of Rules in the Modal System S4 and the Intuitionistic Logic.* - Algebra and Logic, V.23 (1984), No 5, 369 - 384 (Engl. Translation).

29. Rybakov V.V. *The Bases for Admissible Rules of Logics S4 and Int.*- Algebra and Logic, V.24, 1985, 55-68 (English translation).

30. Rybakov V.V. *Rules of Inference with Parameters for Intuitionistic logic.*- Journal of Symbolic Logic, Vol. 57, No. 3, 1992, pp. 912 - 923.

31. Rybakov V.V. *Hereditarily Structurally Complete Modal Logics.*- Journal of Symbolic Logic, Vol. 60, No.1, 1995, pp. 266 - 288.

32. Rybakov V.V. *Modal Logics Preserving Admissible for S4 Inference Rules.* - In theProceedings of the conference CSL'94. *LNCS*, No.993 (1995), Springer-Verlag, 512 - 526.

33. Rybakov V.V. *Admissible Logical Inference Rules.* - Studies in Logic and the Found. of Mathematics, Vol. 136, Elsevier Sci. Publ., North-Holland, New-York-Amsterdam, 1997, 617 pp.

34. Rybakov V.V. *Quasi-characteristic Inference Rules.* - In Book: Eds: S.Adian, A.Nerode, *LNCS,* Vol. 1234, , *Springer,* 1997, pp. 333 - 342.

35. Rybakov V.V., Kiyatkin V.R., Oner T., *On Finite Model Property For Admissible Rules.* - Mathematical Logic Quarterly, Vol.45, No 4, 1999, pp. 505-520.

36. Rybakov V.V. Terziler M., Rimazki V. *Basis in Semi-Reduced Form for the Admissible Rules of the Intuitionistic Logic IPC.*- Mathematical Logic Quarterly, Vol.46, No. 2 (2000), pp. 207 - 218.

37. Rybakov V.V. *Construction of an Explicit Basis for Rules Admissible in Modal System S4.* - Mathematical Logic Quarterly, Vol. 47, No. 4 (2001), pp. 441 - 451.

38. Rybakov V.V. *Logical Consecutions in Intransitive Temporal Linear Logic of Finite Intervals. Journal of Logic Computation,* (Oxford Press, London), Vol. 15 No. 5 (2005) pp. 633 -657.

39. Vardi M. *An automata-theoretic approach to linear temporal logic.*- In the Book: Proceedings of the Banff Workshop on Knowledge Acquisition (1994), (Banff'94).

On the Frequency of Letters in Morphic Sequences

Kalle Saari*

Department of Mathematics and Turku Centre for Computer Science,
University of Turku, 20014 Turku, Finland
kasaar@utu.fi

Abstract. A necessary and sufficient criterion for the existence and value of the frequency of a letter in a morphic sequence is given. This is done using a certain incidence matrix associated with the morphic sequence. The characterization gives rise to a simple if-and-only-if condition that all letter frequencies exist.

Keywords: Morphic sequences, HD0L sequences, Frequency of letters, Incidence matrix, Simple generator.

1 Introduction

Consider an infinite sequence $\mathbf{w}$ that consists of finitely many different symbols, or letters. Let w be a finite prefix of $\mathbf{w}$, and let $|w|$ denote its length. Suppose the letter 1 occurs in $\mathbf{w}$. Denoting the number of occurrences of 1 in w by $|w|_1$, we define the *frequency of the letter 1 in* w to be the ratio

$$\frac{|w|_1}{|w|}.$$

Doing this to every finite prefix of $\mathbf{w}$, we obtain an infinite sequence of ratios which may or may not converge. In case it does converge, we define the limit to be the *frequency of the letter* 1 *in* $\mathbf{w}$, and we denote it by ϕ_1. Hence,

$$\phi_1 = \lim_{|w| \to \infty} \frac{|w|_1}{|w|}$$

if the limit exists. In the literature this type of frequency is often called asymptotic or natural frequency.

The sequences we focus here are so-called *morphic sequences* (see the definition below), which in theory of dynamical systems are known as *substitution sequences* and in theory of L-systems as *HD0L words*. We are interested in the existence of the frequency of a letter in a morphic sequence. Michel [6, 7] proved that if a morphic sequence is *primitive*, then the frequencies of all letters in it exists. He also provided an explicit formula to determine the frequency of each

---

* Supported by the Finnish Academy under grant 206039.

D. Grigoriev, J. Harrison, and E.A. Hirsch (Eds.): CSR 2006, LNCS 3967, pp. 334–345, 2006.

letter. There is a recent result by Peter [9] giving a sufficient and necessary condition for the existence of the frequency of a letter in an automatic sequence. If a sequence is binary and a fixed point of a nonidentity morphism, then the frequency of both letters exist [12]. Until now, however, no simple general characterization for the existence of the letter frequencies was known. Our main result is a solution to this problem.

We give a necessary and sufficient condition when the frequency of a given letter in a morphic sequence exists. The condition is given with the help of the incidence matrix of a so-called *simple* generating morphism of the sequence, and it also provides an explicit formula for the frequency. Using the result, we formulate a simple if-and-only-if criterion for the existence of the frequencies of all letters. We also show that the frequency of all letters exists in any sequence generated by a morphism with polynomial growth.

The notion of frequency is naturally extended to factors, i. e., blocks of letters occurring in a sequence. This more general situation has been studied especially in symbolic dynamics, where frequency is considered as a shift invariant probability measure associated with the given sequence, see [11]. Some nice results on this topic can be found in [3, 4].

2   Definitions and Notations

In this section we fix the terminology used in the paper. Let Σ^* be the free monoid over the finite alphabet Σ, and let ϵ denote its identity element. The map $\varphi\colon \Sigma^* \to \Delta^*$ is a *morphism* if $\varphi(uv) = \varphi(u)\varphi(v)$ for all $u, v \in \Sigma^*$. If furthermore $\varphi(a) \in \Delta$ for all $a \in \Sigma$, then φ is called a *coding*. The *incidence matrix* $M(\varphi)$ associated with φ is defined to be

$$M(\varphi) = (m_{d,a})_{d \in \Delta, a \in \Sigma}, \qquad m_{d,a} = |\varphi(a)|_d.$$

To make this definition unambiguous, we assume that the alphabets Σ, Δ are ordered. Especially if $\Sigma = \Delta$, then the ith column and the ith row of $M(\varphi)$ corresponds to the same letter. We let $M_a = (m_{d,a})_{d \in \Delta}$ denote the column of $M(\varphi)$ that corresponds to the letter $a \in \Sigma$.

Let M be a nonnegative square matrix. Let $J = \mathrm{diag}(J_1, \ldots, J_k)$ be the Jordan form of M (see [5]), where $J_1, \ldots, J_k$ are the Jordan blocks of size $m_1, \ldots, m_k$ associated with the eigenvalues $\lambda_1, \ldots, \lambda_k$ of M, respectively; that is,

$$J_i = \begin{pmatrix} \lambda_i & 1 & 0 & \cdots & 0 \\ 0 & \lambda_i & 1 & \cdots & \vdots \\ \vdots & \ddots & \ddots & \ddots & 0 \\ \vdots & \vdots & \ddots & \ddots & 1 \\ 0 & \cdots & \cdots & 0 & \lambda_i \end{pmatrix}_{m_i \times m_i} . \tag{1}$$

Let r denote the Perron–Frobenius eigenvalue of M (see [8]). Besides being an eigenvalue of M, it is also the maximum of absolute values of all eigenvalues of M. We say that J_i is the *dominating Jordan block of J* if

(i) $\lambda_i = r$ and

(ii) if $\lambda_j = r$, then $m_j \leq m_i$;

in other words, there are no larger Jordan blocks in J associated with r.

Let $\mu^*\colon \Sigma^* \to \Sigma^*$ be a morphism. It is *primitive*, if there exists an integer $k \geq 1$ such that, for all $a, b \in \Sigma$, the letter b occurs in $\mu^k(a)$. If the letter $a \in \Sigma$ satisfies the identity $\mu^k(a) = \epsilon$ for some integer $k \geq 1$, it is called *mortal*. Now suppose that $\mu(0) = 0u$ for a letter $0 \in \Sigma$ and for some word $u \in \Sigma^*$ that contains a non-mortal letter. This ensures that $\mu^i(0)$ is a proper prefix of $\mu^{i+1}(0)$ for all $i \geq 0$. Then we say that μ is *prolongable on* 0. Iterating μ on 0, we obtain a sequence of iterates $0, \mu(0), \mu^2(0), \mu^3(0), \ldots$ that converges to the infinite sequence

$$\mu^\omega(0) := 0u\mu(u)\mu^2(u)\mu^3(u)\cdots.$$

Such a sequence is called a *pure morphic sequence*. If $\tau\colon \Sigma^* \to \Gamma^*$ is a coding, then $\mathbf{w} := \tau(\mu^\omega(0))$ is called a *morphic sequence*. The sequence $\mathbf{w}$ is *primitive* is μ is primitive.

The incidence matrix $M(\mu)$ associated with $\mu\colon \Sigma^* \to \Sigma^*$ is a nonnegative square matrix with the property $M(\mu^n) = M(\mu)^n$. We denote the Perron–Frobenius eigenvalue of $M(\mu)$ by $r(\mu)$ and the Jordan form of $M(\mu)$ by $J(\mu)$.

We say that the morphism $\mu\colon \Sigma^* \to \Sigma^*$ is a *simple generator* of the sequence $\mathbf{w} = \tau(\mu^\omega(0))$ if

(S1) each letter of Σ occurs in $\mu(0)$, and

(S2) if λ is an eigenvalue of $M(\mu)$ and $|\lambda| = r(\mu)$, then $\lambda = r(\mu)$.

See [5] for general concepts and results of matrix theory, and [8] for theory on nonnegative matrices. For further information about morphisms and morphic sequences, we refer to [1].

3 Lemmata

In this section we gather some necessary tools to prove the main result of the paper. The following lemma by Frobenius is proved in [8, Theorem III 1.1]:

Lemma 1. *Let A be an irreducible square matrix with the Perron–Frobenius eigenvalue r, and suppose it has exactly h eigenvalues of modulus r; we denote them by $\lambda_1, \lambda_2, \ldots, \lambda_h$. Then the eigenvalues $\lambda_1, \lambda_2, \ldots, \lambda_h$ are the distinct hth roots of r^h.*

Lemma 2. *Any morphic sequence has a simple generator.*

Proof. Let $\mathbf{w} = \tau(\mu^\omega(0))$ be a morphic sequence, where $\mu\colon \Sigma^* \to \Sigma^*$. We construct a simple generator of $\mathbf{w}$ by modifying μ. Let $\Sigma' \subseteq \Sigma$ be the set of letters occurring in $\mu^\omega(0)$. Then $\mu(\Sigma') \subseteq \Sigma'^*$. Since μ is prolongable on 0, there exists an integer $k \geq 1$ such that all the letters of Σ' occur in $\mu^k(0)$. The matrix $M(\mu)$ is similar to a matrix of the form

$$M(\mu) \sim \begin{pmatrix} A_{11} & 0 & \cdots & 0 \\ B_{21} & A_{22} & \cdots & 0 \\ \vdots & & \ddots & \vdots \\ B_{n1} & B_{n2} & \cdots & A_{nn} \end{pmatrix},$$

where every entry A_{ii} on the diagonal in an irreducible square matrix, and the entries above them are 0. Let h_i be the integer in Lemma 1 that corresponds to A_{ii}, and set $h = \mathrm{lcm}(h_1, h_2, \ldots, h_n)$, the least common multiple of h_i s. It follows that all the eigenvalues λ of $M(\mu)$ with $|\lambda| = r(\mu)$ satisfy the equality $\lambda^h = r(\mu)^h$.

We define the morphism $\mu' : \Sigma'^* \to \Sigma'^*$ from the condition $\mu'(a) = \mu^{hk}(a)$ for all $a \in \Sigma'$ and claim that μ' is a simple generator of $\mathbf{w}$:

First, it is clear that $\mathbf{w} = \tau(\mu'^{\omega}(0))$. Secondly, (S1) is satisfied since $\mu^k(0)$ is a prefix of $\mu^{hk}(0)$, and so all the letters of Σ' occur in $\mu'(0)$. Now suppose that λ is an eigenvalue of $M(\mu')$. Since $M(\mu') = M(\mu)^{hk}$, we see that

$$\det\big(M(\mu') - \lambda I\big) = \det\big(M(\mu) - \omega_1 I\big) \cdots \det\big(M(\mu) - \omega_{hk} I\big) = 0,$$

where ω_i s are the hkth roots of λ. Thus λ is an eigenvalue of $M(\mu')$ if and only if some of its hkth roots, say w_i, is an eigenvalue of $M(\mu)$. Using this observation in the case when $\lambda = r(\mu')$, we obtain $r(\mu') = r(\mu)^{hk}$. If $\lambda \neq r(\mu')$ and $|\lambda| = r(\mu')$, then since $\lambda = w_i^{hk}$ and $w_i^h = r(\mu)^h$, we obtain $\lambda = r(\mu)^{hk} = r(\mu')$, a contradiction. Hence μ' satisfies also (S2), and so it is a simple generator of $\mathbf{w}$. This concludes the proof.

From now on, we assume that $\mathbf{w} = \tau(\mu^{\omega}(0))$ is a morphic sequence, where $\mu : \Sigma^* \to \Sigma^*$ is a simple generator, $\tau : \Sigma^* \to \Gamma^*$ is a coding, and $0 \in \Sigma$. We omit redundant use of parenthesis and denote $\tau\mu^n(x) = \tau(\mu^n(x))$ for $x \in \Sigma^*$.

The matrix $E(\mu)$ defined in the next lemma plays an essential role in our further considerations. Recall that the rank of a matrix is the size of a maximal set of linearly independent columns of the matrix.

Lemma 3. *Let $p+1$ be the size of the dominating Jordan block in $J(\mu)$. Then the limit*

$$E(\mu) := \lim_{n \to \infty} \frac{M(\mu)^n}{n^p r(\mu)^n}$$

exists. Furthermore, it has the following properties:

- *The rank of $E(\mu)$ equals the number of occurrences of the dominating Jordan block in $J(\mu)$.*
- *For all letters $a, b \in \Sigma$,*

$$|\mu^n(a)|_b = e_{b,a} n^p r^n + O(n^{p-1} r^n),$$

 where $e_{b,a}$ is the entry of $E(\mu)$ in the place (b, a).
- *The column $E_0 = (e_{a,0})_{a \in \Sigma}$ is nonzero.*

Proof. We denote $r = r(\mu)$ and let t be the number of occurrences of the dominating Jordan block in $J(\mu) = (J_1, \ldots, J_k)$, where J_i is the Jordan block as in (1). It is easily verified that

$$
J_i^n = \begin{pmatrix}
\lambda_i^n & \binom{n}{1}\lambda_i^{n-1} & \cdots & \binom{n}{m_i-1}\lambda_i^{n-m_i+1} \\
0 & \lambda_i^n & \cdots & \binom{n}{m_i-2}\lambda_i^{n-m_i+2} \\
\vdots & \vdots & \ddots & \vdots \\
0 & 0 & \cdots & \lambda_i^n
\end{pmatrix}
\tag{2}
$$

for all $n \geq 0$. We also note that

$$
\binom{n}{p} r^{n-p} = \alpha n^p r^n + O(n^{p-1} r^n),
\tag{3}
$$

where $\alpha = (p!\, r^p)^{-1} > 0$. Since μ is a simple generator, the equality $|\lambda_i| = r$ implies $\lambda_i = r$, and thus if either $\lambda_i \neq r$ or $j < p$, then

$$
\binom{n}{j} \lambda_i^{n-j} = O(n^{p-1} r^n).
\tag{4}
$$

It follows that the limit

$$
D := \lim_{n\to\infty} \frac{J(\mu)^n}{n^p r^n}
$$

exists, and moreover, each occurrence of the dominating Jordan block contributes exactly one positive entry, namely $(p!\, r^p)^{-1}$, in D; all the other entries equal 0. Since the positive entries lie in different columns, the rank of D equals t.

The identity $M(\mu)^n = P J(\mu)^n P^{-1}$ implies that

$$
\lim_{n\to\infty} \frac{M(\mu)^n}{n^p r^n} = P D P^{-1},
\tag{5}
$$

which is, as the limit of a sequence of nonnegative matrices, a nonnegative matrix. We define $E(\mu) = (e_{a,b})_{a,b\in\Sigma}$ to be this limit. Because $E(\mu)$ is similar to D, the rank of $E(\mu)$ equals that of D, that is, t.

Since $M(\mu)^n = P J(\mu)^n P^{-1}$ and $J(\mu)^n = \mathrm{diag}(J_1^n, \ldots, J_k^n)$, it follows from (2) that for all letters $a, b \in \Sigma$, there exist constants $c_{a,b}^{i,j}$, where $1 \leq i \leq k, 0 \leq j < m_i$ such that

$$
|\mu^n(a)|_b = \sum_{1\leq i\leq k} \sum_{0\leq j<m_i} c_{a,b}^{i,j} \binom{n}{j} \lambda_i^{n-j}.
\tag{6}
$$

Without loss of generality, we may assume that the $J_1, J_2, \ldots, J_t$ are the dominating Jordan blocks in $J(\mu)$. Then

$$
|\mu^n(a)|_b = \sum_{1\leq i\leq t} c_{a,b}^{i,p} \binom{n}{p} r^{n-p} + \sum_{1\leq i\leq t} \sum_{0\leq j<p} c_{a,b}^{i,j} \binom{n}{j} r^{n-j}
$$

$$
+ \sum_{t<i\leq k} \sum_{0\leq j<m_i} c_{a,b}^{i,j} \binom{n}{j} \lambda_i^{n-j}.
\tag{7}
$$

Now it follows from (3), (4), and (7) that

$$|\mu^n(a)|_b = \beta_{b,a} n^p r^n + O(n^{p-1} r^n), \tag{8}$$

where $\beta_{b,a} = \alpha \left(\sum_{1 \le i \le t} c_{a,b}^{i,p} \right)$.

On the other hand, Equation (5) implies

$$|\mu^n(a)|_b = e_{b,a} n^p r^n + o(n^p r^n).$$

This and (8) gives $\beta_{b,a} = e_{b,a}$, and so $|\mu^n(a)|_b = e_{b,a} n^p r^n + O(n^{p-1} r^n)$.

As the rank of $E(\mu)$ equals $t \ge 1$, there is a nonzero column in $E(\mu)$, say E_a, where $a \in \Sigma$. Hence $\gamma_a := \sum_{b \in \Sigma} e_{b,a} > 0$. Now $|\mu^n(a)| = \gamma_a n^p r^n + o(n^p r^n)$, and because the letter a occurs in $\mu(0)$, we see that

$$\frac{|\mu^{n+1}(0)|}{(n+1)^p r^{n+1}} \ge \frac{|\mu^n(a)|}{(n+1)^p r^{n+1}} = \frac{\gamma_a n^p r^n + o(n^p r^n)}{(n+1)^p r^{n+1}} \longrightarrow \gamma_a \frac{1}{r} > 0$$

as $n \longrightarrow \infty$. Since $|\mu^n(0)| = \gamma_0 n^p r^n + o(n^p r^n)$, where $\gamma_0 = \sum_{b \in \Sigma} e_{b,0}$, it follows that E_0 cannot be a zero column.

From this lemma we can conclude:

Corollary 1. *Let $p + 1$ be the size of the dominating Jordan block in $J(\mu)$. Denote $W(\tau, \mu) := M(\tau) E(\mu)$, where $E(\mu)$ is as in Lemma 3. Then*

- *For all letters $a \in \Sigma$, $g \in \Gamma$,*

$$|\tau \mu^n(a)|_g = w_{g,a} n^p r^n + O(n^{p-1} r^n),$$

  where $w_{g,a}$ is the entry of $W(\tau, \mu)$ in the place (g, a).
- *The column W_0 is nonzero.*

The factorization of any finite prefix of **w** presented in the next lemma provides us with an important tool.

Lemma 4. *Let $G \ge 1$ be an integer. We define $M_j = \max_{a \in \Sigma} \{ |\tau \mu^j(a)| \}$. The following holds for any letter $a \in \Sigma$ and integer $n \ge 1$: If w is a prefix of $\tau \mu^n(a)$, then w has a factorization of the form*

$$w = \tau \mu^{n_1}(u_1) \tau \mu^{n_2}(u_2) \cdots \tau \mu^{n_r}(u_r) z,$$

where $r \ge 0$, $n > n_1 > n_2 > \cdots > n_r \ge G$, $u_i \in \Sigma^*$ and $|u_i| \le M_1$ for $i = 1, \dots, r$, and $z \in \Sigma^*$, $|z| \le M_G$.

Proof. We prove the claim by induction on $n \ge 1$. If $n \le G$, then for all $a \in \Sigma$, $|w| \le |\tau \mu^n(a)| \le M_G$, so we may choose $r = 0$ and $z = w$. Suppose then that the claim holds for all $1 \le n \le k$, where $k \ge G$, and for all $a \in \Sigma$, and let $n = k + 1$.

Write $\mu(a) = b_1 b_2 \cdots b_t$, where $b_i \in \Sigma$. If $t = 0$, then $w = \tau \mu^n(a) = \epsilon$, so assume $t \ge 1$. Now $w = \tau \mu^k(b_1 b_2 \cdots b_s) w'$ for some $0 \le s < t$ and some prefix w' of $\tau \mu^k(b_{s+1})$. By the induction assumption,

$$w' = \tau \mu^{n_1}(u_1) \tau \mu^{n_2}(u_2) \cdots \tau \mu^{n_r}(u_r) z,$$

where $r \geq 0$, $k > n_1 > n_2 > \cdots > n_r \geq G$, $u_i \in \Sigma^*$ and $|u_i| \leq M_1$ for $i = 1, \ldots, r$, and $|z| \leq M_G$. Denote $u_0 = b_1 b_2 \cdots b_s$, so that

$$w = \tau \mu^k (u_0) \tau \mu^{n_1} (u_1) \tau \mu^{n_2} (u_2) \cdots \tau \mu^{n_r} (u_r) z.$$

Since $n > k$ and $|u_0| \leq M_1$, we see that this factorization is of the correct form.

A proof of the next well-known lemma can be found in [2].

Lemma 5 (Abel's summation formula). *Let $(a_n)_{n \geq 1}$ be a sequence of complex numbers. Let f be a complex valued continuously differentiable function for $x \geq 1$. We define*

$$A(x) = \sum_{1 \leq n \leq x} a_n.$$

Then

$$\sum_{1 \leq n \leq x} a_n f(n) = A(x) f(x) - \int_1^x A(t) f'(t) dt.$$

We use Abel's summation formula to prove the following lemma.

Lemma 6. *Let $r > 1$ be a real number, and let q be any integer. Then*

$$\sum_{1 \leq n \leq x} n^q r^n = \frac{r}{r-1} x^q r^x + o(x^q r^x).$$

Proof. Denote $A(x) = \sum_{1 \leq n \leq x} r^n$, and let $f(x) = x^q$. Observe that $A(x) \leq \frac{r^{x+1}}{r-1}$. Also,

$$\int_1^x t^{q-1} r^t dt = \frac{(-1)^{q-1} (q-1)!}{(\ln r)^q} \left[\sum_{i=0}^{q-1} \frac{(-1)^i (\ln r)^i}{i!} t^i r^t \right]_1^x = o(x^q r^x),$$

which implies

$$\int_1^x A(t) t^{q-1} dt = o(x^q r^x).$$

Now application of Lemma 5 shows that

$$\sum_{1 \leq n \leq x} n^q r^n = r \frac{r^x - 1}{r-1} x^q - q \int_1^x A(t) t^{q-1} dt = \frac{r}{r-1} x^q r^x + o(x^q r^x)$$

4 Main Results

Having established the necessary tools in the previous section, we are ready to present the main results of this paper.

Theorem 1. *Let $\mathbf{w} = \tau \mu^\omega(0)$ be a morphic sequence and μ be its simple generator. If $r(\mu) = 1$ (that is, $|\mu^n(0)|$ has polynomial growth rate) then the frequency of every letter in $\mathbf{w}$ exists.*

Proof. It is enough to show that the frequencies of all letters in $\mu^\omega(0)$ exist. Let $1 \in \Sigma$, and let $E(\mu)$ and p be as in Lemma 3. Since $r(\mu) = 1$, Lemma 3 implies that

$$|\mu^n(0)| = \gamma_0 n^p + O(n^{p-1}), \tag{9}$$

where $\gamma_0 = \sum_{b \in \Sigma} e_{b,0} > 0$. Moreover,

$$\phi_1 := \lim_{n \to \infty} \frac{|\mu^n(0)|_1}{|\mu^n(0)|} = \frac{e_{1,0}}{\sum_{b \in \Sigma} e_{b,0}}.$$

We claim that the frequency of the letter 1 exists and equals ϕ_1.

Denote $f(n) = |\mu^{n+1}(0)| - |\mu^n(0)|$. Equation (9) implies that $f(n) = o(|\mu^n(0)|)$. Let $\delta > 0$ be a real number. Let $G \geq 1$ be an integer such that

$$\left| \frac{|\mu^n(0)|_1}{|\mu^n(0)|} - \phi_1 \right| < \delta$$

for all $n \geq G$. Let w be a prefix of $\mu^\omega(0)$ of length $|\mu^n(0)| \leq |w| < |\mu^{n+1}(0)|$ for some $n \geq G$. Then

$$\left| \frac{|w|_1}{|w|} - \phi_1 \right| \leq \max \left\{ \left| \frac{|w|_1}{|\mu^n(0)|} - \phi_1 \right|, \left| \frac{|w|_1}{|\mu^{n+1}(0)|} - \phi_1 \right| \right\}$$

$$\leq \max \left\{ \left| \frac{|\mu^n(0)|_1}{|\mu^n(0)|} - \phi_1 \right| + \frac{f(n)}{|\mu^n(0)|}, \left| \frac{|\mu^{n+1}(0)|_1}{|\mu^{n+1}(0)|} - \phi_1 \right| + \frac{f(n)}{|\mu^{n+1}(0)|} \right\}$$

$$< \max \left\{ \delta + \frac{f(n)}{|\mu^n(0)|}, \delta + \frac{f(n)}{|\mu^{n+1}(0)|} \right\} \longrightarrow \delta$$

as $|w| \to \infty$. Since δ can be chosen arbitrarily close to 0, it follows that

$$\lim_{|w| \to \infty} \frac{|w|_1}{|w|} = \phi_1,$$

which concludes the proof.

The following theorem is our main result.

Theorem 2. *Let* $\mathbf{w} = \tau\mu^\omega(0)$ *be a morphic sequence, where* $\mu \colon \Sigma^* \to \Sigma^*$ *is a simple generator and* $\tau \colon \Sigma^* \to \Gamma^*$. *Denote* $W(\tau, \mu) = M(\tau)E(\mu)$. *The frequency of the letter* $1 \in \Gamma$ *in* $\mathbf{w}$ *exists if and only if*

$$\frac{w_{1,a}}{\sum_{g \in \Gamma} w_{g,a}} = \frac{w_{1,0}}{\sum_{g \in \Gamma} w_{g,0}} \tag{10}$$

for all nonzero columns W_a *of* $W(\tau, \mu)$. *If it exists, the frequency is the ratio in* (10).

Proof. We denote $r = r(\mu)$. Let us first assume that the Equation (10) holds for all nonzero columns. According to Theorem 1, we may assume that $r > 1$. Let $A \subseteq \Sigma$ denote the set of the letters $a \in \Sigma$ such that W_a is nonzero, that is,

$$A = \Big\{a \in \Sigma \colon \sum_{g \in \Gamma} w_{g,a} > 0\Big\}.$$

Denote $B = \Sigma \setminus A$. First we will show that the contribution of the letters of B in **w** is "sufficiently" small, so that their behavior when iterated do not affect the existence of the frequency.

By Corollary 1,

$$|\tau\mu^n(0)| = \gamma_0 n^p r^n + O(n^{p-1} r^n), \tag{11}$$

where $\gamma_0 = \sum_{g \in \Gamma} w_{g,0} > 0$, and furthermore for all $b \in B$,

$$|\tau\mu^n(b)| = O(n^{p-1} r^n).$$

Thus there exists a real number $\alpha > 0$ such that $|\tau\mu^n(b)| \leq \alpha n^{p-1} r^n$ for all $n \geq 1$ and all $b \in B$. Since $r > 1$, it follows from Lemma 6 that

$$\sum_{i=1}^{n} \sum_{b \in B} |\tau\mu^i(b)| \leq \alpha |B| \sum_{i=1}^{n} i^{p-1} r^i = o(n^p r^n).$$

Hence by Equation (11),

$$\sum_{i=1}^{n} \sum_{b \in B} |\tau\mu^i(b)| = o(|\tau\mu^{n-1}(0)|). \tag{12}$$

Now we are ready to show that the frequency of the letter 1 in **w** exists and equals

$$\phi_1 := \frac{w_{1,0}}{\sum_{g \in \Gamma} w_{g,0}}.$$

Corollary 1 implies that, for all $a \in A$,

$$\frac{|\tau\mu^n(a)|_1}{|\tau\mu^n(a)|} \longrightarrow \frac{w_{1,a}}{\sum_{g \in \Gamma} w_{g,a}}$$

as $n \longrightarrow \infty$, and by the assumption, this limit equals ϕ_1.

Let $\delta > 0$. Let $G \geq 1$ be an integer such that

$$\big| |\tau\mu^n(a)|_1 - \phi_1 |\tau\mu^n(a)| \big| < \delta |\tau\mu^n(a)|$$

for all $n \geq G$ and all $a \in A$.

Let w be a prefix of **w** of length $|\tau\mu^{n-1}(0)| < |w| \leq |\tau\mu^n(0)|$ for some $n \geq G$, and let

$$w = \tau\mu^{n_1}(u_1)\tau\mu^{n_2}(u_2) \cdots \tau\mu^{n_r}(u_r)z,$$

be the factorization as in Lemma 4. Then $n > n_1 > n_2 > \cdots > n_r \geq G$, $|u_i| \leq M_1$ for $i = 1, \ldots, r$, and $|z| \leq M_G$. Let $x_1^{(i)} x_2^{(i)} \cdots x_{s_i}^{(i)}$, where $x_j^{(i)} \in A$, denote the subword of u_i obtained by erasing all the occurrences of letters of B. Similarly, let $y_1^{(i)} y_2^{(i)} \cdots y_{t_i}^{(i)}$, where $y_j^{(i)} \in B$, denote the subword of u_i obtained by erasing all the occurrences of letters of A. Note that $t_i \leq M_1$ for $i = 1, 2, \ldots, r$. Then

$$
\big| |w|_1 - \phi_1 |w| \big| \leq \sum_{i=1}^{r} \sum_{j=1}^{s_i} \big| |\tau\mu^{n_i}(x_j^{(i)})|_1 - \phi_1 |\tau\mu^{n_i}(x_j^{(i)})| \big|
$$

$$
+ \sum_{i=1}^{r} \sum_{j=1}^{t_i} \big| |\tau\mu^{n_i}(y_j^{(i)})|_1 - \phi_1 |\tau\mu^{n_i}(y_j^{(i)})| \big| + \big| |z|_1 - \phi_1 |z| \big|
$$

$$
< \sum_{i=1}^{r} \sum_{j=1}^{s_i} \delta |\tau\mu^{n_i}(x_j^{(i)})| + \sum_{i=1}^{r} \sum_{j=1}^{t_i} |\tau\mu^{n_i}(y_j^{(i)})| + |z|
$$

$$
\leq \delta |w| + M_1 \sum_{k=1}^{n} \sum_{b \in B} |\tau\mu^{k}(b)| + M_G.
$$

The previous inequality and a combination of (12) and $|\tau\mu^{n-1}(0)| < |w|$ gives

$$
\Big| \frac{|w|_1}{|w|} - \phi_1 \Big| < \delta + M_1 \frac{\sum_{k=1}^{n} \sum_{b \in B} |\tau\mu^{k}(b)|}{|w|} + \frac{M_G}{|w|} \longrightarrow \delta,
$$

as $|w| \longrightarrow \infty$. Since $\delta > 0$ was arbitrarily chosen, this implies that

$$
\lim_{|w| \to \infty} \frac{|w|_1}{|w|} \longrightarrow \phi_1.
$$

Hence the frequency of the letter 1 exists and equals the attested value.

Conversely, assume that the frequency of the letter 1 exists. Denote the frequency by ϕ_1, which then has to equal

$$
\lim_{n \to \infty} \frac{|\tau\mu^n(0)|_1}{|\tau\mu^n(0)|} = \frac{w_{1,0}}{\sum_{g \in \Gamma} w_{g,0}}.
$$

To derive a contradiction, suppose there exists a letter $c \in \Sigma$ such that the column W_c is nonzero and

$$
\frac{w_{1,c}}{\sum_{g \in \Gamma} w_{g,c}} \neq \frac{w_{1,0}}{\sum_{g \in \Gamma} w_{g,0}}.
$$

Denoting the ratio on the left by ρ_1, we have $\rho_1 \neq \phi_1$. Since the morphism μ is a simple generator, the letter c occurs in $\mu(0)$, and we can write $\mu(0) = ucv$, where $u, v \in \Sigma^*$. Here $u \neq \epsilon$ because $c \neq 0$. By Corollary 1,

$$
|\tau\mu^n(u)|_1 = \alpha_u \, n^p r^n + o(n^p r^n), \qquad |\tau\mu^n(u)| = \gamma_u \, n^p r^n + o(n^p r^n),
$$

$$
|\tau\mu^n(c)|_1 = \alpha_c \, n^p r^n + o(n^p r^n), \qquad |\tau\mu^n(c)| = \gamma_c \, n^p r^n + o(n^p r^n),
$$

where

$$\alpha_u = \sum_{i=1}^{|u|} w_{1,u_i}, \qquad\qquad \gamma_u = \sum_{i=1}^{|u|} \sum_{g \in \Gamma} w_{g,u_i},$$

$$\alpha_c = w_{1,c}, \qquad\qquad \gamma_c = \sum_{g \in \Gamma} w_{g,c}.$$

Here the symbol u_i denotes the ith letter of u. In particular, $u_1 = 0$, and since W_0 is nonzero, we see that $\gamma_u > 0$. Similarly, $\gamma_c > 0$ since W_c is nonzero.

Since both u and uc are nonempty prefixes of $\mu(0)$, it follows that

$$\frac{\alpha_u}{\gamma_u} = \lim_{n \to \infty} \frac{|\tau \mu^n(u)|_1}{|\tau \mu^n(u)|} = \phi_1$$

and

$$\frac{\alpha_u + \alpha_c}{\gamma_u + \gamma_c} = \lim_{n \to \infty} \frac{|\tau \mu^n(u)|_1 + |\tau \mu^n(c)|_1}{|\tau \mu^n(u)| + |\tau \mu^n(c)|} = \lim_{n \to \infty} \frac{|\tau \mu^n(uc)|_1}{|\tau \mu^n(uc)|} = \phi_1,$$

so that $\alpha_u / \gamma_u = \alpha_c / \gamma_c$. But

$$\frac{\alpha_c}{\gamma_c} = \frac{w_{1,c}}{\sum_{g \in \Gamma} w_{g,c}} = \rho_1,$$

and consequently $\phi_1 = \rho_1$, a contradiction. This concludes the proof.

We finish by stating two if-and-only-if condition as to whether the frequencies of all letters in a morphic sequence exist.

Corollary 2. *Let* $\mathbf{w} = \tau \mu^\omega(0)$ *be a morphic sequence and* μ *its simple generator. The frequencies of all letters in* $\mathbf{w}$ *exist if and only if the rank of* $W(\tau, \mu)$ *equals 1.*

Proof. By Theorem 2, the frequency of all letters exist if and only if

$$\frac{w_{1,a}}{\sum_{g \in \Gamma} w_{g,a}} = \frac{w_{1,0}}{\sum_{g \in \Gamma} w_{g,0}}$$

for all letters $a \in \Sigma$ for which W_a is nonzero and for all $1 \in \Gamma$. This is true if and only if

$$W_a = \frac{\sum_{g \in \Gamma} w_{g,0}}{\sum_{g \in \Gamma} w_{g,a}} W_0$$

for all nonzero columns W_a. Finally, this is equivalent to the condition that the rank of $W(\tau, \mu)$ is 1 (note that W_0 is nonzero). $\qquad\qquad\qquad$

Corollary 3. *Let* $\mathbf{w} = \mu^\omega(0)$ *be a pure morphic sequence and* μ *its simple generator. Then the frequency of all letters in* $\mathbf{w}$ *exists if and only if the number of occurrences of the dominating Jordan block in* $J(\mu)$ *equals 1.*

Proof. Since now $W(\mu) = E(\mu)$, the claim follows from Lemma 3 and Corollary 2. $\qquad\qquad\qquad$

Acknowledgements

Research for this paper was done while the author enjoyed the hospitality of the School of Computing, Waterloo, Canada. During the visit, the author greatly benefited from discussions with Prof. Jeff Shallit.

The comments and remarks by the anonymous reviewers are much appreciated.

References

1. Allouche, J-P.; Shallit, J. Automatic sequences. Theory, applications, generalizations. *Cambridge University Press, Cambridge*, 2003.
2. Apostol, T.M. Introduction to the Analytic Number Theory. *Springer-Verlag, New York-Heidelberg*, 1976.
3. Dekking, F.M. On the Thue-Morse measure. *Acta Univ. Carolin. Math. Phys.* 33 (1992) 35–40.
4. Frid, A. On the frequency of factors in a DOL word. *J. Autom. Lang. Comb.* 3 (1998), no. 1, 29–41. 29–41,
5. Gantmacher, F. R. The theory of matrices. Vols. 1, 2. *Chelsea Publishing Co., New York* 1959.
6. Michel, P. Sur les ensembles minimaux engendrés par les substitutions de longueur non constante. Ph. D. Thesis, Université de Rennes, 1975.
7. Michel, P. Stricte ergodicité d'ensembles minimaux de substitution. In J.-P. Conze and M. S. Keane, editors, *Théorie Ergodique: Actes des Journées Ergodiques, Rennes 1973/1974*, Vol. 532 of Lecture Notes in Mathematics, pp. 189-201. Springer-Verlag, 1976
8. Minc, H. Nonnegative matrices. Wiley-Interscience Series in Discrete Mathematics and Optimization. A Wiley-Interscience Publication. *John Wiley & Sons, Inc., New York*, 1988.
9. Peter, M. The asymptotic distribution of elements in automatic sequences. *Theoretical Computer Science 301, 2003*, 285–312.
10. Pullman, N. J. Matrix theory and its applications. Selected topics. Pure and Applied Mathematics, Vol. 35. *Marcel Dekker, Inc., New York-Basel*, 1976.
11. Queffélec, M. Substitution dynamical systems—spectral analysis. Lecture Notes in Mathematics, 1294. *Springer-Verlag, Berlin*, 1987.
12. Saari, K. On the Frequency of Letters in Pure Binary Morphic Sequences. *Developments in language theory, 2005*, 397–408.

Functional Equations in Shostak Theories

Sergey P. Shlepakov

Moscow State University, Russia

1 Introduction

We consider Shostak theories introduced in [1]. The class of Shostak theories consists of decidable first order equality theories, specified by two algorithms: a canoniser and a solver. A canoniser calculates the normal form of a term. A solver tests whether an equality can be reduced to an equivalent substitution and constructs this substitution when it exists. The examples of Shostak theories are linear arithmetics of integers and rational numbers, theories of lists, arrays, ets. ([2]).

In [1] the combinations of Shostak theories were considered and it was proposed to combine the solvers of components into a joint solver for the combination of theories. It was shown later ([3]) that this is impossible for arbitrary Shostak theories, so some special cases should be considered. One of the most discussed case ([4], [5]) is the combination of a Shostak theory in a signature Ω with a pure theory of equality in a signature Ω', $\Omega \cap \Omega' = \emptyset$. This is the case under consideration.

It can be expressed in some other way. Let us consider the second order language in the same signature Ω, i.e. we extend the language by second order functional variables $G = \{g_i\}$. A set of equations in combined theory became the system of second order equations in Shostak theory. This formulation naturally leads to a different notion of a solution: the corresponding substitution must act on second order variables too.

We express the values for second order functional variables by consistent infinite substitutions. We propose an algorithm, that determines whether a system of equations has a solution in this sense (a unifier), and returns the finite codes for the functions that constitute the solution. We also propose an algorithm that computes the values of such a function given its code and its arguments. It is shown that the solution θ constructed by the first algorithm is the most general one. Namely, every unifier of the system has the form $\lambda\theta$ for some substitution λ. We also prove that the unifiability of a system of second order equations is equivalent to its satisfiability in the standard term model for the second order language. Finally we apply this approach to the validity problem for Horn sentences with second order variables.

Let Ω be a signature that consists of function symbols f_i and constants c_j. Let Var be a countable set of first order variables. By $Tm^{(1)}$ we denote the set of all (first order) terms constructed from Ω and Var.

For a vector of terms $\bar{t} = (t_1, \ldots, t_n)$ and a map $\sigma : Tm^{(1)} \to Tm^{(1)}$ let $\sigma\bar{t} = (\sigma t_1, \ldots, \sigma t_n)$.

D. Grigoriev, J. Harrison, and E.A. Hirsch (Eds.): CSR 2006, LNCS 3967, pp. 346–351, 2006.

A *substitution* is a map $\sigma : Tm^{(1)} \to Tm^{(1)}$ such that $\sigma c_j = c_j$ and $\sigma f_i(\bar{t}) = f_i(\sigma \bar{t})$ hold for all $\bar{t}$. In general, the set

$$Dom(\sigma) = \{\, x \in Var \mid \sigma x \neq x \,\}$$

for a substitution σ may be infinite. A substitution σ is called *finite*, if $Dom(\sigma)$ is finite. We write $\sigma = [t_i/x_i]$ if $\sigma x_i = t_i$ and the sequence $\{x_i\}$ enumerates $Dom(\sigma)$. A finite substitution $\epsilon = [t_i/x_i]$ can be represented by the formula $\bigwedge_{x_i \in Dom(\epsilon)}(x_i = t_i)$ that will be denoted by the same letter ϵ when it does not lead to ambiguity.

For $\sigma = [t_i/x_i]$ let $Var(\sigma) = Dom(\sigma) \cup \bigcup_i Var(t_i)$.

2 Shostak Theories

The definition of Shostak theory involves two algoriths: canoniser and solver.

Definition 1. *A* canoniser π *is a computable map of* $Tm^{(1)}$ *to* $Tm^{(1)}$ *such that*
  1) $\pi x = x$,
  2) $Var(\pi t) \subseteq Var(t)$,
  3) $\pi \sigma = \pi \sigma \pi$,
  4) $\pi t = f(t_1, \ldots, t_n) \Rightarrow \pi t_i = t_i$,
  5) $\pi f(\bar{t}) = \pi f(\pi \bar{t})$.
Here x *is a variable,* $f \in \Omega$, $t, t_i \in Tm^{(1)}$ *and* σ *is a substitution.*

With a canoniser π we associate the first order equality theory T_π (in the signature Ω) with axioms $t = \pi t$, $t \in Tm^{(1)}$.

Lemma 1. *Let* a, b *be a terms.* $T_\pi \vdash a = b$ *iff* $\pi a = \pi b$.

For a finite set $W \subset Var$ and a fomula φ let $(\exists^W)\varphi$ denote the formula $(\exists \bar{x})\varphi$, where $\bar{x}$ is the list of all variables from W.

Definition 2. *Let* $a, b \in Tm^{(1)}$ *and* $U \subset Var^{(1)}$ *be an infinite decidable set of variables such that* $Var(a,b) \cap U = \emptyset$. *A solver is a procedure* $solve(U, a = b)$ *with the following properties. It returns* $\bot$ *when* $T_\pi \nvdash (\exists^{Var(a=b)})a = b$. *Otherwise it returns a finite idempotent substitution* ϵ, *such that:*
  1) $Dom(\epsilon) \subseteq Var(a = b)$,
  2) $W = Var(\epsilon) \setminus Var(a = b) \subseteq U$,
  3) $T_\pi \vdash (\exists^W)\epsilon \leftrightarrow a = b$.

The last condition means that ϵ is a solution of the equation $a = b$. We admit parametric solutions too and U provides "fresh" variable identifiers for them.

A *Shostak theory* is a theory of the form T_π for some canoniser π that has a solver *solve* ([2]).

Example 1. (Convex theory of lists [2]). The signature consists of a function symbols *car* and *cdr*, that return the head and the tail of the list, and a binary function symbol *cons*, that constructs the list given its head and tail.

The canoniser π: apply the following reduction rules

$$cons(car(x), cdr(x)) \rightarrow x, \quad car(cons(x,y)) \rightarrow x, \quad cdr(cons(x,y)) \rightarrow y$$

while it is possible. (The result does not depend on the order of the reductions.)

The solver works as follows. Let the equation be $car(x) = cdr(y)$ and $x, y \notin U$, $u \in U$. It is provable in T_π that

$$car(x) = cdr(y) \Leftrightarrow \exists u \, (x = cons(cdr(y), u)),$$

so $solve(U, car(x) = cdr(y))$ returns $[cons(cdr(y), u)/x]$. Note that the equation has no idempotent solution ϵ with $Var(\epsilon) \subseteq \{x, y\}$.

The standard variable elimination method extends the solver procedure to the case of a finite system of first order equations $S = \{a_i = b_i\}$. The substitution returned by $solve(U, S)$ will also have the properties similar to 1-3 above. We show that this solution will be the most general one in the following sence.

A *unifier* for the first order system $S = \{a_i = b_i\}$ is a finite idempotent substitution σ such that $\pi\sigma a_i = \pi\sigma b_i$. A *most general unifier for S with respect to U* is a unifier σ for S such that for any unifier ϵ for S holds $\pi\epsilon t = \pi\sigma_\epsilon \sigma t$ for some substitution σ_ϵ and all terms t with $Var^{(1)}(t) \cap U = \emptyset$.

Theorem 1. *Let $U \subset Var$ be infinite and $U \cap Var(S) = \emptyset$. If $T_\pi \vdash (\exists^{Var(S)})S$ then there exists a most general unifier for S w.r.t. U.*

3 The Second Order Language

Let Ω be a signature, $Var^{(1)}$ be a set of first order variables, $G = \{ g_1, g_2, \dots \}$ be a set of second order function variables. The set of second order terms is defined by the following grammar

$$Tm^{(2)} ::= Var^{(1)} \mid c_i \mid f_i(Tm^{(2)}, \dots, Tm^{(2)}) \mid g_i(Tm^{(2)}, \dots, Tm^{(2)}).$$

Any term from the set $Var^{(2)} = Var^{(1)} \cup \{ g_i(\bar{t}) \}$ is called *a solvable* ([5]). A second order solvable is a solvable of the form $g_i(\bar{t})$. Let T^+ be a set of all solvables that are subterms of terms from $T \subset Tm^{(2)}$.

We say that the map $\pi : Tm^{(2)} \rightarrow Tm^{(2)}$ is *functionally consistent* if $\pi\bar{t}_1 = \pi\bar{t}_2$ implies $\pi f(\bar{t}_1) = \pi f(\bar{t}_2)$ and $\pi g(\bar{t}_1) = \pi g(\bar{t}_2)$ for all $f \in \Omega$, $g \in G$.

The notion of substitution remains the same as for the first order case. Let α be any map of $Tm^{(2)}$ to $Tm^{(2)}$. We say that a substitution σ is *α-consistent*, if $\alpha\sigma g(\bar{t}) = \alpha\sigma g(\alpha\bar{t})$ holds for every second order solvable $g(\bar{t})$.

4 The Canoniser for $Tm^{(2)}$

Let $\lambda : Tm^{(2)} \rightarrow Tm^{(1)}$ be an infinite invertible substitution that replaces all solvables (of the forms $g_i(\bar{t})$ and $x_i \in Var^{(1)}$) by the first order variables. We also suppose that λ and λ^{-1} are computable. Let π be a canoniser.

Lemma 2. *A map $\varkappa : Tm^{(2)} \to Tm^{(2)}$ and a substitution μ such that $\varkappa t = \lambda^{-1}\pi\lambda\mu t$, $\mu = \{\, g(\varkappa \bar{t}_1)/g(\bar{t}_1) \mid g(\bar{t}_1) \in Var^{(2)} \,\}$ do exist and are unique.*

Lemma 3. *Let $t_1, t_2 \in Tm^{(1)}$. Then $\pi t_1 = \pi t_2 \Leftrightarrow \varkappa t_1 = \varkappa t_2$.*

Theorem 2. *The map $\varkappa$ has the following properties (similar to the properties 1–5 of a canoniser):*

1) $\varkappa x = x$, for $x \in Var^{(1)}$;
2) $Var^{(1)}(\varkappa t) \subseteq Var^{(1)}(t)$, here $Var^{(1)}(t) = Var(t) \cap Var^{(1)}$;
3) $\varkappa \sigma t = \varkappa \sigma \varkappa t$ for any $\varkappa$-consistent substitution σ;
4) if $\varkappa t = f(t_1, \ldots, t_n)$, then $\varkappa t_i = t_i$;
5) $\varkappa f(\bar{t}) = \varkappa f(\varkappa \bar{t})$.

Thus, the computable map $\varkappa$ has essentially the same properties as a canoniser for the first order term. Moreover, the equivalence relation on $Tm^{(1)}$ induced by $\varkappa$ coincides with one induced by π. We will say that $\varkappa$ is a *generalised canoniser*.

5   Solutions for Second Order Equation Systems

Let $S = \{\, a_i = b_i \,\}$, $a_i, b_i \in Tm^{(2)}$ be a finite system of second order equations. Then $\{\, \lambda a_i = \lambda b_i \,\}$ is a first order system, so we can use the first order solver to solve it: $solve(\lambda U, \{\, \lambda a_i = \lambda b_i \,\}) = [t_i/v_i]$. We may try to define the second order solver as follows:

$$solve^{(2)}(U, S) = [\lambda^{-1} t_i / \lambda^{-1} v_i] = \sigma.$$

The substitution σ satisfies the system: $\varkappa \sigma a_i = \varkappa \sigma b_i$. It is idempotent and $Dom(\sigma) \subseteq Var(S)$. But at the same time this $solve^{(2)}$ cannot guarantee the functional consistency condition. For example, it can return a substitution $\sigma = [1/g(x), 2/x, 3/g(2)]$ for which $\varkappa \sigma x = \varkappa \sigma 2$, but $\varkappa \sigma g(x) \neq \varkappa \sigma g(2)$. Moreover, $solve^{(2)}$ returns finite substitutions whereas a nontrivial functionally consistent substitution must be infinite. Below we provide a better way to solve the second order equation systems.

Definition 3. *Let $S = \{\, a_i = b_i \,\}$, $a_i, b_i \in Tm^{(2)}$ be a finite system of second order equations. A unifier of S is an idempotent functionally consistent and $\varkappa$-consistent substitution ρ such that $\varkappa \rho a_i = \varkappa \rho b_i$.*

Let $U \subset Var^{(1)}$ and $(Var(S))^+ \cap U = \emptyset$. A weak most general unifier of S with respect to U (weak mgu of S w.r.t. U) is a unifier ρ such that for any unifier ρ_1 of S there exists substitution ρ_2 for which $\varkappa \rho_1 t = \varkappa \rho_2 \rho t$ holds for all terms t with $(Var(t))^+ \cap U = \emptyset$.

A week mgu ρ of S w.r.t. U is called stable if $\mu \rho$ is a weak mgu of S w.r.t. U too.

6   The Construction of a Unifier

We construct a unifier for S in two steps. At first we construct some finite substitution. Then we extend it using the extension procedure specified below. The result will be an inifinite substitution that will be a stable weak mgu for S. Now we introduce the extension procedure.

Let $W \subset Var^{(2)}$ be finite. We construct an inifinite functionally consistent substitution $\bar{\sigma}$ for which $\sigma v = \bar{\sigma} v$ holds when $v \in W$.

A substitution σ is called $\langle \varkappa, W \rangle$-consistent if for all $g_k(\bar{t}_1), g_k(\bar{t}_2) \in W$

$$\varkappa \sigma \bar{t}_1 = \varkappa \sigma \bar{t}_2 \Rightarrow \sigma g_k(\bar{t}_1) = \sigma g_k(\bar{t}_2),$$
$$\varkappa \bar{t}_1 = \varkappa \bar{t}_2 \Rightarrow \sigma g_k(\bar{t}_1) = \sigma g_k(\bar{t}_2).$$

Lemma 4. *Let* $W \subset Var^{(2)}$ *be finite,* $W = W^+$, σ *be a* $\langle \varkappa, W \rangle$-*consistent substitution, and* $(Var(\sigma))^+ \subseteq W$. *Then*

1) *there exists unique substituition* $\bar{\sigma}$ *such that*
$$\bar{\sigma}x = \quad \sigma x, \qquad x \in Var^{(1)},$$
$$\bar{\sigma}g_k(\bar{t}) = \begin{cases} \sigma g_k(\bar{t}), & g_k(\bar{t}) \in W; \\ \sigma g_k(\bar{t}_1), & g_k(\bar{t}) \notin W \text{ and } \exists g_k(\bar{t}_1) \in W : \varkappa \bar{\sigma} \bar{t} = \varkappa \bar{\sigma} \bar{t}_1; \\ g_k(\varkappa \bar{\sigma} \bar{t}), & g_k(\bar{t}) \notin W \text{ and } \neg \exists g_k(\bar{t}_1) \in W : \varkappa \bar{\sigma} \bar{t} = \varkappa \bar{\sigma} \bar{t}_1; \end{cases}$$

2) $\bar{\sigma}$ *is* $\varkappa$-*consistent;*

3) $\sigma v = \bar{\sigma} v$ *for every* $v \in W$;

4) *if* σ *is idempotent, then* $\bar{\sigma}$ *is an idempotent and functionally consistent substitution;*

5) *if* σ *is idempotent and* θ *is an idempotent, functionally consistent and* $\varkappa$-*consistent substitution such that* $\theta|_W = \sigma|_W$, *then* $\varkappa \theta = \varkappa \theta \bar{\sigma}$.

Let $U \subset Var^{(2)}$ be an infinite decidable set of variables. Now we introduce the unification algorithm $unify(U, S)$ for finite systems $S = \{ a_i = b_i \}$, $a_i, b_i \in Tm^{(2)}$ with $U \cap (Var(S))^+ = \emptyset$. It tests the unifiability of S and in the positive case returns some finite substitution which can be used as a finite core of a unifier.

1. Let $i = 0$, $S_0 = \varkappa S$.
2. If $solve^{(2)}(U, S_i) = \bot$ then terminate and return $\bot$. Else set $\sigma_i = solve^{(2)}(U, S_i)$.
3. If there exists a solvables $g_k(\bar{t}_1), g_k(\bar{t}_2) \in (Var(\varkappa S))^+$ such that $\varkappa \sigma_i \bar{t}_1 = \varkappa \sigma_i \bar{t}_2$, but $\varkappa \sigma_i g_k(\bar{t}_1) \neq \varkappa \sigma_i g_k(\bar{t}_2)$, then set $S_{i+1} = S_i \cup \{g_k(\bar{t}_1) = g_k(\bar{t}_2)\}$, $i = i + 1$, and goto step 2. If there is no such solvables, goto step 4.
4. Terminate with the result $\sigma = [\varkappa t_j / v_j]$ where $[t_j / v_j] = \sigma_i$.

Theorem 3. *1)* $unify(U, S)$ *always terminates.*

2) If S has no unifier then $unify(U, S)$ *returns* $\bot$.

3) If S has a unifier then $unify(U, S)$ *returns a substitution* σ *such that* $Var(\sigma) \setminus Var(\varkappa S) \subset U$.

4) If S has a unifier then the resulting substitution σ *is* $\langle \varkappa, W \rangle$-*consistent for* $W = (Var(\varkappa S))^+ \cup Var^{(1)}(\sigma)$ *and* $\bar{\sigma}$ *is a stable weak mgu for S w.r.t. U.*

7   The Standard Term Model

Here we define a standart term model $I_\varkappa^2$ for the second order theory and show that a system of equations is unifiable iff it is satisfiable in this model. We also construct an operator *mod* which provides the validity test for Horn clauses.

Definition 4. *Let Ω be a signature and G be a set of second oder variables. The* standart term model $I_\varkappa^2$ *(for the second order language) is the model with domain $D_\varkappa^2 = \{\ \varkappa t \mid t \in Tm^{(2)}\ \}$. The constants and the function symbols from Ω are interpreted as follows: $c_j^{I_\varkappa^2} = \varkappa c_j,\ \ f_j^{I_\varkappa^2}(\bar{t}) = \varkappa f_j(\bar{t}).$*

Definition 5. *A system of equations $S = \{a_i = b_i\}$, $a_i, b_i \in Tm^{(2)}$ is satisfiable in a model I if $I \models (\exists^{Var(S)})S$.*

Theorem 4. *A system of equations $S = \{a_i = b_i\}$, $a_i, b_i \in Tm^{(2)}$ is satisfiable in the standard term model $I_\varkappa^2$ iff S has a unifier.*

Now we introduce a computable operation mod_U and demonstrate how it can be used for testing the validity of Horn clauses in the standard term model. Let a unifiable system of equations $S = \{a_i = b_i\}$, $a_i, b_i \in Tm^{(2)}$ and an infinite decidable set $U \subset Var^{(2)}$ be fixed and $Var(S)^+ \cap U = \emptyset$. Suppose ζ be a stable weak mgu of S w.r.t. U (for example, the week mgu from Theorem 3).

Definition 6. $t \bmod_U S = \varkappa \zeta t$.

Theorem 5. *Let $a, b \in Tm^{(2)}$ and $Var(a, b) \cap U = \emptyset$.*

$$I_\varkappa^2 \models (S \to a = b) \quad \Leftrightarrow \quad a \bmod_U S = b \bmod_U S.$$

References

1. Shostak, R.E.: Deciding combinations of theories. Journal of the ACM **31** (1984) 1–12
2. Manna, Z., Zarba, C.G.: Combining decision procedures. In: Formal Methods at the Cross Roads: From Panacea to Foundational Support. Volume 2757 of Lecture Notes in Computer Science., Springer (2003) 381–422
3. Krstić, S., Conchon, S.: Canonization for disjoint unions of theories. In Baader, F., ed.: Proceedings of the 19th International Conference on Automated Deduction (CADE-19). Volume 2741 of Lecture Notes in Computer Science., Miami Beach, FL, USA, Springer Verlag (2003)
4. Ganzinger, H.: Shostak light. In Voronkov, A., ed.: Proceedings of the 18th International Conference on Automated Deduction. Volume 2392 of Lecture Notes in Computer Science., Springer-Verlag (2002) 332–346
5. Rueß, H., Shankar, N.: Deconstructing Shostak. In: Proceedings of the 16th Annual IEEE Symposium on Logic in Computer Science (Boston, Massachusetts, USA), IEEE Computer Society (2001) 19–28

All Semi-local Longest Common Subsequences in Subquadratic Time

Alexander Tiskin

Department of Computer Science, The University of Warwick,
Coventry CV4 7AL, United Kingdom
tiskin@dcs.warwick.ac.uk

Abstract. For two strings a, b of lengths m, n respectively, the longest common subsequence (LCS) problem consists in comparing a and b by computing the length of their LCS. In this paper, we define a generalisation, called "the all semi-local LCS problem", where each string is compared against all substrings of the other string, and all prefixes of each string are compared against all suffixes of the other string. An explicit representation of the output lengths is of size $\Theta\big((m+n)^2\big)$. We show that the output can be represented implicitly by a geometric data structure of size $O(m + n)$, allowing efficient queries of the individual output lengths. The currently best all string-substring LCS algorithm by Alves et al. can be adapted to produce the output in this form. We also develop the first all semi-local LCS algorithm, running in time $o(mn)$ when m and n are reasonably close. Compared to a number of previous results, our approach presents an improvement in algorithm functionality, output representation efficiency, and/or running time.

1 Introduction

Given two strings a, b of lengths m, n respectively, the longest common subsequence (LCS) problem consists in comparing a and b by computing the length of their LCS. In this paper, we define a generalisation, called "the all semi-local LCS problem", where each string is compared against all substrings of the other string, and all prefixes of each string are compared against all suffixes of the other string. The all semi-local LCS problem arises naturally in the context of LCS computations on substrings. It is closely related to local sequence alignment (see e.g. [7, 9]) and to approximate string matching (see e.g. [6, 15]).

A standard approach to string comparison is representing the problem as a dag (directed acyclic graph) of size $\Theta(mn)$ on an $m \times n$ grid of nodes. The basic LCS problem, as well as its many generalisations, can be solved by dynamic programming on this dag in time $O(mn)$ (see e.g. [6, 7, 15, 9]). It is well-known (see e.g. [13, 1] and references therein) that all essential information in the grid dag can in fact be represented by a data structure of size $O(m + n)$. In this paper, we expose a rather surprising (and to the best of our knowledge, previously unnoticed) connection between this linear-size representation of the string comparison dag, and a standard computational geometry problem known as dominance counting.

D. Grigoriev, J. Harrison, and E.A. Hirsch (Eds.): CSR 2006, LNCS 3967, pp. 352–363, 2006.

If the output lengths of the all semi-local LCS problem are represented explicitly, the total size of the output is $\Theta((m+n)^2)$, corresponding to m^2+n^2 possible substrings and $2mn$ possible prefix-suffix pairs. To reduce the storage requirements, we allow the output lengths to be represented implicitly by a smaller data structure that allows efficient retrieval of individual output values. Using previously known linear-size representations of the string comparison dag, retrieval of an individual output length typically requires scanning of at least a constant fraction of the representing data structure, and therefore takes time $O(m + n)$. By exploiting the geometry connection, we show that the output lengths can be represented by a set of $m + n$ grid points. Individual output lengths can be obtained from this representation by dominance counting queries. This leads to a data structure of size $O(m+n)$, that allows to query an individual output length in time $O\left(\frac{\log(m+n)}{\log\log(m+n)}\right)$, using a recent result by JaJa, Mortensen and Shi [8]. The described approach presents a substantial improvement in query efficiency over previous approaches.

It has long been known [14, 5] that the (global) LCS problem can be solved in subquadratic[1] time $O\left(\frac{mn}{\log(m+n)}\right)$ when m and n are reasonably close. Alves et al. [1], based on an idea of Schmidt [17], proposed an all string-substring (i.e. restricted semi-local) LCS algorithm that runs in time $O(mn)$. In this paper, we propose the first all semi-local LCS algorithm, which runs in subquadratic time $O\left(\frac{mn}{\log(m+n)^{1/2}}\right)$ when m and n are reasonably close. This improves on [1] simultaneously in algorithm functionality, output representation efficiency, and running time.

2   Previous Work

Although our generic definition of the all semi-local LCS problem is new, several algorithms dealing with similar problems involving multiple substring comparison have been proposed before. The standard dynamic programming approach can be regarded as comparing all prefixes of each string against all prefixes of the other string. Papers [17, 11, 1] present several variations on the theme of comparing substrings (prefixes, suffixes) of two strings. In [12, 10], the two input strings are revealed character by character. Every new character can be either appended or prepended to the input string. Therefore, the computation is performed essentially on substrings of subsequent inputs. In [13], multiple strings sharing a common substring are compared against a common target string. A common feature in many of these algorithms is the use of linear-sized string comparison dag representation, and a suitable merging procedure that "stitches together" the representations of neighbouring dag blocks to obtain a representation for the blocks' union. As a consequence, such algorithms could be adapted to work with our new, potentially more efficient geometric representation, without any increase in asymptotic time or memory requirements.

---

[1] The term "subquadratic" here and in the paper's title refers to the case $m = n$.

3   Problem Statement and Notation

We consider strings of characters from a fixed finite alphabet, denoting string concatenation by juxtaposition. Given a string, we distinguish between its contiguous *substrings*, and not necessarily contiguous *subsequences*. Special cases of a substring are *a prefix* and *a suffix* of a string. For two strings $a = \alpha_1 \alpha_2 \ldots \alpha_m$ and $b = \beta_1 \beta_2 \ldots \beta_n$ of lengths m, n respectively, the *longest common subsequence (LCS) problem* consists in computing the LCS length of a and b.

We consider a generalisation of the LCS problem, which we call the *all semi-local LCS problem*. It consists in computing the LCS lengths on substrings of a and b as follows:

- the *all string-substring LCS problem*: a against every substring of b;
- the *all prefix-suffix LCS problem*: every prefix of a against every suffix of b;
- symmetrically, the *all substring-string LCS problem* and the *all suffix-prefix LCS problem*, defined as above but with the roles of a and b exchanged.

It turns out that by considering this combination of problems rather than each problem separately, the algorithms can be greatly simplified.

A traditional distinction, especially in computational biology, is between global (full string against full string) and local (all substrings against all substrings) comparison. Our problem lies in between, hence the term "semi-local". Many string comparison algorithms output either a single optimal comparison score across all local comparisons, or a number of local comparison scores that are "sufficiently close" to the globally optimal. In contrast with this approach, we require to output all the locally optimal comparison scores.

In addition to standard integer indices $\mathbb{Z} = \{\ldots, -2, -1, 0, 1, 2, \ldots\}$, we use *odd half-integer*[2] indices $\hat{\mathbb{Z}} = \{\ldots, -\frac{5}{2}, -\frac{3}{2}, -\frac{1}{2}, \frac{1}{2}, \frac{3}{2}, \frac{5}{2}, \ldots\}$. For two numbers i, j, we write $i \trianglelefteq j$ if $j - i \in \{0, 1\}$, and $i \triangleleft j$ if $j - i = 1$. We denote

$$[i : j] = \{i, i + 1, \ldots, j - 1, j\}$$
$$\langle i : j \rangle = \left\{ i + \tfrac{1}{2}, i + \tfrac{3}{2}, \ldots, j - \tfrac{3}{2}, j - \tfrac{1}{2} \right\}$$

4   Problem Analysis

It is well-known that an instance of the LCS problem can be represented by a dag (directed acyclic graph) on an $m \times n$ grid of nodes, where character matches correspond to edges of weight 1, and gaps to edges of weight 0. To describe our algorithms, we need a slightly extended version of this representation, where the representing dag is embedded in an infinite grid dag.

Definition 1. *Let* $m, n \in \mathbb{N}$. *A grid dag* G *is a weighted dag, defined on the set of nodes* $v_{i,j}$, $i \in [0 : m]$, $j \in [0 : n]$. *For all* $i \in [1 : m]$, $j \in [1 : n]$:

---

[2] It would be possible to reformulate all our results using only integers. However, using odd half-integers helps to make the exposition simpler and more elegant.

- *horizontal edge $v_{i,j-1} \rightarrow v_{i,j}$ and vertical edge $v_{i-1,j} \rightarrow v_{i,j}$ are both always present in G and have weight 0;*
- *diagonal edge $v_{i-1,j-1} \rightarrow v_{i,j}$ may or may not be present in G; if present, it has weight 1.*

Given an instance of the all semi-local LCS problem, its *corresponding grid dag* is an $m \times n$ grid dag, where the diagonal edge $v_{i-1,j-1} \rightarrow v_{i,j}$ is present, iff $\alpha_i = \beta_j$.

Common string-substring, suffix-prefix, prefix-suffix, and substring-string subsequences correspond, respectively, to paths of the following form in the grid dag:

$$v_{0,j} \rightsquigarrow v_{m,j'} \quad v_{i,0} \rightsquigarrow v_{m,j'} \quad v_{0,j} \rightsquigarrow v_{i',n} \quad v_{i,0} \rightsquigarrow v_{i',n} \tag{1}$$

where $i, i' \in [0 : m]$, $j, j' \in [0 : n]$. The length of each subsequence is equal to the weight of its corresponding path. The solution to the all semi-local LCS problem is equivalent to finding the weight of a maximum-weight path of each of the four types (1) between every possible pair of endpoints. (Since the graph is acyclic, this is also equivalent to finding the weight of the corresponding minimum-weight path in a grid dag where all the weights are negated.)

Definition 2. *Given an $m \times n$ grid dag G, its extension G^+ is an infinite weighted dag, defined on the set of nodes $v_{i,j}$, $i, j \in \mathbb{Z}$ and containing G as a subgraph. For all $i, j \in \mathbb{Z}$:*

- *horizontal edge $v_{i,j-1} \rightarrow v_{i,j}$ and vertical edge $v_{i-1,j} \rightarrow v_{i,j}$ are both always present in G^+ and have weight 0;*
- *when $i \in [1 : m]$, $j \in [1 : n]$, diagonal edge $v_{i-1,j-1} \rightarrow v_{i,j}$ is present in G^+ iff it is present in G; if present, it has weight 1;*
- *otherwise, diagonal edge $v_{i-1,j-1} \rightarrow v_{i,j}$ is always present in G^+ and has weight 1.*

An infinite dag that is an extension of some (finite) grid dag will be called an *extended grid dag*. When dag G^+ is the extension of dag G, we will say that G is the *core* of G^+. Relative to G^+, we will call the nodes of G *core nodes*.

By using the extended grid dag representation, the four path types (1) can be reduced to a single type, corresponding to the all string-substring (or, symmetrically, substring-string) LCS problem on an extended set of indices.

Definition 3. *Given an $m \times n$ grid dag G, its extended horizontal (respectively, vertical) score matrix is an infinite matrix defined by*

$$A(i,j) = \max weight(v_{0,i} \rightsquigarrow v_{m,j}) \qquad i,j \in \mathbb{Z} \tag{2}$$
$$A^*(i,j) = \max weight(v_{i,0} \rightsquigarrow v_{j,n}) \qquad i,j \in \mathbb{Z} \tag{3}$$

where the maximum is taken across all paths between the given endpoints in the extension G^+. If $i = j$, we have $A(i,j) = 0$. By convention, if $j < i$, then we let $A(i,j) = j - i < 0$.

The maximum path weights for each of the four path types (1) can be obtained from the extended horizontal score matrix (2) as follows:

$$\max weight(v_{0,j} \rightsquigarrow v_{m,j'}) = A(j,j')$$
$$\max weight(v_{i,0} \rightsquigarrow v_{m,j'}) = A(-i,j') - i$$
$$\max weight(v_{0,j} \rightsquigarrow v_{i',n}) = A(j,m+n-i') - m + i'$$
$$\max weight(v_{i,0} \rightsquigarrow v_{i',n}) = A(-i,m+n-i') - m - i + i'$$

where $i,i' \in [0:m]$, $j,j' \in [0:n]$, and the maximum is taken across all paths between the given endpoints. The same maximum path weights can be obtained analogously from the extended vertical score matrix (3).

For most of this section, we will concentrate on the properties of extended horizontal score matrices, referring to them simply as "extended score matrices". By symmetry, extended vertical score matrices will have analogous properties. We assume $i,j \in \mathbb{Z}$, unless indicated otherwise.

Theorem 1. *An extended score matrix has the following properties:*

$$A(i,j) \trianglelefteq A(i-1,j); \tag{4}$$
$$A(i,j) \trianglelefteq A(i,j+1); \tag{5}$$
$$\text{if } A(i,j+1) \triangleleft A(i-1,j+1), \text{ then } A(i,j) \triangleleft A(i-1,j); \tag{6}$$
$$\text{if } A(i-1,j) \triangleleft A(i-1,j+1), \text{ then } A(i,j) \triangleleft A(i,j+1). \tag{7}$$

Proof. A path $v_{0,i-1} \rightsquigarrow v_{m,j}$ can be obtained by first following a horizontal edge of weight 0: $v_{0,i-1} \rightarrow v_{0,i} \rightsquigarrow v_{m,j}$. Therefore, $A(i,j) \leq A(i-1,j)$. On the other hand, any path $v_{0,i-1} \rightsquigarrow v_{m,j}$ consists of a subpath $v_{0,i-1} \rightsquigarrow v_{l,i}$ of weight at most 1, followed by a subpath $v_{l,i} \rightsquigarrow v_{m,j}$. Therefore, $A(i,j) \geq A(i-1,j) - 1$. We thus have (4) and, by symmetry, (5).

A crossing pair of paths $v_{0,i} \rightsquigarrow v_{m,j}$ and $v_{0,i-1} \rightsquigarrow v_{m,j+1}$ can be rearranged into a non-crossing pair of paths $v_{0,i-1} \rightsquigarrow v_{m,j}$ and $v_{0,i} \rightsquigarrow v_{m,j+1}$. Therefore, we have *the Monge property*:

$$A(i,j) + A(i-1,j+1) \leq A(i-1,j) + A(i,j+1)$$

Rearranging the terms

$$A(i-1,j+1) - A(i,j+1) \leq A(i-1,j) - A(i,j)$$

and applying (4), we obtain (6) and, by symmetry, (7). $\square$

The properties of Theorem 1 are symmetric with respect to i and $n-j$. Alves et al. [1] introduce the same properties but do not make the most of their symmetry. We aim to exploit symmetry to the full.

Corollary 1. *An extended score matrix has the following properties:*

if $A(i,j) \triangleleft A(i-1,j)$, then $A(i,j') \triangleleft A(i-1,j')$ for all $j' \leq j$;
if $A(i,j) = A(i-1,j)$, then $A(i,j') = A(i-1,j')$ for all $j' \geq j$;
if $A(i,j) \triangleleft A(i,j+1)$, then $A(i',j) \triangleleft A(i',j+1)$ for all $i' \geq i$;
if $A(i,j) = A(i,j+1)$, then $A(i',j) = A(i',j+1)$ for all $i' \leq i$.

Proof. These are the well-known properties of matrix A and its transpose A^T being *totally monotone*. In both pairs, the properties are each other's contrapositive, and follow immediately from Theorem 1. $\qquad\square$

Informally, Corollary 1 says that the inequality between the corresponding elements in two successive rows (respectively, columns) "propagates to the left (respectively, downwards)", and the equality "propagates to the right (respectively, upwards)". Recall that by convention, $A(i,j) = j - i$ for all index pairs $j < i$. Therefore, we always have an inequality between the corresponding elements in successive rows or columns in the lower triangular part of matrix A. If we fix i and scan the set of indices j from $j = -\infty$ to $j = +\infty$, an inequality may change to an equality at most once. We call such a value of j *critical* for i. Symmetrically, if we fix j and scan the set of indices i from $i = +\infty$ to $j = -\infty$, an inequality may change to an equality at most once, and we can identify values of i that are critical for j. Crucially, for all pairs (i,j), index i will be critical for j if and only if index j is critical for i. This property lies at the core of our method, which is based on the following definition.

Definition 4. *An odd half-integer point* $(i,j) \in \hat{\mathbb{Z}}^2$ *is called A-critical, if*

$$A\left(i + \tfrac{1}{2}, j - \tfrac{1}{2}\right) \vartriangleleft A\left(i - \tfrac{1}{2}, j - \tfrac{1}{2}\right) = A\left(i + \tfrac{1}{2}, j + \tfrac{1}{2}\right) = A\left(i - \tfrac{1}{2}, j + \tfrac{1}{2}\right)$$

In particular, point (i,j) is never A-critical for $i > j$. When $i = j$, point (i,j) is A-critical iff $A\left(i - \tfrac{1}{2}, j + \tfrac{1}{2}\right) = 0$.

Corollary 2. *Let* $i, j \in \hat{\mathbb{Z}}^2$. *For each i (respectively, j), there exists at most one j (respectively, i) such that the point (i,j) is A-critical.*

Proof. By Corollary 1 and Definition 4. $\qquad\square$

We will represent an extended score matrix by its set of critical points. Such a representation is based on the following simple geometric concept.

Definition 5. *Point* (i_0, j_0) *dominates*[3] *point* (i,j), *if* $i_0 < i$ *and* $j < j_0$.

Informally, the dominated point is "below and to the left" of the dominating point in the score matrix[4]. The following theorem shows that the geometric representation of a score matrix is unique, and gives a simple formula for recovering matrix elements.

Theorem 2. *For an arbitrary integer point* $(i_0, j_0) \in \mathbb{Z}^2$, *let* $d_A(i_0, j_0)$ *denote the number of (odd half-integer) A-critical points it dominates. We have*

$$A(i_0, j_0) = j_0 - i_0 - d_A(i_0, j_0)$$

---

[3] The standard definition of dominance requires $i < i_0$ instead of $i_0 < i$. Our definition is more convenient in the context of the LCS problem.

[4] Note that these concepts of "below" and "left" are relative to the score matrix, and have no connection to the "vertical" and "horizontal" directions in the grid dag.

Proof. Induction on $j_0 - i_0$. Denote $d = d_A(i_0, j_0)$.

Induction base. Suppose $i_0 \geq j_0$. Then $d = 0$ and $A(i_0, j_0) = j_0 - i_0$.
Inductive step. Suppose $i_0 < j_0$. Let d' denote the number of critical points in $\langle i_0 : n \rangle \times \langle 0 : j_0 - 1 \rangle$. By the inductive hypothesis, $A(i_0, j_0 - 1) = j_0 - 1 - i_0 - d'$. We have two cases:

1. There is a critical point $(i, j_0 - \frac{1}{2})$ for some $i \in \langle i_0 : n \rangle$. Then $d = d' + 1$ and $A(i_0, j_0) = A(i_0, j_0 - 1) = j_0 - i_0 - d$ by Corollary 1.
2. There is no such critical point. Then $d = d'$ and $A(i_0, j_0) = A(i_0, j_0 - 1) + 1 = j_0 - i_0 - d$ by Corollary 1.

In both cases, the theorem statement holds for $A(i_0, j_0)$.   □

Recall that outside the core, the structure of an extended grid graph is trivial: all possible diagonal edges are present in the non-core subgraph. This gives rise to an additional property: when $i < -m$ or $j > m + n$, point (i, j) is A-critical iff $j - i = m$. We will call such A-critical points *trivial*. It is easy to see that an A-critical point (i, j) is non-trivial, iff either both $v_{0,i-\frac{1}{2}}$ and $v_{0,i+\frac{1}{2}}$, or both $v_{m,j-\frac{1}{2}}$ and $v_{m,j+\frac{1}{2}}$, are core nodes.

Corollary 3. *There are exactly $m + n$ non-trivial A-critical points.*

Proof. We have $A(-m, m + n) = m$. On the other hand, $A(-m, m + n) = 2m + n - d_A(-m, m + n)$ by Theorem 2. Hence, the total number of non-trivial A-critical points is $d_A(-m, m + n) = m + n$.   □

Since only non-trivial critical points need to be represented explicitly, Theorem 2 allows a representation of an extended score matrix by a data structure of size $O(m + n)$. There is a close connection between this representation and the canonical structure of general Monge matrices (see e.g. [4]).

Informally, Theorem 2 says that the value $A(i_0, j_0)$ is determined by the number of A-critical points dominated by (i_0, j_0). This number can be obtained by scanning the set of all non-trivial critical points in time $O(m + n)$. However, much more efficient methods exist when preprocessing of the critical point set is allowed.

The dominance relation between points is a well-studied topic in computational geometry. The following theorems are derived from two relevant geometric results, one classical and one recent.

Theorem 3. *For an extended score matrix A, there exists a data structure which*

- *has size $O\big((m + n) \log(m + n)\big)$;*
- *can be built in time $O\big((m + n) \log(m + n)\big)$, given the set of all non-trivial A-critical points;*
- *allows to query an individual element of A in time $O\big(\log(m + n)^2\big)$.*

Proof. The structure in question is a 2D range tree [3] (see also [16]), built on the set of non-trivial critical points. There are $m + n$ non-trivial critical points, hence the total number of nodes in the tree is $O((m + n) \log(m + n))$. A dominance query on the set of non-trivial critical points can be answered by accessing $O(\log(m + n)^2)$ of the nodes. A dominance query on the set of trivial critical points can be answered by a simple constant-time index calculation (note that the result of such a query can only be non-zero when the query point lies outside the core subgraph of the extended grid dag). The sum of the two dominance queries above provides the total number of critical points dominated by the query point (i_0, j_0). The value $A(i_0, j_0)$ can now be obtained by Theorem 2.   □

Theorem 4. *For an extended score matrix A, there exists a data structure which*

- *has size $O(m + n)$;*
- *allows to query an individual element of A in time $O\left(\frac{\log(m+n)}{\log\log(m+n)}\right)$.*

Proof. As above, but the range tree is replaced by the asymptotically more efficient data structure of [8].   □

While the data structure used in Theorem 4 provides better asymptotics, the range tree used in Theorem 3 is simpler, requires a less powerful computation model, and is more likely to be practical.

We conclude this section by formulating yet another previously unexploited symmetry of the all semi-local LCS problem, which will also become a key ingredient of our algorithm. This time, we consider both the horizontal score matrix A as in (2), and the vertical score matrix A^* as in (3). We show a simple one-to-one correspondence between the geometric representations of A and A^*, allowing us to switch easily between these representations.

Theorem 5. *Point (i, j) is A-critical, iff point $(-i, m + n - j)$ is A^*-critical.*

Proof. Straightforward case analysis based on Definition 4.   □

5   The Algorithm

We now describe an efficient algorithm for the all semi-local LCS problem. We follow a divide-and-conquer approach, which refines the framework for the string-substring LCS problem developed in [17, 1].

Strings a, b are recursively partitioned into substrings. Without loss of generality, consider a partitioning $a = a_1 a_2$ into a concatenation of two substrings of length m_1, m_2, where $m_1 + m_2 = m$. Let A, B, C denote the extended score matrices for the all semi-local LCS problems comparing respectively a_1, a_2, a against b. In every recursive call our goal is, given matrices A, B, to compute matrix C efficiently. We call this procedure *merging*. Trivially, merging can be performed in time $O((m + n)^3)$ by standard matrix multiplication over the (max, +)-semiring. By exploiting the Monge property of the matrices, the time complexity of merging can be easily reduced to $O((m + n)^2)$, which is optimal if the matrices are represented explicitly. We show that a further reduction in the time complexity of merging is possible, by using the data representation and algorithmic ideas introduced in Section 4.

By Theorem 2, matrices A, B, C can each be represented by the sets of respectively $m_1 + n$, $m_2 + n$, $m + n$ non-trivial critical points. Alves et al. [1] use a similar representation; however, for their algorithm, n critical points per matrix are sufficient. They describe a merging procedure for the special case $m_1 = 1$ (or $m_2 = 1$), that runs in time $O(n)$. On the basis of this procedure, they develop a string-substring LCS algorithm that runs in time $O(mn)$, and produces a data structure of size $O(n)$, which can be easily converted into the critical point set for the output matrix. By adding a post-processing phase based on Theorems 3, 4, this algorithm can be adapted to produce a query-efficient output data structure.

Our new algorithm is based on a novel merging procedure, which works for arbitrary values m_1, m_2.

Lemma 1. *Given subproblems with extended score matrices A, B, C as described above, the sets of A- and B-critical points can be merged into the set of C-critical points in time $O(m + n^{1.5})$ and memory $O(m + n)$.*

Proof. Our goal is to compute the set of all non-trivial C-critical points. Without loss of generality, we may assume that $2m_1 = 2m_2 = m$, and that n is a power of 2 (otherwise, appropriate padding can be applied to the input). We will compute non-trivial C-critical points in two stages:

1. points $(i, k) \in \langle -m, -\frac{m}{2} \rangle \times \langle 0, \frac{m}{2} + n \rangle \cup \langle -\frac{m}{2}, n \rangle \times \langle \frac{m}{2} + n, m + n \rangle$;
2. points $(i, k) \in \langle -\frac{m}{2}, n \rangle \times \langle 0, \frac{m}{2} + n \rangle$.

It is easy to see that every non-trivial C-critical point (i, j) is computed in either the first or the second stage. Informally, each C-critical point in the first stage is obtained as a direct combination of an A-critical and a B-critical point, exactly one of which is trivial. All A-critical and B-critical points remaining in the second stage are non-trivial, and determine collectively the remaining C-critical points. However, in the second stage the direct one-to-one relationship between C-critical points and pairs of A- and B-critical points need not hold.

We now give a formal description of both stages of the algorithm.

First stage. Let $i \in \langle -m, -\frac{m}{2} \rangle$, $j = i + \frac{m}{2}$. Recall that (i, j) is a trivial A-critical point. It is straightforward to check that for all k, (i, k) is C-critical, iff (j, k) is B-critical. Therefore, all $m/2$ C-critical points in $\langle -m, -\frac{m}{2} \rangle \times \langle 0, \frac{m}{2} + n \rangle$ can be found in time $O(m + n)$. Analogously, all $m/2$ C-critical points in $\langle -\frac{m}{2}, n \rangle \times \langle \frac{m}{2} + n, m + n \rangle$ can also be found in time $O(m + n)$. The overall memory cost of the first stage is $O(m + n)$.

Second stage. First, we simplify the problem by eliminating all half-integer indices that correspond to critical points considered in the first stage. We then proceed by partitioning the resulting square index range recursively into regular half-sized square blocks. For each block, we establish the number of C-critical points contained in it, and perform the recursive partitioning of the block as long as this number is greater than 0. (The details are omitted due to space restrictions. They will appear in the full version of the paper.)

In summary, the first stage takes time and memory $O(m + n)$. The second stage takes time and memory $O(m + n)$ for index elimination and renumbering, and then time $O(n^{1.5})$ and memory $O(n)$ for the recursion. Therefore, we have the total time and memory cost as claimed.   □

From now on, we assume without loss of generality that $n \leq m$. We will also assume that m and n are reasonably close, so that $(\log m)^c \leq n$ for some constant c, specified separately for each algorithm. First, we describe a simple algorithm running in overall time $O(mn)$, and then we modify it to achieve running time $o(mn)$.

Algorithm 1 (All semi-local LCS, basic version).
Input: strings a, b of length m, n, respectively; we assume $\log m \leq n \leq m$.
Output: extended score matrix on strings a, b, represented by $m+n$ non-trivial critical points.
Description. The computation proceeds recursively, partitioning the longer of the two current strings into a concatenation of two strings of equal length (within ± 1 if string length is odd). Given a current partitioning, the corresponding sets of critical points are merged by Lemma 1. Note that we now have two nested recursions: the main recursion of the algorithm, and the inner recursion of Lemma 1.

In the process of main recursion, the algorithm may (and typically will, as long as the current values of m and n are sufficiently close) alternate between partitioning string a and string b. Therefore, we will need to convert the geometric representation of a horizontal score matrix into a vertical one, and vice versa. This can be easily achieved by Theorem 5.

The base of the main recursion is $m = n = 1$.
Cost analysis. Consider the main recursion tree. The computational work in the top $\log(m/n)$ levels of the tree is at most $\log(m/n) \cdot O(m) + (m/n) \cdot O(n^{1.5}) = O(mn)$. The computational work in the remaining $2 \log n$ levels of the tree is dominated by the bottom level, which consists of $O(mn)$ instances of merging score matrices of size $O(1)$. Therefore, the total computation cost is $O(mn)$.

The main recursion tree can be evaluated depth-first, so that the overall memory cost is dominated by the top level of the main recursion, running in memory $O(n)$.   □

The above algorithm can now be easily modified to achieve the claimed subquadratic running time, using an idea originating in [2] and subsequently applied to string comparison by [14].

Algorithm 2 (All semi-local LCS, full version).
Input, output: as in Algorithm 1; we assume $(\log m)^{5/2} \leq n \leq m$.
Description. Consider an all semi-local LCS problem on strings of size $t = \frac{1}{2} \cdot \log_{\sigma} m$, where σ is the size of the alphabet. All possible instances of this problem are precomputed by Algorithm 1 (or by the algorithm of [1]). After

that, the computation proceeds as in Algorithm 1. However, the main recursion is cut off at the level where block size reaches t, and the precomputed values are used as the recursion base.

Cost analysis. In the precomputation stage, there are σ^{2t} problem instances, each of which costs $O(t^2)$. Therefore, the total cost of the precomputation is $\sigma^{2t} \cdot O(t^2) = \frac{1}{4} \cdot m(\log_\sigma m)^2 = O\left(\frac{mn}{\log^{1/2}(m+n)}\right)$.

Consider the main recursion tree. The computational work in the top $\log(m/n)$ levels of the tree is at most $\log(m/n) \cdot O(m) + (m/n) \cdot O(n^{1.5}) = O\left(\frac{mn}{\log^{1.5} m}\right) + O\left(\frac{mn}{\log^{5/4} m}\right) = o\left(\frac{mn}{\log^{1/2}(m+n)}\right)$. The computational work in the remaining $2\log(n/t)$ levels of the tree is dominated by the cut-off level, which consists of $O(mn/t^2)$ instances of merging score matrices of size $O(t)$. Therefore, the total computation cost is $mn/t^2 \cdot O(t^{1.5}) = O\left(\frac{mn}{t^{1/2}}\right) = O\left(\frac{mn}{\log^{1/2}(m+n)}\right)$. □

6   Conclusions

We have presented a new approach to the all semi-local LCS problem. Our approach results in a significantly improved output representation, and yields the first subquadratic algorithm for the problem, with running time $O\left(\frac{mn}{\log^{1/2}(m+n)}\right)$ when m and n are reasonably close.

An immediate open question is whether the time efficiency of our algorithm can be improved even further, e.g. to match the (global) LCS algorithms of [14, 5] with running time $O\left(\frac{mn}{\log(m+n)}\right)$. Currently, our algorithm assumes constant alphabet size; it may be possible to remove this assumption by the technique of [5].

Another interesting question is whether our algorithm can be adapted to more general string comparison. The *edit distance problem* concerns a minimum-cost transformation between two strings, with given costs for character insertion, deletion and substitution. The LCS problem is equivalent to the edit distance problem with insertion/deletion cost 1 and substitution cost 2 or greater. By a constant-factor blow-up of the grid dag, our algorithm can solve the all semi-local edit distances problem, where the insertion, deletion and substitution edit costs are any constant rationals. It remains an open question whether this can be extended to arbitrary real costs, or to sequence alignment with non-linear gap penalties.

Finally, our technique appears general enough to be able to find applications beyond semi-local comparison. In particular, could it be applied in some form to the biologically important case of fully local (i.e. every substring against every substring) comparison?

Acknowledgement

The author is partially supported by a Royal Society Conference Grant.

References

1. C. E. R. Alves, E. N. Cáceres, and S. W. Song. An all-substrings common subsequence algorithm. *Electronic Notes in Discrete Mathematics*, 19:133–139, 2005.
2. V. L. Arlazarov, E. A. Dinic, M. A. Kronrod, and I. A. Faradzev. On economical construction of the transitive closure of an oriented graph. *Soviet Mathematical Doklady*, 11:1209–1210, 1970.
3. J. L. Bentley. Multidimensional divide-and-conquer. *Communications of the ACM*, 23(4):214–229, 1980.
4. R. E. Burkard, B. Klinz, and R. Rudolf. Perspectives of Monge properties in optimization. *Discrete Applied Mathematics*, 70:95–161, 1996.
5. M. Crochemore, G. M. Landau, and M. Ziv-Ukelson. A subquadratic sequence alignment algorithm for unrestricted score matrices. *SIAM Journal on Computing*, 32(6):1654–1673, 2003.
6. M. Crochemore and W. Rytter. *Text Algorithms*. Oxford University Press, 1994.
7. D. Gusfield. *Algorithms on Strings, Trees, and Sequences: Computer Science and Computational Biology*. Cambridge University Press, 1997.
8. J. JaJa, C. Mortensen, and Q. Shi. Space-efficient and fast algorithms for multidimensional dominance reporting and counting. In *Proceedings of the 15th ISAAC*, volume 3341 of *Lecture Notes in Computer Science*, pages 558–568, 2004.
9. N. C. Jones and P. A. Pevzner. *An introduction to bioinformatics algorithms*. Computational Molecular Biology. The MIT Press, 2004.
10. S.-R. Kim and K. Park. A dynamic edit distance table. *Journal of Discrete Algorithms*, 2:303–312, 2004.
11. G. M. Landau, E. Myers, and M. Ziv-Ukelson. Two algorithms for LCS consecutive suffix alignment. In *Proceedings of the 15th CPM*, volume 3109 of *Lecture Notes in Computer Science*, pages 173–193, 2004.
12. G. M. Landau, E. W. Myers, and J. P. Schmidt. Incremental string comparison. *SIAM Journal on Computing*, 27(2):557–582, 1998.
13. G. M. Landau and M. Ziv-Ukelson. On the common substring alignment problem. *Journal of Algorithms*, 41:338–359, 2001.
14. W. J. Masek and M. S. Paterson. A faster algorithm computing string edit distances. *Journal of Computer and System Sciences*, 20:18–31, 1980.
15. G. Navarro. A guided tour to approximate string matching. *ACM Computing Surveys*, 33(1):31–88, 2001.
16. F. P. Preparata and M. I. Shamos. *Computational Geometry: An Introduction*. Texts and Monographs in Computer Science. Springer, 1985.
17. J. P. Schmidt. All highest scoring paths in weighted grid graphs and their application to finding all approximate repeats in strings. *SIAM Journal on Computing*, 27(4):972–992, 1998.

Non-approximability of the Randomness Deficiency Function

Michael A. Ustinov[*]

Moscow State University
ustinov@mccme.ru.

Abstract. Let x be a binary string of length n. Consider the set P_x of all pairs of integers (a, b) such that the randomness deficiency of x in a finite set S of Kolmogorov complexity at most a is at most b. The paper [4] proves that there is no algorithm that for every given x upper semicomputes the minimal deficiency function $\beta_x(a) = \min\{b \mid (a, b) \in P_x\}$ with precision $n/\log^4 n$. We strengthen this result in two respects. First, we improve the precision to $n/4$. Second, we show that there is no algorithm that for every given x enumerates a set at distance at most $n/4$ from P_x, which is easier than to upper semicompute the minimal deficiency function of x with the same accuracy.

1 Introduction

We first recall the basic notions of Kolmogorov complexity and randomness deficiency. Let $\Xi = \{0, 1\}^*$ be the set of all binary strings. The length of $x \in \Xi$ is denoted by $l(x)$. A *decompressor* is a partial computable function D mapping $\Xi \times \Xi$ to Ξ.

The complexity of x conditional to y with respect to D is defined as

$$K_D(x|y) = \min \{l(p) \mid D(p, y) = x\}$$

(the minimum of the empty set is defined as $+\infty$). A decompressor U is called *universal* if for every other decompressor D there is c such that for all $x, y \in \Xi$ it holds $K_U(x|y) \leq K_D(x|y) + c$. Fix a universal decompressor and define the Kolmogorov complexity of a binary string x conditional to a binary string y by $K(x|y) = K_U(x|y)$. This definition and a proof of existence of universal decompressors was given in [2], see also the textbook [3]. The (unconditional) Kolmogorov complexity of x is defined as $K(x) = K(x|\text{empty string})$.

In this paper we need to define Kolmogorov complexity also for finite subsets of Ξ. To this end fix a computable injective mapping $S \mapsto [S]$ of the set of finite subsets of Ξ to Ξ and let $K(S) = K([S])$. The Kolmogorov complexity of x conditional to a set S is defined as $K(x|S) = K(x|[S])$.

---

[*] The work was supported by the RFBR grant 06-01-00122-a and NSH grant 358.2003.1.

D. Grigoriev, J. Harrison, and E.A. Hirsch (Eds.): CSR 2006, LNCS 3967, pp. 364–368, 2006.

The randomness deficiency of x in a finite set $S \subset \Xi$ (where $x \in S$) was defined by Kolmogorov as

$$\delta(x|S) = \log|S| - K(x|S).$$

The randomness deficiency measures how unlikely looks the hypothesis "x was obtained by a random sampling in S". For example, if someone claims that he has tossed a fair coin $n = 1000$ times and the result is $x_1 = 00\ldots0$ (n tails) or $x_2 = 0101\ldots01$ we will hardly believe him. The randomness deficiency of both strings x_1, x_2 in the set $S = \{0,1\}^n$ is close to n, as both $K(x_1|S)$ and $K(x_2|S)$ are small. However the string x_3 with no regularities ($K(x_3|n)$ is close to n) does not look that suspicious and we have $\delta(x_3|S) \approx n - n = 0$.

2   Our Results and Related Works

In the algorithmic statistics initiated by Kolmogorov and developed in [1, 4] we want for given data $x \in \Xi$ and given complexity level α to find an explanation for x, that is, to find a finite set $S \ni x$ of complexity at most α such that the randomness deficiency of x in S is small. The minimal randomness deficiency function of x measures the best randomness deficiency we can achieve at each complexity level α:

$$\beta_x(\alpha) = \min_S \{\delta(x|S) : S \ni x, \ K(S) \le \alpha\}$$

The *profile* of x is defined as

$$P(x) = \{\langle a, b\rangle \mid \exists S : S \ni x, \ K(S) \le a, \ \delta(x|S) \le b\}$$

The profile of x is determined uniquely by the minimal randomness deficiency function and vice versa.

Fix α. How can we find a set $S \ni x$ with small $\delta(x|S)$? The following idea is inspired by the Maximal likelihood principle. Find a set $S \ni x$ of complexity at most α of minimal possible cardinality. Following this idea, Kolmogorov has defined the *structure function* of x:

$$h_x(\alpha) = \min_S \{\log|S| : S \ni x, \ K(S) \le \alpha\}$$

The paper [4] proves that this method indeed works: every set $S \ni x$ of complexity at most $K(S) \le \alpha$ of minimal cardinality (among sets of complexity at most α) minimizes also the randomness deficiency of x in sets of complexity at most $\alpha - O(\log n)$, where $n = l(x)$. Formally, write $f(n) = O(g(n))$ if $|f(n)| \le cg(n) + C$, where c ia an absolute constant and C depends on the universal decompressor in the definition of $K(x|y)$. The paper [4] proves that for some function $\delta' = O(\log n)$ and for every x of length n and set $S \ni x$ with $K(S) \le \alpha + \delta'$ it holds:

$$\delta(x|S) < \log|S| + K(S) - K(x) + \delta' \quad (1)$$

$$\beta_x(\alpha) > h_x(\alpha + \delta') + \alpha - K(x) - \delta'. \quad (2)$$

The first inequality implies that if S witnesses $h_x(\alpha + \delta') = i$ then $\delta(x|S) \le i + \alpha - K(x) + 2\delta'$. The second inequality implies that $\beta_x(\alpha) > i + \alpha - K(x) - \delta'$ thus $\delta(x|S)$ is only $3\delta'$ apart from $\beta_x(\alpha)$.

This result implies that the set

$$Q(x) = \{\langle a, b + a - K(x) \rangle \mid \exists S : S \ni x, \ K(S) \le a, \ \log|S| \le b\}$$

is $O(\log n)$ close to the profile of x (we say that P is ε close to Q if for all $(a, b) \in P$ there is $(a', b') \in Q$ with $|a - a'| \le \varepsilon$, $|b - b'| \le \varepsilon$, and vice versa). Note that there is a (non-halting) algorithm A that given x and $K(x)$ enumerates the set $Q(x)$ (prints all the pairs in $Q(x)$). A natural question is whether there is an algorithm that enumerates a set which is $O(\log n)$ close to $P(x)$ given only x (and not $K(x)$)? The main result of the present paper is the negative answer to this question (Theorem 1 below): given only x it is impossible to enumerate a set that is $n/4 - O(\log n)$ close to $P(x)$.

A result from [4] states that the minimal deficiency function $\beta_x(\alpha)$ is not upper-semicomputable with accuracy $n/\log^4 n$. This means that there is no (non-halting) algorithm that given x and α enumerates a set A of naturals such that $\beta_x(\alpha) < \min A < \beta_x(\alpha) + n/\log^4 n$. Note that this result does not imply anything about approximating the profile of x. Indeed, as it is proven in [4], the function $\beta_x(\alpha)$ can decrease much (say by $n/2$) when α increases only a little (by $O(\log n)$). Therefore there is no simple way to transform an algorithm approximating the set $P(x)$ to an algorithm upper semicomputing β_x: a set Q can approximate $P(x)$ but the set $\{b \mid (a, b) \in Q\}$ can be far apart from $\{b \mid (a, b) \in P(x)\}$ (for some a).

Our second result shows that it is impossible to approximate the value of $\beta_x(\alpha)$ in a point α that is not very close to 0 or n. Let $\alpha(n)$ be a computable function such that $n/10 < \alpha(n) < 9n/10$ (say). Then there is no algorithm that given x and α enumerates a set A whose minimal element is at distance at most $n/30$ from a point in the range $[\beta_x(\alpha(n)); \beta_x(\alpha(n)) - n/30)]$. This follows from Theorem 2 below.

3   Theorems and Proofs

Theorem 1. *There is a function $\delta = O(\log n)$ with the following property. There is no algorithm that for every string x of length n enumerates a set of pairs of naturals $P' = P'(x)$ that is $n/4 - \delta$ close to $P(x)$.*

Proof. We first prove that there is no algorithm enumerating a set P' that is $n/6 - \delta$ close to $P(x)$. Then using the inequality (2), we show how to modify the proof to obtain the theorem. Assume that there is an algorithm that for every given x enumerates a set P' at distance $n/6 - \delta$ from $P(x)$, where the value of δ will be chosen later. Then run the following algorithm A. Its input is the length n and the number of finite subsets of Ξ of Kolmogorov complexity at most $n/3 - \delta$.

We first enumerate finite subsets of Ξ of complexity at most $n/3 - \delta$ until we find the given number of them. Then we find all the strings x of length n that do not belong to any set of complexity at most $n/3 - \delta$ and cardinality at most $2^{2n/3}$. Then for every such x we run the algorithm enumerating a set $P'(x)$ at distance at most $n/6 - \delta$ from $P(x)$. If for some x we find a pair $\langle a', b' \rangle \in P'(x)$ with $a' < n/6$, $b' < n/6$, we halt and output that x. (End of algorithm A.)

We claim that (1) the algorithm will eventually halt and (2) the complexity of the output string x is more than $n/3$. The claim (1) means that there is a string x that does not belong to any set of complexity at most $n/3 - \delta$ and cardinality at most $2^{2n/3}$ such that the set $P'(x)$ has a pair $\langle a', b' \rangle$ with $a' \leq n/6$, $b' < n/6$. Indeed let x be a string of length n with $K(x|n) \geq n$ (simple counting shows that there is such x). If x belonged to a set S of complexity at most $n/3 - \delta$ and cardinality at most $2^{2n/3}$ then we could describe x by its index in S and by S. The total length of this description would be $n/3 - \delta + 2n/3 + O(\log n)$ (we need extra $O(\log n)$ bits to separate the description of S from the binary index of x in S). If δ is large enough the total length of this description would be less than n. For S equal to the set of all strings of length n we have

$$\delta(x|S) = \log|S| - K(x|S) \leq n - K(x|n) + O(1) \leq O(1).$$

This proves that the profile of x has the pair $\langle O(1), O(1) \rangle$ and the set $P'(x)$ the pair $\langle a', b' \rangle$ with $a' \leq n/6 - \delta + O(1) < n/6$, $b' \leq n/6 - \delta + O(1) < n/6$.

To prove the claim (2) note that the profile of the x output by A has the pair (a, b) with $a, b < n/3 - \delta$. That is, there is a set $S \ni x$ of complexity at most $n/3 - \delta$ with $\delta(x|S) \leq n/3 - \delta$. As x is outside all sets of complexity at most $n/3 - \delta$ and cardinality at most $2^{2n/3}$, we know that $\log|S| \geq 2n/3$ Therefore

$$K(x) \geq K(x|S) - O(1) = \log|S| - \delta(x|S) - O(1) \geq 2n/3 - n/3 + \delta - O(1)$$
$$= n/3 + \delta - O(1).$$

Thus we have shown that the algorithm outputs a string x of complexity at least $n/3 + \delta - O(1)$ on input of complexity at most $n/3 - \delta + O(\log n)$. For large enough $\delta = O(\log n)$ this is a contradiction.

Let us prove now the theorem in the original formulation. Consider the same algorithm as above but this time look for a string outside sets of complexity at most $n/2 - \delta$ and cardinality at most $2^{n/2}$ such that $P'(x)$ has a pair $\langle a', b' \rangle$ with $a' < n/4$, $b' < n/4$. Just as above, we can prove that the algorithm terminates. To lower bound the complexity of its output x we use the inequality (2) for $\alpha = n/2 - \delta - \delta'$. We have:

$$K(x) \geq h_x(\alpha + \delta') + \alpha - \delta' - \beta_x(\alpha) \geq n/2 + n/2 - \delta - 2\delta' - n/2 + \delta = n/2 - 2\delta'$$

Thus the algorithm on the input of complexity $n/2 - \delta + O(\log n)$ outputs a string x of complexity at least $n/2 - 2\delta'$. Recall that $\delta' = O(\log n)$. Hence for some $\delta = O(\log n)$ we obtain a contradiction.   □

The next theorem is proved by essentially the same arguments.

Theorem 2. *There is a function $\delta = O(\log n)$ with the following property. Then there is no algorithm that for every n, every string x of length n and every α and ε such that $2\varepsilon \le \alpha \le n - 2\varepsilon - \delta$ enumerate a set $T = T(x)$ of natural numbers whose minimal element is in the range*

$$\beta_x(\alpha) < \min T < \beta_x(\alpha - \varepsilon) + \varepsilon.$$

Proof. Assume that such algorithm and ε, α exist. Consider the following algorithm A. As input it receives n and the number of all finite subsets of Ξ of complexity at most $\alpha + \delta'$, where δ' is the function from (2). We find all x outside sets of complexity at most $\alpha + \delta'$ and cardinality at most $2^{n-\alpha-\varepsilon-\delta'}$. For every such x we enumerate the set $T(x)$, and when we find an x with $\min T(x) < \varepsilon$ we halt and output that x. The complexity of that x is at least

$$h_x(\alpha + \delta') + \alpha - \delta' - \beta_x(\alpha) > n - 2\varepsilon - \delta'.$$

As the input has complexity at most $\alpha + \delta' + O(\log n)$ we obtain $\alpha + \delta' + O(\log n) > n - 2\varepsilon - \delta'$. For some $\delta = O(\log n)$ this inequality contradicts the condition $\alpha < n - 2\varepsilon - \delta$.

It remains to show that there is x of length n outside all sets of complexity at most $\alpha + \delta'$ and cardinality at most $2^{n-\alpha-\delta'-\varepsilon}$ with $\min T(x) < \varepsilon$. As in the previous theorem consider an x with $K(x|n) \ge n$. It is outside all sets of complexity at most $\alpha + \delta'$ and cardinality at most $2^{n-\alpha-\delta'-\varepsilon}$, as otherwise we could describe it in $n - \varepsilon + O(\log n) < n$ bits. As the randomness deficiency of x in the set S of all strings of length n is $O(1)$ we have $\beta_x(\alpha - \varepsilon) = O(1)$ thus $\min T(x) < \beta_x(\alpha - \varepsilon) + \varepsilon \le \varepsilon$. $\qquad\square$

Acknowledgement

The author is sincerely grateful to his scientific advisor N. Vereshchagin for drawing his attention to the problem and helpful discussions.

References

1. P. Gács, J. Tromp, P.M.B. Vitányi. Algorithmic statistics, *IEEE Trans. Inform. Th.*, 47:6(2001), 2443–2463.
2. A.N. Kolmogorov, Three approaches to the quantitative definition of information, *Problems Inform. Transmission* 1:1 (1965) 1–7.
3. M.Li and P.M.B.Vitanyi. An Intoduction to Kolmogorov Complexity and its Applications, Springer Verlag, New York, 2nd Edition, 1997
4. N.Vereshchagin and P.Vitanyi. Kolmogorov's Structure Functions with an Application to the Foundations of Model Selection. IEEE Transactions on Information Theory 50:12 (2004) 3265-3290. Preliminary version: Proc. 47th IEEE Symp. Found. Comput. Sci., 2002, 751–760.

Multi-agent Explicit Knowledge

Tatiana Yavorskaya (Sidon)

Department of Mathematical Logic and Theory of Algorithms,
Faculty of Mechanics and Mathematics,
Moscow State University, Moscow 119992, Russia
tanya@lpcs.math.msu.su

Abstract. Logic of proofs LP, introduced by S. Artemov, originally designed for describing properties of formal proofs, now became a basis for the theory of knowledge with justification. So far, in epistemic systems with justification the corresponding "evidence part", even for multi-agent systems, consisted of a single explicit evidence logic. In this paper we introduce logics describing two interacting explicit evidence systems. We find an appropriate formalization of the intended semantics and prove the completeness of these logics with respect to both symbolic and arithmetical models. Also, we find the forgetful projections for the logics with two proof predicates which are extensions of the bimodal logic $S4^2$.

1 Introduction

The Logic of Proofs LP introduced by S. Artemov in 1995 (see the detailed description in [1, 2]) was originally designed to express in logic the notion of a proof. It is formulated in the propositional language enriched by new atoms $[\![t]\!]F$ with the intended meaning *"t is a proof of F"*. Proofs are represented by *proof terms* constructed from *proof variables* and *proof constants* by means of three elementary computable operations: binary $\cdot$, $+$ and unary $!$ specified by the axioms

$$[\![t]\!](A \to B) \to ([\![s]\!]A \to [\![t \cdot s]\!]B) \qquad \textit{application}$$
$$[\![t]\!]A \to [\![t + s]\!]A, \quad [\![s]\!]A \to [\![t + s]\!]A \qquad \textit{nondeterministic choice}$$
$$[\![t]\!]A \to [\![!t]\!][\![t]\!]A \qquad \textit{positive proof checker}$$

LP is axiomatized over propositional calculus by the above axioms and the principle

$$[\![t]\!]A \to A \qquad \textit{weak reflexivity}$$

The rules of inference are *modus ponens* and *axiom necessitation rule*. The latter allows to specify proof constants as proofs of the concrete axioms

$$\frac{}{[\![a]\!]A}, \qquad \text{where } a \text{ is an axiom constant, A is an axiom of LP.}$$

The intended semantics for LP is given by formal proofs in Peano Arithnmetic PA: proof variables are interpreted by codes of PA-derivations, $[\![t]\!]F$ stands for

D. Grigoriev, J. Harrison, and E.A. Hirsch (Eds.): CSR 2006, LNCS 3967, pp. 369–380, 2006.
© Springer-Verlag Berlin Heidelberg 2006

the arithmetical proof predicate "t is a proof of F". It is proven in [2] that LP is arithmetically complete with respect to the class of all proof systems. Furthermore, LP suffices to realize Gödel's provability logic S4 and thus provides S4 and intuitionistic logic with the exact provability semantics.

In [3] it was suggested to treat $[\![t]\!]F$ as a new type of knowledge operator called *evidence–based knowledge* with the meaning "t is an evidence for F." Evidence based knowledge (EBK) systems are obtained by augmenting a multi–agent logic of knowledge with a system of evidence assertions $[\![t]\!]F$. Three main cases of EBK–systems were introduced in [3] in which the base knowledge logic is T_n, $S4_n$ or $S5_n$. The evidence part for all of them consists of a single logic of proofs LP.

In this paper we study multiple interacting EBK–systems, namely, we study logics that describe the behavior of two reasoning agents $\mathcal{P}_1$ and $\mathcal{P}_2$ which somehow communicate to each other. For simplicity, let us think about a reasoning agent as a proof system, then evidences are proofs in this system. We develop a language with two proof operators $[\![\cdot]\!]_1(\cdot)$ and $[\![\cdot]\!]_2(\cdot)$ representing proof predicates for $\mathcal{P}_1$ and $\mathcal{P}_2$. In general, proofs of these two systems are distinct, so proof terms for a proof system $\mathcal{P}_i$ ($i = 1, 2$) are constructed from its own atomic proofs represented by proof variables p_k^i and proof constants c_k^i. We suppose that both $\mathcal{P}_1$ and $\mathcal{P}_2$ has all the power of LP, so we reserve a copy of LP–operations $\times_i$, $+_i$ and $!_i$ for application, nondeterministic choice and positive proof checker in $\mathcal{P}_i$ ($i = 1, 2$).

For the minimal logic of two proof systems denoted by LP^2 we assume that there is no communication between them, except that all axioms are common knowledge, so we extend the axiom necessitation rule and allow it to derive all the formulas

$$[\![c_{j_1}^{k_1}]\!]_{k_1}[\![c_{j_2}^{k_2}]\!]_{k_2} \ldots [\![c_{j_n}^{k_n}]\!]_{k_n} A, \text{ where all } k_i \in \{1, 2\}, \ A \text{ is an axiom.}$$

Going further, we may assume that the two systems $\mathcal{P}_1$ and $\mathcal{P}_2$ are allowed to communicate, that is, one of the proof systems is able to derive something about the other one. We study two types of communications.

Proof checking. We assume that $\mathcal{P}_2$ can verify all proofs of $\mathcal{P}_1$ and introduce a unary operation $!_1^2$ specified by the axiom

$$[\![t]\!]_1 A \rightarrow [\![!_1^2 t]\!]_2 [\![t]\!]_1 A.$$

Further, we can consider the case when both $\mathcal{P}_1$ and $\mathcal{P}_2$ are able to verify each other, then we add the dual operation $!_2^1$ with the specification

$$[\![t]\!]_2 A \rightarrow [\![!_2^1 t]\!]_1 [\![t]\!]_2 A.$$

The resulting logics are denoted by $\mathsf{LP}_!^2$ and $\mathsf{LP}_{!!}^2$ respectively.

Proof embedding. Here we suppose that all proofs of $\mathcal{P}_1$ can be converted to $\mathcal{P}_2$–proofs; this is done by the operation $\uparrow_1^2$ specified by the principle

$$[\![t]\!]_1 A \rightarrow [\![\uparrow_1^2 t]\!]_2 A.$$

If $\mathcal{P}_1$ can also imitate $\mathcal{P}_2$–proof, we add a converse operation $\uparrow_2^1$ with the specification

$$[\![t]\!]_2 A \rightarrow [\![\uparrow_2^1 t]\!]_1 A.$$

We denote the resulting logics by $\mathsf{LP}_\uparrow^2$ and $\mathsf{LP}_{\uparrow\uparrow}^2$.

In this paper for all the logics L mentioned above we do the following:

- describe symbolic semantics and prove completeness of L;
- find the forgetful projection of L, i.e. a bimodal logic obtained from L by replacing all occurrences of $[\![t]\!]_i$ by $\square_i$ for $i = 1, 2$;
- describe arithmetical interpretation and prove completeness of L.

The structure of the paper is the following. In section 2 we give a precise description of the language and the logics we are dealing with. In section 3 the modal counterparts of all the described logics are found. It turned out that the forgetful projections of $\mathsf{LP}_\uparrow^2$ and $\mathsf{LP}_{\uparrow\uparrow}^2$ coincide with the projections of $\mathsf{LP}_\uparrow^2$ and $\mathsf{LP}_{\uparrow\uparrow}^2$ respectively. Section 4 is devoted to symbolic and arithmetical semantics.

2 Explicit Evidence Logics for Two Agents: Definitions

Definition 1. *The minimal language L of the bimodal explicit evidence logic is denoted by* LP^2. *It contains*

- *propositional variables* $SVar = \{S_1, S_2, \ldots\}$;
- *two disjoint sets of proof variables* $PVar^i = \{p_1^i, p_2^i, \ldots\}$ *and two disjoint sets of proof constants* $\{c_1^i, c_2^i, \ldots\}$ *where* $i = 1, 2$;
- *two copies of every operation on proofs from* LP: *binary* $\times_1$, $+_1$, $\times_2$, $+_2$ *and unary* $!_1$ *and* $!_2$;
- *Boolean connectives and two operational symbols* $[\![\cdot]\!]_1(\cdot)$ *and* $[\![\cdot]\!]_2(\cdot)$ *of the type* proof $\rightarrow$ (proposition $\rightarrow$ proposition)

We also consider extensions of L. The first option is to add one or both of the unary functional symbols $!_1^2$ *and* $!_2^1$; *we denote the result by* $\mathsf{LP}_!^2$, $\mathsf{LP}_{!!}^2$ *respectively. Another option is to add one or both of the unary functional symbols* $\uparrow_1^2$ *and* $\uparrow_2^1$; *the result is denoted by* $\mathsf{LP}_\uparrow^2$, $\mathsf{LP}_{\uparrow\uparrow}^2$ *respectively.*

For every language L from the definition above we define two sets of terms $Tm_i(L)$, $(i = 1, 2)$. For $L = \mathsf{LP}^2$ the set $Tm_i(L)$ consists of all terms constructed from variables and constants labelled with sup-i by operations labelled by i. Namely, for $i = 1, 2$, every proof variable p_j^i or proof constant c_j^i is an element of $Tm_i(L)$ and if $t, s \in Tm_i(L)$, then $t \times_i s$, $t +_i s$ and $!_i t$ belong to $Tm_i(L)$ too. For the extensions of the minimal language we add the following clauses to the definition of terms

- for $L = \mathsf{LP}_!^2$, if $t \in Tm_1(L)$ then $!_1^2 t \in Tm_2(L)$;
- for $L = \mathsf{LP}_{!!}^2$, if $t \in Tm_1(L)$ then $!_1^2 t \in Tm_2(L)$ and if $t \in Tm_2(L)$ then $!_2^1 t \in Tm_1(L)$;

- for $L = \mathsf{LP}^2_\uparrow$, if $t \in Tm_1(L)$ then $\uparrow^2_1 t \in Tm_2(L)$;
- for $L = \mathsf{LP}^2_{\uparrow\uparrow}$, if $t \in Tm_1(L)$ then $\uparrow^2_1 t \in Tm_2(L)$ and if $t \in Tm_2(L)$ then $\uparrow^1_2 t \in Tm_1(L)$.

Formulas of the language L are constructed from sentence variables by boolean connectives and according to the rule: for $i = 1, 2$ if $t \in Tm_i(L)$ and F is a formula of L then $[\![t]\!]_i F$ is a formula of L too. The set of all formulas is denoted by $Fm(L)$. Formulas of the form and $[\![t]\!]_i F$ are called *q-atomic*, the set of such formulas is denoted by $QFm_i(L)$. We write $QFm(L)$ for $QFm_1(L) \cup QFm_2(L)$.

Operations on proofs are specified by the following formulas (t, s are terms, A, B are formulas):

$$
\begin{array}{ll}
\mathrm{Ax}(\times_i) & [\![t]\!]_i(A \to B) \to ([\![s]\!]_i A \to [\![t \times_i s]\!]_i B) \\
\mathrm{Ax}(+_i) & [\![t]\!]_i A \to [\![t +_i s]\!]_i A, \quad [\![s]\!]_i A \to [\![t +_i s]\!]_i A \\
\mathrm{Ax}(!_i) & [\![t]\!]_i A \to [\![!_i t]\!]_i [\![t]\!]_i A \\
\mathrm{Ax}(!^2_1) & [\![t]\!]_1 A \to [\![!^2_1 t]\!]_2 [\![t]\!]_1 A \\
\mathrm{Ax}(!^1_2) & [\![t]\!]_2 A \to [\![!^1_2 t]\!]_1 [\![t]\!]_2 A \\
\mathrm{Ax}(\uparrow^2_1) & [\![t]\!]_1 A \to [\![\uparrow^2_1 t]\!]_2 A \\
\mathrm{Ax}(\uparrow^1_2) & [\![t]\!]_2 A \to [\![\uparrow^1_2 t]\!]_1 A
\end{array}
$$

Definition 2. *For every language L from Definition 1 we define the corresponding bimodal logic of proofs L. It is axiomatized by the following schemas:*

$A0$ *classical propositional axioms*
$A1$ $[\![t]\!]_i A \to A, \quad i = 1, 2$
$A2...$ *axioms for all operations of L.*

The rules of inference are modus ponens and axiom necessitation rule

$$
[\![c^{k_1}_{j_1}]\!]_{k_1} [\![c^{k_2}_{j_2}]\!]_{k_2} \dots [\![c^{k_n}_{j_n}]\!]_{k_n} A, \; \text{where all } k_i \in \{1, 2\}, \; A \; \text{is an axiom.}
$$

Informally speaking, the language LP^2 describes the structure which contains objects of three types: *propositions* represented by formulas, *proofs*$_1$ and *proofs*$_2$ represented by proof terms. We suppose that there are two proof systems $\mathcal{P}_1$ and $\mathcal{P}_2$; the system $\mathcal{P}_i$ tries to find $t \in proofs_i$ for $A \in propositions$. The structure is supplied with two proof predicates $[\![t]\!]_1 A$ and $[\![t]\!]_2 A$, which correspond to $\mathcal{P}_1$ and $\mathcal{P}_2$. Both proof predicates are supposed to be recursive. For every $p \in proofs_i$ the set of propositions proved by p in $\mathcal{P}_i$ is finite and the function that maps proofs to the corresponding sets is total recursive.

Both proof systems $\mathcal{P}_1$ and $\mathcal{P}_2$ are supplied with operations on proofs taken from LP, thus, they are capable of internalizing there own proofs. The minimal language LP^2 corresponds to the situation when two proof systems do not communicate. The only information about $\mathcal{P}_1$ which is available to $\mathcal{P}_2$ and vise versa is transferred via the axiom necessitation rule. For example, the second proof system knows that $!_1$ is a proof checker of the first one since we can derive $[\![c^2]\!]_2([\![t]\!]_1 A \to [\![!t]\!]_1 [\![t]\!]_1 A)$. Externally we can prove that something is provable in $\mathcal{P}_1$ iff it is provable in $\mathcal{P}_2$, that is, the following two assertions are equivalent:

there exists a term $t \in Tm_1(\mathsf{LP}^2)$ such that $\mathsf{LP}^2 \vdash [\![t]\!]_1 A$

and

there exists a term $s \in Tm_2(\mathsf{LP}^2)$ such that $\mathsf{LP}^2 \vdash [\![s]\!]_2 A$

However, this fact cannot be derived in LP^2, that is, there is no term $t \in Tm_2(\mathsf{LP}^2)$ such that $\mathsf{LP}^2 \vdash [\![p^1]\!]_1 S \to [\![t]\!]_2 S$ (this fact easily can be proven using symbolic semantics from section 4). So, neither $\mathcal{P}_1$ nor $\mathcal{P}_2$ is able to formalize or proof the equivalence just mentioned.

The communication between the two proof systems becomes possible in the extensions of LP^2. In $\mathsf{LP}_!^2$ and $\mathsf{LP}_\uparrow^2$ it is one-way: $\mathcal{P}_2$ can derive some facts about $\mathcal{P}_1$. In $\mathsf{LP}_{!!}^2$ in $\mathsf{LP}_{\uparrow\uparrow}^2$ information can be transferred symmetrically both-ways. Operations $!_1^2$ and $!_2^1$ are proof checkers. $\mathsf{LP}_!^2$ corresponds to the case when $\mathcal{P}_2$ is able to check proofs of $\mathcal{P}_1$; in $\mathsf{LP}_{!!}^2$ we suppose that both of $\mathcal{P}_i$ can proof-check each other. Operations $\uparrow_1^2$ or $\uparrow_2^1$ appear if one of the systems can prove everything that the other one can.

Operations $!_1^2$ and $\uparrow_1^2$ can imitate each other in the following sense.

Lemma 1. *1. For every term $t \in Tm_1(\mathsf{LP}_!^2)$ and formula $F \in Fm(\mathsf{LP}_!^2)$, there is a term $s \in Tm_2(\mathsf{LP}_!^2)$ such that $\mathsf{LP}_!^2 \vdash [\![t]\!]_1 F \to [\![s]\!]_2 F$.*

2. For every term $t \in Tm_1(\mathsf{LP}_\uparrow^2)$ and formula $F \in Fm(\mathsf{LP}_\uparrow^2)$, there is a term $s \in Tm_2(\mathsf{LP}_\uparrow^2 0$ such that $\mathsf{LP}_\uparrow^2 \vdash [\![t]\!]_1 F \to [\![s]\!]_2 [\![t]\!]_1 F$.

Proof. 1. Derive in $\mathsf{LP}_!^2$

$$[\![t]\!]_1 F \to [\![!_1^2 t]\!]_2 [\![t]\!]_1 F$$
$$[\![c^2]\!]_2([\![t]\!]_1 F \to F)$$
$$[\![t]\!]_1 F \to [\![c^2 \times_2 (!_1^2 t)]\!]_2 F$$
$$\text{take } s = c^2 \times_2 (!_1^2 t)$$

2. Derive in $\mathsf{LP}_\uparrow^2$

$$[\![t]\!]_1 F \to [\![!_1 t]\!]_1 [\![t]\!]_1 F$$
$$[\![!_1 t]\!]_1 [\![t]\!]_1 F \to [\![\uparrow_1^2 !_1 t]\!]_2 [\![t]\!]_1 F$$
$$[\![t]\!]_1 F \to [\![\uparrow_1^2 (!_1 t)]\!]_2 [\![t]\!]_1 F$$
$$\text{take } s = \uparrow_1^2 (!_1 t).$$

Lemma 2 (Internalization property). *Let L be one of the logics from Definition 2. If $L \vdash F$, then for $i = 1, 2$ there exists a term t_i constructed from constants with the help of operations $\times_i$ and $!_i$ such that $L \vdash [\![t_i]\!]_i F$.*

Proof. Standard induction on derivation of F.

Lemma 3. *Let L be one of the logics from definition 2. For $i = 1, 2$ let δ_i be a $\wedge, \vee$−combination of q-atoms from $QFm_i(L)$. Let δ stand for a $\wedge, \vee$−combination of q-atoms from $QFm_1(L) \cup QFm_2(L)$.*

1. There exists a term t_i such that $L \vdash \delta_i \to [\![t_i]\!]_i \delta_i$.

2. If L contains either $!_1^2$ or $\uparrow_1^2$ then there exists a term $t \in Tm_2(L)$ such that $L \vdash \delta \to [\![t]\!]_2 \delta$.

3. If L contains either $!_2^1$ or $\uparrow_2^1$ then there exists a term $t \in Tm_1(L)$ such that $L \vdash \delta \to [\![t]\!]_1 \delta$.

Proof. 1. Induction on the construction of δ_i. If $\delta_i = [\![t]\!]_i F$ then apply $Ax(!_i)$ to obtain $L \vdash \delta_i \to [\![!_i t]\!]_i \delta_i$. If $\delta_i = \alpha_i \wedge \beta_i$ or $\delta_i = \alpha_i \vee \beta_i$ then, by the induction hypothesis, there exist terms u and v such that $L \vdash \alpha_i \to [\![u]\!]_i \alpha_i$ and $L \vdash \beta_i \to$

$[\![v]\!]_i\beta_i$. By the axiom necessitation rule, $L \vdash [\![c^i]\!]_i(\alpha_i \to (\beta_i \to (\alpha_i \wedge \beta_i)))$. Using $\mathrm{Ax}(\times_i)$, we derive

$$L \vdash \alpha_i \wedge \beta_i \to [\![c^i \times_i u \times_i v]\!]_i(\alpha_i \wedge \beta_i).$$

By axiom necessitation we also have $L \vdash [\![c_1^i]\!]_i(\alpha_i \to \alpha_i \vee \beta_i)$ and $L \vdash [\![c_2^i]\!]_i(\beta_i \to \alpha_i \vee \beta_i)$. Hence $L \vdash \alpha_i \to [\![c_1^i \times_i u]\!]_i(\alpha_i \vee \beta_i)$ and $L \vdash \beta_i \to [\![c_2^i \times_i v]\!]_i(\alpha_i \vee \beta_i)$. Therefore $L \vdash \alpha_i \vee \beta_i \to [\![(c_1^i \times_i u) +_i (c_2^i \times_i v)]\!]_i(\alpha_i \vee \beta_i)$.

2. The induction step is similar to the previous case. For the induction base now we should consider two options $[\![t]\!]_1 F$ and $[\![s]\!]_2 F$ instead of one. The second option is treated similarly with the previous case. For $\delta = [\![t]\!]_1 F$ we have $\mathsf{LP}_!^2 \vdash [\![t]\!]_1 F \to [\![!_1^2 t]\!]_2 [\![t]\!]_1 F$. In $\mathsf{LP}_\uparrow^2$ we reason as follows: $\mathsf{LP}_\uparrow^2 \vdash [\![t]\!]_1 F \to [\![!_1 t]\!]_1 [\![t]\!]_1 F$ and $\mathsf{LP}_\uparrow^2 \vdash [\![!_1 t]\!]_1 \delta_{1,2} \to [\![\uparrow_1^2 !_1 t]\!]\delta_{1,2}$. Hence $\mathsf{LP}_\uparrow^2 \vdash \delta_{1,2} \to [\![\uparrow_1^2 !_1 t]\!]\delta_{1,2}$.

3. Similar to 2.

3 Realization of Bimodal Logics

In [2] it is proven that LP is able to realize all derivations in the modal logic S4, namely, if A is a theorem of S4 then there is an assignment of LP–terms to all occurrences of $\Box$'s in A such that the resulting formula is a theorem in LP. In this section we describe the modal counterparts of the logics LP^2, $\mathsf{LP}_!^2$ and $\mathsf{LP}_\uparrow^2$.

We need the bimodal logic $\mathsf{S4}^2$ and its extension $\mathsf{S4}_{\mathsf{mon}}^2$. $\mathsf{S4}^2$ is given by the following axioms and rules of inference: for $i = 1, 2$,

A1 propositional tautologies
A2 $\Box_i A \to A$
A3 $\Box_i(A \to B) \to (\Box_i A \to \Box_i B)$
A4 $\Box_i A \to \Box_i \Box_i A$
R1 Modus Ponens: $A, A \to B \vdash B$
R2 Necessitation: if $\vdash A$ then $\vdash \Box_i A$.

$\mathsf{S4}_{\mathsf{mon}}^2$ is an extension of $\mathsf{S4}^2$ by the principle

A5 $\Box_1 F \to \Box_2 F$.

We prove that the analog of the realization theorem for S4 and LP holds for the following pairs of logics: $\mathsf{S4}^2$ and LP^2, $\mathsf{S4}_{\mathsf{mon}}^2$ and $\mathsf{LP}_!^2$, $\mathsf{S4}_{\mathsf{mon}}^2$ and $\mathsf{LP}_\uparrow^2$. We need the following definition.

Definition 3. *Let L be one of the languages from definition 1. Suppose that A is a formula with two modalities. A realization of A in the language L is a formula $A^r \in Fm(L)$ which is obtained from A by substitution of terms from $Tm_i(L)$ for all occurrences of $\Box_i$ in A. A realization is normal if all negative occurrences of modalities are assigned proof variables.*

Theorem 1. *1. $\mathsf{S4}^2 \vdash A$ iff there exists a normal realization r in the language LP^2 such that $\mathsf{LP}^2 \vdash A^r$.*

2. For $L \in \{\mathsf{LP}_!^2, \mathsf{LP}_\uparrow^2\}$, $\mathsf{S4}_{\mathsf{mon}}^2 \vdash A$ iff there exists a normal realization r in the language L such that $L \vdash A^r$.

The proof of this theorem goes along the lines of the proof of realization of S4 in LP (sf. [2]). First of all, we need the normalized Gentzen-style versions of $S4^2$ and $S4^2_{mon}$. Sequential calculus for $S4^2$, denoted by $GS4^2$, has the same axioms and rules as sequential calculus for classical propositional logic plus four modal rules (two for each modality):

$$(\text{Left}\square_i) \quad \frac{A, \Gamma \Rightarrow \Delta}{\square_i A, \Gamma \Rightarrow \Delta} \qquad (\text{Right}\square_i) \quad \frac{\square_i \Gamma \Rightarrow A}{\square_i \Gamma \Rightarrow \square_i A} \qquad (i = 1, 2).$$

In the Gentzen-style version of $S4^2_{mon}$ denoted by $GS4^2_{mon}$ the rule $(\text{Right}\square_2)$ is replaced by a stronger version

$$\frac{\square_1 \Gamma_1, \square_2 \Gamma_2 \Rightarrow A}{\square_1 \Gamma_1, \square_2 \Gamma_2 \Rightarrow \square_2 A}.$$

Theorem 2. *For a logic $L \in \{S4^2, S4^2_{mon}\}$ the following connection between L and its Gentzen-style version $\mathcal{G}$ holds:*

$$\mathcal{G} \vdash \Gamma \Rightarrow \Delta \ \text{ iff } \ L \vdash \bigwedge \Gamma \to \bigvee \Delta.$$

Theorem 3. *Any logic $\mathcal{G} \in \{GS4^2, GS4^2_{mon}\}$ enjoys cut-elimination: if $\mathcal{G} \vdash \Gamma \Rightarrow \Delta$ then $\Gamma \Rightarrow \Delta$ can be derived in $\mathcal{G}$ without using of the Cut-rule.*

Lemma 4. *1. $GS4^2 \vdash \Gamma \Rightarrow \Delta$ iff there exists a normal realization r such that $LP^2 \vdash (\bigwedge \Gamma \to \bigvee \Delta)^r$.*
   2. For $L \in \{LP^2_!, LP^2_\uparrow\}$, $GS4^2_{mon} \vdash \Gamma \Rightarrow \Delta$ iff there exists a normal realization r in the language L such that $L \vdash (\bigwedge \Gamma \to \bigvee \Delta)^r$.

Proof. Similar to the proof of the realization theorem for LP. Goes by induction on the cut-free proof of $\Gamma \Rightarrow \Delta$. Uses Internalization and δ-completeness.

4   Symbolic and Arithmetical Semantics

Models of multi-agent logics of explicit knowledge below are natural generalizations of Mkrtychev models for LP (cf. [6]).

Definition 4. *Let L be any language from definition 1. An L–model $\mathcal{M} = (\#, v)$ consists of two objects*

  – $\#$ is a mapping from proof terms of L to sets of formulas of L, called an evidence function;
  – v is a truth evaluation of sentence variables.

For every functional symbol from L the evidence function $\#$ should satisfy the corresponding closure condition from the list given below: suppose that t, s are in $Tm_i(L)$, $i = 1, 2$

- if $(A \to B) \in \#(t)$, $A \in \#(s)$ then $B \in \#(t \times_i s)$;
- if $A \in \#(t)$ then $A \in \#(t +_i s)$ and $A \in \#(s +_i t)$;
- if $A \in \#(t)$ then $[\![t]\!]_i A \in \#(!_i t)$;
- if $A \in \#(u)$ and $u \in Tm_1(L)$ then $[\![u]\!]_1 A \in \#(!_1^2 u)$;
- if $A \in \#(v)$ and $v \in Tm_2(L)$ then $[\![v]\!]_2 A \in \#(!_2^1 v)$;
- if $A \in \#(u)$ and $u \in Tm_1(L)$ then $A \in \#(\uparrow_1^2 u)$;
- if $A \in \#(v)$ and $v \in Tm_2(L)$ then $A \in \#(\uparrow_2^1 v)$.

Definition of the truth relation $\mathcal{M} \models A$ is inductive: for propositional variables $\mathcal{M} \models S$ iff $v(S) = true$, $\models$ commutes with Boolean connectives and for $t_i \in Tm_i$

$$\mathcal{M} \models [\![t]\!]_i A \;\rightleftharpoons\; A \in \#(t) \text{ and } \mathcal{M} \models A.$$

A model $\mathcal{M} = (\#, v)$ is called *finitely generated* (or f.g. for short) if

- for every term t the set $\#(t)$ is finite; the set $\{p \in PVar | \#(p) \neq \emptyset\}$ is finite;
- the set of terms, for which the converse of the conditions on the evidence function does not hold, is finite;
- the set $\{S \in SVar \mid v(S) = true\}$ is finite.

Definition 5. *For any logic L from definition 2 a constant specification CS is any finite set of formulas derived by the axiom necessitation rule. We say that $L \vdash A$ meeting CS if all axiom necessitation rules in the derivation of A introduce formulas from CS. We say that an L–model $\mathcal{M}$ meets CS if $\mathcal{M} \models (\bigwedge CS)$.*

Theorem 4. *Let L be any logic from definition 2.*
1. If $L \vdash A$ meeting CS then for every L-model $\mathcal{M}$ meeting CS one has $\mathcal{M} \models A$.
2. If $L \nvdash A$ meeting CS then there exists a f.g. L-model $\mathcal{M}$ meeting CS such that $\mathcal{M} \nvDash A$.

Proof. We give the sketch of the proof for $L = \mathsf{LP}^2$; for the remaining systems the proof differs in saturation and completion algorithms (see below) to which the cases corresponding to the additional operations should be added. It is enough to consider the case $CS = \emptyset$; the general case can be reduces to this one by the deduction theorem which holds in all logics L. We omit all the proofs of technical lemmas.

Soundness can be easily proven by induction on the derivation of A. In order to prove completeness suppose that $\mathsf{LP}^2 \nvdash A$. We will construct a finitely generated model $\mathcal{M} = (\#, v)$ such that $\mathcal{M} \nvDash A$.

Step 1: Saturation algorithm. It constructs a finite set of formulas $Sat(A)$ which is called an *adequate set*. We need the following definition: the *complexity of a proof term t* denoted by $|t|$ is the length of the longest branch in the tree representing this term. The saturation algorithm works as follows:

1. Initialization. Put $Sat_0(A) := SubFm(A)$. Calculate the maximal complexity of terms which occur in A; let N denote the result.

2. For every $l = 1, \ldots, N + 1$ we calculate the set $Sat_l(A)$ as follows.
   - Initially $Sat_l(A) := Sat_{l-1}(A)$.
   - if $[\![t]\!]_i(A \to B), [\![s]\!]_i A \in Sat_{l-1}(A)$ then extend $Sat_l(A)$ by $[\![t \times_i s]\!]_i B$;
   - if $[\![t]\!]_i A \in Sat_{l-1}(A)$ and $|s| \leq l$ then extend $Sat_l(A)$ by $[\![t +_i s]\!]_i A$ and $[\![s +_i t]\!]_i A$;
   - if $[\![t]\!]_i A \in Sat_{l-1}(A)$ then extend $Sat_l(A)$ by $[\![!_i t]\!]_i [\![t]\!]_i A$.
3. Put $Sat(A) := Sat_{N+1}(A)$.

Lemma 5. *(Properties of adequate sets.) For every* $l = 0, \ldots, N + 1$,

1. $Sat_l(A)$ *is closed under subformulas, that is,* $SubFm(Sat_l(A)) \subseteq Sat_l(A)$.
2. *If* $G \in Sat_{l+1}(A) \setminus Sat_l(A)$ *then* G *has the form* $[\![t]\!]E$ *and* $|t| \geq l + 1$.
3. *If* $[\![t]\!]_i(F \to G), [\![s]\!]_i F \in Sat(A)$ *and* $|t \times_i s| \leq N$ *then* $[\![t \times_i s]\!]_i G \in Sat(A)$.
   If $[\![t]\!]_i G \in Sat(A)$ *and* $|t +_i s| \leq N$, *then* $[\![t +_i s]\!]_i G, [\![s +_i t]\!]_i G \in Sat(A)$.
   If $[\![t]\!]_i G \in Sat(A)$ *and* $|!_i t| \leq N$ *then* $[\![!_i t]\!]_i [\![t]\!]_i G \in Sat(A)$.

Proof. Joint induction on l.

Step 2. Now we describe a translation of the language LP^2 into the pure propositional language. For every q-atom $[\![t]\!]_i B \in Sat(A)$ we reserve a fresh propositional variable $S_{t,i,B}$. For every formula G whose all q-atomic subformulas belong to $Sat(A)$ by G' we denote the result of substitution of all outermost occurrences of q-atomic subformulas in G by the corresponding propositional variables. Namely, we define G' by induction on the construction of G in the following way: for propositional variables $S' \rightleftharpoons S$; $(\cdot)'$ commutes with boolean connectives and $([\![t]\!]_i B)' \rightleftharpoons S_{t,i,B}$.

Let $Ax(A)$ stand for the conjunction of all substitutional instances of axioms A1–A4 whose all q-atomic subformulas are from $Sat(A)$. Put $A_p \rightleftharpoons (Ax(A) \to A)'$. Since $\mathsf{LP}^2 \nvdash A$ we conclude that A_p is not provable in propositional logic (otherwise after the reverse substitution of $[\![t]\!]_i B$ for $S_{t,i,B}$ in the derivation of A_p in propositional calculus we get $\mathsf{LP}^2 \vdash Ax(A) \to A$, hence $\mathsf{LP}^2 \vdash A$). Therefore, there exists an evaluation w of propositional letters from A_p by (*true*, *false*) such that $w(A_p) = false$. Define

$$\Gamma_0 \rightleftharpoons \{B \in Sat(A) \mid w(B') = true\},$$
$$\Delta_0 \rightleftharpoons \{B \in Sat(A) \mid w(B') = false\}.$$

Lemma 6. *The sets* Γ_0 *and* Δ_0 *has the following properties:*

1. $\Gamma_0 \cap \Delta_0 = \emptyset$.
2. *If* $[\![t]\!]E \in \Gamma_0$ *then* $E \in \Gamma_0$.
3. *If* $[\![t]\!]_i(F \to G), [\![s]\!]_i F \in \Gamma_0$ *and* $|t \times_i s| \leq N$ *then* $[\![t \times_i s]\!]_i G \in \Gamma_0$.
   If $[\![t]\!]_i G \in \Gamma_0$ *and* $|t +_i s| \leq N$ *then* $[\![t +_i s]\!]_i G \in \Gamma_0$ *and* $[\![s +_i t]\!]_i G \in \Gamma_0$.
   If $[\![t]\!]_i G \in \Gamma_0$ *and* $|!_i t| \leq N$ *then* $[\![!_i t]\!]_i [\![t]\!]_i G \in \Gamma_0$.

Step 3. Completion algorithm. It goes through infinite number of iterations; the l-th iteration produces the set Γ_l which is finite. Start with Γ_0. For every $l = 1, 2, \ldots$ on the l-th iteration construct the set Γ_l as follows

- Initially $\Gamma_l := \Gamma_{l-1}$.
- if $[\![t]\!]_i(A \to B), [\![s]\!]_i A \in \Gamma_{l-1}$ then extend Γ_l by $[\![t \times_i s]\!]B$;
- if $[\![t]\!]_i A \in \Gamma_{l-1}$ and $|s| \le l$ then extend Γ_l by $[\![t +_i s]\!]_i A$ and $[\![s +_i t]\!]_i A$;
- if $[\![t]\!]_i A \in \Gamma_{l-1}$ then extend Γ_l by $[\![!_i t]\!]_i [\![t]\!]_l A$;
- Go to the next l.

Put $\Gamma := \bigcup_l \Gamma_l$.

Lemma 7. *For every* $l = 0, 1, 2, \ldots,$

1. *The set* Γ_l *is finite and* $\Gamma_l \cap \Delta = \emptyset$
2. *If* $E \in \Gamma_{l+1} \setminus \Gamma_l$ *then* E *is of the form* $[\![t]\!]G$ *and* $|t| \ge N + l + 1$.
3. $\Gamma_l \cup \Delta_0$ *is closed under subformulas, that is,* $SubFm(\Gamma_l \cup \Delta) \subseteq \Gamma_l \cup \Delta$.
4. *If* $[\![t]\!]_i(F \to G), [\![s]\!]_i F \in \Gamma$ *then* $[\![t \times_i s]\!]_i G \in \Gamma$. *If* $[\![t]\!]_i G \in \Gamma$ *then* $[\![t +_i s]\!]_i G \in \Gamma$ *and* $[\![s +_i t]\!]_i G \in \Gamma$. *If* $[\![t]\!]_i G \in \Gamma$ *then* $[\![!_i t]\!]_i [\![t]\!]_i G \in \Gamma$.
5. *For every term* t *the set* $I(t) = \{E \mid [\![t]\!]E \in \Gamma\}$ *is finite and the function* $t \mapsto I(t)$ *is primitive recursive.*

Proof. Induction on l.

Step 4. For every $t \in Tm_i$ and $S \in SVar$ put

$$\#(t) \rightleftharpoons \{E \mid [\![t]\!]E \in \Gamma\} \qquad v(S) \rightleftharpoons w(S).$$

Lemma 8. *For every formula* G *one has*

$$G \in \Gamma \Rightarrow \mathcal{M} \models G;$$
$$G \in \Delta \Rightarrow \mathcal{M} \not\models G.$$

Proof. Induction on G. We use lemma 7.

From lemmas 8 and 7 it follows that $\mathcal{M}$ is a finitely generated model for LP^2. Since $w(A') = false$ we conclude $A \in \Delta$, hence $\mathcal{M} \not\models A$. This completes the proof of the theorem.

Corollary 1. LP^2 *is decidable.*

Epistemic semantics for LP^2 is given by the following natural generalization of Fitting models (cf. [4]). For any language L from definition 1 one could define a *Fitting model* as follows. An L–model $\mathcal{M} = (W, R_1, R_2, \mathcal{E}, v)$ has the following parameters

- a nonempty set of possible worlds W;
- two reflexive transitive accessibility relations on W denoted by R_1, R_2
- an evidence function $\mathcal{E}$ which maps $W \times Tm(L)$ to sets of formulas of L,
- for every $x \in W$ a truth evaluation $v(x)$ maps propositional variables to $\{true, false\}$.

We require that for every node $x \in W$ the restriction of $\mathcal{E}$ to x satisfies all the conditions for $\#$ and $\mathcal{E}$ is monotone in the following sense: for $i = 1, 2$ if xR_iy and $t \in Tm_i(L)$ then $\mathcal{E}(x, t) \subseteq \mathcal{E}(y, t)$.

The truth relation for every node $x \in W$ is defined in the standard way; we put $\mathcal{M}, x \models [\![t]\!]_i F$ iff $F \in \mathcal{E}(x, t)$ and $\mathcal{M}, y \models F$ for every $y \in W$ such that xR_iy.

Note that a model in the sense of definition 4 is a Fitting model, namely, take W a singleton set and R_1, R_2 total relations on W. It is easy to prove that all the logics considered in this paper are sound and complete with respect to the models just described. In particular, the completeness with respect to Fitting semantics follows from Theorem 4 and the fact that aforementioned Mkrychev models are singleton versions of the corresponding Fitting models.

Now let us describe the interpretation of bimodal logics of proofs in Peano Arithmetic PA (the definition of PA and related topics can be found in [7]).

Definition 6. *A normal proof predicate* Prf *is an arithmetical provably* Δ_1 *formula satisfying the following conditions:*
1) for every arithmetical formula φ PA $\vdash \varphi$ *iff there exists a natural number* n *such that* $Prf(n, \lceil \varphi \rceil)$;
2) for every n *the set* $Th(n) \rightleftharpoons \{\varphi \mid Prf(n, \lceil \varphi \rceil)\}$ *is finite and the function* $n \mapsto Th(n)$ *is total recursive;*
3) for every finite set of arithmetical theorems Γ *there exists a natural number* n *such that* $\Gamma \subseteq Th(n)$.

Lemma 9. *Let* L *be a language from definition 1. For every pair of normal proof predicates* Prf_1, Prf_2 *and every operation of* L *there exist a total recursive function which satisfies the corresponding axiom. For example, there exists a function* app_i *such that for all natural numbers* k, n *for all arithmetical sentences* φ, ψ

$$\text{PA} \vdash Prf_i(k, \lceil \varphi \rightarrow \psi \rceil) \rightarrow (Prf_i(n, \lceil \varphi \rceil) \rightarrow Prf_i(app_i(k, n), \lceil \psi \rceil))$$

Definition 7. *Let* L *be one of the languages from definition 2. An arithmetical interpretation* $* = (Prf_1, Prf_2, (\cdot)^*)$ *for the language* L *has the following parameters:*

- *two normal proof predicate* Prf_1 *and* Prf_2;
- *total recursive functions for operations of* L *which satisfy lemma 9*
- *an evaluation* $(\cdot)^*$ *that assigns natural numbers to proof variables and arithmetical sentences to propositional variables.*

Arithmetical evaluation $(\cdot)^*$ *can be extended to all* LP^2 *terms and formulas in the following way. It commutes with the Boolean connectives and*

$$([\![t]\!]_i A)^* \rightleftharpoons \exists x \, (i = i \wedge x = \lceil A^* \rceil \wedge Prf_i(t^*, x)).$$

Note that PA $\vdash ([\![p]\!]A)^* \leftrightarrow Prf(p^*, \lceil A^* \rceil)$. The reasons why we interpret the proof predicates in the more sophisticated way is that it makes the following problem decidable: being given Prf_i, an arithmetical formula φ and an $\mathcal{L}$-formula F, decide whether there exists $(\cdot)^*$, such that $F^* = \varphi$. If such $*$ exists then it is unique.

Theorem 5. *[Arithmetical soundness and completeness]*
For every LP^2 *formula A the following three propositions are equivalent:*
   1) $\mathsf{LP}^2 \vdash A$;
   2) for every interpretation $*$, $\mathsf{PA} \vdash A^*$;
   3) for every interpretation $*$, A^* is true.

References

1. S. Artemov, *Uniform provability realization of intuitionistic logic, modality and λ-terms*, Electronic Notes in Theoretical Computer Science 23 (1999).
2. S. Artemov, *Explicit provability and constructive semantics*, Bulletin of Symbolic Logic 7 (2001), 1–36.
3. S. Artemov, *Evidence–based common knowledge*, Technical Report TR–2004018, CUNY Ph.D. Program in Computer Science, 2005.
4. M. Fitting, *The Logic of Proofs, Semantically*, Annals of Pure and Applied Logic, vol. 132 (2005), no. 1, pp. 1-25.
5. R. Kuznets, *On the complexity of explicit modal logics*, Computer Science Logic 2000, Lecture Notes in Computer Science 1862 (2000), 371–383.
6. A. Mkrtychev, *Models for the Logic of Proofs*, Lecture Notes in Computer Science, v. 1234 (1997), *Logical Foundations of Computer Science '97, Yaroslavl'*, pp. 266–275.
7. C. Smoryński, *Self-reference and Modal Logic*, Springer, New York, 1985.

Polarized Subtyping for Sized Types

Andreas Abel*

Department of Computer Science, University of Munich,
Oettingenstr.67, D-80538 München, Germany
abel@tcs.ifi.lmu.de

Abstract. We present an algorithm for deciding polarized higher-order subtyping without bounded quantification. Constructors are identified not only modulo β, but also η. We give a direct proof of completeness, without constructing a model or establishing a strong normalization theorem. Inductive and coinductive types are enriched with a notion of size and the subtyping calculus is extended to account for the arising inclusions between the sized types.

1 Introduction

Polarized kinding and subtyping has recently received interest in two contexts. First, in the analysis of container types in object-oriented programming languages [12]. If List A is a functional (meaning: read-only) collection of objects of type A and A is a subtype (subclass) of B then List A should be a subtype of List B. However, for read-write collections, as for instance Array, such a subtyping relation is unsound[1], hence these two collection constructors must be kept apart. The conventional modeling language for object types, System $\mathsf{F}^\omega_\le$, does not distinguish List and Array in their kind—both map types to types, thus, have kind $* \to *$. To record subtyping properties in the kind of constructors, polarities were added by Cardelli, Pierce, Steffen [25], and Duggan and Compagnoni [12]. Now, the type constructor List gets kind $* \xrightarrow{+} *$, meaning that it is a monotone (or covariant) type-valued function, whereas Array gets kind $* \xrightarrow{\circ} *$, meaning that Array is neither co- nor contravariant or its variance is unknown to the type system.

Another application of polarized kinding is normalizing languages[2] with recursive datatypes. It is well-known that if a data type definition has a negative recursive occurrence, a looping term can be constructed by just using the constructors and destructors of this data type, without actually requiring recursion on the level of programs [21]. Negative occurrences can be excluded by polarized kinding [2]—a recursive type μF is only admitted if $F : * \xrightarrow{+} *$.

A promising way to formulate a normalizing language is by using *sized types*. Hughes, Pareto, and Sabry [19] have presented such a language, which can be used, e. g., as a basis for embedded programming. It features sized first-order parametric data

---

* Research supported by the coordination action *TYPES* (510996) and thematic network *Applied Semantics II* (IST-2001-38957) of the European Union and the project *Cover* of the Swedish Foundation of Strategic Research (SSF).

[1] Nevertheless, such a subtyping rule has been added for arrays in Java.

[2] In a normalizing language, each program is terminating.

D. Grigoriev, J. Harrison, and E.A. Hirsch (Eds.): CSR 2006, LNCS 3967, pp. 381–392, 2006.

types, where the size parameter induces a natural subtyping relation. Independently, Barthe et. al. [6] have arrived at a similar system, which is intended as the core of a theorem prover language. Both systems, however, fail to treat higher-order and hetero-geneous (or nested) data types which have received growing interest in the functional programming community [5, 7, 18, 22, 3].

In order to extend the sized type system to such higher-order constructions, we need to handle polarized higher-order subtyping! Steffen [25] has already defined the necessary concepts and an algorithm that decides this kind of subtyping. But because he features also bounded quantification, his completeness proof for the algorithm is long and complicated. In this article, I present a different subtyping algorithm, without bounded quantification, but instead fitted to the needs of sized types, and prove it sound and complete in a rather straightforward manner.

Main technical contribution. I define a polarized higher-order subtyping algorithm which respects not only β but also η-equality and computes the normal form of the considered type constructors incrementally. A novelty is the succinct and direct proof of completeness, which relies neither on a normalization theorem nor a model construction. Instead, a lexicographic induction on kinds and derivations is used.

Organization. In Section 2, we recapitulate the polarized version of F^ω defined in a previous article [2] and extend it by declarative subtyping. A subtyping algorithm is added and proven sound in Section 3. In Section 4, we prove completeness of the algorithmic equality. The extension to sized types is presented in Section 5, and we close with a discussion of related work.

Preliminaries. The reader should be familiar with higher-order polymorphism and subtyping. Pierce's book [23] provides an excellent introduction.

Judgements. In the following, we summarize the inductively defined judgements in this article.

$\Gamma \vdash F : \kappa$	constructor F has kind κ in context Γ
$\Gamma \vdash F = F' : \kappa$	F and F' of kind κ are $\beta\eta$-equal
$\Gamma \vdash F \leq F' : \kappa$	F is a higher-order subtype of F'
$F \searrow W$	F has weak head normal form W
$\Gamma \vdash_a W \leq^q W' : \kappa$	algorithmic subtyping

When we write $\mathcal{D} :: J$, we mean that judgement J has derivation $\mathcal{D}$. Then, $|\mathcal{D}|$ denotes the height of this derivation.

2   Polarized System F^ω

In this section, we present a polarized version of F^ω. This is essentially Fix^ω [2] without fixed-points, but with the additional polarity $\top$. A technical difference is that Fix^ω uses Church-style (kind-annotated) constructors whereas we use Curry-style (domain-free) constructors. However, all result of this paper apply also to the Church style.

$p, q \in$ Pol		$p \le q$	pq

$p, q ::= \circ$ non-variant
     $\mid +$ covariant
     $\mid -$ contravariant
     $\mid \top$ invariant

$p \le q$ lattice:

$$\top$$
$$+ \quad\quad -$$
$$\circ$$

pq	$\circ$	$+$	$-$	$\top$
$\circ$	$\circ$	$\circ$	$\circ$	$\top$
$+$	$\circ$	$+$	$-$	$\top$
$-$	$\circ$	$-$	$+$	$\top$
$\top$	$\top$	$\top$	$\top$	$\top$

Fig. 1. Polarities: definition, ordering, composition

2.1 Polarities

We aim to distinguish constructors with regard to their *monotonicity* or *variance*. For instance, the product constructor $\times$ is *monotone* or *covariant* in both of its arguments. If one enlarges the type A or B, more terms inhabit $A \times B$. The opposite behavior is called *antitone* or *contravariant*. Two more scenarios are possible: the value $F\,A$ does not change when we modify A. Then F is called *constant* or *invariant*. Finally, a function F might not exhibit a uniform behavior, it might grow or shrink with its argument, or we just do not know how F behaves. This is the general case, we call it *non-variant*. Each of the behaviors is called a *polarity* and abbreviated by one of the four symbols displayed in Fig. 1.

The polarities are related: Since "non-variant" just means we do not have any information about the function, and we can always disregard our knowledge about variance, each function is non-variant. The inclusion order between the four sets of in-, co-, contra-, and non-variant functions induces a partial *information order* $\le$ on Pol. The smaller a set is, the more information it carries. Hence $\circ \le p$, $p \le \top$, and $p \le p$ for all p. This makes Pol a bounded 4-element lattice as visualized in Fig. 1.

Polarity of composed functions. Let F, G be two functions such that the composition $F \circ G$ is well-defined. If F has polarity p and G has polarity q, we denote the polarity of the composed function $F \circ G$ by pq. It is clear that polarity composition is monotone: if one gets more information about F or G, certainly one cannot have *less* information about $F \circ G$. Then, if one of the functions is constant, so is their composition. Otherwise, if one of them is non-variant, the same holds for the composition. In the remaining cases, the composition is covariant if F and G have the same variance, otherwise it is contravariant. This yields the multiplication table in Fig. 1. Polarity composition, as function composition, is associative. It is even commutative, but not *a priori*, since function composition is not commutative.

Inverse application of polarities. If $f(y) = py$ is the function which composes a polarity with p, what would be its inverse $g(x) = p^{-1}x$? It is possible to define g in such a way that f and g form a Galois connection, i.e.,

$$p^{-1}x \le y \iff x \le py.$$

It is not hard to see that $+^{-1}x = x$, $-^{-1}x = -x$, $\top^{-1}x = \circ$, $\circ^{-1}\circ = \circ$ and $\circ^{-1}x' = \top$ (for $x' \ne \circ$). As for every Galois connection, it holds that $p^{-1}py \le y$ and $x \le pp^{-1}x$, and both f and g are monotone.

2.2   Kinds, Constructors, and Kinding

Fig. 2 lists *kinds*, which are generated by the base kind $*$ of types and polarized function space, and by *(type) constructors*, which are untyped lambda-terms over some constructor constants C. As usual, $\lambda X F$ binds variable X in F. We identify constructors under α-equivalence, i. e., under renaming of bound variables. $\mathsf{FV}(F)$ shall denote the set of free variables of constructor F.

The *rank* $\mathsf{rk}(\kappa) \in \mathbb{N}$ of a kind κ is defined recursively by $\mathsf{rk}(*) = 0$ and $\mathsf{rk}(p\kappa \to \kappa') = \max(\mathsf{rk}(\kappa) + 1, \mathsf{rk}(\kappa'))$.

Polarized contexts. A polarized context Γ fixes a polarity p and a kind κ for each free variable X of a constructor F. If $p = +$, then X may only appear positively in F; this ensures that $\lambda X F$ is a monotone function. Similarly, if $p = -$, then X may only occur negatively, and if $p = \circ$, then X may appear in both positive and negative positions. A variable labeled with $\top$ may only appear in arguments of an invariant function. The domain $\mathsf{dom}(\Gamma)$ is the set of constructor variables Γ mentions. As usual, each variable can appear in the context only once.

We say context Γ' is more *liberal* than context Γ, written $\Gamma' \leq \Gamma$, iff

$$(X : p\kappa) \in \Gamma \text{ implies } (X : p'\kappa) \in \Gamma' \text{ for some } p' \leq p.$$

In particular, Γ' may declare more variables than Γ and assign weaker polarities to them. The intuition is that all constructors which are well-kinded in Γ are also well-kinded in a more permissive context Γ'.

The *application* $p\Gamma$ of a polarity p to a context Γ is defined as pointwise application, i. e., if $(X : q\kappa) \in \Gamma$, then $(X : (pq)\kappa) \in p\Gamma$. *Inverse application* $p^{-1}\Gamma$ is defined analogously. Together, they form a Galois connection, i. e., for all Γ and Γ',

$$p^{-1}\Gamma \leq \Gamma' \iff \Gamma \leq p\Gamma'.$$

Kinding. We introduce a judgement $\Gamma \vdash F : \kappa$ which combines the usual notions of well-kindedness and positive and negative occurrences of type variables. A candidate for the application rule is

$$\frac{\Gamma \vdash F : p\kappa \to \kappa' \qquad \Gamma' \vdash G : \kappa}{\Gamma \vdash F\,G : \kappa'} \; \Gamma \leq p\Gamma'.$$

The side condition is motivated by polarity composition. Consider the case that $X \notin \mathsf{FV}(F)$. If G is viewed as a function of X, then $F\,G$ is the composition of F and G. Now if G is q-variant in X, then $F\,G$ is pq-variant in X. This means that all q-variant variables of Γ' must appear in Γ with a polarity of at most pq. Now if $X \in \mathsf{FV}(F)$, it could be that it is actually declared in Γ with a polarity smaller than pq. Also, variables which are not free in G are not affected by the application $F\,G$, hence they can carry the same polarity in $F\,G$ as in F. Together this motivates the condition $\Gamma \leq p\Gamma'$.

Since $p^{-1}\Gamma$ is the most liberal context which satisfies the side condition, we can safely replace Γ' by $p^{-1}\Gamma$ in the above rule. Hence, we arrive at the kinding rules as given in Fig. 2. Although these rules are not fully deterministic, they can easily be turned into a bidirectional kind checking algorithm for constructors in β-normal form. Kinding enjoys the usual properties of weakening, strengthening, and substitution.

Syntactic categories.

$$X, Y, Z \qquad\qquad\qquad\qquad\qquad\qquad \text{(type) constructor variable}$$
$$C \qquad\qquad ::= \; \to \; | \; \forall_\kappa \qquad\qquad\qquad \text{(type) constructor constant}$$
$$A, B, F, G, H, I, J ::= C \; | \; X \; | \; \lambda X F \; | \; F\, G \qquad \text{(type) constructor}$$
$$\kappa \qquad\qquad ::= * \; | \; p\kappa \to \kappa' \qquad\qquad\qquad \text{kind}$$
$$\Gamma \qquad\qquad ::= \diamond \; | \; \Gamma, X : p\kappa \qquad\qquad \text{polarized context}$$

The signature Σ assigns kinds to constants ($\kappa \xrightarrow{p} \kappa'$ means $p\kappa \to \kappa'$).

$$\to \; : * \xrightarrow{-} * \xrightarrow{+} * \qquad\qquad \text{function space}$$
$$\forall_\kappa : (\kappa \xrightarrow{\circ} *) \xrightarrow{+} * \qquad \text{quantification}$$

Kinding $\Gamma \vdash F : \kappa$.

$$\text{KIND-C} \; \frac{C : \kappa \in \Sigma}{\Gamma \vdash C : \kappa} \qquad \text{KIND-VAR} \; \frac{X : p\kappa \in \Gamma \qquad p \le +}{\Gamma \vdash X : \kappa}$$

$$\text{KIND-}\lambda \; \frac{\Gamma, X : p\kappa \vdash F : \kappa'}{\Gamma \vdash \lambda X F : p\kappa \to \kappa'} \qquad \text{KIND-APP} \; \frac{\Gamma \vdash F : p\kappa \to \kappa' \qquad p^{-1}\Gamma \vdash G : \kappa}{\Gamma \vdash F\, G : \kappa'}$$

Polarized equality $\Gamma \vdash F = F' : \kappa$: Symmetry (EQ-SYM), transitivity (EQ-TRANS), and:

$$\text{EQ-}\beta \; \frac{\Gamma, X : p\kappa \vdash F : \kappa' \qquad p^{-1}\Gamma \vdash G : \kappa}{\Gamma \vdash (\lambda X F)\, G = [G/X]F : \kappa'} \qquad \text{EQ-}\eta \; \frac{\Gamma \vdash F : p\kappa \to \kappa'}{\Gamma \vdash (\lambda X.\, F X) = F : p\kappa \to \kappa'}$$

$$\text{EQ-}\top \; \frac{\Gamma \vdash F : \top\kappa \to \kappa' \qquad \top^{-1}\Gamma \vdash G : \kappa \qquad \top^{-1}\Gamma \vdash G' : \kappa}{\Gamma \vdash F\, G = F\, G' : \kappa'}$$

$$\text{EQ-VAR} \; \frac{X : p\kappa \in \Gamma \qquad p \le +}{\Gamma \vdash X = X : \kappa} \qquad \text{EQ-}\lambda \; \frac{\Gamma, X : p\kappa \vdash F = F' : \kappa'}{\Gamma \vdash \lambda X F = \lambda X F' : p\kappa \to \kappa'}$$

$$\text{EQ-C} \; \frac{C : \kappa \in \Sigma}{\Gamma \vdash C = C : \kappa} \qquad \text{EQ-APP} \; \frac{\Gamma \vdash F = F' : p\kappa \to \kappa' \qquad p^{-1}\Gamma \vdash G = G' : \kappa}{\Gamma \vdash F\, G = F'\, G' : \kappa'}$$

Polarized subtyping $\Gamma \vdash F \le F' : \kappa$: Transitivity (LEQ-TRANS) and:

$$\text{LEQ-REFL} \; \frac{\Gamma \vdash F = F' : \kappa}{\Gamma \vdash F \le F' : \kappa} \qquad \text{LEQ-ANTISYM} \; \frac{\Gamma \vdash F \le F' : \kappa \qquad \Gamma \vdash F' \le F : \kappa}{\Gamma \vdash F = F' : \kappa}$$

$$\text{LEQ-}\lambda \; \frac{\Gamma, X : p\kappa \vdash F \le F' : \kappa'}{\Gamma \vdash \lambda X F \le \lambda X F' : p\kappa \to \kappa'} \qquad \text{LEQ-APP} \; \frac{\Gamma \vdash F \le F' : p\kappa \to \kappa' \qquad p^{-1}\Gamma \vdash G : \kappa}{\Gamma \vdash F\, G \le F'\, G : \kappa'}$$

$$\text{LEQ-APP+} \; \frac{\Gamma \vdash F : +\kappa \to \kappa' \qquad \Gamma \vdash G \le G' : \kappa}{\Gamma \vdash F\, G \le F\, G' : \kappa'}$$

$$\text{LEQ-APP-} \; \frac{\Gamma \vdash F : -\kappa \to \kappa' \qquad -\Gamma \vdash G' \le G : \kappa}{\Gamma \vdash F\, G \le F\, G' : \kappa'}$$

Fig. 2. $\mathsf{F}_{\widehat{\omega}}$: Kinds and constructors

2.3 Equality and Subtyping

Constructor equality is given by judgement $\Gamma \vdash F = F' : \kappa$. In contrast to most presentations of System F^ω, we consider constructors equivalent modulo β (EQ-β) *and* η (EQ-η), see Fig. 2. There is a third axiom, EQ-$\top$, that states that invariant functions F yield equal results if applied to arbitrary constructors G, G' of the right kind. This axiom can only be formulated for *kinded* equality; it is already present in Steffen's thesis [25, p. 74, rule E-APP$\circ$]. A rule for reflexivity is not included since it is admissible.

Simultaneously with equality, we define higher-order polarized subtyping $\Gamma \vdash F \leq F' : \kappa$ (see Fig. 2). Rule LEQ-REFL includes the subtyping axioms for variables and constants as special cases. Reflexivity and transitivity together ensure that subtyping is compatible with equality. The antisymmetry rule LEQ-ANTISYM potentially enlarges our notion of equality.

There are two kinds of congruence rules for application: one kind states that if functions F and F' are in the subtyping relation, so are their values $F G$ and $F' G$ at a certain argument G. The other kind of rules concern the opposite case: If F is a function and two arguments G and G' are in a subtyping relation, so are the values $F G$ and $F G'$ of the function at these arguments. However, such a relation can only exist if F is covariant or contravariant, or, of course, invariant.

3 Algorithmic Polarized Subtyping

In this section, we present an algorithm for deciding whether two well-kinded constructors are equal or related by subtyping. The algorithm is an adaption of Coquand's $\beta\eta$-equality test [11] to the needs of subtyping and polarities. The idea is to first weak-head normalize the constructors under consideration and then compare their head symbols. If they are related, one continues to recursively compare the subcomponents, otherwise, subtyping fails. First, we define weak head evaluation (see Fig. 3).

For the subtyping algorithm, note that at any point during subtyping checking we may require kinding information. For example, consider checking $X G \leq X G'$. If X is covariant, we need to continue with $G \leq G'$, but if X is contravariant, the next step would be checking $G' \leq G$. Hence, the algorithm needs both context Γ and kind κ of the two considered constructors as additional input. The general form of algorithmic subtyping $\Gamma \vdash_{\mathsf{a}} F \leq^q F' : \kappa$ is defined as $F \searrow W$ and $F' \searrow W'$ and $\Gamma \vdash_{\mathsf{a}} W \leq^q W' : \kappa$, where the last judgement is defined in Fig. 3. The polarity q codes the relation that we seek to establish between W and W': If $q = \circ$, we expect them to be equal, if $q = +$, we expect $W \leq W'$, and if $q = -$, then the other way round. Finally if $q = \top$, then W and W' need not be related, and the algorithm succeeds immediately (rule AL-$\top$). If rule AL-$\top$ is given priority over the other rules, then the relation above is deterministic and can be directly implemented as an algorithm (apply the rules backwards).

Theorem 1 (Soundness of algorithmic subtyping). *Let* $\Gamma \vdash F, F' : \kappa$.

1. *If* $F \searrow W$ *then* $\Gamma \vdash F = W : \kappa$.
2. *If* $\Gamma \vdash_{\mathsf{a}} F \leq^\circ F' : \kappa$ *then* $\Gamma \vdash F = F' : \kappa$.
3. *If* $\Gamma \vdash_{\mathsf{a}} F \leq^+ F' : \kappa$ *then* $\Gamma \vdash F \leq F' : \kappa$.
4. *If* $\Gamma \vdash_{\mathsf{a}} F \leq^- F' : \kappa$ *then* $\Gamma \vdash F' \leq F : \kappa$.

Weak head normal forms $W \in \mathsf{Val}$.

$$\mathsf{Ne} \ni N ::= C \mid X \mid N\,G \qquad \text{neutral constructors}$$
$$\mathsf{Val} \ni W ::= N \mid \lambda X F \qquad \text{weak head values}$$

Weak head evaluation $F \searrow W$ (big-step call-by-name operational semantics).

$$\text{EVAL-C} \ \overline{C \searrow C} \qquad \text{EVAL-VAR} \ \overline{X \searrow X} \qquad \text{EVAL-LAM} \ \overline{\lambda X F \searrow \lambda X F}$$

$$\text{EVAL-APP-NE} \ \frac{F \searrow N}{F\,G \searrow N\,G} \qquad \text{EVAL-APP-}\beta \ \frac{F \searrow \lambda X F' \qquad [G/X]F' \searrow W}{F\,G \searrow W}$$

Algorithmic subtyping.

$$\Gamma \vdash_{\mathsf{a}} F \leq^q F' : \kappa \quad \Longleftrightarrow \quad F \searrow W \text{ and } F' \searrow W' \text{ and } \Gamma \vdash_{\mathsf{a}} W \leq^q W' : \kappa,$$

Algorithmic subtyping $\Gamma \vdash_{\mathsf{a}} W \leq^q W' : \kappa$ for weak head values.

$$\text{AL-}\top \ \overline{\Gamma \vdash_{\mathsf{a}} W \leq^\top W' : \kappa}$$

$$\text{AL-VAR} \ \frac{(X : p\kappa) \in \Gamma \qquad p \leq +}{\Gamma \vdash_{\mathsf{a}} X \leq^q X : \kappa} \qquad \text{AL-}\lambda \ \frac{\Gamma, X : p\kappa \vdash_{\mathsf{a}} F \leq^q F' : \kappa'}{\Gamma \vdash_{\mathsf{a}} \lambda X F \leq^q \lambda X F' : p\kappa \to \kappa'}$$

$$\text{AL-C} \ \frac{(C : \kappa) \in \Sigma}{\Gamma \vdash_{\mathsf{a}} C \leq^q C : \kappa} \qquad \text{AL-APP-NE} \ \frac{\Gamma \vdash_{\mathsf{a}} N \leq^q N' : p\kappa \to \kappa' \quad p^{-1}\Gamma \vdash_{\mathsf{a}} G \leq^{pq} G' : \kappa}{\Gamma \vdash_{\mathsf{a}} N\,G \leq^q N'\,G' : \kappa'}$$

$$X \notin \mathsf{FV}(N) : \quad \text{AL-}\eta\text{-L} \ \frac{\Gamma, X : p\kappa \vdash_{\mathsf{a}} F \leq^q N\,X : \kappa'}{\Gamma \vdash_{\mathsf{a}} \lambda X F \leq^q N : p\kappa \to \kappa'} \qquad \text{AL-}\eta\text{-R} \ \frac{\Gamma, X : p\kappa \vdash_{\mathsf{a}} N\,X \leq^q F : \kappa'}{\Gamma \vdash_{\mathsf{a}} N \leq^q \lambda X F : p\kappa \to \kappa'}$$

Fig. 3. Algorithmic subtyping

4 Completeness

While soundness of the algorithmic subtyping/equality is easy to show, the opposite direction, completeness, is usually hard and requires either the construction of a model [10, 17] or strong normalization for constructors [24, 25, 16]. We will require neither.

Algorithmic subtyping is *cut-free* in a twofold sense: First, a rule for transitivity is missing (this is the cut on the level of subtyping). Its admissibility can often be shown directly by induction on the derivations [24, 10]—so also in our case. The second kind of cut is on the level of kinds: Kinds can be viewed as propositions in minimal logic and constructors as their proof terms, and an application which introduces a redex is a cut in natural deduction. Algorithmic subtyping lacks a general rule for application; its admissibility corresponds to the property of normalization or cut admissibility, resp. We manage to show the admissibility of application directly by a lexicographic induction of the kind of the argument part and the derivation length of the function part. This way, we save ourselves considerable work, and completeness is relatively straightforward.

Lemma 1 (Weakening). *If $\mathcal{D} :: \Gamma \vdash_{\mathsf{a}} F \leq^q F' : \kappa$ and both $\Gamma' \leq \Gamma$ and $q \leq q'$ then $\mathcal{D}' :: \Gamma' \vdash_{\mathsf{a}} F \leq^{q'} F' : \kappa$ and derivation $\mathcal{D}'$ has the same height as $\mathcal{D}$.*

Lemma 2 (Swap). *If* $\Gamma \vdash_a F \leq^q F' : \kappa$ *then* $\Gamma \vdash_a F' \leq^{-q} F : \kappa$.

Antisymmetry of algorithmic subtyping is straightforward in our case since our judgement is deterministic—it is more difficult in the presence of bounded quantification [9].

Lemma 3 (Antisymmetry). *If* $\Gamma \vdash_a F \leq^q F' : \kappa$ *and* $\Gamma \vdash_a F \leq^{q'} F' : \kappa$ *then* $\Gamma \vdash_a F \leq^{\min\{q,q'\}} F' : \kappa$.

Lemma 4 (Transitivity). *If* $\Gamma \vdash_a F_1 \leq^q F_2 : \kappa$ *and* $\Gamma \vdash_a F_2 \leq^{q'} F_3 : \kappa$ *then* $\Gamma \vdash_a F_1 \leq^{\max(q,q')} F_3 : \kappa$.

Transitivity is basically proven by induction on the sum of the lengths of the two given derivations. For the η-rules to go through we need strengthen the induction hypothesis a bit; alternatively, one can use a different measure [16].

The next lemma states that the η-rules can be extended beyond neutral constructors. It can be proven directly:

Lemma 5 (Generalizing the η-rules).
If $\Gamma, Y : p\kappa \vdash_a FY \leq^q F' : \kappa'$ *and* $Y \notin \mathsf{FV}(F)$ *then* $\Gamma \vdash_a F \leq^q \lambda Y F' : p\kappa \to \kappa'$. *If* $\Gamma, Y : p\kappa \vdash_a F' \leq^q FY : \kappa'$ *and* $Y \notin \mathsf{FV}(F)$ *then* $\Gamma \vdash_a \lambda Y F' \leq^q F : p\kappa \to \kappa'$.

Now we come to the main lemma:

Lemma 6 (Substitution and application). *Let* $\Gamma \leq p\Delta$ *and* $\Delta \vdash_a G \leq^{pq} G' : \kappa$.

1. *If* $\mathcal{D} :: \Gamma, X : p\kappa \vdash_a F \leq^q F' : \kappa'$ *then* $\Gamma \vdash_a H \leq^q H' : \kappa'$ *for* $H \equiv [G/X]F$ *and* $H' \equiv [G'/X]F'$ *and either both* H *and* H' *are neutral or* $\mathsf{rk}(\kappa') \leq \mathsf{rk}(\kappa)$.
2. *If* $\mathcal{D} :: \Gamma \vdash_a F \leq^q F' : p\kappa \to \kappa'$ *then* $\Gamma \vdash_a FG \leq^q F'G' : \kappa'$.

Both propositions are proven simultaneously by lexicographic induction on $(\mathsf{rk}(\kappa), |\mathcal{D}|)$. This works because the constructor language is essentially the simply-typed λ-calculus (STL), which has a small proof-theoretical strength. The idea is taken from Joachimski and Matthes' proof of weak normalization for the STL [20] which I have formalized in Twelf [1]. The argument goes probably back to Anne Troelstra, it is implicit in Girard's combinatorial weak normalization proof [13, Ch. 4.3] and has been reused by Watkins et. al. [26] and Adams [4, p. 65ff] to define a logical framework based solely on normal terms.

Theorem 2 (Completeness).

1. *If* $\mathcal{D} :: \Gamma \vdash F : \kappa$ *then* $\Gamma \vdash_a F \leq^q F : \kappa$.
2. *If* $\mathcal{D} :: \Gamma \vdash F = F' : \kappa$ *then* $\Gamma \vdash_a F \leq^\circ F' : \kappa$.
3. *If* $\mathcal{D} :: \Gamma \vdash F \leq F' : \kappa$ *then* $\Gamma \vdash_a F \leq^+ F' : \kappa$.

Proof. Simultaneously by induction on $\mathcal{D}$. In the difficult cases of β-reduction and application, use Lemma 6. In case of EQ-η, use Lemma 5. In the cases of transitivity and antisymmetry, use Lemmata 4 and 3. For LEQ-REFL, apply Lemma 1.

Now we have a sound and complete subtyping algorithm, but we have nothing yet to get it started. Since there are no subtyping assumptions or basic subtyping relations (like Nat $\leq$ Real), two constructors related by subtyping are already equal. In the next section, we will extend the subtyping relation to make it more meaningful.

5 Extension to Sized Types

We introduce a new base kind ord. Kinds that do not mention ord are called *pure kinds* from here and denoted by κ_*. The signature is extended as described in Fig. 4. The first argument to μ_κ and ν_κ shall be written as superscript.

In this signature we can, for instance, model lists of length $< n$ as $\mathsf{List}^a A := \mu_*^a \lambda X. 1 + A \times X$ where $a = \mathsf{s}(\mathsf{s}\ldots(\mathsf{s}\,0))$ (n times s). Streams that have a depth of at least n are represented as $\mathsf{Stream}^a A := \nu_*^a \lambda X. A \times X$. The type $\mathsf{List}^\infty A$ contains lists of arbitrary length, and $\mathsf{Stream}^\infty A$ productive streams (which never run out of elements). We call such types with an ordinal index *sized*. Naturally, lists are covariant in their size argument and streams are contravariant (each stream which produces at least $n + 1$ elements produces of course also at least n elements).

Barthe et. al. [6] define a calculus $\lambda^{\widehat{}}$ with sized inductive and coinductive types in which all recursive functions are terminating and all streams are productive. Their sizes

---

Extension of the signature Σ:

$+$	$: * \xrightarrow{+} * \xrightarrow{+} *$	disjoint sum	$1\ :*$	unit type
$\times$	$: * \xrightarrow{+} * \xrightarrow{+} *$	cartesian product	$0\ : \mathsf{ord}$	ordinal zero
μ_{κ_*}	$: \mathsf{ord} \xrightarrow{+} (\kappa_* \xrightarrow{+} \kappa_*) \xrightarrow{+} \kappa_*$	inductive types	$\mathsf{s}\ : \mathsf{ord} \xrightarrow{+} \mathsf{ord}$	successor
ν_{κ_*}	$: \mathsf{ord} \xrightarrow{-} (\kappa_* \xrightarrow{+} \kappa_*) \xrightarrow{+} \kappa_*$	coinductive types	$\infty : \mathsf{ord}$	infinity

Extension of equality and subtyping.

$$\text{EQ-}\infty \; \frac{}{\Gamma \vdash \mathsf{s}\infty = \infty : \mathsf{ord}}$$

$$\text{LEQ-0} \; \frac{\Gamma \vdash a : \mathsf{ord}}{\Gamma \vdash 0 \leq a : \mathsf{ord}} \qquad \text{LEQ-S} \; \frac{\Gamma \vdash a : \mathsf{ord}}{\Gamma \vdash a \leq \mathsf{s}a : \mathsf{ord}} \qquad \text{LEQ-}\infty \; \frac{\Gamma \vdash a : \mathsf{ord}}{\Gamma \vdash a \leq \infty : \mathsf{ord}}$$

Weak head normal forms.

$$\mathsf{Ne} \ni N ::= C \mid X \mid N\,G \mid \mathsf{s}\,N \quad (C \notin \{\mathsf{s}, \infty\}) \qquad \text{neutral constructors}$$
$$\mathsf{Val} \ni W ::= N \mid \lambda XF \mid \mathsf{s} \mid \infty \qquad \text{weak head values}$$

Extension of weak head evaluation and algorithmic subtyping.

$$\text{EVAL-S} \; \frac{F \searrow \mathsf{s} \quad G \searrow \infty}{F\,G \searrow \infty} \qquad \text{EVAL-APP-S} \; \frac{F \searrow \mathsf{s} \quad G \searrow N}{F\,G \searrow \mathsf{s}\,N}$$

$$\text{AL-0-L} \; \frac{}{\Gamma \vdash_{\mathsf{a}} 0 \leq^+ W : \mathsf{ord}} \qquad \text{AL-0-R} \; \frac{}{\Gamma \vdash_{\mathsf{a}} W \leq^- 0 : \mathsf{ord}}$$

$$\text{AL-}\infty\text{-R} \; \frac{}{\Gamma \vdash_{\mathsf{a}} W \leq^+ \infty : \mathsf{ord}} \qquad \text{AL-}\infty\text{-L} \; \frac{}{\Gamma \vdash_{\mathsf{a}} \infty \leq^- W : \mathsf{ord}}$$

$$\text{AL-S-R} \; \frac{\Gamma \vdash N_1 \leq^+ N_2}{\Gamma \vdash_{\mathsf{a}} N_1 \leq^+ \mathsf{s}\,N_2 : \mathsf{ord}} \; N_1 \not\equiv \mathsf{s}\,N \qquad \text{AL-S-L} \; \frac{\Gamma \vdash N_1 \leq^- N_2}{\Gamma \vdash_{\mathsf{a}} \mathsf{s}\,N_1 \leq^- N_2 : \mathsf{ord}} \; N_2 \not\equiv \mathsf{s}\,N$$

Fig. 4. Extensions for sized types

are given by *stage expressions* which can be modelled by constructors of kind ord in our setting. Sized data types are introduced by a set of data constructors—we can define them using μ and ν. The normalization property is ensured by a restricted typing rule for recursion; in our notation it reads

$$\imath : \mathrm{ord}, f : \mu_*^\imath F \to G\,\imath \vdash e : \mu_*^{\mathsf{s}\imath} F \to G\,(\mathsf{s}\,\imath) \qquad F : * \xrightarrow{+} * \qquad G : \mathrm{ord} \xrightarrow{+} *$$
$$(\mathsf{letrec}\ f = e) : \forall_{\mathrm{ord}}\lambda\imath.\ \mu_*^\imath F \to G\,\imath$$

Similarly, corecursive functions are introduced by

$$\imath : \mathrm{ord}, f : G\,\imath \to \nu_*^\imath F \vdash e : G\,(\mathsf{s}\,\imath) \to \nu_*^{\mathsf{s}\imath} F \qquad F : * \xrightarrow{+} * \qquad G : \mathrm{ord} \xrightarrow{-} *$$
$$(^{\mathrm{co}}\mathsf{letrec}\ f = e) : \forall_{\mathrm{ord}}\lambda\imath.\ G\,\imath \to \nu_*^\imath F$$

We have reduced the stage expressions of $\widehat{\lambda}$ to just constructors of a special kind and model the inclusion between sized types of different stages simply by variance. Lifting the restriction of $\widehat{\lambda}$ that type constructors must be monotone in all arguments comes at no cost in our formulation: We can define the type of A-labeled, B-branching trees as $\mathsf{Tree}^a\,A\,B = \mu_*^a\lambda X.\,1 + A \times (B \to X)$, where now $\mathsf{Tree} : \mathrm{ord} \xrightarrow{+} * \xrightarrow{+} * \xrightarrow{-} *$.

Higher-order subtyping becomes really necessary when we allow *higher-kinded* inductive types. These are necessary to model heterogeneous (also called nested) datatypes, as for instance powerlists: $\mathsf{PList} := \lambda a.\ \mu_{+*\to*}^a \lambda X \lambda A.\ A + X\,(A \times A) : \mathrm{ord} \xrightarrow{+} * \xrightarrow{+} *$. More examples for such datatypes can be found in the literature [22, 5, 7, 18, 3].

Having the machinery of higher-order subtyping running, the extensions needed for sized types are minimal. They are summarized in Fig. 4

Theorem 3. *Algorithmic subtyping for the extended system is still sound and complete.*

6   Conclusion and Related Work

We have presented algorithmic subtyping for polarized F^ω, without bounded quantification, but with rules for η-equality. The algorithm is economic since computes the β-normal form incrementally, just enough to continue with the subtyping test. Due to the trick with the lexicographic induction on kinds and derivations, its completeness proof is quite short compared to completeness proofs of related systems in the literature [25, 10]. However, it is unclear whether the proof scales to bounded quantification— this is worthwhile investigating in the future. Another extension is subkinding induced by the polarities [25], but no big difficulties are to be expected from this side.

Related work. The inspiration for the algorithmic subtyping judgement presented here came from Coquand's $\beta\eta$-conversion algorithm [11] and the idea for the crucial substitution and application lemma (6) from Joachimski and Matthes' proof of weak normalization for the simply-typed λ-calculus [20]. Both Coquand's algorithm and Joachimski and Matthes' characterization of weakly normalizing terms bear strong resemblances to Goguen's typed operational semantics [14, 15].

Our algorithmic subtyping is closely related to Compagnoni and Goguen's weakhead subtyping [8, 9, 10], however, they are additionally dealing with bounded quantification and require neutral constructors to be fully β-normalized. They do not, however, treat η-equality and polarities.

Pierce and Steffen [24] show decidability of higher-order subtyping with bounded quantification. Their calculus of strong cut-free subtyping likens our subtyping algorithm, only that they fully β-normalize the compared constructors and do not treat η. Steffen [25] extends this work to polarities; in his first formulation, kinding and subtyping are mutually dependent. He resolves this issue by introducing a judgement for variable occurrence. Matthes and I [2] have independently of Steffen developed a polarized version of F^ω which unifies variable occurrence and kinding through polarized contexts. Duggan and Compagnoni investigate subtyping for polarized object type constructors [12]. The system is similar to Steffen's, albeit has no constructor-level λ-abstraction, hence there is no need to care for η.

Acknowledgments. Thanks to Thierry Coquand and Ralph Matthes for inspiring discussions and to my former colleagues at Chalmers, especially the bosses of my project, for giving me time to work on my PhD thesis. Thanks to Healfdene Goguen and the anonymous referees for reading the draft and giving helpful and encouraging comments. Thanks to Freiric Barral for proof reading the final version.

References

1. A. Abel. Weak normalization for the simply-typed lambda-calculus in Twelf. In *Logical Frameworks and Metalanguages (LFM 04)*. IJCAR, Cork, Ireland, 2004.
2. A. Abel and R. Matthes. Fixed points of type constructors and primitive recursion. In *Computer Science Logic, CSL'04*, vol. 3210 of *LNCS*, pp. 190–204. Springer, 2004.
3. A. Abel, R. Matthes, and T. Uustalu. Iteration schemes for higher-order and nested datatypes. *Theoretical Computer Science*, 333(1–2):3–66, 2005.
4. R. Adams. *A Modular Hierarchy of Logical Frameworks*. Ph.D. thesis, University of Manchester, 2005.
5. T. Altenkirch and B. Reus. Monadic presentations of lambda terms using generalized inductive types. In J. Flum and M. Rodríguez-Artalejo, eds., *Computer Science Logic, CSL '99, Madrid, Spain, September, 1999*, vol. 1683 of *LNCS*, pp. 453–468. Springer, 1999.
6. G. Barthe, M. J. Frade, E. Giménez, L. Pinto, and T. Uustalu. Type-based termination of recursive definitions. *Mathematical Structures in Computer Science*, 14(1):1–45, 2004.
7. R. S. Bird and R. Paterson. De Bruijn notation as a nested datatype. *J. of Funct. Program.*, 9(1):77–91, 1999.
8. A. Compagnoni and H. Goguen. Anti-symmetry of higher-order subtyping and equality by subtyping. *Submitted*, 2005.
9. A. B. Compagnoni and H. Goguen. Anti-symmetry of higher-order subtyping. In J. Flum and M. Rodríguez-Artalejo, eds., *Computer Science Logic, CSL '99, Madrid, Spain, September, 1999*, vol. 1683 of *LNCS*, pp. 420–438. Springer, 1999.
10. A. B. Compagnoni and H. Goguen. Typed operational semantics for higher-order subtyping. *Inf. Comput.*, 184(2):242–297, 2003.
11. T. Coquand. An algorithm for testing conversion in type theory. In G. Huet and G. Plotkin, eds., *Logical Frameworks*, pp. 255–279. Cambridge University Press, 1991.
12. D. Duggan and A. Compagnoni. Subtyping for object type constructors, 1998. Presented at FOOL 6.
13. J.-Y. Girard, Y. Lafont, and P. Taylor. *Proofs and Types*, vol. 7 of *Cambridge Tracts in Theoretical Computer Science*. Cambridge University Press, 1989.

14. H. Goguen. Typed operational semantics. In M. Deziani-Ciancaglini and G. D. Plotkin, eds., *Typed Lambda Calculi and Applications (TLCA 1995)*, vol. 902 of *LNCS*, pp. 186–200. Springer, 1995.

15. H. Goguen. Soundness of the logical framework for its typed operational semantics. In J.-Y. Girard, ed., *Typed Lambda Calculi and Applications, TLCA 1999*, vol. 1581 of *LNCS*. Springer, L'Aquila, Italy, 1999.

16. H. Goguen. Justifying algorithms for $\beta\eta$ conversion. In V. Sassone, ed., *Foundations of Software Science and Computational Structures, FOSSACS 2005, Edinburgh, UK, April 2005*, vol. 3441 of *LNCS*, pp. 410–424. Springer, 2005.

17. R. Harper and F. Pfenning. On equivalence and canonical forms in the LF type theory. *ACM Transactions on Computational Logic*, 6(1):61–101, 2005.

18. R. Hinze. Generalizing generalized tries. *J. of Funct. Program.*, 10(4):327–351, 2000.

19. J. Hughes, L. Pareto, and A. Sabry. Proving the correctness of reactive systems using sized types. In *Symposium on Principles of Programming Languages*, pp. 410–423. 1996.

20. F. Joachimski and R. Matthes. Short proofs of normalization. *Archive of Mathematical Logic*, 42(1):59–87, 2003.

21. N. P. Mendler. Recursive types and type constraints in second-order lambda calculus. In *Logic in Computer Science (LICS'87), Ithaca, N.Y.*, pp. 30–36. IEEE Computer Society Press, 1987.

22. C. Okasaki. From Fast Exponentiation to Square Matrices: An Adventure in Types. In *International Conference on Functional Programming*, pp. 28–35. 1999.

23. B. C. Pierce. *Types and Programming Languages*. MIT Press, 2002.

24. B. C. Pierce and M. Steffen. Higher order subtyping. *Theor. Comput. Sci.*, 176(1,2):235–282, 1997.

25. M. Steffen. *Polarized Higher-Order Subtyping*. Ph.D. thesis, Technische Fakultät, Universität Erlangen, 1998.

26. K. Watkins, I. Cervesato, F. Pfenning, and D. Walker. A concurrent logical framework I: Judgements and properties. Tech. rep., School of Computer Science, Carnegie Mellon University, Pittsburgh, 2003.

Neural-Network Based Physical Fields Modeling Techniques

Konstantin Bournayev

Belgorod Shukhov State Technological University

Abstract. The possibility of solving elliptic and parabolic partial differential equations by using cellular neural networks with specific structure is investigated. The method of solving varialble coefficients parabolic PDEs is proposed. Issues of cellular neural network stability are examined.

1   The Problem

Physical fields models are widely used in practical tasks. The problem of non-destructive construction diagnostic or the process of the hydrodynamic shock wave expansion in could be taken as an examples. Strictly speaking, almost all problems related to the thermal conductivity, diffusion, convection and electromagnetism use more or less approximate stationary or non-stationary physical field model.

By physics definition field is an assignment of a quantity to every point in space (or more generally, spacetime). The stationary fields — like those describing the process of diffusion, laminar fluid flow or electromagnetic field around stationary charged object — may be described by elliptic Laplace/Poisson equations, while the non-stationary fields — for example, describing shock waves or thermal impulses — may be described by parabolic parabolic differential equations (PDEs).

Let us take a sample region $\Omega \subset E^n$, where Γ is a boundary of Ω. An arbitrary statical field in this region may be defined by an elliptic PDE (using the Laplace operator):

$$\Delta F = g \tag{1}$$

and a set of boundary conditions. The most often met boundary condition type may be described with the following equations:

$$F(\boldsymbol{x})|_\Gamma = \nu(\boldsymbol{x}) \tag{2}$$

$$\left.\frac{\partial F(\boldsymbol{x})}{\partial \boldsymbol{n}}\right| = \nu(\boldsymbol{x}) \tag{3}$$

where $\boldsymbol{n}$ is a vector normal to region boundary.

Often he region Ω have "uncomfortable", in partucular, non-rectanle shape.

In its turn, the model of non-stationery field may be described the similar way. In addition to boundary conditions mentioned above, the time-dependent conditions should be taken into account in this case.

D. Grigoriev, J. Harrison, and E.A. Hirsch (Eds.): CSR 2006, LNCS 3967, pp. 393–402, 2006.

Note that domain the problem is being solved on can be either infinite of finite and can have an arbitrary shape. Such boundary value problems could be extremely resource-consuming. As most real objects are non-homogeneous and anisotropic, their models will result in PDEs with variable coefficients, which make solution process even more complex. In is almost impossible to solve such problems without software and hardware specifically designed for these particular tasks.

2   Existing Approaches

Existing methods of solving the described problem can be divided into two groups: analytical and numeric ones.

Analytical methods provide the exact analytical solution. On the other side, they cannot be applied to the arbitrary task, as these methods impose may constraints on the type of boundary conditions and/or region shape. In rare cases the problem can be transformed to the form required by the particular analytic method using the artificial approaches, but this transformation cannot be performed in generic case. Thus, application of analytic methods to the practical problems is almost impossible.

The second group is numerical methods. The common approach used by numerical methods is the replacing of the continuous problem by a discrete problem, usually by estimating the values of the decision function in the finite set of discretization nodes. The most widely used numerical methods of solving PDEs are the finite differences method and finite elements method. Other numerical methods — like particle-in-cell method and its derivatives — have the smaller area of application and are specific to concrete tasks. Also, many particular implementations of these methods impose constraints on the region shape and boundary condition types.

In spite of the fact the there's a whole range of more or less task-specific numeric methods, most of them come to giant systems of algebraic equations, containing thousands of variables and taking a vast amount of computational resources to be solved. The solution of typical task having the discretization mesh may take up to several hours on the modern average-class workstation; the would require the specialized hardware. Besides, an issue of parallelizing calculations is raised. As most equation system solution algorithms are non-parallel in their nature, thus this issue is of vital importance. There are some techniques allowing to parallelize such kind of calculations, but unfortunately they usually scale bad and may be applied only to specific matrix types. Therefore, there's a need in development and/or enhancement of methods permitting to solve the PDEs which would use the existing processing powers more effectively.

Several last years brought publications referring to neural networks as an approach to boundary-value problems solution, nevertheless, these publications are not numerous and describe only several particular problems. The question of interrelationships between the net topology and solution precision is explored insufficiently. Also, no concrete designs of hardware or software neural network implementations were described. This leads to conclusion that the theory of solving PDEs with the help of neural networks is still an actual unresolved problem.

At the moment, a three main techniques of solving PDEs with the help of neural networks may be marked out.

The first one uses the approach similar to the FEM, in particular, the PDE solution approximation with the set of simple basis functions, like a linear B-splines. These splines can be derived by the superposition of piecewise linear activation functions. Thus, the solution of a differential equation using linear B-splines can be directly mapped on the architecture of a feedforward neural network. Unfortunately, this method scales bad, as the number of nodes in the network depends on the number of basis functions; thus, as the size of discretization grid grows, the network size grows too, making the process of learning slow and ineffective for large grids. The same problem arise for the multidimensional problems. [1, 2]

The second approach bases on the fact that the the solution of a linear system of algebraic equations may be mapped onto the architecture of a Hopfield neural network. The minimization of the network's energy function gives the solutions of the mapped system. Thus, after the PDE's domain have been discretized, the resulting algebraic system could be solved using the neural network. Unfortunately, this approach is characterized by a long net learning time; the nets are problem-specific, so the network cannot be applied to another task after it learned to solve one. [3, 4]

The third one is closely related to analytical methods; it provides the differentiable solution using the neural-networks of specific structure. Unlike two approaches described above, this one provides the approximate analytical solution in form of superposition of two functions, one of which — satisfies the boundary conditions and another — is reduced to zero on the region's boundary and partially defined by the neural network . The neural network is a standard three-layer perceptron; the learning set is a set of apriori-known points of the problem solution. This method is rather fast by itself, but the proposed method of building the non neural-network-defined part of composite solution requires the rectangular shape of the PDE domain, which is not possible for all practical tasks. Besides, the process of perceptron learning requires the presence of more rough numerical solution of the problem, assuming that one of the classical methods should be used together with proposed. [5]

Also, besides the mentioned publications, publications examining the possibility of cellular (or "locally connected") neural networks application to the mathematical problems solution, in particular, for solving linear algebraic equation systems and PDEs. Nevertheless, no concrete theories or algorithms were offered in these works. [6, 7]

3   The Proposed Approach

A hypothesis is made that an technique exists allowing to build a cellular neural network for a arbitrary boundary-value problem; the stable state of the network may be directly mapped to the problem solution.

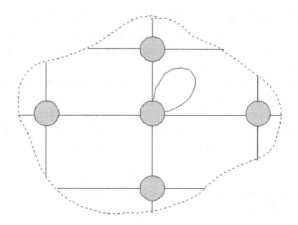

Fig. 1. A simple two-dimensional cellular network structure

In general case a cellular neural network is a multidimensional (possibly irregular) mesh, containing neurons in the mesh nodes. Unlike most other network topologies, neurons are connected only to their neighbors, contained in some area with small radius value. Most links are bidirectional; the neuron activation functions are simple, often linear. The advantage of this structure compared to other widely used network topologies is its simplicity and small number of links; this reduces the volume of computation resources required by the network, simplifies the parallelization of the calculation process and make the neural networks hardware implementation simple, cheaper and more reliable.

This approach is inspired by approach used in analog computers; on the other side, anaolog computer have a limited area of use. The proposed methodic should permit using modern widely-available digital electronic devices with similar methodics while retaining all analog modelling merits.

The proposed approach assumes that every solution domain discretization node is mapped to one network neuron. The neuron output value is treated as an approximate value of the PDE solution in mapped node. As number of links depends linearly on the neurons count, combinatorial explosion will not take place as the number of nodes increase. Provided that link weights will be chosen so that summary output of all neurons remains constant in case the network contains no neurons representing the boundary conditions, it can be proved that the cellular-network based model correctly represents the behavior of the physical systems, in particular, the law of energy conservation.

4   Solving Elliptic PDE Boundary-Value Problems

Let's examine the possibility of solving the elliptic PDEs with the cellular neural networks containing the only one layer. The structure of the network may be described by the following rules and definitions:

- the network is a n-dimensional rectangular grid; every neuron is mapped to discretization mesh node in the solution domain Ω;
- $L(\boldsymbol{X})$ is a number of neuron having an n-dimensional coordinate vector $\boldsymbol{X}$;
- f_i is an output value of the i-th neuron;
- designate the number of neuron preceding the i-th neuron by the j-th coordinate as $P(i,j)$, and the following one — as $S(i,j)$.
- assign the feedback link the weight $w_{ii} = b$, where $b \in (0,1)$ is an user-defined coefficient; it can be proved that the value of b will not affect the network stable state;
- assign all other links the weight

$$w_{ij} = \left\{ \begin{array}{c} \frac{1-b}{2n}, i \in N(j) \\ 0 \end{array} \right. \tag{4}$$

where N is a set of all neighbors of i-th neuron, n — a number of dimension of domain Ω;
- all neurons not representing boundary conditions use simplest linear activation functions $f(x) = x$
- boundary conditions having the form $F(\boldsymbol{x})|_\Gamma = \nu(\boldsymbol{x})$ are represented by neurons having constant activation functions

$$f_i(\boldsymbol{x}) = \nu(\boldsymbol{x}_i) = \nu_i = const \tag{5}$$

- boundary conditions having the form $\left. \frac{\partial F(\boldsymbol{x})}{\partial \boldsymbol{n}} \right|_\Gamma$ are represented by neurons having linear activation functions

$$f_i(x) = x + \nu(\boldsymbol{x}_i) = x + \nu_i \tag{6}$$

Note that the well-posed problem would have $\oint_\Gamma \frac{\partial F(\boldsymbol{x})}{\partial \boldsymbol{n}} = 0$, thus we may assume $\sum_i \nu_i = 0$

Note that network stable state will satisfy the following equation:

$$f_i = w_{ii} f_i + \sum_{j \in N(i)} (w_{ji} f_i) = w_{ii} f_i + \sum_{d=1}^{n} \left(w_{(P(i,d)i)} f_{P(i,d)} + w_{S(i,d),i} f_{S(i,d)} \right), \forall i \tag{7}$$

To prove that output values in stable state will satisfy the finite-difference form of the elliptic PDE transform this equation:

$$cf_i = bf_i + \sum_{d=1}^{n} \left(\frac{1-b}{2n} f_{P(i,d)} + \frac{1-b}{2n} f_{S(i,d)} \right) \tag{8}$$

$$nf_i = \sum_{d=1}^{n} \left(\frac{f_{P(i,d)} + f_{S(i,d)}}{2} \right) \tag{9}$$

$$\sum_{d=1}^{n} \left(\frac{f_{P(i,d)} - 2f_i + f_{S(i,d)}}{2} \right) \approx \Delta F = 0 \tag{10}$$

It is obvious that the last equation is equivalent to the simplest second-order finite-difference representation of the Laplace equation.

This result can be easily generalized on the higher-precision finite difference approximation using the network having the radius of the "neighborhood area" greater than 1. Note that the radius affects both approximation precision and network complexity, thus increasing the need in computational resources.

Let's examine the network stability issue. Network may not converge to the stable state due the following two reasons: positive feedbacks causing the neuron output values to grow indefinitely or network oscillation. Output values of the network is limited due the following reasons:

- if no boundary-conditions given (network corresponds to the fully isolated system) then the summary output is constant:

$$\sum_{j=1}^{N} f_i^{(t+1)} = \sum_{j=1}^{N}\sum_{i=1}^{N} f_i^{(t)} w_{ij} = \sum_{i=1}^{N}\sum_{j=1}^{N} f_i^{(t)} w_{ij} = \sum_{i=1}^{N} f_i^{(t)} = const \quad (11)$$

- network containing boundary conditions of type $\frac{\partial F(\boldsymbol{x})}{\partial \boldsymbol{n}}|_\Gamma = \nu(\boldsymbol{x})$:

$$\sum_{j=1}^{N} f_j^{(t+1)} = \sum_{j=1}^{N}\left(\sum_{i=1}^{N}\left(f_i^{(t)} w_{ij}\right) + \nu_j\right) = \sum_{i=1}^{N}\sum_{j=1}^{N} f_i^{(t)} w_{ij} + \sum_i \nu_i \quad (12)$$

using the restriction on the sum of ν_i, the following expression could be written:

$$\sum_{i=1}^{N}\sum_{j=1}^{N} f_i^{(t)} w_{ij} + \sum_j \nu_j = \sum_{i=0}^{N}\sum_{j=1}^{N} f_i^{(t)} w_{ij} = const \quad (13)$$

- network containing boundary conditions of type $F(\boldsymbol{x})|_\Gamma = \nu(\boldsymbol{x})$ (k designates the number of "internal" neurons, which do not define the boundary conditions).

$$\sum_{i=1}^{k} f_i^{(t+1)} = \sum_{i=1}^{k}\left(\left(\sum_{j=1}^{k} w_{ji} f_j^{(t)}\right) + \left(\sum_{j=k+1}^{N} w_{ji} f_i^{(t)}\right)\right) = \quad (14)$$

$$\sum_{i=1}^{k}\sum_{j=1}^{k} w_{ji} f_j^{(t)} + C = \sum_{j=1}^{k}\left(\sum_{i=1}^{k} w_{ji}\right) f_j^{(t)} + C$$

Designate $\sum_{i=1}^{k} w_{ji}$ as α_i; note that as $K < N$, then $\forall i : \alpha_i < 1$. Hence the system converges to the state $\sum_{i=1}^{N} (1 - \alpha_i) F_i^{(t)} = C$.

Let's show that network cannot oscillate: prove that no set of sequential states $A_1, \ldots, A_n$ exists such as :

$$A_{i+1} = T(A_i), \forall i \in (1, n-1), A_1 = T(A_n) \quad (15)$$

(T stands for one network state change; in our case when neurons have linear activation functions $T(X) = WX + G$)

If the set $A_1, \ldots, A_n$ exists, then

$$
\begin{cases}
A_1 = W A_n + G \\
\ldots \\
A_n = W A_{n-1} + G
\end{cases}
\tag{16}
$$

Consequently:

$$
A_1 = W^n A_1 + \left(\sum_{i=0}^{n-1} W^i \right) G
\tag{17}
$$

and

$$
(E - W^n) A_1 = \left(\sum_{i=0}^{n-1} W^i \right) G
\tag{18}
$$

Note that with chosen link weights the equality $W^n = E$ never comes true. Then, solving the simple matrix equation , we come to

$$
A_1 = \frac{\left(\sum_{i=1}^{n-1} W^i \right) G}{(E - W) \sum_{i=0}^{n-1} W^i} = \frac{G}{1 - W}
\tag{19}
$$

But in this case the following expression will be true:

$$
A_2 = W A_1 + G = W \frac{G}{1 - W} + G = \frac{G}{1 - W} = A_1
\tag{20}
$$

resulting in $A_1 = A_i, \forall i$. Thus, no set of distinct states $A_1, \ldots, A_n$ exists.

Hence, a conclusion can be made that the stable condition of the net is accessible, single and does match the approximate finite-difference approximation of the PDE problem.

5   Solving Parabolic PDE Boundary-Value Problems

The structure of the network may be described by the following rules and definitions:

- the network consists of several layers; every layer is a n-dimensional rectangular grid; every neuron is mapped to discretization mesh node in the solution domain Ω;
- $L(X)$ is a number of neuron having an n-dimensional coordinate vector X in a layer t;
- values f_i, $P(i,j)$ and $S(i,j)$ have the same meaning as for elliptic PDEs;
- variable coefficient values $k(X, t)$ are referred as $k_{L(X,t)}$;
- an auxiliary value is introduced:

$$
\hat{k}_{L(X,t)} = \frac{\tau}{h^2} \sum_{i=1}^{n} \left(\frac{2k_{L(P(X,i),t)}}{k_{L(P(X,i),t)} + k_{L(X,t)}} + \frac{2k_{L(S(X,i),t)}}{k_{L(S(X,i),t)} + k_{L(X,t)}} \right)
\tag{21}
$$

note that if $k = const$, $\hat{k} = \frac{\tau}{h^2} 2n$

– assign the feedback link the weight

$$w_{ii} = b - \frac{1-b}{k_i \hat{k}_i} \qquad (22)$$

where $b \in (0,1)$ is an user-defined coefficient;
– assign all other links between neurons inside the same layer the weight

$$w_{ij} = \begin{cases} \frac{1-b}{k_j} \frac{2k_i}{k_j+k_i}, & i \in N(j) \\ 0 \end{cases} \qquad (23)$$

– assign links between layers the weight:

$$w_{ij} = \begin{cases} w_{L(\boldsymbol{X},i),L(\boldsymbol{Y},i+1)} = 0, \ \boldsymbol{X} \neq \boldsymbol{Y}, \\ w_{L(\boldsymbol{X},i),L(\boldsymbol{Y},i+1)} = \frac{1-b}{k_{L(\boldsymbol{X},i)}\hat{k}_{L(\boldsymbol{X},i)}} \end{cases} \qquad (24)$$

– neuron activation functions are selected using the same rules as for elliptic PDE.

Note that network stable state will satisfy the following equation:

$$cf_{L(\boldsymbol{X},t)} = w_{L(\boldsymbol{X},t),L(\boldsymbol{X},t)}f_{L(\boldsymbol{X},t)} + \quad (25)$$

$$+ \sum_{i=1}^{n} \left(w_{L(P(\boldsymbol{X},i),t),L(\boldsymbol{X},t)}f_{L(P(\boldsymbol{X},i),t)} + w_{L(S(\boldsymbol{X},i),t),L(\boldsymbol{X},t)}f_{L(S(\boldsymbol{X},i),t)} \right) +$$

$$+ w_{L(\boldsymbol{X},t-1),L(\boldsymbol{X},t)}f_{L(\boldsymbol{X},t-1)}$$

Under the condition $k = const$ the selected model satisfies the finite-difference approximation of the parabolic PDE with constant coefficients. The proof is simple; the equation describing the stable state can be transformed the following way:

$$cccf_{L(\boldsymbol{X},t)} = \left(b - \frac{1-b}{2kn(\tau/h^2)} \right) f_{L(\boldsymbol{X},t)} + \quad (26)$$

$$+ \frac{1-b}{2n(\tau/h^2)} \sum_{i=1}^{n} \left(f_{L(P(\boldsymbol{X},i),t)} + f_{L(S(\boldsymbol{X},i),t)} \right) + \frac{1-b}{2kn(\tau/h^2)} f_L \boldsymbol{X}, t-1$$

$$\left(2kn\frac{\tau}{h^2}(1-b) + (1-b) \right) f_{L(\boldsymbol{X},t)} = \quad (27)$$

$$= k(1-b) \sum_{i=1}^{n} \left(f_{L(P(\boldsymbol{X},i),t)} + f_{S(P(\boldsymbol{X},i),t)} + (1-b)f_{L(\boldsymbol{X},t-1)} \right)$$

$$f_{L(\boldsymbol{X},t)} - f_{L(\boldsymbol{X},t-1)} = k \left(\frac{\tau}{h^2} \right) \sum_{i=1}^{n} \left(f_{L(P(\boldsymbol{X},i),t)} - 2f_{L(\boldsymbol{X},t)} + f_{L(S(\boldsymbol{X},i),t)} \right) \quad (28)$$

the latter expression is obviously equivalent to the finite-difference form.

Under the condition $k \neq const$ the selected model satisfies the finite-difference approximation of the parabolic PDE with variable coefficients. In this case the stable state equation will take the following form:

$$cccf_{L(\boldsymbol{X},t)} = \left(b - \frac{1-b}{k\hat{k}}\right) f_{L(\boldsymbol{X},t)} + \qquad (29)$$

$$+\frac{1-b}{\hat{k}} \sum_{i=1}^{n} \left(\frac{2k_{L(P(\boldsymbol{X},i),t)}}{k_{L(\boldsymbol{X},t)} + k_{L(P(\boldsymbol{X},i),t)}} + \frac{2k_{L(S(\boldsymbol{X},i),t)}}{k_{L(\boldsymbol{X},t)} + k_{L(S(\boldsymbol{X},i),t)}}\right) +$$

$$\frac{1-b}{k\hat{k}} f_{L(\boldsymbol{X},t-1)}$$

After similar transformation it comes to:

$$f_{L(\boldsymbol{X},t)} - f_{L(\boldsymbol{X},t-1)} = k\hat{k}f_{L(\boldsymbol{X},t)} + \qquad (30)$$

$$+k_{L(\boldsymbol{X},t)} \sum_{i=1}^{n} \left(\frac{2k_{L(P(\boldsymbol{X},i),t)}}{k_{L(\boldsymbol{X},t)} + k_{L(P(\boldsymbol{X},i),t)}} f_{L(P(\boldsymbol{X},i),t)} +\right.$$

$$\left. + \frac{2k_{L(S(\boldsymbol{X},i),t)}}{k_{L(\boldsymbol{X},t)} + k_{L(S(\boldsymbol{X},i),t)}} f_{L(S(\boldsymbol{X},i),t)} +\right)$$

After regrouping items it take the form equivalent to finite-difference representation:

$$f_{L(\boldsymbol{X},t)} - f_{L(\boldsymbol{X},t-1)} = \sum_{i=1}^{n} \left(\frac{2k_{L(\boldsymbol{X},t)}k_{L(P(\boldsymbol{X},i),t)}}{k_{L(\boldsymbol{X},t)} + k_{L(P(\boldsymbol{X},i),t)}} \left(f_{L(P(\boldsymbol{X},i),t)} - f_{L(\boldsymbol{X},t)}\right) +\right.(31)$$

$$\left.\frac{2k_{L(\boldsymbol{X},t)}k_{L(S(\boldsymbol{X},i),t)}}{k_{L(\boldsymbol{X},t)} + k_{L(S(\boldsymbol{X},i),t)}} \left(f_{L(S(\boldsymbol{X},i),t)} - f_{L(\boldsymbol{X},t)}\right) +\right)$$

The proof of network stability is analogous to the presented in the section devoted to elliptic PDEs.

6   Conclusion

Hence, a conclusion can be made that it is possible to build a neural network which does not require learning and is able to get the approximate numeric solution of boundary-value elliptic or parabolic PDE problem on the arbitrary finite domain in a finite time. The precision of such solution is comparable to the solutions base on widely used numerical method, while neural-network-based approach allows to use parallel calculations more effectively.

References

1. A.J.Meade, Jr and A.A.Fernandez. The numerical solution of linear ordinary differential equations by feedforward neural networks // Mathematical Computer Modeling. 1994. V.19. No. 12. Pp. 1-25.
2. A.J.Meade, Jr and A.A.Fernandez. Solution of Nonlinear Ordinary Differential Equations by Feedforward Neural networks // Mathematical Computer Modeling. 1994. V.20. No. 9, Pp. 19-44.

3. H.Lee and I.Kang. Neural algorithms for solving differential equations // Journal of Computational Physics. 1990. V. 91. Pp.110-117.
4. L.Wang and J.M.Mendel. Structured trainable networks for matrix algebra // IEEE Int. Joint Conference on Neural Networks. 1990. V. 2. Pp. 125-128.
5. I.E. Lagaris, A. Likas and D.I. Fotiadis. Artificial Neural Networks for Solving Ordinary and Partial Differential Equations // IEEE Trans. on Neural Networks. 1998. V. 4. P. 987-1000.
6. Luigi Fortuna, Paolo Arena, David Balya, Akos Zarandy. Cellular Neural Networks: A Paradigm for Nonlinear Spatio-Temporal Processing // Circuits and Systems Magazine, IEEE, 2001. V. 1, N 4, Pp. 6-21.
7. Csaba Rekeczky , Istvan Szatmari, Peter Foldesy and Tamas Roska. Analogic Cellular PDE Machines // Analogical and Neural Computing Systems Laboratory. Computer and Automation Research Institute. Hungarian Academy of Sciences, Budapest, Hungary. Tech. Report, 2004. Pp. 1-6.

Approximate Methods for Constrained Total Variation Minimization*

Xiaogang Dong and Ilya Pollak**

School of Electrical and Computer Engineering, Purdue University,
465 Northwestern Ave., West Lafayette, IN 47907, USA
{dongx, ipollak}@ecn.purdue.edu
Tel.: +1-765-494-3465, +1-765-494-5916; Fax: +1-765-494-3358

Abstract. Constrained total variation minimization and related convex optimization problems have applications in many areas of image processing and computer vision such as image reconstruction, enhancement, noise removal, and segmentation. We propose a new method to approximately solve this problem. Numerical experiments show that this method gets close to the globally optimal solution, and is 15-100 times faster for typical images than a state-of-the-art interior point method. Our method's denoising performance is comparable to that of a state-of-the-art noise removal method of [4]. Our work extends our previously published algorithm for solving the constrained total variation minimization problem for 1D signals [13] which was shown to produce the globally optimal solution exactly in $O(N \log N)$ time where N is the number of data points.

1 Problem Statement

Suppose the observed data $\mathbf{f}$ is a noisy measurement of an unknown image $\mathbf{g}$:

$$\mathbf{f} = \mathbf{g} + \mathbf{w},$$

where $\mathbf{w}$ is zero-mean additive white noise with variance σ^2. It was proposed in [16] to recover an estimate $\hat{\mathbf{g}}$ of $\mathbf{g}$ by solving the following problem of constrained minimization of the total variation (TV):

$$\hat{\mathbf{g}} = \arg \min_{\mathbf{g}:\, \|\mathbf{f}-\mathbf{g}\| \leq \sigma} \mathrm{TV}(\mathbf{g}). \tag{1}$$

A number of more effective noise removal paradigms have since been developed [15,10]. However, problem (1) and related variational and PDE-based methods have been successfully used in a variety of other application areas such as tomographic reconstruction [8, 17], deblurring [3, 6, 12, 20], and segmentation [14, 9]. This motivates continued interest in problem (1) as well as the need to develop fast algorithms for solving it.

---

* This work was supported in part by the National Science Foundation (NSF) through CAREER award CCR-0093105 and through grant IIS-0329156.
** Corresponding author.

D. Grigoriev, J. Harrison, and E.A. Hirsch (Eds.): CSR 2006, LNCS 3967, pp. 403–414, 2006.

The original formulation of [16] treated continuous-space images for which the total variation is defined as follows:

$$TV(\mathbf{g}) = \int |\nabla \mathbf{g}|.$$

A numerical procedure was developed in [16] based on discretizing the corresponding Euler-Lagrange equations. Since then, many other numerical schemes have been proposed to approximately solve this and related optimization problems [3, 4, 5, 6, 11, 12, 14, 19, 20].

We consider a discrete formulation of the problem in which $\mathbf{f}$ and $\mathbf{g}$ are both images defined on an undirected graph $\mathcal{G} = (\mathcal{N}, \mathcal{L})$ where $\mathcal{N}$ is a set of nodes (pixels) and $\mathcal{L}$ is the set of links which define the neighborhood structure of the graph. We use the following definition of the total variation for such discrete images:

$$TV(\mathbf{g}) \stackrel{\triangle}{=} \sum_{\{m,n\} \in \mathcal{L}} |g_m - g_n|. \tag{2}$$

The norm we use in Eq. (1) is the ℓ_2 norm:

$$\|\mathbf{f} - \mathbf{g}\| = \sqrt{\sum_{n \in \mathcal{N}} (f_n - g_n)^2}.$$

The optimization problem (1) can then be cast as a second-order cone program (SOCP), i.e., the minimization of a linear function over a Cartesian product of quadratic cones [2, 1, 18]. The globally optimal solution to this SOCP problem can be obtained with interior point methods whose computational complexity is $O(N^2 \log \varepsilon^{-1})$ where $N = |\mathcal{N}|$ is the number of pixels, and ε is a precision parameter [18, 11].

The main contribution of the present paper is the development of a suboptimal algorithm which we empirically show to be about 15-100 times faster than a state-of-the-art interior point method, for typical natural images. This algorithm is also empirically shown to achieve values of $TV(\mathbf{x})$ which are quite close to the globally optimal ones achieved by SOCP. Moreover, the images recovered by the new method and via SOCP are visually very similar. We also experimentally evaluate our algorithm's performance as a denoising method, using the algorithm of [4] as a benchmark. We show that the two algorithms perform comparably. A specialization of this algorithm to 1D discrete signals was proposed and analyzed in [13]. This specialization was shown to exactly solve problem (1) in 1D, in $O(N \log N)$ time, and with $O(N)$ memory complexity. Its variants which have the same complexity were shown in [13] to exactly solve two related discrete 1D problems, specifically, the Lagrangian version of problem (1) and the minimization of $\|\mathbf{f} - \mathbf{g}\|$ subject to a constraint on $TV(\mathbf{g})$.

We note in addition that, if the graph $\mathcal{G}$ is a regular rectangular grid, then each pixel n can be represented by its horizontal and vertical coordinates i and j. In this case, another possible discretization of the TV is:

$$TV'(\mathbf{g}) \stackrel{\triangle}{=} \sum_{i,j} \sqrt{(g_{i+1,j} - g_{i,j})^2 + (g_{i,j+1} - g_{i,j})^2}. \tag{3}$$

With this definition of the discrete TV, the optimization problem (1) can also be cast as an SOCP problem, as shown in [11]. It can therefore also be solved using interior point methods for SOCP.

2   Notation

We define a real-valued image on an arbitrary finite set $\mathcal{N}$ of points as any function which assigns a real number to every point in $\mathcal{N}$. For the algorithm described in Section 3, it is important to define adjacency relationships on the points in $\mathcal{N}$, and therefore we assume that $\mathcal{N}$ is the set of nodes of an undirected graph $\mathcal{G} = (\mathcal{N}, \mathcal{L})$ where the set $\mathcal{L}$ of *links* consists of unordered pairs of distinct nodes. If $\{m, n\}$ is a link, we say that the nodes m and n are *neighbors*. For example, $\mathcal{G}$ could be a finite 2D rectangular grid where each node has four neighbors: east, west, north, and south, as in Fig. 1. We say that $R \subset \mathcal{N}$ is a *connected set* if, for any two nodes $m, n \in R$, there exists a *path* between m and n which lies entirely within R—i.e., if there exists a sequence of links of the form $\{m, m_1\}, \{m_1, m_2\}, \ldots, \{m_{k-1}, m_k\}, \{m_k, n\}$ with $m_1, m_2, \ldots, m_k \in R$.

If $N = |\mathcal{N}|$ is the total number of nodes, then an image $\mathbf{u}$ on $\mathcal{N}$ can be thought of as an N-dimensional vector: $\mathbf{u} \in \mathbb{R}^N$. We say that a set $\mathcal{S} = \{R_1, \ldots, R_I\}$ is a *segmentation* of an image $\mathbf{u}$ if:

- every R_i is a connected set of nodes;
- the image intensity within every R_i is constant, $u_m = u_n$ for all $m, n \in R_i$ for $i = 1, \ldots, I$;
- $R_1, \ldots, R_I$ are pairwise disjoint sets whose union is $\mathcal{N}$.

We then say that R_i is a *region* of $\mathcal{S}$. When convenient, we also say in this case that R_i is a region of $\mathbf{u}$. For example, two segmentations of the image in Fig. 1(a) are shown in Figs. 1(b,c). We use μ_R to denote the common intensity within region R and $|R|$ to denote the number of pixels in R.

Given a segmentation $\mathcal{S}$ of an image $\mathbf{u}$, two regions $R, R' \in \mathcal{S}$ are called *neighbors* if there exist two nodes $m \in R$, $n \in R'$ which are neighbors, i.e., such that $\{m, n\} \in \mathcal{L}$. The *multiplicity* $\lambda_{R,R'}$ of two neighbor regions R and R' is the

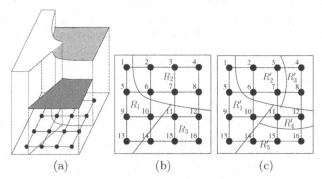

Fig. 1. (a) Image $\mathbf{u}$. (b) Segmentation $\mathcal{S}$. (c) Segmentation $\mathcal{S}'$.

length of the boundary between them—i.e., the number of links $\{m, n\}$ such that $m \in R$ and $n \in R'$. We let NBR-PRS$_{\mathcal{S}}$ be the set of all pairs of neighbor regions in $\mathcal{S}$, and we let NBRS$_{\mathcal{S}}(R)$ be the set of all regions in $\mathcal{S}$ that are neighbors of a region $R \in \mathcal{S}$. For example, in the segmentation $\mathcal{S}$ of Fig. 1(b), each region is a neighbor of the other two, and the multiplicities are $\lambda_{R_1,R_2} = 3$, $\lambda_{R_1,R_3} = 3$, and $\lambda_{R_2,R_3} = 2$. In the segmentation $\mathcal{S}'$ of Fig. 1(c), NBRS$_{\mathcal{S}'}(R_1') = \{R_2', R_4', R_5'\}$.

3   Algorithm Description

We define a dynamical system which generates a family of images $\{\mathbf{u}(t)\}_{t=0}^{\infty}$, parameterized by a *time parameter* t. We suppose that the initial data for this system is the observed noisy image, $\mathbf{u}(t = 0) = \mathbf{f}$. We let $\mathcal{S}(t)$ be the segmentation of $\mathbf{u}(t)$ such that $\mu_R(t) \neq \mu_{R'}(t)$ for any pair of neighbor regions $R, R' \in \mathcal{S}(t)$. The output of our algorithm is the image $\mathbf{u}(t^*)$ at such time t^* that $\|\mathbf{f} - \mathbf{u}(t^*)\| = \sigma$. The basic reason for the fact that our algorithm is fast is that it does not explicitly compute the solution $\mathbf{u}(t)$ for any $t \neq t^*$. The basic reason for the fact that it achieves values of the TV which are close to globally optimal ones, is the fact that the underlying dynamical system is based on the gradient descent for the TV, as we presently explain. The remainder of the section is devoted to the description of the dynamical system and our algorithm for computing $\mathbf{u}(t^*)$.

We first rewrite Eq. (2) for any image $\mathbf{u}(t)$, as follows, using the notation introduced in the previous section:

$$\text{TV}(\mathbf{u}(t)) = \sum_{\{R,R'\} \in \text{NBR-PRS}_{\mathcal{S}(t)}} \lambda_{R,R'} \cdot |\mu_R(t) - \mu_{R'}(t)|.$$

It is shown in [9] that if the gradient is taken in the space of all images which are piecewise constant on $\mathcal{S}(t)$, then the gradient descent for the TV is given by

$$\dot{\mu}_R(t) = \frac{1}{|R|} \sum_{R' \in \text{NBRS}_{\mathcal{S}(t)}(R)} \lambda_{R,R'} \cdot \text{sgn}(\mu_{R'}(t) - \mu_R(t)), \qquad (4)$$

where $\mu_R(t)$ denotes the intensity within region R of image $\mathbf{u}(t)$. This equation is valid as long as $\mathcal{S}(t) = \text{const}$, i.e., as long as $\mu_R(t) \neq \mu_{R'}(t)$ for every pair of neighbor regions R and R'. As soon as $\mu_R(t)$ and $\mu_{R'}(t)$ become equal for some pair of neighbor regions R and R', their respective rates of evolution $\dot{\mu}_R(t)$ and $\dot{\mu}_{R'}(t)$ become undefined since the right-hand side of Eq. (4) undergoes a discontinuity in this case. To handle this scenario, we supplement Eq. (4) with the following rule.

Region merging: How and when. Suppose that for some time instant $t = \tau_{R,R'}$ we have: $R, R' \in \mathcal{S}(\tau_{R,R'}^-)$ and $\mu_R(\tau_{R,R'}^-) = \mu_{R'}(\tau_{R,R'}^-)$. Then we merge R and R' into a new region $R \cup R'$, with the same intensity:

$$\mathcal{S}(\tau_{R,R'}^+) = \mathcal{S}(\tau_{R,R'}^-) \backslash \{R, R'\} \cup \{R \cup R'\},$$

$$\mu_{R \cup R'}(\tau_{R,R'}^+) = \mu_R(\tau_{R,R'}^-) = \mu_{R'}(\tau_{R,R'}^-).$$

In addition, as shown through numerical experiments in the next section, it may sometimes be beneficial to split a region into two different regions. We postpone until later the discussion of when our algorithm decides to split a region. Once it does, region splitting occurs as follows.

Region splitting: How. Splitting of region R into R' and $R\backslash R'$ at some time instant $t = \tau_R$ means that new regions R' and $R\backslash R'$ are formed, and that they have the same intensity as R:

$$\mathcal{S}(\tau_R^+) = \mathcal{S}(\tau_R^-)\backslash\{R\} \cup \{R, R\backslash R'\},$$
$$\mu_{R'}(\tau_R^+) = \mu_{R\backslash R'}(\tau_R^+) = \mu_R(\tau_R^-)$$

Time instants when regions are merged and split are called *event times*. Given a time T and the corresponding segmentation $\mathcal{S}(T)$, we define the *birth time* b_R for every region $R \in \mathcal{S}(T)$ as $b_R \stackrel{\triangle}{=} \sup\{t < T : R \notin \mathcal{S}(t)\}$. Similarly the *death time* d_R of R is defined as $d_R \stackrel{\triangle}{=} \inf\{t > T : R \notin \mathcal{S}(t)\}$ Note that b_R can be either zero, or a time instant when two regions get merged to form R, or a time instant when R is formed as a result of splitting another region into two; d_R can be either a time instant when R is merged with a neighbor to form another region, or a time instant when R is split into two other regions.

Let

$$\beta_R = \sum_{R' \in \mathrm{NBRS}_{\mathcal{S}(b_R)}(R)} \lambda_{R,R'} \cdot \mathrm{sgn}(\mu_{R'}(b_R^+) - \mu_R(b_R^+)), \tag{5}$$

$$v_R = \beta_R/|R|. \tag{6}$$

Note that our split and merge rules are such that the intensity of every pixel is a continuous function of t. The continuity property means that the righthand side of Eq. (4) is constant for $b_R < t < d_R$, and is equal to v_R:

$$\dot{\mu}_R(t) = v_R \qquad \text{for } b_R < t < d_R. \tag{7}$$

For each region $R \in \mathcal{S}(t)$, we therefore have:

$$\mu_R(t) = \mu_R(b_R) + (t - b_R) \cdot v_R, \qquad \text{for } b_R < t < d_R. \tag{8}$$

Let $\bar{u}_R^0 \stackrel{\triangle}{=} \dfrac{1}{|R|} \sum_{n \in R} f_n$ be the average of the initial data $\mathbf{f}$ over the set R. For the regions R with $b_R = 0$, we have $\mu_R(b_R) = \bar{u}_R^0$, and therefore the following holds:

$$\mu_R(t) = \bar{u}_R^0 + t \cdot v_R, \qquad \text{for } b_R < t < d_R. \tag{9}$$

In order for our algorithm to be fast, it is important that Eq. (9) hold not only for regions with $b_R = 0$, but also for all other regions, at all times t. Proposition 1 below relies on this property. It is straightforward to show that, if this equation holds for every region for all times $t < \tau_{R,R'}$, and if regions R and R' get merged at the time instant $t = \tau_{R,R'}$, then Eq. (9) will also hold for the new region $R \cup R'$. We are now finally in a position to state our strategy for splitting regions.

Region splitting: When. We use two criteria to determine whether a region R is to be split into two regions R' and $R\backslash R'$. First, we check whether this split is consistent with the dynamics (8)—in other words, we check that R and $R\backslash R'$ will not be merged back together immediately after they are split. Second, we determine if there exists a time instant τ_R at which a split can be performed in such a way that, on the one hand, the intensity of every pixel is a continuous function of time, and, on the other hand, Eq. (9) is satisfied for the new regions. Note that, since there are $O(2^{|R|})$ possible two-way splits of R, searching over all possible splits is not computationally feasible. Instead, we only search over a small number of possible splits, namely, only horizontal and vertical splits that result in at least one of R', $R\backslash R'$ being a rectangle. This can be efficiently accomplished through an algorithm that walks around the boundary of R.

We now describe the termination of the algorithm. It is based on the following proposition, which can be proved using Eq. (9).

Proposition 1. *Let* $\alpha(t) = \|\mathbf{u}(t) - \mathbf{u}(0)\|^2$. *Then, for* $t \in [0, \infty)$, $\alpha(t)$ *is a monotonically increasing function of time, which changes continuously from 0 to* $\|\mathbf{u}(0) - \overline{\mathbf{u}(0)}\|^2$ *where* $\overline{\mathbf{u}(0)}$ *is the constant image whose every pixel is equal to the average intensity of the initial data* $\mathbf{u}(0)$. *It is a differentiable function of time except at the times of merges and splits, and its rate of change is:*

$$\dot{\alpha}(t) = 2t \sum_{R \in \mathcal{S}(t)} \frac{\beta_R^2}{|R|}.$$

The algorithm starts by checking whether $\|\mathbf{f} - \bar{\mathbf{f}}\|^2 \le \sigma^2$. If this is true, we stop the algorithm and use $\bar{\mathbf{f}}$ as the output. Otherwise, we initialize $\alpha(0) = 0$. Given $\alpha(\tau_l)$ at the current event time τ_l, we use the following equation, which is derived from Proposition 1, to calculate $\alpha(\tau_{l+1})$ at the next event time τ_{l+1}:

$$\alpha(\tau_{l+1}) = \alpha(\tau_l) + (\tau_{l+1}^2 - \tau_l^2) \sum_{R \in \mathcal{S}(\tau_l)} \frac{\beta_R^2}{|R|}.$$

The algorithm keeps running until $\alpha(\tau_{l+1}) > \sigma^2$ for a certain l. Then our algorithm's termination time t^* can be calculated as follows:

$$t^* = \sqrt{\tau_l^2 + \frac{\sigma^2 - \alpha(\tau_l)}{\sum_{R \in \mathcal{S}(\tau_l)} \frac{\beta_R^2}{|R|}}}.$$

We then use Eq. (9) to calculate $\mu_R(t^*)$ for each $R \in \mathcal{S}(t^*)$, and output $\mathbf{u}(t^*)$.

Putting everything together, we have the following outline of the algorithm.

1. **Initialize**

   If $\bar{\mathbf{f}}$ satisfies the constraint, output $\bar{\mathbf{f}}$ and terminate. Otherwise, initialize t_0 to be zero, $\mathbf{u}(t_0)$ to be the data $\mathbf{f}$, the initial segmentation to consist of singleton regions, and the neighborhood structure to be the standard four-neighbor grid. Initialize the parameters $|R|$, b_R, β_R, $\mu_R(t_0)$, and $\lambda_{R,R'}$ according to the definitions above.

2. **Find Potential Merge Times**
   Assuming that the intensity $\mu_R(t)$ of every region evolves according to Eq. (9), find, for every pair of neighbor regions R and R', the time $\tau_{R,R'}$ which is the earliest time when these two regions have equal intensities.

3. **Construct the Event Heap**
   Store all potential merge events on a binary min-heap [7] sorted according to the merge times.

4. **Merge, Split, or Stop**
   Extract the root event from the heap. Calculate $\alpha(\tau)$ where τ is the event time. If $\alpha(\tau) \geq \sigma$, go to Step 7.
   if the event is a merge event
        merge;
   if the event is a split event
        split;

5. **Update the Heap**
   Decide whether the newly formed regions may be split. If so, add the corresponding split events to the event heap. Add the merge events for the newly formed regions to the event heap. Remove all the events involving the discarded regions from the heap.

6. **Iterate**
   Go to Step 4.

7. **Output**
   Calculate t^*; calculate and output $\mathbf{u}(t^*)$.

4   Comparison to SOCP

Given a noise-free image $\mathbf{g}$ with dynamic range $[0,1]$, we generate noisy images $\mathbf{f}$ by adding white Gaussian noise with standard deviation σ ranging from 0.01 to 0.30. The correct value of σ is used in the simulations both for our

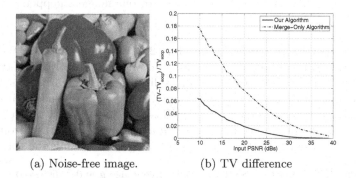

(a) Noise-free image.   (b) TV difference

Fig. 2. (a) Noise-free "peppers" image. (b) Percentage difference between the TV for our split-and-merge algorithm and the optimal TV obtained via SOCP, for many different input PSNR levels (solid line); percentage difference between the TV for our merge only algorithm and the optimal TV (dashdot).

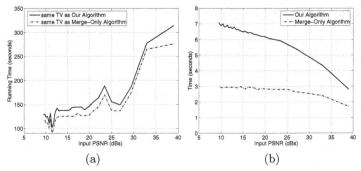

Fig. 3. (a) Running time for SOCP; (b) running time for our algorithms

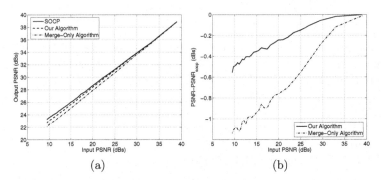

Fig. 4. (a) Output PSNR as a function of the input PSNR. (b) Output PSNR difference between our algorithm and SOCP.

algorithm and for the SOCP-based algorithm. We use the MOSEK software as the solver for SOCP. It implements a state-of-the-art interior point method [2]. Running the comparative experiments on 12 different natural images yields very similar results; we only provide the results for one image, "peppers," shown in Fig. 2(a).

SOCP solver converges to the globally optimal solution whereas this is not necessarily the case for our algorithm. Fig. 2(b) shows how close our algorithm gets to the globally optimal value for the total variation, for a range of typical input PSNRs.[1] As shown in the figure, our split-and-merge algorithm (solid lines) essentially finds the globally optimal solution at high PSNRs and is within 7% of the globally optimal solution at low PSNRs. Note also that the merge-only version of our algorithm (dashed lines) is about a factor of three farther from the optimal total variation than the split-and-merge algorithm. While the optimization

---

[1] The definition we use for PSNR for a noise-free image $\mathbf{g}$ and a distorted image $\hat{\mathbf{g}}$ is:

$$PSNR = 10 \log \left(\frac{(\max \mathbf{g})^2 \cdot N}{\|\hat{\mathbf{g}} - \mathbf{g}\|^2} \right)$$

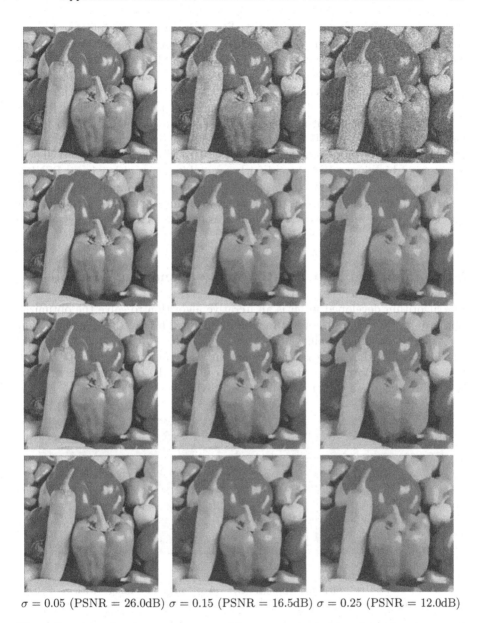

$\sigma = 0.05$ (PSNR $= 26.0$dB) $\sigma = 0.15$ (PSNR $= 16.5$dB) $\sigma = 0.25$ (PSNR $= 12.0$dB)

Fig. 5. Denoising for the peppers image. First row: noisy images. Second row: restored via our split-and-merge algorithm. Third row: restored via our merge-only algorithm. Fourth row: restored via SOCP.

performance of the split-and-merge algorithm is very similar to that of the SOCP, their running times are drastically different. In order to make the comparison of the running times as fair to the SOCP method as possible, we only calculate its running time to get to the total variation achieved by our algorithm, rather than the running time until convergence.[2] As Fig. 3 shows, the running time for our split-and-merge algorithm is 15 to 100 times lower than that for the interior point method. Note also that the merge-only version of our algorithm is about twice as fast as the split-and-merge, and that a similar comparison to SOCP shows that the merge-only algorithm is 30 to 150 times faster than the SOCP method.

The visual appearance of the output images and the corresponding PSNRs is very similar for the split-and-merge algorithm and the SOCP interior point method. The output PSNR is displayed in Fig. 4(a) as a function of the input PSNR. The PSNR differences displayed in Fig. 4(b) reveal virtually identical performance at high input PSNRs and fairly small differences at low input PSNRs. Moreover, the visual quality of the output images is very similar, as evidenced by Fig. 5. The visual quality of the output images for the merge-only algorithm is very similar as well; however, these images have lower PSNRs, especially for low input PSNRs.

5   Comparison to a Related Image Denoising Method [4]

The algorithm developed in [4] is an iterative algorithm to solve the optimization problem (1) where the total variation is defined by Eq. (3). The algorithm converges to the unique solution of (1). Since the algorithm in [4] and our split-and-merge algorithm address different optimization problems, we compare their noise removal performance. Specifically, we compare the running times of these two algorithms when they reach the same PSNR.

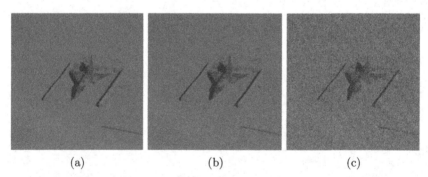

(a)   (b)   (c)

Fig. 6. (a) Noise-free "airplane" image. (b) Noisy "airplane" image with $\sigma = 0.05$ (PSNR = 26.0dB). (c) Noisy "airplane" image with $\sigma = 0.15$ (PSNR = 16.5dB).

The setup of our experiments is similar to the previous section. The noise-free images have the dynamic range $[0, 1]$. The noisy images are generated by adding white Gaussian noise with various standard deviations σ. The correct value of σ is assumed to be known in the simulations. We first apply our split-and-merge algorithm to noisy images and calculate the running times and the PSNRs of

---

[2] Note also that presolving and matrix reordering are not counted towards the running time of SOCP since these procedures are reusable for images with the same size [11].

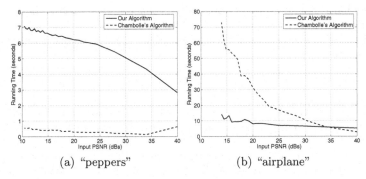

(a) "peppers"   (b) "airplane"

Fig. 7. Running times for our split-and-merge algorithm (solid lines) and the algorithm of [4] (dashed lines) to reach the same levels of PSNR

the denoised images. We then rerun each simulation using the algorithm of [4], until the same PSNR is obtained. The results for two images, for a range of input PSNRs, are shown in Fig. 7. For the "peppers" image, the algorithm of [4] is faster than our algorithm, by a factor of 4-33. For the "airplane" image, the two algorithms perform similarly at high PSNRs whereas our algorithm is up to about 6 times faster at low PSNRs. Since our algorithm is a multiscale segmentation process which works with flat regions, it is better adapted to images such as "airplane" which have large homogeneous regions.

References

1. F. Alizadeh and D. Goldfarb. Second-order cone programming. *Math. Program.*, 95(1):3–51, Jan. 2003.
2. E. D. Andersen, C. Roos, and T. Terlaky. On implementing a primal-dual interior-point method for conic quadratic optimization. *Math. Program.*, 95(2):249–277, Feb. 2003.
3. P. Blomgren and T. F. Chan. Modular solvers for image restoration problems using the discrepancy principle. *Numer. Linear Algebra Appl.*, 9:347–358, 2002.
4. A. Chambolle. An algorithm for total variation minimization and applications. *J. Math. Imaging Vis.*, 20:89–97, 2004.
5. T. F. Chan, G. H. Golub, and P. Mulet. A nonlinear primal-dual method for total variation-based image restoration. *SIAM J. Sci. Comput.*, 20(6):1964–1977, 1999.
6. P. L. Combettes and J. Luo. An adaptive level set method for nondifferentiable constrained image recovery. *IEEE Trans. Image Proc.*, 11(11):1295–1304, Nov. 2002.
7. T. H. Cormen, C. E. Leiserson, R. L. Rivest, and C. Stein. *Introduction to Algorithms.* MIT Press, Cambridge, MA, second edition, 2001.
8. D. C. Dobson and F. Santosa. An image-enhancement technique for electrical impedance tomography. *Inverse Problems*, 10:317–334, 1994.
9. X. Dong and I. Pollak. Multiscale segmentation with vector-valued nonlinear diffusions on arbitrary graphs. *IEEE Trans. Image Proc.* To appear.
10. A. Foi, V. Katkovnik, and K. Egiazarian. Pointwise shape-adaptive DCT as an overcomplete denoising tool. In *Proc. Int. TICSP Workshop on Spectral Methods and Multirate Signal Processing, SMMSP 2005*, Riga, Latvia, June 2005.

11. D. Goldfarb and W. Yin. Second-order cone programming methods for total variation-based image restoration. *SIAM J. Sci. Comput.*, 27(2):622–645, 2005.

12. Y. Li and F. Santosa. A computational algorithm for minimizing total variation in image restoration. *IEEE Trans. Image Proc.*, 5(6):987–995, June 1996.

13. I. Pollak, A. S. Willsky, and Y. Huang. Nonlinear evolution equations as fast and exact solvers of estimation problems. *IEEE Trans. Signal Proc.*, 53(2):484–498, Feb. 2005.

14. I. Pollak, A. S. Willsky, and H. Krim. Image segmentation and edge enhancement with stabilized inverse diffusion equations. *IEEE Trans. Image Proc.*, 9(2):256–266, Feb. 2000.

15. J. Portilla, V. Strela, M. J. Wainwright, and E. P. Simoncelli. Image denoising using scale mixtures of gaussians in the wavelet domain. *IEEE Trans. Image Proc.*, 12(11):1338–1351, Nov. 2003.

16. L. I. Rudin, S. Osher, and E. Fatemi. Nonlinear total variation based noise removal algorithms. *Physica D*, 60:259–268, 1992.

17. K. Sauer and C. Bouman. Bayesian estimation of transmission tomograms using segmentation based optimization. *IEEE Trans. on Nuclear Science*, 39(4): 1144–1152, 1992.

18. T. Tsuchiya. A convergence analysis of the scaling-invariant primal-dual path-following algorithms for second-order cone programming. *Optimization Methods and Software*, 11(12):141–182, 2000.

19. C. R. Vogel and M. E. Oman. Iterative methods for total variation denoising. *SIAM J. Sci. Comput*, 17(1):227–238, Jan. 1996.

20. C. R. Vogel and M. E. Oman. Fast, robust total variation-based reconstruction of noisy, blurred images. *IEEE Trans. Image Proc.*, 7(6), June 1998.

Dynamic Isoline Extraction
for Visualization of Streaming Data

Dina Goldin* and Huayan Gao

University of Connecticut, Storrs, CT, USA
{dqg, ghy}@engr.uconn.edu

Abstract. Queries over streaming data offer the potential to provide timely information for modern database applications, such as sensor networks and web
services. Isoline-based visualization of streaming data has the potential to be of
great use in such applications. Dynamic (real-time) isoline extraction from the
streaming data is needed in order to fully harvest that potential, allowing the users
to see in real time the patterns and trends – both spatial and temporal – inherent
in such data. This is the goal of this paper.

Our approach to isoline extraction is based on *data terrains*, triangulated irregular networks (TINs) where the coordinates of the vertices corresponds to
locations of data sources, and the height corresponds to their readings. We dynamically maintain such a data terrain for the streaming data. Furthermore, we
dynamically maintain an isoline (contour) map over this dynamic data network.
The user has the option of continuously viewing either the current shaded triangulation of the data terrain, or the current isoline map, or an overlay of both.

For large networks, we assume that complete recomputation of either the data
terrain or the isoline map at every epoch is impractical. If n is the number of
data sources in the network, time complexity per epoch should be $O(\log n)$ to
achieve real-time performance. To achieve this time complexity, our algorithms
are based on efficient dynamic data structures that are continuously updated rather
than recomputed. Specifically, we use a *doubly-balanced interval tree*, a new data
structure where both the tree and the edge sets of each node are balanced.

As far as we know, no one has applied TINs for data terrain visualization
before this work. Our dynamic isoline computation algorithm is also new. Experimental results confirm both the efficiency and the scalability of our approach.

1 Introduction

Queries over streaming data offer the potential to provide timely information for modern database applications, such as sensor networks and web services. Isoline-based visualization of streaming data has the potential to be of great use in such applications.
Isoline (contour) maps is particularly informative if the streaming data values are related
to phenomena that tend to be continuous for any given location, such as temperature,
pressure or rainfall in a sensor network.

Dynamic (real-time) isoline extraction from the streaming data is needed in order
to allow the users to see in real time the patterns and trends – both spatial and temporal
– inherent in such data. Such isoline extraction is the goal of this paper.

---

* Supported by NSF award 0545489.

D. Grigoriev, J. Harrison, and E.A. Hirsch (Eds.): CSR 2006, LNCS 3967, pp. 415–426, 2006.

Our approach to isoline extraction is based on *data terrains*, triangulated irregular networks (TINs) where the (x, y)-coordinates of the vertices corresponds to locations of data sources, and the z-coordinate corresponds to their readings. Efficient algorithms, especially when implemented in hardware, allow for fast shading of TINs, which are three-dimensional. By combining shading with user-driven rotation and zooming, data terrains provide a very user-friendly way to visualize data networks.

While the rendering of static TINs is a well-researched problem, we are concerned with *dynamic* networks, where data sources may change their readings over time; they may also join the network, or leave the network. We dynamically maintain a data terrain for the streaming data from such a network of data sources.

Furthermore, we dynamically maintain an isoline (contour) map over this dynamic data network. Isolines consist of points of equal value; they are most commonly used to map mountainous geography. The isoline map can be displayed in isolation, or over-layed on the underlying TIN, providing the user with a visualization that is both highly descriptive and very intuitive.

For large networks, we assume that complete recomputation of either the data terrain or the isoline map at every epoch is impractical. If n is the number of data sources in the network, time complexity per epoch should be $O(log\ n)$ to achieve real-time performance. To achieve this time complexity, our algorithms are based on efficient dynamic data structures that are continuously updated rather than recomputed. Specifically, we use a *doubly-balanced interval tree*, a new data structure where both the tree and the edge sets of each node are balanced.

Dynamic isoline maps have been proposed before in the context of sensor networks [7, 11]. However, as far as we know, no one has applied TINs for this purpose before this work. Our dynamic isoline computation algorithm is also new. As a result, earlier approaches produce isoline maps that are in both more costly and less accurate.

We have implemented the data structures and algorithms proposed in the paper. The user has the option of continuously viewing either the current shaded triangulation, or the current isoline map, or an overlay of both. Experimental results, simulating a large network of randomly distributed data sources, confirm both the efficiency and the scalability of our approach.

Overview. We describe data terrains in section 2, and discuss the algorithms for their computation and dynamic maintenance. In section 3, we give an algorithm for computing isoline maps over the data terrain, as well as their dynamic maintenance. In section 4 we present our implementation of isoline-based visualization. Related work is discussed in section 5, and we conclude in section 6.

2   Data Terrains

Our notion of a data terrain is closely related to the notion of a geographic terrain, commonly used in Geographic Information Systems (GIS). Geographic terrains represent elevations of sites and are static.

There are two main approaches to represent terrains in GIS. One is *Digital Elevation Models* (DEM), representing it as gridded data within some predefined intervals, which is volume-based and regular. DEMs are typically used in raster surface models. Due to the regularity of DEMs, they are not appropriate for networks of streaming data sources, whose locations are not assumed to be regular. The other representation is *Triangulated Irregular Networks* (TIN). The vertices of a TIN, sometimes called *sites*, are distributed irregularly and stored with their location (x, y) as well as their height value z as vector data (x, y, z); TINs are typically used in vector data models. For a detailed survey of terrain algorithms, including TINs, see [17].

In this paper, we chose to use TINs to represent the state of a data network. TINs are good for visualization, because they can be efficiently shaded to highlight the 3D aspects of the data. The (x, y) coordinates of the TIN's vertices corresponds to the locations of data sources, and the z coordinates corresponds to their current value (that is being visualized). We refer to this representation as *data terrains*.

We initially construct the data terrain with the typical algorithm for TIN construction, as in [8, 9] in $O(n \log n)$ time. Since the construction of TIN is only dependent on the *location* of its sites, the topology of the TIN does not change with the change of the data values. The only possibility for a TIN to change is when a new data source joins the network, or when some data source leaves the network, e.g. due to power loss. In the following, we describe the algorithm for updating the TIN when this happens.

Insertion. When a new data source is added to the network, we need to add the corresponding vertex to the data terrain. It would be the same algorithm as when building a new data terrain, since it is an incremental algorithm. As discussed in [9], the worst case of the time performance for site insertion could be $O(n)$. Note that we assume that data sources are not inserted often, much less frequently than their values are updated, giving us *amortized* performance of $O(\log n)$ for this operation.

Deletion. When a data source leaves the network, we need a local updating algorithm to maintain our dynamic TIN. Basically, this is the inverse of the incremental insertion algorithm, but in practice there are a variety of specific considerations. [10] first described a deletion algorithm in detail, but unfortunately, it had mistakes. The algorithm was corrected in [5], and further improved in [14]. The performance for deletion algorithm is $O(k \log k)$ where k is the number of neighbors of the polygon incident the vertex to be deleted. Figure 1 illustrates how one site is removed from a TIN.

Efficient algorithms, especially when implemented in hardware, allow for fast shading of data terrains. By combining shading with user-driven rotation and zooming, data terrains provide a very user-friendly way to visualize data networks.

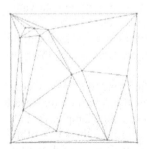 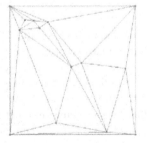

Fig. 1. Deleting one site from TIN: (a) before deletion; (b) after deletion

3   Dynamic Isoline Extraction

In this section, we describe how to extract an isoline map from a data terrain for a dynamic data network.

3.1   Interval Tree

One naive way to extract an isoline map at a given height h directly from the data terrain is to traverse all the triangles in the data terrain, intersect each one with the plane $z = h$, and return all the resulting segments. $O(n)$ time is needed for this brute-force approach, where n is the number of sites. We use *interval trees* [6] to obtain a more efficient solution. For every edge in the TIN, this tree contains an interval corresponding to the edge's z-span.

Interval trees are special binary trees; besides the key which is called a *split value*, each node also contains an *interval list*. Given the set of intervals for the z-spans of a TIN's edges, an (unbalanced) interval tree is constructed as follows.

1. Choose a split value s for the root. This value may be determined by the first inserted interval. For example, if (a, b) is the first inserted interval, then the split value will be $(a + b)/2$.
2. Use s to partition the intervals into three subsets, I_{left}, I, I_{right}. Any interval (a, b) is in I if $a \leq s \leq b$; it is in I_{left} if $b < s$; and it is in I_{right} if $a > s$.
3. Store the intervals in I at the root; they are organized into two sorted lists. One is the *left list*, where all the intervals are sorted in increasing order of a; the other is the *right list*, where all the intervals are sorted by decreasing order of b.
4. Recurse on I_{left} and I_{right}, creating the left and right subtrees, respectively.

Next, we doubly balance the tree; we use AVL trees [1] for this purpose. The first AVL tree is for the interval tree itself, the other is for the edge lists stored at the nodes of the interval tree. This enables us to provide quick updates to the tree (section 3.2).

Figure 2 gives an example of a TIN, composed of 7 sites and 19 edges, and the corresponding interval tree. More details can be found in [16], which uses an interval tree as the data structure to conveniently retrieve isoline segments. But they only consider static TIN, while we will give an algorithm for dynamic TIN in section 3.2.

We now describe the algorithm for using an interval tree T to create an isoline at value v. It is a recursive algorithm that begins with the root node. Let the split value be s. If $v < s$ then we will do the following two things. First search the left list of the root, and then search the left subtree of the root recursively. If $v > s$ then we will do the similar two things. First search the right list of the root, and then search the right subtree of the root recursively. We stop at the leaf node.

We finish by describing the details of the algorithm for querying the matched edge list in the left list or right list of the interval node, mentioned briefly before. Recall that we stored the edge list as an AVL tree. Take the left list as an example, we only consider the left evaluation of the smaller point of the edge. Let it be the key k of the AVL node. We search the AVL tree recursively. If $v < k$, then we search the left subtree recursively. If $v > k$, which means that all the edges in the left subtree are the

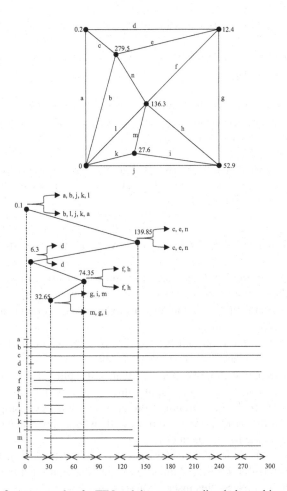

Fig. 2. An example of a TIN and the corresponding balanced interval tree

matched edges, output them and search the right subtree recursively. Querying the right list of the interval node is symmetric.

Note that an interval tree can also be constructed for *triangles*, rather than *edges*, of the TIN, since a z-span can just as easily be defined for triangles. We can quickly find those triangles that intersect with the plane $z = h$, and avoid considering others. One such algorithm is given in [16].

In our work, we found it more convenient to use edges to compute isolines from the TIN instead of triangles. Note that this kind of substitution does not affect the efficiency, because of the following fact: if there are n vertices in the TIN, the number of edges and the number of triangles are each $O(n)$ [13]. Let n_b denote the number of sites on the boundary of the TIN, and n_i be the number of sites in the interior; the total number of sites is $n = n_b + n_i$. The number of edges is $n_e = 2n_b + 3(n_i - 1) <= 3n$. The number of triangles, let be n_t, would be $n_t = 2n - 6$ when $n > 3$. Therefore, both the edge-based and the triangle-based interval trees allow for a more efficient algorithm to get the isolines from TIN than the naive one described at the beginning of the section.

3.2 Dynamic Interval Tree

In our setting, the data values at the sources change as time passes. Since our interval tree is built up on the edges of the TIN, and the z-span of the edge is dependent on the values at the sites adjacent to the edge, a change in these values will necessitate a change to the interval tree.

We begin with a built TIN and a constructed interval tree, as described above. In the following, we give a detailed description of the algorithm to update the tree after some data source s changes its value from v_0 to v.

1. We use the TIN to find all edges incident with the data source s. Since the TIN contains an incidence list L for each vertex, we can find these edges in constant time $O(1)$.
2. For every edge e in the list L, we need to update its position in the interval tree. Suppose that the original z-span for e is z_e, and the new one is z'_e.
3. Run a binary search from the root to find the node x which contains the interval z_e. This is done in $O(\log n)$ time.
4. Delete z_e from both the right and the left lists of x. Since both of these lists are implemented as a standard AVL tree, the performance is $O(\log n)$.
5. Look for the node y that should contain z'_e, that is, its split value overlaps z'_e. First, check whether z'_e overlaps with the split value of x, in which case we need look no further. Otherwise, begin searching from the root of the tree, comparing z'_e with the split value of the node, until we find the node whose split value overlaps z'_e or reach a leaf. The time for this is in $O(\log n)$.
6. If we found y, then insert z'_e into the right and the left lists of y. Both lists are implemented with AVL trees, and the size of each list is at most the total number of the edges in the TIN. So this insert should be in $O(\log n)$.
7. If we have reached a leaf without finding y, we insert a new leaf into the interval tree to store the new interval. Its interval lists will contain just z'_e, and its split value will be the midpoint of z'_e; this is in $O(1)$.

Recall that we are using balanced (AVL) trees both for the interval tree, and for the interval lists within each node of the interval tree. To keep the trees balanced, all insertions and deletions are followed by a *rebalance* operation. There exists the rebalance algorithm for AVL trees in $O(\log n)$ time [19], and it is easy to see that double balancing does not increase the time complexity. An alternative method is *relaxed AVL tree* [12]. Instead of rebalancing the tree at every update, we relax the restriction and accumulate a greater difference is heights before we need to adjust the height of the AVL tree.

Figure 3 illustrates an update to the interval tree of Figure 2 when the reading of data source s changes from 136.3 to 170.4. Note that all edges incident on s need to be checked. In this figure, we need to update the position of intervals f, h, l, m, n in the interval tree.

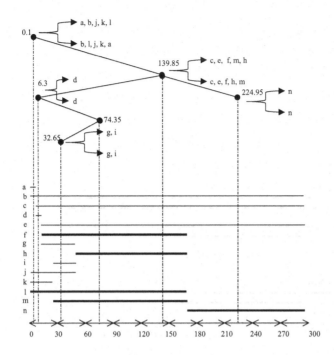

Fig. 3. An update to the interval tree

This tree is highly unbalanced; it is a snapshot after the insertion but before rebalancing. Figure 4 shows the result after balancing the interval tree. To insert n into the interval tree, we needed to insert a new node to store n. After rebalancing, there are no intervals in the node whose split value is 74.35; this decreases the height of the interval tree by 1, so we delete this node. The time complexity of this algorithm is $O(log\ n)$.

Note that our algorithm, while sometimes adding new nodes (with singleton interval lists), does not delete old nodes when their interval lists become empty. We determined experimentally that there was no benefit in doing so. Since the data readings move up and down (rather than monotonically increasing or decreasing), the empty node is very likely to be used up at some point; it turns out that keeping it around for this eventuality is more time efficient than deleting it right away.

To complete the performance analysis of the interval tree update algorithm, we need to know how many edges are incident on a given site s, since each of these k edges needs to be updated separately. It can be proved that k is never more than 6. We already know the number of edges is $n_e <= 3n$, where n is the number of sites, one per data source. Since each edge is incident on two sites, clearly we have: $k = \frac{2*n_e}{n} <= 6$. We confirmed this experimentally, measuring the average of k; we found that it was never more than 6, not growing as the number of sites increased. Therefore, we assume that k is $O(1)$.

For each edge, we took its z-span interval from its old position and found an appropriate new position to insert the interval, rebalancing when needed. Each of these operations is in $O(log\ n)$ time. Therefore, the overall algorithm is $O(log\ n)$.

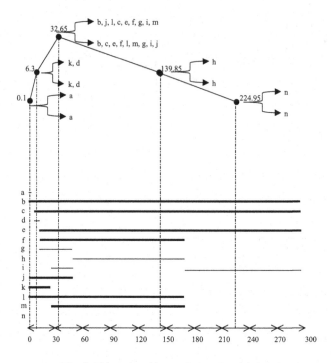

Fig. 4. The updated interval tree after rebalancing

4 Performance

4.1 Assumptions

Our visualization algorithm relies on continuously updating the TIN and the corresponding interval tree. The updates are triggered by changes to data values, insertions (new data sources), or deletions (loss of a data source). We use the *amortized* approach to complexity analysis, assuming that changes to data values happen much more often than either insertions or deletions.

4.2 Experiments

Our implementation of the visualization of data terrain uses *OpenGL* [15], a software interface to graphics hardware. In our simulation, we used GNU C++ with OpenGL library under Linux platform to render the data terrain and isoline maps. We implemented isoline extraction from data terrains using the algorithm described above. In our experiment, we started by generating the initial data terrain from the initial values at data sources, using the algorithm in section 2. Then we constructed our interval tree using the algorithm in section 3.

For the initial values of our data, we used the actual data which describes the terrain around the University of Connecticut. As shown in figure 5, we simulated a network of 257 data sources deployed around UConn which were in the region $(0,0) -$ $(9600, 10115)$. The data readings were the local height at that coordinate, which ranged

Fig. 5. Data terrain of UConn. (a) before change; (b) after change.

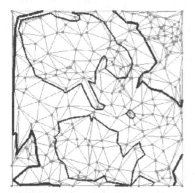 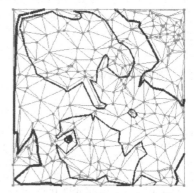

Fig. 6. Isolines from TIN, (a) before and (b) after change

from 0 to 420 (feet). Figure 5 shows the shaded data terrain (a) before and (b) after a site n lower left quadrant changed its value, from 350 to 149.49. One can clearly see the difference in the shape of the two data terrains.

Our data stream consisted of changes to the readings of one site (chosen at random) at a time. As we processed the stream, we updated the TIN and the interval trees. Figure 6 illustrates how the isolines can be affected by a change to the reading at a single site. It shows the TIN overlayed with the isolines; the thick line represents the isoline value 200, and the thin line represents the isoline value 300. When we change the reading at one site (colored as black) from 350 in (a) to 149.49 in (b), it is apparent how the isolines changes accordingly.

We measured the performance of our update algorithms described in sections 2 and 3, plotting time performance against the number of data sources in the network, n. We varied n from 50 to 2500 in 50 unit intervals: $\{50, 100, 150, \ldots, 2450, 2500\}$. Given n data sources distributed randomly in a region $(0,0) - (300, 300)$, we chose one at random and changed its reading, updating everything. We repeated this 100 times, getting the cumulative time for each n in microseconds. Figure 7 (a) shows the plot we obtained from our experiments.

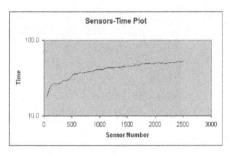
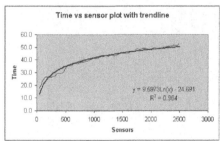

Fig. 7. Time performance vs. number of sites, (a) without and (b) with the logarithmic trendline

The logarithmic trendline through this plot has the function $y = 9.5973(ln\ x) - 24.691$, and the value of .964 for R^2 as shown in Figure 7 (b) . This value of R^2 shows that there is a 96.4% reliability of the relationship between the plot and the trendline. From this picture, we can see that our algorithm is logarithmic and scalable. This confirms our analysis in section 3.

5 Related Work

The visualization of data terrains involves much knowledge in computer graphics. A good review of rendering techniques such as transformations, shading, interpolation, texture mapping, ray tracing, etc., as well as the mathematics theory behind it, can be found in [18, 2]. One of the popular rendering libraries is OpenGL. It is said to be industry standard; it is stable, reliable, portable, extensible, scalable and easy to use. Documents are available from *http://www.opengl.org*. [15], written by the OpenGL Architecture Review Board, is the most authoritative one. In our simulation, we used GNU C++ with OpenGL under Linux to render our data terrain and our isoline map.

Interval tree first was propose by Edelsbrunner [6] to efficiently retrieve intervals of real lines that contain a given query value. Cignoni et. al. [4] uses interval tree as the data structure to extract isosurfaces. Chiang [3] describes how to extract isosurface from volumetric data using interval tree.

Van Kreveld [16] uses the interval tree as the data structure to extract isolines from TINs by associating each triangle with the intervals of the elevation it spans. There are several differences between that algorithm and ours. Their intervals are based on z-spans of triangles rather than edges. They do not use balanced trees as we do. Their algorithm is not dynamic (the update operations are not defined). Finally, as far as we know, their algorithm was never implemented.

To our knowledge, no one has described an algorithm to extract isolines efficiently and dynamically from data terrains as we have done. Related work in the sensor network community has tried to extract isolines directly from sensor readings, using in-network protocols. For example, in [11], an isobar computation from sensor networks is performed as a form of aggregation. This work assumes that every sensor is in a rectangular grid and merge these grids in-network. The communication between the sensors is based on a tree. The general process is that each sensor gets the isobar map from its children, combines its own information, and sends the isobar map up to its parent. Finally, the root aggregates the isobar maps. So they need some polygon operations such as intersect and union; the time complexity is $O(n \log n)$, where n is the number of edges in the polygon. Whenever there are some sensors changing their reading, we need to compute the isobar again, so this approach is not efficient in a dynamic real-time setting.

6   Conclusions

Isoline-based visualization of streaming data has the potential to be of great use in modern database applications, such as sensor networks and web services. This paper was concerned with dynamic (real-time) isoline extraction from the streaming data, so as to allowing the users to see in real time the patterns and trends – both spatial and temporal – inherent in such data.

Our approach to isoline extraction was based on *data terrains*, triangulated irregular networks (TINs) where the coordinates of the vertices corresponds to locations of data sources, and the height corresponds to their readings. We dynamically maintained such a data terrain for the streaming data. Furthermore, we dynamically maintained an isoline (contour) map over this dynamic data network.

For large networks, we assumed that complete recomputation of either the data terrain or the isoline map at every epoch is impractical. If n is the number of data sources in the network, time complexity per epoch should be $O(\log n)$ to achieve real-time performance. To achieve this time complexity, our algorithms are based on efficient dynamic data structures that are continuously updated rather than recomputed. Specifically, we used a *doubly-balanced interval tree*, a new data structure where both the tree and the edge sets of each node are balanced.

As far as we know, no one has applied TINs for data terrain visualization before this work. Our dynamic isoline computation algorithm is also new. Experimental results confirm both the efficiency and the scalability of our approach. All our implementation was in GNU C++ with OpenGL library under Linux.

References

1. G.M. Adelson-Velskii and E.M. Landis. An algorithm for the organization of information. *In Soviet Math. Doclady 3*, pages 1259–1263, 1962.
2. Edward Angel. *Interactive Computer Graphics: A Top-Down Approach with OpenGL*. Pearson Addison-Wesley, Jul. 2002.
3. Yi-Jen Chiang and Claudio T. Silva. I/O Optimal Isosurface Extraction. *IEEE Visualization '97*, pages 293–250, 1997.
4. Paolo Cignoni, Paola Marino, Claudio Montani, Enrico Puppo, and Roberto Scopigno. Speeding up Isosurface Extraction Using Interval Tree. *IEEE Trans. on Visualization and Computer Graphics*, 3, Apr.-Jun. 1997.
5. Olivier Devillers. On Deletion in Delaunay Triangulations. *Proc. 15th Annual Symp. on Computational Geometry*, pages 181–188, Jun. 1999.
6. Edelsbrunner. Dynamic Data Structure for Orthogonal Intersection Queries. *Tech. Rep. F59, Inst. Informationsverarb. Tech. Univ. Graz, Graz, Austria*, 1980.

7. Deborah Estrin. Embedded Networked Sensing for Environmental Monitoring: Applications and Challenges. *DIALM-POMC Joint Workshop on Found. of Mobile Computing, San Diego, CA*, Sep. 2003.

8. Michael Garland and Paul S. Heckbert. Fast Polygonal Approximations of Terrains and Height Fields. *Tech. Rep. CMU-CS-95-181, Carnegie Mellon Univ.*, Sep. 1995.

9. Leonidas Guibas and Jorge Stolfi. Primitives for the Manipulation of General Subdivisions and the Computation of Vorono Diagrams. *ACM Trans. on Graphics*, 4(2):74–123, 1985.

10. Martin Heller. Triangulation Algorithms for Adaptive Terrain Modeling. *Proc. 4th Int'l Symp. of Spatial Data Handling*, pages 163–174, 1990.

11. Joseph M. Hellerstein, Wei Hong, Samuel Madden, and Kyle Stanek. Beyond Average: Towards Sophisticated Sensing with Queries. *2nd Int'l Workshop on Information Proc. in Sensor Networks (IPSN '03)*, Mar. 2003.

12. Kim S. Larsen, Eljas Soisalon-Soininen, and Peter Widmayer. Relaxed Balance through Standard Rotations. *Workshop on Alg. and Data Structures*, 1997.

13. Charles L. Lawson. Software for C^1 Surface Interpolation. *In John R. Rice, ed., Mathematical Software III, Academic Press, NY*, pages 161–194, 1977.

14. Mir Abolfazl Mostafavi, Christopher Gold, and Maciej Dakowicz. Delete and Insert Operations in Voronoi / Delaunay Methods and Applications. *Computers & Geosciences*, 29(4):523–530, May 2003.

15. Dave Shreiner, Mason Woo, Jackie Neider, and Tom Davis. *OpenGL Programming Guide: The Official Guide to Learning OpenGL, Version 1.4, 4th edition*. Addison-Wesley, Nov. 2003.

16. Marc van Kreveld. Efficient Methods for Isoline Extraction from a Digital Elevation Model Based on Triangulated Irregular Networks. *In 6th Int'l Symp. on Spatial Data Handling Proc.*, pages 835–847, 1994.

17. Marc van Kreveld. Digital Elevation Models and TIN Algorithms. *Algorithmic Found. of Geographic Information Systems in LNCS (tutorials),Springer-Verlag, Berlin*, 1340:37–78, 1997.

18. Alan H. Watt. *3D Computer Graphics*. Addison-Wesley, Dec. 1999.

19. Mark Allen Weiss. *Data Structures and Algorithm Analysis in C*. Addison-Wesley, Jul 1997.

Improved Technique of IP Address Fragmentation Strategies for DoS Attack Traceback

Byung-Ryong Kim[1], and Ki-Chang Kim[2]

[1] School of Computer Science and Engineering, Inha Univ., 253, YongHyun-Dong,
Nam-Ku, Incheon, 402-751, Korea
doolyn@inha.ac.kr
[2] School of Information and Communication Engineering, Inha Univ.,
253, YongHyun-Dong, Nam-Ku, Incheon, 402-751, Korea
kichang@inha.ac.kr

Abstract. Defending against denial-of-service(DoS) attacks is one of the hardest security problems on the Internet today. One difficulty to thwart these attacks is totrace the source of the attacks because they often use incorrect, or spoofed IP source addresses to disguise the true origin Traceback mechanisms are a critical part of the defense against IP spoofing and DoS attacks, as well as being of forensic value to law enforcement. Currently proposed IP traceback mechanisms are inadequate to address the traceback. problem for the following reasons: they require DoS victims to gather thousands of packets to reconstruct a single attack path; they do not scale to large scale Distributed DoS attacks; and they do not support incremental deployment. This study suggests to find the attack origin through MAC address marking of the attack origin. It is based on an IP trace algorithm, called Marking Algorithm. It modifies the Marking Algorithm so that we can convey the MAC address of the intervening routers, and as a result it can trace the exact IP address of the original attacker. To improve the detection time, our algorithm also contains a technique to improve the packet arrival rate. By adjusting marking probability according to the distance from the packet origin, we were able to decrease the number of needed packets to traceback the IP address.

1 Introduction

The most typical form of the malicious internet attacks interfering with company activities is DoS(Denial-of-Service) attack. DoS attack causes damages to prevent the victim system from conducting normal service and make the system down by sending infinite malicious packets to systems using more than one attacking system. Likewise DoS attack can not only temporarily cause denial of service after simply paralyzing server but also terribly cause the complete loss of credit which can be the most important part of internet service for internet service providers or corporate. Because there are no suitable measures to cope with the attacks, there just is no other way except reactivating paralyzed server or blocking attacks with IDS(Intrusion Detection System). If attacker's correct location cannot be found further attacks may be attempted and in this way it will be exposed to potential risks.

In an effort to detect location of attacker, IP traceback was approached through studies but because attacker spoofs its own IP there has been limit in tracing back IP.

D. Grigoriev, J. Harrison, and E.A. Hirsch (Eds.): CSR 2006, LNCS 3967, pp. 427–437, 2006.

Attackers can hide its location spoofing the IP employing the point that unlike other attack forms DoS attack does not require creditable connection between the attack system and the victim system and even packet having spoofed IP has enough effect on DoS attack.

Accordingly other than the technique to use source IP address recorded at packet, technique to detect attacker's location should be studied and marking type IP trace-back technique was proposed as an alternative measure. It detects the attack path by having every router on network mark its own IP address to packet passing through the router and using information on the victim system. This can detect to the first router where packet passes through out of the victim place but the disadvantage is that it cannot find the source location of the practical attack.

Therefore in the study we understand how to find attack path in current marking type IP traceback techniques and furthermore propose how to detect the true attack source with MAC address by marking MAC address of attack source.

For this each router marks the MAC address of front end together with its own IP address and it is proposed as follows; mechanism to perform integrity test on MAC address just as the integrity test performed on IP address in the victim system, and minimize the number of packets needed to find out attack path, which was revealed as weakness.

This paper is composed of the following parts. We review background information and highlight the main challenges of the IP traceback approach in section 2. In Section 3, we give an overview of address fragmentation techniques for the purposes of packet marking for internet traceback. Section 4 shows the experimental results and in final chapter 5 conclusion and prospects for further study are described.

2 Related Works

This section briefly introduces several previously proposed techniques to IP trace back the origins of attack. The importance of IP traceback has prompted many researchers to work on this topic [1], [2], [3], [5], [6], [7], [9], [10], [11]. We review these efforts in chronological order.

Burch and Cheswick introduce the concept of network traceback. They identify attack paths by selectively flooding network links and monitoring the changes caused in attack traffic [3].

Savage et al. propose the Fragment Marking Scheme(FMS) for IP traceback [10]. They suggest that routers probabilistically mark the 16 bit IP identification field, and that the receiver reconstructs the IP addresses of routers on the attack path using these markings. Bellovin et al. develop iTrace [2]. In iTrace, routers probabilistically send a message to either the source or destination IP address of a packet, indicating the IP address of the router. This approach does not alter packets in-flight and victims can also detect attackers that use reflectors to hide their presence [8], however, it does generate additional traffic.

Goodrich presents a marking scheme that marks nodes instead of links into packets [5]. Because this approach does not use a distance field, it has issues with attack graph reconstruction and does not scale to a large number of attackers.

Dean et al. suggest algebraic traceback, an algorithm to encode a router's IP address as a polynomial in the IP identification field [4]. We show in the next subsection that it does not scale to large number of attackers. Adler presents a theoretical analysis of traceback, presenting a one-bit marking scheme [1]. This work is primarily of theoretical interest, and does not scale to large numbers of attackers.

Snoeren et al. propose SPIE, a mechanism using router state to track the path of a single packet [10]. The main advantage of SPIE is that it enables a victim to trace back a single packet by querying the router state of upstream routers, however, it does require routers to keep a large amount of state. Li et al. have further developed their approach, lowering the required router state, at the expense of a large communication overhead for traceback [7].

In this mechanism, referred to as Fragment Marking Scheme, router marks that packet appointed as given probability, p or less, not all the packet, passes through the router. Namely it marks the router IP to source field for additional data to show path of packet which is selected with probability of p or less at each router and distance information, the number of hops is set to 0. When the probability to select the packet at the next router through which the packet passes is p or more, the last router marked at the packet is the IP of the router. The record on both the source and last IP undergoes the following; when passing through the next router, if probability is p or more, the number of hops is added by the number of routers passed through from the source router and if less than p, this record is disregarded and source is recorded again. This series of processes are reliable to find out the source of the real packet although modification is made to the source address where packet was generated on purpose by increasing the random rate. This mechanism has come from the idea that the part to record the data is the identification field marking the packet's identity for the separation of packet's IP header and the rate to use the part is statistically 0.25%.

In current marking algorithm, router data is sent to identification field of packet's IP header, router data is sent in slice and marking process at router is processed as probability by sampling basis. In order for router to mark its IP address to packet, identification field(16 bit) of IP header is used.

Using the router's IP address, R and bit algorithm to the IP address 64 bit R is created by bitinterleaving the hash(R) value and o(displacement data, the random number)th slice is loaded to packet after the value is divided into 8 slices. So distance data, the slice of IP address, and the displacement data indicating the slice's location are marked. The packets marking the IP address of passed router are composed by distance data at the victim system and if the IP address obtained from the combination is found to be right router's IP address, it will be recorded to path tree. So attack path can be the path tree composed through the above described processes.

Current marking algorithm to the first router of attack path can be found . Although attack paths are found and using the paths measures can be devised to cope with DoS attack, source of attack cannot be found. To the first router through which packet passed it is possible to trace but there is no way to trace attack source here. Also due to the algorithm's characteristic that the first data is lost(marking the router of new router) in the middle by each router till packet arrives victim systems, it is relatively low for packet having the first router data to arrive to the victim system. The more the number of hops from attacker to the victim system, the smaller the arrival rate of the packet.

3 Low-Cost Mechanism to Correctly Detect Attack Source

Router's location is marked by marking the router's IP address to packet in existing marking algorithms, but because only the router data on path(IP address) is included in packet, it is traceable only to the first router on path. Hence it is not possible to trace the location of attack source in the existing mechanisms.

In this paper we intend to mark not only router's IP address but also the MAC address of front end(the previous router or attack source). While there is advantage in marking MAC address there also is problem. The problem is that it can send 32 bit IP address and 48 bit MAC address in same way. In case of IP address, making 32 bit hash value, 64 bit is made adding the hash value to the existing IP address, transmission is performed in 8 slices and integrity can be checked in reverse process. But in case that transmission is done in 8 slices after making 48 bit MAC address into 64 bit, each router should process two stages of marking its IP address and the front end's MAC address and one router leads to consuming two distance data to discern each stage. Because of this the number of hops to be traced back is limited to 16 which is the half in the existing mechanism and which is a great obstacle to the original aim to trace attack source. In this way new mechanism is required to allow to mark MAC address and at the same time not to decrease the traceback scope under current mechanism.

To load router's IP and MAC address data together to identification field as in marking algorithm, check the integrity on the two data in the victim system, and maintain the number of hops of attack path that was possible to trace in current mechanism, this paper proposes that IP and MAC addresses are transmitted in slices after making the addresses into combination of 56 bit slices. Here mechanism to check the integrity should be added in making 56 bit combination. For this identification field of IP header is to be recombined. Proposed recombination technique is to combine 16 bit of identification field into 3 bit displacement data field, 6 bit distance data field, and 7 bit slice data field. This can be illustrated as Fig 1.

Offset information	Distance information	Fragment information
0 2 3	8 9	15

Fig. 1. Identification field configuration of proposed IP header

Mechanism to create IP slice and MAC slice to be loaded to slice data field of IP header identification field and check the integrity is proposed as follows.

3.1 Create Slice Combination

Unlike in current marking algorithms that 64 bit IP-slice combination is made by adding hash 32 bit to 32 bit IP address. in this paper 56 bit IP-slice combination is composed and technique to check the integrity of IP address is also to be included. 56 bit IP-slice combination is composed by adding 24 bit hash value to 32 bit IP address. An example is shown on Fig 2 to illustrate the technique. Algorithm 1 to show process to create IP-slice combination is described with example of Fig 2.

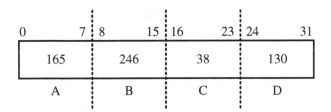

Fig. 2. IP address example

Algorithm 1. IP-slice combination creation algorithm.
Input : 32 bit IP address, Output: 56 bit IP-slice combination.

- 1) Divide 32 bit IP address(R) into 4 parts and call it A, B, C, D each.
- 2) Perform XOR($\oplus$) over the former two 8 bits(A, B) and the latter two 8 bits(C, D) and make two new 8 bits(E, F) and then perform XOR to the two(E, F), create one new 8 bit and the value is called G.

$$E = A \oplus B$$
$$F = C \oplus D$$
$$G = E \oplus F$$

24 bit(H) is created by combining newly created three 8 bits.
$$H = E + F + G$$

- 3) 56 bit is created adding 32 bit R and the new 24 bit H.(Separating 4 bit at R and 3 bit at H from the front, adding the two makes 7 bit. Repeat the process 8 times then 56 bit R' is created.) Here the process to create the 56 bit R' is defined as BitInterleave1. Divide the created R' into 8 slices and load them to packet according to probability.

In the victim system a complete IP address is combined after recombination over 8 packets having the same value of distance data, and here it should be checked whether the combined IP address is correct or incorrect mixed with other slice in the process. Algorithm 2, the checking stage, is the reverse process of the above proposed Algorithm 1.

Algorithm 2. Integrity Check Algorithm of IP address.
Input: 56 bit IP-slice combination, Output: 32 bit IP address.

- 1) BitDeinterleave and combine the 5, 6, 7 multiple bit of 56 bit's IP-slice combination(R') and make 24 bit and this is called H.
- 2) Divide this 24 bit by 3 and call E, F, G each and where $G' = E \oplus F$, perform process 3. Where $G \neq E \oplus F$, delete R' and re-preform with new slice combination from process 1.
- 3) Combine created three 8 bits and call it H.
$$H = E + F + G$$
- 4) Here 32 bit excluding R' at H becomes IP address.
- 5) Divide R' by 4 and perform XOR over the former two 8 bits(A, B) and the later two 8 bits (C, D) each. After making two new 8 bits(E', F') perform XOR over these two 8 bits, create one new 8 bit and call the value G'.

$$E' = A \oplus B$$
$$F' = C \oplus D$$
$$G' = E \oplus F$$

- 6) After combining three newly created 8 bits create 24 bit(H').

$$H' = E' + F' + G'$$

Where $H \neq H'$, accept the R' is correctly transmitted combined value in correct order, and where $H \neq H'$, delete R' and perform from process 1 with new combination slice.

3.2 Create IP-Slice Combination and Check the Integrity

As each router on attack path marks its own IP address to packet, if it marks the MAC address of front end also, it needs to check the integrity of the transmitted MAC address as the integrity of IP address transmitted to packet is checked in the victim system. In addition by having the intermediate router mark the MAC address of the front end, practically each router consumes two distance data(its own IP address - distance data 0, MAC address of the front end - distance data 1). This causes a new problem of decreasing the number of traceable hops under the current mechanism to the half.

To fill out this problem, in this study adjusting the configuration of identification field and making distance data field 6 bit, the number of traceable hops can be maintained as in current marking algorithm by composing slice data field with 7 bit. Also the problem of decreasing slice data field by 1 bit was settled by proposing a technique to check integrity after adjusting the existing 64 bit slice combination into 56 bit. Fig 3 is an example to illustrate this technique.

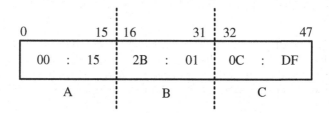

Fig. 3. MAC address example

Algorithm 3, an algorithm to show the process to create MAC-Slice combination, explains the process with the example of Fig 3.

Algorithm 3. MAC-slice combination creation.
Input: 48 bit MAC address, Output: 56 bit MAC-slice combination.

- 1) Divide 48 bit MAC address (M) by 6, make it 6 parts, and call it A, B, C, D, E, F each.
- 2) XOR each 8 bit, create new 8 bit and call this value G.

$$G = A \oplus B \oplus C \oplus D \oplus F$$

- 3) BitInterleave each bit of new 8 bit, G, to the 7th multiple place of MAC address, create 56 bit, M. Here define the process of creating M as BitInterleave2.

Divide the created M by 8 slices, and load them to packet according to probability. In the victim system, complete MAC address is combined through recombination process over 8 packets having the same distance data. Here it should be checked whether the combined MAC address is right or wrong mixed with other slice in the middle. Algorithm 4 is reverse process of the above proposed Algorithm 3.

Algorithm 4. Integrity check algorithm of MAC address.
Input: 56 bit MAC-slice combination, Output: 48 bit MAC address.

- 1) BitDeinterleave2 the 7th multiple digit bit from M' of combined 56 bit, make 8 bit after combination, and call it G.
- 2) Divide 48 bit (MAC address, M') excluding bits obtained from BitDeinterleave2 in 1) by 6 parts and call it A, B, C, D, E, F each.
- 3) Compare result value of $A \oplus B \oplus C \oplus D \oplus E \oplus F$ with G.
- 4) Where $A \oplus B \oplus C \oplus D \oplus E \oplus F = G$, accept it as combined value in right order through correct transmission.
  Where $A \oplus B \oplus C \oplus D \oplus E \oplus F \neq G$, delete it since M is not right value.

3.3 Technique to Use Function to Weighing Probability p

Fig 4 shows that router data saved at packet is newly created or changed in the course that packet starts sending after router's marking its data to the packet at router R_1 and arrives to Router R_5 thru each router and also shows each packet discerned by router data saved at such packet. In Fig 4 d_0 deletes the previous data loaded on packet at router and newly marks its own IP address, d1 performs XOR and marks its IP address to the former router's data loaded on packet, and values after d_2 just increase and send packet's distance data only with the former data maintained.

Fig 4 shows that router data saved at packet is newly created or changed in the course that packet starts sending after router's marking its data to the packet at router R_1 and arrives to Router R_5 thru each router and also shows each packet discerned by router data saved at such packet. In Fig 4 d_0 deletes the previous data loaded on packet at router and newly marks its own IP address, d_1 performs XOR and marks its IP address to the former router's data loaded on packet, and values after d_2 just increase and send packet's distance data only with the former data maintained.

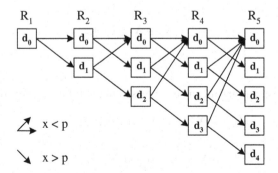

Fig. 4. Packet flow and data in marking algorithm

In Fig 4 arrow means to send to the next router applying probability p, for example if probability p applies to d_0 of R_1, the value is d_0 and d_1 of R_2. Here if applying probability produces random number and $x<p$, follow process(1) in Fig 4, if packet distance data is 0 and $x>p$, follow process(2), and if greater than 0 and $x>p$, follow process(3)

It can be described in the following equation.

$$R_2[d_0] = R_1[d_0] \times p \tag{1}$$

$$R_2[d_1] = R_1[d_0] \times (1-p) \tag{2}$$

In current mechanism, probability p is fixed at every router. Now by giving weight to the probability by distance, data loss and arrival rate is to be improved. It is helpless to lose data owned whenever passing through router but it is necessary to minimize time to spend in finding attack path by having packet with distant data arrive at the victim system as much as possible.

Value to indicate the number of packets arriving at each router, X, is defined below.

Definition. Where number of hops is n, the number of packets arriving at router with packet, whose distance is i, is expressed in X_i^{n+1}

Also function j to probability p is defined in the following equation to weigh by distance data.

$$f_i = \frac{1}{(i+1)\times 2} \ (i=\text{distance}) \tag{3}$$

With the two above definitions value to hop can be expressed in the equation below.

Where $n=0$,

$$X_0^1 = X_0^0 \cdot f_0$$
$$X_1^1 = X_0^0 \cdot (1-f_0)$$

Where $n=1$,

$$X_0^2 = X_0^1 \cdot f_0 + X_1^1 \cdot f_1$$
$$X_1^2 = X_0^1 \cdot (1-f_0)$$
$$X_2^2 = X_1^1 \cdot (1-f_1)$$

Combining above equations it can be expressed in the following function equations.

$$X_0^{n+1} = X_0^n \cdot f_0 + X_1^n \cdot f_1 ... X_n^n \cdot f_n \tag{4}$$

$$X_i^{n+1} = X_{i-1}^n \cdot f_{i-1} \ \text{for } i=1, \dots, n+1 \tag{5}$$

Without setting the probability fixed as expressed above, if weighing by packet's distance data, whenever passing through router the packet's arrival rate with the previous data, which was greatly diminished, will be largely improved. In chapter 4 the test results on the improvement of packet arrival rate and the decrease of deviation will be presented.

4 Performance Evaluation and Test Results

Fig 5 shows the flow of packets arriving from attack source A to the victim system V through intermediate routers R_3, R_4, R_5 using the proposed algorithm. Intermediate routers mark their IP addresses and MAC address of the front end to packet passing through it. Unlike the existing marking algorithms because it should mark MAC address as well router should generate random number y and if y<0, distance data is 0 and it marks its IP address, and if y≥0, the distance data is 1 and performs XOR and marks MAC address of the front end and the IP address.

With slices of the packets accepted from Fig 5, after combining table by slice with the same distance data and combining slices with the same distance data it can be shown by distance data. Performing XOR over slice combination(R_5) of the first node(distance data 0) and slice combination($R_5 \oplus M_4$) of the second node(distance data 1), the result can be found as follows.

$$R_5 \oplus (R_5 \oplus M_4) = M_4$$

And performing XOR slice combination(R_5) of the first node(distance data 0) and slice combination($R_5 \oplus R_4$) of the third node(distance data 2), the result can be found as follows.

$$R_5 \oplus (R_5 \oplus R_4) = R_4$$

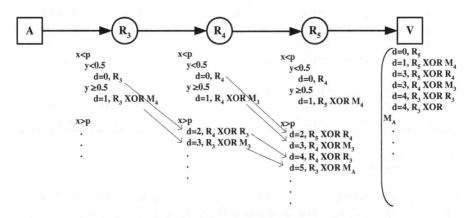

Fig. 5. Detection of attack source using proposed algorithm

Through the above two processes IP address and MAC address of router R_4 can be obtained. Applying the mentioned processes to each node of Table 1 will give IP address and MAC address of each router on attack path, and finally the MAC address of attack source too. Mechanism to weigh router's probability p by packet's distance data was proposed above. To confirm whether equation 4 is optimized or not, function equation to probability p is changed to the following function having two coefficients a, b.

$$f_i = \frac{1}{ai+b} \quad (i=\text{distance data}, \ a \geq 0, b \succ 0) \tag{6}$$

As Hop increases the arrival rate of packet arriving to remote router diminishes to half each. With some distance data, packet's arrival rate is changed by the following equation.

$$R_i = \frac{100}{2^i} \quad (i=\text{distance data}) \tag{7}$$

To sum up this, it can be found the more hops increase and the bigger distance data gets, the dramatically smaller packet arrival rate. As shown in the results packet arrival rate has been improved by flexibly weighing p by distance data rather than by fixing the probability p.

With equation 6, conversing the coefficient not exceeding $0 \leq a \leq 10$, $1 \leq b \leq 10$, calculate the deviation of packet arrival rate arriving to router.

Table 1 shows the deviation of packet arrival rate where hop=31, and the deviation increases proportionally to the two coefficients a, b excluding two cases where $a=0$ or $b=1$.

Table 1. Table 4. Packet's arrival rate by hop(p flexible)

a \ b	1	2	3	...	5	6
0	17.399264	9.71602	7.26184	...	2.96899	2.66695
1	17.399264	3.99885	2.61227	...	3.22738	3.55309
2	17.399264	1.97575	2.48060	...	5.36424	5.69557
3	17.399264	2.73581	3.86648	...	6.89371	7.19553
4	17.399264	3.69190	5.00034	...	8.00487	8.28221
5	17.399264	4.45516	5.87619	...	8.84917	9.10777
6	17.399264	5.05487	6.55819	...	9.51436	9.75853
7	17.399264	5.53446	7.10851	...	10.05332	10.68621
8	17.399264	5.92565	7.55987	...	10.49975	10.73368
9	17.399264	6.25043	7.93658	...	10.87615	11.09287
10	17.399264	6.52423	8.25569	...	11.19818	11.40900

5 Conclusion

Defending against denial-of-service(DoS) attacks is one of the hardest security problems on the Internet today. One difficulty to thwart these attacks is totrace the source of the attacks because they often use incorrect, or spoofed IP source addresses to disguise the true origin Traceback mechanisms are a critical part of the defense against IP spoofing and DoS attacks, as well as being of forensic value to law enforcement. Currently proposed IP traceback mechanisms are inadequate to address the traceback. problem for the following reasons: they require DoS victims to gather thousands of packets to reconstruct a single attack path; they do not scale to large scale Distributed DoS attacks; and they do not support incremental deployment. This study suggests to find the attack origin through MAC address marking of the attack origin. It is based on an IP trace algorithm, called Marking Algorithm. It modifies the Marking Algorithm so that we can convey the MAC address of the intervening routers, and as a result it can trace the exact IP address of the original attacker. To improve the detection time, our algorithm also contains a technique to improve the packet arrival rate. By adjusting marking probability according to the distance from the packet origin, we were able to decrease the number of needed packets to traceback the IP address.

Acknowledgements

This work was supported by INHA UNIVERSITY Research Grant.

References

1. Micah Adler. Tradeoffs in probabilistic packet marking for IP traceback. In Proceedings of 34th ACM Symposium on Theory of Computing (STOC), 2002.
2. S. Bellovin, M. Leech, and T. Taylor. The ICMP traceback message. Internet-Draft, draft-ietf-itrace-01.txt, October 2001. Work in progress, available at ftp://ftp.ietf.org/internet-drafts/draft-ietf-itrace-01.txt.
3. Hal Burch and Bill Cheswick. Tracing anonymous packets to their approximate source. Unpublished paper, December 1999.
4. Drew Dean, Matt Franklin, and Adam Stubblefield. An algebraic approach to IP traceback. ACM Transactions on Information and System Security, May 2002.
5. Michael Goodrich. Efficient packet marking for large-scale IP traceback. In Proceedings of the 9th ACM Conference on Computer and Communications Security, pages 117.126, November 2001.
6. Heejo Lee and Kihong Park. On the effectiveness of probabilistic packet marking for IP traceback under denial of service attack. In Proceedings IEEE Infocomm 2001, April 2001.
7. J. Li, M. Sung, J. Xu, and L. Li. Large-scale IP traceback in high-speed Internet: Practical techniques and theoretical foundation. In Proceedings of the IEEE Symposium on Security and Privacy, May 2004.
8. Vern Paxson. An analysis of using reflectors for distributed denial-of-service attacks. Computer Communication Review, 31(3), July 2001.
9. Stefan Savage, David Wetherall, Anna Karlin, and Tom Anderson. Practical network support for IP traceback. In Proceedings of ACM SIGCOMM 2000, August 2000.
10. Alex C. Snoeren, Craig Partridge, Luis A. Sanchez, Christine E. Jones, Fabrice Tchakountio, Stephen T. Kent, and W. Timothy Strayer. Hash-based IP traceback. In Proceedings of ACM SIGCOMM 2001, pages 3.14, August 2001.
11. Dawn Song and Adrian Perrig. Advanced and authenticated marking schemes for IP traceback. In Proceedings IEEE Infocomm 2001, April 2001.

Performance Comparison Between Backpropagation, Neuro-Fuzzy Network, and SVM

Yong-Guk Kim[1], Min-Soo Jang[2], Kyoung-Sic Cho[1], and Gwi-Tae Park[2]

[1] School of Computer Engineering, Sejong University, Seoul, Korea
ykim@sejong.ac.kr
[2] Dept. of Electrical Engineering, Korea University, Seoul, Korea
gtpark@elec.korea.ac.kr

Abstract. In this study, we compare the performance of well-known neural networks, namely, back-propagation (BP) algorithm, Neuro-Fuzzy network and Support Vector Machine (SVM) using the standard three database sets: Wisconsin breast cancer, Iris and wine data. Since such database have been useful for evaluating performance of a group of machine learning algorithms, a series of experiments have been carried out for three algorithms using the cross validation method. Results suggest that SVM outperforms the others and the Neuro-Fuzzy network is better than the BP algorithm for this data set.

1 Introduction

Although machine learning has a long history, recently it becomes an important research topic, partly because it has been used extensively in medical area. In particular, a group of machine learning algorithms have been used in classifying the diseases and symptoms. There are many algorithms available for such purpose. Among them, three machine learning algorithms are used extensively: artificial neural network, Neuro-Fuzzy network and SVM. However, a group of researchers prefer one algorithm to others, and the other group dose so the other algorithm. It appears that the preference is subjective. Given such situation, we have thought that it could be worth to compare the performances of those algorithms. The present study aims to evaluate such algorithms using the standard database sets, available in the public domain.

The rest of this paper is organized as follows. Section 2 introduces the structure and training method of the artificial neural network and Section 3 describes the structure of the Neuro-Fuzzy network. In section 4, the basic principles of SVM such as maximum margin, optimization and slack variable are described. Experimental results and discussion are given in Section 5 and some conclusions and discussion are given in Section 6.

2 Artificial Neural Networks

The BackPropagation (BP) algorithm, a gradient descent method for training the weights in a multilayer neural network, is a specific technique for the training of neural network [1]. The backpropagation process consists of two passes through the

D. Grigoriev, J. Harrison, and E.A. Hirsch (Eds.): CSR 2006, LNCS 3967, pp. 438–446, 2006.

different layers of the network, a forward pass and a backward pass. In the forward pass, an input vector is applied to the input nodes of the network, and its effect propagates through the network layer by layer. Finally, a set of output is produced as the actual response of the network. During the forward pass, the network weights are all fixed. During the backward pass, on the other hand, the network weights are all adjusted in accordance with the error-correction rule. Specifically, the actual response of the network is subtracted from a desired output to produce an error signal. This error signal is the propagated backward through the network, against the direction of synaptic connections. The network weights are adjusted to make the actual response of the network move close to the desired response. Fig. 1 shows a diagram of typical multilayer neural network.

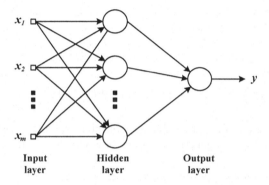

x_1
x_2
x_m

| Input layer | Hidden layer | Output layer |

y

Fig. 1. Diagram for an Artificial Neural Network

3 Neuro-Fuzzy Network

Self-Adaptive Neuro-Fuzzy Inference System (SANFIS) is capable of self-adapting and self-organizing its internal structure to acquire a parsimonious rule-base for interpreting the embedded knowledge of a system from the given training data set [2]. SANFIS can have three types of IF-THEN rule structures as shown in Fig. 2,

$$\text{Rule } j: \text{IF } x_1 \text{ is } A_1^j \text{ and } \dots \text{ and } x_n \text{ is } A_n^j , \tag{1}$$

$$\text{THEN } y_1 \text{ is } f_1^j \text{ and } \dots \text{ and } y_m \text{ is } f_m^j .$$

$$\text{Where } f_k^j = \begin{cases} B_k^j & \text{(type I)} \\ \theta_k^j & \text{(type II)} \\ b_{0k}^j + b_{1k}^j x_1 + \dots + b_{nk}^j x_n & \text{(type III)} \end{cases} \tag{2}$$

where $j = 1, 2, \dots, J$, $x_i (i = 1, 2, \dots, n)$,and $y_k (k = 1, 2, \dots, m)$ are the input and output variables, respectively, and A_i^j are the input fuzzy term sets. B_k^j , θ_k^j , and $b_{0k}^j + b_{1k}^j x_1 + \dots + b_{nk}^j x_n$ are output fuzzy term sets (type I), singleton constituents

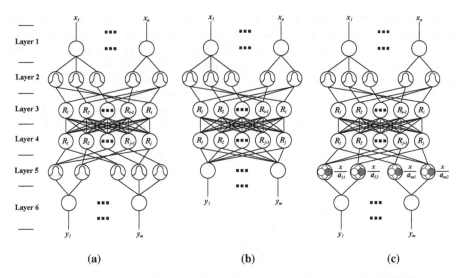

Fig. 2. Structures of SANFIS (a) SANFIS I, (b) SANFIS II, and (c) SANFIS III

(type II), and function of linear combination of input variables (type III, TSK model), respectively. Fig. 2 shows generic structures of SANFIS type I, type II, and type III, respectively.

The SANFIS learning algorithm consists of two components: 1) an MCA clustering algorithm that identifies a parsimonious internal structure in the sense that the number of clusters (fuzzy rules) is equal (or close) to the true number of clusters in a given training data set, and 2) a fast recursive linear/nonlinear least-squares optimization algorithm that is utilized to accelerate the learning convergence and fine tune the link weights of the whole system to achieve a better performance.

4   Support Vector Machines

4.1   Linear Separable Case

The fundamental idea of SVM is to construct a hyperplane as the decision line, which separates the positive (+1) classes from the negative (-1) ones with the largest margin [1, 3, 9]. In a binary classification problem, let us consider the training sample $\{(x_i, d_i)\}_{i=1}^{N}$, where x_i is the input pattern for the i-th sample and d_i is the corresponding desired response (target output) with subset $d \in \{-1, +1\}$. The equation of a hyperplane that does the separation is

$$w^T x + b = 0 \qquad (3)$$

where x is an input vector, w is an adjustable weight vector, and b is a bias. Fig. 3 shows an optimal hyper plane for the linearly separable case and margin, γ. The aim of the SVM classifier is to maximize the margin.

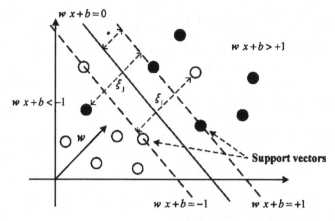

Fig. 3 An optimal hyperplane for the linearly separable case with the maximum margin and slack variables, ξ_i

The margin, distance of the nearest point to the typical hyper plane, equals to $1/\|w\|$. So, the problem turns into a quadratic programming:

$$\text{minimize} \quad \frac{1}{2}\|w\|^2$$

$$\text{s.t.} \quad y_i\left(w \cdot x_i + b\right) \geq 1 \quad , i = 1, 2, 3, \cdots, N \tag{4}$$

Problems of this kind are dealt with by introducing Lagrange multipliers $\alpha_i \geq 0, i = 1, \ldots, n$ and a Lagrangian

$$L(w, b, \alpha) = \frac{1}{2}\|w\|^2 - \sum_{i=1}^{n} \alpha_i \left(y_i \left(w \cdot x_i + b \right) - 1 \right) \tag{5}$$

The Lagrangian L has to be minimized with respect to the variables w and b and maximized with respect to the dual variables α_i (in other words, a saddle point has to be found). The statement that at the saddle point, the derivatives of L with respect to the variables must vanish,

$$\frac{\partial}{\partial b} L(w, b, \alpha) = 0 \quad \text{and} \quad \frac{\partial}{\partial w} L(w, b, \alpha) = 0 \tag{6}$$

Leads to

$$\sum_{i=1}^{n} \alpha_i y_i = 0 \quad \text{and} \quad w = \sum_{i=1}^{n} \alpha_i y_i x_i \tag{7}$$

By substituting Eq. 7 into the Lagrangian (Eq. 5), one eliminates the variables w and b, arriving at the so-called dual quadratic optimization problem, which is the problem that one usually solves in practice:

$$\max_{\alpha} \quad \sum_{i=1}^{n} \alpha_i - \frac{1}{2}\sum_{i,j=1}^{n} \alpha_i \alpha_j y_i y_j x_i^T x_j$$

$$\text{subject to } \alpha_i \geq 0, i = 1,\dots,n, \quad \sum_{i=1}^{n} \alpha_i y_i = 0 \tag{8}$$

Thus, by solving the problem, one obtains the optimized solution.

$$w^* = \sum_{i=1}^{n} \alpha_i^* y_i x_i \tag{9}$$

$$b^* = y_i - w^* \cdot x_i \tag{10}$$

The decision function of classification is

$$f(x) = \text{sgn}\left(\sum_{i=1}^{n} y_i \alpha_i^* (x \cdot x_i) + b^* \right) \tag{11}$$

4.2 Nonlinearly Separable Case

In practice, a separating hyperplane may not exist because the outliers as shown in Fig. 3. In that case, it can be an over-fitting problem. Slack variable ξ_i is introduced to relax such situation and it gives a certain tolerance as follows:

$$y_i (w \cdot x_i + b) \geq 1 - \xi_i \quad , i = 1,2,3,\cdots,N \tag{12}$$

A classifier that generalizes well is then found by controlling both the weight vector and the sum of the slack variables $\sum_i \xi_i$ minimize the cost function:

$$\Phi(w,\xi) = \frac{1}{2}\|w\|^2 + C\sum_{i=1}^{n} \xi_i \tag{13}$$

where C is a user-specified positive parameter. Using the method of Lagrange multipliers and proceeding in a manner similar to that described above, we may lead to the problem of maximizing (Eq. 8), subject to the constraints

$$0 \leq \alpha_i \leq C, i = 1,\dots,n, \quad \text{and } \sum_{i=1}^{n} \alpha_i y_i = 0 \tag{14}$$

This way, the influence of the outliers gets limited.

The other important property of SVM is that it can use diverse kernels in dealing with the input vectors. The kernels have non-linear decision boundaries, it is possible to separate the positive set from the negative one in the complex cases as shown in Fig. 4.

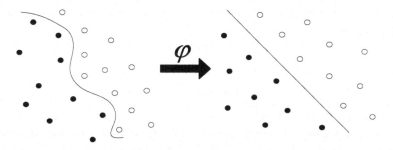

Fig. 4. Nonlinear mapping from the input space to the feature space

Let $\varphi(\cdot)$ denote a nonlinear transformation from the input space to the feature space and m denote the dimension of the feature space. Given such a nonlinear transformation, we may define a hyperplane as follows:

$$\sum_{j=1}^{m} w_j \varphi_j(x) + b = 0 \qquad (15)$$

We may simplify matters by writing

$$\sum_{j=0}^{m} w_j \varphi_j(x) = 0 \qquad (16)$$

where it is assumed that $\varphi_0(x) = 1$ for all x, so that w_0 denotes the bias b. Adapting Eq. 7 to this case, we may write

$$w = \sum_{i=1}^{n} \alpha_i y_i \varphi(x_i) \qquad (17)$$

Substituting Eq. 17 in Eq. 16, we may define the decision surface computed in the feature space as:

$$\sum_{i=1}^{n} \alpha_i y_i \varphi^T(x_i) \varphi(x) = 0 \qquad (18)$$

The term $\varphi^T(x_i)\varphi(x)$ represents the inner product of two vectors induced in the feature space by the input vector x and the input pattern x_i pertaining to the i-th example. We may therefore introduce the inner-product kernel denoted by $K(x, x_i)$ and defined by

$$K(x, x_i) = \varphi^T(x_i)\varphi(x) \qquad (19)$$

The optimal hyperplane is now defined by

$$\sum_{i=1}^{n} \alpha_i y_i K(x, x_i) = 0 \qquad (20)$$

Generally, several kernels such as polynomial, Gaussian and sigmoid kernels are used for applications.

5 Experiment and Performance Comparison

We have compared the performance of the neural network method, the Neuro-Fuzzy method and the SVM method using three well known data sets in the classification task - the iris data, the Wisconsin breast cancer data, and the wine classification data. These three data sets are available from the University of California, Irvine, via an anonymous ftp [5].

5.1 Iris Data

The Fisher-Anderson iris data consist of four input parameters, sepal length (sl), sepal width (sw), petal length (pl), and petal width (pw), on 150 specimens of iris plant. Three species of iris are involved, Iris Sestosa, Iris Versiolor and Iris Virginica, and each species contains 50 instances. To compare the three methods, we randomly split the data into training set (50% of the data) and a test set (50%). The data set was normalized to the range [-1, 1]. We used the public domain implementation of **libSVM** [4] and predefined kernel, linear kernel, polynomial kernel, and radial basis function kernel. It has two type of SVM: C-SVM and Nu-SVM. And, the slack ξ_i variable is implemented as the C value. For instance, when C is infinite, ξ_i equals to zero.

The experiment was accomplished 10 times and average results are shown in table 1. Table 1 indicates that the best result was obtained when C-SVM is combined with the RBF kernel. Performance of SANFIS was lower that that of SVM, and BP was the worst.

Table 1. Performance comparison for the Iris data

Type / Kernel	C-SVM C = 1	C-SVM C = 1000	Nu-SVM	SANFIS	BP
Linear	96.459%	96.459%	96.188%	97.20 ~ 97.47% [2]	95.33%
Polynomial	97.622%	98.556%	91.263%		
RBF	95.776%	98.658%	95.916%		

5.2 Breast Cancer Data

The Wisconsin Breast Cancer Diagnostic data consist of nine input parameters, clump thickness, uniformity of cell size, uniformity of cell shape, marginal adhesion, single epithelial cell size, bare nuclei, bland chromatin, normal nucleoli, and mitoses, on 699 patterns. Two output classes are adopted, benign and malignant. 458 patterns are in the benign class and the other 241 patterns are in the malignant class. We used '0' value for missed values contained in 16 patterns. As prior experiment of the iris data, we randomly split the data into training set (50% of the data) and a test set (50%). The data was normalized to the range [-1, 1]. Result suggests that RBF is better than

Table 2. Performance comparison for the breast cancer data

Type Kernel	C-SVM		Nu-SVM	SANFIS	BP
	C = 1	C = 1000			
Linear	96.859%	97.475%	94.703%	96.07 ~	
Polynomial	97.474%	99.667%	96.369%	96.30%	96.28%
RBF	97.155%	100%	96.317%	[2]	

the other SVM kernels as shown in table 2. And performances of SANFIS and BP were similar and yet lower than that of SVM.

5.3 Wine Classification Data

The wine classification data set contains 178 wines that are brewed in the same region of Italy but derived from three different cultivars. Each data consist of 13 parameters, alcohol, malic acid, ash, alkalinity of ash, magnesium, total phenols, flavonoids, non-flavonoid phenols, proanthocyanins, color intensity, hue, OD280/OD315 of diluted wines and praline. The data set was classified three classes. As prior experiments, we randomly split the data into training set (50% of the data) and a test set (50%). The data was normalized to the range [-1, 1]. Result summarized in table 3 shows that SVMs again outperform the other neural networks.

Table 3. Performance comparison for the wine data

Type Kernel	C-SVM		Nu-SVM	SANFIS	BP
	C = 1	C = 1000			
Linear	98.461%	100%	96.375%	98.876%	
Polynomial	96.946%	100%	97.399%	[2]	90.44%
RBF	98.437%	100%	97.776%		

6 Conclusions and Discussion

The present paper compares the performance of well-known machine learning algorithms, namely, back-propagation (BP) algorithm, Neuro-Fuzzy network and Support Vector Machine (SVM) using the standard database sets: Wisconsin breast cancer, Iris and wine data. Our results suggest that SVM outperforms the others and the Neuro-Fuzzy network is better than the BP algorithm. Given the fact that BP typically consists of three layers and the Neuro-Fuzzy network has multi-layer, it is surprising that the performance of SVM is better than the others, since an SVM is a two-layer network, suggesting that the number of layers is not a crucial factor. Our experience also indicates that SVM converges reliably compared to the other algorithms. Therefore, we conclude that the learning principle of an algorithm is important than the structure of it.

References

1. S. Haykin, "Newral Network", Prentice Hall, 1999
2. J. S. Wang and G. C. Lee, "Self-Adaptive Neuro-Fuzzy Inference Systems for Classification Applications", IEEE Trans. on Fuzzy System, vol. 10, no. 6, pp. 790-802, 2002
3. V. Vapnik, "The Nature of Statistical Learning Theory", Springer-Verlag, NY, USA, pp. 45-98, 1995
4. LibSVM. http://www.csie.ntu.edu.tw/~cjlin/libsvm/index.html
5. UCI Database. ftp://ftp.ics.uci.edu/pub/machine-learning-databases
6. P.K. Simpson, "Fuzzy Min-Max Neural Network – Part O: Classification", IEEE Trans. on Neural Networks, vol. 3, pp. 776-786, 1992
7. B. C. Lovel and A.P. Bradley, "The Multiscale Classifier", IEEE Trans. on Pattern Anal. Machine Intell., vol. 18, pp. 124-137, 1996
8. A. L. Corcoran and S. Sen, "Using Real-Valued Genetic Algorithms to Evolve Rule Sets for Classification", in Proc. of 1st IEEE conf. Evolutionary Computation, pp. 120-124, 1994
9. B. Scholkopf, A.J. Smola, "Learning with Kernels", The MIT Press, 2002

Evolutionary Multi-objective Optimisation by Diversity Control

Pasan Kulvanit[1], Theera Piroonratana[2],
Nachol Chaiyaratana[2], and Djitt Laowattana[1]

[1] Institute of Field Robotics,
King Mongkut's University of Technology Thonburi, Bangkok 10140, Thailand
{pasan, djitt}@fibo.kmutt.ac.th
[2] Research and Development Center for Intelligent Systems,
King Mongkut's Institute of Technology North Bangkok, Bangkok 10800, Thailand
theerapi@yahoo.com, nchl@kmitnb.ac.th

Abstract. This paper presents an improved multi-objective diversity control oriented genetic algorithm (MODCGA-II). The performance comparison between the MODCGA-II, a non-dominated sorting genetic algorithm II (NSGA-II) and an improved strength Pareto evolutionary algorithm (SPEA-II) is carried out where different two- and three-objective benchmark problems with specific multi-objective characteristics are used. The results indicate that the two-objective MODCGA-II solutions are better than the solutions generated by the NSGA-II and SPEA-II in terms of the closeness to the true Pareto optimal solutions and the uniformity of solution distribution along the Pareto front. In contrast, the NSGA-II in overall produces the best solutions in three-objective problems. As a result, the limitation of the proposed algorithm is identified.

1 Introduction

It is undeniable that a major factor that contributes to the success of genetic algorithms is the parallel search mechanism embedded in the algorithm itself. However, this does not prevent the occurrence of premature convergence in the situation when the similarity among individuals in the population becomes too high. As a result, the prevention of a premature convergence must also be considered during the genetic algorithm design. One of the direct approaches for achieving the necessary prevention is to maintain population diversity [1].

Various strategies can be used to maintain or increase the population diversity. Nonetheless, a modification on the selection operation has received much attention. For instance, Mori et al. [2] has introduced a notion of thermodynamical genetic algorithm where the survival of individuals is regulated by means of monitoring the free energy within the population. The modification on the selection operation can also be done in the cross-generational sense [3, 4, 5]. Whitley [3] has proposed a GENITOR system where offspring generated by standard operators are chosen for replacing parents based upon the ranks of the individuals. In contrast to Whitley [3], Eshelman [4] recommends the application of mating

D. Grigoriev, J. Harrison, and E.A. Hirsch (Eds.): CSR 2006, LNCS 3967, pp. 447–456, 2006.
© Springer-Verlag Berlin Heidelberg 2006

restriction while Shimodaira [5] suggests the use of variable-rate mutation as a means to create offspring. Then a cross-generational survival selection is carried out using a standard fitness-based selection technique in both cases.

In addition to the early works described above, another genetic algorithm that utilises standard crossover and mutation operators has been specifically developed by Shimodaira [6] to handle the issue of population diversity; this algorithm is called a diversity control oriented genetic algorithm or DCGA. During the cross-generational survival selection in the DCGA, duplicated individuals in the merged population containing both parent and offspring individuals are first eliminated. The remaining individuals are then sorted according to their fitness values in descending order. Following that the best individual from the remaining individuals is determined and kept for passing onto the next generation. Then either a cross-generational deterministic survival selection (CDSS) method or a cross-generational probabilistic survival selection (CPSS) method is applied in the top-down fashion to the remaining non-elite individuals in the sorted array. In the case of the CDSS, the remaining non-elite individuals with high fitness values will have a higher chance of being selected since they reside in the top part of the array and hence have a higher selection priority than individuals with low fitness values. In contrast, a survival probability value is assigned to each non-elite individual according to its similarity to the best individual in the case of the CPSS. If the genomic structure of the individual interested is very close to that of the best individual, the survival probability assigned to this individual will be close to zero. On the other hand, if the chromosome structure of this individual is quite different from that of the best individual, its survival probability will be close to one. Each individual will then be selected according to the assigned survival probability where the survival selection of the sorted non-elite individuals is still carried out in the top-down manner. The DCGA has been successfully benchmarked in various continuous test problems [7].

With a minor modification, the DCGA can also be used in multi-objective optimisation. One possible approach for achieving this is to integrate the DCGA with other genetic algorithms that are specifically designed for multi-objective optimisation such as a multi-objective genetic algorithm or MOGA [8]. Such approach has been investigated by Sangkawelert and Chaiyaratana [9] where the inclusion of cross-generational survival selection with the multi-objective genetic algorithm is equivalent to the use of elitism, which is proven to be crucial to the success of various multi-objective algorithms including a non-dominated sorting genetic algorithm II or NSGA-II [10] and an improved strength Pareto evolutionary algorithm or SPEA-II [11]. In addition, the similarity measurement between the non-elite individual and the elite individual required by the diversity control operator is still carried out in the genotypic space. The resulting combined algorithm, which can be uniquely referred to as a multi-objective diversity control oriented genetic algorithm or MODCGA has been successfully tested using a two-objective benchmark suite [12]. Although some insights into the behaviour of the MODCGA have been gained, further studies can be made. In multi-objective optimisation the trade-off surface, which is the direct result from the spread of solutions, is generally defined in objective space. This means that diversity control can also be achieved by considering the similarity between objective vectors of the individuals. Moreover, in the initial study the multi-objective benchmark problems contain only two objectives. The performance study where benchmark problems contain a higher number of objectives [13] should also be investigated.

The organisation of this paper is as follows. In section 2, the explanation of the improved multi-objective diversity control oriented genetic algorithm or MODCGA-II is given. In section 3, the multi-objective benchmark problems and performance evaluation criteria are explained. Next, the multi-objective benchmarking results of the MODCGA-II, NSGA-II and SPEA-II are illustrated and discussed in section 4. Finally, the conclusions are drawn in section 5.

2   MODCGA-II

The MODCGA-II functions by seeking to optimise the components of a vector-valued objective function where the desired solutions are members of the Pareto optimal set. A solution is said to be Pareto optimal if no improvement can be achieved in one objective that does not lead to degradation in at least one of the remaining objectives. Hence, one solution is better than or dominates another solution if and only if there is an improvement in at least one objective without the sacrifice in the other objectives. Similar to the MOGA [8], the rank of an individual is given by the number of solutions in the set that dominate the candidate individual. Non-dominated individuals will posses the highest rank while dominated individuals will have lower ranks. However, the comparison will be made among individuals in the merged population, which is the result from combining parent and offspring populations together. Since the best individuals in the multi-objective context are the non-dominated individuals, when the CPSS method is used there will be more than one survival probability value that can be assigned to each dominated individual. In this study, the lowest value in the probability value set is chosen for each dominated individual. After the survival selection routine is completed and the fitness values have been interpolated onto the individuals according to their ranks, standard genetic operations including fitness sharing [8], fitness-based selection, crossover and mutation can then be applied to the population. In this work, the similarity measurement will be conducted in objective space; two advantages are gained through this modification. Firstly, since the aim of multi-objective optimisation is to obtain multiple solutions at which together produce a trade-off objective surface that represents a Pareto front, diversity control in objective space would directly enforce this aim. Secondly, a diversity control operator that is designed for use in objective space would be independent of the chromosome encoding scheme utilised. With the modification described above, in the CPSS scheme the survival probability (p_s) will be given by

$$p_s = \{(1 - c)d/d_{max} + c\}^\alpha \tag{1}$$

where d is the distance between the interested individual and a non-dominated individual in objective space, d_{max} is the maximum distance between two individuals in the population, c is the shape coefficient and α is the exponent coefficient. The formula in equation (1) is adapted from the original genotypic operator [6].

In addition to the modification on the diversity control operation, the use of a preserved non-dominated solution archive is included in the MODCGA-II. Basically, the parent individuals will be picked from a population which includes both individuals obtained after the diversity control and that from the archive. Each time that a new population is created after the diversity control operation, non-dominated solutions within the archive will be updated. If the solution that survives the diversity control operation is neither dominated by any solutions in the archive nor a duplicate of a solution in the archive, then this solution will be added to the archive. At the same time, if the solution that survives the diversity control operation dominates any existing solution in the archive, the dominated solution will be expunged from the archive. In order to maintain the diversity within the preserved non-dominated solution archive, k-nearest neighbour clustering technique [11] is used to regulate the size of the archive.

3   Multi-objective Problems and Performance Criteria

The MODCGA-II will be benchmarked using six optimisation test cases developed by Deb et al. [13]. The problems DTLZ1–DTLZ6 are scalable minimisation problems with n decision variables and m objectives. In this paper, two-objective problems with 11 decision variables and three-objective problems with 12 decision variables are investigated. DTLZ1 has a linear Pareto front and contains multiple local fronts. DTLZ2 has a spherical Pareto front. DTLZ3 and DTLZ4 also have spherical Pareto fronts where DTLZ3 contains multiple local fronts while the DTLZ4 solutions are non-uniformly distributed in the search space. DTLZ5 has a curve Pareto front. DTLZ6 also has a curve Pareto front but the problem contains multiple local fronts.

Zitzler et al. [12] suggest that to assess the optimality of non-dominated solutions identified by a multi-objective optimisation algorithm, these solutions should be compared among themselves and with the true Pareto optimal solutions. Two corresponding measurement criteria are considered: the average distance between the non-dominated solutions to the Pareto optimal solutions (M_1) and the distribution of the non-dominated solutions (M_2). These criteria are calculated from the objective vectors of the solutions obtained. A low M_1 value implies that the solutions are close to the true Pareto optimal solutions. In addition, when two solution sets have similar M_1 indices, the set with a higher M_2 value would have a better distribution.

4   Results and Discussions

In this section, the results from using the MODCGA-II to solve test problems DTLZ1–DTLZ6 will be presented. The results will be benchmarked against that obtained from the non-dominated sorting genetic algorithm II or NSGA-II [10] and the improved strength Pareto evolutionary algorithm or SPEA-II [11] where the executable codes for the implementation of both algorithms are obtained directly from A Platform and Programming Language Independent Interface for Search Algorithms (PISA) web site (http://www.tik.ee.ethz.ch/pisa). Both CDSS and CPSS techniques are utilised in the implementation of the MODCGA-II. The diversity control study will be conducted with other genetic parameters remain fixed throughout the trial. The parameter setting for the MODCGA-II, NSGA-II and SPEA-II that is used in all problems is displayed in Table 1.

Table 1. Parameter setting for the MODCGA-II, NSGA-II and SPEA-II

Parameter	Value and Setting
Chromosome coding	Real-value representation
Fitness sharing	Triangular sharing function (MODCGA-II only)
Fitness assignment	Linear fitness interpolation (MODCGA-II only)
Selection method	Stochastic universal sampling (MODCGA-II) or tournament selection (NSGA-II and SPEA-II)
Crossover method	SBX recombination with probability = 1.0 [14]
Mutation method	Variable-wise polynomial mutation with probability = 1/number of decision variables [14]
Population size	100
Archive size	100 (MODCGA-II and SPEA-II only)
Number of generations	300 (MODCGA-II) or 600 (NSGA-II and SPEA-II)
Number of repeated runs	30

Five values of the shape coefficient (c)—0.00, 0.25, 0.50, 0.75 and 1.00—and six values of the exponent coefficient (α)—0.00, 0.20, 0.40, 0.60, 0.80 and 1.00—are used to create 30 different diversity control settings for the MODCGA-II. From equation (1), the settings of $c = 1.00$ and $\alpha = 0.00$ are for the implementation of the CDSS technique since the survival probability of each dominated individual is equal to one. For each setting, the MODCGA-II runs for the DTLZ1–DTLZ6 problems with two and three objectives are repeated 30 times. The M_1 and M_2 performance indices from each run are subsequently obtained and the average values of the two indices calculated from all two-objective problems and that from all three-objective problems are displayed in the form of contour plots in Fig. 1. The M_2 index is calculated using the neighbourhood parameter $\sigma = 0.488$ where the parameter is set using the extent of the true Pareto front in the objective space as the guideline. In addition, the M_2 index has been normalised by the maximum attainable number of non-dominated individuals from a single run. From Fig. 1, it is noticeable that diversity control settings considered have a small effect in three-objective problems while a significant performance variation can be detected in two-objective problems. For the case of two-objective problems, the region where the M_1 index has a small value coincides with the area where the M_2 index is small. At the same time the region where the M_1 index is high is also in the vicinity of the area where the M_2 index has a large value. In a successful multi-objective search, the M_1 index should be as small as possible. Although a large M_2 index usually signifies a good solution distribution, the interpretation of the M_2 result must always be done while taken the M_1 index into consideration. This is because in the case where the solutions are further away from the true Pareto optimal solutions, the obtained value of the M_2 index is generally high since each solution would also be far apart from one another. In other words, the M_2 index has a lesser priority than the M_1 index and should be considered only when the obtained values of the M_1 index from two different algorithms or algorithm settings are close to one another. Using the above argument, multiple settings of the c and α values in Fig. 1 can be used to achieve low M_1 indices. In the current investigation, the setting where $c = 0.75$ and $\alpha = 0.2$ is chosen as the candidate setting that represents the diversity control that leads to a low M_1 value.

The search performance of the MODCGA-II with $c = 0.75$ and $\alpha = 0.2$ will be compared with that from the NSGA-II and SPEA-II. As stated in Table 1, each algorithm run will be repeated 30 times where the M_1 and normalised M_2 indices are subsequently calculated for each repeated run. The performance of the MODCGA-II, NSGA-II and SPEA-II in terms of the average and standard deviation of the M_1 and normalised M_2 indices on the two-objective and three-objective DTLZ1–DTLZ6 problems is summarised in Tables 2 and 3, respectively. In Tables 2 and 3, the neighbourhood parameter (σ) for the calculation of M_2 indices for all test problems is also set to 0.488.

Table 2. Summary of the MODCGA-II, NSGA-II and SPEA-II performances on the two-objective DTLZ1–DTLZ6 problems

Problem Index		MODCGA-II		NSGA-II		SPEA-II	
		Average	S.D.	Average	S.D.	Average	S.D.
DTLZ1	M_1	3.1157	4.7837	11.9186	5.1490	12.9616	5.2649
	M_2	0.4326	0.2942	0.6391	0.0474	0.7810	0.0547
DTLZ2	M_1	0.0030	0.0008	0.0148	0.0088	0.0190	0.0096
	M_2	0.5039	0.0439	0.5672	0.0310	0.5053	0.0494
DTLZ3	M_1	22.2335	18.1880	78.6069	24.8055	88.4823	22.4487
	M_2	0.5642	0.3941	0.6119	0.0814	0.7463	0.0745
DTLZ4	M_1	0.0023	0.0018	0.0238	0.0138	0.0252	0.0104
	M_2	0.3353	0.2111	0.2871	0.2353	0.3457	0.2417
DTLZ5	M_1	0.0030	0.0006	0.0148	0.0088	0.0175	0.0079
	M_2	0.4972	0.0495	0.5672	0.0310	0.5026	0.0501
DTLZ6	M_1	1.0199	0.3685	6.4295	0.3509	6.4986	0.3355
	M_2	0.8044	0.0603	0.7157	0.0493	0.8946	0.0166

Firstly, consider the results from the two-objective test problems, which are displayed in Table 2. In terms of the average distance from the non-dominated solutions identified to the true Pareto front or the M_1 criterion, the MODCGA-II posses the highest performance in all six test problems. Nonetheless, the

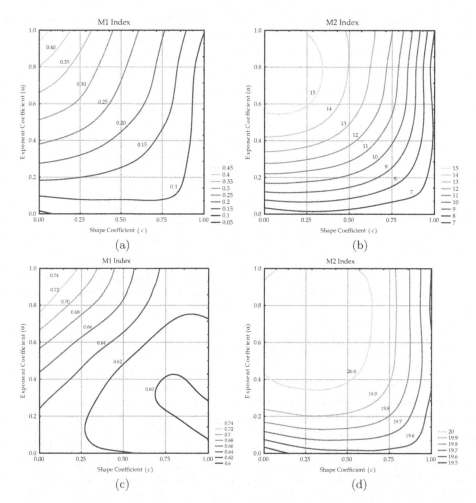

Fig. 1. Average values of M_1 and M_2 indices from all multi-objective problems for each diversity control setting (a) M_1 index from two-objective problems (b) M_2 index from two-objective problems (c) M_1 index from three-objective problems (d) M_2 index from three-objective problems

MODCGA-II is unable to identify the true Pareto optimal solutions in the DTLZ3 and DTLZ6 problems. These two problems are difficult to solve in the sense that they contains multiple local Pareto fronts. Although the DTLZ1 problem also contains numerous local Pareto fronts, the majority of results from all 30 MODCGA-II runs indicate that the MODCGA-II is capable solving this problem. This means that the shape of Pareto front in two-objective problems can also affect the algorithm performance since the DTLZ1 problem has a straight-line Pareto front while the DTLZ3 and DTLZ6 problems have curve Pareto fronts. The M_1 index also reveals that the performance of NSGA-II and SPEA-II are very similar in all six problems. Since the M_1 indices from both algorithms are

Table 3. Summary of the MODCGA-II, NSGA-II and SPEA-II performances on the three-objective DTLZ1–DTLZ6 problems

Problem Index		MODCGA-II		NSGA-II		SPEA-II	
		Average	S.D.	Average	S.D.	Average	S.D.
DTLZ1	M_1	24.8583	8.3026	7.3259	3.1478	12.2895	3.9570
	M_2	0.9923	0.0130	0.8762	0.0900	0.4542	0.0283
DTLZ2	M_1	0.0473	0.0069	0.0119	0.0095	0.0120	0.0069
	M_2	0.7281	0.0395	0.6776	0.0262	0.3502	0.0227
DTLZ3	M_1	282.8405	73.2361	67.5248	18.6048	89.5977	31.0686
	M_2	1.0000	0.0000	0.8333	0.0880	0.4575	0.0229
DTLZ4	M_1	0.0519	0.0240	0.0192	0.0097	0.0257	0.0153
	M_2	0.7229	0.0724	0.4371	0.1772	0.2714	0.0941
DTLZ5	M_1	0.0088	0.0024	0.0138	0.0119	0.0197	0.0088
	M_2	0.4916	0.0527	0.4061	0.0733	0.2080	0.0346
DTLZ6	M_1	3.9762	0.7985	5.8548	0.3046	6.3882	0.3039
	M_2	0.9847	0.0119	0.9842	0.0014	0.4999	0.0005

quite close, a further inspection on the M_2 indices can be easily made. Again, the M_2 indices from the NSGA-II and SPEA-II are also very close to one another. This leads to the conclusion that for the two-objective problems, the capability of both the NSGA-II and SPEA-II is pretty much the same.

Moving onto the results from the three-objective test problems, which are displayed in Table 3. In terms of the M_1 criterion, the NSGA-II posses the highest performance in the DTLZ1–DTLZ4 problems while the MODCGA-II is the best algorithm for the DTLZ5 and DTLZ6 problems. However, the DTLZ1, DTLZ3 and DTLZ6 problems cannot be solved using either the MODCGA-II or NSGA-II. These three problems contain multiple local Pareto front and hence can be classified as difficult problems. It is also noticeable that in contrast to the two-objective case, the shape of the Pareto front has no effect on the algorithm's ability to identify the correct solutions. By comparing the two-objective results with the three-objective results, it can be seen that there is deterioration in performances in the MODCGA-II and SPEA-II while the performance of the NSGA-II remains unchanged. The deterioration of the SPEA-II performance is detectable only in terms of the solution distribution (M_2 index) and not in terms of the closeness of solutions to the true Pareto front (M_1 index). As a result, the SPEA-II solutions now have a worse distribution than that from the NSGA-II while the solutions from both algorithms are at a similar distance from the true Pareto front. On the other hand, the performance degradation in the MODCGA-II is highest when the problem involves multiple local Pareto fronts. Nonetheless, even with the performance reduction, the MODCGA-II is still better than the NSGA-II at solving the DTLZ5 and DTLZ6 problems where the Pareto fronts can be visually displayed as two-dimensional curves. However, the difference in the search performances of the two algorithms in the DTLZ5 and DTLZ6 problems is not as evident as that in the DTLZ1–DTLZ4 problems. The M_2 indices confirm that the distributions of solutions from the MODCGA-II and NSGA-II are very similar in the last two problems. Using both performance indices and algorithm performance deterioration, it can be concluded that in overall the best algorithm for the three-objective test problems is the NSGA-II.

5 Conclusions

In this paper, an improved multi-objective diversity control oriented genetic algorithm or MODCGA-II is presented. The proposed algorithm differs from the original MODCGA [9] in the sense that the MODCGA-II performs diversity control via similarity measurement in objective space, which makes the diversity control operation becomes independent from the chromosome encoding scheme, and the use of a preserved non-dominated solution archive is also included. The MODCGA-II has been successfully tested on six scalable multi-objective benchmark problems [13]. The criteria used to assess the algorithm performance include the closeness of non-dominated solutions to the true Pareto front and the distribution of the solutions across the front [12]. The analysis indicates that the MODCGA-II with the CPSS technique where $c = 0.75$ and $\alpha = 0.20$ can produce non-dominated solutions that are better than that generated by the non-dominated sorting genetic algorithm II or NSGA-II [10] and the improved strength Pareto evolutionary algorithm or SPEA-II [11] in two-objective benchmark problems. On the other hand, when the number of objectives increases to three, the MODCGA-II performance is worse than that of the NSGA-II and the limitation of the proposed algorithm is hence identified.

Acknowledgements

This work was supported by the Thailand Research Fund through the Royal Golden Jubilee Ph.D. Program (Grant No. PHD/1.M.KT.44/C.1) and the Research Career Development Grant (Grant No. RSA4880001).

References

1. Mauldin, M.L.: Maintaining diversity in genetic search. In: Proceedings of the National Conference on Artificial Intelligence, Austin, TX (1984) 247–250
2. Mori, N., Yoshida, J., Tamaki, H., Kita, H., Nishikawa, Y.: A thermodynamical selection rule for the genetic algorithm. In: Proceedings of the Second IEEE International Conference on Evolutionary Computation, Perth, WA (1995) 188–192
3. Whitley, D.: The GENITOR algorithm and selection pressure: Why rank-based allocation of reproduction trials is best. In: Proceedings of the Third International Conference on Genetic Algorithms, Fairfax, VA (1989) 116–121
4. Eshelman, L.J.: The CHC adaptive search algorithm: How to have safe search when engaging in nontraditional genetic recombination. In: Rawlins, G.J.E. (ed.): Foundations of Genetic Algorithms, Vol. 1. Morgan Kaufmann, San Mateo, CA (1991) 265–283

5. Shimodaira, H.: A new genetic algorithm using large mutation rates and population-elitist selection (GALME). In: Proceedings of the Eighth IEEE International Conference on Tools with Artificial Intelligence, Toulouse, France (1996) 25–32

6. Shimodaira, H.: DCGA: A diversity control oriented genetic algorithm. In: Proceedings of the Second International Conference on Genetic Algorithms in Engineering Systems: Innovations and Applications, Glasgow, UK (1997) 444–449

7. Shimodaira, H.: A diversity-control-oriented genetic algorithm (DCGA): Performance in function optimization. In: Proceedings of the 2001 Congress on Evolutionary Computation, Seoul, Korea (2001) 44–51

8. Fonseca, C.M., Fleming, P.J.: Multiobjective optimization and multiple constraint handling with evolutionary algorithms–Part 1: A unified formulation. IEEE Transactions on Systems, Man, and Cybernetics–Part A: Systems and Humans **28**(1) (1998) 26–37

9. Sangkawelert, N., Chaiyaratana, N.: Diversity control in a multi-objective genetic algorithm. In: Proceedings of the 2003 Congress on Evolutionary Computation, Canberra, Australia (2003) 2704–2711

10. Deb, K., Pratap, A., Agarwal, S., Meyarivan, T.: A fast and elitist multiobjective genetic algorithm: NSGA-II. IEEE Transactions on Evolutionary Computation **6**(2) (2002) 182–197

11. Zitzler, E., Laumanns, M., Thiele, L.: SPEA2: Improving the strength Pareto evolutionary algorithm for multiobjective optimization. In: Giannakoglou, K., Tsahalis, D., Periaux, J., Papailiou, K., Fogarty, T. (eds.): Evolutionary Methods for Design, Optimisation and Control. International Center for Numerical Methods in Engineering (CIMNE), Barcelona, Spain (2002) 95–100

12. Zitzler, E., Deb, K., Thiele, L.: Comparison of multiobjective evolutionary algorithms: Empirical results. Evolutionary Computation **8**(2) (2000) 173–195

13. Deb, K., Thiele, L., Laumanns, M., Zitzler, E.: Scalable test problems for evolutionary multi-objective optimization. In: Abraham, A., Jain, L.C., Goldberg, R. (eds.): Evolutionary Multiobjective Optimization: Theoretical Advances and Applications. Springer, Berlin, Germany (2005)

14. Deb, K.: Multi-objective Optimization Using Evolutionary Algorithms. Wiley, Chichester, UK (2001)

3D Facial Recognition Using Eigenface and Cascade Fuzzy Neural Networks: Normalized Facial Image Approach

Yeung-Hak Lee and Chang-Wook Han

School of Electrical Engineering and Computer Science, Yeungnam University,
214-1, Dae-dong, Gyongsan, Gyongbuk, 712-749 South Korea
{annaturu, cwhan}@yumail.ac.kr

Abstract. The depth information in the face represents personal features in detail. In particular, the surface curvatures extracted from the face contain the most important personal facial information. The principal component analysis using the surface curvature reduces the data dimensions with less degradation of original information, and the proposed 3D face recognition algorithm collaborated into them. The recognition for the eigenface referred from the maximum and minimum curvatures is performed. To classify the faces, the cascade architectures of fuzzy neural networks, which can guarantee a high recognition rate as well as parsimonious knowledge base, are considered. Experimental results on a 46 person data set of 3D images demonstrate the effectiveness of the proposed method.

1 Introduction

Today's computer environments are changing because of the development of intelligent interface and multimedia. To recognize the user automatically, people have researched various recognition methods using biometric information – fingerprint, face, iris, voice, vein, etc [1]. In a biometric identification system, the face recognition is a challenging area of research, next to fingerprinting, because it is a no-touch style. For visible spectrum imaging, there have been many studies reported in literature [2]. But the method has been found to be limited in their application. It is influenced by lighting illuminance and encounters difficulties when the face is angled away from the camera. These factors cause low recognition. To solve these problems a computer company has developed a 3D face recognition system [2][3]. To obtain a 3D face, this method uses stereo matching, laser scanner, etc. Stereo matching extracts 3D information from the disparity of 2 pictures which are taken by 2 cameras. Even though it can extract 3D information from near and far away, it has many difficulties in practical use because of its low precision. 3D laser scanners extract more accurate depth information about the face, and because it uses a filter and a laser, it has the advantage of not being influence by the lighting illuminance when it is angled away from the camera. A laser scanner can measure the distance, therefore, a 3D face image can be reduced by a scaling effect that is caused by the distance between the face and the camera [4][5].

D. Grigoriev, J. Harrison, and E.A. Hirsch (Eds.): CSR 2006, LNCS 3967, pp. 457–465, 2006.

Broadly speaking the two ways to establish recognition employs the face feature based approach and the area based approach [5-8]. A feature based approach uses feature vectors which are extracted from within the image as a recognition parameter. An area based approach extracts a special area from the face and recognizes it using the relationship and minimum sum of squared difference. Face recognition research usually uses 2 dimensional images. Recently, the 3D system becomes cheaper, smaller and faster to process than it used to be. Thus the use of 3D face image is now being more readily researched [3][9-12]. Many researchers have used 3D face recognition using differential geometry tools for the computation of curvature [9]. Hiromi et al. [10] treated 3D shape recognition problem of rigid free-form surfaces. Each face in the input images and model database is represented as an Extended Gaussian Image (EGI), constructed by mapping principal curvatures and their directions. Gordon [11] presented a study of face recognition based on depth and curvature features. To find face specific descriptors, he used the curvatures of the face. Comparison of the two faces was made based on the relationship between the spacing of the features. Lee and Milios [13] extracted the convex regions of the face by segmenting the range of the images based on the sign of the mean and Gaussian curvature at each point. For each of these convex regions, the Extended Gaussian Image (EGI) was extracted and then used to match the facial features of the two face images.

One of the most successful techniques of face recognition as statistical method is principal component analysis (PCA), and specifically eigenfaces [14][15]. In this paper, we introduce a novel face recognition for eigenfaces using the curvature that well presenting personal characteristics and reducing dimensional spaces. Moreover, the normalized facial images are considered to improve the recognition rate.

Neural networks (NNs) have been successfully applied to face recognition problems [16]. However, the complexity of the NNs increases exponentially with the parameter values, i.e. input number, output number, hidden neuron number, etc., and becomes unmanageable [17]. To overcome this curse of dimensionality, the cascade architectures of fuzzy neural networks (CAFNNs), constructed by the memetic algorithms (hybrid genetic algorithms) [18], are applied to this problem.

2 Face Normalization

The nose is protruded shape and located in the middle of the face. So it can be used as the reference point, firstly we tried to find the nose tip using the iterative selection method, after extraction of the face from the 3D face image [20]. And in face recognition, we have to consider the obtained face posture. Face recognition systems suffer drastic losses in performance when the face is not correctly oriented. The normalization process proposed here is a sequential procedure that aims to put the face shapes in a standard spatial position. The processing sequence is panning, rotation and tilting [21].

3 Surface Curvatures

For each data point on the facial surface, the principal, Gaussian and mean curvatures are calculated and the signs of those (positive, negative and zero) are used to determine the surface type at every point. The $z(x, y)$ image represents a surface where the individual Z-values are surface depth information. Here, x and y is the two spatial co-ordinates. We now closely follow the formalism introduced by Peet and Sahota [19], and specify any point on the surface by its position vector:

$$R(x, y) = xi + yj + z(x, y)k \tag{1}$$

The first fundamental form of the surface is the expression for the element of arc length of curves on the surface which pass through the point under consideration. It is given by:

$$I = ds^2 = dR \cdot dR = Edx^2 + 2Fdxdy + Gdy^2 \tag{2}$$

where

$$E = 1 + \left(\frac{\partial z}{\partial x}\right)^2, \quad F = \frac{\partial z}{\partial x}\frac{\partial z}{\partial y}, \quad G = 1 + \left(\frac{\partial z}{\partial y}\right)^2 \tag{3}$$

The second fundamental form arises from the curvature of these curves at the point of interest and in the given direction:

$$II = edx^2 + 2fdxdy + gdy^2 \tag{4}$$

where

$$e = \frac{\partial^2 z}{\partial x^2}\Delta, \quad f = \frac{\partial^2 z}{\partial x \partial y}\Delta, \quad g = \frac{\partial^2 z}{\partial y^2}\Delta \tag{5}$$

and

$$\Delta = (EG - F^2)^{-1/2} \tag{6}$$

Casting the above expression into matrix form with;

$$V = \begin{pmatrix} dx \\ dy \end{pmatrix}, \quad A = \begin{pmatrix} E & F \\ F & G \end{pmatrix}, \quad B = \begin{pmatrix} e & f \\ f & g \end{pmatrix} \tag{7}$$

the two fundamental forms become:

$$I = V'AV \quad I = V'BV \tag{8}$$

Then the curvature of the surface in the direction defined by V is given by:

$$k = \frac{V'BV}{V'AV} \tag{9}$$

Extreme values of k are given by the solution to the eigenvalue problem:

$$(B - kA)V = 0 \tag{10}$$

or

$$\begin{vmatrix} e - kE & f - kF \\ f - kF & g - kG \end{vmatrix} = 0 \tag{11}$$

which gives the following expressions for k_1 and k_2, the minimum and maximum curvatures, respectively:

$$k_1 = \{gE - 2Ff + Ge - [(gE + Ge - 2Ff)^2$$
$$-4(eg - f^2)(EG - F^2)]^{1/2}\}/ 2(EG - F^2) \tag{12}$$

$$k_2 = \{gE - 2Ff + Ge + [(gE + Ge - 2Ff)^2$$
$$-4(eg - f^2)(EG - F^2)]^{1/2}\}/ 2(EG - F^2) \tag{13}$$

Here we have ignored the directional information related to k_1 and k_2, and chosen k_2 to be the larger of the two. For the present work, however, this has not been done. The two quantities, k_1 and k_2, are invariant under rigid motions of the surface. This is a desirable property for us since the cell nuclei have no predefined orientation on the slide (the x – y plane).

The Gaussian curvature K and the mean curvature M is defined by

$$K = k_1 k_2 , \quad M = (k_1 k_2)/ 2 \tag{14}$$

which gives k_1 and k_2, the minimum and maximum curvatures, respectively. It turns out that the principal curvatures, k_1 and k_2, and Gaussian are best suited to the detailed characterization for the facial surface, as illustrated in Fig. 1. For the simple facet model of second order polynomial of the form, i.e. an 3x3 window implementation in our range images, the local region around the surface is approximated by a quadric

$$z(x, y) = a_{00} + a_{10}x + a_{01}y + a_{01}y + a_{20}x^2 + a_{02}y^2 + a_{11}xy \tag{15}$$

and the practical calculation of principal and Gaussian curvatures is extremely simple.

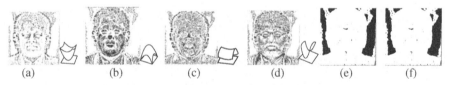

(a)   (b)   (c)   (d)   (e)   (f)

Fig. 1. Six possible surface type according to the sign of principal curvatures for the face surface; (a) concave (pit), (b) convex (peak), (c) convex saddle, (d) concave saddle, (e) minimal surface, (f) plane

4   Eigenface

4.1   Computing Eigenfaces [14]

Consider face images of size N x N, extracted contour line value. These images can be thought as a vector of dimension N^2, or a point in N^2 – dimensional space. A set of images, therefore, corresponds to a set of points in this high dimensional space. Since facial images are similar in structure, these points will not be randomly distributed, and therefore can be described by a lower dimensional subspace. Principal component analysis gives the basis vectors for this subspace. Each basis vector is of length N^2, and is the eigenvector of covariance matrix corresponding to the original face images. Let Γ_1, Γ_2, ... , Γ_M be the training set of face images. The average face is defined by

$$\Psi = \frac{1}{M}\sum_{n=1}^{M}\Gamma_n \qquad (16)$$

Each face differs from the average face by the vector $\Phi_i = \Gamma_n - \Psi$. The covariance matrix

$$C = \frac{1}{M}\sum_{n=1}^{M}\Phi_n\Phi_n^T \qquad (17)$$

has a dimension of N^2 x N^2 . Determining the eigenvectors of C for typical size of N is intractable task. Once the eigenfaces are created, identification becomes a pattern recognition task. Fortunately, we determine the eigenvectors by solving an M x M matrix instead.

4.2 Identification

The eigenfaces span an M-dimensional subspace of the original N^2 image space. The M significant eigenvectors are chosen as those with the largest corresponding eigen-values. A test face image Γ is projected into face space by the following operation: $\omega_n = u_n^T(\Gamma - \Psi)$, for n=1, ..., M, where u_n is the eigenvectors for C. The weights ω_n from a vector $\Omega^T = [\omega_1\ \omega_2 ... \omega_{M'}]$ which describes the contribution of each eigenface in representing the input face image. This vector can then be used to fit the test image to a predefined face class. A simple technique is to use the Euclidian distance $\varepsilon_n = \|\Omega - \Omega_n\|$, where Ω_n describes the nth face class. In this paper, we used the cascade architectures of fuzzy neural networks to compare with the distance as described next chapter.

5   Cascade Architectures of Fuzzy Neural Networks (CAFNNs)

As originally introduced in [17], the structure of CAFNNs is the cascade combination of the logic processors (LPs) which consist of fuzzy neurons. The LP, described in

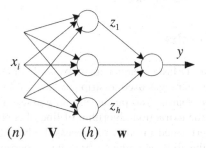

Fig. 2. Architecture of the LP regarded as a generic processing unit

Fig. 2, is a basic two-level construct formed by a collection of "*h*" AND neurons whose results of computing are then processed by a single OR neuron located in the output layer. Because of the location of the AND neurons, we will be referring to them as a hidden layer of the LP.

LPs are basic functional modules of the network that are combined into a cascaded structure. The essence of this architecture is to stack the LPs one on another. This results in a certain sequence of input variables. To assure that the resulting network is homogeneous, we use LPs with only two inputs, as shown in Fig. 3. In this sense, with "*n*" input variables, we end up with $(n-1)$ LPs being used in the network. Each LP is fully described by a set of the connections (V and w). To emphasize the cascade-type of architecture of the network, we index each LP by referring to its connections as $V[ii]$ and $w[ii]$ with "ii' being an index of the LP in the cascade sequence.

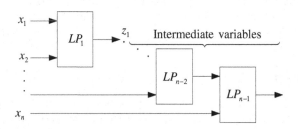

Fig. 3. A cascaded network realized as a nested collection of LPs

The sequence of relevant input subset and the connections were optimized by memetic algorithms in [18] to construct parsimonious knowledge base, but accurate one. As illustrated in [18], the memetic algorithms are more effective than the optimization scenario in [17]. Therefore, the optimization scenario in [18] will be considered in this approach. For more details about CAFNNs and its optimization, please refer to [17][18].

To apply the CAFNNs to classification problems, the output (class) should be fuzzified as binary. For example, if we assume that there are 5 classes (5 persons) in the data sets, the number of output crisp set should be 5 that are distributed uniformly. If the person belongs to the 2nd-class, the Boolean output can be discretized as "0 1 0 0 0". In this classification problem, the winner-take-all method is used to decide the class of the testing data set. This means that the testing data are classified as the class which has the biggest membership degree.

6   Experimental Results

In this study, we used a 3D laser scanner made by a 4D culture to obtain a 3D face image. First, a laser line beam was used to strip the face for 3 seconds, thereby obtaining a laser profile image, that is, 180 pieces and no glasses. The obtained image size was extracted by using the extraction algorithm of line of center, which is 640 x 480. Next, calibration was performed in order to process the height value, resampling and interpolation. Finally, the 3D face images for this experiment were extracted, at

320x320. A database is used to compare the different strategies and is composed of 92 images (two images of 46 persons). Of the two pictures available, the second photos were taken at a time interval of 30 minutes.

From these 3D face images, finding the nose tip point, using contour line threshold values (for which the fiducial point is nose tip), we extract images around the nose area. To perform recognition experiments for extracted area we first need to create two sets of images, i.e. training and testing. For each of the two views, 46 normal-expression images were used as the training set. Training images were used to generate an orthogonal basis, as described in section 3, into which each 3D image in training data set is projected in section 4. Testing images are a set of 3D images extracted local area we wish to identify.

Table 1. The comparison of the recognition rate (%)

		Best1	Best5	Best10	Best15
k_1	CAFNN(normalized)	64.5	78.4	89.6	95.9
	CAFNN	56.2	73.9	85.2	90.5
	k-NN	42.9	57.1	66.7	66.7
k_2	CAFNN (normalized)	68.1	86.4	90.1	96.3
	CAFNN	63.7	80.3	85.8	90.1
	k-NN	61.9	78.5	83.3	88.1

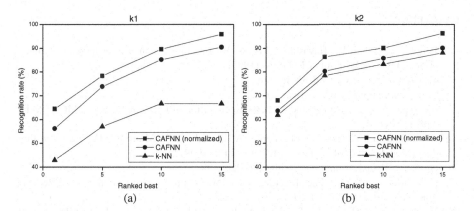

Fig. 4. The recognition results for each recognition method: (a) k_1, (b) k_2

Once the data sets have been extracted with the aid of eigenface, the development procedure of the CAFNNs should be followed for the face recognition. The used parameter values are the same as [18]. Since a genetic algorithm is a stochastic optimization method, ten times independent simulations were performed to compare the results with the conventional classification methods, as described in Table 1 and Fig. 4. In

Table 1 and Fig. 4, the results of the CAFNN are averaged over ten times independent simulations, and subsequently compared with the results of the conventional method (k-nearest neighborhood: k-NN). Also, the normalized facial images were considered to generate the curvature-based data set. As can be seen from Table 1 and Fig. 4, the recognition rate is improved by using normalized facial images.

7   Conclusions

The surface curvatures extracted from the face contain the most important personal facial information. We have introduced, in this paper, a new practical implementation of a person verification system using the local shape of 3D face images based on eigenfaces and CAFNNs. The underlying motivations for our approach originate from the observation that the curvature of face has different characteristic for each person. We found the exact nose tip point by using an iterative selection method. The low-dimensional eigenfaces represented were robust for the local area of the face. The normalized facial images were also considered to improve the recognition rate. To classify the faces, the CAFNNs were used. The CAFNNs have reduced the dimensionality problem by selecting the most relevant input subspaces too. Experimental results on a group of face images (92 images) demonstrated that our approach produces excellent recognition results for the local eigenfaces.

From the experimental results, we proved that the process of face recognition may use low dimension, less parameters, calculations and less same person images (used only two) than earlier suggested. We consider that there are many future experiments that could be done to extend this study.

References

1. Jain, L. C., Halici, U., Hayashi, I., Lee, S. B.: Intelligent biometric techniques in finger-print and face recognition. CRC Press (1999)
2. 4D Culture. http://www.4dculture.com
3. Cyberware. http://www.cyberware.com
4. Chellapa, R., et al.: Human and Machine Recognition of Faces: A Survey. UMCP CS-TR-3399 (1994)
5. Hallinan, P. L., Gordon, G. G., Yuille, A. L., Giblin, P., Mumford, D.: Two and three dimensional pattern of the face. A K Peters Ltd. (1999)
6. Grob, M.: Visual computing. Springer Verlag (1994)
7. Nikolaidis, A., Pitas, I.: Facial feature extraction and pose determination. Pattern Recognition, Vol. 33 (2000) 1783-1791
8. Moghaddam, B., Jebara, T., Pentland, A.: Bayesian face recognition. Pattern Recognition, Vol. 33 (2000) 1771-1782
9. Chua, C. S., Han, F., Ho, Y. K.: 3D Human Face Recognition Using Point Signature. Proc. of the 4th ICAFGR (2000)
10. Tanaka, H. T., Ikeda, M., Chiaki, H.: Curvature-based face surface recognition using spherial correlation. Proc. of the 3rd IEEE Int. Conf. on Automatic Face and Gesture Recognition (1998) 372-377

11. Gordon, G. G.: Face Recognition based on depth and curvature feature. Proc. of the IEEE Computer Society Conf. on Computer Vision and Pattern Recognition (1992) 808-810
12. Chellapa, R., Wilson, C. L., Sirohey, S.: Human and machine recognition of faces: A survey. Proceedings of the IEEE, Vol. 83, No. 5 (1995) 705-740
13. Lee, J. C., Milios, E.: Matching range image of human faces. Proc. of the 3rd Int. Conf. on Computer Vision (1990) 722-726
14. Turk, M., Pentland, A.: Eigenfaces for Recognition. Journal of Cognitive Neuroscience, Vol. 3, No. 1 (1991) 71-86
15. Hesher, C., Srivastava, A., Erlebacher, G.: Principal Component Analysis of Range Images for Facial Recognition. Proc. of CISST (2002)
16. Zhao, Z. Q., Huang, D. S., Sun, B. Y.: Human face recognition based on multi-features using neural networks committee. Pattern Recognition Letters, Vol. 25 (2004) 1351-1358
17. Pedrycz, W., Reformat, M., Han, C. W.: Cascade architectures of fuzzy neural networks. Fuzzy Optimization and Decision Making, Vol. 3 (2004) 5-37
18. Han, C. W., Pedrycz, W.: A new genetic optimization method and its applications. submitted to International Journal of Approximate Reasoning
19. Peet, F. G., Sahota, T. S.: Surface Curvature as a Measure of Image Texture. IEEE Trans. PAMI, Vol. 7, No. 6 (1985) 734-738
20. Lee, Y., Park, G., Shim, J., Yi, T.: Face Recognition from 3D Face Profile using Hausdorff Distance. Proc. of PRIA-6-2002 (2002)
21. Lee, Y.: 3D Face Recognition Using Longitudinal Section and Transection. Proc. of DICTA-2003 (2003)

A New Scaling Kernel-Based Fuzzy System with Low Computational Complexity

Xiaojun Liu, Jie Yang, Hongbin Shen, and Xiangyang Wang

Institute of Image Processing & Pattern Recognition, Shanghai Jiaotong University,
Shanghai, P.R. China 200240
pi4414@sjtu.edu.cn

Abstract. The approximation capability of fuzzy systems heavily depends on the shapes of the chosen fuzzy membership functions. When fuzzy systems are applied in adaptive control, computational complexity and generalization capability are another two important indexes we must consider. Inspired by the conclusion drawn by S.Mitaim and B.Kosko and wavelet analysis and SVM, the scaling kernel-based fuzzy system SKFS(Scaling Kernel-based Fuzzy System) is presented as a new simplified fuzzy system in this paper, based on Sinc x membership functions. SKFS can approximate any function in $L_2(R)$, with much less computational complexity than classical fuzzy systems. Compared with another simplified fuzzy system GKFS(Gaussian Kernel-based Fuzzy System) using Gaussian membership functions, SKFS has a better approximation and generalization capabilities, especially in the coexistence of linearity and nonlinearity. Therefore, SKFS is very suitable for fuzzy control. Finally, several experiment results are used to demonstrate the effectiveness of the new simplified fuzzy system SKFS.

1 Introduction

In the last decade, fuzzy system has been successfully applied in many fields, such as fuzzy modeling, function approximation, pattern recognition and adaptive control[13]. Classical fuzzy systems and their numerous variations were proposed [6,13]. The basic reason for the success of fuzzy system originates from the fact that it can effectively integrate data with expert knowledge in the unified framework, and furthermore, it is also a universal approximator[6,10,12,14], that is to say, it can approximate any continuous function with any given accuracy.

Generally speaking, a fuzzy system consists of 4 main parts: fuzzification, a rule base, an inference engine and a defuzzifier. When applying fuzzy systems to fuzzy modeling and adaptive control, the computational complexity is an important factor we often have to consider, so we hope to build such a fuzzy system that it can not only be computed with less complexity, but also hold the property of a universal approximator. In [4,13], based on Gaussian membership functions and the famous TSK fuzzy-rule forms, the authors presented a simplified fuzzy system (We call it GKFS here). A remarkable feature of such a fuzzy system is that it avoids defuzzification computation, therefore, the fuzzy system can

D. Grigoriev, J. Harrison, and E.A. Hirsch (Eds.): CSR 2006, LNCS 3967, pp. 466–474, 2006.

be computed more quickly than the corresponding classical fuzzy system, and moreover, we can prove that the fuzzy system still keeps a universal approximator based on Weirstone theorem, and so it is comparatively suitable for fuzzy modeling and adaptive control.

Although the universal approximation is kept theoretically, due to the coexistence of linearity and nonlinearity such as piecewise functions in fuzzy adaptive control, GKFS and current fuzzy systems sometimes become inefficient. Thus, a challenging problem appears whether we can design such a fuzzy system that it can *effectively* approximate *any* function in $L_2(R)$ with the same computational complexity as GKFS.

In [11,12], based on the experimental results obtained from more than 100 different types of membership functions, B.Kosko and B.M.Novakovic claimed that the classical fuzzy system with Sinc x(or the corresponding cosx type) membership functions has better approximation capability than fuzzy system with other types of membership functions. In [12], the authors wrote: *"frequent winning status of the sinc set in the simulations shows that: This seems to be the first time anyone has used the sinc function as a fuzzy set and yet such sets may well have improved the performance of many real fuzzy systems."* and *"the success of the sinc set and the hyperbolic-tangent bell curve further suggest that the familiar Gaussian or Cauchy or other familiar unimodal curves will not emerge as optimal set functions in other searches."* In fact, it is well known that sinc x is a scaling kernel function in wavelet analysis, and this fact implies that we may perhaps build another new kind of simplified fuzzy system with the help of the Sinc x functions.

In this paper, we will present a new Scaling Kernel-based fuzzy system (SKFS), and this new fuzzy system preserves the advantage of the less computational complexity like GKFS, however, SKFS can universally approximate any function and have better approximation and generalization capabilities. Such merits of SKFS make it more feasible than the simplified fuzzy system GKFS.

This paper is organized as follows. In Section 2, we propose the Scaling Kernel-based fuzzy system (SKFS) and build the link between wavelet analysis, SVM and SKFS. The learning algorithms of SKFS are discussed in Section 3. Experimental results are demonstrated in Section 4. Section 5 concludes this paper.

2   Scaling Kernel-Based Fuzzy System SKFS

In [4,13], the authors proposed the simplified fuzzy system GKFS independently. GKFS takes the TSK fuzzy-rule forms as follows.

$$R^i : if \ x_1 \ is \ A_{i1} \ ,..., x_n \ is \ A_{in} \ then \ \overline{y}_i = p_{i0} + p_{i1}x_1 + ... + p_{in}x_n \qquad (1)$$

where $x_1, x_2, ..., x_n$ are input variables of the fuzzy system, $\overline{y}_i$ is the output variable of the ith fuzzy rule, $i = 1, 2, ..., M$ (M is the total number of fuzzy rules).The Gaussian membership function A_{ij} is defined as

$$A_{ij}(x_j) = e^{-(\frac{x_j - m_{ij}}{\sigma_{ij}})^2} \qquad (2)$$

where $j = 1, 2, \ldots, n$, m_{ij}, σ_{ij} are free adjustable parameters. Obviously $\overline{y}_i$ is the linear combination of input variables $x_1, x_2, \ldots, x_n$ and $p_{i0}, p_{i1}, \ldots, p_{in}$ are the corresponding coefficients.

GKFS does not use the defuzzification computation, and it directly defines its output as

$$y = \sum_{i=1}^{M} \overline{y}_i \prod_{j=1}^{n} A_{ij}(x_j) \tag{3}$$

We call (3) the simplified fuzzy system GKFS (Gaussian Kernel-based fuzzy system). The authors[4,13] have proved that GKFS is also a universal approximator. It should be pointed out that the output of the corresponding TSK fuzzy system is

$$y' = \sum_{i=1}^{M} \overline{y}_i \prod_{j=1}^{n} A_{ij}(x_j) / \sum_{i=1}^{M} \prod_{j=1}^{n} A_{ij}(x_j) \tag{4}$$

where (4) uses the defuzzification procedure. Obviously, (3) has much less computational complexity than (4).

Recently, based on the experimental results obtained from more than 100 different types of membership functions, B.Kosko et al claimed Sinc x is the type of membership function that could make the corresponding classical fuzzy system have the best approximation capability. Inspired by this conclusion, we redefine the membership function A_{ij} of of (2) as the form of Sinc x, i.e.

$$A_{ij}(x_j) = sin(b_{ij}(x_j - m_{ij}))/(b_{ij}(x_j - m_{ij})) \tag{5}$$

where b, m_{ij} are free parameters, so, the corresponding (3) immediately draws a new version of the simplified fuzzy system and we call it Scaling Kernel-based fuzzy system(SKFS), owing to the fact that Sinc x is often used as the scaling kernel function in wavelet analysis. Note, as pointed out in [12], Sinc x can sometimes take negative values, so when we use this membership function, we may use a flag to guarantee that the denominator of the corresponding Sinc x is a positive value.

Accordingly, whether SKFS is a universal approximator or not will directly affect its applicable value. In fact, (5) is a scaling function, which plays an important role in wavelet analysis[1,2,7]. Thus we can study SKFS with the help of wavelet analysis, and then obtain the stronger approximation result, i.e. it can universally approximate any function in $L_2(R)$.

Remark. Many researchers have revealed the equivalent and complementary relationships among neural networks, fuzzy system and wavelet analysis. In [7], Y.Yu and S.Tan discussed the complementary and equivalence relationships between fuzzy system and wavelet neural network through B-spline functions. Although the study for SKFS here is related to these fruitful results of other scholars, it is different. The principle difference lies in that we introduce the concept of scaling functions of wavelet analysis instead of B-Spline functions and avoid computing defuzzification simultaneously. Thus, SKFS can universally approximate any function with much less computational complexity.

In wavelet analysis, scaling space, scaling functions, sub-wavelet space and sub-wavelet function are very important concepts[2]. When dimension n is large enough, the square and integral space $L_2(R)$ is approximately equal to V_N. So we can decompose $L_2(R)$ space as follows: $L_2(R) = V_N = V_{N-1} \oplus W_{N-1} = V_{N-2} \oplus W_{N-2} \oplus W_{N-1} = \ldots = V_{-\infty} \oplus W_{-\infty} \oplus \ldots \oplus W_{-1} \oplus W_0 \oplus W_1 \oplus \ldots \oplus W_{N-2} \oplus W_{N-1}$.

Hence, if we take the scaling functions in V_N space as the basis functions, in terms of the properties of scaling functions, we can derive a complete base V_N through translation operations. Thus, any function in this space can be represented as the linear combination of scaling functions. In other words, the linear combination of a complete base can approximate any function in this space within *any* given accuracy.

If we denote the 1-dimentional scaling function as $\phi(x)$, then a special form of the separable multi-dimensional scaling function can be written as

$$\phi(\overline{x}) = \prod_{i=1}^{n} \phi(x^i) \qquad (6)$$

where $\overline{x} = (x_1, x_2, \ldots, x_n) \in R^n$, and each 1-dimensional scaling function satisfies

$$\int_{-\infty}^{\infty} \phi(x)dx = 1 \qquad (7)$$

So we can construct such a scaling kernel function as

$$K(\overline{x}, \overline{x}') = \prod_{i=1}^{n} \phi(x^i - x'^i) \qquad (8)$$

If we take $\phi(x) = Sinc(x) = sin(bx)/bx$, then the corresponding scaling kernel function is

$$K(\overline{x}, \overline{x}') = \prod_{i=1}^{n} \sin(b_i(x^i - x'^i))/b_i(x^i - x'^i) \qquad (9)$$

where b is an adjustable parameter. In fact, the scaling kernel function (9) is a multi-dimensional scaling function with translation terms. So, according to wavelet analysis[8], if we take

$$f(x) = \sum_{i=1}^{M} \omega_i K(\overline{x}, \overline{m}_i) = \sum_{i=1}^{M} \omega_i \prod_{j=1}^{n} \sin(b_i(x_j - m_{ij}))/b_i(x_j - m_{ij}) \qquad (10)$$

then we can surely build such a $f(x)$ that it can approximate any function in $L_2(R)$ space within *any* given accuracy.

If we substitute $\overline{y}_i = p_{i0}$ for $\overline{y}_i = p_{i0} + p_{i1}x_1 + \ldots + p_{in}x_n$ in SKFS, then we obtain SKFS0. Easily, we can derive the output of SKFS0 as

$$y = \sum_{i=1}^{M} p_{i0} \prod_{j=1}^{n} A_{ij}(x_j) = \sum_{i=1}^{M} p_{i0} \prod_{j=1}^{n} \sin(b_{ij}(x_j - m_{ij}))/b_{ij}(x_j - m_{ij}) \qquad (11)$$

Obviously, SKFS0 is a special case of SKFS. Comparing (10) with (11), in terms of wavelet analysis, if M is large enough, then SKFS0 can universally approximate any function in $L_2(R)$ space. That is to say, we derive the following Theorem 1.

Theorem 1. *Scaling kernel-based fuzzy system SKFS0 is a universal approximator for any function in $L_2(R)$.*

Theorem 1 means the equivalence between SKFS0 and the corresponding wavelet analysis and thus SKFS0 preserves the multiresolution capability of wavelet analysis. This capability will lead SKFS0 to better approximation accuracy.

In fact, SKFS fuzzy system is an extension of SKFS0 based on TSK model by substituting p_{i0} with $p_{i0} + p_{i1}x_1 + \ldots + p_{in}x_n$. Similar to [4,5], we immediately have the following Theorem 2.

Theorem 2. *Scaling kernel-based fuzzy system SKFS is a universal approximator for any function in $L_2(R)$.*

In terms of [4,5], we can easily know that SKFS has better approximation capability than SKFS0 since SKFS has the higher order sugeno consequents than SKFS0.

Similarly, in terms of [10], both SKFS and SKFS0 are smooth approximators for any continuous function in $L_2(R)$.

Comparing with the simplified fuzzy system GKFS , the new SKFS has the following merits:

- SKFS can universally approximate any function in $L_2(R)$ while GKFS can only universally approximate any continuous function.
- The computation of the membership functions of Sinc x and exp is relatively easy and simple, and their corresponding simplified fuzzy system structures are the same. But, as pointed out by B.Kosko et al, the Sinc x membership function will bring better approximation capability, and our experimental results here will support the B.Kosko's conclusion again.
- From another point of view, if we set $M = K$, $m_{ij} = \overline{x}_i^l$, $b_{ij} = b$, where K is the total number of samples, $\overline{x}_i^l$ denotes the ith component of the lth sample, then then SKFS0's output, i.e. (11) can be written as (12), i.e.

$$y = \sum_{i=1}^{K} p_{i0} \prod_{j=1}^{n} \sin(b(x_j - \overline{x}_j))/b(x_j - \overline{x}_j) \qquad (12)$$

In terms of [8], (12) becomes a scaling kernel function support vector machines, since Sinc(x) can satisfy the second condition of support vector kernels[8]. As well, for GKFS, if we take R^i : *if x_1 is $A_{i1}, \ldots, x_n$ is A_{in} then $\overline{y}_i = p_{i0}$*, where $i = 1, 2, \ldots, K$, and

$$A_{ij}(x_j) = e^{-(\frac{x_j - \overline{x}_j^l}{\sigma})^2} \qquad (13)$$

the output of the corresponding GKFS is

$$y = \sum_{i=1}^{K} \bar{y}_i \prod_{j=1}^{n} e^{-(\frac{x_j - \bar{x}_j^l}{\sigma})^2} \tag{14}$$

Similarly, in terms of [8,9], (14) turns into a Gaussian kernel function support vector machine. Because Gaussian kernel function only satisfies Mercer condition, the Gaussian kernel function support vector machine represented by (14) can not approximate any function in $L_2(R)$ space (Note: Weirstone theorem only guarantees that (14) can universally approximate any continuous function). However, based on the theory of support vector machine and wavelet analysis, (12) can approximate any function in $L_2(R)$. So, to some extent, the above discussions have explained the reasons that SKFS has better approximation capability than GKFS. Our experiments will also verify this assertion.

3   SKFS's Learning Algorithm

With the gradient method, we can derive the learning algorithm of SKFS. Suppose we have K samples: $(\bar{x}_1^l, \bar{x}_2^l, \ldots, \bar{x}_n^l; \bar{y}^l)$, where $l = 1, 2, \ldots, K$, $\bar{x}_1^l, \bar{x}_2^l, \ldots, \bar{x}_n^l$ is the inputs of the lth sample, $\bar{y}^l$ is the real output of the lth sample.

We define the error function as

$$e_l = \frac{1}{2}(\bar{y}^l - y^l)^2 \tag{15}$$

where y^l is the output of the fuzzy system SKFS, then, the update equations of $m_{ij}(t)$ and $p_{ij}(t)$ are as follows.

$$m_{ij}(t+1) = m_{ij}(t) - \alpha \frac{\partial e_l}{\partial m_{ij}} \tag{16}$$

$$p_{ij}(t+1) = p_{ij}(t) - \alpha \frac{\partial e_l}{\partial p_{ij}} \tag{17}$$

$$b_{ij}(t+1) = b_{ij}(t) - \alpha \frac{\partial e_l}{\partial b_{ij}} \tag{18}$$

where

$$\frac{\partial e_l}{\partial m_{ij}} = (y^l - \bar{y}^l)\frac{\partial y^l}{\partial m_{ij}} = (y^l - \bar{y}^l)\bar{y}_i \frac{\partial \prod_{j=1}^{n} A_{ij}(\bar{x}_j^l)}{\partial m_{ij}}$$

$$= (y^l - \bar{y}^l)\bar{y}_i A_{i1}(\bar{x}_1^l)A_{i2}(\bar{x}_2^l)...A_{ij-1}(\bar{x}_{j-1}^l)(\bar{x}_n^l) \times \frac{\partial \frac{(b(\bar{x}_j - m_{ij}))}{b(\bar{x}_j - m_{ij})}}{\partial m_{ij}} A_{ij+1}(\bar{x}_{j+1}^l)...A_{in}$$

$$= (y^l - \bar{y}^l)\bar{y}_i \prod_{j=1}^{n} A_{ij}(\bar{x}_j^l) \left[\frac{1}{\bar{x}_j - m_{ij}} - b \times ctg(b(\bar{x}_j^l - m_{ij})) \right] \tag{19}$$

and

$$\frac{\partial e_l}{\partial p_{i0}} = (y^l - \bar{y}^l)\frac{\partial y^l}{\partial p_{i0}} = (y^l - \bar{y}^l)\frac{\partial(\sum_{i=1}^{M} \bar{y}_i \prod_{j=1}^{n} A_{ij}(\bar{x}_j^l))}{\partial p_{i0}}$$

$$= (y^l - \bar{y}^l)\prod_{j=1}^{n} A_{ij}(\bar{x}_j^l) \tag{20}$$

$$\frac{\partial e_l}{\partial p_{ij}} = (y^l - \bar{y}^l)\frac{\partial y^l}{\partial p_{ij}} = (y^l - \bar{y}^l)\frac{\partial(\sum\limits_{i=1}^{M} \bar{y}_i \prod\limits_{j=1}^{n} A_{ij}(\bar{x}_j^l))}{\partial p_{ij}}$$
$$= (y^l - \bar{y}^l)\bar{x}_j^l \prod_{j=1}^{n} A_{ij}(\bar{x}_j^l) \tag{21}$$

and

$$\frac{\partial e_l}{\partial b_{ij}} = (y^l - \bar{y}^l)\frac{\partial y^l}{\partial b_{ij}} = (y^l - \bar{y}^l)\bar{y}_i\frac{\partial \prod\limits_{j=1}^{n} A_{ij}(\bar{x}_j^l)}{\partial b_{ij}}$$
$$= (y^l - \bar{y}^l)\bar{y}_i \prod_{j=1}^{n} A_{ij}(\bar{x}_j^l) \left[(\bar{x}_j - m_{ij}) \times ctg(b_{ij}(\bar{x}_j^l - m_{ij})) - \frac{1}{b_{ij}}\right] \tag{22}$$

Where α is the learning rate, $i = 1, 2, \ldots, M$ (is the number of rules used), $j = 1, 2, \ldots, n$.

Similarly, we can also derive the update equations of GKFS, which are omitted for simplicity.

4   Experimental Results

In this section, we take several benchmark examples[1,2,3] to test the approximation and/or generalization performance of SKFS and compare it with GKFS.

In order to compare the performance of GKFS with SKFS, we take the measure in[1,2] as the performance index

$$J_i = \sqrt{\frac{\sum\limits_{l=1}^{K} (\hat{y}_l - y_l^d)^2}{\sum\limits_{l=1}^{K} (y_l^d - \bar{y})^2}} \quad with \ \bar{y} = \frac{1}{K}\sum_{l=1}^{K} y_l^d \tag{23}$$

where K is the total numbers of samples, y_l^d denotes the SKFS or GKFS's predicted/desired output of the lth sample, $\hat{y}_l$ denotes the real output of the lth sample .

Example 1. Approximation of Piecewise Function: The piecewise function is continuous and analyzable. And we use this function to compare the performances of SKFS and GKFS when they are used to approximate a function in the coexistence of linearity and nonlinearity. The piecewise function is defined as follows:

$$f(x) = \begin{cases} -2.186x - 12.864 & -10 \leq x < -2 \\ 4.246x & -2 \leq x < 0 \\ 10e^{-0.05x-0.5}\sin[(0.03x + 0.7)x] & 0 \leq x \leq 10 \end{cases} \tag{24}$$

In this example, we sampled 200 points distributed randomly over [-10,10] as the training data, another 200 points as the checking data. Table 1 shows the better approximation/generalization performances of SKFS according to the performance index defined above.

Table 1. Comparison of SKFS and GKFS for Example 1

Method	Iterative Numbers	Number of Parameters	Performance Index J in training case	performance Index J in checking case
SKFS	3000	2	0.051207	0.070493
GKFS	4000	2	0.109445	0.289166

From Table 1 we can easily find that SKFS has better approximation /generalization performance than that of GKFS. One of the main reasons is that SKFS borrows the multiresolution capability of the scaling functions and wavelet analysis to reach better performance while GKFS cannot.

Example 2. Predicting Chaotic Time Series: A benchmark problem is to predict future values of Mackey-Glass time series, which is a differential delay equation defined as follows.

$$\dot{x}(t) = \frac{0.2x(t - \tau)}{1 + x^{10}(t - \tau)} - 0.1x(t) \tag{25}$$

This is a non-periodic and non-convergent time series that is very sensitive to initial conditions, let $\tau = 17$ and $x(0) = 1.2$.(We assume $x(t) = 0$,where $t < 0$)

Similar to Jang[15], we extracted 1000 input-output data pairs x^d, y^d from the Mackey-Glass time series(where $\tau = 17$ and $x(0) = 1.2$ with the following format:

$$x^d = [x(t - 18), x(t - 12), x(t - 6), x(t)], y^d = x(t + 6) \tag{26}$$

where $t = 118$ to 1117. Also, we use the first 500 pairs as the training samples, while the remaining 500 pairs as the checking samples.

Table 2 shows the comparison of approximation/generalization performance of SKFS and GKFS.

Table 2. Comparison of SKFS and GKFS for Example 2

Method	Iterative Numbers	Number of Parameters	Performance Index J in training case	performance Index J in checking case
SKFS	1000	3	0.125460	0.132116
GKFS	1000	3	0.140374	0.149399

In summary, the above experimental results demonstrate that SKFS has better approximation and generalization capabilities than GKFS. Since SKFS has the same simple structure as GKFS, so it is much more suitable for fuzzy adaptive control.

5 Conclusions and Future Work

In this paper, inspired by B.Kosko et al, we have presented a new simplified fuzzy system SKFS. Because SKFS avoids the defuzzification computation, it's computational complexity is lower than other similar fuzzy systems. Numerical examples also demonstrated the better approximation accuracy of SKFS.

Some open problems await us to explore in the future. For example, we can integrate SKFS with adaptive control to study the stability of control system. Another interesting problem worthy to be further studied is how to combine SKFS with CMAC(Cerebellar Model Articulation Controller) neural network to build a new scaling kernel-based fuzzy CMAC.

References

1. D.W.C.Ho, P.A.Zhang, J.H.Xu, Fuzzy wavelet networks for function learning. IEEE Trans. Fuzzy Systems,9(2001) 200-211
2. Q.Zhang, A.Benveniste, Wavelet networks, IEEE Trans. Neural Networks,3 (1992) 889-898
3. J.Chen, D.D.Bruns, WaveMRX neural network development for system identification using a systematic design synthesis, Ind.Eng. Chem.Res,34(1995) 4420-4435
4. M.Alata, C.Y.Su, K.Demirli, Adaptive control of a class of nonlinear systems with a fast-order parameterized Sugeno fuzzy approximator, IEEE Trans. Systems, Man and Cybernetics(part C),31(2001) 410-419
5. K.Demirli, P.Mufhukumran, fuzzy system identification with high order subtractive clustering, J.Intell. Fuzzy Systems, 9 (2001) 129-158
6. L.X.Wang and J.M.Mendel, Fuzzy basis functions, universal approximation, and OLS, IEEE Trans. Neural networks,3 (1992) 807-814
7. Y.Yu,S.H.Tan, Complementarity and equivalence relationships between convex fuzzy systems with symmetry restrictions and wavelets, Fuzzy sets and systems,101 (1999) 423-438
8. C.J.C.Burges, Geometry and invariance in kernel based methods, in Advance in Kernel Methods-Support vector learning, Cambridge, MA:MIT Press,(1999) 89-116
9. V.Vapnik, The nature of statistical learning theory, New York: Springer Verlag,1995
10. M.Landajo, M.J.Rio, R.Perez, Anote on smooth approximation capabilities of fuzzy systems, IEEE Trans. Fuzzy systems,9 (2001) 229-237
11. B.M.Novakovic, Fuzzy logic control synthesis without any rule base, IEEE Trans. Systems, Man and Cybernetics(part B),29 (1999) 459-466
12. S.Mitaim,B.Kosko, The shape of fuzzy sets in adaptive function approximation, IEEE Trans. Fuzzy systems,9 (2001) 637-655
13. Wang Shitong, fuzzy systems, fuzzy neural networks and programming, Publishing House of Science and Technologies of Shanghai, Shanghai 1998
14. Wang Shitong, Fuzzy system and CMAC network with B-spline membership/basis functions are smooth approximators, Int.J.Soft computing,7(2003) 566-573
15. J.S.R.Jang, C.T.Sun, E.Mizutani, Neuro-Fuzzy and Soft Computing, Englewood Clitts,N: Prentice-Hall,1997

Bulk Synchronous Parallel ML: Semantics and Implementation of the Parallel Juxtaposition

F. Loulergue[1], R. Benheddi[1], F. Gava[2], and D. Louis-Régis[1]

[1] Laboratoire d'Informatique, Fondamentale d'Orléans, Université d'Orléans, France
{floulerg, rbenhedd}@univ-orleans.fr
[2] Laboratory of Algorithms, Complexity and Logic,
University Paris XII, France
gava@univ-paris12.fr

1   Introduction

The design of parallel programs and parallel programming languages is a trade-off. On one hand the programs should be efficient. But the efficiency should not come at the price of non portability and unpredictability of performances. The portability of code is needed to allow code reuse on a wide variety of architectures and to allow the existence of legacy code. The predictability of performances is needed to guarantee that the efficiency will always be achieved, whatever is the used architecture.

Another very important characteristic of parallel programs is the complexity of their semantics. Deadlocks and indeterminism often hinder the practical use of parallelism by a large number of users. To avoid these undesirable properties, a trade-off has to be made between the expressiveness of the language and its structure which could decrease the expressiveness.

Bulk Synchronous Parallelism [22,20] (BSP) is a model of computation which offers a high degree of abstraction like PRAM models but yet a realistic cost model based on a structured parallelism: deadlocks are avoided and indeterminism is limited to very specific cases in the BSPlib library [13]. BSP programs are portable across many parallel architectures.

Over the past decade, Bulk Synchronous Parallelism (and the Coarse-Grained Multicomputer or CGM which can be seen as a special case of the BSP model) have been used for a large variety of applications. It is to notice that "A comparison of the proceedings of the eminent conference in the field, the ACM Symposium on Parallel Algorithms and Architectures, between the late eighties and the time from the mid nineties to today reveals a startling change in research focus. Today, the majority of research in parallel algorithms is within the coarse-grained, BSP style, domain" [8].

Our research aims at combining the BSP model with functional programming. We obtained the Bulk Synchronous Parallel ML language (BSML) based on a *confluent* extension of the λ-calculus. Thus BSML is deadlock free and deterministic. Being a high-level language, programs are easier to write, to reuse and to compose. It is even possible to *certify* the correctness of BSML programs [9] with the help of the Coq proof assistant [2]. The performance prediction of BSML

D. Grigoriev, J. Harrison, and E.A. Hirsch (Eds.): CSR 2006, LNCS 3967, pp. 475–486, 2006.

programs is possible. BSML has been extended in many ways throughout the years and the papers related to this research are available at http://bsml.free.fr.

One direction for the extension of BSML was to offer new primitives for the programming of divide-and-conquer Bulk Synchronous Parallel algorithms. Two new primitives have been designed :

- the parallel superposition [17,10] which creates two parallel threads whose communication and synchronization phases are fused ;
- the parallel juxtaposition [16] which divides the parallel machine in two independent sub-machines while preserving the Bulk Synchronous Parallel model.

[16] presents the programming model of BSML with juxtaposition. This model presents a global view to the programmer, easier to understand than what actually happens when a BSML program is run on a parallel machine. Nevertheless to implement BSML with juxtaposition we need a *distributed semantics* (section 3) which specifies the execution model i.e. what actually happens on a parallel machine. Using this specification we implemented (section 4) the juxtaposition using the parallel superposition and *imperative features*.

We begin the paper with an overview of BSML with juxtaposition (section 2). Related work and conclusions end the paper (sections 5 and 6).

2   Bulk Synchronous Parallel ML with Juxtaposition: An Overview

There is currently no implementation of a full BSML language but rather a partial implementation as a library for Objective Caml language [14,6]. BSML follows the Bulk Synchronous Parallel (BSP) model which offers a model of architecture, a model of execution and a cost model.

A BSP computer contains a set of uniform *processor-memory* pairs, a *communication network* allowing inter-processor delivery of messages and a *global synchronization unit* which executes collective requests for a *synchronization barrier* (for the sake of conciseness, we refer to [3] for more details). In this model, a parallel computation is divided in *super-steps*, at the end of which a the routing of the messages and barrier synchronization are performed. Hereafter all requests for data which have been posted during a preceding super-step are fulfilled.

The performance of the machine is characterized by 3 parameters expressed as multiples of the local processing speed r: p is the number of processor-memory pairs, l is the time required for a global synchronization and g is the time for collectively delivering a 1-relation (communication phase where every processor receives/sends at most one word). The network can deliver an h-relation in time $g \times h$ for any arity h. The execution time of a super-step is thus the sum of the maximal local processing time, of the data delivery time and of the global synchronization time. The execution time of a program is the sum of the execution time of its super-steps.

BSML does not rely on SPMD programming. Programs are usual "sequential" Objective Caml programs but work on a parallel data structure. Some of the advantages is a simpler semantics and a better readability: the execution order follows (or at least the results is such as the execution order seems to follow) the reading order.

The core of the BSMLlib library is based on the following elements:

bsp_p: unit→int
mkpar: (int →α) →α **par**
apply: (α →β) **par** →α **par** →β **par**
type α option = None | Some **of** α
put: (int→α option) **par** →(int→α option) **par**
proj: α option **par** →int →α option

It gives access to the BSP parameters of the underling architecture. In particular, **bsp_p**() is p, the *static* number of processes. There is an abstract polymorphic type α **par** which represents the type of p-wide parallel vectors of objects of type α one per process. The nesting of **par** types is prohibited. Our type system enforces this restriction [11].

The BSML parallel constructs operate on parallel vectors. Those parallel vectors are created by **mkpar** so that (**mkpar** f) stores (f i) on process i for i between 0 and $(p-1)$. We usually write f as **fun** pid→e to show that the expression e may be different on each processor. This expression e is said to be *local*. The expression (**mkpar** f) is a parallel object and it is said to be *global*.

A BSP algorithm is expressed as a combination of asynchronous local computations and phases of global communication with global synchronization.

Asynchronous phases are programmed with **mkpar** and **apply**. The expression (**apply** (**mkpar** f) (**mkpar** e)) stores ((f i)(e i)) on process i.

Let consider the following expression:

let vf = **mkpar**(**fun** i→(+) i) **and** vv = **mkpar**(**fun** i→2∗i+1) **in**
apply vf vv

The two parallel vectors are respectively equivalent to:

$$\left|\text{fun x→x+0}\right|\text{fun x→x+1}\right| \cdots \left|\text{fun x→x+(p−1)}\right| \text{ and } \boxed{0}\boxed{3}\ \cdots\ \boxed{2 \times (p-1)+1}$$

The expression **apply** vf vv is then evaluated to:

$$\boxed{0}\boxed{4}\ \cdots\ \boxed{2 \times (p-1)+2}$$

Readers familiar with BSPlib [13] will observe that we ignore the distinction between a communication request and its realization at the barrier. The communication and synchronization phases are expressed by **put**. Consider the expression:

$$\text{put}(\text{mkpar}(\text{fun } i→fs_i)) \tag{1}$$

To send a value v from process j to process i, the function fs_j at process j must be such that (fs_j i) evaluates to Some v. To send no value from process j to process i, (fs_j i) must evaluate to None. Expression (1) evaluates to a parallel vector containing a function fd_i of delivered messages on every process. At process i, $(\text{fd}_i \; j)$ evaluates to None if process j sent no message to process i or evaluates to Some v if process j sent the value v to the process i.

BSML also contains a synchronous projection operation **proj** whose detailed presentation is omitted here. It is necessary of express algorithms like:

Repeat Parallel Iteration **Until** Max of local errors $< \epsilon$

The projection should not be evaluated inside the scope of a **mkpar**. This is enforced by our type system [11].

To evaluate two parallel programs on the same machine, one can partition it into two sub-machines and evaluate each program independently on each partition. Nevertheless in this case the BSP cost model is lost since for example a global synchronization of each sub-machine would no more cost L. To preserve the BSP model, which is the best solution [12], synchronization barriers need to remain global for the whole machine. In a first definition of parallel composition [15], it was possible to compose two programs whose evaluations need the same number of super-steps. It is of course restrictive and the programmer was responsible to write programs which fulfill this constraint. A new version called *parallel juxtaposition* removes this constraint [16]. It is the version that we present in this paper.

Consider the expression (**juxta** $m \; E_1 \; E_2$). It means that the m first processors will evaluate E_1 and the others will evaluate E_2. From the point of view of E_1 the network will have m processors named $0, \ldots, m-1$. From the point of view of E_2 the network will have $p - m$ processors (where p is the number of processors of the current network) named $0, \ldots, (p - m - 1)$ (processor m is renamed 0, *etc.*). The value of **bsp_p**() is also changed. Otherwise the evaluation of the expressions is the same, on each sub-machine, than without parallel juxtaposition, but the evaluation of **put** and **at** need the whole machine for the global synchronization. A problem occurs when the evaluation of E_1 and the evaluation of E_2 need a different number of super-steps. That is why a new primitive is necessary. The **sync** primitive is a loop of synchronization barrier calls. It loops until a synchronization barrier call is made by **sync** on the whole machine.

In case of the evaluation of E_1 needs one more super-step than the evaluation of E_2, the evaluation of (**sync** (**juxta** $m \; E_1 \; E_2$)) can be described as follows:

- at the beginning, each synchronization barrier request for the evaluation of E_1 matches a synchronization barrier request for the evaluation of E_2;
- then the evaluation of E_2 ends. E_2 requests one more synchronization barrier for its last super-step. The second sub-machine has finished the evaluation of E_2 so it evaluates **sync**: the synchronization barrier request of **sync** will match the request of the first sub-machine;
- each sub-machine has finished the evaluation of its expressions and they both request a synchronization barrier from a **sync**. As this request concerns the whole machine the evaluation of **sync** ends.

Evaluation result of a parallel juxtaposition is a parallel vector:

$$(\textbf{juxta } m \; \langle \, v_0 \, , \ldots, \, v_{m-1} \, \rangle \; \langle \, v'_0 \, , \ldots, \, v'_{p-1-m} \, \rangle) = \langle v_0, \ldots, v_{m-1}, v'_0, \ldots, v'_{p-1-m} \rangle$$

From the functional point of view, the **sync** function is identity.

In the BSML library, the fact that Objective Caml is a language with a weak call-by-value evaluation strategy must be taken into account. To avoid the evaluation of the two last arguments of the function **juxta** and the argument of the function **sync**, these arguments should be functions:

juxta: int $\rightarrow$(unit $\rightarrow\alpha$ **par**) $\rightarrow$(unit $\rightarrow\alpha$ **par**) $\rightarrow\alpha$ **par**
sync: (unit$\rightarrow\alpha$ **par**) $\rightarrow\alpha$ **par**

The following example is a divide-and-conquer version of the scan program which is defined by scan $\oplus \langle v_0, \ldots, v_{p-1} \rangle = \langle v_0, \ldots, v_0 \oplus v_1 \oplus \ldots \oplus v_{p-1} \rangle$ where $\oplus$ is an associative binary operation.

```
let rec scan op vec =
 if bsp_p()=1 then vec
 else
  let mid=bsp_p()/2 in
  let vec'=juxta mid (fun ()→scan op vec) (fun ()→scan op vec) in
  let msg vec=apply (mkpar(fun i v→
   if i=mid−1
   then fun dst→ if dst>=mid then Some v else None
   else fun dst→ None)) vec
  and parop=parfun2(fun x y→match x with None→y|Some v→op v y)in
  parop (apply(put(msg vec'))(mkpar(fun i→mid−1))) vec'
```

The juxtaposition divides the network into two parts the scan is recursively applied to each part. The value held by the last processor of the first part is broadcast to all the processors of the second part, then this value and the value held locally are combined by the operator op on each processor of the second part.

To use this function at top-level, it must be put into a **sync** primitive:

$$(\mathbf{sync} \ (\mathbf{fun} \ () \rightarrow \text{scan} \ (+) \ \text{this}))$$

3   Distributed Semantics

High level semantics corresponds to the programming model. Distributed semantics corresponds to the execution model. In the former, all the parallel operations seem synchronous, even those which do not need communication. In the latter, the operations without communication are asynchronous and the operations with communications are synchronous.

The distributed evaluation can be defined in two steps:

1. local reduction (performed by one process) ;
2. global reduction of distributed terms which allows the evaluation of communications.

480 F. Loulergue et al.

3.1 Syntax

We consider here only a small subset of the Ocaml language with our parallel primitives. We first consider the *flat* part of the language, i.e. without parallel juxtaposition :

$$
\begin{array}{llll}
e & ::= & x & \text{(variable)} \\
& | & c & \text{(constant)} \\
& | & \textbf{bsp\_p} & \text{BSP parameter p} \\
& | & (\textbf{fun } x \rightarrow e) & \text{(abstraction)} \\
& | & op & \text{(operator)} \\
& | & (e\ e) & \text{(application)} \\
& | & (\textbf{let } x = e \textbf{ in } e) & \text{(binding)} \\
& | & (\textbf{if } e \textbf{ then } e \textbf{ else } e) & \text{(conditional)} \\
& | & (\textbf{mkpar } e) & \text{(parallel vector)} \\
& | & (\textbf{apply } e\ e) & \text{(parallel application)} \\
& | & (\textbf{get } e\ e) & \text{(communication primitive)} \\
& | & (\textbf{if } e \textbf{ at } e \textbf{ then } e \textbf{ else } e) & \text{(global conditional)} \\
& | & \langle e \rangle & \text{(enumerated parallel vector)} \\
& | & (\textbf{sync } e') & \text{(sync primitive)}
\end{array}
$$

The use of the juxtaposition is only allowed in the scope of a **sync** primitive :

$$
\begin{aligned}
e' ::= &\ x \ | \ c \ | \ \textbf{bsp\_p} \ | \ (\textbf{fun } x \rightarrow e') \ | \ op \ | \ (e'\ e') \ | \ (\textbf{let } x = e' \textbf{ in } e') \\
& | \ (\textbf{if } e' \textbf{ then } e' \textbf{ else } e') \ | \ (\textbf{mkpar } e') \ | \ (\textbf{apply } e'\ e') \ | \ (\textbf{get } e'\ e') \\
& | \ (\textbf{if } e' \textbf{ at } e' \textbf{ then } e' \textbf{ else } e') \ | \ \langle e' \rangle \ | \ (\textbf{juxta } m\ e'\ e') \ | \ \|e'\|
\end{aligned}
$$

For the sake of conciseness, we use the **get** and **if at** constructs instead of the more general **put** and **proj** functions. There is no fundamental differences, but the semantics is simpler. We also omit in the remaining of the paper to distinguish expressions e and e'. Most of the rules are valid for both. We also omit a simple type system (with explicit typing of variables with two possible annotations: local or global) which allows to avoid the nesting of parallel values.

The user is not supposed to write enumerated parallel vectors $\langle e \rangle$. These expressions are created during the evaluation of a **mkpar** expressions. $\|e'\|$ indicates that the expression e is a branch of a juxtaposition.

Values are given by the following grammar :
$$
v ::= (\textbf{fun } x \rightarrow e) \ | \ c \ | \ op \ | \ \langle v \rangle
$$

3.2 Local Reduction

The distributed evaluation is an SPMD semantics. Each processor will evaluate one copy of the BSML program. As long as the expression is not an expression which requires communications, the evaluation can proceed asynchronously on each processor.

When the juxtaposition primitive is used two sub-machines are considered. For a given process it means that the process identifier and the number of processes can change. Nevertheless these values are constant for the actual parallel machine. Thus we choose to put these parameters on the arrow. $\longrightarrow_p^i$ is the local reduction at processor whose absolute process identifier is i on a parallel machine of p processors. The absolute process identifier of the first process and number of processes of the sub-machine a process belong to are stored in two stacks : $\mathcal{E}^f$ and $\mathcal{E}^p$.

The location reduction is a relation on tuples of one expression, and two stacks. It is defined by the rules of figure 1 (the set of rules for predefined sequential operators is omitted[1]) plus the following contexts and context rule :

$$
\begin{aligned}
\Gamma := {} & [] \mid (\Gamma\ e) \mid (v\ \Gamma) \mid (\textbf{let } x = \Gamma \textbf{ in } e) \mid (\textbf{if } \Gamma \textbf{ then } e \textbf{ else } e) \\
& \mid (\textbf{mkpar } \Gamma) \mid (\textbf{apply } \Gamma\ e) \mid (\textbf{apply } v\ \Gamma) \mid (\textbf{get } \Gamma\ e) \mid (\textbf{get } v\ \Gamma) \\
& \mid (\textbf{if } e \textbf{ at } \Gamma \textbf{ then } e \textbf{ else } e) \mid (\textbf{if } \Gamma \textbf{ at } v \textbf{ then } e \textbf{ else } e) \\
& \mid (\textbf{juxta } \Gamma\ e\ e) \mid \langle \Gamma \rangle \mid \|\Gamma\| \mid (\textbf{sync } \Gamma)
\end{aligned}
$$

$$
\begin{aligned}
(e_1\ \mathcal{E}_1^f,\ \mathcal{E}_1^p) & \longrightarrow_p^i (e_2,\ \mathcal{E}_2^f,\ \mathcal{E}_2^p) \\
(\Gamma(e_1),\ \mathcal{E}_1^f,\ \mathcal{E}_1^p) & \longrightarrow_p^i (\Gamma(e_2),\ \mathcal{E}_2^f,\ \mathcal{E}_2^p)
\end{aligned}
\tag{2}
$$

The four first rules of figure 1 are usual rules of a functional language. Rule (7) returns the head of the stack of number of processors. For the stacks we use "::" for adding a value to the stack. The function h is defined by $\mathrm{h}(v :: \mathcal{E}) = v$. If the stack is empty, then if it is the $\mathcal{E}^f$ then the hd function returns 0, if it is the $\mathcal{E}^p$ stack the hd function return the value if p given by the $\longrightarrow_p^i$ arrow.

The two next rules formalize the informal semantics of the BSML primitives **mkpar** and **apply**, but as opposed as section 2, we consider here only what happens at process i. For example for rule (8), the processor i has (in the current sub-machine) the process identifier $i - \mathrm{h}(\mathcal{E}^f)$. Thus for it evaluating **mkpar** f is evaluating $(\mathrm{f}\ (i - \mathrm{h}(\mathcal{E}^f)))$.

The three last rules are devoted to the juxtaposition :

- the two first are used to choose which branch is evaluated by the given processor, depending on its identifier. New values of the process identifier of the first processor and the number of processor of the sub-machine are push on top of the respective stacks.
- the last one is used to restore the values of the process identifier of the first processor and the number of processor of the larger machine at the end of the evaluation of the branch.

3.3   Global Reduction

The global reduction $\rightarrow$ concerns the whole parallel machine. A distributed expression is thus p tuples manipulated by the local reduction. We use the following syntax for distributed expressions : $\left\langle (e_0, \mathcal{E}_0^f, \mathcal{E}_0^p), \ldots, (e_{p-1}, \mathcal{E}_{p-1}^f, \mathcal{E}_{p-1}^p) \right\rangle$

---

[1] This set includes rules for the **fix** operator used for recursion.

$$(\ ((\textbf{fun} \ x \to e) \ v), \ \mathcal{E}^f, \ \mathcal{E}^p \ ) \longrightarrow^i_p (\ e[x \leftarrow v], \ \mathcal{E}^f, \ \mathcal{E}^p \ ) \tag{3}$$

$$(\ (\textbf{let} \ x = v \ \textbf{in} \ e), \ \mathcal{E}^f, \ \mathcal{E}^p \ ) \longrightarrow^i_p (\ e[x \leftarrow v], \ \mathcal{E}^f, \ \mathcal{E}^p \ ) \tag{4}$$

$$(\ (\textbf{if true then} \ e_1 \ \textbf{else} \ e_2), \ \mathcal{E}^f, \ \mathcal{E}^p \ ) \longrightarrow^i_p (\ e_1, \ \mathcal{E}^f, \ \mathcal{E}^p \ ) \tag{5}$$

$$(\ (\textbf{if false then} \ e_1 \ \textbf{else} \ e_2), \ \mathcal{E}^f, \ \mathcal{E}^p \ ) \longrightarrow^i_p (\ e_1, \ \mathcal{E}^f, \ \mathcal{E}^p \ ) \tag{6}$$

$$(\ \textbf{bsp\_p}, \ \mathcal{E}^f, \ \mathcal{E}^p \ ) \longrightarrow^i_p (\ h(\mathcal{E}^p), \ \mathcal{E}^f, \ \mathcal{E}^p \ ) \tag{7}$$

$$(\ (\textbf{mkpar} \ v), \ \mathcal{E}^f, \ \mathcal{E}^p \ ) \longrightarrow^i_p (\ (v \ (i - h(\mathcal{E}^f))), \ \mathcal{E}^f, \ \mathcal{E}^p \ ) \tag{8}$$

$$(\ (\textbf{apply} \ \langle v_1 \rangle \ \langle v_2 \rangle), \ \mathcal{E}^f, \ \mathcal{E}^p \ ) \longrightarrow^i_p (\ \langle (v_1 \ v_2) \rangle, \ \mathcal{E}^f, \ \mathcal{E}^p \ ) \tag{9}$$

$$(\ (\textbf{juxta} \ m \ e_1 \ e_2), \ \mathcal{E}^f, \ \mathcal{E}^p \ ) \longrightarrow^i_p \begin{array}{l} (\ \|e_1\|, \ h(\mathcal{E}^f) :: \mathcal{E}^f, \ m :: \mathcal{E}^p \ ) \\ \text{if } 0 \leq m < h(\mathcal{E}^p) \\ \text{and } (i - h(\mathcal{E}^f)) < m \end{array} \tag{10}$$

$$(\ (\textbf{juxta} \ m \ e_1 \ e_2), \ \mathcal{E}^f, \ \mathcal{E}^p \ ) \longrightarrow^i_p \begin{array}{l} (\ \|e_2\|, \ (h(\mathcal{E}^f){+}m) :: \mathcal{E}^f, \ (h(\mathcal{E}^p){-}m) :: \mathcal{E}^p \ ) \\ \text{if } 0 \leq m < h(\mathcal{E}^p) \\ \text{and } (i - h(\mathcal{E}^f)) \geq m \end{array} \tag{11}$$

$$(\ \|v\|, \ f :: \mathcal{E}^f, \ p' :: \mathcal{E}^p \ ) \longrightarrow^i_p (\ v, \ \mathcal{E}^f, \ \mathcal{E}^p \ ) \tag{12}$$

Fig. 1. Local reduction

The first rule takes into account the local reduction :

$$\frac{(e_i, \mathcal{E}^f, \mathcal{E}^p) \longrightarrow^i_p (e'_i, \mathcal{E}'^f_i, \mathcal{E}'^p_i)}{\langle \ldots, (e_i, \mathcal{E}^f_i, \mathcal{E}^p_i), \ldots \rangle \to \langle \ldots, (e'_i, \mathcal{E}'^f_i, \mathcal{E}'^p_i), \ldots \rangle} \tag{13}$$

The second one is used for communications and synchronisation. The p processors are partitioned into $1 \leq k \leq p$ parts, each part containing one of more successive processor. Two processors belongs to the same part n ($1 \leq n \leq k$) if the values at the top of their $\mathcal{E}^f$ stacks are equal. In this case they also have the same value on top of $\mathcal{E}^p$ which we note p_n.

We note $(e_{n,i}, \mathcal{E}^f_{n,i}, \mathcal{E}^p_{n,i})$ the process i *in the nth part*. This processor has process identifier $i + h(\mathcal{E}^f)$ in the whole parallel machine. We want the reduction :

$$\left\langle \begin{array}{l} (e_{1,0}, \mathcal{E}^f_{1,0}, \mathcal{E}^p_{1,0}), \ldots, (e_{1,p_1-1}, \mathcal{E}^f_{1,p_1-1}, \mathcal{E}^p_{1,p_1-1}), \\ (e_{2,0}, \mathcal{E}^f_{2,0}, \mathcal{E}^p_{2,0}), \ldots, (e_{k,p_k-1}, \mathcal{E}^f_{k,p_k-1}, \mathcal{E}^p_{k,p_k-1}) \end{array} \right\rangle$$

$$\to$$

$$\left\langle \begin{array}{l} (e'_{1,0}, \mathcal{E}'^f_{1,0}, \mathcal{E}'^p_{1,0}), \ldots, (e'_{1,p_0-1}, \mathcal{E}'^f_{1,p_0-1}, \mathcal{E}'^p_{1,p_0-1}), \\ (e'_{2,0}, \mathcal{E}'^f_{2,0}, \mathcal{E}'^p_{2,0}), \ldots, (e'_{k,p_k-1}, \mathcal{E}'^f_{k,p_k-1}, \mathcal{E}'^p_{k,p_k-1}) \end{array} \right\rangle$$

We have either :

– all processors are evaluating a **sync**, i.e. $\forall n.\forall i.(1 \leq n \leq k) \& (0 \leq i < p_n) \Rightarrow e_{n,i} = \Gamma(\textbf{sync} \ v_{n,i})$ then $\forall n.\forall i.(1 \leq n \leq k) \& (0 \leq i < p_n) \Rightarrow e'_{n,i} = \Gamma(v_{n,i})$
– at least one part is evaluating a primitive of communication. For each part we have to evaluate the corresponding primitive. For all n such that $1 \leq n \leq k$, we have either :

get: $\begin{cases} \text{we have} \quad \forall i.0 \le i < p_n \Rightarrow e_{n,i} = \Gamma(\text{get } \langle v_i \rangle \langle n_i \rangle) \\ \text{then} \qquad \forall i.0 \le i < p_n \Rightarrow e'_{n,i} = \Gamma(\langle v_{n_i} \rangle) \end{cases}$

if at: we have $\forall i.0 \le i < p_n \Rightarrow e_{n,i} = \Gamma(\text{if } \langle v_i \rangle \text{ at } \langle m \rangle \text{ then } e_i^1 \text{ else } e_i^2)$
then:
- if $0 \le m < p_n$ and $v_m = \textbf{true}$ then $\forall i.0 \le i < p_n \Rightarrow e'_{n,i} = \Gamma(e_i^1)$.
- if $0 \le m < p_n$ and $v_m = \textbf{false}$ then $\forall i.0 \le i < p_n \Rightarrow e'_{n,i} = \Gamma(e_i^2)$.

sync: $\begin{cases} \text{we have} \quad \forall i.0 \le i < p_n \Rightarrow e_{n,i} = \Gamma(\textbf{sync } v_i) \\ \text{then} \qquad \forall i.0 \le i < p_n \Rightarrow e'_{n,i} = \Gamma(\textbf{sync } v_i) \end{cases}$

Theorem 1. $\rightarrow$ *is confluent.*

4 Implementation

The semantics given in the previous section is a specification for a distributed SPMD implementation. The current implementation of BSML is modular [18]. The module of primitives is a function which takes as argument a module of lower-level communications. Several such modules are provided and built on top of MPI, PVM, PUB, and also directly TCP/IP.

There are two implementations of the parallel juxtaposition. One which needs to extend the lower-level module interface. This solution adds very little sequential computation overhead. The drawback is that each lower-level module should be modified.

The second one implements the juxtaposition using the superposition. The advantage is that the implementation of the superposition is independent from the lower-level module. Moreover the implementation of the juxtaposition using the superposition is quite simple. and it is moreover very close to the semantics. This implementation adds more useless sequential computations, but there is no need for the **sync** additional synchronization barrier.

The superposition [16] allows two parallel expressions to be concurrently evaluated by two parallel threads running on the whole parallel machine. The results is a pair of parallel vectors of size p. The outline of the implementation of **juxta** m e1 e2 is as follows, where two stacks – one for the current number of processors p of the current machine and one for the number (in the previous machine) of the first processor f of the current machine – are used. The implementation o evaluate :

1. check if m is not greater than the number of processors
2. push the current values of f and p on the stacks
3. superpose the following expressions:
   (a) set p to m and evaluate e1()
   (b) set f to $m + f$ and p to $p - m$ and evaluate e2()
   the pair (va,vb) of parallel vectors is obtained
4. restore the values of f and p by popping them from the stacks
5. return a parallel vector in which:
   - the m first values come from va (from index f to $f + m - 1$)
   - the $p - m$ next values come from vb (from index $f + m$ to $f + p - 1$)

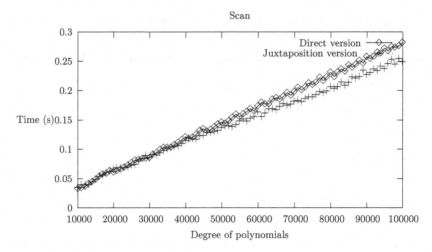

Fig. 2. Experiments with the scan programs

We did some experiments with the scan program shown in section 2. We run it on parallel vectors of polynomial, the operation being the addition. The tests were done on a 10 nodes Pentium IV cluster with Giga-bit Ethernet network (figure 2).

We compared the version with juxtaposition which requires $\log p$ super-steps with a direct version without superposition in 1 super-step. In the latter case, p polynomials are received by the last processor. In the former case, at each step, a processor receives at most 2 polynomial. As the polynomial are quite big, the $\log p$ version performs better than the direct version. Of course for smaller polynomials the direct version is better. It also depends on the BSP parameters of the parallel machine.

5   Related Work

[21] presents another way to divide-and-conquer in the framework of an object-oriented language. There is no formal semantics and no implementation from now on. The proposed operation is similar to the parallel *superposition*, several BSP threads use the whole network. The same author advocates in [19] a new extension of the BSP model in order to ease the programming of divide-and-conquer BSP algorithms. It adds another level to the BSP model with new parameters to describe the parallel machine.

[23] is an algorithmic skeletons language based on the BSP model and offers divide-and-conquer skeletons. Nevertheless, the cost model is not really the BSP model but the D-BSP model [7] which allows subset synchronization. We follow [12] to reject such a possibility.

In the BSPlib library [13] subset synchronization is not allowed as explained in [20]. The PUB library [4] is another implementation of the BSPlib standard proposal. It offers additional features with respect to the standard which follows the BSP* model [1] and the D-BSP model [7]. Minimum spanning trees nested BSP algorithms [5] have been implemented using these features.

6   Conclusion and Future Work

We have presented a distributed semantics – which formalizes the execution model – and an implementation of the parallel juxtaposition primitive for Bulk Synchronous Parallel ML. This primitive allows to write parallel divide-and-conquer BSP algorithms.

The programming model, and its formalization, of BSML with juxtaposition has been presented in a previous paper [16]. We need now to prove that the programming model and the execution model are equivalent i.e. that their formalizations are equivalent semantics.

The parallel juxtaposition is in fact an imperative extension of BSML. It has for example a side effect on the number **bsp_p**() of processors of the current parallel machine. Thus the method presented in [9] used to prove the correctness of BSML programs with the Coq proof assistant cannot be used. Another direction of research is thus to provide a transformation from a program with parallel juxtaposition to an *equivalent pure functional* program without parallel juxtaposition. The equivalence is in this case a semantic equivalence, the performance of the two programs being different. This transformation should also be proved correct. The correctness of the original program can then be ensured by proving, using Coq, the correctness of the transformed pure functional program.

Acknowledgments. The authors wish to thank the anonymous referee for their comments. This work is supported by the "ACI Jeunes chercheurs" program from the French ministry of research under the project "Programmation parallèle certifiée" PROPAC (http://wwwpropac.free.fr).

References

1. W. Bäumker, A. adn Dittrich and F. Meyer auf der Heide. Truly efficient parallel algorithms: c-optimal multisearch for an extension of the BSP model. In 3rd *European Symposium on Algorithms (ESA)*, pages 17–30, 1995.
2. Y. Bertot and P. Castéran. *Interactive Theorem Proving and Program Development*. Springer, 2004.
3. R. Bisseling. *Parallel Scientific Computation. A structured approach using BSP and MPI*. Oxford University Press, 2004.
4. O. Bonorden, B. Juurlink, I. von Otte, and O. Rieping. The Paderborn University BSP (PUB) library. *Parallel Computing*, 29(2):187–207, 2003.
5. O. Bonorden, F. Meyer auf der Heide, and R. Wanka. Composition of Efficient Nested BSP Algorithms: Minimum Spanning Tree Computation as an Instructive Example. In *Proceedings of PDPTA*, 2002.
6. E. Chailloux, P. Manoury, and B. Pagano. *Développement d'applications avec Objective Caml*. O'Reilly France, 2000.

7. P. de la Torre and C. P. Kruskal. Submachine locality in the bulk synchronous setting. In L. Bougé et al., eds., *Euro-Par'96*, LNCS 1123–1124, Springer, 1996.
8. F. Dehne. Special issue on coarse-grained parallel algorithms. *Algorithmica*, 14: 173–421, 1999.
9. F. Gava. Formal Proofs of Functional BSP Programs. *Parallel Processing Letters*, 13(3):365–376, 2003.
10. F. Gava. *Approches fonctionnelles de la programmation parallèle et des méta-ordinateurs. Sémantiques, implantations et certification.* PhD thesis, University Paris Val-de-Marne, LACL, 2005.
11. F. Gava and F. Loulergue. A Static Analysis for Bulk Synchronous Parallel ML to Avoid Parallel Nesting. *Future Generation Computer Systems*, 21(5):665–671, 2005.
12. G. Hains. Subset synchronization in BSP computing. In H.R.Arabnia, ed., *Proceedings of PDPTA*, vol. I, pages 242–246, CSREA Press, 1998.
13. J.M.D. Hill, W.F. McColl, and al. BSPlib: The BSP Programming Library. *Parallel Computing*, 24:1947–1980, 1998.
14. X. Leroy, D. Doligez, J. Garrigue, D. Rémy, and J. Vouillon. The Objective Caml System release 3.09, 2005. web pages at www.ocaml.org.
15. F. Loulergue. Parallel Composition and Bulk Synchronous Parallel Functional Programming. In S. Gilmore, ed., *Trends in Functional Programming, Volume 2*, pages 77–88. Intellect Books, 2001.
16. F. Loulergue. Parallel Juxtaposition for Bulk Synchronous Parallel ML. In H. Kosch et al., eds., *Euro-Par 2003*, LNCS 2790, pages 781–788, Springer, 2003.
17. F. Loulergue. Parallel Superposition for Bulk Synchronous Parallel ML. In Peter M. A. Sloot et al., eds., *Proceedings of ICCS 2003, Part I*, LNCS 2659, pages 223–232, Springer, 2003.
18. F. Loulergue, F. Gava, and D. Billiet. BSML: Modular Implementation and Performance Prediction. In Vaidy S. Sunderam et al., eds., *Proceedings of ICCS 2005, Part II*, LNCS 3515, pages 1046–1054, Springer, 2005.
19. J. M. R. Martin and A. Tiskin. BSP modelling a two-tiered parallel architectures. In B. M. Cook, ed., *WoTUG'99*, pages 47–55, 1999.
20. D. B. Skillicorn, J. M. D. Hill, and W. F. McColl. Questions and Answers about BSP. *Scientific Programming*, 6(3):249–274, 1997.
21. A. Tiskin. A New Way to Divide and Conquer. *Parallel Processing Letters*, (4), 2001.
22. Leslie G Valiant. A bridging model for parallel computation. *Communications of the ACM*, 33(8):103–111, August 1990.
23. A. Zavanella. *Skeletons and BSP : Performance Portability for Parallel Programming.* PhD thesis, Universita degli studi di Pisa, 1999.

A Shortest Path Algorithm Based on Limited Search Heuristics

Feng Lu[1] and Poh-Chin Lai[2]

[1] State Key Laboratory of Resources and Environmental Information System,
The Institute of Geographical Sciences and Natural Resources Research,
Chinese Academy of Sciences, Beijing 100101, P.R. China
luf@lreis.ac.cn
[2] Department of Geography, The University of Hong Kong,
Hong Kong SAR, P.R. China

Abstract. Dijkstra's algorithm is arguably the most popular computational solution to finding single source shortest paths. Increasing complexity of road networks, however, has posed serious performance challenge. While heuristic procedures based on geometric constructs of the networks would appear to improve performance, the fallacy of depreciated accuracy has been an obstacle to the wider application of heuristics in the search for shortest paths. The authors presented a shortest path algorithm that employs limited search heuristics guided by spatial arrangement of networks. The algorithm was tested for its efficiency and accuracy in finding one-to-one and one-to-all shortest paths among systematically sampled nodes on a selection of real-world networks of various complexity and connectivity. Our algorithm was shown to outperform other theoretically optimal solutions to the shortest path problem and with only little accuracy lost. More importantly, the confidence and accuracy levels were both controllable and predictable.

Keywords: shortest path algorithm, road network, heuristic.

1 Introduction

A variety of the shortest path algorithm has been presented and implemented in the past decades. Literature [1][2][3] made detailed evaluation and comparison of those most popular shortest path algorithms. It was argued that no single algorithm consistently outperformed all others over the various classes of simulated networks[2]. Their analyses used randomly generated networks whose connectivity patterns rarely reflected the situation faced by transportation analysts [4]. Moreover, the results of finding the shortest paths between two nearby locations from a transportation network of sparse connection, the theoretical worse-case scenario for the shortest path algorithm, usually mismatched the computational efficiency in practice [2][5]. Literature [6] conducted a lot of experiments with the most popular shortest path algorithms on a lot of real road networks and identified some algorithms they argued most suitable for road networks.

Shortest path algorithms traditionally concerned topological or phase characteristics of a network and neglect the spatial or proximity characteristics. Such a viewpoint has

D. Grigoriev, J. Harrison, and E.A. Hirsch (Eds.): CSR 2006, LNCS 3967, pp. 487–497, 2006.
© Springer-Verlag Berlin Heidelberg 2006

resulted in a looping and radial approach to path searching which cannot prevent redundant searching even in situations when the destination nodes have been located. Some researchers have argued that the locational information (i.e. relative positions) of nodes can be employed in heuristics to inform the searching process [4][7][8]. Heuristics have been widely used in optimum path algorithms [7][8][9][10].

Most heuristics such as A* algorithm utilized local controlling strategies to guide the node search until the destination was found. They were efficient in locating approximately optimal one-to-one shortest paths. The localized approach could not handle one-to-some or some-to-one shortest path problems because of inability to treat many nodes simultaneously.

Our method makes use of geographical proximity in a transport network in our search heuristics. We would argue that the optimal path algorithms and heuristic strategies can be integrated to make a trade-off between efficiency and theoretical exactness of shortest paths, which is especially important for large-scale web applications or real time vehicular path finding. This paper discusses the search heuristics built upon geographic proximity between spatial objects and highlights advantages of the integration. Our method was tested for its efficiency and accuracy in solving shortest paths, using ten real-world networks of different complexity and connectedness.

2 Real-World Networks in the Study

We downloaded from the Internet ten real-world networks in the United States (Table 1) representing road structures of varied complexity (http://www.fhwa.dot.gov/ and http://www.bts.gov). The networks were arranged in order of complexity (as indicated by increasing number of nodes and arcs from top to bottom in Table 1) and the density of network data at both ends was illustrated in Figure 1. All but the last two network data in Table 1 (Alabama, Georgia, Pennsylvania, New York, Texas, California,

Table 1. Characteristics of real-world networks in the study

Network data	Number of nodes	Number of arcs	Arc/Node ratio	Max arc length	Mean arc length	STDEV of arc length
Alabama	952	1482	1.557	0.5960	0.1071	0.1102
Georgia	2308	3692	1.600	0.4783	0.0703	0.0793
Pennsylvania	2640	4183	1.584	0.6987	0.0551	0.0732
New York	3579	5693	1.591	0.7063	0.0476	0.0643
Texas	3812	6340	1.663	1.402	0.0858	0.1383
California	5636	9361	1.661	1.5300	0.0414	0.0911
Northeast USA	14009	19788	1.413	1.114	0.0558	0.0799
All of USA	26322	44077	1.675	2.3320	0.1429	0.1674
Utah Detailed	72558	100533	1.386	0.6953	0.0127	0.0159
Alabama Detailed	154517	199656	1.292	0.1936	0.0074	0.0069

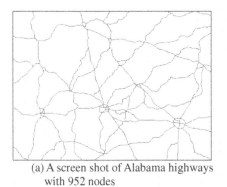

| (a) A screen shot of Alabama highways with 952 nodes | (b) A screen shot of Alabama roadways with 154,517 nodes |

Fig. 1. Network data of different complexity

northeast USA, and all of USA) contained state and inter-state highways. The last two network data included street networks in Utah and Alabama respectively. The network data were stored in ArcGIS shapefile format and recorded as pairs of longitude and latitude readings in decimal degrees.

The network data were edited to topological correctness by removing pseudo nodes and resolving connectivity. They were then transformed and and exported into our custom-built ASCII files to store geometrical and topological information on arcs and nodes.

3 Search Heuristics Based Upon Geographic Proximity

An ellipse based strategy is utilized to limit the searching within a local area generated with the source and destination nodes. Detailed description of the strategy can be found in [11][12]. The approximate maximum distance or cost M_D for traveling on a road network can be determined if the origin and the destination can be identified. A zone can then be defined to limit the search for a solution. Nodes that lied beyond the limit would not be considered in the search for the shortest path solution. Given that the Euclidean distance from a node N to the origin S and to the destination T is $|SN|$ and $|NT|$ respectively, as in Figure 2, the limiting condition can be defined as $|SN|+|NT| \leq M_D$. The critical points of N and N' will circumscribe an ellipse focused at S and T with a major axis of M_D [12]. The key to the approach was to fix a reasonable major axis M_D, representing the maximum endurable cost to transverse a pair of points. Because the cost factor between two nodes was considered to vary proportionally with distance, the maximum endurable cost could be derived from node locations.

3.1 Correlation Between Shortest Path and Euclidean Distance in Road Networks

First, 1000 sample nodes were extracted systematically from each network to reconstitute sets A and B, each containing 500 nodes. In the case of Alabama, all 952 nodes were used (Table 1). Every node in A or B would be regarded in turn as origins and destinations between which the shortest paths must be determined. In other words, a total of 500x500 shortest paths (or 476x476 for Alabama) would be required for each

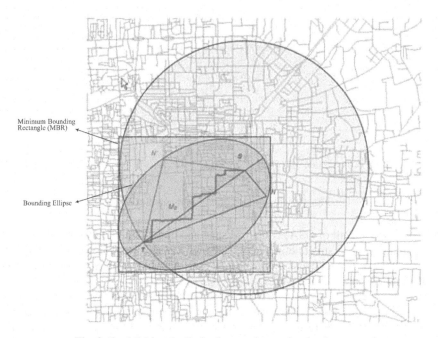

Minimum Bounding
Rectangle (MBR)

Bounding Ellipse

Fig. 2. Establishing the limit of a search area for the shortest path

network. A systematical sampling method makes sure that both of the sampled origin and destination nodes are evenly distributed in the networks, and thereby make the statistical analysis on the samples more representative.

Suppose that the Euclidean distance between the node pair was e_{ab} and the shortest path distance was p_{ab}, a set R of ratios $r_{ab} = p_{ab}/e_{ab}$ could be computed for each sample set. The ratio r compares the shortest path against the Euclidean distance. In general, we could establish a real number τ as the threshold value for the elements in R at a stated confidence level. A τ value at 95 percent would imply 95 percent confidence that the shortest path for a pair of nodes could be found within the extent built with the τ value. A larger τ would occur if the network exhibited connectivity disrupted by terrain features, such as physical presence of rivers or mountain ranges.

A plot of the shortest path against the Euclidean distances for sample nodes in the ten networks was illustrated in Figure 3. Our computation for the shortest paths used Dijkstra's algorithm implemented with a quad-heap structure. Given that the shortest distance between a node pair could never be shorter than their Euclidean distance, we could infer that a concentration of points along the 45° trend line would indicate a close to perfect match between the two distances. Conversely, dispersal from the 45° trend line would imply the presence of longer and meandering shortest paths indicating the lack of direct connectivity between pairs of points.

The τ values in our study were derived statistically at 95 percent confidence level and the 95 percent line for each network ($y = \tau \cdot x$) was also shown in Figure 3. The τ values ranged from 1.238 (Texas) to 1.585 (Utah detailed) which appeared quite consistent across the networks. Texas highway network embedded with uniformly distributed nodes and radial roads yielded the smallest τ thus signifying more compact

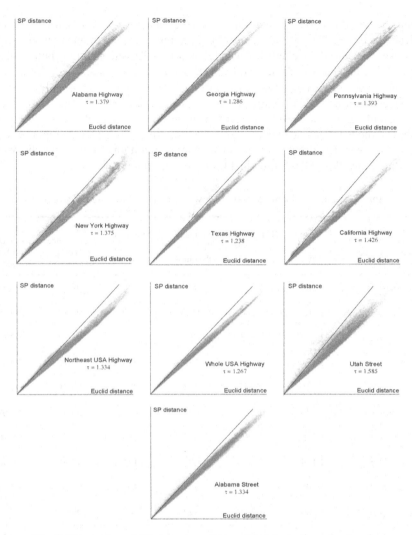

Fig. 3. The τ values at 95 percent confidence level for each road network

connectivity. In contrast, Utah detailed network had the largest τ as evident from the presence of physical separation by mountain ranges. The relationship between r, τ, and the networks will be explored further after the mathematical explanation below.

3.2 Numerical Premise

Using τ as a product coefficient, we could derive the major axis M_D for the ellipse that encircled a pair of origin and destination nodes. However, the computationally intensive node-in-ellipse search would offset benefits brought about by limiting the search extent, particularly when the destination and the origin were further apart. To make up for the computational deficiency of ellipses, we proposed the use of a minimum bounding rectangle (MBR) to limit our search. The use of MBR was computationally superior to the ellipse method, and provided an enlarged searching extent which would enhance the possiblility finding a solution in one round of search.

Table 2 presented the average r-values against the 95 percent τ threshold of the ten networks. Other than California and Utah Detailed, the τ thresholds were less than 120 percent of average r-values. We would attribute a larger τ for California because of its elongated shape and Utah Detailed for its disrupted landscape. In other words, there is great likelihood (i.e. at least 95 percent of the time) that we could locate the shortest paths (SP) between any node pairs within the bounding ellipse. Our attempt to compare success-rates of the elliptical versus rectangular search limits showed that the MBR is superior to the ellipse; the former had fewer SPs beyond bound.

Table 2. τ values and comparison between elliptical and rectangular search areas

Network data (based on 1000 sampled nodes in each network)	Average ratio $r_{ab} = p_{ab}/e_{ab}$	Threshold τ (at 95 % confidence)	Number of SP beyond the ellipse	Number of SP beyond the MBR	Real confidence with ellipse search (%)	Real confidence with MBR search (%)
Alabama	1.201	1.379	1588	840	99.30	99.63
Georgia	1.162	1.286	1013	619	99.59	99.75
Pennsylvania	1.179	1.393	761	453	99.70	99.82
New York	1.191	1.375	750	406	99.70	99.84
Texas	1.128	1.238	1220	369	99.51	99.85
California	1.170	1.426	1163	802	99.53	99.68
Northeast USA	1.166	1.334	438	250	99.82	99.90
All of USA	1.139	1.267	226	109	99.91	99.96
Utah Detailed	1.308	1.585	717	405	99.71	99.84
Alabama Detailed	1.209	1.334	822	549	99.67	99.78
Average	1.185	1.362	-	-	99.64	99.81

Although the 95 percent τ threshold was set to limit the searching area (generally as depicted in Figure 4a), our empirical findings proved that the use of either the ellipse or the MBR to limit a search area in the computation of shortest paths was better than 95 percent. We computed evenly distributed 500*500 (476*476 for Alabama) shortest paths using the Dijkstra's algorithm implemented with quad-heap. We then computed the number and percentage of shortest paths that could be located within the ellipse (as in Figure 4b) and within the MBR (as in Figure 4c). The numbers of shortest paths that went beyond the ellipse or the MBR were then derived.

The last two columns in Table 2 recorded that more than 99 percent of the shortest paths (average 99.64 percent for ellipse and 99.81 percent for MBR) could be identified within the limited search areas. Figures 4b and 4c explained why. r_{pq} in Figure 4b was

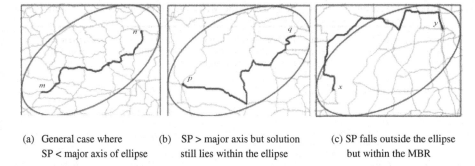

(a) General case where   (b) SP > major axis but solution   (c) SP falls outside the ellipse
     SP < major axis of ellipse   still lies within the ellipse   but within the MBR

Fig. 4. Comparing shortest path and search areas limited by ellipse and the MBR

10.49 percent bigger than the 95 percent τ threshold but the shortest path remained within the ellipse. Figure 4c showed that the ratio r_{xy} exceeded τ by 28.49 percent but the shortest path was still within the MBR. The higher confidence levels in both cases meant that the limited search was extremely effective because solutions to shortest paths (including cases similar to Figures 4b and 4c) could be found 99 percent of the time in one round of search. In other words, only a very small percentage of the shortest paths (an average of 0.36 percent for ellipse and 0.19 percent for MBR) would be found beyond the search areas. These special cases would occur when no path between a pair of nodes exists within the search limits and such a condition would, in turn, invoke an additional routine to conduct a liberal search for the shortest path.

3.3 Establishing Accuracy and Computational Efficiency

While our method seemed effective, it was still possible that some near-shortest (or optimum) paths found within the limit of the ellipse or MBR might not be the real shortest paths. We therefore repeated the experiment using the same set of data to undertake a liberal computation for the shortest paths without limiting the search area. The corresponding results were listed in Table 3 which showed that very few shortest paths (i.e. an average of 0.074 percent) would be mis-calculated when MBR was used to limit the search area. It also showed that the use of MBR would result in locating near-shortest paths with slightly longer lengths (i.e. an average of 7.65 percent longer than the actual shortest paths). The results were encouraging as they indicated that the MBR would not cause excessive loss in the computational accuracy except in the case of California whose network was deficient in connectivity. Our next task was to assess the computational efficiency in locating shortest paths with and without the use of MBR to limit the search area.

Assuming that all sample nodes in our networks were distributed uniformly throughout, a liberal search for the shortest path would begin at the source and spread radially outward until all node pairs were processed. Our schematic showed that the search area limited with MBR and the radial search would overlap and possess an area ratio to the circle as $(arctg \ (\dfrac{\sqrt{\tau^2 - 1}}{\sqrt{5 - \tau^2}}) + (\tau - 1 + \dfrac{\sqrt{5 - \tau^2}}{2}) * \dfrac{\sqrt{\tau^2 - 1}}{2}) / \pi$. With $\tau < 2$, the area ratio will less than 0.747. We recorded the computational time for resolving the one-to-one shortest paths with the 1000 sample nodes (all 952 nodes

Table 3. Possibility for finding SPs beyond the MBR but with near-SPs found within it

Network data	Number of shortest paths beyond the MBR but with near-shortest paths found within its bounds	Possibility (%)	Difference in mean path lengths (%)
Alabama	159	0.070	13.60
Georgia	269	0.108	6.94
Pennsylvania	172	0.069	9.10
New York	53	0.021	8.06
Texas	151	0.060	3.29
California	517	0.207	17.74
Northeast USA	193	0.077	7.46
All of USA	175	0.070	6.28
Utah Detailed	74	0.030	1.50
Alabama Detailed	66	0.026	2.56
Average	-	0.074	7.65

Table 4. Computational efficiency between limited search with MBR and liberal search

Network data	Average CPU time for one-to-one SP calculation with MBR search (s)	Average CPU time for one-to-one SP calculation with liberal search (s)	Time saving with MBR search (%)
Alabama	0.00020	0.00028	29.25
Georgia	0.00045	0.00068	34.41
Pennsylvania	0.00062	0.00074	16.44
New York	0.00077	0.00102	24.18
Texas	0.00072	0.00115	37.64
California	0.00150	0.00177	15.35
Northeast USA	0.00423	0.00630	32.83
All of USA	0.00941	0.01416	33.51
Utah Detailed	0.02850	0.03737	23.74
Alabama Detailed	0.05543	0.08624	35.73
Average	-	-	26.68

for Alabama) using the MBR to limit the search area on the one hand and by means of liberal search on the other. Total 250,000 shortest paths were calculated for each network (226,576 paths for Alabama). The results were listed in Table 4. It showed that the computational times with MBR were shorter than those of liberal search on all counts. Indeed, an average 26.68 percent saving in computational time was realized with only minimal loss in accuracy (i.e. 0.074 percent of the optimum paths would be 7.65 percent longer than the actual shortest paths on average, as shown in Table 3).

3.4 Special Considerations

It should be noted that networks with different spatial configurations would yield different results. Figure 5 showed three city road networks of unique spatial arrangement: (a) a strongly connected network without dangles or dead-end roads; (b) a network containing many cul-de-sacs; and (c) a network comprising of some sub-networks connected at several junctures across a river.

Samples of 500*500 node pairs were extracted from each of the networks in Figure 5 and the statistics for shortest path computation were summarized in Table 5. Table 5

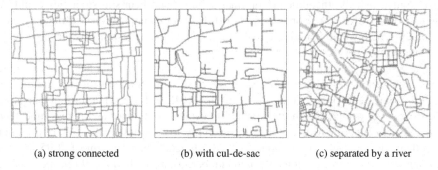

| (a) strong connected | (b) with cul-de-sac | (c) separated by a river |

Fig. 5. Special cases for shortest path heuristics

Table 5. Efficiency comparison of ellipse MBR and free search for SPs on city road networks

City road networks	Average ratio $r_{ab} = p_{ab}/e_{ab}$	Threshold τ (at 95 % confidence)	Average computational time for MBR search (s)	Average computational time for liberal search (s)	Time saving with MBR search (%)
Network A (4892 nodes, 7924 arcs)	1.275	1.441	0.00108	0.00155	30.32
Network B (11893 nodes, 16689 arcs)	1.308	1.584	0.00359	0.00476	24.58
Network C (12803 nodes, 18107 arcs)	1.381	2.019	0.00493	0.00537	8.28

(such as $\tau > 2$ in Figure 5c) was not as good as earlier results. While the MBR in showed that the performance of MBR search for networks with larger τ thresholds comparison with the liberal search method managed to yield time saving in this case, the margin of difference was relatively small. This observation alerted the need for some adjustments in our approach for cases with sub-networks as in Figure 5c. For example, decomposition by graph method using the divide-and-conquer strategy to solve for the shortest paths would be appropriate when the source and destination nodes were in separate sub-networks. The MBR heuristic would still apply in the search for shortest paths for node pairs within the same sub-network.

4 Conclusion

The single-source shortest path algorithm, as one of the most natural problems in network optimization, has generated much research from a variety of disciplines such as computer science, logistics, and geographical information systems. Among the many algorithms presented, the Dijkstra's algorithm was by far the most mature and robust in practice, especially for one-to-one or one-to-some solutions to the shortest paths. We have argued that the theoretically optimal solution and heuristic strategies could be integrated to offer a reasonable trade-off between computational efficiency and accuracy, which would benefit the realization of shortest path solutions for large-scale web applications or real-time vehicular navigation.

The geographical proximity of road networks provides excellent heuristics to guide the search for one-to-one or one-to-some shortest paths. Such a heuristic can greatly limit the search extent for a quick solution. The MBR method we put forth in this paper allows users to resolve the shortest paths for a stated confidence and in less time. The MBR is also superior to the ellipse in resolving the shortest paths. While a higher confidence level would bring about reduced efficiency, our experiment at 95 percent confidence showed better than 95 percent performance. In fact, 99.81 percent of one-to-one shortest paths could be found within the MBR along with 26.68 percent time saving on average, with little loss in accuracy as reflected by an average of only 0.074 percent of the optimal paths exceeding 7.65 percent of the lengths of the actual shortest paths.

In short, the MBR heuristics provides an attractive alternative to finding one-to-one or one-to-some (some-to-one) shortest paths in less time and with little loss in accuracy. Employing the heuristics with a stated confidence makes the approach holistic. We believe that further integration with local heuristics, such as decomposition by graph method, would render the search for shortest paths more efficient and practical.

Acknowledgments

This research was supported by the National Natural Science Foundation of China under grant No. 40201043 and Knowledge Innovation Project of CAS under grant No. CXIOG-D04-02.

References

1. Deo N. and Pang C.Y. Shortest-path algorithms: taxonomy and annotation, Networks, 4(1984) 275-323.
2. Cherkassky B.V., Goldberg A.V. and Radzik T. Shortest paths algorithms: theory and experimental evaluation, Mathematical Programming, 73(1996) 129-174.
3. Pallottino S., Scutellà M. G., Shortest path algorithms in transportation models: classical and innovative aspects, in Equilibrium and advanced transportation modeling, Marcotte P. and Nguyen S. (Editors), Kluwer, Norwell, MA, (1998) 245-281
4. Miller H.J. and Shaw S.L. Geographic Information Systems for Transportation: Principles and Applications. New York: Oxford University Press (2001).
5. Goldberg A.V. and Tarjan R.E. Expected performance of Dijkstra's shortest path algorithm, Technical Report No. PRINCETONCS//TR-530-96, Princeton University (1996).
6. Zhan F.B. and Noon C.E. Shortest path algorithms: an evaluation using real road networks, Transportation Science, 32(1998) 65-73.
7. Zhao Y.L. Vehicle Location and Navigation Systems. Boston: Artech House Publishers (1997).
8. Nilsson. N.J. Artificial Intelligence: A New Synthesis. San Francisco: Morgan Kaufmann Publishers (1998).
9. Fisher P. F. A primer of geographic search using artificial intelligence, Computers and Geosciences, 16(1990) 753-776.
10. Holzer M., Schulz F., and Willhalm T., Combining speed-up techniques for shortest-path computations, Lecture Notes in Computer Science, 3059((2004) 269-284
11. Nordbeck, S. and Rystedt, B., Computer cartography --- range map, BIT, 9(1969) 157-166.
12. Feng Lu, Yanning Guan, An optimum vehicular path solution with multi-heuristics, Lecture Notes in Computer Science, 3039(2004) 964-971

A New Hybrid Directory Scheme for Shared Memory Multi-processors

Guoteng Pan, Lunguo Xie, Qiang Dou, and Erhua He

School of Computer Science,
National University of Defense Technology,
Changsha 410073, P.R. China
gtpan@nudt.edu.cn

Abstract. It is increasingly popular to adopt DSM systems to maximize parallelism beyond the limits of SMP. Wherein, a proper cache coherence scheme should be efficient at the memory overhead for maintaining the directories because of its significant impact on overall performance of the system. In this paper, we propose a new hybrid directory scheme which reduces the memory overhead of the directory and improves the memory access time by using our new hybrid directory scheme. We evaluate the performance of our proposed scheme by running six applications on an execution-driven simulator (RSIM). The simulation results show that the performance of a system with hybrid directory can achieve close to that of a multiprocessor with bit-vector directory.

1 Introduction

In implementing Distributed-Shared Memory (DSM) systems, it is rather difficult to maintain cache coherence efficiently, which not only determines the correctness of the system, but also has a significant impact on system performance. A variety of mechanisms have been proposed for solving the cache coherence problem.

Snooping protocol [1] is a popular cache coherence protocol which is designed based on a shared bus connecting the processors. But bus is a kind of broadcast medium, its scalability is limited. When the number of processors increases beyond 32, the shared bus becomes a bottleneck for such situations.

Directory-based protocols were proposed for medium-scale and large-scale multiprocessors [2, 3]. Its basic idea is keeping a directory entry for each memory line, all of which consist of the state of the memory line and a sharing code indicating the processors that contain a copy of the line in their local caches. By examining a directory entry, a processor can determine the other processors sharing the memory line that it wishes to read, write or send invalidate messages to as required. Directory- based protocols scale better than snooping protocols because they do not rely on a shared bus to exchange coherence information. Many commercial multiprocessor systems implement directory-based coherence such as SGI Origin 2000 [4] and SGI Origin 3000[5].

D. Grigoriev, J. Harrison, and E.A. Hirsch (Eds.): CSR 2006, LNCS 3967, pp. 498–504, 2006.
© Springer-Verlag Berlin Heidelberg 2006

Directory maintains information on the sharing of cache lines in the system. Directory organization defines the storage structure that makes up the directory and the nature of the information stored within it. The memory overhead of directory has significant impact on the system performance and the choice of the directory structure is important for implementing high performance systems.

In this paper, we propose a new directory organization combining bit-vector scheme with limited pointers scheme. There are two directories in a nodes local memory, one is organized as the limited pointers scheme that keeps a directory entry for each memory line and the other is organized as bit-vector scheme that has only a few entries, not all memory lines have associated entry.

The rest of the paper is organized as followings. The next section gives an overview of the related work. In the third section, we introduce the hybrid directory scheme. The fourth section shows the performance evaluation and the simulation results. Finally we conclude this paper in section 5.

2 Related Work

In this section, we will briefly review some directory-based cache coherence schemes, most of these schemes attempt to address the issue of memory overhead of the directory.

In bit-vector scheme [3], for each memory line in main memory, the bit-vector protocol keeps a directory entry for that line, and in each entry every processor or node has a presence bit associated with each memory block indicating whether they have a copy of the block. Because of the huge memory overhead, the bit-vector directory is mainly used in small number of multiprocessors.

In the limited pointers directory scheme [6], the directory entry uses L pointers for the first L nodes caching the block and each pointer represents a nodes ID. When a request of read miss arrives at the home node and the directory finds that no more free pointer is available in the associated entry to record the node sending the request, this situation is called directory entry overflow. To tackle this problem, a pointer will be randomly selected and freed by sending invalidate message to the node that it represents. Under this situation, system performance may be poor.

Acacio from Spain proposed multilayer clustering as an effective approach [7] to reduce the directory-entry width. The evaluation results for 64 processors shows that this approach can drastically reduce the memory overhead, while suffering performance degradation similar to previous compressed schemes.

Tree-based cache coherence protocol [8] was proposed by Chang from Texas A&M University, which was a hybrid of the limited directory and the linked list scheme. By utilizing a limited number of pointers in the directory, the proposed scheme connects the nodes caching a shared block in a tree fashion. In addition to the low communication overhead, the proposed scheme also contains the advantages of the existing bit-map and tree-based linked list protocols, namely, scalable memory requirement and logarithmic invalidation latency.

3   Hybrid Directory

By comparing several directory-based schemes (bit-vector, coarse vector and limited pointers), we have learned that these schemes all have advantages and disadvantages, some have large memory overhead, and some have lower time-efficiency or result in many meaningless messages.

For large-scale multiprocessors, some of these directory-based schemes can work well, but for very large-scale systems, they will limit the scalability of the system and do not scale well. To achieve higher performance in cc-NUMA systems, a new directory organization must be put forward.

We can decrease the memory overhead of the directory in two ways: one is reducing the directory width as is adopted in coarse vector and limited pointers; the other is reducing the directory height as is used in sparse directory.

In practice, the average number of the copy of the memory line is small. Research has shown that when the proportion of the shared read is less than 70more than 5 nodes local cache contain the given memory line between two writes [9]. According to this, we propose a new hybrid directory scheme that combines the bit- vector scheme and the limited-pointers scheme.

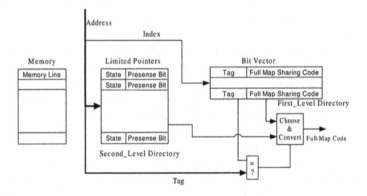

Fig. 1. The Organization of Hybrid Directory

Figure 1 depicts the hybrid organization of limited pointers and bit-vectors directories.

In our proposal, each node must maintain two kinds of directories:

1. The first-level directory using bit-vector scheme is organized as a cache, it only contains small number of entries and can be implemented in SRAM. Every directory entry consists of a presence bit vector and the tag information indicating whether it is hit or not. Because the main access manner of the directory is read- modify-write (RMW), we adopt write-back policy for directory cache. One item in the first-level directory writes back to the second-level directory only when it is replaced.

2. The limited pointers scheme is used in the second-level directory and the memory blocks all have a corresponding item in the directory. Every directory entry consists of a state bit and a presence bit vector, indicating the state of directory entry and the sharing information of the block copy respectively.

4   Performance Evaluation

4.1   Simulation Methodology

To evaluate the performance impact of hybrid directory on the application performance of CC-NUMA multiprocessors, we use a modified version of the Rice Simulator for ILP Multiprocessors (RSIM) which is an execution driven simulator for shared memory multiprocessors. Table 1 gives some main parameters of the simulated system in our simulation.

We have selected several numerical applications to investigate the performance benifits of our proposed scheme. These applications are QS and SOR from RSIM, Water from SPLASH benchmark suite [10] and FFT, Radix and LU from SPLASH-2 benchmark suite [11]. The input date sizes are shown in Table 2.

Table 1. Simulation parameters

The Simulated System (64 Processors)			
Processor		Memory	
Speed	200MHz	Interleaving	4
Issue	4	Access Time	18 cycles
Cache		Network	
L1 Cache	16KB, Direct-mapped	Topology	2D Mesh
Access Time	1 cycle	Router Cycle	3
L2 Cache	64KB, 4-way set associative	Flit Length	8B
Access Time	5 cycles	Flit Delay	4 cycles
Block Size	64B	Link Width	64B
Internal Bus			
Bus Cycle	3	Bus Width	32b
Hybrid Directory			
1- level Access Time 8 cycles		2-level Access Time 20 cycles	

Table 2. Applications Workload

Applications	Size
QS	16384
SOR	12864
Water	512 molecules
FFT	65536
Radix	524288 keys
LU	256256, block size = 8

4.2   Base Simulation Results

In this subsection, we present and analyze the results obtained by simulations running on two kinds of systems: base and two-level directory. We choose 64 processors for our modeled system, and the base system does not use the hybrid directory. For the first level directory (directory cache), there are 1K items which use direct-mapped policy and every entry has 6 pointers in the sencond level directory.

Figure 2 shows the improvements on execution time for each application by using the hybrid directory scheme. As it was expected, some degradation occurs

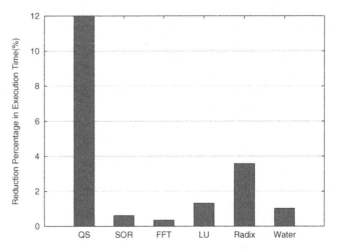

Fig. 2. Application Execution Time Improvements

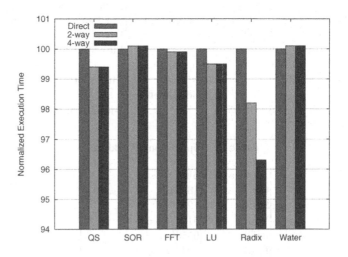

Fig. 3. Impact of Set Associative on Execution Time

when the hybrid directory is used and the degradation depends on the sharing patterns of these applications at a certain extent. The degradation for some appliactions is very small and can be negligible (the slowdown for FFT is only 0.35% and 0.61% for SOR). QS achieves the hightest reduction in execution time (12.9% slowdown). On the other hand, the performance of Water, LU and Radix are also degraded but not so significantly (1.03%, 1.32% and 3.57%, respectively).

In order to determine the effect of set associative on application performance, we vary the set associative of the directory cache from 1-way (direct-mapped) to 4-way and compare the performance. Figure 3 shows the impact of the set associative on the overall execution time. As the set associative is increased, we find that for most of the applications, the execution time decreases except SOR and Water. The set associative has obvious effect on the application performance for Radix, but for other applications, the performance with 2-way set associative is similar to that with 4-way set associative.

5   Conclusions

In this paper, we presented a new hybrid directory scheme which combines the bit- vector scheme (organized as a directory cache) and the limited-pointers scheme, to reduce the directory memory overhead and improve the performance of CC-NUMA multiprocessors.

We evaluate the performance of our proposed scheme by running six applications on a modified version of the Rice Simulator for ILP Multiprocessors (RSIM). The simulation results show that the performance of a system with hybrid directory can achieve close to that of a multiprocessor with bit-vector directory, even more better for some of the six applications.

Acknowledgement

This paper is supported by the National High-Tech 863 Project of China under grant No.2002AA104510.

References

1. S. J. Eggers and R. H. Katz. Evaluating the Performance of Four Snooping Cache Coherency Protocols. In Proc. of the 16th Annual International Symposium on Computer Architecture, 1989
2. Daniel Lenoski and James Laudon. The directory-based cache coherence protocol for the DASH multiprocessor. In Proc. of the 17th annual international symposium on Computer Architecture, 1990
3. L. M. Censier and P. Feautrier. A New Solution to Coherence Problems in Multi-cache Systems, IEEE Transactions on Computers, vol. C-27, no. 12, 1112-1118, 1978
4. J. Laudon and D. Lenoski. The SGI Origin: A ccNUMA Highly Scalable Server. In Proceedings of the 24th international Symposium on Computer Architecture, 1997

5. Silicon   Graphics   Inc.   http://www.sgi.com/products/servers/origin/3000/overview.html

6. Anant Agarwal and Richard Simoni. An Evaluation of Directory Schemes for Cache Coherence. In Proc. of the 15th Znt. Sym. On Computer Architecture, pp. 280-289, 1988

7. Acacio, et al. A New Scalable Directory Architecture for Large-Scale Multiprocessors. In Proceedings of the 7th International Symposium on High-Performance Computer Architecture, 97-106, 2001

8. Chang and Bhuyan. An Efficient Hybrid Cache Coherence Protocol for Shared Memory Multiprocessors. IEEE Transactions on Computers, vol. 48, no. 3, 352-360, 1999

9. Rangyu Deng and Lunguo Xie. Sharing Pattern of Shared Data in Multiprocessors. Computer Engineering and Science, vol.20, no A1, 66-69, 1998

10. J.P. Singh, et al. SPLASH: Stanford Parallel Applications for Shared-Memory. Computer Architecture News, vol. 20, 5-44, 1992

11. S.C.Woo, et al. SPLASH-2 Programs: Characterization and Methodological Considerations. In Proc. of the 22nd Intl Symposium on Computer Architecture, 24-36, 1995

Manipulator Path Planning in 3-Dimensional Space

Dmitry Pavlov

Saint-Petersburg State Polytechnical University, Saint-Petersburg, Russia
dmitry.pavlov@gmail.com

Abstract. Present paper is aimed to work out efficient algorithm of multi-chain manipulator path planning in 3D space with static polygonal obstacles. The resulting solution is based on navigational maps approach. Using this approach, manipulator features are considered as intellectual agent, and reachability information is stored in compact form. This enables fast adaptation to arbitrary parameters of manipulator and workspace. The paper describes two algorithms: (i) a local walkthrough with obstacle avoidance, and (ii) incremental navigational map building, performed at running stage. Both algorithms make an extensive use of the specific features of the problem. Working simultaneously, they allow real-time manipulator path planning, as well as self-learning in idle mode. Algorithms are implemented as a demonstration program.

1   Introduction and Acknowledgements

The problem of navigation, being currently very popular in robotics, is traditionally related to artificial intelligence. In most cases, human can easily figure out a path of a rigid body or a system of rigid bodies, with presence of obstacles, while computers solve this problem much more difficultly [1].

The particular problem that we are facing is the path planning of multi-chain manipulator, moving in an ambient space with static obstacles. Human cannot easily imagine manipulator motion (due to large amount of degrees of freedom and heavy robot's shape inconsistency). So, in this case the computer can take precedence over a human.

This paper describes an approach based on navigational maps, which has never been used for manipulator path plainning. The research process was supervised by S. Zhukov, and I am very grateful for his insights and guidance.

2   Problem Definition

2.1   Terminology

We define *workspace* (WS) as a 3-dimensional space with static polygonal obstacles, and *N-linked robot manipulator* as a construction of N polygonal links, joined in chain, each rotating around an axis fixed in the coordinate system

D. Grigoriev, J. Harrison, and E.A. Hirsch (Eds.): CSR 2006, LNCS 3967, pp. 505–513, 2006.

Fig. 1. A screenshot of a demonstration program

of the "parent" link. The rotation axis of the first link is fixed in workspace coordinate system. Each of manipulator's links has one degree of freedom.

We define *configuration* of our manipulator as a tuple of values in range $[0, 2\pi)$, which completely determines manipulator's position. Thus, a configuration space (CS) is $[0, 2\pi)^N$. We call the configuration *acceptable* if each of the manipulator links have no collision with static obstacles and with each other (except the neighbor links).

2.2   Informal Problem Definition

The main purpose of our work is to create a prototype of a navigational system, intended to execute a multiple path planning queries. For fast adaptation, we prohibit long preprocessing of WS, but accept learning, i.e. saving intermediate results, obtained during the execution, for to handle the consequent queries faster. Well trained navigational system does all the necessary computations close to real-time.

Our problem is a rough simplification of a real robot navigation problem: (a) we ignore any kind of kinematical constraints (so instead of a physical problem we have pure geometrical one), and (b) we do not provide optimal trajectory.

While manipulator's position is flexible, obstacles are fixed. When any obstacle changes its position, navigational system might reset all collected data.

2.3   Formal Problem Definition

Our task is to prototype a teachable navigation system for execution of multiple path planning queries (i.e. building trajectories). System might to:

- Provide fast adaptation to arbitrary parameters of manipulator and workspace
- Accumulate intermediate computational results
- Do self-learning in "idle mode"

Path planning query looks like this: given configurations θ_{start} and θ_{goal}, build a continuous function $\theta(t)$, satisfying the following conditions:

- $\theta(0) = \theta_{\text{start}}$
- $\theta(t_0) = \theta_{\text{goal}}$
- For any t in $[0, t_0]$, $\theta(t)$ is acceptable manipulator configuration.

3   Navigational Map

3.1   Principles

Given approach is based on navigational maps proposed by Zhukov [2], and we will use terminology from his paper. In the sequel we denote by *intellectual agent* a moving robot; by a *walkthrough algorithm* the one used by agent for walking to current (local) target, by an accessibility zone a set of configurations, joint by some criterion of accessibility; by the *accessibility graph* the one constructed by accessibility relation on zones; by navigational map a combination of walkthrough algorithm and accessibility graph.

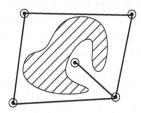

Fig. 2. Simple navigational map (for 2D point considered as an intellectual agent. Wlkthrough algorithm is "go straight". Accessibility graph contains 5 nodes.

In our problem, by an agent configuration we denote the set of numbers in $[0, 2\pi)$, representing manipulator's links rotation angles. So, every node in accessibility graph represents some unique manipulator configuration. The edge between two nodes is present if and only if a (local) walkthrough between their configurations is possible. The accessibility graph is not oriented, since a walkthrough algorithm is symmetric by definition.

The planning query is formed as a pair of configurations: starting and target. The task is to find the trajectory (path in CS) between them.

3.2   Path Planning

When starting and target manipulator positions are given, they are associated with some nodes of the existing navigational map by a virtual walkthrough. If no nodes are associated with position, it will be added to map as is.

While there is no path on navigational map between nodes associated with starting and ending points, the map is updated, as shown in the previous paragraph. That process can loop to infinity, for example if in fact there is no any path! We cannot confidently detect the absence of the path, as it is theoretically unsolvable problem. We can only interrupt the process after some time limit.

As the path on the accessibility graph is found, it is easy to construct manipulator's trajectory from it. The trajectory, obtained by a local walkthrough from starting configuration to the one of associated map node, is appended to the chain of trajectories of walkthroughs between map nodes, till the one associated with target configuration, and finally from the last node to the target configuration.

3.3   Building of the Navigational Map

Presented approach is designed for building navigational map on the fly – that means, than no navigational map is given at the very beginning, and it has to be constructed during path planning. Obviously, the algorithm has to be focused on given query, i.e. to build only those parts of navigational map that can help one to complete current planning task. Of course some heuristics should be used. There is some similarity to the widely used A* path planning algorithm. Its main idea is first to explore the areas that will most likely lead us to success. However, presented algorithm can work in idle mode (without planning query). In both modes (idle and focused), map building is incremental (and all obtained results are to be used during the execution of future queries).

Navigational map building procedure takes into consideration a very important manipulator's feature - its "modularity", i.e. the dependency of each manipulator's link on all previous links and independence on all the subsequent links. Simultaneously with construction of navigational map for given (say N-linked) manipulator, the navigational maps for 1-,2-,...(N-1)- linked manipulators are constructed. Those "reduced" manipulators are obtained from given one by removing links "from top to down", with preservation of links closer to manipulator's base. As the map building is incremental and performed "on the fly", all maps except the first are empty after initialization. The first map (for the one-link manipulator) is constructed trivially, in a very short time.

The increment of navigational map starts only when the existing map does not succeed to complete current query. The increment of the N-th map is performed by analyzing the (N-1)-th map, and can invoke its increment, which can stimulate the increment if (N-2)-th map, and so on. The first trivial map never gets updated.

4   Local Walkthrough Algorithm

4.1   Idea

It is well understandable that the simplier walkthrough algorithm we have, the more complex navigational map we have to build, and vice versa. Zhukov [2] has noticed that the quite simple walkthrough algorithm is the best choice in the sense of overall performance: big navigational map is more preferable than slow and unpredictable walkthrough procedure. So, we also use the very light one, without backtracking and complicated heuristics.

4.2   Description of Algorithm

Algorithm of an adaptive coordinate descent is developed as a simplified (and very undemanding to resources) analog of GoCDT [2] for manipulator. Algorithm is based on the principle of getting manipulator position to the desired one by coordinate descent method [3], with adaptive step selection and without back-offs. On each iteration only one manipulator's chain is rotated – the chain whose rotation angle has the biggest difference with desired one. The step (angle increment) changes adaptively, depending on collision detection: if the collision after changing angle with current step occurs, we decrease the step until given threshold. If no step in any direction can be done, walkthrough fails. Since AdapticeCD is not symmetric, the walkthrough algorithm is taken as combination of 2 walkthroughs (to and from the goal position), and considered to be successful if either of them succeeded.

Algorithm 1. Local walkthrough algorithm

```
AdaptiveCD (goal_state[])
    scale ← 1
    old_path_length ← ∞
    iter ← 0
    while scale > σ₀ and iter < max_iters do
        path_length← distance to goal configuration
        if |path_length − old_path_length| > δ₀ then
            scale ← min(1, scale * 2) {increase step}
        else {have not come nearer}
            scale ← scale/2 {decrease step}
        end if
        if path_length < threshold then
            return true{goal reached}
        end if
        for i from 1 to N do {by all chains}
            delta[i] ← (goal_state[i] − current_state[i]) * scale
            if rotation of i-th link by angle delta[i] causes collision then
                delta[i] ← 0
            end if
        end for
        i_max ← index of maximal absolute value in delta
        Rotate link i_max by angle delta[i_max]
        old_path_length ← path_length
        iter ← iter + 1
    end while
    return false {step value ot iterations exhausted; goal not reached}
```

The max_iters limitation is not obligatory, because algorithm definitely stops after at most

$$\frac{\sum_{i=1}^{N}(|\text{goal\_state}[i] - \text{start\_state}[i]|)}{\delta_0}$$

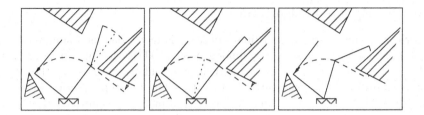

Fig. 3. AdaptiveCD algorithm illustration (for 2-dimensional manipulator)

iterations. The artificial constraint of iterations amount is supplied for saving computation time (as it is said before, the walkthrough should not be a "bottleneck" of the whole navigation process). We use $\sigma_0 = 0.125$ and `max_iters` = 30, and `AdaptiveCD` usually succeeds after 5-10 iterations, or fails after 10-15 iterations.

5   Incremental Update of the Navigational Map

5.1   Basic Principles

Firstly, it is necessary to mention that was described above: navigational map update procedure can work in "focused mode" to complete current query, as well as randomly improve navigational map for future use. One procedure's call increases the amount of nodes by 1.

We denote by *virtual edge* a pair of nodes of the accessibility graph, which are not connected, but can possibly be connected. Focused map update is based on finding the virtual edge, which is most perspective for connecting the "start" and "goal" connected components of the accessibility graph. Our algorithm selects the best virtual edge as the shortest one, adjacent to some vertex from the "start" or "goal" component. Once the best virtual edge is selected, procedure tries to generate new node "closer" to it. We use Euclid metrics of distance between nodes in CS. Another metrics can be found in [4].

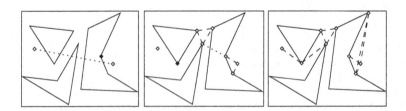

Fig. 4. A navigational map for point in 2-dimensional space can be represented by part of the visibility graph. Dashed lines denote edges of accessibility graph, pointed line denotes virtual edge.

5.2   Manipulator-Specific Features

For the update of k-th navigational map, one node from (k-1)-th navigational map is selected. Before it is selected, the update of (k-1)-th map can be initiated. Let us call opening the procedure of generation nodes k-th map nodes from (k-1)-th map nodes. So, the node selected from previous map is *opened*. Each node

Algorithm 2. Updating k-th navigatiohal map

```
UpdateNavigMap (k)
   while time frame not exhausted do
      node ← FindOrCreate(k - 1)
      if node = nil then
         return false {no unrevealed nodes}
      end if
      if OpenNode(node) = true then
         return true
      end if
   end while
```

Algorithm 3. Finding the unrevealed node to open (or create new) in k-th navigational map

```
FindOrCreate (k)
   if no unrevealed nodes then
      if k < 2 or UpdateNavigMap(k) = false then
         return nil
      end if
   end if
   if k > 1 and Random(0, 1) > 0.5 then
      UpdateNavigMap(k)
   end if
   node ← optimal unrevealed node
   return node
```

Algorithm 4. Opening node

```
OpenNode(node)
   newnodes ← list of new nodes, generated from node
   if newnodes = ∅ then
      return false
   end if
   for each node in newnodes do
      add node to navigational map
      if k = N then {for "main" map only}
         Perform virtual walkthrough from node to all the components of the accessibility
         graph; add the new edges on success
      end if
   end for
```

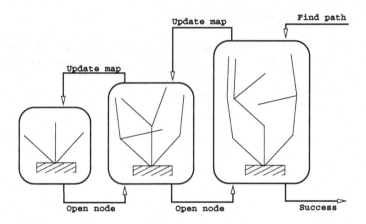

Fig. 5. A simplified example of navigational maps for 3-linked 2-dimensional manipulator, and the workflow of path planning procedure

Table 1. Minimal, maximal and average time taken by path planning procedure. Two tables are represening two independent environments (i.e. geometrical parameters of the manipulator and static obstacles.)

N	t_{min}	t_{max}	$\bar{t}$	$\|V\|$
4	0.16	3.41	0.66	27
5	0.26	18.12	2.54	43
6	0.89	360.4	38.51	209

N	t_{min}	t_{max}	$\bar{t}$	$\|V\|$
4	0.17	6.46	2.20	40
5	0.26	7.64	2.35	46
6	0.92	380.5	43.18	350

can be opened only once. The "opening" procedure copies all (k-1) angle values from the selected node to each of the generated nodes. Thus, the position of the first (k-1) manipulator links is unchanged. The angles, which determine a position of k-th manipulator link, are selected in some discrete subset of $[0, 2\pi)$, according to collision conditions.

6   Results and Conclusion

All described algorithms were implemented as a demonstration program, which visualizes manipulator path planning on a very complex (for this problem) obstacles. After a short period of learning (a few queries) path planning is performed in real time. But, the tests are showing that the increase of manipulator's links amount causes exponential growth of the navigational map.

The presented technique is powerful and can be more extended more by many features such as moving obstacles, building optimal (by some real-conditioned criteria [5]) trajectory, navigational map reduction without loss of accessibility information, and more.

References

1. Latombe, J.-C.: Robot Motion Planning. Kluwer Academic Publishers, 1991.
2. Zhukov, S: Accessibility Knowledge Representation. Programming and Computer Software, vol. 3, Moskow, RAS, 1999.
3. Baxter, B.: Fast numerical Methods for Inverse Kinematics. IK Courses '2000. Department of CS. The university of North Carolina. http://www.cs.unc.edu/~baxter/courses/290/html
4. Kuffer, J.J.: Effective Sampling and Distance Metrics for 3D Rigid Body Path Planning, Proc. IEEE Int'l Conf. on Robotics and Automation (ICRA'2004), New Orleans, April 2004.
5. Demyanov, A.: Lower semicontinuity of some functionals under the PDE constraints: a quasiconvex pair. Proceedings of Steklov Mathematical Institute vol. 318 (2004) pp. 100-119.

Least Likely to Use: A New Page Replacement Strategy for Improving Database Management System Response Time

Rodolfo A. Pazos R.[1], Joaquín Pérez O.[1], José A. Martínez F.[1,2],
Juan J. González B.[2], and Mirna P. Ponce F.[2]

[1] Centro Nacional de Investigación y Desarrollo Tecnológico,
Cuernavaca, Morelos, Mexico
{pazos, jperez,jam}@cenidet.edu.mx
[2] Instituto Tecnológico de Cd. Madero, Cd. Madero, Tamps., Mexico
jjgonzalezbarbosa@hotmail.com, mirna_poncef@yahoo.com.mx

Abstract. Since operating systems (OSs) file systems are designed for a wide variety of applications, their performance may become suboptimal when the workload has a large proportion of certain atypical applications, such as a database management system (DBMS). Consequently most DBMS manufacturers have implemented their own file manager relegating the OS file system. This paper describes a novel page replacement strategy (Least Likely to Use) for buffer management in DBMSs, which takes advantage of very valuable information from the DBMS query planner. This strategy was implemented on an experimental DBMS and compared with other replacement strategies (LRU, Q2 and LIRS) which are used in OSs and DBMSs. The experimental results show that the proposed strategy yields an improvement in response time for most types of queries and attains a maximum of 97-284% improvement for some cases.

1 Introduction

In the setting of a modern operating system (OS) a database management system (DBMS) is treated like a normal application program, which may deteriorate DBMS performance. This is due to the pagination of its code and data by the OS memory management system, which is designed for general purposes and so its algorithms pursue to satisfy a wide variety of applications[12].

It is known in the database (DB) community that buffer management plays a key role in providing an efficient access to data resident on disk and optimal use of main memory [2]. Since for a DBMS it is possible to know the access pattern to data when processing queries, then there is more certainty in deciding *which data to keep in main memory and which replacement policy to use*, thus resulting in a much lower response time [2]. If the DBMS query optimizer and the buffer manager worked together and not independently, the query optimizer could provide useful information to the buffer manager; since for most queries it can determine the data access sequence for answering the query [7], and therefore hand it over to the buffer manager for taking advantage of this information.

D. Grigoriev, J. Harrison, and E.A. Hirsch (Eds.): CSR 2006, LNCS 3967, pp. 514–523, 2006.

Since accessing a DB page in disk is much slower that accessing a DB page in the buffer, the main goal of a DBMS buffer manager is to minimize page I/O. For this reason devising better page replacement algorithms is very important for the overall performance of DBMSs [6]. Despite the development of many replacement policies, it is expected that more will be proposed or the existing ones will be improved [2, 11]; however, without prefetching its benefits can be limited [3].

The goal of this investigation is to demonstrate that the approach proposed in this work (Least Likely to Use) improves the performance of DBMSs buffer manager. This investigation has four features that set it apart from the rest:

- The evaluation was carried out implementing several algorithms (including ours) on an experimental DBMS, instead of using simulation.
- By using implementation we were able to focus our evaluation on the ultimate performance criterion: response time; as opposed to the evaluations carried out in other investigations which use page hit ratio.
- The experiments were designed to evaluate the algorithms performance for different types of queries.
- The proposed approach combines and synchronizes a new page replacement algorithm and prefetching.

It has been claimed that implementation-based evaluation yields more realistic results [11]. One of the reasons is that when using simulation one might inadvertently ignore implementation details that are essential for the DBMS to generate correct query results. If the overlooked detail requires much processing time in a real implementation, then the simulation results will not be realistic since the ignored time might offset any potential time gain. Conversely, when implementation based evaluation is used, one cannot miss any essential detail, for otherwise query results will not be correct. For example, we were able to carry out experiments involving queries whose processing needs database indexes, whose simulation is extremely difficult to carry out.

2   Related Works

A large number of works on replacement policies have been carried out to date (most in the OS area and much fewer in the DBMS area). This section includes a brief description of five of the most relevant replacement algorithms, two of which (2Q and LIRS) were selected for comparison against our proposed algorithm. Unfortunately there exist too few works that integrate and synchronize prefetching and buffer management, which is of great importance for the performance of the entire system.

In [10] a new page replacement algorithm, Time of Next Reference Predictor (TNRP), is proposed, which is based on predicting the time each page will be referenced again and thus replacing the page with the largest predicted time for its next reference. According to the tests conducted, they claim that their algorithm outperforms LRU, by up to 25-30% in some cases and over 100% in one of the cases. The best performance was obtained when the number of page frames is 4 as compared with 8 and 16, which were also used in their tests. A major difference between this work and ours is that they used simulation for evaluation, while we implemented the algorithms, which yields more realistic results. Like most work on replacement policies their algorithm was designed for OSs, while ours is for DBMSs.

In [9] the authors that devised 2Q mention that it performs similarly as LRU/2 (in fact, usually slightly better). Additionally, they affirm that 2Q yields an average improvement of 5-10% in hit ratio with respect to LRU for a large variety of applications and buffer sizes, with little or no tuning needed. However, in [8] its authors state that two pre-determined 2Q parameters (which control the sizes of the A1in y A1out queues) need to be carefully tuned and are sensitive to the workload type. 2Q was selected for comparison vs. our algorithm (LLU), and the comparative results are presented in section 5.

In [8] a replacement algorithm was proposed, called LIRS, which has a complexity similar to that of LRU, using the distance between the last and the penultimate page reference to estimate the probability of the page being referenced again. The authors divide the referenced blocks into two sets (like 2Q), one of them with 1% of the buffer size (suggested by the authors), which contains the pages that have been accessed again. According to tests conducted by the authors, they claim that LIRS has better performance than LRU and 2Q. LIRS was also selected for comparison, and the comparative results are presented in section 5.

In [4] prefetching is used to assist caching by predicting future memory references. The authors present a new mechanism which does not require additional tables or training time for predicting the data access pattern. In their project only data are prefetched, and one of the problems consists of discerning between address and data for carrying out prefetching. In our project prefetching is combined with buffer replacement policies using information provided by the query optimizer.

3   Architecture of the DBMS Used for Testing Algorithms

Currently there exists an experimental distributed database management system (called SiMBaDD), developed by us. In order to test the proposed replacement algorithm, SiMBaDD was used for implementing on it the buffer management module (Figure 1). The reason for selecting this DBMS is that the source code is available and it was easier for us to add and modify software modules.

It is important to point out that the *Buffer Manager* includes sub-modules for different page replacement algorithms (LRU, 2Q, LIRS and LLU), which permits to evaluate the DBMS performance using any given replacement algorithm. It is worth mentioning that the DBMS allows processing queries using indexes, which permits to conduct experiments for cases 3, 5, 6, 7 and 8, described in subsection 4.1.

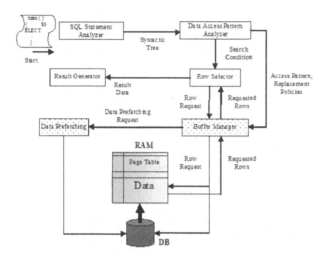

Fig. 1. Architecture experimental distributed database management system

4   Description of the LLU Page Replacement Policy

The Least Likely to Use (LLU) page replacement policy uses two parameters: usage value and replacement policy mark. The usage value is an estimation of the probability of a given page to be used again. The usage value of any page lays in the range from 0 to 10, where 0 indicates that the page will not be used, 10 specifies that it will be used again, and an intermediate value means that there is a chance that it will be used. The usage policy mark is set (to MRU or APU) according to the most convenient replacement policy for the type of the query that requested the page (according to the query classification described in subsection 4.1).

Upon reception of a query, the DBMS performs a lexical and syntactic parsing of the query (*SQL Statement Analyzer* module). Then if the query is correctly formulated, it is interpreted for selecting the table rows that satisfy the search condition (*Row Selector module*). Finally the DBMS formats the final results of the query (*Result Generator module*). In the original version of the DBMS, the *Row Selector* module requested table rows directly to the OS as needed, without intervention of the *Buffer Manager* and *Data Prefetching* modules (boxes with crisscross filling), which were added in the new version.

One of the main components of the proposed architecture is the *Data Access Pattern Analyzer*, which is in charge of identifying the type of query in order to determine the data access pattern and the ad-hoc replacement policy for such query (using the algorithm described in section 4.1). This permits to prefetch rows (*Data Prefetching* module) and to select the most convenient rows for replacement (*Buffer Manager* module), and consequently, to achieve the desired speed up. When the *Buffer Manager* receives a row request, the *Row Searcher* sub-module (Figure 2) finds out if the row is located in disk or main memory. If

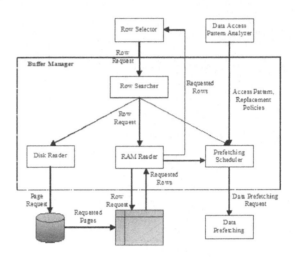

Fig. 2. Buffer Manager module

the row is in main memory, it requests the row to the *RAM Reader*, otherwise
it requests the row to the *Disk Reader*, which loads the row into main memory.
Afterwards, the requested row is delivered to the *Row Selector*.

The LLU policy takes first into consideration the replacement of the page
with the lowest usage value. When there exist several pages with the same lowest
usage value, the victim page is selected according to the usage policy mark; e.g.,
the most recently used page is selected among all the pages with an MRU mark
and the same lowest usage value.

Each time a page is transferred from disk to main memory, the *Buffer Manager* assigns to it a usage value and a replacement policy mark, which are based
on information supplied by the *Data Access Pattern Analyzer*. Whenever a page
has to be replaced, the *Buffer Manger* will use the LLU replacement policy,
described in previous paragraphs. The LLU policy selects the page that is least
likely to be used again. This decision is based primarily on the usage value of
the candidate pages and secondarily on the policy mark.

4.1 Query Classification for Deciding LLU Parameters and Prefetching

One of the most important modules of the proposed architecture is the *Data
Access Pattern Analyzer*, whose task consists of determining when and what data
to prefetch and what data to replace when processing a SELECT statement of
SQL, by using the information provided by the query optimizer. The following
paragraphs describe how to decide whether to use prefetching and to determine
the usage value and the most convenient replacement policy (replacement policy
mark) for each case of the SELECT statement.

1. **Query involving one table, without WHERE clause (with and without index).** Since the information of all the table rows has to be delivered, it is convenient to use prefetching to move to main memory the rows (stored in pages) that will be used in the near future. This speeds up data access because the system anticipates row loading so that they will already be in main memory when they are required. For this case it is convenient to use the APU (Any Previously Used) replacement algorithm, because the rows already read will not be used again and therefore it is immaterial which row to remove. (Note: in this case the usage value of the rows is set to 0.)

2. **Query involving only one table, with simple WHERE clause without index.** When the column appearing in the WHERE clause does not have an index, it is necessary to check each table row to determine if it satisfies the search condition (WHERE clause); therefore this case is treated just as case 1.

3. **Query involving only one table, with simple WHERE clause with index.** First it is necessary to determine if it is convenient to use the index, if so the nodes of the index tree are loaded as they are needed. In this case it is not possible to benefit from prefetching, because in order to prefetch any tree node it is necessary to know its address. However it would be convenient to prefetch the root node, since this node is indispensable to initiate the search. Once the leaf node sought is found, it is possible to access the table row (page). Subsequently, the adjacent nodes of the leaf (to the left or right) have to be read as long as the WHERE clause is satisfied (e.g., column > value), which as a result may deliver one, several or no rows. It is convenient to use the MRU (Most Recently Used) replacement algorithm for the tree nodes, in order to keep in the buffer the first accessed nodes (those closest to the root), which are most likely to be used by another query. The APU replacement algorithm can be used for managing pages that contain table rows. (Note: since once a certain row is used it could happen that some other row in the same page could be requested later, thus the usage value of the page is set to 5.)

4. **Query involving only one table, with complex WHERE clause without index.** When there is no index all the table rows have to be inspected for finding those that fulfill the search condition; therefore this case is treated exactly as case 1.

5. **Query involving only one table, with complex WHERE clause with index.** The WHERE clause has to be decomposed into sub-queries, and if some of these has an index it is used as long as it is useful; if there exist several indices, it is necessary to make an analysis to select the most beneficial. If there exists an index and it is convenient to use it, the process continues just as in case 3. It is necessary to note that the rows retrieved using the index have to be checked to find out if they fulfill the other conditions (sub-queries). If no index can be used, this case will be treated just as case 1.

6. **Query involving two tables, without WHERE clause (with and without index).** Since a cartesian product has to be performed, an index is

not useful for this. For performing the cartesian product one of the tables has to be read sequentially only once, thus the rows already read will not be used again and therefore it is advisable to use the APU replacement algorithm. Concurrently, the rows of the other table are read sequentially, but these rows have to be accessed a number of times proportional to the cardinality of the first table; in this case it is convenient to use the MRU (most recently used) replacement algorithm, which according to the literature is the most efficient for improving the response time in joins. (Note: in this case the usage value of the first table rows is set to 0, while for the second table rows is set to 10.)

7. **Query involving two tables, with simple WHERE clause (with and without index).** It is convenient to create an index for the small table (the table with the smallest cardinality) if it has no useful index; in parallel the rows of the large table (with the largest cardinality) have to be prefetched; this is because accessing these rows using prefetching is faster than accessing them using an index. If the operator appearing in the clause is $>$, $> =$, $<$, $< =$ or $=$, the process must proceed as follows: read a row of the large table, next obtain the data of the column that participates in the WHERE clause, then using this data look for the first leaf in the index tree of the small table that satisfies the WHERE clause similarly to case 3; from this leaf scan to the left or right for finding the tree leaves that satisfy the clause according to the operator type; this process must be carried out for each row of the large table. The replacement algorithm that should be used for the pages that contain rows of the small table is APU, because they have the same probability of being accessed again; while for the large table the APU algorithm should be used, since each of these rows will not be used again. (Note: in this case the usage value of the small table rows is set to 5, while for the rows of the large table is set to 0.) If the operator were $<>$, the process would proceed similarly to case 6.

8. **Query involving two tables, with complex WHERE clause (with and without index).** The query has to be decomposed into sub-queries, treating this case just as the previous one. It is necessary to mention that for some clauses a cartesian product has to be performed, so this case should be treated exactly as case 6.

9. **Query involving three or more tables (with and without WHERE clause and with and without index).** First two tables have to be processed, then the resulting table and one of the remaining tables are processed similarly, and so on for the rest of the tables. Each pair of tables should be processed as described in cases 6, 7 or 8 depending on the existence or inexistence of a WHERE clause, a simple clause or a complex clause.

5   Experimental Results

The replacement policy proposed in this work was compared with the traditional LRU replacement policy, as well as newer policies such as 2Q and LIRS.

The reasons for selecting 2Q and LIRS for comparison is that the first was designed specifically for DBMSs and the second has a good performance with respect to other policies according to its authors' claims. All of the policies were implemented on the DBMS, whose architecture is depicted in Figure 1. In particular 2Q and LIRS were implemented according the descriptions given in [9] and [8].

The DBMS used for conducting the experiments was implemented in standard C and runs on the Fedora version of the Linux operating system. The data and queries used for tests were obtained from the Open Source Database Benchmark [1]. The four tables used in the experiments (*tenpct, uniques, updates* and *hundred*) have 100,000 rows each and occupy approximately 15MB in hard disk. The experiments were conducted on a computer with a Centrino CPU at 1.6 MHz with 512 MB of main memory.

Table 1 shows the results of the tests conducted for queries of type 2 (using prefetching). For each replacement policy a series of runs were carried out for each of three different buffer sizes: 1,000, 2,000 and 3,000 pages (with 4,096KB each page). Each series consisted of 10-12 runs, and for each series the average query processing time (expressed in seconds) was recorded and shown in the table.

Table 1. Case 2: query involving one table, with simple WHERE clause without index

Number of Pages	LRU	LIRS	2Q	LLU
1000	3.348	2.849	2.034	**1.166**
2000	4.323	2.192	2.802	**1.008**
3000	4.850	1.583	3.428	**0.979**

Table 2. Case 3: query involving one table, with simple WHERE clause with index

Number of Pages	LRU	LIRS	2Q	LLU
1000	0.678	**0.617**	0.629	0.657
2000	0.660	**0.653**	0.678	0.672
3000	**0.636**	0.653	0.656	0.648

Table 2 presents the results obtained for queries of type 3 (using prefetching). In this case LLU is outperformed sometimes by LIRS and sometimes by LRU; notice, however, that the difference in performance is small. Table 3 presents the results for case 7, which show that the LLU replacement policy yields an average improvement of 284%, 140% and 97% in response time with respect to LRU, LIRS and 2Q respectively. Additionally, the last four columns of Table 3 show the results without prefetching for all the algorithms; as expected, the use of prefetching *generally reduces* the response time for all the algorithms.

Table 3. Case 7: Query involving two tables, with simple WHERE clause

Number of Pages	With Prefetching				Without Prefetching			
	LRU	LIRS	2Q	LLU	LRU	LIRS	2Q	LLU
1000	24.758	27.204	27.227	**13.420**	49.268	15.422	31.754	17.394
2000	64.447	46.825	26.542	**15.394**	71.007	30.928	35.179	15.189
3000	76.435	29.296	31.127	**14.283**	44.483	31.665	48.984	17.905

Experiments similar to the previous ones were conducted for cases 1, 4, 5, 6 and 8. For case 6 smaller tables were used because the cartesian product of 100,000-rows tables took too much time and disk space. No experiments were carried out for case 9 since it is similar to cases 6, 7 and 8.

As expected, the performance of each policy varies from case to case, such that it happens that for one case one policy may outperform another, while for another case the latter may outperform the first. It is worth mentioning that in all cases except two (cases 3 and 5) LLU outperformed all the other policies.

6   Final Remarks

In the area of optimization of access strategies there exist some open research issues. Though much work has been carried out for attempting to obtain or predict the access pattern to data, mainly in the OS area, which is useful for a mixture of a wide variety of applications; however, improving the prediction of the access pattern for a mixture of applications is a difficult task. Fortunately, predicting the access pattern in DBMSs can be more precise, since the computer works with a limited number of operation types and the page reference patterns exhibited by most of these operations are regular and predictable.

The proposed algorithm (LLU) takes advantage of this behavior, and in order to assess its performance it was evaluated with respect to algorithms that have been used or proposed for OSs and DBMSs, such as LRU, Q2 and LIRS. The results showed that for most cases LLU outperforms the other algorithms.

This investigation has several features that set it apart from the rest: the evaluation was carried out by implementing different replacement algorithms on an experimental DBMS, the criteria used for comparing algorithms was response time, and the experiments were designed to evaluate the algorithms performance for different types of queries. It has been claimed that implementation-based evaluation yields more realistic results, since when using simulation one might overlook details that are essential for the DBMS to generate correct query results.

It is important to remark that the replacement algorithms by themselves are not enough, since they have to be combined and synchronized with prefetching to improve their performance [3], as it has been shown for OSs, which constitutes a complex problem [5]. To date several investigators are aiming their research at combining two or more techniques to improve the performance of data management. However, few researchers have intimately synchronized prefetching with buffer management.

References

1. Andy Riebs and Compaq Computer Corporation. The Open Source Database Benchmark (2005). http://osdb.sourceforge.net/
2. Bressan, S., Goh, C.L., Ooi, B. C., Tan, K.: A Framework for Modeling Buffer Replacement Strategies. Proc. Ninth International Conference on Information and Knowledge Management. McLean, Virginia, USA (2000) 62-69
3. Butt, A.R., Gniady, C., Hu, Y.C.: The Performance Impact of Kernel Prefetching on Buffer Cache Replacement Algorithms. Proc. ACM SIGMETRICS, Banff, Canada (2005) 157-168
4. Cooksey, R.: Content-Sensitive Data Prefetching. PhD dissertation, University of Colorado, Colorado, USA (2002)
5. Dan, A., Yu, P.S., Chung, J.: Characterization of Database Access Pattern for Analytic Prediction of Buffer Hit Probability. VLDB Journal, Vol. 4, No. 1 (1995) 127-154
6. Effelsberg, W.: Principles of Database Buffer Management. ACM Transactions on Database Systems, Vol. 9, No. 4 (1984) 560-595
7. Jeon, S.H., Noh, S.H.: A Database Disk Buffer Management Algorithm Based on Prefetching. Proc. ACM CIKM, Bethesda, USA. (1998) 167-174
8. Jiang, S., Zhang, X.: LIRS: An Efficient Low Inter-Reference Recency Set Replacement Policy to Improve Buffer Cache Performance. Proc. ACM SIGMETRICS, Marina del Rey, USA (2002) 31-42
9. Johnson, T., Shasha, D.: 2Q: A Low Overhead High Performance Buffer Management Replacement Algorithm. Proc. VLDB, Santiago de Chile, Chile (1994) 439-450
10. Juurlink, B.: Approximating the Optimal Replacement Algorithm. Proc. Conf. Computing Frontiers, Ischia, Italy (2004) 313-319
11. Smaragdakis, Y.,: General Adaptive Replacement Policies. Proc. Fourth International Symposium on Memory Management ISMM, Vancouver, Canada. (2004) 108-119
12. Traub, O., Lloyd, A., Ledlie, J.: Application-Specific File Caching for Unmodified Applications and Kernels. Technical Report, MIT, Massachusetts, USA (2002)

Nonlinear Visualization of Incomplete Data Sets

Sergiy Popov

Senior Research Scientist, Control Systems Research Laboratory,
Kharkiv National University of Radio Electronics,
14 Lenin av., Kharkiv, 61166, Ukraine
Serge.Popov@ieee.org

Abstract. Visualization of large-scale data inherently requires dimensionality reduction to 1D, 2D, or 3D space. Autoassociative neural networks with bottleneck layer are commonly used as a nonlinear dimensionality reduction technique. However, many real-world problems suffer from incomplete data sets, i.e. some values may be missing. Common methods dealing with missing data include deletion of all cases with missing values from the data set or replacement with mean or "normal" values for specific variables. Such methods are appropriate when just a few values are missing. But in the case when a substantial portion of data is missing, these methods may significantly bias the results of modeling. To overcome this difficulty, we propose a modified learning procedure for the autoassociative neural network that directly takes into account missing values. The outputs of the trained network may be used for substitution of the missing values in the original data set.

1 Introduction

When a scientist or an engineer faces a new problem, the first steps towards its solution are to understand what is given and which aspects of this are most important. Since humans perceive most of the information in the course of their life in a visual form, it is preferable to present this new problem also in some kind of visual form: directly or through some analogy, i.e. visualize it. It is quite easy to visualize structures or logical relationships by means of flow charts and block diagrams. But when we come to data sets describing quantitative characteristics of objects or their relationships, the problems of dealing with high dimensionality arises.

People inherently are ably to think only in 1D, 2D, and 3D spaces. On the other hand, most real-world scientific and engineering problems deal with tens to thousands of dimensions. Thus, presenting (visualizing) high-dimensional data in a low-dimensional space requires dimensionality reduction. It is a technique intended to cut the number of dimensions while preserving maximum useful information in the data set.

Some of well-known dimensionality reduction methods are the following:

- principal component analysis (PCA) [4, 9];
- principal curves [3, 5];
- multidimensional scaling [7, 11];
- autoassociative (bottleneck) artificial neural networks (AANN) [6, 8].

D. Grigoriev, J. Harrison, and E.A. Hirsch (Eds.): CSR 2006, LNCS 3967, pp. 524–533, 2006.
© Springer-Verlag Berlin Heidelberg 2006

These and many other methods work seamlessly on complete data sets, when all numerical values are present, but most of them cannot be applied to data sets with missing values. The essence of the problem lies in the mathematics: for the formulas to be computed, all included variables must take some exact numerical values. When the value is missing, the formula cannot be computed at all, or it must be modified to omit this value. When missing values are positioned randomly in the data set, formulas cannot be modified to handle all possible situations, thus a way is sought to fill the missing values with some numerical values.

There are several simple methods to fill missing values:

1. When the number of samples with missing measurements is very small, discard these whole samples.
2. Replace missing values with mean or some "normal" (tolerable) value for this parameter.
3. If it is appropriate for the problem at hand, interpolate the value from the neighboring cells.

These methods have common drawbacks:

1. Missing data replacement (imputation) leads to biased estimates.
2. When the "restored" data set is presented to a dimensionality reduction algorithm, it does not "know" which values are true and which are replaced, and thus they have the same ranking in terms of their information load. On the other hand, dimensionality reduction implies information loss, so it is preferable to keep as much as possible information from the true data and completely ignore all missing data.

Thus it is desirable to develop an algorithm that would explicitly handle missing data, eliminating the abovementioned drawbacks. For PCA such an algorithm exists, it is a well-known expectation maximization (EM) algorithm [10]. However, PCA provides only linear projection, and more efficient results can be obtained by employing nonlinear dimensionality reduction techniques.

Autoassociative (bottleneck) neural networks can be seen as a generalization of PCA to the nonlinear case. It is proven [1, 2] that if only linear activation functions are used and the network is optimally trained, it performs exactly the same projection as PCA. In this paper, we propose modifications to standard learning procedures for AANN, which allow direct handling of missing data, extract most information from the present data, and estimate missing values from the low-dimensional representation of the data set.

The paper is organized as follows: section 2 gives a short overview of standard AANN architecture and learning algorithms; in section 3, the proposed modifications are presented; section 4 supports theoretical findings with experimental evidence; and finally, conclusions are made on the basis of the obtained results.

The following notation is adopted:

X – $N{\times}D$ matrix containing the data set, where
N – number of samples,
D – original (high) dimensionality of the data;
$x(k)$ – k-th row of X, i.e. one sample;

$y(k)$ – low-dimensional representation of $x(k)$;

$\hat{x}(k)$ – reconstruction of $x(k)$, obtained from $y(k)$.

2 Autoassociative Neural Networks

AANN is a kind of feedforward neural network with multiple hidden layers. Depending on the architecture and activation functions used, AANN can perform linear or nonlinear mapping. General AANN architecture is presented in fig. 1.

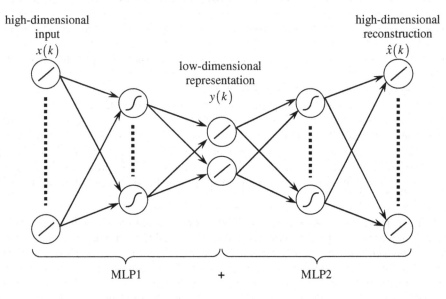

Fig. 1. General AANN architecture

It consists of input and output layers (with linear activation functions) with the number of neurons equal to the original dimensionality of the data. The first and the third hidden layers (with nonlinear activation functions) contain equal number of neurons, which is chosen according to the problem at hand. The second hidden layer (with linear activation functions) is the "bottleneck" layer, whose number of neurons is equal to the target low dimensionality. The outputs of the network are extracted from this layer. Such a network can be considered as two parts: input and the first two hidden layers form a multilayer perceptron (MLP1) with one hidden layer that performs nonlinear mapping $x(k) \Rightarrow y(k)$; the second and third hidden layers with the output layer form the second multilayer perceptron (MLP2) that solves the reverse problem of reconstructing the original data $y(k) \Rightarrow \hat{x}(k)$. The idea is to "squeeze" high-dimensional data through a low-dimensional "bottleneck" (the second hidden layer) so that the reconstruction $\hat{x}(k)$ is as close to the original data $x(k)$ as possible, i.e. maximum of information is retained. Thus the network inputs are also used as the learning targets.

To achieve this goal, the network is trained in the supervised mode with respect to the following criterion

$$E = \sum_{k=1}^{N} \left\| x(k) - \hat{x}(k) \right\|^2 . \tag{1}$$

If only linear mapping is required, the first and the third hidden layers (with nonlinear activation functions) are unnecessary, and the general architecture can be simplified (fig. 2).

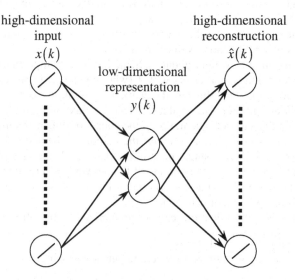

Fig. 2. Simplified AANN architecture for linear mapping

AANN can be trained with any learning algorithm suitable for feedforward neural networks. Most of them are based on the backpropagation procedure to calculate error gradients for hidden layers. Since for our further consideration, the choice of a particular learning algorithm does not matter, we will not focus on this issue.

3   Modified Learning Procedure

Now consider the case of missing values. The goal is to eliminate their influence on the network output and weights update.

The network outputs are formed by feeding forward the inputs through the network layers. The only layer that directly receives the inputs $x(k)$ is the first hidden layer. The weighted inputs are accumulated to form the neurons' activations as follows

$$a_j = \sum_{i=1}^{D} w_{ji} x_i \tag{2}$$

where a_j – activation of the j-th neuron, x_i – the i-th input, w_{ji} – the corresponding synaptic weight.

If the input is missing, it is natural to exclude the corresponding term from summation, which is equivalent to setting the corresponding x_i to zero. In this way, the missing value does not influence the sum (the neuron activation), hence, the neuron output, hence, the network output.

The network learning is basically an optimization procedure performed with respect to criterion (1). Ideally, the absolute value of E can drop to 0 if the network perfectly reconstructs its inputs. In reality it is always above 0, and the goal of learning is to minimize it by adjusting the synaptic weights of the network. Obviously, higher errors at particular network outputs lead to bigger adjustment of weights. On the contrary, zero errors lead to no adjustment. Thus, to eliminate the influence of missing values on the network learning, it is necessary to zero out errors at those outputs, where the target values are missing.

Such a modification to weight update scheme has a very important advantage: it is equivalent to weighting output errors with 1 (when the target is present) and 0 (when the target is missing), thus shifting the learning "attention" only to real data and completely discarding missing values. Hence, maximum retention of useful information from the data set is achieved in the course of dimensionality reduction.

When the learning procedure converges, the outputs of the network may be used to replace the corresponding missing values in the data set. If the task of missing value restoration is the primary one, the outputs of the network, where the target values are missing, may be fed back to the corresponding inputs. This will lead to an iterative process of missing value reconstruction.

Thus, the proposed modifications can be summarized in the following two rules:

1. Replace missing inputs with zeros.
2. Replace learning errors with zeros where targets are missing.

4   Experimental Results

The proposed approach was applied to real-world problems of biomedical data visualization. The first data set contains blood tests (35 parameters) for 26 patients taken before and after treatment (total of 52 samples). Dimensionality reduction is performed from 35D to 2D. To compare different algorithms, 20% of data is randomly discarded, so we have both full and incomplete data sets.

The reference visualization is obtained by applying standard PCA to the full data set (fig. 3). Each line represents one patient, the end with a solid circle corresponds to the blood test taken before treatment, and the end with an empty circle corresponds to the blood test taken after treatment.

The incomplete data set is visualized using three different approaches: fig. 4 – missing data is replaced with mean values for the corresponding parameter and then standard PCA is applied; fig. 5 – EM algorithm is applied directly to the incomplete data set; fig. 6 – the proposed modified AANN method is applied directly to the incomplete data set. Linear AANN (35-2-35 architecture) was used, because we are comparing to linear PCA methods.

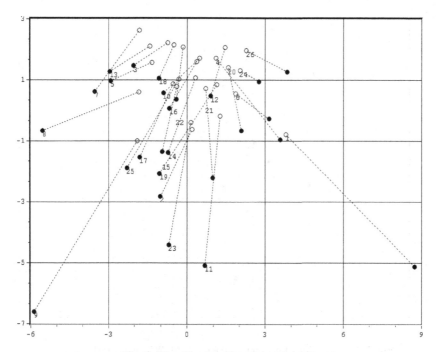

Fig. 3. Visualization of complete data set

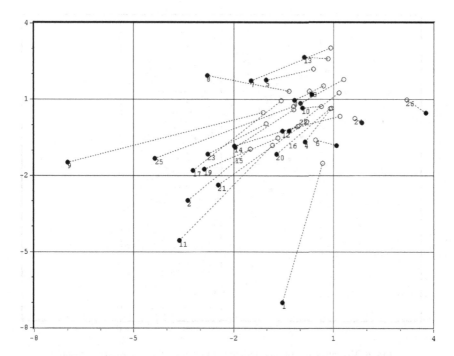

Fig. 4. Visualization of incomplete data set with PCA (missing values are replaced by means)

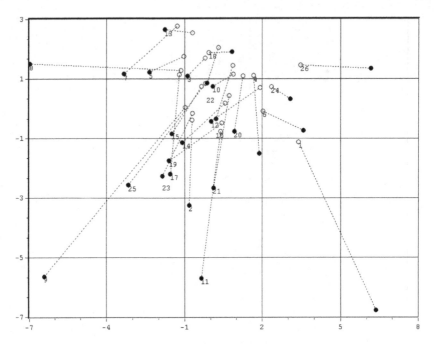

Fig. 5. Visualization of incomplete data set with PCA using EM algorithm

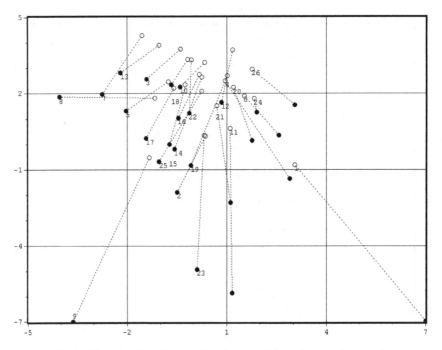

Fig. 6. Visualization of incomplete data set with the proposed approach

Visual analysis of the obtained visualizations reveals the following:

1. Replacing missing data with means leaded to a severe distortion of the data cloud shape and interrelations between data points. This could lead to wrong conclusions if this visualization was used for decision making or express diagnostics.
2. EM algorithm performed much better than the previous method. The overall shape is only slightly distorted, but interrelations between data points are still wrong in many cases and are closer to the result of the previous method than to the reference.
3. The proposed approach yielded the best visualization in terms of its closeness to the reference PCA visualization of the complete data set. The shape is preserved almost precisely and data points interrelations are only slightly distorted.

The second data set helps demonstrate the full power of the proposed approach. It contains results of 135 biomedical tests for 202 cases. The hard part is that 70% of data is missing because only a part of the total number of tests was performed for each particular case. Two visualizations were obtained: linear (fig. 7), using 135-2-135 AANN architecture; and nonlinear (fig. 8), using 135-25-2-25-135 AANN architecture. The nonlinear visualization offered a much better insight into the data, clearly showing two clusters, which are not visible in the linear visualization but are indeed present in the data.

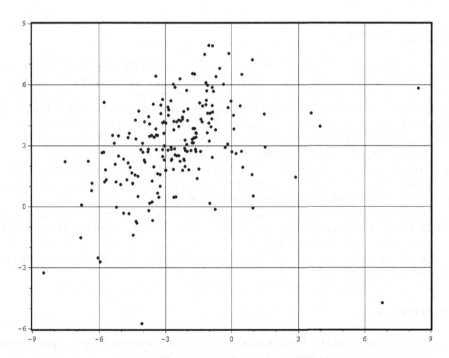

Fig. 7. Linear visualization of the second data set (70% of missing values)

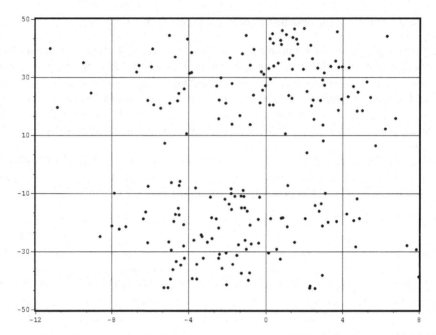

Fig. 8. Nonlinear visualization of the second data set (70% of missing values)

5 Conclusions

In this paper, we proposed modifications to AANN learning procedures that allow direct handling of data sets with missing values. One of its most important advantages for dimensionality reduction is the ability to preserve most of information from the present data while completely ignoring missing values that is very useful when a substantial portion of a data set is missing. This approach can be used for both linear and nonlinear dimensionality reduction and does not depend on the network architecture and learning algorithm, i.e. any supervised learning can be applied, any type of neural network can be used that performs weighted summation of inputs. The proposed approach can be easily generalized to other types of feedforward neural networks that are trained by supervised learning algorithms.

Comparison of experimental results have shown the superiority of the proposed method over other approaches to missing data handling, in particular, replacement of missing values by mean values, and the expectation maximization algorithm. Nonlinear visualization can offer additional insights that was demonstrated on the second incomplete data set.

References

1. Baldi, P. and Hornik, K.: Neural networks and principal component analysis: learning from examples without local minima, Neural Networks 2, (1989) 53-58.
2. Bourlard, H. and Kamp, Y.: Auto-association by multilayer perceptrons and singular value decomposition, Biological Cybernetics 59, (1988) 291-294.

3. Hastie, T. and Stuetzle, W.: Principal curves, Journal of the American Statistical Association 84, (1989) 502-516.
4. Jolliffe, I.T.: Principal component analysis, Springer, New-York. (1986)
5. Kegl, B., Krzyzak, A., Linder, T. and Zeger, K.: Learning and design of principal curves, IEEE Transactions on Pattern Analysis and Machine Intelligence, 22, (2000) 281-297.
6. Kramer, M.A.: Nonlinear principal component analysis using autoassociative neural networks, Journal of the American Institute of Chemical Engineers 37, (1991) 233-243.
7. Kruskal, J.B. and Wish, M.: Multidimensional Scaling, Sage Publications, Newbury Park. (1978)
8. Oja, E.: Data compression, feature extraction, and autoassociation in feedforward neural networks, in Kohonen, T., Makisara, M., Simula, O. and Kangas J. (eds.), Proceedings of the International Conference on Artificial Neural Networks, 1, (1991) 737-745.
9. Pearson, K.: On lines and planes of closest fit to systems of points in space, The London, Edinburgh and Dublin Philosophical Magazine and Journal of Sciences 6, (1901) 559-572.
10. Roweis, S.: EM algorithm for PCA and SPCA, Neural Information Processing Systems 10, (1997) 626-632.
11. Torgerson, W.S.: Multidimensional scaling: I. Theory and method, Psychometrika 17, (1952) 401-419.

A Review of Race Detection Mechanisms

Aoun Raza

Institute of Software Technology, University of Stuttgart,
Universitaetsstr. 38, 70569 Stuttgart, Germany
raza@informatik.uni-stuttgart.de

Abstract. This survey examines research in the area of race detection techniques. Diverse flavors of on-the-fly, ahead-of-time and post-mortem techniques are covered. This survey tries to present advantages and limitations exhibited by different race detection techniques.

1 Introduction

A race condition occurs when shared data is read and written by different processes without prior synchronization. To use shared resources correctly, parallel programming requires mechanisms for determining when a program is free from race conditions, and for assisting programmers in locating race conditions when they occur. Accurate locating requires that only those races that are direct manifestations of program bugs be reported, and not those that may be artifacts of other races or imprecise run-time traces. The common techniques for detection are based on dynamic or static analysis of programs. Compared to static analysis, dynamic analysis has the advantage that noise and undecideability is less of concern: by definition, an executed path is possible and at runtime all values can be determined. However, dynamic analysis has a cost in terms of time overhead and space, and does not consider unrealized paths as well. In contrast, static techniques can analyze programs regardless of their input, execution path and all flows through a program. There has been a lot of research carried out defining different techniques and building tools for race detection. As mentioned above, tools are based on either dynamic or static analysis. Some tools show better detection performance but high overhead in terms of time and space, other tools need user annotations and mechanisms to reduce the number of spuriously detected races. Tools based on static analysis sometimes lack accurate control- or data-flow information. Some tools also need precise points-to analysis as well.

2 Background

Since the detection of race condition in parallel programs is notoriously difficult, a large community has focused on this issue. In fact, it is quite difficult to detect such problems by manually testing the programs. Additionally, most of the existing concurrent software systems are written in C. Therefore, the need of an efficient mechanism for detecting parallel program anomalies is always present. As a consequence of detection

D. Grigoriev, J. Harrison, and E.A. Hirsch (Eds.): CSR 2006, LNCS 3967, pp. 534–543, 2006.
© Springer-Verlag Berlin Heidelberg 2006

difficulties the tools for automatic detection are extremely valuable. Hence, there has been a substantial amount of past work in building tools for analysis and detection of data races [1, 2, 3, 4]. These tools are either based on verification of access event ordering or verifying a lock discipline [22]. This means, if there is no unordered access to a shared variable such that at least one access is writing, the program is free from race conflicts. Similarly, if the accesses to shared variables in a program obey a locking discipline, then program is race free. In the traditional manner, the research can be categorized as on-the-fly, ahead-of-time, and post-mortem techniques. These techniques exhibit different strengths for race detection in programs. The ahead-of-time approaches encompass those detection techniques that apply static analysis and compile-time heuristics while on-the-fly approaches are dynamic in nature. The post-mortem techniques are a combination of static and dynamic techniques. The next sections provide an overview of techniques in each of these categories. Section 3 will discuss on-the-fly techniques and section 4 focuses on ahead-of-time techniques. Finally, section 5 presents the post-mortem based analysis techniques for race detection. At the end, we will summarize our study of different race detection techniques.

3 On-the-Fly Race Detection Techniques

On-the-fly race detection techniques are based on dynamic program analysis. Therefore, on-the-fly analyses operate at run time, visit only feasible paths, and have accurate views of the values of shared data and of other resource state. However, due to their dynamic nature, they impose a heavy computational overhead, making it time-consuming to run test cases and impossible on programs that have strict timing requirements. The term high overhead means that, while, in theory, on-the-fly tools can compute arbitrarily precise information, in practice they are limited to what can be computed efficiently both in time and space. Additionally, it is very difficult or even impossible to elicit race conditions by on-the-fly techniques, due to the non-determinism introduced by schedulers [20]. Furthermore, their reliance on invasive instrumentation typically precludes their use on low-level code such as OS kernels, device drivers and complex embedded systems. Finally, on-the-fly tools can find errors only on executed paths, which depend on input to the system. This not only makes dynamic analysis difficult but also sometimes impossible. Therefore, it is desirable to have a detection mechanism that can find races on a certain input with a single program execution, i.e., has the *Single Input, Single Execution* (SISE) property [21]. Nevertheless, the SISE property can be violated for programs that have internal non-determinism [22]. Thus, the complete test of such system is generally not possible. Unfortunately, the number of feasible paths can grow exponentially with the size of code [2]. This means that, in practice, testing can only exercise a tiny fraction of all feasible paths, leaving large systems with a residue of errors that could take weeks of execution to manifest. In some systems it is even worse, i.e., in a operating system some code might never run. The bulk of such code resides in device drivers, and only a small fraction of these drivers can be tested at a typical site, since there are usually a small number of installed devices.

One of the well known dynamic data race detectors is eraser [3], which uses binary rewriting techniques to monitor every shared-memory reference and to verify that consistent locking behavior is observed. The core idea of eraser is the "Lockset algorithm". For each shared variable, eraser maintains a set of candidate locks that protected it for all reads and writes as the program executes. Before any access to shared memory by a thread, eraser checks if it has obtained the required lockset. If the lockset becomes empty, it indicates that there is no lock to protect the variable and a warning will be given. To handle the situation when the above general discipline is violated but the situation is free from data races, such as variable initialization, shared read on data, single-writer, multiple-reader locks, the lockset algorithm is modified and extended. Eraser goes slightly beyond the work of Dinning and Schonberg [21], which is based on the traditional *happens-before* mechanism. The *happens-before* relation was originally defined by Lamport [6] and determines where conflicting memory accesses from different threads are separated by synchronization events. Unfortunately, *happens-before* is difficult to implement efficiently because it requires per-thread information about the concurrent access to shared-memory locations. Most importantly, the effectiveness of tools based on *happens-before* is highly dependent on the interleaving produced by the scheduler. Therefore, eraser's approach of enforcing a locking discipline is simpler and more thorough at catching races than the approach based on *happens-before*. Eraser is limited in that it can only process mutex synchronization operations. It fails when other synchronization primitives are built on top of it. Eraser takes an unmodified program binary as input and adds instrumentation to produce a new binary that is functionally identical, but includes calls to eraser. This makes the applications typically slow down by a factor of 10 to 30. Furthermore, the modifications can change the order in which threads are scheduled and can affect behavior of time sensitive applications.

Christoph and Gross's object race detection [22] greatly improves on eraser's performance by applying escape analysis to filter out non-data race statements. Object race detection extends eraser's ownership model and detects data races at the object level instead of at the level of each memory location. The design goal is to carry out expensive lock set operations for only those object which are shared. The overhead ranges from 16% to 129% which is obviously better than with eraser. However, their coarse granularity of data race detection leads to the reporting of many data races which are not true, i.e., the reported races do not indicate unordered concurrent access to shared state. Choi et al. present a novel approach to dynamic data race detection which is both more efficient and more precise than previously described object race detection with a runtime overhead in the range of 13% to 42% [4]. The key idea is the weaker-than relation, which is used to identify memory accesses that are provably redundant from the view point of data race detection. Another source of reduction in overhead is that this approach does not report all access pairs that participate in data races, but instead guarantees that at least one access is reported for each distinct memory location involved in a data race. This approach results in runtime overhead ranging from 13% to 42%, which is well below the runtime overhead of previous approaches with comparable precision. This performance improvement is the result of a unique combination of complementary static and dynamic optimisation techniques. The static analysis phase is used to compute a conservative set of statements that are identified as potentially participating in data races. Then these statements are used to detect on-the-fly data races. Several optimisations are applied to identify and discard redundant access events that do not contain new information. However, the weaker-than relation makes it difficult to handle the situation where the ownership state changes dynamically.

Another solution to locate race conditions dynamically in explicitly parallel message-passing programs is discussed in [9]. The authors classify races into two categories, namely non-artifact and possible artifact races that are caused by former ones. They argue that only races guaranteed to be non-artifact should be reported. Additionally, they argue that accurate detection using a pure on-the-fly algorithm requires space proportional to the length of the execution, an impractical requirement for long program runs. To address this problem, they present a two-pass on-the-fly algorithm that requires space independent of the execution length. The first pass is an approximation of an on-the-fly algorithm that determines whether any races occurred, but does not pinpoint their locations. More specifically, it locates the second message that races (first participator in a race) towards the racing receive (second participator) in each process, but does not locate the racing receive. The second pass is a re-execution on the program on the same input. It performs an accurate detection of non-artifact races. The authors claim that space usage does not grow with the length of the execution, and, even though the non-deterministic re-execution may differ from the original run, non-artifacts are still guaranteed to be detected. The hybrid 2-pass technique has been implemented under PVM 3.3.6 on an Ethernet-connected network of Sparc-Station 10s. Several message-passing programs were compiled with a library that performs the pass I (race detection) and pass II (race location). Experiment data show that about 50% of possible non-artifact races can only be labeled "Tangled," because the race checking system is unable to determine whether the races are non-artifact. In most cases the average slowdown was under 3%, which is an advantage. The other benefit of the algorithm is accurately reported non-artifact races. However, the instrumentation process of the mechanism requires re-compilation of the program. Furthermore, the 2-pass algorithm only locates the "first" non-artifact race in each process during each run. To locate subsequent non-artifact races, the "first" non-artifact race needs to be fixed and the program needs to be re-executed.

In the chain of on-the-fly techniques, Choi and Min present a mechanism called race frontier to debug data races in the execution of parallel programs [18]. The key idea is to identify a set of detected data races whose execution histories, including undetected race events preceding them, can be reproduced. The presented technique shows how to extend the mechanism from the case of two processes and a single shared variable to handle the general case of an arbitrary number of shared variables. Generally, dynamic race detection requires keeping the history of all accesses that have a potential for a data race, incurring potentially unbounded space and time overhead. A solution to this problem is to limit the number of entries in the access history of each shared variable and only report the latest entries involved in a data race. However, the reported data race is not guaranteed to occur when the program is re-executed. In their approach, Choi and Min ensure the reproduction of not only the detected data race but also of all the data races that were undetected because of the limited entries kept in the access history. The effect is the same as reproducing complete data race histories from the abridged data race history collected during program execution, allowing well-known methods for debugging sequential programs to be applied during re-execution. The race frontier is developed as an extension of *Partial Order Execution Graphs* (POEG) [19]. The PEOG is based on Lamport's *happens-before* relationship captured by imposing a partial memory access order on the operations (read or write) performed in the execution instance (event). However, the mechanism cannot locate the exact position of an unprotected (unreported) data race, which may be the cause of a following detected (reported) data race. Additionally, no experimental results have been reported.

4 Ahead-of-Time Race Detection Techniques

On the other end of the spectrum are static techniques. The static techniques have less precise local information; still they can provide significant advantages for large code bases. Unlike a dynamic approach, a static technique does not require code execution. As a consequence it is very advantageous for operating systems where a large part of code never executes. Static approaches immediately find races in obscure code paths that are difficult to reach with testing. The static techniques exhibiting linear nature can also do analysis impractical at runtime. Many static techniques are based on strong type checking. Boyapati and Rinard have introduced a new static type system for multithreaded programs to prevent both data races and deadlocks [11]. Their research is based on the premise that well-typed programs are guaranteed to be free of these kinds of errors. The proposed type system allows programmers to specify the locking discipline in their programs in the form of type declarations. The system also allows programmers to partition the locks into a fixed number of equivalence classes, and to use a recursive tree-based data structure to describe the partial order among the equivalence classes. Additionally, the system allows mutations to the data structure that change the partial order at runtime; the type checker statically verifies that the mutations do not introduce cycles in the partial order, and that the changing of the partial order does not lead to deadlocks. The system uses a variant of ownership types to prevent data races and deadlocks. Ownership types provide a statically enforceable way of specifying object encapsulation. Moreover, ownership types are useful for preventing data races and deadlocks because the lock that protects an object can also protect its encapsulated objects. The system has been implemented as a JVM-compatible prototype, which translates well-typed programs into byte codes that can run on regular JVMs. The implementation handles all features of the Java programming language. Besides the efficiency and effectiveness, the system has some limitations; it only supports Java programs. Furthermore, it requires proper type annotations, either inferred by the type systems or manually inserted by programmers in source code. To support such kinds of type systems, a specially designed compiler or mid-layer translator is needed.

Flanagan and Qadeer [10] presented an idea based on the observation that the absence of race conditions is neither necessary nor sufficient to ensure the absence of errors due to unexpected thread interactions. They propose a stronger non-interference property, namely the atomicity of code blocks, and they present a type system for specifying and verifying such *atomicity* properties. The type system allows statement blocks and functions to be annotated with the key word *atomic*. The type system guarantees that for any arbitrary interleaved program execution there is a corresponding execution with equivalent behavior in which the instructions of each atomic block executed by a thread are not interleaved with instructions from other threads. This property allows programmers to reason about the behavior of well-typed programs at a higher level of abstraction, where each atomic block is executed "in one step," thus significantly simplifying both formal and informal reasoning. The idea is formalized in terms of CAT, a small, imperative, multithreaded language with high-order functions. The proof of the correctness of the type system is based on the reduction theorem of Cohen and Lamport. The benefits of the type system are illustrated with an application to the *java.util.Vector* library class. The advantage of this mechanism is that by reducing atomic code blocks into "single step" the work of reasoning about the interactions between threads can be greatly reduced. However, the static type system needs to be used together with other race detection tools, and cannot guarantee the absence of synchronization errors such as race conditions by itself. Additionally, it requires the use of a special programming language, or the modification of current language tools to implement such a type system.

Another static analysis system for detecting race conditions in Java programs is discussed in [1]. The analysis supports the lock-based synchronization discipline by tracking the protecting lock for each shared field in the program and verifies that the appropriate lock is held whenever a shared field is accessed. The reasoning and performed checks are expressed as an extension of a race-free Java type system. The extended type system is capable of capturing many common synchronization patterns, which includes classes with internal synchronization, classes that require client-side synchronization, and thread-local classes. Mechanisms are provided for escaping the type system in places where it proves to be too restrictive, or where a particular race condition is considered benign. The implementation is done as a race condition checker, called *rccjava*. It has been tested on a variety of java programs totaling over 40,000 lines of code. The additional type of information required by *rccjava* is embedded in Java comments to preserve compatibility with existing Java tools. *rccjava* relies on the programmer to manually insert annotations into source code, which incurs a burden of about 20 additional type annotations per 1000 lines of code. A number of races have been found in the standard Java libraries and other tested programs. This system presents an effective model of static race analysis, which has been integrated with other techniques such as race detection in large programs [7]. The system is a better candidate for real-time race detection than using event ordering due to its efficiency in time and space. However, it requires the programmer to manually insert annotations. Additionally, it needs to access the source codes of tested programs. There is no time and space performance data available.

An extended *rccjava* approach is defined by Flanagen and Freud in [7]. They have improved *rccjava* to be used on large and realistic programs, including an annotation inference system and a user interface to help programmers understand warnings generated by the tool. To achieve practical analysis of large programs, an annotation assistant, called Houdini/rcc based on the Houdini annotation inference architecture [8], was developed to automatically insert annotations into analyzed programs. Additionally a number of techniques were added to reduce false alarms caused by the automatic annotations. The improved *rccjava* provides meaningful information about potential races to programmers through a simple interface. Furthermore, it can cluster race conditions together according to their probable cause so that related race conditions can be dealt with as a single unit. Still, *rccjava* is limited mainly to support lock-based synchronization operations. Additional annotation rules need to be added to deal with other types of synchronization idioms. The tool needs to access the source code of Java programs. Furthermore, some techniques used to reduce false alarms generated by the annotation interference system are unsound due to the employed approximations.

Other well known static race detection approaches are the Warlock tool [14] for finding races in C programs, the Extended Static Checking [15] (ESC) and ESC/Java [16] tools for Modula-3 and Java respectively, which use theorem proving to find errors. Burrows and Leino [17] have since extended ESC/Java to check for stronger properties than only reporting unprotected variables accesses. Unfortunately, because of lack of precision at compile time, both Warlock and ESC make heavy use of annotations to inject knowledge into the analysis and to reduce the number of false positives. Anecdotally this caused problems when applying Warlock to large code bases; sophisticated code requires many annotations just to suppress spurious errors.

Engler and Ashcraft define another static tool, RacerX, based on the lockset approach [3]. RacerX uses flow sensitive, inter-procedural analysis to detect both race conditions and deadlocks. It aggressively infers checking information such as which locks protect which operations, which code contexts are multithreaded, and which shared accesses are dangerous. It tracks a set of code features which it uses to sort errors from most to least severe. At a high level, checking a system with RacerX involves targeting to system-specific locking functions and extracting a control-flow graph from the checked system, which is used for further analysis to find races and deadlocks. RacerX uses novel techniques such as multithreading inference and belief analysis [23] to counter the impact of analysis mistakes. The tool requires 2-14 minutes to analyze a 1.8 million line system. The tool has been applied to Linux, FreeBSD and a large commercial code base and has found serious errors in all of them. Nevertheless, the tool has several limitations. Firstly, it lacks good pointer analysis; secondly, it does only simple function pointer resolution. Finally, [3] notes speed limitations for the analysis of OS code with a propensity for functions invoking huge code parts, each time with different locks.

5   Post-mortem Race Detection Techniques

Post-mortem techniques [5] analyze log or trace data after the program has executed in a manner similar to dynamic techniques. The post-mortem techniques are the combination of static and dynamic techniques. While post-mortem analyses can affect performance less than dynamic analyses, they suffer from the same limitation as dynamic techniques in that they can only find errors along executed paths. The solutions based on the post-mortem techniques collect information at compile time, and then analyze the re-execution of the program based on the collected information. A system which uses post-mortem techniques for cyclic debugging of non-deterministic parallel programs was presented by Ronse and Bosschere [12]. The system, RecPlay, traces a program execution and stores the information in trace files; then this information is used to guide a faithful re-execution. It runs a race detector as watchdog during replay without changing the behavior of the execution. RecPlay can only correctly replay programs that are free of data races. Once the data race occurs, the replayed execution stops and the user is notified. After that, there is no guarantee for a correct re-execution. To detect the data race, during recording phase the tool only records the synchronization operations by storing the timestamp increments in each thread. During race detection time (replay time), RecPlay traces data access operations by collecting memory reference information; then it detects conflicting memory references in concurrent segments by using a logical vector clock and clock snooping. Finally, it identifies the instructions that cause the data race. The data race detection is based on the *happens-before* relation. Several programs running on Solaris have been tested by RecPlay. The RecPlay is completely independent of any compiler or programming language, and it does not require recompilation or re-linking. However, as it is based on the *happens-before* relation, it can only detect data races that appear in a particular execution. Additionally, the average overhead for replay is 91% and automatic race detection slows down the program execution about 36 times. Further, the solution only runs on Solaris.

Another record replay tool for Java, Déjà Vu, which provides deterministic replay of a program's execution, was presented by Choi and Srinivasan [13]. The mechanism introduces the concept of logical thread schedule, which is a sequence of intervals of critical events wherein each interval corresponds to the critical and non-critical events executing consecutively in a specific thread. Déjà Vu is independent of the underlying thread scheduler (either an operating system or a user-level thread scheduler). It records all critical events, i.e., all synchronization events and the shared variable accesses by capturing logical thread schedule intervals. To identify the schedule intervals, one single global clock and a local clock for each thread are used. All critical events are traced by updating the global clock and assigning the global lock value to the local clock. When the thread is scheduled out, the global clock continues to tick and the local clock pauses. At the start of a replay, Déjà Vu reads the thread schedule information from a file created at the end of the recording. When a thread is created and starts its execution, it receives an ordered list of its logical thread schedule intervals. When a critical event execution is reached, it will wait until the global clock value becomes the same as the local value (read from record file). After execution, it will update the global clock. To implement the record/replay mechanism, Sun Microsystems' Java Virtual Machine has been modified. Several programs such as Chaos and MTD have been tested by using Déjà Vu. The observed execution time overhead is from 17% to 87%. Its implementation by modifying JVM instead of the operating system makes it a portable tool for Java applications across different platforms. The techniques of handling Java synchronization operations can be extended to general multithreaded programming systems with similar synchronization primitives. However, Déjà Vu only can deterministically replay the non-deterministic execution behavior due to thread and related concurrent constructs such as synchronization primitives. Window events, input/outputs, and system calls have not been taken care of. Actually, this is a common dilemma existing in record/replay systems: replaying a faithful execution requires recording as many non-deterministic events as possible; but on the other hand, recording all kinds of non-deterministic events is an extremely challenging task, sometimes may be infeasible and may incur intolerable overhead during the recording phase.

6 Summary

We have discussed several race detection techniques having the flavor of on-the-fly, ahead-of-time and post-mortem approaches. We notice that the majority of existing research use either static or dynamic approaches underneath. Further, some techniques have been developed by hybridization of static and dynamic approaches. This study described advantages and limitation of different techniques. The techniques which are dynamic in nature have the advantage of visiting only feasible paths and have accurate views of interactions. However, they impose high overhead in terms of time and space. Additionally, on-the-fly mechanisms face difficulties in race detection if the program contains internal non-determinism. In contrast, static techniques have no limitation in terms of time and space. They can provide significant advantages for large code bases. Static techniques can also analyze those program parts which might never execute. The only constraint could be the lack of precise information. Therefore, static techniques require algorithms to reduce the reported false-positives [2]. The hybrid techniques combine the best part of static and dynamic techniques, but still lack performance due to large program traces. Lastly, the post-mortem techniques can only find errors along executed paths. In conclusion, we expect to see significant development in the area of race detection as programs are increasingly based on multithreaded design.

Acknowledgements

I would like to thank Prof. Dr. Erhard Ploedereder for his revisions and insightful comments on this paper.

References

[1] C. Flanagan, S.N. Freund. Type-Based Race Detection for Java. In Proceedings of the ACM SIGPLAN 2000 Conference on Programming Language Design and Implementation, 2000, Vancouver, British Columbia, Canada, pages 219-232.

[2] D. Engler and K. Ashcraft. RacerX: Effective, Static Detection of Race Conditions and Deadlocks. In Proceedings of the 19th ACM Symposium on Operating Systems Principles, October 2003, NY USA.

[3] S. Savage, M. Burrows, G. Nelson, P. Sobalvarro, and T. Anderson. Eraser: A Dynamic Data Race Detector for Multi-Threaded Programs. ACM Transactions on Computer Systems, 1997.

[4] J. D. Choi, K. Lee, A. Loginov, R. O'Callahan, V. Sarkar, and M. Sridharan. Efficient and precise datarace detection for multithreaded object-oriented programs. In Proceedings of the ACM SIGPLAN 2002 Conference on Programming Language Design and Implementation (PLDI'02), pages 258–269, June 2002.

[5] D.P. Helmbold and C.E. McDowell. A taxonomy of race detection algorithms. Technical Report UCSC-CRL-94-35, 1994.

[6] L. Lamport. Time clocks and ordering of events in a distributed system. Communications of the ACM (CACM), Vol.21, No.7, pp:558-565, 1978.

[7] C. Flanagan and S. Freund. Detecting race conditions in large programs. In Workshop on Program Analysis for Software Tools and Engineering (PASTE 2001), pages 90–96, June 2001.

[8] C. Flanagan, R. Joshi, and K. R. M. Leino. Annotation inference for modular checkers. Information Processing Letters Volume 77, Issue 2-4 (February 2001) Pages: 97 - 108 Year of Publication: 2001.

[9] R.H.B. Netzer, T.W. Brennan, S.K Damodaran-Kamal. Debugging race conditions in message-passing programs, Proceeding of the SIGMETRICS Symposium on Parallel and Distributed Tools, January 1996.

[10] C. Flanagan and S. Qadeer. Types for atomicity. In Proceedings of the 2003 ACM SIGPLAN international Workshop on Types in Languages Design and Implementation, Jan 2003, pages 1-12.

[11] C. Boyapati and M. Rinard. A parameterised type system for race-free java programs. In ACM Conference on Object-Oriented Programming Systems Languages and Applications, 2001, pages 56-69.

[12] M. Ronse and K. De Bosschere. RecPlay : A Fully Integrated Practical Record/Replay System. In ACM Transactions on Computer Systems, Vol.17, No.2, May 1999, pages 133-132.

[13] J. D. Choi and H. Srinivasan. Deterministic Replay of Java Multithreaded Applications. In Proceedings of SIGMETRICS Symposium on Parallel and Distributed Tools (SPDT-98), pages 48-49, August 1998.

[14] N. Sterling. Warlock: A static data race analysis tool. In Proceedings of the 1993 USENIX Winter Technical Conference, pages 97-106, 1993.

[15] D. Detlefs, K. R. M. Leino, G. Nelson, and J. Saxe. Extended static checking, COMPAQ SRC Research Report 159, Dec 1998.

[16] K. R. M. Leino, G. Nelson, J. B. Saxe. ESC/Java user's manual. Technical note 2000-002, Compaq Systems Research Center, Oct. 2000.

[17] M. Burrows and K. R. M. Leino. Finding stale-value errors in concurrent programs. Technical Report SRC-TN-2002-004, Compaq Systems Research Center, May 2002.

[18] J.D. Choi and S. L. Min. Race Frontier: reproducing data races in parallel programs debugging. In Proceedings of the third ACM SIGPLAN Symposium on Principles & practice of parallel programming, April 1991, pages 145-154

[19] A. Dinning and E. Schonberg. An empirical comparison of monitoring algorithms for access anomaly detection. In Proceedings of the second ACM SIGPLAN Symposium on Principles & Practice of Parallel Programming, pp.1-10, March 14-16, 1990, Seattle, Washington, United States.

[20] C. von Praun, T. Gross. Static conflict analysis for multi-threaded object-oriented programs. In Proceedings of the ACM Conference on Programming Language Design and Implementation, pages 115-128, 2003.

[21] A. Dinning, E. Schonberg. Detecting access anomalies in programs with critical sections. In Proceedings of the 1991 ACM/ONR Workshop on Parallel and Distributed Debugging, December 1991, Pages 85-96.

[22] C. von Praun and Thomas Gross. Object-Race Detection, Conference on Object-Oriented Programming, Systems, Languages, and Applications (OOPSLA), October 2001.

[23] D. Engler, D. Chen, S. Hallem, A. Chou, and B. Chelf. Bugs as deviant behavior: A general approach to inferring errors in systems code. In Proceedings of the Eighteenth ACM Symposium on Operating Systems Principles, 2001.

[24] P. Emrath and D. Padua. Automatic detection of non-determinacy in parallel programs. In Proceedings of the ACM Workshop on Parallel and Distributed Debugging, pages 89-99, Madison, Wisconsin, Jan. 1989.

Fuzzy-Q Knowledge Sharing Techniques with Expertness Measures: Comparison and Analysis

Panrasee Ritthipravat[1], Thavida Maneewarn[1],
Jeremy Wyatt[2], and Djitt Laowattana[1]

[1] Institute of FIeld roBOtics (FIBO), King Mongkut's University of Technology,
Thonburi, 91 Suksawasd 48, Bangmod, Bangkok, Thailand
{pan, praew, djitt}@fibo.kmutt.ac.th
[2] School of Computer Science, University of Birmingham,
Birmingham, B15 2TT, United Kingdom
jlw@cs.bham.ac.uk

Abstract. Four knowledge sharing techniques based on fuzzy-Q learning are investigated in this paper. These knowledge sharing techniques are 'Shared Memory', 'Adaptive Weighted Strategy Sharing', 'Exploration Guided Method', and 'Greatest Mass Method'. Different robot expertness measures are applied to these knowledge sharing techniques in order to improve learning performance. We proposed a new robot expertness measure based on regret evaluation. The regret takes uncertainty bounds of two best actions, i.e. greedy action and the second best action, into account. Simulations were performed to compare the effectiveness of the three expertness measures i.e. expertness based on accumulated rewards, on average move and on regret measure, when applied to different sharing techniques. Our proposed measure resulted in better performance than the other expertness measures. Analysis and comparison of different knowledge sharing techniques are also provided herein.

1 Introduction

Reinforcement learning notoriously requires a long learning period, particularly when applied with a complicated task. Additionally, it is difficult for a robot to explore huge state and action spaces in a short time. To alleviate these problems, multiple mobile robots have been served to learn a task by exploring different part of state and action spaces simultaneously. During the learning period, they may share the knowledge they have learnt. Unfortunately, most reinforcement learning techniques require auxiliary methods to integrate external knowledge sources into the robot's knowledge. In general, knowledge gained from one robot could be different from that of the others, even if the robots have the same mechanism and learn the same task. This happens because the robots have different experiences and properties. Therefore, knowledge sharing among robots is one of the most challenging topics in robotic research.

Knowledge sharing among reinforcement learning robots has been extensively studied in order to utilize and gain benefit from multiple knowledge sources. The

D. Grigoriev, J. Harrison, and E.A. Hirsch (Eds.): CSR 2006, LNCS 3967, pp. 544–554, 2006.

robot may use external knowledge sources gained from other robots or human to improve the robot's learning performances. Moreover it can use previously learned knowledge for speeding up learning in the new task. Previous research of knowledge sharing can be classified into two groups: direct and indirect methods.

1. Direct Method

The Direct method focuses on directly integrating all available sources of shared knowledge into robot's knowledge. Various techniques were proposed. They are the 'Policy Averaging' [1] which all policies are averaged into the new knowledge. The 'Weighted Strategy Sharing: WSS' [2], [3], [4] is the method in which weights were assigned to all knowledge sources according to the robot expertise or compatibility of agent state spaces [5] and then summed into the new knowledge. The 'Same-policy' [1] is the method in which all agents used and updated the same policy.

2. Indirect Method

In the indirect method, external knowledge sources will be used to guide robot's decision making. These external knowledge will not be integrated into robot's learning directly. Most works used shared knowledge to bias action selection. The robot selects an action according to the resulting probabilities. Techniques in this group are Skill Advice Guided Exploration (SAGE)' [6], [7], 'Supervised Reinforcement Learning (SRL)' [8] etc.

To study knowledge sharing among real robots, the following problems are taken into consideration. They are continuous state and action spaces, generalization, learning time constraints, uncertainty and the imprecision of sensing and actuation in real robots [9]. A fuzzy Q-learning method is a promising technique due to its abilities to deal with most problems and also nonlinearities and unknown dynamics of the system. Therefore, sharing knowledge among fuzzy Q-learning robots is our interest. Shared knowledge becomes state-action values in each fuzzy rule. Additionally, robot expertness measures are investigated and used to improve learning performance of several knowledge sharing techniques. In the next section, knowledge sharing algorithms studied in this paper will be presented.

2　Knowledge Sharing Techniques

Knowledge sharing techniques investigated in this paper can be summarized as follows.

2.1　Shared Memory: SM

This technique is inspired from the 'Same-policy' technique [1]. After interaction with an environment, the robots use and update the same set of state-action values. Since all robots have the same brain, each individual robot's experiences directly affects the overall robots' decision making. For learning a task with n robots, action values will be updated n times in each iteration. Learning should be faster than individual learning since n robots explore various states simultaneously.

2.2   Adaptive Weighted Strategy Sharing: Adaptive WSS

This strategy is developed from the Weighted Strategy Sharing (WSS) method proposed by Ahmadabadi's team [2], [3], [4]. In the WSS method, learning is composed of two phases: individual learning and cooperative learning phases. For the individual learning phase, all learners will learn a task separately. At a certain period, named cooperative time, learning will be switched to the cooperative learning phase which allows the robots to share the learned state-action values, $Q(s, a)$. In this phase, the action values of all robots will be weighted and summed as the new knowledge for every robot as shown in Eq. 1

$$Q^{new}(s, a) = \sum_{m=1}^{n} W_m Q^m(s, a),$$ (1)

where $Q^{new}(s, a)$ is a new set of state-action values initialized for all n sharing robots. Superscript m indicates the mth robot's. W_m is weight calculated from robots' expertise as presented in Eq. 2

$$W_m = \frac{expertness_m}{\sum_{p=1}^{n} expertness_p}, \qquad (m = 1, \ldots, n)$$ (2)

where $expertness_m$ is the mth robot expertness value. Therefore, at the end of cooperative learning phase, all robots have homogeneous set of state-action values. The individual learning phase will then continue thereafter. These phases will switch back and forth at every cooperative time. The cooperative time is set at every predefined end of learning trials. In [10], we have shown that the WSS method does not support asynchronous knowledge sharing among robots. The adaptive version was proposed to solve such problem and it showed that learning performance could be improved if weights were properly assigned. For the Adaptive WSS, each robot learns a task independently and it is able to make a decision whether to share knowledge with the other $n - 1$ robots by itself. The robot is presumed to perceive all the other robots' knowledge and their expertness values at any time t. At the end of robot learning trial, the robot will assign weights to all sources of shared knowledge as computed from Eq. 2. Difference between the robot's weight, W_i, and that of the other robots, W_j where $\underset{j \neq i}{j \in} n - 1$ will be employed to determine probability of sharing as shown in Fig. 1.

In Fig. 1, two thresholds Th_1 and Th_2 will be set. In this paper, they are 0.1 and 0.5 respectively. If the difference is less than Th_1, sharing will not be occurred. In contrary, if the difference is higher than Th_2, sharing will be arisen with probability 1. Doing in this manner, each robot is able to determine which robot should the knowledge be obtained from. Once the sharing robots have been

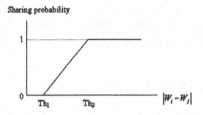

Fig. 1. Sharing probability

determined, a new knowledge can be obtained from Eq. 1. After the sharing is performed, all sharing robots will have the same level of expertise. Therefore the new expertness value for all sharing robots can be computed from an average of the sharing robot expertness values. The robot will update its knowledge and its expertness value immediately while the other sharing robots which are learning a task will keep their new knowledge and their expertness value in their memories. Once a robot's learning trial is finished, they will be employed for the next learning trial.

2.3   Exploration Guided Method: EGM

From the above techniques, other robots' experiences directly affect the robot's state-action values. What the robot has learned may be drastically changed by the other robots' knowledge. This may cause loss of useful information which has been learned from the robots' past experiences [11], [12]. One of the powerful techniques, which allows integrating external knowledge sources into a reinforcement learning agent, is the use of the SAGE framework [6], [7]. In this methodology, action selection probabilities of advice policies and that of the robot were weighted and summed. The robot selected an action according to a resulting probability mass function (PMF). In their papers, weights were determined from two indices, i.e. exploration cost and impatience measurement. Unfortunately, there was no clarification on how to set suitable parameters in each task. In this paper, an algorithm named Exploration Guided Method (EGM) developed from the SAGE framework will be presented. At a given state s, PMF over actions of the mth knowledge source, $Pr_m(s, a)$, will be generated for all n available knowledge sources. These action selection PMFs will be weighted and summed as shown below

$$f(s, a) = \sum_{m=1}^{n} W_m Pr_m(s, a),$$ (3)

where W_m is weight for the mth knowledge source which can be calculated from Eq. 2. The resulting PMF $f(s, a)$ will then be mapped into action selection PMF, $Pr(s, a)$ as follows:

$$Pr(s, a) = \frac{f(s, a)}{\sum_{a' \in A} f(s, a')}.$$ (4)

The robot will select an action according to the action selection PMF as presented in Eq. 4. In this paper, the action selection probabilities in each fuzzy rule will be formed. For each rule, the local action will then be selected by the use of ϵ-greedy method.

2.4 Greatest Mass Method: GMM

The greatest mass method was presented by [13]. The original work studied the learning algorithm based on a modular architecture. Each module learned different subtasks. At a given state s, an arbitrator would select an action, a, which maximized a summation of state-action values, being estimated from all n modules as:

$$a = \arg \max_{a' \in A}\Big(Q_1(s,a') + Q_2(s,a') + \cdots + Q_n(s,a')\Big). \tag{5}$$

In Eq. 5, the selected action is the one that gets a common agreement from all modules to receive the highest return after taking such action. To apply the greatest mass method to our study, each module represents individual source of shared knowledge. The robots are forced to learn the same task. At a certain state s, a chosen action is the one which maximizes the resulting PMF calculated from Eq. 3. Therefore, the taken action is selected from

$$a = \arg \max_{a' \in A}\Big(\sum_{m=1}^{n} W_m Pr_m(s,a')\Big). \tag{6}$$

3   Measure of Expertness

From the techniques presented above, weights play an important role in knowledge sharing among robots. Weight will be used not only to determine whether the knowledge from the sources should be used but also how much the knowledge from the source should contribute to the new knowledge. The weight can be determined from the robot expertness measure which indicates the performance of its current policy. Two approaches of expertness measures previously proposed by Ahmadabadi's team were the measure based on accumulated rewards and on average move. For the first approach, two values are examined. They are the Normal (Nrm) and the Gradient (G). The Nrm takes accumulated rewards received since learning begins into account. Therefore, it can be biased from long history of experiences. The G is used to alleviate this problem. It takes accumulated rewards since each individual learning phase has begun. However, both the Nrm and the G suffer when they have negative value. The higher negative value could have two possible meanings, either it has sufficiently learned to indicate which actions should not be executed or it is exploring improper actions. In the second case, the use of these measures can degrade the robot learning performance. In this paper, the G value will be only applied with the technique that the cooperative time is explicitly determined, i.e. the Adaptive WSS. The Nrm value will be employed with the EGM and GMM based techniques.

Another expertness measure is the Average Move (AM). The AM takes an average number of moves that the robot executed before achieving the goal into consideration. The lower number of moves that the robot has done, the higher expertness value is. However, when the robot randomly explores an environment, the AM cannot be used to represent the robot expertness measure.

In this paper, we proposed a new measure of expertise based on regret evaluation. The regret measure is formed from the uncertainty bounds of the two best actions, i.e. the greedy action and the second best action. Bounds of both actions will be compared. If the lower limit of the bound of the greedy action is higher than the upper limit of the bound of the second best action, it is more likely that the greedy action is the best action. The regret measure given state s at time $t + 1$ is calculated from

$$regret(s_{t+1}) = -(lb(Q(s_{t+1}, a_1)) - ub(Q(s_{t+1}, a_2))), \tag{7}$$

where $lb(Q(s_{t+1}, a_1))$ is the lower limit of estimated state-action value given state s at time $t + 1$ of the greedy action. $ub(Q(s_{t+1}, a_2))$ is the upper limit of approximated state-action value given state s at time $t + 1$ of the second best action. They are approximated from past state-action values sampled from time $t - k + 1$ to t as: $\{Q_T(s_T, a)\}_{T=t-k+1}^{t} = \{Q_{t-k+1}(s_{t-k+1}, a), \ldots, Q_t(s_t, a)\}$ where k is a number of samples. Normal distribution in each state-action value is assumed. Mapping the regret measure into the expertise value given state s at time t can be defined as follows:

$$expertness_m(s_t) = 1 - \frac{1}{1 + exp(-b * regret(s_t))}. \tag{8}$$

From Eq. 8, the regret value is mapped into a flipped sigmoid function ranges between [0 1]. b is slope of the mapped function. The large negative regret measure causes the expertness value to approach one.

4   Simulation

Thirteen techniques were used to test the performance of our proposed regret measure within the scope of two problems: knowledge sharing among robots that learn a task from scratch and that relearn the transferred knowledge. Each robot learns its task by the use of Fuzzy Q-learning technique following unmodified version presented in [9]. The tested techniques are 1.)'SP'; Separate learning or learning without sharing knowledge. 2.) 'SM': Shared Memory. 3.) 'AdpWSS': Adaptive WSS with G expertness measure 4.) 'AdpWSSAM': Adaptive WSS with AM measure 5.) 'AdpWSSR': Adaptive WSS with regret evaluation 6.) 'EGM': Exploration Guided Method without weight assignment 7.) 'EGMN': EGM with Nrm measure 8.) 'EGMAM': EGM + AM 9.) 'EGMR': EGM + regret 10.) 'GMM': Greatest Mass Method without weight assignment 11.) 'GMMN': GMM + Nrm 12.) 'GMMAM': GMM + AM 13.) 'GMMR': GMM + regret.

For the first problem, an intelligent goal capturing behaviour is simulated. The goal will avoid being captured once it realizes that the distance between the goal and the robot is smaller than 30 cm and the orientation of the goal w.r.t the robot is in between -45° and 45°. The goal will run in a perpendicular direction to the robot's heading direction. 1000 trials are tested for a single run. Learning rate is varied from 0.1 to 0.3 for the first nine techniques and from 0.01 to 0.2 for the left ones. The best results are selected and compared.

For the second problem, two robots relearn knowledge gained from static obstacle avoidance to dynamic obstacle avoidance behaviour. Two robots learn to approach a goal which has two opponents move in opposite direction. Learning rate is varied from 0.05 to 0.5 for all techniques. The best results are selected and compared. The discount factor for both problems is 0.9. Parameters were tuned for each algorithm by hand as presented in [10]. The accumulated reward averaged over trial and collision rate in each run are recorded. Their average value over 50 runs are used to compare the learning performance as in the previous problem. From the simulation results, accumulated rewards and collision rates of these behaviours are summarized in Tables 1 and 2. Robots' path and accumulated rewards averaged over trial of the first and the second problems are shown in Figs. 2 and 3 respectively.

In Fig. 2(a), two robots learn to capture their goals in separated environment. The robots are presented by the bigger circles with lines indicated their heading direction. In Fig. 3(a), two robots learn to move to the goal while avoiding collision with dynamic obstacles in the same environment. Two opponents move in opposite direction presented by arrow lines as to obstruct the robots. The robots have to avoid both opponent robots and their teammate.

The performance of knowledge sharing is indicated by the accumulated rewards as presented in Figs. 2(b) and 3(b). In these figures, the best learning rate is presented after the algorithm's name. For example, SP:0.05 indicates the separate learning with the best learning rate is set at 0.05. The simulation results showed that, the AdpWSSR gave the best performance for the first problem. In the second problem, the EGMR is the best one. As seen in Figs. 2(b) and 3(b), the SM was better than the AdpWSS and the AdpWSSAM. However the SM converged to suboptimal accumulated rewards in both problems. For the

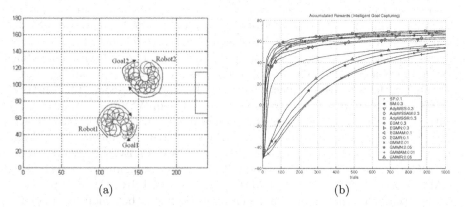

Fig. 2. Intelligent goal capturing (a) robots' path (b) accumulated rewards

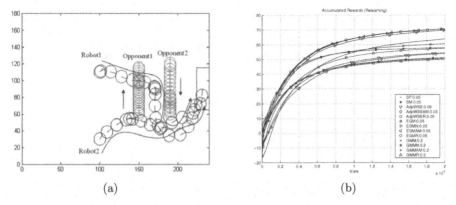

Fig. 3. Dynamic obstacle avoidance (a) robots' path (b) accumulated rewards

GMM based methods, the GMMR gave the best performances compared to the other GMM based techniques in the first problem. In the second problem, the GMM outperformed the other GMM based techniques. Additionally, from the simulation results, the separate learning outperformed various knowledge sharing techniques in the second problem once learning rate is set as small value. In the simulations, the best learning rate for the SP is 0.05. Small learning rate implies that the state-action values were gradually changed. Therefore, knowledge gained from static obstacle avoidance behaviour was good and it required slow adjustment to form the dynamic obstacle avoidance behaviour. It is worth noting that after knowledge sharing was begun, the average of robots' learning performance of the AdpWSS based algorithm was slightly decreased. This was not surprised since the robot's knowledge was averaged with the others' shared knowledge.

5 Discussion

As seen in the simulation results, our proposed measure gave the best results compared to the other measures when applied to the majority of knowledge sharing algorithms. Each knowledge sharing technique has different advantages and disadvantages which can be summarized as follows:

For the SM technique, n robots have the same brain. They use and update the same memory. The learning is fast since action values will be updated n times in each iteration. Additionally, it is simple for implementation and it requires less storage memory compared to the other investigated knowledge sharing techniques. However, its problem arises when learning is encountering certain local minimum. It is difficult for the robots to get out from the situation. This happens because the robots use the same decision making policy. It is difficult to achieve different solutions from the group's judgment. Moreover, it is difficult to integrate other forms of shared knowledge source into the robots' knowledge.

In the Adaptive WSS based algorithm, it uses weighted average of each knowledge source. To see advantages of this algorithm, simple examples are given as

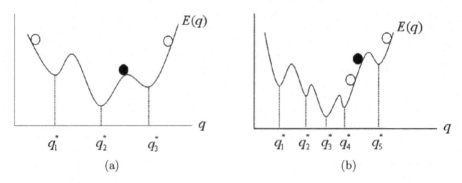

Fig. 4. Two simple examples of the Adaptive WSS mechanism

shown in Figs. 4(a) and 4(b). In those figures, q is a parameter vector used for state-action value approximation. $E(q)$ represents approximation error. q_k^* is local optimum estimated value at the k optimum point. Learning is performed in the manner of minimizing error. Two robots' current knowledge are presented by two white circles. Once knowledge is shared, the new knowledge will be generated and can be shown by the black circle.

In Fig. 4(a), there exist three optimum points. Once sharing is arisen, the new knowledge moved out of the local optimum points to the global one. In Fig. 4(b), the new knowledge can also move out of the local minimum point corresponding to the q_5^* to the next lower point. In those examples, we can see that the Adaptive WSS and the WSS can improve learning performance by changing the minimum points or disturbing the robots' knowledge to move out of the local optimums. However, all sources of shared knowledge must have the same state-action representation.

For the EGM based method, it does not integrate the shared knowledge into the robot's knowledge directly. This allows simple integrating external knowledge sources which may have different representation. However, there are various ways

Table 1. Simulation results (Intelligent Goal Capturing)

Table 1.1 General methods

Methods	Accum. Rewards.
SP	61.6995
SM	**66.1877**

Table 1.2 The AdpWSS based methods

Methods	Accum. Rewards.
AdpWSS	62.4671
AdpWSSAM	63.9752
AdpWSSR	**70.3805**

Table 1.3 The EGM based methods

Methods	Accum. Rewards.
EGM	67.8646
EGMN	66.5634
EGMAM	69.0422
EGMR	**69.5178**

Table 1.4 The GMM based methods

Methods	Accum. Rewards.
GMM	53.8878
GMMN	53.5770
GMMAM	51.2447
GMMR	**57.2711**

Table 2. Simulation results (Relearning)

Table 2.1 General methods

Methods	Accum. Rewards.	Coll. rate
SP	**63.9507**	**0.0956**
SM	57.9447	0.0985

Table 2.2 The AdpWSS based methods

Methods	Accum. Rewards.	Coll. rate
AdpWSS	54.3682	0.1022
AdpWSSAM	53.9923	0.1032
AdpWSSR	**57.5093**	**0.0913**

Table 2.3 The EGM based methods

Methods	Accum. Rewards.	Coll. rate
EGM	70.8249	**0.0940**
EGMN	69.8329	0.0947
EGMAM	70.6941	0.0942
EGMR	**70.8393**	0.0945

Table 2.4 The GMM based methods

Methods	Accum. Rewards.	Coll. rate
GMM	**60.7095**	**0.0428**
GMMN	51.1320	0.0447
GMMAM	50.6641	0.0443
GMMR	50.0231	0.0446

of generating probability mass function. One may suffer from the difficulty of temperature setting if the Boltzmann distribution is used.

In the last technique, the GMM was slightly better than the separate learning. This happens because there is no exploration in this algorithm. The action is selected from the maximum of resulting PMFs. No exploration causes the robots to get stuck by taking suboptimal actions. Though the low collision rate was observed, the simulation showed that the robots rarely moved through the obstacles to the goal. They decided to move out of the field instead for avoiding the collision.

6 Conclusion

In this paper, various knowledge sharing algorithms on fuzzy-Q learning architecture were studied. A new expertness measure based on regret evaluation was proposed. Simulation results showed that our proposed measure applying with the investigated algorithms gave the best performances comparing to the use of other previously proposed measures. The expertness measure based on regret evaluation better represents the robot expertise compared to the other measures. Additionally, it can be applied with variety of knowledge sharing algorithms. Among the investigated knowledge sharing techniques, the AdpWSSR gave the best performance when knowledge was shared among the robots that learn a task from scratch. The EGMR gave the best performance when the knowledge sharing was performed among robots that relearn the transferred knowledge.

Acknowledgement

This work was supported by the Thailand Research Fund through the Royal Golden Jubilee Ph.D Program (Grant No. PHD/1.M.KT.43/C.2). Matlab software was supported by TechSource System (Thailand).

References

1. Tan, M.: Multi-agent reinforcement learning: Independent vs cooperative agents. In: Proc. 10th Int. Conf. Machine Learning (1993)
2. Ahmadabadi, M.N., Asadpour, M.: Expertness Based Cooperative Q-Learning. IEEE Trans. SMC.–Part B. **32**(1) (2002) 66–76
3. Ahmadabadi, M.N., Asadpour, M., Khodanbakhsh, S.H., Nakano, E.: Expertness Measuring in Cooperative Learning. IEEE Int. Conf. on IROS. (2000) 2261–2267
4. Ahmadabadi, M.N., Asadpour, M., Nakano, E.: Cooperative Q-learning: The Knowledge Sharing Issue. J. Adv. Robotics. **15**(8) (2002) 815–832.
5. Bitaghsir, A.A., Moghimi, A., Lesani, M., Keramati, M.M., Ahmadabadi, M.N., Arabi, B.N.: Successful Cooperation between Heterogeneous Fuzzy Q-learning Agents. IEEE Int. Conf. SMC. (2004)
6. Dixon, K.R., Malak, R.J., Khosla, P.K.: Incorporating Prior Knowledge and Previously Learned Information into Reinforcement Learning Agents. Tech. report. Institute for Complex Engineered Systems, Carnegie Mellon University (2000)
7. Malak, R.J. and Khosla, P.K.,: A framework for the Adaptive Transfer of Robot Skill Knowledge Using Reinforcement Learning Agents. In: Proc. IEEE Int. Conf. ICRA. (2001)
8. Moreno, D.L., Regueiro, C.B., Iglesias, R., Barro, S.: Using Prior Knowledge to Improve Reinforcement Learning in Mobile Robotics. Proc. Towards Autonomous Robotics Systems. Univ. of Essex, UK (2004)
9. Ritthipravat, P., Maneewarn, T., Laowattana, D., Wyatt, J.: A Modified Approach to Fuzzy Q-Learning for Mobile Robots. In: Proc. IEEE Int. Conf. SMC. (2004)
10. Ritthipravat, P., Maneewarn, T., Wyatt, J., Laowattana, D.: Comparison and Analysis of Expertness Measure in Knowledge Sharing among Fuzzy Q-learning Robots. In: Proc. The nineteenth International Conference on Industrial, Engineering and other Applications of Applied Intelligent Systems (IEA/AIE), June 27-30. (2006)
11. Peterson, T.S., Owens, N.E. and Carroll, J.L.: Towards Automatic Shaping in Robot Navigation. In: Proc. IEEE Int. Conf. ICRA. (2001)
12. Carroll, J.L., Peterson, T.S. and Owens, N.E.: Memory-guided Exploration in Reinforcement Learning. In: Proc. IEEE Int. Conf. IJCNN. (2001)
13. Whitehead, S., Karlsson J. and Karlsson, J.: Learning Multiple Goal Behavior via Task Decomposition and Dynamic Policy Merging. Robot Learning, Connell, J.H. and Mahadevan, S. (Eds.), Kluwer Academic Publishers, Norwell, MA. (1993)

Explaining Symbolic Trajectory Evaluation by Giving It a Faithful Semantics

Jan-Willem Roorda and Koen Claessen

Chalmers University of Technology, Sweden
{jwr, koen}@chalmers.se

Abstract. Symbolic Trajectory Evaluation (STE) is a formal verification technique for hardware. The current STE semantics is not faithful to the proving power of existing STE tools, which obscures the STE theory unnecessarily. In this paper, we present a new *closure semantics* for STE which does match the proving power of STE model-checkers, and makes STE easier to understand.

1 Introduction

The rapid growth in hardware complexity has lead to a need for *formal verification* of hardware designs to prevent bugs from entering the final silicon. *Model-checking* is a verification method in which a model of a system is checked against a *property*, describing the desired behaviour of the system over time. An exhaustive search through the model determines whether the property holds or not. Today, all major hardware companies use model-checkers in order to reduce the number of bugs in their designs.

Symbolic Trajectory Evaluation (STE) [8] is a model-checking technique for hardware. STE uses *abstraction*, meaning that details of the circuit behaviour are removed from the circuit model. This improves the capacity limits of the method, but has as down-side that certain properties cannot be proved if the wrong abstraction is chosen. To be able to reason about STE verification without having to bother with the implementation details of STE model-checkers a *semantics* for STE is used. Unfortunately, as we argue in this paper, the semantics currently described in the STE-literature is not *faithful* to the proving power of STE model-checking algorithms, that is, STE model-checkers can actually prove more properties than the STE-semantics predicts. Therefore, in this paper, we give an alternative semantics for STE, called the *closure semantics*. The closure semantics is faithful to the proving power of STE model-checkers and makes understanding STE easier.

Introduction to STE. STE is a model-checking technique based on *simulation*. STE combines three-valued simulation (using the standard values 0 and 1 together with the extra value X, "don't know") with symbolic simulation (using symbolic expressions to drive inputs). STE is able to verify properties of circuits containing large data paths that are beyond the reach of traditional symbolic model checking [1, 7, 8]. Most implementations of STE use a canonical representation of propositional logic formulas called Binary Decision Diagrams (BDDs) to represent values of nodes during simulation.

Consider the circuit the in Fig. 2. In the figure, p, q, r, s, u, v, and out are node names. The nodes are connected via *logical gates* and wires. The only gates in this circuit are

D. Grigoriev, J. Harrison, and E.A. Hirsch (Eds.): CSR 2006, LNCS 3967, pp. 555–566, 2006.
© Springer-Verlag Berlin Heidelberg 2006

x	$\neg x$	x	y	$x \& y$	x	y	$x + y$
0	1	0	0	0	0	0	0
1	0	0	1	0	0	1	1
X	X	1	0	0	1	0	1
		1	1	1	1	1	1
		X	0	0	X	0	X
		0	X	0	0	X	X
		X	1	X	X	1	1
		1	X	X	1	X	1
		X	X	X	X	X	X

Fig. 1. Three-valued extensions of the gates   **Fig. 2.** An example circuit

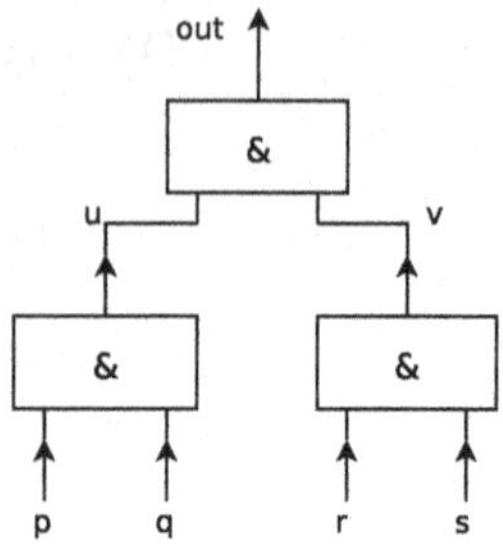

AND-gates, depicted by the symbol $\&$. The circuit implements a 4-ary AND-gate by means of three 2-ary AND-gates.

In STE, circuit specifications are *assertions* of the form $A \implies C$. Here, A is called the *antecedent* and C the *consequent*. For example, an STE-assertion for the circuit in Fig. 2 is:

$$(\text{p is } a) \ \textbf{and} \ (\text{s is } \neg a) \implies (\text{out is } 0) \tag{1}$$

Here a is a *symbolic variable*, which can take on the value 0 or 1, and p, s and out are *node names*. The assertion states that when node p has value a, and node s has value $\neg a$, then node out must have value 0.

STE Model-Checking. An STE-model checker performs a three-valued symbolic simulation run of the circuit. The antecedent of the assertion drives the simulation by providing symbolic values of nodes, the consequent specifies the conditions that should result. When an antecedent does not specify a value for a particular input, this input receives value X.

During simulation, values of nodes are represented by *symbolic three-valued expressions*. Three-valued symbolic expressions are expressions that given a valuation of the symbolic variables to Boolean values evaluate to 0,1 or X, using the truth-tables in Figure 1. Usually, STE-model checking algorithms implement these three-valued symbolic expressions canonically by a pair of BDDs using a *dual-rail encoding*, which represents a three-valued value by two Boolean values.

Example 1. For assertion (1) three-valued symbolic simulation works as follows. The antecedent specifies that node p has symbolic value a, so the simulated value for this node is a. Furthermore, the antecedent specifies that node s has symbolic value $\neg a$, so the simulated value for this node is $\neg a$. However, the antecedent does not specify the value of input nodes q and r, therefore these nodes receive the unknown value X in the simulation. As node u is the output of the AND-gate with inputs p and q, this node receives value $(a \& X)$. In a similar fashion, node v receives the value $(X \& \neg a)$. Finally, node out is the output of the AND-gate with inputs u and v, so node out is represented by the expression $((a \ \& \ X) \ \& \ (X \ \& \ \neg a))$, which is simplified to the expression 0.

During simulation, the STE model-checker checks, for each node in the consequent, whether its simulated value meets the required value, as specified by the consequent. In this case, the simulated value of node out is 0, and the required value is 0. Therefore, the property is proved.   □

STE Abstractions. The power of STE comes from its use of *abstraction*. Three abstractions are used: (1) the value X can be used to abstract from a specific Boolean value of a circuit node, (2) in the simulation, information is only propagated forwards through the circuit (i.e. from inputs to outputs of gates) and through time (i.e. from time t to time $t + 1$), (3) the circuit model does not contain an initial state for registers, in effect removing the concept of reachable states. Below, (1) and (2) are discussed in more detail.

Three-Valued Abstraction. The three-valued abstraction is induced by the antecedent of the assertion; when the antecedent does not specify a value for a certain node, the value of the node is abstracted away by using the unknown value X. Because of this abstraction, the values of circuit nodes during simulation can be represented by BDDs in terms of the symbolic variables occurring in the assertion.

For instance, in the example above, the values of node q and r were abstracted away from, and only one BDD-variable was needed to perform the verification. Without abstraction two more variables would have been needed to represent the values of nodes p and q. Of course, for this example the efficiency gain is not very impressive. But for many applications the three-valued abstraction is essential.

For instance, consider the verification of the correct behaviour of the read action of a memory of 2^{16} memory locations of each 8 bits. In conventional BDD-based model checking, BDDs with 2^{19} BDD-variables are needed for this verification, this leads immediately to a BDD-blow up, making verification impossible. In STE, however, it is possible to only use variables for the address and the contents of the memory location being read, and to abstract away from the values in all other memory locations. Therefore, in STE based verification of the property, the BDDs contain only 24 variables, and verification goes through without running out of memory.

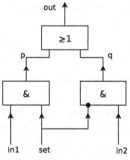

Fig. 3. A multiplexer

The drawback of this abstraction is the *information loss* inherent to the use of three-valued logic. This information loss is illustrated in the following example.

Example 2. Consider the circuit given in Fig. 3. The circuit consists of two AND-gates and an OR-gate, depicted by ≥ 1. The circuit implements a *multiplexer*: if set is 1, input in_1 is routed to the output out, if set is 0, input in_2.

An STE-assertion for this circuit is:

$$(in_1 \text{ is } a) \text{ and } (in_2 \text{ is } a) \Longrightarrow (out \text{ is } a) \qquad (2)$$

The assertion states that when nodes in_1 and in_2 both have value a, then node out has value a as well. It is easy to see that this assertion is true when no abstraction is used. However, the reader can verify that symbolic three-valued simulation calculates the expression $(a \ \& \ X) + (\neg X \ \& \ a)$ for node out, which is simplified to $a \ \& \ X$. This expression is not equivalent to required value a: when $a = 1$ the simulated expression has value X. Therefore, the property is not proved by STE.

Such an information loss can be repaired by introducing extra symbolic variables. In this case, by driving input node set by a symbolic variable b. This yields the following assertion.

$$(\text{in}_1 \text{ is } a) \textbf{ and } (\text{in}_2 \text{ is } a) \textbf{ and } (\text{set is } b) \implies (\text{out is } a) \tag{3}$$

For this assertion, three-valued symbolic simulation calculates the expression $(a \mathbin{\&} b) + (\neg b \mathbin{\&} a)$ for node out, which is simplified to a.   □

Forwards Abstraction. The STE-simulator performs *forwards simulation*. That means that information is only propagated forwards through the circuit (i.e. from inputs to outputs of gates) and through time (i.e. from time t to time $t + 1$).

Example 3. Consider the following assertion for the multiplexer circuit.

$$\text{p is } 1 \implies \text{set is } 1 \tag{4}$$

It is easy to see that the assertion is true when no abstraction is used: if the output of AND-gate has value 1 then both of the inputs to the gate should have value 1 as well. This assertion can, however, not be proved with STE.

Symbolic three-valued simulation proceeds as follows: The assertion does not contain any assumptions on the node set, it will therefore receive value X in the simulation. Node p receives value 1 in the simulation, but as this node is the output of AND-gate with node set as input, this does not influence the value of node set. Therefore, after simulation, node set has value X and not the required value 1.   □

Semantics for STE. The above examples illustrate that it is non-trivial to decide whether an assertion can be proved with an STE-model checker. In the examples, detailed knowledge about the implementation of STE-model checkers was used to do so.

To be able to reason about STE verification, without having to bother with the implementation details of STE model-checkers a *semantics* for STE is used. A semantics for STE consists of a formal theory that decides whether an STE-assertion is *true* or *false* for a particular circuit. Ideally, such a semantics deems a property to be true if-and-only-if the property can be proved by an STE-model checker. So the semantics should, for instance, state that assertions (1) and (3) are true, but that assertions (2) and (4) are false. A semantics for STE was first described in [8] by Seger and Bryant. Later, a simplified and easier to understand semantics was given in [4] by Melham and Jones. Unfortunately, neither of these semantics matches the proving power of currently available STE model checkers. The problem is that they

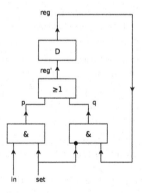

Fig. 4. A memory cell

p   = in AND set
q   = ¬set AND reg
reg' = p OR q

Fig. 5. Its netlist

cannot deal with *combinational* properties (properties ranging over one single point in time). All such properties are deemed to be false by the semantics. Of course, STE model-checkers can deal with such properties. For example, the STE model-checker in Intel's in-house verification toolkit Forte [3] easily proves assertions (1) and (3). However, the current semantics deems these properties to be false. This situation makes it harder to understand STE, and we believe it to be both unnecessary and undesirable.

Contribution. In this paper, we introduce a new semantics for STE, called the *closure semantics*, that is faithful to the proving power of STE algorithms. We believe this semantics makes STE easier to understand and provides a solid basis for STE-theory.

2   Y-Semantics

In this section we describe the semantics commonly given in the STE-literature [8, 4, 2] which we, in this paper, informally call the Y-semantics. Furthermore, we explain why this semantics is not faithful to the proving power of STE model-checkers. Before doing so, we first introduce the concept of *states*.

n	$s(n)$
in	1
set	1
other	X

n	$Y(s)(n)$
reg	1
other	X

Fig. 6.   Example states

Values and States.   For technical reasons, besides the three values $0, 1$ and X we have already discussed, a fourth value T (called the *over-constrained value*) is used in STE simulation. The value T represents a clash; it is the resulting value of a node that is required to have both the value 0 and 1 during simulation. The three-valued gate-definitions in Fig. 1 are extended to deal with this fourth value in such a way that whenever at least one of their inputs is T, their output is also T.

A *circuit state*, written s : **State**, is a function from the set of nodes of circuit to the values $\{0, 1, X, T\}$. For instance, consider the memory cell circuit in Fig. 4. The state s in Fig. 6 gives value 1 to nodes in and set and value X to all other nodes.

Next-State Functions.   In the Y-semantics, circuits are modelled by their *next-state functions*, written Y : **State** $\rightarrow$ **State**. Given a circuit state, a next-state function calculates the state of the circuit in the next moment in time.

Example 4. Consider the memory cell circuit given in Fig. 4. The circuit consists of two AND-gates, an OR-gate, and a delay element, depicted by the letter D. The delay-element has output node reg and input node reg$'$. The value of the output of the delay element at time $t + 1$ is the value of its input at time t. A next state function for the memory cell given in Fig. 4 is:

n	$Y(s)(n)$
in	X
set	X
p	X
q	X
reg$'$	X
reg	$(s(\mathsf{in})\ \&\ s(\mathsf{set})) + (\neg s(\mathsf{set})\ \&\ s(\mathsf{reg}))$

Note that the next-state function yields the unknown value X for all nodes that depend on the values of the input nodes. The reason is that the next-state function cannot predict the value of the input nodes, it predicts only the values of nodes that depend on the *previous* values of other nodes. For example, when given state s (given in Fig. 6), the next-state function yields the state $Y(s)$ (also given in Fig. 6) for the next moment in time in which only node reg has value 1 and all other nodes have value X. The intuition is that when at time t nodes in and set have value 1, then at time $t + 1$ node reg is required to have value 1 as well.   □

We argue that using next-state functions is problematic in defining a proper semantics for STE. The problem is that next-state functions only express a relation between nodes in *successive* points in time, while ignoring the relation between nodes in the circuit at the *same* time point. Therefore, a semantics based on next-state functions cannot deal with assertions that express a relation between circuit nodes at the same time-point.

Before making this statement more precise, we formally define the Y-semantics. But first, we introduce the concepts of *sequences*, *trajectories*, and *assertions*.

Sequences. A *sequence* $\sigma : \mathbb{N} \to$ **State** is a function from a point in time to a circuit state, describing the behaviour of a circuit over time. The set of all sequences σ is written **Seq**. A *three-valued sequence* is a sequence that does not assign the value T to any node at any time.

Trajectories. Given a circuit, a trajectory of the circuit is a sequence that meets the constraints of the circuit c, taking the STE-abstractions into account. In the Y-semantics, trajectories are defined in terms of next-state functions.

To define trajectories, we first need to introduce the information order $\leq$ on the values $0, 1, \mathsf{X}$, and T. The unknown value X contains the least information, so $\mathsf{X} \leq 0$, $\mathsf{X} \leq 1, \mathsf{X} \leq \mathsf{T}$, while 0 and 1 are incomparable. The over-constrained value T contains most information, so $0 \leq \mathsf{T}$ and $1 \leq \mathsf{T}$. If $v \leq w$ it is said that v is *weaker* than w. The information order $\leq$ is extended to states as follows: state s_1 is weaker than state s_2, written $s_1 \leq s_2$, iff. for every node n, $s_1(n) \leq s_2(n)$.

Given a circuit c with a next-state function Y. A sequence τ is a *Y-trajectory* of Y iff for all $t \in \mathbb{N}$:

$$Y(\tau(t)) \leq \tau(t + 1)$$

The intuition behind this definition is that in a trajectory all information that can be derived from time t should be propagated to time $t + 1$.

Example 5. For the next-state function given in Example 4, the following sequence is a trajectory:

n	$t = 0$	$t = 1$	$t > 1$
in	1	X	X
set	1	X	X
reg	X	1	X
p	X	X	X
other	X	X	X

Notice that although the nodes in and set have value 1 at time 0, node p does not receive value 1. The reason is that in Y-trajectories there is no propagation of information between nodes at the same point; information is only propagated from a time point to the next time point.

STE-Assertions. have the form $A \implies C$. Here A and C are formulas in *Trajectory Evaluation Logic* (TEL). The only variables in the logic are time-independent Boolean variables taken from the set V of *symbolic variables*. The language is given by the following grammar:

$$f ::= n \text{ is } 0 \mid n \text{ is } 1 \mid f_1 \text{ and } f_2 \mid P \rightarrow f \mid \mathbf{N} f$$

where n is a circuit node and P is a Boolean propositional formula over the set of symbolic variables V. The operator **is** is used to make a statement about the Boolean value of a particular node in the circuit, **and** is conjunction, $\rightarrow$ is used to make conditional statements, and $\mathbf{N}$ is the next time operator. Note that symbolic variables only occur in the Boolean propositional expressions on the left-hand side of an implication. The notation n **is** P, where P is a Boolean symbolic expression over the set of symbolic variables V, is used to abbreviate the formula: $(\neg P \rightarrow n \text{ is } 0) \text{ and } (P \rightarrow n \text{ is } 1)$.

The meaning of a TEL formula is defined by a satisfaction relation that relates valuations of the symbolic variables and sequences to TEL formulas. Here, the following notation is used: The time shifting operator σ^1 is defined by $\sigma^1(t)(n) = \sigma(t+1)(n)$. Standard propositional satisfiability is denoted by $\models_{\text{Prop}}$. Satisfaction of a trajectory evaluation logic formula f, by a sequence $\sigma \in \mathbf{Seq}$, and a valuation $\phi : V \rightarrow \{0,1\}$ (written $\phi, \sigma \models f$) is defined by

$$
\begin{aligned}
\phi, \sigma \models n \text{ is } b &\equiv \sigma(0)(n) = b \; , \; b \in \{0,1\} \\
\phi, \sigma \models f_1 \text{ and } f_2 &\equiv \phi, \sigma \models f_1 \text{ and } \phi, \sigma \models f_2 \\
\phi, \sigma \models P \rightarrow f &\equiv \phi \models_{\text{Prop}} P \text{ implies } \phi, \sigma \models f \\
\phi, \sigma \models \mathbf{N} f &\equiv \phi, \sigma^1 \models f
\end{aligned}
$$

Y-Semantics. The Y-semantics is defined as follows: A circuit with next-state function Y *satisfies* a trajectory assertion $A \implies C$ iff for every valuation $\phi : V \rightarrow \{0,1\}$ of the symbolic variables and for every three-valued trajectory τ of Y, it holds that

$$\phi, \tau \models A \implies \phi, \tau \models C.$$

The following example illustrates that the Y-semantics cannot be used to reason about STE assertions that specify a combinational property of a circuit.

Example 6. Consider the following STE assertion for the memory cell:

$$(\text{in is } 1) \text{ and } (\text{set is } 1) \implies (\text{p is } 1)$$

This assertion does not hold in the Y-semantics. The trajectory given in Example 5 is a counter example. Of course, this simple combinational property can be proved by an STE model-checker. $\square$

3   Closure Semantics

The previous section showed that the Y-semantics is not faithful to the proving power of STE model checkers. The problem lies in the next-state function which only propagates information from one time-point to the next, and thereby ignores propagation of information between nodes in the same time-point.

In this section we introduce a new semantics for STE called the *closure semantics*. In this semantics, *closure functions* are used as circuit models. The idea is that a closure function, written F : **State** $\rightarrow$ **State** takes as input a state of the circuit, and calculates all information about the circuit state at the *same* point in time that can be derived by propagating the information in the input state in a *forwards* fashion. Later, trajectories are defined using closure functions.

Before we can give an example of a closure function, we need to define the *least upper bound* operator. The least-upper bound operator, written $\sqcup$, given two values taken out of $\{0, 1, X, T\}$ calculates their least upper bound w.r.t. the information order $\leq$. So, for instance $0 \sqcup X = 0$, $X \sqcup 1 = 1$, but also $T \sqcup 0 = T$ and $T \sqcup X = T$, and in particular $0 \sqcup 1 = T$.

Example 7. The closure function for a circuit consisting of a single AND-gate with inputs p and q, and output r is given in Fig. 7. Here, The least upper bound operator in the expression for $F(s)(r)$ combines the value of r in the given state s, and the value for r that can be derived from the values of p and q, being $s(\mathsf{p})$ & $s(\mathsf{q})$.

n	$F(s)(n)$
p	$s(\mathsf{p})$
q	$s(\mathsf{q})$
r	$(s(\mathsf{p})$ & $s(\mathsf{q})) \sqcup s(\mathsf{r})$

Fig. 7. Closure function for the AND-gate

A state s : $\{\mathsf{p}, \mathsf{q}, \mathsf{r}\} \rightarrow \mathbb{V}$ can be written as a vector $s(\mathsf{p}), s(\mathsf{q}), s(\mathsf{r})$. For example, the state that assigns the value 1 to p and q and the value X to node r is written as 11X. Applying the closure function to the state 11X yields 111. The reason is that when both inputs to the AND-gate have value 1, then by forwards propagation of information, also the output has value 1. Applying the closure function to state 1XX yields 1XX. The reason is that the output of the AND-gate is unknown when one input has value 1 and the other value X. The *forwards* nature of simulation becomes clear when the closure function is applied to state XX1, resulting in XX1. Although the inputs to the AND-gate must have value 1 when the output of the gate has value 1, this cannot be derived by forwards propagation.

A final example shows how the over-constrained value T can arise. Applying the closure function to state 0X1 yields 0XT. The reason is that the input state gives node r value 1 and node p value 0. From p having value 0 it can be derived by forwards propagation that r has value 0, therefore r receives the over-constrained value T. □

Before we describe how a closure function for an arbitrary circuit can be defined, we define the concept of *netlists* for describing circuits. Here, a netlist is an acyclic list of definitions describing the relations between the values of the nodes. For example, the netlist of the memory cell circuit in Fig. 4 is given in Fig. 5. Inverters are not modelled explicitly in our netlists, instead they occur implicitly for each mention of the negation operator ¬ on the inputs of the gates. Delay elements are not mentioned explicitly in the netlist either. Instead, for a delay element with output node n in the circuit, the input of the delay element is node n' which is mentioned in the netlist. So, from the netlist in Fig. 5 it can

n	$F(s)(n)$
in	$s(\mathsf{in})$
set	$s(\mathsf{set})$
p	$(s(\mathsf{in})$ & $s(\mathsf{set})) \sqcup s(\mathsf{p})$
q	$(\neg s(\mathsf{set})$ & $s(\mathsf{reg})) \sqcup s(\mathsf{q})$
reg'	$(F(s)(\mathsf{p}) + F(s)(\mathsf{q})) \sqcup s(\mathsf{reg'})$
reg	$s(\mathsf{reg})$

Fig. 8. Closure function for the memory cell

be derived that the node reg is the output of a delay element with input reg′. For simplicity, we only allow AND-gates and OR-gates in netlists. It is, however, straightforward to extend this notion of netlists to include more operations.

Given the netlist of a circuit c, the *induced closure function* for the circuit, written F_c, can easily be constructed by interpreting each definition in the netlist as a three-valued gate. Given a state s, for every circuit input n, the value of the node is given by $s(n)$ Also, for every output n of a delay element, the value of the node is given by $s(n)$. Otherwise, if n is the output of an AND-gate with input nodes p and q, and values for p and q are already calculated, the value of node n is the least upper bound of $s(n)$ and the three-valued conjunction of the values for p and q. In a similar way, values for the outputs of OR-gates are defined. This definition is well-defined because netlists are acyclic by definition.

Example 8. The induced closure function for the memory cell circuit in Fig. 4 is given in Fig. 8. Consider the state s_1 given in Fig. 9. Applying the closure function to this state yields the state $F(s_1)$. Node p receives value 1 as it is the output of an AND-gate with inputs in and set which both have value 1 in the input state. In the same way, node q receives value 0. Node reg′ receives value 1, because it is the output of an OR-gate with input nodes p and q. Finally, node reg receives value X as its value depends on the *previous* value of node reg′, its value cannot be determined from the current values of the other nodes.

n	$s_1(n)$
in	1
set	1
other	X

n	$F(s_1)(n)$
in	1
set	1
p	1
q	0
reg′	1
reg	X

n	$s_2(n)$
p	1
other	X

n	$F(s_2)(n)$
p	1
reg′	1
other	X

Fig. 9. Example states

When an internal node is given a value in the input state, the least upper bound of this value and the value derived by forwards propagation is used. For instance, driving only node p with value 1 yields state s_2 given in Fig. 9. Applying the closure function to this state yields state $F(s_2)$. Note that there is no backwards information flow: the closure function does not demand that nodes set and in have value 1, though they are the input nodes to an AND-gate whose output node has received value 1.   □

Closure functions should meet several requirements:

– Closure functions are required to be *monotonic*, that is, for all states s_1, s_2: $s_1 \leq s_2$ implies $F(s_1) \leq F(s_2)$. This means that a more specified input state cannot lead to a less specified result. The reason is that given a more specified input state, more information about the state of the circuit can be derived.
– Closure functions are required to be *idempotent*, that is, for every state s: $F(F(s)) = F(s)$. This means that repeated application of the closure function has the same result as applying the function once. The reason is that the closure function should derive all information about the circuit state in one go.
– Finally, we require that closure functions are *extensive*, that is, for every state s: $s \leq F(s)$. This means that the application of a closure function to a circuit state should yield a state as least as specified as the input state. The reason is that the closure function is required not to loose any information.

n	$\sigma(0)$	$\sigma(1)$	$\sigma(t), t>1$
in	1	X	X
set	1	0	X
other	X	X	X

n	$F^{\to}(\sigma)(0)$	$F^{\to}(\sigma)(1)$	$F^{\to}(\sigma)(2)$	$F^{\to}(\sigma)(t), t>2$
in	1	X	X	X
set	1	0	X	X
p	1	0	X	X
q	0	1	X	X
reg'	1	1	X	X
reg	X	1	1	X

Fig. 10. Sequence σ **Fig. 11.** Sequence $F^{\to}(\sigma)$

The induced closure function for a circuit is by construction monotonic, idempotent and extensive.

Closure over Time. In STE, a circuit is simulated over multiple time steps. During simulation, information is propagated forwards through the circuit and through time, from each time step t to time step $t+1$.

To model this forwards propagation of information through time, a *closure function over time*, notation $F^{\to} : \mathbf{Seq} \to \mathbf{Seq}$, is used. Given a sequence, the closure function over time calculates all information that be can derived from that sequence by forwards propagation. Recall that for every delay element with output node n the input to the delay element is node n'. Therefore, the value of node n' at time t is propagated to node n at time $t+1$ in the forwards closure function over time.

Below, the closure function over time is defined as a function of a closure function. First, an example is given.

Example 9. For the memory cell, consider the closure function given in Fig. 8, and the sequence σ, given in Fig. 10. The sequence $F^{\to}(\sigma)$ at time 0 depends only on the sequence σ at time 0, and is computed by applying the closure function F to $\sigma(0)$. See Fig. 11.

The sequence $F^{\to}(\sigma)$ gives node reg' value 1 at time 0. Node reg' is the input to the delay element with output node reg. The value 1 should be propagated from reg' at time 0, to reg at time 1. Therefore, when calculating the state $F^{\to}(\sigma)(1)$, the node values given by the state $\sigma(1)$ and the value of node reg propagated from the value of node reg' at time 0 are combined. Let us call the state that combines these node values $\sigma'(1)$:

$$\sigma'(1)(\text{reg}) = \sigma(1)(\text{reg}) \sqcup F^{\to}(\sigma)(0)(\text{reg}')$$
$$\sigma'(1)(n) = \sigma(1)(n) \qquad\qquad , n \neq \text{reg}$$

In this case $\sigma'(1)$ is given by: $\sigma'(1)(\text{set}) = 0$, $\sigma'(1)(\text{reg}) = 1$, and $\sigma'(1)(n) = $ X for each other node n. Applying the forwards closure F to $\sigma'(1)$ yields state $F^{\to}(\sigma)(1)$ given in Fig. 11. The value of $F^{\to}(\sigma)(2)$ is given by applying F to $\sigma'(2)$, where $\sigma'(2)$ is calculated in a similar fashion as $\sigma'(1)$, and is given by: $\sigma'(2)(\text{reg}) = 1$, and $\sigma'(2)(n) = $ X for all other nodes n. Repeating this procedure gives the complete sequence $F^{\to}(\sigma)$ given in Fig. 11.   □

Given a closure function F for a circuit with has a set of outputs of delay-elements $\mathcal{S}$, the *closure function over time*, written $F^{\to} : \mathbf{Seq} \to \mathbf{Seq}$, is inductively defined by:

$$F^{\to}(\sigma)(0) = F(\sigma(0))$$
$$F^{\to}(\sigma)(t+1) = F(\sigma'(t+1))$$

where
$$\sigma'(t+1)(n) = \begin{cases} \sigma(t+1)(n) \sqcup F^{\rightarrow}(\sigma)(t)(n'), & n \in \mathcal{S} \\ \sigma(t+1)(n), & \text{otherwise} \end{cases}$$

The function $F^{\rightarrow}$ inherits the properties of being monotonic, idempotent and extensive from F.

Trajectories. A trajectory is defined as a sequence in which no more information can be derived by forwards propagation. That is, a sequence τ is a trajectory of a closure function when it is a fixed-point of the closure function over time. So, a sequence τ is a *trajectory* of F iff $\tau = F^{\rightarrow}(\tau)$.

Closure Semantics of STE. Using the definition of trajectories of a circuit, we can now define the closure semantics. A circuit c with closure function F *satisfies* a trajectory assertion $A \implies C$ iff for every valuation $\phi : V \rightarrow \{0, 1\}$ of the symbolic variables, and for every three-valued trajectory τ of F, it holds that:

$$\phi, \tau \models A \implies \phi, \tau \models C.$$

Example 10. The reader can verify that assertions (1) and (3), given in the introduction in the paper, are true in this semantics, but assertions (2) and (4) are false.

For instance, for assertion (1), a case distinction can be made on the value of the symbolic variable a. If $\phi(a) = 0$ then in every trajectory τ that satisfies A, node p has value 0, so by forwards propagation also node out has value 0, so the trajectory also satisfies C. By similar reasoning it can be derived that for $\phi(a) = 1$, every trajectory that satisfies A also satisfies C.

For assertion (2), a counter-example is the trajectory that gives both in_1 and in_2 value 1, and all other nodes value X. $\square$

Cautious Semantics. In our proposed semantics, as well as in the Y-semantics, an assertion can be true even if for a particular valuation ϕ of the symbolic variables there are no three-valued trajectories of the circuit that satisfy the antecedent. This is illustrated in the example below.

Example 11. For an AND-gate with inputs in_1 and in_2, and output out, the assertion

$$(\text{out is } 1) \textbf{ and } (\text{in}_1 \text{ is } a) \textbf{ and } (\text{in}_2 \text{ is } b) \implies (\text{in}_1 \text{ is } 1) \textbf{ and } (\text{in}_2 \text{ is } 1)$$

is true in the closure semantics. For valuations that give at least one of the symbolic variables a and b the value 0, there are no three-valued trajectories that meet the antecedent: there are no three-valued trajectories in which at least one of the inputs of the AND-gate (nodes in_1 and in_2) has value 0, while the output (node out) has value 1. Only for the valuation that gives both the symbolic variables value 1, there exists a three-valued trajectory that satisfies the antecedent. As this trajectory satisfies the consequent as well, the assertion is true in the closure semantics. $\square$

The default approach taken in Intel's in-house verification toolkit Forte is to demand that for each valuation of the symbolic variables there exists at least one three-valued trajectory that satisfies the antecedent. In order to warn the user of a possible mistake, Forte reports an 'antecedent-failure' when the top-value $\top$ is required to satisfy the antecedent. We have formalised this approach in the *cautious closure semantics* of STE. A circuit with closure function F *cautiously satisfies* a trajectory assertion $A \implies C$ iff *both* F satisfies $A \implies C$ *and* for every valuation ϕ of the symbolic variables there exists a three-valued trajectory τ such that $\phi, \tau \models A$.

4   Conclusion

We have introduced a new semantics for STE, called the *closure semantics*, that is faithful to the proving power of STE algorithms. We believe this semantics makes STE easier to understand and provides a solid basis for STE-theory.

The semantics presented here is closely related to BDD-based model-checking algorithms for STE. In such algorithms, a simulator calculates a weakest trajectory satisfying the antecedent. The theory of STE guarantees that only this trajectory need be considered to check whether an STE-assertions holds. The calculation done by the simulator is essentially the same as the calculation done by the closure function of the circuit; the difference is that the simulator works on symbolic values, whereas closure functions work on scalar values.

In a recent paper [6] we have presented a new SAT-based algorithm for STE, together with a *stability semantics* for STE. The stability semantics describes a set of constraints on sequences such that the constraints are satisfiable if-and-only-if the sequence is a trajectory. These constraints can be directly translated to SAT-clauses. The stability semantics is equivalent to the closure semantics [5]. But, as the stability semantics is more closely related to SAT-based STE model-checkers, the closure semantics presented in this paper is better suited for explaining BDD-based STE.

Acknowledgements. Thanks to Mary Sheeran for commenting on earlier drafts of this paper. Thanks to Tom Melham for contributing to the discussions on the semantics of STE. This research is supported by the department of Computing Science and Engineering of Chalmers University of Technology and by the Swedish Research Council (Vetenskapsrådet).

References

1. M. Aagaard, R. B. Jones, T. F. Melham, J. W. O'Leary, and C.-J. H. Seger. A methodology for large-scale hardware verification. In *FMCAD*, 2000.
2. C.-T. Chou. The Mathematical Foundation of Symbolic Trajectory Evaluation. In *Computer Aided Verification (CAV)*, 1999.
3. FORTE. http://www.intel.com/software/products/opensource/tools1/verification.
4. T. F. Melham and R. B. Jones. Abstraction by symbolic indexing transformations. In *Formal Methods in Computer-Aided Design FMCAD*, volume 2517 of *LNCS*, 2002.
5. J.-W. Roorda. Symbolic trajectory evaluation using a satisfiability solver. Licentiate thesis, Computing Science, Chalmers University of Technology, 2005.
6. J.-W. Roorda and K. Claessen. A new SAT-based Algorithm for Symbolic Trajectory Evaluation. In *Correct Hardware Design and Verification Methods (CHARME)*, 2005.
7. T. Schubert. High level formal verification of next-generation microprocessors. In *Proceedings of the 40th conference on Design automation*. ACM Press, 2003.
8. C.-J. H. Seger and R. E. Bryant. Formal verification by symbolic evaluation of partially-ordered trajectories. *Formal Methods in System Design*, 6(2), 1995.

Analytic Modeling of Channel Traffic in n-Cubes

Hamid Sarbazi-Azad, Hamid Mahini, and Ahmad Patooghy

IPM School of Computer Science, and Sharif University of Technology,
Tehran, Iran
azad@ipm.ir, azad@sharif.edu

Abstract. Many studies have shown that the imbalance of network channel traffic is of critical effect on the overall performance of multicomputer systems. In this paper, we analytically model the traffic rate crossing the network channels of a hypercube network under different working conditions. The effect of different parameters on the shaping of non-uniformity and traffic imbalance over network channels, are considered and analytical models for each case are proposed.

1 Introduction

The *routing algorithm* indicates the next channel to be used at each intermediate node. That channel may be selected from a set of possible choices and according to the size of this set, the routing algorithm may be divided into three categories: *deterministic, partially adaptive, and fully adaptive* [6, 16].

Deterministic routing assigns a single path to each source and destination node (the size of the mentioned set is one in this category) resulting in a simple router implementation. However, under deterministic routing, a message cannot use alternative paths to avoid congested channels along its route and therefore the low network performance is inevitable. The XY and e-cube routing algorithms are the most known routing algorithm of this category in meshes and hypercubes [5].

Fully adaptive routing has been suggested to overcome this limitation by enabling messages to explore all available paths (the above mentioned set has its maximum possible size) and consequently offers the potential for making better use of network resources. But these algorithms imply more router complexity for deadlock-freedom. An example of a fully adaptive routing algorithm is Duato's [7] routing algorithm. Hop-based routing [3] and Linder-Harden's [13] algorithm are also adaptive routing algorithms proposed for the mesh and hypercube networks.

Partially adaptive routing algorithms try to combine the advantages of the two other categories to produce a routing with limited adaptivity and establish a balance between performance and router complexity. They allow selecting a path from a subset of all possible paths. In fact, these algorithms limit the size of the set of possible choices. Turn model based algorithms and planar adaptive routing algorithm are the most important partially adaptive algorithms for the mesh and hypercube networks [8].

The performance of the network is mainly determined by the three characteristics of interconnection networks mentioned above (topology, switching method, and routing algorithm). However, traffic distribution of the workload is another important

D. Grigoriev, J. Harrison, and E.A. Hirsch (Eds.): CSR 2006, LNCS 3967, pp. 567–579, 2006.
© Springer-Verlag Berlin Heidelberg 2006

factor in determining the overall performance of the system. Although this factor is not considered a network characteristic and is determined by the applications being executed on the machine, it has great impact on performance. The three abovementioned factors are also influential on the traffic pattern [16].

Numerous studies have shown that with load balanced channel traffic greater network performance can be expected. This infers that overall performance is also affected by traffic pattern [14]. In this paper, we analytically model the effect of some important factors that cause imbalance in the channel traffic rates. This study focuses on the hypercube network for the sake of presentation and derives some analytic models to predict traffic in the network. In particular we model the effect of destination address distribution and routing algorithm on network channel traffic.

2 Preliminaries

This section describes the network and node structure in a hypercube along with e-cube [6] deterministic routing algorithm with virtual channels.

A n-dimensional hypercube can be modeled as a graph H_n (V, E), with the node set V (H_n) and edge set E (H_n), where $|V|=2^n$ and $|E|=n2^n$ nodes. The 2^n nodes are distinctly addressed by n-bit binary numbers, with values from 0 to 2^n-1. Each node has link at n dimensions, ranging from 1 (lowest dimension) to n (highest dimension), connecting to n neighbours. An edge connecting nodes $X= x_n x_{n-1} \dots x_1$ and $Y= y_n y_{n-1} \dots y_1$ is said to be at dimension j or to the jth dimensional edge if their binary addresses $x_n x_{n-1} \dots x_1$ and $y_n y_{n-1} \dots y_1$ differ at bit position j only, i.e. $x_j \neq y_j$. An edge in H_n can also be represented by an n-character string with one hyphen (-) and $n-1$ binary symbols $\{0, 1\}$. For example in a H_4, the string 00-1 denotes the edge connecting nodes 0001 and 0011.

Each node consists of a processing element (PE) and router. The PE contains a processor and some local memory. The router has $(n+1)$ input and $(n+1)$ output channels. A node is connected to its neighbouring nodes through n inputs and n output channels. The remaining channels are used by the PE to inject/eject messages to/from the network respectively. Messages generated by the PE are transferred to the router through the injection channel. Messages at the destination are transferred to the local PE through the ejection channel. The router contains flit buffers for incoming virtual channels. The input and output channels are connected by a $(n+1)$-way crossbar switch which can simultaneously connect multiple input to multiple output channels in the absence of channel contention [2].

Many routing algorithms have been proposed for hypercubes. Given a message with source and destination addresses $x= x_n x_{n-1} \dots x_1$ and $y= y_n y_{n-1} \dots y_1$ in an n-dimensional hypercube, the current router (the router which message is in) chooses the next channel to be taken by the message as follows. If the current router address $C= c_n c_{n-1} \dots c_1$, the router calculate bit pattern $B = C \oplus y$ (bit-wise exclusive-or of C and y) the set of channels which can be used to take the message closer to the destination node, are those positions in the pattern B which are "1" [6].

Choosing one of them in a fixed manner (say the first least significant "1"), the routing algorithm will be *deterministic* and referred to as e-cube routing [1]. Choosing a subset of this set will result in partially adaptive routing. For example, p-cube [8] divides this set into two subsets: one set including those positions that are "0" in C and "1" in y, and the other set including those positions that are "1" in C and "0" in y. According to e-cube routing, a message in its first step can take any of channels belonging to the first set and when all channels in the first set are all passed, it can choose any of the channels in the second set. Choosing any channel from the main set in any order can result in a fully adaptive routing algorithm. Linder-Harden's, Duato's, and hop-based routing algorithms are examples of fully adaptive routing algorithms. However, making the algorithm deadlock free requires some careful consideration in order to avoid cyclic dependencies which may result in deadlock. We usually use virtual channels to develop such deadlock avoidance treatments.

3 Traffic Pattern and Load Distribution

Uniform Traffic Pattern Analysis: With uniform network traffic, all network nodes (except the source node) can be the destination of a message with equal probability. In this case, the number of nodes which are i hops away from a given node in an n-dimensional hypercube is $\binom{n}{i}$. Given that a uniform message can be destined to the other network nodes with equal probability, the probability that a uniform message generated at a given source node makes i hops to reach its destination, P_{u_i}, can be given by [1, 4]

$$P_{u_i} = \frac{\binom{n}{i}}{N-1}.$$
(1)

where N is the number of nodes in the network $N=2^n$. The average number of hops that a uniform message makes across the network is given by [16]

$$d_u = \sum_{i=1}^{n} i P_{u_i}.$$
(2)

Now, we can estimate the message arrival rate over each channel as [15]

$$\lambda_{channel} = \frac{N \lambda d_u}{nN} = \frac{\lambda d_u}{n}.$$
(3)

Note that the above rate is valid if the routing algorithm can distribute the traffic evenly over network channels. The e-cube deterministic routing and the fully adaptive

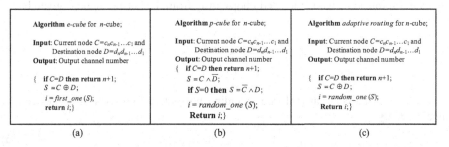

Algorithm *e-cube* for *n*-cube;	Algorithm *p-cube* for *n*-cube;	Algorithm *adaptive routing* for *n*-cube;
Input: Current node $C=c_n c_{n-1}\dots c_1$ and Destination node $D=d_n d_{n-1}\dots d_1$ **Output:** Output channel number { **if** $C=D$ **then return** $n+1$; $S = C \oplus D$; $i = first\_one\ (S)$; **return** i;}	**Input:** Current node $C=c_n c_{n-1}\dots c_1$ and Destination node $D=d_n d_{n-1}\dots d_1$ **Output:** Output channel number { **if** $C=D$ **then return** $n+1$; $S = C \wedge \overline{D}$; **if** $S=0$ **then** $S = \overline{C} \wedge D$; $i = random\_one\ (S)$; **Return** i;}	**Input:** Current node $C=c_n c_{n-1}\dots c_1$ and Destination node $D=d_n d_{n-1}\dots d_1$ **Output:** Output channel number { **if** $C=D$ **then return** $n+1$; $S = C \oplus D$; $i = random\_one\ (S)$; **Return** i;}
(a)	(b)	(c)

Fig. 1. Different routing algorithms in the nD hypercube; a) Deterministic, b) Partially adaptive, c) Fully adaptive

routing algorithms shown in Figure 1 can balance the traffic over network channels [1, 9, 10, 11, 12] while p-cube partially adaptive routing algorithm can not [6] as will be shown in the next section.

3.1 Non-uniform Traffic Pattern Analysis

The model used here to create non-uniform traffic pattern has been widely used in the literature. According to this model a node can partly create non-uniform traffic. To this end, we consider each node to generate non-uniform traffic with a probability of x, where $0 \le x \le 1$, and creates messages with uniform destination distribution with a probability 1-x. Thus, $x = 0$ corresponds to a uniform traffic while $x=1$ indicates a complete non-uniform traffic defined by the non-uniform traffic generator used [16].

In what follows, we consider traffic analysis for three different popular traffic patterns used in the literature: *hotspot*, *bit-reversal*, and *matrix-transpose*.

Hotspot traffic pattern: According to the hotspot traffic distribution a node can create a message to the hotspot node with rate h [14]. It is easy to see that a node may create an i-hop hotspot node with the probability of

$$P_{h_i} = \frac{\binom{n}{i}}{N-1}.$$

(4)

Thus, the average number of hops that a hotspot message may traverse in the network is equal to that for the uniform traffic. Therefore, hotspot traffic does not change the average distance a message (hotspot or uniform) may take to reach its destination. However, it can make unbalanced traffic over network channel as the channels closer to the hotspot node will receive more messages than others.

Consider a channel that is j hops, $1 \le j \le n$, away from the hot-spot node. The probability that a hot-spot message has used this channel during its network journey can be derived as follows. Consider the set J of all the channels located j hops away from the hotspot node. The number of source nodes for which an element of J can act as an intermediate channel to reach the hotspot node is $N - \sum_{k=0}^{j-1}\binom{n}{k}$. Since there are $(n-j+1)\binom{n}{j-1}$ such intermediate channels (or elements in the set J), the hotspot message arrival rate at a channel located j hops away from the hotspot node is given be

$$\lambda_{h,channel_j} = \frac{\lambda h \left[N - \sum_{k=0}^{j-1}\binom{n}{k} \right]}{(n-j+1)\binom{n}{j-1}}.$$

(5)

The overall message arrival rate including hotspot and uniform messages can be then given by

$$\lambda_{channel_j} = (1-h)\lambda_{channel} + \lambda_{h,channel_j}.$$

(6)

where $\lambda_{channel}$ is given by equation (3).

Bit-reversal traffic pattern: Many applications, such as FFT, may produce bit-reversal non-uniform traffic pattern in the network [6, 16]. According to this traffic pattern, a source node $x_1 x_2 \cdots x_n$ sends message to node $B(x_1 x_2 \cdots x_n) = x_n x_{n-1} \cdots x_1$. Let us now calculate the probability, P_{β_i}, that a generated bit-reversal message makes i hops to cross the network.

Let $x = x_1 x_2 \cdots x_n$ be the source address and $B(x) = x'_1 x'_2 \cdots x'_n$ be the destination address for the bit-reversal message. When n is even every single bit difference between the first half parts of the source and destination addresses, $x_1 x_2 \cdots x_{n/2}$ and $x'_1 x'_2 \cdots x'_{n/2}$, results in another bit difference in the next halves, $x_{n/2} x_{n/2+1} \cdots x_n$ and $x'_{n/2} x'_{n/2+1} \cdots x'_n$. Thus, the probability that a bit-reversal message makes i hops, where i is odd, is zero. Let us calculate the number of possible ways that the source and destination nodes of the bit-reversal message are located i hops away from each other (i= 0, 2, 4, ..., n). To do so, we can simply consider the first half of the address patterns of the source and destination nodes and calculate the number of possible combinations that $x_1 x_2 \cdots x_{n/2}$ and $x'_1 x'_2 \cdots x'_{n/2}$ are different in j bit positions for j= 0, 1, 2,..., n/2. A bit in the address $x_1 x_2 \cdots x_{n/2}$ with its corresponding bit in the address $x'_1 x'_2 \cdots x'_{n/2}$ make up four combinations, 00, 01, 10 and 11. In two combinations, 00 and 11, those two bits are equal while in the other two combinations, 01 and 10, they are different. Therefore, the number of possible combinations that $x_1 x_2 \cdots x_{n/2}$ and $x'_1 x'_2 \cdots x'_{n/2}$ are different in exactly j bits is $\binom{n/2}{j} 2^{n/2-j} 2^j$ (j= 0, 1, 2, ..., n/2), and consequently the number of possible combinations that $x_1 x_2 \cdots x_n$ and $x'_1 x'_2 \cdots x'_n$ are different in exactly i bits (i= 0, 2, 4, ..., n) is given by

$$ n_{\beta_i} = \binom{\frac{n}{2}}{\frac{i}{2}} 2^{\frac{n-i}{2}} 2^{\frac{i}{2}} = 2^{\frac{n}{2}} \binom{\frac{n}{2}}{\frac{i}{2}} \quad (i = 0, 2, 4, ..., n). \tag{7} $$

When n is odd we can apply the above derivation for the bit patterns $x_1 x_2 \cdots x_{(n-1)/2} x_{(n+1)/2+1} \cdots x_n$ and $x'_1 x'_2 \cdots x'_{(n-1)/2} x'_{(n+1)/2+1} \cdots x'_n$. Note that the bit $x_{(n+1)/2}$ in x is equal to the bit $x'_{(n+1)/2}$ in $B(x)$. Therefore, the number of combinations where the bit patterns x and $B(x)$ are different in i bits (given by equation 4) is doubled considering the two possible values 0 and 1 for $x_{(n+1)/2}$ when n is odd. Combining these two cases (odd and even n) will result in a general equation for the number of possible combination that x and $B(x)$ are different in exactly i bits (i= 0, 1,..., n) as

$$ n_{\beta_i} = \begin{cases} \binom{\lfloor \frac{n}{2} \rfloor}{\frac{i}{2}} 2^{\lfloor \frac{n}{2} \rfloor + (n \bmod 2)} & \text{if } i \text{ is even} \\ 0 & \text{otherwise} \end{cases} \tag{8} $$

Thus, the probability that a bit-reversal message makes i hops to reach its destination can be written as

$$P_{\beta_i} = \frac{n_{\beta_i}}{N - n_{\beta_0}} = \begin{cases} \dfrac{\binom{\lfloor \frac{n}{2} \rfloor}{\frac{i}{2}}}{2^{\lfloor \frac{n}{2} \rfloor - (n \bmod 2)}}, & \text{if } i \text{ is even}. \\ 0, & \text{otherwise} \end{cases} \qquad (9)$$

The average number of hops that a bit-reversal message makes across the network is given by

$$d_{bit-reversed} = \sum_{i=1}^{n} i P_{\beta_i} . \qquad (10)$$

Using a similar model to combine hotspot and uniform traffic, we can assume that the generation rate of bit-reversal messages at a node is β and thus the rate for generating uniform messages is $1-\beta$. Therefore, the average distance that a message takes in the network can be given as

$$d_\beta = \beta d_{bit-reversed} + (1-\beta) d_u . \qquad (11)$$

Matrix-transpose traffic pattern: Many applications, such as matrix problems and signal processing, may produce matrix-transpose non-uniform traffic pattern in the network [6, 16]. According to this traffic pattern a source node $x_1 x_2 \cdots x_n$ sends messages to node $B(x_1 x_2 \cdots x_n) = x_{n/2} \cdots x_1 x_n \cdots x_{n/2+1}$ (when n is even) or $x_{(n-1)/2} \cdots x_1 x_n \cdots x_{(n-1)/2+1}$ (when n is odd). Let us now calculate the probability, P_{β_i}, that a generated bit-reversal message makes i hops to cross the network.

Let us now calculate the probability, P_{m_i}, that a newly-generated matrix-transpose message makes i hops to cross the network. Examining the address patterns generated by matrix-transpose permutations reveals that this probability has to be calculated in different ways for odd and even values of n (the dimensionality of the hypercube). Let $x = x_1 x_2 \cdots x_n$ be the source address and $M(x) = x'_1 x'_2 \cdots x'_n$ be the destination address for a matrix-transpose message. When n is even, every single bit difference between the first $n/2$ bit positions of the source and destination addresses, $x_1 x_2 \cdots x_{n/2}$ and $x'_1 x'_2 \cdots x'_{n/2}$, results in another bit difference in the remaining $n/2$ bit positions of addresses, $x_{n/2+1} x_{n/2+2} \cdots x_n$ and $x'_{n/2+1} x'_{n/2+2} \cdots x'_n$. Therefore, the probability that a matrix-transpose message makes i hops is zero when i is odd. Let us determine the number of possible cases where the source and destination of a matrix-transpose message are located i hops away from each other ($i = 0, 2, 4, \ldots, n$). This can be done by simply considering only the first $n/2$ bit positions in the source and destination addresses, and thus enumerating the number of combinations where $x_1 x_2 \cdots x_{n/2}$ and $x'_1 x'_2 \cdots x'_{n/2}$ are different in exactly j bit positions ($j = 0, 1, 2, \ldots, n/2$). Any bit in the address pattern $x_1 x_2 \cdots x_{n/2}$ with the corresponding bit in the pattern $x'_1 x'_2 \cdots x'_{n/2}$ make up four combinations, which are 00, 01, 10 and 11. In two combinations, 00 and 11, those two bits are equal while in the other two combinations, 01 and 10, they are different.

Therefore, the number of possible combinations that result in the patterns $x_1 x_2 \cdots x_{n/2}$ and $x'_1 x'_2 \cdots x'_{n/2}$ being different in exactly j bits is $\binom{n/2}{j} 2^{n/2-j} 2^j$ ($j = 0, 1, 2, \ldots, n/2$). So, the number of possible combinations where $x_1 x_2 \cdots x_n$ and $x'_1 x'_2 \cdots x'_n$ are different in exactly i bits ($i = 0, 2, 4, \ldots, n$) is given by

$$n_{m_i,even} = \binom{\frac{n}{2}}{\frac{i}{2}} 2^{\frac{n-i}{2}} 2^{\frac{i}{2}} = 2^{\frac{n}{2}} \binom{\frac{n}{2}}{\frac{i}{2}}, \quad i = 0, 2, 4, \ldots, n. \tag{12}$$

Consider the case where n is odd. Examining the address patterns of the source $x_1 x_2 \cdots x_n$ and destination $x'_1 x'_2 \cdots x'_n$ generated by the matrix-transpose permutation shows that finding the number of combinations where these address patterns are different in exactly i bits ($i = 0, 1, 2, \ldots, n$) is equivalent to the problem of finding the number of ways that i bits can be placed on a "fictive" circle such that no two adjacent bits on the circle be equal. It is can easily be checked that when i is odd there is no way to place i bits on the circle where no two adjacent bits are equal. When i is even, two configurations meet the desired condition. These i bits can be selected from n bits in $\binom{n}{i}$ different combinations resulting in a total of

$$n_{m_i,odd} = 2\binom{n}{i}, \quad i = 0, 2, 4, \ldots, n\text{-}1. \tag{13}$$

combinations where the address patterns $x_1 x_2 \cdots x_n$ and $x'_1 x'_2 \cdots x'_n$ are different exactly in i bits. Combining equations (4) and (5) gives a general equation for the number of possible combinations that result in the address patterns $x_1 x_2 \cdots x_n$ and $x'_1 x'_2 \cdots x'_n$ being different in exactly i bits ($i = 0, 1, \ldots, n$) as

$$n_{m_i} = \begin{cases} 0 & \text{if } i \text{ is odd} \\ 2^{\frac{n}{2}} \binom{\frac{n}{2}}{\frac{i}{2}} & \text{if } (i \text{ is even)and}(n \text{ is even)} \\ 2\binom{n}{i} & \text{if } (i \text{ is even)and}(n \text{ is odd)} \end{cases} \tag{14}$$

Thus, the probability that a matrix-transpose message makes i hops to reach its destination can be written as

$$P_{m_i} = \frac{n_{m_i}}{N - n_{m_0}} = \begin{cases} 0, & \text{if } i \text{ is odd} \\ \dfrac{\binom{\frac{n}{2}}{\frac{i}{2}}}{2^{\frac{n}{2}} - 1}, & \text{if } (i \text{ is even)and}(n \text{ is even)} \\ \dfrac{\binom{n}{i}}{2^{n-1} - 1}, & \text{if } (i \text{ is even)and}(n \text{ is odd)} \end{cases} \tag{15}$$

The average number of hops that a matrix-transpose message makes across the network is given by

$$d_{matrix-transpose} = \sum_{i=1}^{n} iP_{m_i} .$$
(16)

Again assuming that the generation rate of bit-reversal messages at a node is m and thus the rate for generating uniform messages is 1-m. Therefore, the average distance that a message takes in the network can be given as

$$d_m = md_{matrix-transpose} + (1-m)d_u .$$
(17)

4 The Effect Routing Algorithm

In this section, we show that even with uniform traffic pattern a routing algorithm may result in unbalanced traffic rate over network channels. With uniform traffic and the use of e-cube or fully adaptive routing the message arrival rate over network channels is balanced.

Let us now analyze the traffic rate over network channels when p-cube partially adaptive routing is used. The way in which this algorithm partitions channels and uses them in two steps, causes traffic load to become heavier in some corners of the network.

Assume the source node x and destination node y as shown below. Let us now calculate the probability that a message passes node a when going from x to y. It is easy to see that the number of combinations to have t bits of l zero bits equal to one is given by $\binom{l}{t}$. Out of these combinations only one pattern corresponds to node a as shown below.

$$
\begin{array}{ll}
x & 00\cdots00\cdots00\cdots01111\cdots111 \\
\downarrow & \\
a & 00\cdots00\cdots01\cdots11111\cdots111 \\
\downarrow & \\
y & 00\cdots01\cdots11\cdots10\cdots01\cdots11
\end{array}
$$

Thus, the probability to pass node a when traversing the network from node x to node y is given by

$$P_{\underset{x \to y}{a}} = \frac{1}{\binom{l}{t}} .$$
(18)

Assuming a hypercube, H_n, with p-cube routing, we now calculate the traffic arrival rate, $\lambda_{<a,b>}$, over channel $<a,b>$, where $a = \underbrace{00\cdots0}_{n-m}\underbrace{11\cdots1}_{m}$, and $b = \underbrace{00\cdots0}_{n-m-1}\underbrace{11\cdots1}_{m+1} .$

Note that all possible source and destination nodes, x and y, for which a and b are passed have the address form shown below

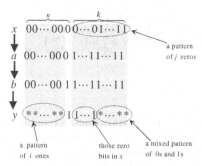

So, if x and y can be a source and destination node with a possible path between them including channel $<a, b>$, the k-th bit of x must be 0 while it is 1 in y. Note that bits $1, 2, \ldots, k\text{-}1$ in x can not be 1, otherwise channel $<a, b>$ can not be passed. This is because if bit t, $t < k$, in x be 1, then the t-th bit in y can not be 1 as this does not result in passing channel $<a, b>$; also the t-th bit in y can not be 0, since we are still in the first phase of routing (note that still k-th bit has not changed) and again channel $<a, b>$ can not be passed.

Also note that none of the two corresponding bits in x and y in the last m-bit part can be 0 at the same time, otherwise channel $<a, b>$ can not be passed.

Considering the above conditions we can see that x and y have the following attributes:
- i bits in the first $(k\text{-}1)$-bit part of y are equal to 1,
- j bits in the last m-bit part of x equals 0 while they are 1 in y,
- the remaining $m\text{-}j$ bits of the last m-bit part are either 1 in x and 0 in y, or 1 in both x and y.

It is easy to see that the number of cases for which x and y fulfill the above conditions is given by $\binom{k}{i}\binom{m-1}{j}2^{m-j-1}$. The probability that a message from x to y passes node a along its path is given by equation (18) as $1/\binom{i+j+1}{j}$, and the probability that it passes channel $<a, b>$ after node a is equal to $1/(i+1)$. Thus, the probability that a message from x to y passes channel $<a, b>$ can be given by

$$P_{\substack{<a,b> \\ x \to y}} = \frac{1}{\binom{i+j+1}{j}(i+1)} \cdot \tag{19}$$

Recalling that the message generation rate of each node, e.g. node x, is λ, and that a message can be destined to any other $N = 2^{n+m} - 1$ nodes, and summing up all the possible combination of i's and j's (in above discussion), we can calculate the message arrival rate over channel $<a, b>$ as

$$\lambda_{<a,b>} = \frac{\lambda}{2^n - 1} \sum_{i=0}^{n-m-1} \sum_{j=0}^{m} \frac{\binom{n-1}{i}\binom{m}{j}2^{m-j}}{(i+j+1)\binom{i+j}{i}} \cdot \tag{20}$$

Using a similar analysis, we can derive an expression to calculate the traffic arrival rate on channel $< a,b >$, where $a = \underbrace{00 \cdots 0}_{n-m}\underbrace{1 1 \cdots 1}_{m}$, and $b = \underbrace{00 \cdots 0}_{n-m+1}\underbrace{1 1 \cdots 1}_{m-1}$, as

$$\lambda_{<a,b>} = \frac{\lambda}{2^n - 1} \sum_{i=0}^{n-m} \sum_{j=0}^{m-1} \frac{\binom{n}{i} \binom{m-1}{j} 2^{m-j-1}}{(i+j+1) \binom{i+j}{i}} . \tag{21}$$

5   Discussions

In this section, we utilize the analytical expressions derived in previous sections. The network configuration used for our analysis is a 10-dimensional hypercube, for the sake of our presentation. Nodes generate messages using a Poisson model with an average message generation of $\lambda = 0.01$. However, the conclusions made for each case are valid for other network configurations and message generation rates at each node.

Figure 2 shows the message arrival rate over different channels of a 10-D hypercube for different hotspot traffic portions $h = 0$ (defining a pure uniform traffic), 0.2, 0.4, 0.6, 0.8 and 1.0 (corresponding to pure hotspot traffic). The obvious non-uniformity in message arrival rate over network channels (as can be seen in the figure) may result in significant performance degradation compared to the uniform traffic. As shown in this figure, the closer the channel is to the hotspot node, the higher the message arrival rate is. However, the arrival rate over the channel located one hop away from the hotspot node is extremely higher than the rest of channels (those located some hops away from the hotspot node). In addition, the message arrival rates for channels located 3 hops away, or more, from the hotspot node are of little difference.

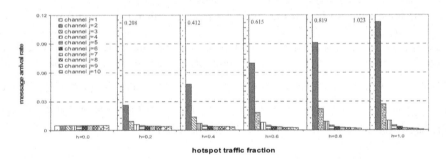

Fig. 2. The message arrival rate over different channels in a 10-cube for different hotspot rates

Figure 3 shows the average number of hops that a message takes in different sized hypercubes under different bit-reversal traffic portions $\beta = 0$ (as the representative of pure uniform traffic pattern), 0.2, 0.4, 0.6, 0.8, 1.0 (as the pure bit-reversal traffic) when $\lambda = 0.01$. It can be seen in this figure that the average path length decreases in proportion with the increase in the bit-reversal traffic ratio β. The average path length, however, increases when network size (n) increases for different bit-reversal traffic portions β.

In Figure 4, the average number of hops that a message takes in different sized hypercubes under different matrix-transpose traffic portions $m = 0$ (as the representative of pure uniform traffic pattern), 0.2, 0.4, 0.6, 0.8, 1.0 (defining a pure matrix-transpose traffic) is shown when $\lambda = 0.01$. It can be seen in the figure, the average path length decreases proportionally when the matrix-transpose traffic portion increases. The interesting thing about the effect of matrix-transpose traffic pattern is the effect of network dimensionality.

It can be seen from the figure than for low matrix-transpose rates the average path length increases when network size increase. The opposite, however, is observed for high matrix-transpose traffic rates where the average path length decreases when network dimensionality increases. This is different from that observed under the bit-reversal traffic pattern.

Figure 5 shows the message arrival rate over network channels in the 10-D hypercube when $\lambda = 0.01$ and P-cube routing algorithm is used for uniform traffic pattern.

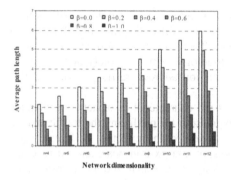

Fig. 3. The average path length messages take in a 10-cube for different bit-reversal traffic rates

Fig. 4. The average path length messages take in a 10-cube for different matrix-transpose traffic rates

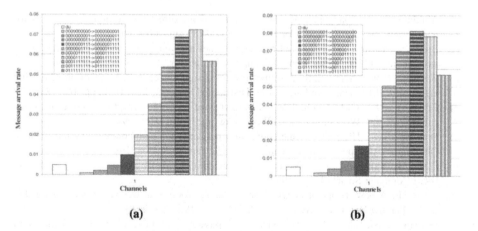

Fig. 5. The message arrival rate on different channels in a 10-cube using p-cube routing; a) negative channels, b) positive channels

As can be seen in the figure, although the traffic pattern used is uniform, the message arrival rates over different channels are very different. For the sake of comparison, the white bar shows the rate when e-cube or full-adaptive routing algorithms are used. In that case, the traffic rate over network channels is distributed evenly. The address pattern of nodes indicating the channel) is small and it gradually increases when the channel weight increases. This continues until the channel Hamming weight is approximately 8 (or 9) after which there is a little decrease again on channel arrival rate. Note that each bar represents the message arrival rate over a group of channels with equal Hamming weight.

6   Conclusions

The performance of an interconnection network mainly depends on the distribution of channel traffic. In most studies, it is shown when the traffic is evenly distributed over network channels that performance increases. There are many sources of network channel load imbalance, including network topology (which is solely dependent on the definition of the network), message destination address traffic pattern, and the routing algorithm employed.

In this paper, we have analytically studied the characteristics of network channel load and have derived some mathematical expression for predicting the traffic arrival rate over different network channels. We considered the effect of different well-known traffic patterns including uniform, hotspot, matrix-transpose, and bit-reversal traffic patterns on channel traffic rate. In addition, we have analyzed p-cube routing (a partially adaptive routing algorithm in hypercubes) and have shown that even under uniform traffic, the traffic load on different channels may vary.

References

1. Abraham, S., Padmanabhan, K.,: Performance of the direct binary n-cube networks for multiprocessors. IEEE Trans. Computers, 37(7), 1989, pp. 1000-1011.
2. Boura, Y.: Design and analysis of routing schemes and routers for wormhole-routed mesh architectures, Ph.D. Thesis, Dept of Computer Sci. & Eng., Penn. State Univ., 1995.
3. Boppana, R.V., Chalasani, S.: A framework for designing deadlock-free wormhole routing algorithms. IEEE Trans. on Parallel and Distributed Systems, 7(2), 1996, pp. 169-183.
4. Boura, Y., Das, C.R., Jacob, T.M.: A performance model for adaptive routing in hypercubes. Proc. Int. Workshop Parallel Processing, pp. 11-16, Dec. 1994.
5. Dandamudi, S.: Hierarchical interconnection networks for multicomputer systems. Ph.D. Dissertation, Computer Science, Univ. Saskatchewan, Saskatoon, 1988.
6. Duato, J., Yalamanchili, S., Ni, L.: Interconnection networks: An engineering approach. IEEE Computer Society Press, 1997.
7. Duato, J.: A new theory of deadlock-free adaptive routing in wormhole networks. IEEE Trans. Parallel Distributed Systems, Vol. 4, No. 12, 1993, pp. 1320-1331.
8. Glass, C.J., Ni, L.M.: The turn model for adaptive routing. Proc. of 19th Int. Symposium Computer Architecture, 1992, pp. 278-287.
9. Guan, W.J., Tsai, W.K., Blough, D.: An analytical model for wormhole routing in multicomputer interconnection networks. Proc. Int. Conference Parallel Processing, 1993, pp. 650-654.

10. Hady, F.T., Menezes, B.L.: The performance of crossbar-based binary hypercubes. IEEE Trans. Computers, 44(10), 1995, pp. 1208-1215.
11. Horng, M., Kleinrock, L.: On the performance of a deadlock-free routing algorithm for boolean n-cube interconnection networks with finite buffers. Proc. Int. Conference Parallel Processing, 1991, pp. III-228-235.
12. Kim, J., Das, C.R.: Hypercube communication delay with wormhole routing. IEEE Trans. Computers, 43(7), 1994, pp. 806-814.
13. Linder, D.H., Harden, J.C.: An adaptive and fault tolerant wormhole routing strategy for k-ary n-cubes. IEEE Trans. on Computers, 40(1,) 1991, pp. 2-12.
14. Pfister, G.J., Norton, V.A.: Hot-spot contention and combining in multistage interconnection networks. IEEE Trans. Computers, 34(10), pp. 943-948, 1985.
15. Sarbazi-Azad, H.: A mathematical model of deterministic wormhole routing in hypercube multicomputers using virtual channels. Journal of Applied Mathematical Modeling, 27(12), 2003, pp. 943-953.
16. Sarbazi-Azad, H.: Performance analysis of multicomputer interconnection networks. PhD Thesis, Glasgow University, Glasgow, UK, 2002.

Capturing an Intruder in the Pyramid

Pooya Shareghi, Navid Imani, and Hamid Sarbazi-Azad

IPM School of Computer Science, and Sharif University of Technology, Tehran, Iran
Azad@ipm.ir, Azad@Sharif.edu

Abstract. In this paper, we envision a solution for the problem of capturing an intruder in one of the most popular interconnection topologies, namely the pyramid. A set of agents collaborate to capture a hostile intruder in the network. While the agents can move in the network one hop at a time, the intruder is assumed to be arbitrarily fast, i.e. it can traverse any number of nodes contiguously as far as there are no agents in those nodes. Here we consider a new version of the problem where each agent can replicate new agents when needed, i.e. the algorithm starts with a single agent and new agents are created on demand. In particular, we propose two different algorithms on the pyramid network and we will later discuss about the merits of each algorithm based on some performance criteria.

1 Introduction

In networked environments that support software migration, a possible threat is the existence of a hostile software agent called the *intruder* that can move from node to node and possibly perform harmful attacks to its host. The intruder can be any kind of malicious software, ranging from viruses and spywares to malfunctioning programs. To cope with the intruder, some special mobile agents, i.e. the *searcher agents* are deployed in the network. They normally start from a homebase and move within the network according to a predefined strategy to find and neutralize the intruder. On the other hand, the intruder moves arbitrarily fast, and it is aware of the position of other agents, thus if feasible, it finds an escape path. The intruder is captured when it cannot find any undefended neighboring node to escape to. We consider that the agents (both types) only reside at nodes, not links, and they do not jump through the network; i.e. they can only move from their current position to a neighboring node through the link between the two nodes.

The concept of *Intrusion Detection* is extensively studied in the past [1], [2], [3], [4], [5], [17], and [18]. Most of the previous works have assumed that in the beginning, all the searcher agents are present at the homebase, and they move to other nodes whenever needed [5]. In our model, however, the searcher agents are capable of replicating other agents, thus the agents are created on demand at run-time. Moreover, the searcher agents though collaborative are completely autonomous. While the recent assumptions are quite realistic in the software world, they decrease the overall steps taken by the agents and hence the network traffic. In the above problem, it is crucial to define an optimal strategy with respect to both the *number of steps* taken to find the intruder and the *number of agents* needed.

D. Grigoriev, J. Harrison, and E.A. Hirsch (Eds.): CSR 2006, LNCS 3967, pp. 580–590, 2006.
© Springer-Verlag Berlin Heidelberg 2006

The described scenario is also known as the *contiguous search problem* in which agents cannot be removed from the network, and clear links must form a connected sub-network at any time, providing safety of movements [2]. This new problem is NP-complete in general [16].

In fact, intrusion detection is a variant of the more general problem of *Graph Searching*, which is well studied in the discrete mathematics [6], [7], [8], [9], and [10]. The graph search algorithms studied in the literature vary in the agents' level of environmental information. In most of the models, it is assumed that the agents, i.e. the searchers and the intruder, have a comprehensive knowledge of their environment and can see and act according to the movement of other participants. This model is known as the *basic pursuit-evasion* model with complete information or BPE for short [9].

In order to narrow the gap between our model and the problem in the real world, we discuss the worst case where the collaborative agents are only aware of the topology of the network, and are blind in the sense that they can see neither the intruder nor the other collaborators. Intruder on the other hand, is assumed to be arbitrarily fast and can track the movement of the searchers and is intelligent enough to select the best possible move in order to evade the searchers. Although the intruder is able to cause damage to its host, it cannot imperil the other agents.

Beside network security, the graph-searching problem has many other applications, including pursuit-evasion problems in a labyrinth, decontamination problems in a system of tunnels, and mobile computing problems in which agents or robots are looking for a hostile intruder. Moreover, the graph-searching problem also arises in VLSI design through its equivalence with the gate-matrix layout problem [2].

In this paper, we demonstrate two different cleaning strategies applied to one of the most common interconnection topologies namely the pyramid. The pyramid has been one of the most important network topologies as it has been used as both hardware architecture and software structure for parallel and network-based computing, image processing and machine vision [12], [13], [14], and [15].

2 Definitions and Notations

2.1 The Pyramid

A pyramid network (or a pyramid for short) is a hierarchical structure based on meshes. An $a \times b$ mesh (Fig. 4a in the appendix), $M_{a,b}$, is a set of nodes $V(M_{a,b}) = \{(x,y) \mid 1 \le x \le a, 1 \le y \le b\}$ where nodes (x_1, y_1) and (x_2, y_2) are connected by an edge iff $|x_1 - x_2| + |y_1 - y_2| = 1$ [11].

A *pyramid* of n levels (Fig. 4b in the appendix), denoted as P_n, consists of a set of nodes $V(P_n) = \{(k,x,y) \mid 0 \le k \le n, 1 \le x, y \le 2^k\}$. A node $(k,x,y) \in V(P_n)$ is said to be a node at level k. All the nodes in level k form a $2^k \times 2^k$ mesh, $M_{2^k \times 2^k}$, which with a slight abuse of notation, we denote as M^k. Hence, there are $N = \sum_{k=0}^{n} 4^k = (4^{n+1} - 1)/3$ nodes in P_n. A node $(k,x,y) \in V(P_n)$ is connected, within the mesh at level k, to four nodes $(k, x-1, y)$ if $x>1$, $(k, x, y-1)$ if $y>1$, $(k, x+1, y)$ if $x<2^k$, $(k, x, y+1)$ if $y<2^k$. It is also connected to nodes $(k+1, 2x-1, 2y)$, $(k+1, 2x, 2y-1)$, $(k+1, 2x-1, 2y-1)$, and $(k+1, 2x, 2y)$ for $0 \le k < n$ in the next level, $k+1$, and to node $(k-1, \frac{x-1}{2}+1, \frac{y-1}{2}+1)$ in level $k-1$ [11]. We use term k-*neighbor* to refer to every neighbor of a node $u=(k_u, x_u, y_u)$ that has the same x and y value than u but a k value equal to k_u+1 or k_u-1. In a similar manner we use terms x-*neighbor* and y-*neighbor*. The apex node in P_n is the node $(0,1,1)$ which can alternatively be denoted as P_n ▲. The corner nodes of P_n are $(n,1,1)$, $(n,2^n,1)$, $(n,2^n,2^n)$ and $(n,1,2^n)$ or P_n ▼, P_n ◥, P_n ◢, and P_n ◣, respectively.

2.2 Basic Graph Searching Definitions

Given a pyramid network that accommodates an intruder in one of its nodes, the problem is to find a strategy according to which the network reaches a state that all of nodes are clean.

At any instant of time, any node u in the pyramid could be:

- *Contaminated*, when capable of harboring an intruder,
- *Guarded*, when an agent resides at node u,
- *Clean*, when all neighboring nodes are clean or guarded, and an agent has already passed node u.

On the other hand, during each search step, a searcher agent is capable of doing one the following actions:

- It can move to any of its neighboring nodes.
- It can wait until a specific condition is met.
- The agent can produce a new searcher agent and inject it to a neighboring node.

Simultaneously, a searcher agent can sense the state of neighboring nodes, i.e. if they are clean/defended or contaminated. To achieve this capability, we consider a tiny memory space in every node (almost equal to the degree of that node) as a bulletin board where information about neighbors is stored. In addition, when a searcher clears a node, it sends a single bit message to all neighboring nodes to indicate the new state of the current node. The above consideration obviates the necessity of a synchronizer to pace the agents while cleaning the network.

Definition 1. The *search number* of a graph G according to a model M for searching G is the smallest k such that k searchers can capture any intruder lurked in G using model M, and is denoted as $sn_M(G) = k$.

Definition 2 [16]. A search strategy is a sequence of search steps that will clear an initially contaminated graph.

According to definition 1, a search strategy uses optimal amount of agents for a given graph G if it needs only $sn_M(G)$ searcher agents for capturing an intruder in graph G. Throughout this paper, we use the terms strategy and algorithm, interchangeably.

Definition 3 [5]. A node search strategy is called *contiguous* and *monotone* where:
1. the removal of agents is not allowed,
2. at any step of the search strategy, the set of clean nodes forms a connected subnetwork,
3. a clean node cannot be recontaminated.

The following notations will be used throughout the paper.

- G_t: a graph G referred to at time instant t.
- $|G|$: size (or order) of a graph G (this notation is also used to denote the size of a set G).
- $d_G(u,v)$: distance between two vertices u and v, in G, i.e. the length of a shortest path between the nodes u and v of G.
- $diam(G)$: diameter of a graph G, i.e. the maximum $d_G(u,v)$ for all u and v in G.
- $A^G = \{a_1^G, a_2^G, ..., a_k^G\}$: the set of searcher agents a_1^G, a_2^G, ..., a_k^G employed to decontaminate a graph G.
- C_t^G : the set of clean nodes in a graph G_t.
- I_t^G : the set of nodes in a graph G_t that are not clean (i.e. can accommodate the intruder), $I_t^G = G_t \setminus C_t^G$.

Definition 4. The *Clean Territory* of a graph G_t is an induced subgraph CT_t^G , where C_t^G induces CT_t^G in G_t.

As the clean nodes do not get contaminated again and the agents can only move to their neighboring nodes, one may simply infer that in a contiguous and monotone node search strategy the clean territory, CT_t^G , is a connected graph at any instant of time.

Definition 5. The *Intruder Territory* of a graph G_t is the induced subgraph IT_t^G , where $I^G(t)$ induces IT_t^G in G_t.

To hinder the intruder from recontaminating the clean territory, any efficient algorithm should move the agents in such a way that they form an impenetrable frontier at any intermediate stage.

Definition 6. The *Battle Front* of a graph G_t is the set of nodes like v, in the clean territory, that has a neighboring node in the intruder territory. It is denoted as $BF_t^G = \{v \mid v \in C_t^G, \exists u \in I_t^G :<u,v> \in E(G_t)\}$.

In order to decontaminate a network we need a strategy that is able to neutralize the intruder after a finite number of steps. Thus, a more formal definition of *strategy* is given below.

Definition 7. A strategy for capturing an intruder in a graph G, when given an intruder and the initial searcher at any two arbitrary nodes, is capable of cleaning the

network after a finite number of steps, T; i.e. $\exists T : \forall t \geq T$, $CT_t^G = G$, where $t=i$ is the time instant of the ith step.
Here we define the lower limit of the steps taken during any algorithm A applied to an arbitrary graph G that starts from a single vertex and must span all vertices of G. Using the following lemma and definition we can evaluate optimality of our algorithms.

Lemma 1. Let A be an arbitrary algorithm that must be applied to a given graph G and must span all of its vertices starting from a single vertex. The lower limit of the steps taken during algorithm A is equal to $\operatorname{diam}(G)$.

Proof. Since the farthest distance between any two vertices in a graph like G is equal to its diameter, thus for spanning all nodes of G one must take $\operatorname{diam}(G)$ steps to reach the farthest node from the initial vertex, when following shortest paths in G.

Definition 8. Any spanning algorithm A applied to a given graph G is said to be *execution-time-optimal* if its execution time complexity is some linear order of $\operatorname{diam}(G)$.

3   Straight Chain Movement (SCM) Strategy

In this section, we propose our first strategy for capturing the intruder in the pyramid, which comprises of two phases:

1. creating straight chains or arrays of searcher agents in each level of the pyramid graph, and
2. moving the chains in a manner that no clear node becomes recontaminated.

3.1   Creating Chains

In this phase, we make a chain of searcher agents in the x dimension, with $y=1$, for every mesh of the pyramid. Without loss of generality, we assume that in the beginning there is just a single searcher agent residing at node $P_n \blacktriangledown =(n,1,1)$. Since the diameter of the pyramid is $2n$, if the searcher agent was elsewhere in the pyramid, we could move it with at most $2n$ moves to the desired location.

In the first step, $t=1$, the initial agent $a_{n,1}$ spawns a new agent $a_{n-1,1}$ and injects it into the neighboring corner node of the upper level mesh, i.e. $(n-1,1,1)$. In time instant $t=2$, agent $a_{n-1,1}$ creates a new agent $a_{n-2,1}$ and sends it to the neighboring node $(n-2,1,1)$.

From time instant $t=2$ up to $t=2^n+n-n=2^n$, any agent $a_{n,t-1}$ creates a new agent $a_{n,t}$ and sends it to the neighboring node $(n,t,1)$. The same procedure is done in the upper level mesh from $t=3$ up to $t=2^{n-1}+n-(n-1)=2^{n-1}+1$. Thus, as a general rule, when agent $a_{k,1}$ enters the mesh of level $k>0$, in the first step, $t=n-k+1$, it creates a new agent and sends it to the mesh at level $k-1$, and then from $t=n-k+2$ up to $t=2^k+n-k$, any agent $a_{k,t+k-n-1}$ creates a new agent $a_{k,t+k-n}$ and sends it to node $(k,t+k-n,1)$. In this way, the apex node will have a resident searcher agent after $t=k$.

Lemma 2. The whole chain making phase takes 2^n steps, and a total amount of $\sum_{i=0}^{n} 2^i = 2^{n+1} - 1$ agents will be created.

Fig. 1 demonstrates steps taken during the chain making phase in P_4.

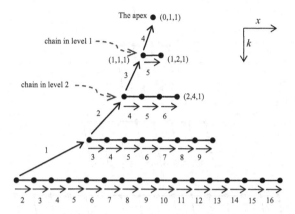

- Links in y and k dimensions are not shown.
- Numbers around the arrow represent the time of *Create_Inject* step.

Fig. 1. Creating Chains in xk-plane, with $y=1$

3.2 Moving Chains

The chains created in the previous phase, together form the battle front. In this phase, no new agents are created and we only move the battlefront along y direction, in such a way that the intruder would not be able to penetrate the already cleaned territory. The agents are identical and autonomous. All the agents behave according to the following simple rule:

- If all k-neighbor nodes of the current node are clean, move to the next y-neighbor node that is contaminated.
- Otherwise wait.

The chains are initially at nodes with $y=1$. Each chain at any level k moves for the first time at $t=n-k+1$. Then, it moves again every 2^{n-k} steps. The total amount of steps needed to accomplish this phase is 2^n-1, which we derive formally in the following lemma.

Lemma 3. The chain-moving phase takes 2^n-1 steps on an n level pyramid to complete.

Proof. Each chain in any level k will move for the last time at $t=n-k+1+(2^k-2)\times 2^{n-k}$. The maximum value for the given formula is achieved when $k=n$. Thus, the whole chain-moving phase takes 2^n-1 steps.

Fig. 2 demonstrates the order in which chains are moved in P_4.

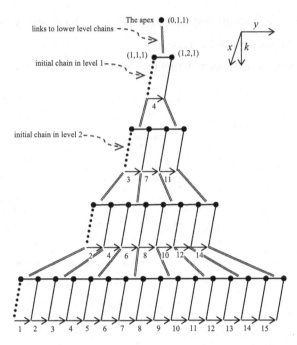

- Bold points represent first node of each chain.
- Black inclined lines represent chains in x dimension, with the leftmost doted ones being the chains created in the previous phase.
- Gray double lines represent links in k dimension.
- Some links in x and y dimensions are not shown.
- Numbers below the arrows represent the time of each *Chain_Move* step.

Fig. 2. Moving chains along dimension y

3.3 Complexity of the SCM Strategy

Theorem 1. The total number of agents needed to clean an n-level pyramid using the SCM strategy is $2^{n+1}-1$, or $O\left(\sqrt{|P_n|}\right)$, which is better than any linear algorithm with respect to the order of the pyramid.

Proof. Since, no agent is created in the second phase, the total number of agents needed is equal to the amount of agents created in the first phase. Thus, according to lemma 2 we need $2^{n+1}-1$ agents.

Theorem 2. Cleaning the whole pyramid takes exactly $3\times2^n-1$, or roughly $O\left(\sqrt{|P_n|}\right)$, steps using the SCM strategy.

Proof. According to lemmas 2 and 3, the first and second phases take 2^n and $2^{n+1}-1$ steps, respectively. Thus, the whole strategy will need roughly 3×2^n steps to clean the network. On the other hand, the pyramid has an order of roughly $4^{n+1}/3$. Therefore, the complexity of the algorithm is about $O\left(\sqrt{|P_n|}\right)$, or exactly $3\times2^n-1$.

We now prove that the proposed strategy is correct in the sense of contiguity and monotony assumptions given in definition 3.

Theorem 3. During the SCM strategy, no node will be recontaminated, and after $3\times2^n-1$ steps the whole network will be clean.

Proof. The pyramid topology, in the kx-plane (or alternatively ky-plane), is like a complete binary tree having the apex node as the root where every two successive nodes in the same level are connected. The intuition behind the SCM strategy is to construct the mentioned semi complete binary tree graph in kx-plane where $y=1$, and then pivoting the tree along dimension y, with the apex node being the pivot point. In this way, obviously there would be no way out for the intruder in any intermediate stage. In addition, according to theorem 2 the network will be clean after $3\times2^n-1$ steps.

Lemma 4. During the SCM strategy, when the agent at a node $u=(k_u,x_u,y_u)$ migrates to node (k_u,x_u,y_u+1), all k-neighbors of u and all its y-neighbors with $y<y_u$ are clean.

Regarding the proof of theorem 3, the set of clean nodes forms a connected subnetwork at any step of the search strategy. On the other hand, according to lemma 4, the algorithm moves the agents in such a way that the clean territory is always expanding toward $y=2^n$; hence, no node can be recontaminated. In addition, the proposed strategy does not allow removing of agents from the network. Therefore, one can derive the following theorem.

Theorem 4. The SCM strategy is contiguous and monotonic.

4   Hierarchical Migration (HM) Strategy

4.1   The Algorithm

The SCM strategy proposes an algorithm for capturing the intruder with an acceptable amount of searcher agents. However, the number of agents is not the sole effective factor in determining the performance of an algorithm for capturing an intruder. The other factors may encompass the total number of steps that the algorithm takes to complete, i.e. execution time and the total number of walks, which determines the traffic of agents on the network. Our second strategy tries to minimize the execution time of the algorithm but this swiftness is provided at the cost of more agents in the network. The *Hierarchical Migration* strategy for capturing the intruder in the pyramid is very fast, since it takes only $4n$ steps. However, it needs more agents than the previous algorithm. It begins with an initial agent located at the apex node. The initial agent replicates three agents and injects them into 3 nodes out of the 4 contaminated lower level neighbors. Later, the initial agent moves to the remaining contaminated lower level neighbor. If every agent does the same, the network will be cleared after a finite number of steps. The whole algorithm can be summarized into the following steps. Each agent at a node like (k,x,y) does the following.

1. create a new agent and inject it to node $(k+1,2x-1,2y-1)$,
2. create a new agent and inject it to node $(k+1,2x-1,2y)$,
3. create a new agent and inject it to node $(k+1,2x,2y)$,
4. Move to node $(k+1,2x,2y-1)$.

Fig. 3 demonstrates the order in which nodes are cleaned in P_4.

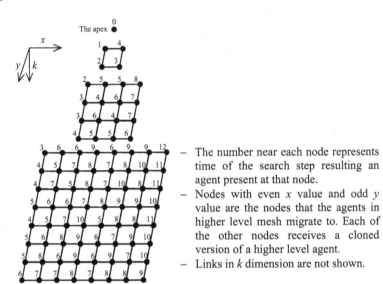

- The number near each node represents time of the search step resulting an agent present at that node.
- Nodes with even x value and odd y value are the nodes that the agents in higher level mesh migrate to. Each of the other nodes receives a cloned version of a higher level agent.
- Links in k dimension are not shown.

Fig. 3. Hierarchical Migration

4.2   Complexity of the HM Strategy

Theorem 5. The total number of agents needed to clean an n-level pyramid using the HM strategy is 2^{2n}, or $O\left(|P_n|\right)$, which is of linear order with respect to the order of the pyramid.

Proof. The intuition behind the algorithm is to decontaminate the pyramid level by level, where for decontaminating a mesh of level k we need $2^k \times 2^k$ agents one for each node. Thus, to decontaminate the last level pyramid, 2^{2n} agents are essential, which is of the same order than the size of the pyramid, $(4^{n+1}-1)/3$.

Theorem 6. Cleaning the whole pyramid takes exactly $4n$, or roughly $O\left(\log|P_n|\right)$, steps using the HM strategy.

Proof. Since every agent present at level k involves in four search steps, and agents act independently and in parallel, thus it takes four steps for migrating from level k to $k+1$. Hence, it takes $4n$ steps to reach the last level mesh, which is of the logarithmic order.

Theorem 7. The HM strategy is *execution-time-optimal*.

Proof. Since the algorithm is of complexity $O\left(\log|P_n|\right)$, which is of linear order of network diameter, according to definition 8 it is execution-time-optimal.

Lemma 5. During the HM strategy, when the agent at node $u=(k,x,y)$ migrates to node $(k+1,2x,2y-1)$, all neighbors of u at level $k-1$ are clean.

Regarding the proof of theorem 5, the set of clean nodes forms a connected subnetwork at any step of the search strategy. On the other hand, according to lemma 5, the algorithm creates and moves the agents in such a way that the clean territory is always expanding downward, i.e. toward the base $k=n$; hence, no node can be recontaminated. In addition, the proposed strategy does not allow removing of agents from the network. Therefore, one can derive the following theorem.

Theorem 8. The HM strategy is contiguous and monotonic.

5 Conclusions

In this paper, we proposed two classes of algorithms for capturing an intruder in the pyramid interconnection network, namely the Straight Chain Movement (SCM) and Hierarchical Migration (HM). The SCM method is interesting in the sense that it uses a very low amount of agents for capturing the intruder in the graph, yet it may take much time to do so. The HM strategy, on the other hand, presents an execution-time-optimal algorithm, which is of linear order of network diameter.

The results of this paper are interesting in the sense that they can be generalized in order to be employed in similar hierarchical interconnection topologies. For example, the HM strategy can be easily adapted to a model of client server systems, where some clients are attached to a server and the server itself is a member of a network of servers which itself is controlled and managed by an ultimate server.

References

1. N. Foukia,J. G. Hulaas and J. Harms. "Intrusion Detection with Mobile Agents". *Proc. of the 11th Annual Conference of the Internet Society (INET '01)*, 2001.
2. L. Barrière, P. Flocchini, P. Fraigniaud, and N. Santoro, "Capture of an Intruder by Mobile Agents", *Proc. of SPAA'02*, Winnipeg, Manitoba, Canada, 2002.
3. M. Asaka, S. Okazawa, A. Taguchi, and S. Goto. "A method of tracing intruders by use of mobile agents", *Proc. of the 9th Annual Conference of the Internet Society (INET'99)*, 1999.
4. G. G. Helmer, J. S. K. Wong, V. Honavar, and L. Miller. "Intelligent agents for intrusion detection", *IEEE Information Technology Conference*, 1998, pp. 121-124.
5. P. Flocchini, M. J. Huang and F. L. Luccio, "Contiguous Search in the Hyperdiamond for Capturing an Intruder", *Proc. 19th IEEE International Parallel and Distributed Processing Symposium (IPDPS)*, Denver, Colorado, 2005.
6. R. Breisch. "An intuitive approach to speleology", *Southwestern Cavers* 6, 1967, pp. 72-78.
7. T. D. Parsons, "Pursuit-evasion in a graph", *Theory and Applications of Graphs*, Springer, Berlin, 1976, pp.426–441.
8. T. D. Parsons, "The search number of a connected graph", *Proc. Ninth Southeastern Conf. Combinatorics, Graph Theory and Computing*, Congress. Num. XXI, Winnipeg, 1978, pp. 549–554.
9. B. Alspach, "Searching and Sweeping Graphs: A Brief Survey", September 2004, accessible at ww.dmi.unict.it/combinatorics04/documenti%20pdf/alspach.pdf.

10. D. Bienstock and P. Seymour, "Monotonicity in graph searching", *Journal of Algorithms* 12, 1991, pp. 239–245.
11. F. Cao, D.F. Hsu, "Fault-tolerance properties of pyramid networks", *IEEE Transactions on Computers*, Vol. 48, No. 1, 1999, pp. 88-93.
12. H. Sarbazi-Azad, M. Ould-Khaoua, and L. M. Mackenzie, "Algorithmic construction of Hamiltonians in pyramids", *Information Processing Letters*, Vol. 80, Elsevier, 2001, pp. 75–79.
13. A. Dingle, H. Sudborough, "Simulation of binary trees and x-trees on pyramid networks", *Proc. IEEE Symp. on Parallel & Distributed Processing*, May 1992, pp. 220-229.
14. F.T. Leighton, "Introduction to parallel algorithms and architectures: arrays, trees, hypercubes", Morgan Kaufmann, 1992.
15. J. F. Jenq, and S. Sahni, "Image Shrinking and Expanding on a Pyramid", *IEEE Transactions On Parallel and Distributed Systems*, Vol. 4, No. 11, Nov. 1993, pp. 1291-1296.
16. N. Megiddo, S. Hakimi, M. Garey, D. Johnson and C. Papadimitriou. "The complexity of searching a graph", *Journal of the ACM*, Vol. 35, No. 1, 1988, pp.18-44.
17. M. Demirbas, A. Arora, and M. Gouda, "A pursuer-evader game for sensor networks", In *Proceedings of the Sixth Symposium on Self- Stabilizing Systems*, 2003, San Francisco, CA, USA, June 24-25, LNCS 2704, pp 1-16.
18. P. Yospanya, B. Laekhanukit, D. Nanongkai, and J. Fakcharoenphol, "Detecting and cleaning intruders in sensor networks", *Proceedings of the National Comp. Sci. and Eng. Conf. (NCSEC'04)*.

Appendix

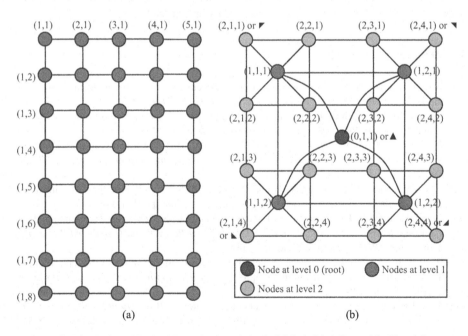

(a)   (b)

Fig. 4. Mesh and Pyramid topologies; a) An 8x5 Mesh, b) A Pyramid of level 2

Speech Enhancement in Short-Wave Channel Based on Empirical Mode Decomposition

Li-Ran Shen, Qing-Bo Yin, Xue-Yao Li, and Hui-Qiang Wang

The College Of Computer Science and Technology,
Harbin Engineering University,
NO.145 Nantong Street, Nangang District, Harbin, China

Abstract. A novel speech enhancement method based on empirical mode decomposition is proposed. The method is a fully data driven approach. Noisy speech signal is decomposed adaptively into oscillatory components called Intrinsic Mode Functions (IMFs) using a process called sifting. The empirical mode decomposition denoising involves thresholding each IMFs. A nonlinear function is introduced for amplitude thresholding. And then reconstructs the estimated speech signal using the processed IMFs. The experimental results show significant improvement in output SNR and quality as compared to recently reported results.

1 Introduction

In short-wave channel communication there is a great deal of interferential noise exist-ing in the surrounding environment, transmitting media, electronic communication device and other speakers' sound, etc. common in most practical situations. In general, the addition of noise reduces intelligibility and degrades the performance of digital voice processors used for applications such as speech compression and recognition. Therefore, the problem of removing the uncorrelated noise component from the noisy speech signal, i.e., speech enhancement, has received considerable attention. In speech communication over the short-wave channel the purpose is to elevate the objective quality of speech signal and the intelligibility of noisy speech in order to reduce the listener fatigue. There have been numerous studies on the enhancement of the noisy speech signal. Many different types of speech enhancement algorithms have been proposed and tested [1, 2, 3, 4, 6]. Spectral subtraction is a traditional method of speech enhancement [6]. The major drawback of this method is the remaining musical noise. Additionally a drawback of speech enhancement methods is the distortion of the useful signal. The resolution is the compromise between signal distortion and residual noise. Though this problem is well known, the study results indicate that both of these cannot be minimized simultaneously. Minimum mean square error (MMSE) [3] estimates on speech spectrum have been proposed. And Ephraimand Van Trees proposed a signal-subspace-based spectral domain algorithm, which controls the energy of residual noise in a certain threshold while minimizing the signal distortion. Hence the probability of noise perception can

D. Grigoriev, J. Harrison, and E.A. Hirsch (Eds.): CSR 2006, LNCS 3967, pp. 591–599, 2006.

be minimized. The drawback of this method is that it deals only with white noise. EMD theory is the newly developed time-frequency analysis technology and is especially of interest in non-stationary signals such as water, sonar, seismic signal, etc [7, 8, 9, 10].

EMD is a technique that has been designed primarily for obtaining AM-FM type representations in the case of signal, which are oscillatory (possibly non-stationary or generated by a non-linear system), in automatic, full data-driven way. Summarily, the main point of EMD is to consider oscillatory signals at the level of their local oscillations and to formalize the idea that: "signal=fast oscillations superimposed to slow oscillations," and to iterate on the slow oscillations component considered as a new signal.

The application of EMD has many important problems, the first one is end effect [7] and the second is the IMF criterion [7].

In this paper, a novel stop criterion was used to IMF criterion. And regression support vector machine was used to solve the end effect problem. At last, the new technique is utilized to decompose speech signals. A soft thresholding is used to denoise every de-composed component. And then reconstruct signal to achieve speech enhancement.

2   Empirical Mode Decomposition

The empirical mode decomposition(EMD) was first introduced by Huang et al. [7]. The principle of this technique is to decompose adaptively a given signal into oscillating components. These components are called intrinsic mode functions (IMFs) and are obtained from the signal by means of an algorithm, called sifting. It is a fully data driven method.

2.1   The Algorithm of EMD

The algorithm to create IMFs is elegant and simple. Firstly, the local extremes in the time series data $X(t)$ are identified, and then all the local maxims are connected by a cubic spline line $U_X(t)$, known as the upper envelope of the data set. Then, we repeat the procedure for the local minima to produce the lower envelope, $L_X(t)$.

Their mean $m_1(t)$ is given by:

$$m_1(t) = \frac{L_X(t) + U_X(t)}{2} \tag{1}$$

It is a running mean. We note that both envelopes should cover by construction all the data between them.

Then we subtract the running mean $m_1(t)$, from the original data $X(t)$, and we get the first component, $h_1(t)$, i.e.:

$$h_1(t) = X(t) - m_1(t) \tag{2}$$

To check if $h_1(t)$ is an IMF, we demand the following conditions: (i)$h_1(t)$ should be free of riding waves i.e. The rest component should not display under-shots or over-shots riding on the data and producing local extremes without zero crossing. (ii) To display symmetry of the upper and lower envelops with respect to zero. (iii) Obviously the number of zero crossing and extremes should be the same in both functions.

The sifting process has to be repeated as many times as it is required to reduce the extracted signal to an IMF. In the subsequent sifting process steps, $h_1(t)$ is treated as the data, then:

$$h_{11}(t) = h_1(t) - m_{11}(t) \tag{3}$$

If the function $h_{11}(t)$ does not satisfy criteria (i)-(iii), then the sifting process continues up to k times, h_{1k}, until some acceptable tolerance is reached:

$$h_{1k}(t) = h_{1(k-1)}(t) - m_{1k}(t) \tag{4}$$

The resulting time series is the first IMF, and then it is designated as $C_1(t) = h_{1k}$. The first IMF component from the data contains the highest oscillation frequencies found in the original data $X(t)$.

The first IMF is subtracted from the original data, and this difference, is called a residue $r_1(t)$ by:

$$r_1(t) = X(t) - C_1(t) \tag{5}$$

The residue $r_1(t)$ is taken as if it was the original data and we apply to it again the sifting process. The process of finding more intrinsic modes $C_j(t)$ continues until the last mode is found. The final residue will be a constant or a monotonic function; in this last case it will be the general trend of the data.

$$X(t) = \sum_{j=1}^{n} C_j(t) - r_n(t) \tag{6}$$

Thus, one achieves a decomposition of the data into n-empirical IMF modes, plus a residue, $r_n(t)$, which can be either the mean trend or a constant. We must point out that this method do not requires a mean or zero reference, and only needs the locations of the local extremes.

2.2   Maximum Power Entropy Used to IMF Stop Criterion

EMD decomposes the signal according to the signal itself and the number of the IMFs is finite. IMFs reflect the intrinsic and reality information of the analyzed signal. Therefore, EMD method is a self-adaptive signal-processing method that is suitable for the analysis of non-linear and non-stationary process.

But if the decomposition is not correct, Each IMF component does not reflect the real physical processing. In addition, to obtain the IMF components with EMD method, it is necessary to confirm the IMF criterion. Because EMD method is a 'sifting' process, to decompose one IMF component needs many times 'sifting'.

In this paper, we take the EMD as a system. And the aim of 'sifting' is to make the system stationary. If the entropy is the same between the sifting, we take the sifting processing should to be stopped.

Before sifting we take the system state is random. And the entropy is:

$$
\begin{aligned}
H_c[p(x), X] &= -\int_{-\infty}^{\infty} p(x) \log_2 p(x) dx \\
&= -\int_{-\infty}^{\infty} p(x) \log_2 \frac{1}{\sqrt{2\pi\sigma^2}} e^{-\frac{(x-m)^2}{2\sigma^2}} dx \\
&= -\int_{-\infty}^{\infty} p(x)(-\log_2 \sqrt{2\pi\sigma^2}) dx + \int_{-\infty}^{\infty} p(x)(\log_2 e)[\frac{(x-m)^2}{2\sigma^2}] \\
&= \log_2 \sqrt{2\pi\sigma^2} + \frac{1}{2} \log_2 e \\
&= \frac{1}{2} \log_2(2\pi e \sigma^2)
\end{aligned}
\tag{7}
$$

After the sifting the system entropy is:

$$
\begin{aligned}
H_c[q(x), X] &= -\int_{-\infty}^{\infty} q(x) \log_2 q(x) dx \\
&= \int_{-\infty}^{\infty} q(x) \log_2[\frac{1}{q(x)} \cdot \frac{p(x)}{p(x)}] dx \\
&= -\int_{-\infty}^{\infty} q(x) \log_2 p(x) dx + \int_{-\infty}^{\infty} q(x) \log_2[\frac{p(x)}{q(x)}] dx \\
&\leq \frac{1}{2} \log_2(2\pi e \sigma^2) = H_c[p(x), X]
\end{aligned}
\tag{8}
$$

Now we can use the difference entropy to decision whether the sifting processing should to be stopped.

$$
\begin{aligned}
I_{p,q} &= H_c[p(x), X] - H_c[q(x), X] \\
&= \frac{1}{2} \log_2(2\pi e P) - \frac{1}{2} \log_2(2\pi e \overline{P}) \\
&= \frac{1}{2} \log_2 \frac{P}{\overline{P}}
\end{aligned}
\tag{9}
$$

2.3   Regression Support Vector Machine Used to End Effect

To solve the end effects of EMD, the regression support vector machines are used to predict the signals before the signal is decomposed by EMD method.

Regression Support Vector Machines (SVRs) [11] are non-parametric predictors designed to provide good generalization performance even on small training sets. A SVM maps (through f) input (real-valued) feature vectors ($x \in X$ with labels $y \in Y$) into a (much) higher dimensional feature space ($z \in Z$) through some nonlinear mapping (something that captures the nonlinearity of the true decision boundary). In a feature space, we can classify the labeled feature vectors (z_i, y_i) using hyper-planes:

$$y_i[< z_i, w > +b] \geq \varepsilon \tag{10}$$

And minimize the functional $\varphi(w) = \frac{1}{2} < w, w >$. The solution to this quadratic program can be obtained from the saddle point of the Lagrangian:

$$L(w, b, \alpha) = \frac{1}{2} < w, w > -\Sigma\alpha_i(y_i[< z_i, w >] - \varepsilon) \tag{11}$$

$$w^* = \Sigma y_i \alpha_i^* z_i, \alpha_i^* \geq 0 \tag{12}$$

The input feature vectors that have positive α_i^* are called Support Vectors $S = \{z_i | \alpha_i^* > 0\}$ and because of the Karush-Kuhn-Tucker optimality conditions, the optimal weight can be expressed in terms of the Support Vectors alone.

This determination of w fixes the optimal separating hyper-plane. The above step has the daunting task of transforming all the input raw features x_i into the corresponding z_i and carrying out the computations in the higher dimensional space Z. This can be avoided by finding a symmetric and positive semi-definite function call the Kernel function between pairs of $x \in X$ and $z \in Z$.

$$K : X \times X \rightarrow R, K(a, b) = K(b, a), a, b \in X \tag{13}$$

Then, by a theorem of Mercer, a transformation $f : X \rightarrow Z$ is induced for which,

$$K(a, b) = < f(a), f(b) > \tag{14}$$

Then the above Lagrangian optimization problem gets transformed to the maximization of the following function:

$$W(\alpha) = \Sigma\alpha_i - \frac{1}{2}\Sigma\alpha_i\alpha_j y_i y_j K(x_i, x_j) \tag{15}$$

$$w^* = \Sigma y_i \alpha_i^* z_i, \alpha_i^* \geq 0 \tag{16}$$

The set of hyper-planes considered in the higher dimensional space Z have a small estimated VC dimension. That is the main reason for the good generalization performance of SVRs.

3   Speech Enhancement Based on EMD

The objective of this research is to extract the clean speech $x(n)$ from degraded speech $y(n)$ by a new soft thresholding criterion.

$$y(n) = x(n) + w(n) \tag{17}$$

Where $w(n)$ is an additive noise.

Because the EMD transform satisfy to addition principle. So we can obtain the following equal.

$$imf_i(n) = imfx_i(n) + imfw_i(n) \tag{18}$$

Where $imf_i(n)$, $imfx_i(n)$ and $imfw_i(n)$ are the ith IMF of $y(n)$, $x(n)$ and $w(n)$ respectively.

Let $im\hat{f}_i(n)$ be an estimation of $imfx_i(n)$ based on the observation noisy speech signal $imf_i(n)$. And then we can use the $im\hat{f}_i(n)$ to reconstruct the speech signal. The following is the method that how to estimate $im\hat{f}_i(n)$ using soft threshold.

Transform domain speech enhancement methods used amplitude subtraction based soft thresholding defined by:

$$im\hat{f}_i(n) = \begin{cases} sgn(imf_i(n))[imf_i(n) - \frac{\lambda_i}{\exp(\frac{imf_i(n)-\lambda_i}{N})}], & |imf_i(n)| < \lambda_i \\ 0, & else. \end{cases} \tag{19}$$

Where the parameter λ_i is

$$\lambda_i = \frac{3\hat{\sigma}_i}{\log(i+1)} \tag{20}$$

$\hat{\sigma}_i$ is the standard deviation of the $imf_i(n)$

$$\hat{\sigma}_i = (\frac{1}{N-1} \sum_{j=1}^{N} (imf_i(j) - \overline{imf_i})^2)^{1/2} \tag{21}$$

Where $\overline{imf_i}$ is the mean of $imf_i(n)$

$$\overline{imf_i} = \frac{1}{N} \sum_{j=1}^{N} imf_i(j) \tag{22}$$

Note that the noise is traditionally characterized by high frequency. The energy will often be concentrated on the high frequency temporal modes(the first IMFs) and decrease towards the coarser ones.

The relationship between SNR, the level of IMF with $\hat{\sigma}_i$ of an arbitrarily selected speech and a noise of the noise-92 database is shown in Fig.1. It is apparent from Fig.1 that alone with the increase of SNR or decomposed layer of IMF, the standard deviation $\hat{\sigma}_i$ make no difference to decrease. So $\hat{\sigma}_i$ is an effective and adaptive threshold.

At last, we can construct the denoised speech signal

$$\hat{x}_i(n) = \sum_{j=1}^{k} im\hat{f}_i(n) + r_n(n) \tag{23}$$

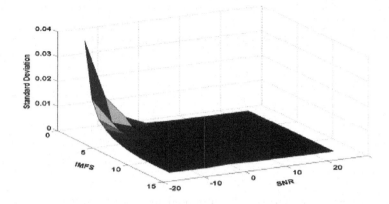

Fig. 1. The relationship between layer, SNR and $\hat{\sigma}_i$

4   Experiments and Analysis

4.1   The Source of Experiment Data

The experiment data come from two parts. The first one is standard noise data-base and speech database. It is used to test the proposed method using different noise types and different SNRs. The second come from real short-wave speech signal records on the spot.

4.2   The Results of Experiments and Analysis

Using the noise database NOISEX-92, different type noise are added to the same pure speech signal. The enhanced SNR is shown in table 1. Because the different of noise, the enhancement effect are different. From the table 1, we can see the proposed method can effectively reduce the pink and white noise. For the babble noise the proposed method performs not so well.

The second database comes from the real records on the spot, which include many kinds of languages such as China, English, Japanese and so on. Each signal length of 8 frequency bands is 5 minutes, and the sample rate is 11025Hz.Because we can not get the clean speech signal, in this part we use subjective score for the proposed method. Two methods are used to test. The first one is the proposed method, and the other is the classical method spectral subtraction.

Table 1. The enhancement results of different noise with different SNR

Noise	Pink	Factory	Airforce	Babble	White
-5	3.1	1.7	1.3	-1.2	5.2
0	9.2	5.4	4.5	2.5	12.7
5	15.2	11.9	11.2	8.2	18.0

Table 2. The tests result of real short-wave speech signal

Frequency band	Proposed method	SS
1	4.2	3.1
2	4.6	3.3
3	4.5	3.2
4	3.7	2.1
5	5	4.2
6	4.8	4.4
7	3.4	2.1
8	4.9	4.7

From table 2, we can see the proposed method can efficiently remove the noise. Because the different noise is in different frequency band, the effects of enhancement are different also. In some band the enhancement effects of the proposed method performed over the traditional method SS. And in the other frequency band the enhancement effects almost equal to the method SS. But it is worth to point out that the proposed method didn't produce music noise.

5 Conclusions

In this paper, a novel method for speech enhancement was proposed. The enhancement scheme based on EMD is simple and full data-driven. It is fully automated de-noising approach. The primary experiments show that the proposed method can efficiently remove the non-stationary noise. It is important for the short-wave communication, because it can elevate the objective quality of speech signal and the intelligibility of noisy speech in order to reduce the listener fatigue.

References

1. Martin, R.:Noise Power Spectral Density Estimation Based on Optimal Smoothing and Minimum Statistics. IEEE Trans on Speech and Audio Processing,Vol.9(2001)504-512
2. Ephraim. Y,Malah. D.: Speech enhancement using a minimum mean-square error short-time spectral amplitude estimator. IEEE Transactions on Acoustics, Speech and Signal Processing, Vol.32 (1984) 1109-1121
3. Zheng W.T, Cao Z.H.: Speech enhancement based on MMSE-STSA estimation and resid-ual noise reduction,1991 IEEE Region 10 International Conference on EC3-Energy, Computer, Communication and Control SystemsVol.3(1991) 265-268
4. Liu Zhibin, Xu Naiping.: Speech enhancement based on minimum mean-square error short-time spectral estimation and its realization, IEEE International conference on intelli-gent processing system, Vol.28 (1997) 1794-1797
5. Lim ,J.S, Oppenheim. A.V.: Enhancement and bandwidth compression of noisy speech. Proc. of the IEEE, Vol.67 (1979) 1586-1604

6. Goh.Z, Tan.K and Tan.T.: Postprocessing method for suppressing musical noise generated by spectral subtraction. IEEE Trans. Speech Audio Procs, Vol. 6(1998) 287-292

7. He. C and Zweig.Z.: Adaptive two-band spectral subtraction with multi-window spectral estimation. ICASSP, Vol.2(1999) 793-796

8. Huang. N.E.: The Empirical Mode Decomposition and the Hilbert Spectrum for Nonlinear and Non-stationary Time Series Analysis, J. Proc. R. Soc. Lond. A,Vol.454(1998) 903-995

9. Huang. W,Shen.Z,Huang.N.E,Fung. Y.C.: Engineering Analysis of Biological Variables: an Example of Blood Pressure over 1 Day. Proc. Natl. Acad. Sci. USA, Vol.95(1998) 4816-4821

10. Huang. W,Shen.Z,Huang.N.E,Fung.: Nonlinear Indicial Response of Complex Nonsta-tionary Oscillations as Pulmonary Pretension Responding to Step Hypoxia. Proc. Natl. Acad. Sci, USA, Vol.96(1999) 1833-1839

11. V. Vapnik, The Nature of Statistical Learning Theory, Springer Verlag, New York, 1995.

Extended Resolution Proofs for Conjoining BDDs

Carsten Sinz and Armin Biere

Institute for Formal Models and Verification,
Johannes Kepler University Linz, Austria
{carsten.sinz, armin.biere}@jku.at

Abstract. We present a method to convert the construction of binary decision diagrams (BDDs) into extended resolution proofs. Besides in proof checking, proofs are fundamental to many applications and our results allow the use of BDDs instead—or in combination with—established proof generation techniques, based for instance on clause learning. We have implemented a proof generator for propositional logic formulae in conjunctive normal form, called EBDDRES. We present details of our implementation and also report on experimental results. To our knowledge this is the first step towards a practical application of extended resolution.

1 Introduction

Propositional logic decision procedures [1, 2, 3, 4, 5, 6] lie at the heart of many applications in hard- and software verification, artificial intelligence and automatic theorem proving [7, 8, 9, 10, 11, 12]. They have been used to successfully solve problems of considerable size. In many practical applications, however, it is not sufficient to obtain a yes/no answer from the decision procedure. Either a model, representing a sample solution, or a justification, why the formula possesses none is required. So, e.g. in declarative modeling or product configuration [9, 10] an inconsistent specification given by a customer corresponds to an unsatisfiable problem instance. To guide the customer in correcting his specification a justification why it was erroneous can be of great help. In the context of model checking proofs are used, e.g., for abstraction refinement [11], or approximative image computations through interpolants [13]. In general, proofs are also important for certification through proof checking [14].

Using BDDs for SAT is an active research area [15, 16, 17, 18, 19, 20]. It turns out that BDD [21] and search based techniques [2] are complementary [22, 23]. There are instances for which one works better than the other. Therefore, combinations have been proposed [16, 17, 20] to obtain the benefits of both, usually in the form of using BDDs for preprocessing. However, in all these approaches where BDDs have been used, proof generation has not been possible so far.

In our approach, conjunction is the only BDD operation considered to date. Therefore our solver is far less powerful than more sophisticated BDD-based SAT solvers [15, 16, 17, 18, 19, 20]. In particular, we currently cannot handle existential quantification. However, our focus is on proof generation, which none of the other approaches currently supports. We also conjecture that similar ideas as presented in this paper can be used to produce proofs for BDD-based existential quantification and other BDD

D. Grigoriev, J. Harrison, and E.A. Hirsch (Eds.): CSR 2006, LNCS 3967, pp. 600–611, 2006.

operations. This will eventually allow us to generate proofs for all the mentioned approaches.

We have chosen extended resolution as a formalism to express our proofs, as it is on the one hand a very powerful proof system equivalent in strength to extended Frege systems [24], and on the other hand similar to the well-known resolution calculus [25]. Despite its strength, it still offers simple proof checking: after adding a check to avoid cyclical definitions, an ordinary proof checker for resolution can be used.

Starting with [26], extended resolution has been mainly a subject of theoretical studies [27, 24]. In practical applications it did not play an important role so far. This may be due to the fact that direct generation of (short) extended resolution proofs is very hard, as there is not much guidance on how to use the *extension rule*. However, when "proofs" are generated by another means (by BDD computations in our case), extended resolution turns out to be a convenient formalism to concisely express proofs. We expect that a wide spectrum of different propositional decision procedures can be integrated into a common proof language and proof verification system using extended resolution.

The rest of this paper is organized as follows: First, we give short introductions to extended resolution and BDDs. Then we present our method to construct extended resolution proofs out of BDD constructions. Thereafter, we portray details of our implementation EBDDRES and show experimental results obtained with it. Finally, we conclude and give possible directions for future work.

2 Theoretical Background

In this paper we are mainly dealing with propositional logic formulae in conjunctive normal form (CNF). A formula F (over a set of variables V) in CNF is a conjunction of clauses, where each clause is a disjunction of literals. A literal is either a variable or its negation. We use capital letters ($C, D, \ldots$) to denote clauses and small-case letters to denote variables ($a, b, c, \ldots x, y, \ldots$) and literals ($l, l_1, l_2, \ldots$). Instead of writing a clause $C = (l_1 \vee \cdots \vee l_k)$ as a disjunction, we alternatively write it as a set of literals, i.e. as $\{l_1, \ldots, l_k\}$, or in an abbreviated form as $(l_1 \ldots l_k)$. For a literal l, we write $\bar{l}$ for its complement, i.e $\bar{x} = \neg x$ and $\overline{\neg x} = x$ for a variable x.

2.1 Extended Resolution

Extended resolution (ER) was proposed by Tseitin [26] as an extension of the resolution calculus [25]. The resolution calculus consists of a single inference rule,[1]

$$\frac{C \,\dot{\cup}\, \{l\} \quad \{\bar{l}\} \,\dot{\cup}\, D}{C \cup D}$$

and is used to refute propositional logic formulae in CNF. Here C and D are arbitrary clauses and l is a literal. A refutation proof is achieved, when the empty clause (denoted by $\square$) can be derived by a series of resolution rule applications. Extended

---

[1] By $\dot{\cup}$ we denote the disjoint union operation, i.e. $A \dot{\cup} B$ is the same as $A \cup B$ with the additional restriction that $A \cap B = \emptyset$.

resolution adds an *extension rule* to the resolution calculus, which allows introduction of definitions (in the form of additional clauses) and new (defined) variables into the proof. The additional clauses must stem out of the CNF conversion of definitions of the form $x \leftrightarrow F$, where F is an arbitrary formula and x is a new variable, i.e. a variable neither occurring in the formula we want to refute or in previous definitions nor in F. In this paper—besides introducing two variables for the Boolean constants—we only define new variables for if-then-else (*ITE*) constructs written as $x \;?\; a : b$ (for variables x, a, b), which is an abbreviation for $(x \rightarrow a) \wedge (\neg x \rightarrow b)$. So introduction of a new variable w as an abbreviation for $ITE(x, a, b) = x \;?\; a : b$ is reflected by the rule

$$\overline{(\bar{w}\bar{x}a)(\bar{w}xb)(w\bar{x}\bar{a})(wx\bar{b})}$$

which has no premises, and introduces four new clauses at once for the given instance of the *ITE* construct. We have used the abbreviated notation for clauses here that leaves out disjunction symbols. Concatenation of clauses is assumed to denote the conjunction of these. It should also be noted that the extended clause set produced by applications of the extension rule is only equisatisfiable to the original clause set, but not equivalent.

The interest in extended resolution stems from the fact that no super-polynomial lower bound is known for extended resolution [24] and that it is comparable in strength to the most powerful proof systems (extended Frege) for propositional logic [24]. This also means that for formulae which are hard for resolution (e.g. Haken's pigeon-hole formulae [28]) short ER proofs exist [27]. Moreover, as it is an extension of the resolution calculus, ER is a natural candidate for a common proof system integrating different propositional decision procedures like the resolution-based DPLL algorithm [1, 2] .

2.2   Binary Decision Diagrams

Binary Decision Diagrams (BDDs) were proposed by Bryant [21] to compactly represent Boolean functions as DAGs (directed acyclic graphs). In their most common form as reduced ordered BDDs (that we also adhere to in this paper) they offer the advantage that each Boolean function is uniquely represented by a BDD, and thus all semantically equivalent formulae share the same BDD. BDDs are based on the Shannon expansion

$$f = ITE(x, f_1, f_0) = (x \rightarrow f_1) \wedge (\neg x \rightarrow f_0) \;,$$

decomposing f into its *co-factors* f_0 and f_1 (w.r.t variable x). The co-factor f_0 (resp. f_1) is obtained by setting variable x to false (resp. true) in formula f and subsequent simplification. By repeatedly applying Shannon expansion to a formula selecting splitting variables according to a global variable ordering until no more variables are left (ending in terminal nodes 0 and 1), its BDD representation is obtained (resembling a decision tree). Merging equivalent nodes (i.e. same variable and co-factors) and deleting nodes with coinciding co-factors results in reduced ordered BDDs. Fig. 1 shows the BDD representation of formula $f = x \vee (y \wedge \neg z)$.

To generate BDDs for (complex) formulae, instead of performing Shannon decomposition and building them *top-down*, they are typically built *bottom-up* starting with basic BDDs for variables or literals, and then constructing more complex BDDs by using BDD operations (e.g., *BDD-and*, *BDD-or*) for logical connectives. As we will

Algorithm 1. BDD-and(a, b)

1: if $a = 0$ or $b = 0$ then return 0
2: if $a = 1$ then return b else if $b = 1$ then return a
3: (x, a_0, a_1) = decompose(a); (y, b_0, b_1) = decompose(b)
4: if $x < y$ then return new-node(y, BDD-and(a, b_0), BDD-and(a, b_1))
5: if $x = y$ then return new-node(x, BDD-and(a_0, b_0), BDD-and(a_1, b_1))
6: if $x > y$ then return new-node(x, BDD-and(a_0, b), BDD-and(a_1, b))

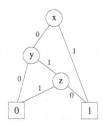

Fig. 1. BDD representation of formula $x \lor (y \land \neg z)$ using variable ordering $x > y > z$

need it in due course, we give the *BDD-and* algorithm explicitly (Algorithm 1). Here, *decompose* breaks down a non-terminal BDD node into its constituent components, i.e. its variable and cofactors. The function *new-node* constructs a new BDD node if it is not already present, and otherwise returns the already existent node. The comparisons in steps 4 to 6 are based on the global BDD variable order.

3 Proof Construction

We assume that we are given a formula F in CNF for which we want to construct an ER proof that shows unsatisfiability of F (i.e., we show that $\neg F$ is a tautology). Instead of trying to derive such a proof directly in the ER calculus—which could be quite hard, as there are myriads of ways to introduce new definitions—we first construct a BDD equivalent to formula F and then extract an ER proof out of this BDD construction. The BDD for formula F is built gradually (bottom-up) by conjunctively adding more and more clauses to an initial BDD representing the Boolean constant true.

Starting with an ordered set of clauses $S = (C_1, \ldots, C_m)$ for formula $F_S = C_1 \land \cdots \land C_m$ (which we want to proof unsatisfiable), we thus first build a BDD c_i for each clause C_i. Then we construct intermediate BDDs h_i corresponding to partial conjunctions $C_1 \land \cdots \land C_i$, until, by computing h_m, we have reached a BDD for the whole formula. These intermediate BDDs can be computed recursively by the equations

$$h_2 \leftrightarrow c_1 \land c_2 \quad \text{and} \quad h_i \leftrightarrow h_{i-1} \land c_i \quad \text{for } 3 \leq i \leq m .$$

If h_m is the BDD consisting only of the 0-node, we know that formula F_S is unsatisfiable and we can start building an ER proof.

The method to construct the ER proof works by first introducing new propositional variables (thus ER is required), one for each node of each BDD that occurs during the construction process, i.e. for all c_i and h_i. New variables are introduced based on the Shannon expansion of a BDD node: for an internal node f containing variable x and having child nodes f_1 and f_0, a new variable (which we also call f), defined by

$$f \leftrightarrow (x \ ? \ f_1 : f_0) \qquad (\bar{f}\bar{x}f_1)(\bar{f}xf_0)(f\bar{x}\bar{f}_1)(fx\bar{f}_0)$$

is introduced. We have also given the clausal representation of the definition on the right. Terminal nodes are represented by additional variables n_0 and n_1 defined by

$$n_0 \leftrightarrow 0 \qquad (\bar{n}_0) \qquad \text{and} \qquad n_1 \leftrightarrow 1 \qquad (n_1) \ .$$

Note that introducing new variables in this way does not produce cyclic definitions, as the BDDs themselves are acyclic. So by introducing variables bottom-up from the leaves of the BDD up to the top node, we have an admissible ordering for applying the extension rule.

With these definitions, we can give an outline of the ER proof we want to generate. It consists of three parts: first, we derive unit clauses (c_i) for the variables corresponding to the top BDD nodes of each clause. Then out of the recursive runs of each *BDD-and*-operation we build proofs for the conjunctions $h_{i-1} \wedge c_i \leftrightarrow h_i$ (in fact, only the implication from left to right is required). And finally, we combine these parts into a proof for h_m. If h_m is the variable representing the zero node, i.e. $h_m = n_0$, we can derive the empty clause by another single resolution step with the defining clause for n_0. We thus have to generate ER proofs for all of the following:

$$
\begin{aligned}
& S \vdash c_i && \text{for all } 1 \leq i \leq m && \text{(ER-1)} \\
& S \vdash c_1 \wedge c_2 \rightarrow h_2 && && \text{(ER-2a)} \\
& S \vdash h_{i-1} \wedge c_i \rightarrow h_i && \text{for all } 3 \leq i \leq m && \text{(ER-2b)} \\
& S \vdash h_m && && \text{(ER-3)}
\end{aligned}
$$

For a proof of (ER-1) for some i assume that clause $D = C_i$ consists of the literals $(l_1, \ldots, l_k)$. We assume literals to be ordered decreasingly according to the BDD's global variable ordering. Then the newly introduced variables for the nodes of the BDD representation of clause D are $d_j \leftrightarrow (l_j \ ? \ n_1 : d_{j+1})$ if l_j is positive, and $d_j \leftrightarrow (\bar{l}_j \ ? \ d_{j+1} : n_1)$ if l_j is negative. We identify d_{k+1} with n_0, and c_i with d_1 here. These definitions induce—among others—the clauses $(d_j \bar{l}_j \bar{n}_1)$ and $(d_j l_j \bar{d}_{j+1})$ for all $1 \leq i \leq k$. We therefore obtain the following ER proof for (d_1): First, we derive $(d_k l_1 \ldots l_{k-1} \bar{n}_1)$ by resolving $(l_1 \ldots l_k)$ with $(d_k \bar{l}_k \bar{n}_1)$. Then, iteratively for $j = k$ down to $j = 2$, we derive $(d_{j-1} l_1 \ldots l_{j-2} \bar{n}_1)$ from $(d_j l_1 \ldots l_{j-1} \bar{n}_1)$ by

$$
\frac{(d_j l_1 \ldots l_{j-1} \bar{n}_1) \quad (d_{j-1} l_{j-1} \bar{d}_j)}{\dfrac{(d_{j-1} l_1 \ldots l_{j-1} \bar{n}_1) \quad (d_{j-1} \bar{l}_{j-1} \bar{n}_1)}{(d_{j-1} l_1 \ldots l_{j-2} \bar{n}_1)}}
$$

And finally, we derive the desired (d_1) by resolving $(d_1 \bar{n}_1)$ with (n_1).

For (ER-2a) and (ER-2b), assume that we want to generate a proof for the general implication $f \wedge g \rightarrow h$ (i.e. for the clause $(\bar{f}\bar{g}h)$), where we have h computed as the conjunction of f and g by the *BDD-and* operation. Definitions for the BDD nodes resp. variables are given by the Shannon expansion of f, g and h as

$$f \leftrightarrow (x ? f_1 : f_0) \qquad (\bar{f}\bar{x}f_1)(\bar{f}xf_0)(f\bar{x}\bar{f}_1)(fx\bar{f}_0)$$
$$g \leftrightarrow (x ? g_1 : g_0) \qquad (\bar{g}\bar{x}g_1)(\bar{g}xg_0)(g\bar{x}\bar{g}_1)(gx\bar{g}_0)$$
$$h \leftrightarrow (x ? h_1 : h_0) \qquad (\bar{h}\bar{x}h_1)(\bar{h}xh_0)(h\bar{x}\bar{h}_1)(hx\bar{h}_0) \ .$$

We can now recursively construct an ER proof for the implication $f \wedge g \rightarrow h$, where in the recursion step we assume that proofs for both $f_0 \wedge g_0 \rightarrow h_0$ and $f_1 \wedge g_1 \rightarrow h_1$ are already given. We thus obtain:

$$
\cfrac{
\cfrac{
\cfrac{
\cfrac{(\bar{f}xf_0) \quad (\bar{f}_0\bar{g}_0h_0)}{(\bar{f}x\bar{g}_0h_0)} \quad (\bar{g}xg_0)
}{(\bar{f}\bar{g}xh_0)} \quad (hx\bar{h}_0)
}{(\bar{f}\bar{g}hx)}
\quad
\cfrac{
\cfrac{
\cfrac{(\bar{f}_1\bar{g}_1h_1) \quad (\bar{f}\bar{x}f_1)}{(\bar{f}\bar{x}\bar{g}_1h_1)} \quad (\bar{g}\bar{x}g_1)
}{(\bar{f}\bar{g}\bar{x}h_1)} \quad (h\bar{x}\bar{h}_1)
}{(\bar{f}\bar{g}h\bar{x})}
}{(\bar{f}\bar{g}h)}
$$

The recursive process stops when we arrive at the leave nodes resp. the base case of the recursive *BDD-and* algorithm. This algorithm stops, if either of its arguments is a Boolean constant, or when both arguments are the same. In each case we obtain tautological clauses, like e.g. $(\bar{n}_0\bar{g}n_0)$ for f being the 0-BDD, or $(\bar{f}f)$ for computing the conjunction of two identical arguments. We call this a *trivial step* of the *BDD-and* algorithm. Moreover, we may stop the recursion, if f implies g (or vice versa), as we then have generated the tautological clause $(\bar{f}gf)$. This case we call a *redundant step* of the algorithm. We will show below how such tautological clauses can be avoided or eliminated in the ER proof.

For a proof of (ER-3) we just have to combine parts one and two: We can resolve the derived unit clauses (c_1) and (c_2) with the implication $(\bar{c}_1\bar{c}_2h_2)$ to produce a proof of (h_2), and then continue iteratively using (ER-2b) to derive further unit clauses (h_i) for all $3 \leq i \leq m$. Finally, (h_m) is resolved with $(\bar{n}_0)$ to produce the empty clause. This completes the whole ER proof for unsatisfiability of F_S.

3.1 Avoiding Tautological Clauses

The proofs (ER-2a) and (ER-2b) presented above may contain tautological clauses that are introduced by equivalent BDD nodes. So if, e.g., $f_0 = h_0$, the recursive proof step starts with a tautological clause in the left branch. We now give a detailed construction that avoids such clauses completely, starting with some definitions:

Definition 1 (Line, Step). *A triplet of BDD nodes (f, g, h) such that h is the BDD node representing $f \wedge g$ is called a (BDD-and) step. A step is called trivial if f or g are Boolean constants or if $f = g$. A non-trivial step is also called a (cache) line. Steps and lines are also identified with clauses, where a line (f, g, h) corresponds to the clause $(\bar{f}\bar{g}h)$. We identify nodes with ER variables here.*

Definition 2 (Redundancy). *A line* $L = (f, g, h)$ *is called redundant if* $f = h$ *or* $g = h$, *otherwise it is called irredundant. The notion of redundancy also carries over to the clause* $(\bar{f}\bar{g}h)$ *corresponding to line* L.

When we have reached an irredundant step $(\bar{f}\bar{g}h)$, we can check whether the co-factor clauses of the assumptions $(\bar{f}_0\bar{g}_0 h_0)$ and $(\bar{f}_1\bar{g}_1 h_1)$ of the step are redundant. If this is the case, the proof has to be simplified and recursion stops (in all but one case) at the redundant step. We now give simplified proofs that contain no tautological clauses for the recursion step of the proofs (ER-2a) and (ER-2b). In what follows, we call the sub-proof of $(\bar{f}\bar{g}hx)$ out of $(\bar{f}_0\bar{g}_0 h_0)$ the left branch and the sub-proof of $(\bar{f}\bar{g}h\bar{x})$ out of $(\bar{f}_1\bar{g}_1 h_1)$ the right branch of the recursive proof step.

R1. If $f_0 = h_0$, we obtain a proof for the left branch (and analogously for $g_0 = h_0$ and for $f_1 = h_1$ or $g_1 = h_1$ on the right branch) by resolving $(hx\bar{h}_0)$ and $(\bar{f}xf_0)$ to produce $(\bar{f}hx)$. Although we have proved a stronger claim on the left branch in this case, it cannot happen that g also disappears on the right branch, as this would only be possible if $f_1 = h_1$. But then $f = h$ would also hold and the step $(\bar{f}\bar{g}h)$ would already be redundant, contradicting our assumption.

T1. If $f_0 = g_0$ (this is not a tautological case, however) then $h_0 = f_0 = g_0$ also holds, so that we arrive at the case above (and similarly for $f_1 = g_1$). We can even choose which of the definitions (either for f or for g) we want to use.

T2. If $f_0 = 1$ we obtain $h_0 = g_0$ and we can use the proof given under (R1) for the left branch (similar for $g_0 = 1$, $f_1 = 1$, and $g_1 = 1$). If $f_0 = 0$, we can use the definition $(\bar{f}xn_0)$ of f and $(\bar{n}_0)$ of 0 to derive the stronger $(\bar{f}x)$. It cannot happen that $f_1 = 0$ at the same time (as then the step would be trivial), so the only possibility where we are really left with a stronger clause than the desired $(\bar{f}\bar{g}h)$ occurs when $f_0 = 0$ and $g_1 = 0$ (or $f_1 = g_0 = 0$). Then we have $h = 0$ and we can derive $(\bar{f}\bar{g})$. In this case we just proceed as in case (H0) below.

H0. If $h = 0$ we let $h_0 = h_1 = 0$ and recursively generate sub-proofs skipping the definition of h by rule (X1) below.

H1. The case $h = 1$ could only happen if $f = g = 1$ would also hold. But then the step would be redundant. If $h_0 = 1$ we derive the stronger (hx) by resolving $(hx\bar{n}_1)$ with (n_1), and similar for $h_1 = 1$. It cannot happen that we have $h_0 = h_1 = 1$ at the same time, as this would imply $h = 1$. Thus, on the other branch we always obtain a clause including $\bar{f}$ and $\bar{g}$ and therefore the finally resulting clause is always $(\bar{f}\bar{g}h)$.

X1. If the decision variable x does not occur in one or several of the BDDs f, g, or h (i.e., for example, if $f = f_0 = f_1$) the respective resolution step(s) involving $\bar{f}$, $\bar{g}$, or h, can just be skipped.

Note that in all degenerate cases besides cases (T2) (only for $f_0 = g_1 = 0$ or $f_1 = g_0 = 0$) and (H0) the proof stops immediately and no recursive descent towards the leaves of the BDD is necessary. If the last proof step in any of the sub-proofs of (ER-2a) or (ER-2b) results in a redundant step, the proof of part (ER-3) contains tautological clauses and must be simplified. So assume in (ER-2a) that the last step is redundant, i.e. in $(\bar{c}_1\bar{c}_2 h_2)$ either $c_1 = h_2$ or $c_2 = h_2$ holds. In both cases the *BDD-and* computation is not needed and we can skip the resolution proof for h_2 and use clause c_1 or c_2, whichever is equivalent to h_2, in the subsequent proof. The same holds for (ER-2b), if the last proof step is redundant and we thus have $h_{i-1} = h_i$ or $c_i = h_i$. Again, we can drop the *BDD-and* computation and skip the resolution proof for h_i out of h_{i-1} and c_i, but instead using h_{i-1} or c_i directly.

A further reduction in proof size could be achieved by simplifying definitions introduced for BDD nodes with constants 0 and 1 as co-factors. So instead of using $f \leftrightarrow (x ? n_1 : f_0)$ as definition for a node f with constant 1 as its first co-factor, we could use the simplified definition $f \leftrightarrow (\bar{x} \rightarrow f_0)$ which would result in only three simpler clauses $(\bar{f} x f_0)$, $(f \bar{x})$, and $(f \bar{f_0})$.

4 Implementation and Experimental Results

We have implemented our approach in the SAT solver EBDDRES. It takes as input a CNF in DIMACS format and computes the conjunction of the clauses after transforming them into BDDs. The result is either the constant zero BDD, in which case the formula is unsatisfiable, or a non-zero BDD. In the latter case a satisfying assignment is generated by traversing a path from the root of the final BDD to the constant one leaf.

In addition to solving the SAT problem, a proof trace can be generated. If the formula is unsatisfiable the empty clause is derived. Otherwise a unit clause can be deduced. It contains a single variable which represents the root node of the final BDD. The trace format is similar to the trace format used for ZCHAFF [14] or MINISAT [6]. In particular, we do not dump individual resolution steps, but combine piecewise regular input resolution steps into *chains*, called *trivial* resolution steps in [29]. Each chain has a set of *antecedent* clauses and one *resolvent*. The antecedents are treated as input clauses in the regular input resolution proof of the resolvent. Our trace checker is able to infer the correct resolution order for the antecedents by unit propagation after assuming the negation of the resolvent clause. Original clauses and those used for defining new variables in the extended resolution proof are marked by an empty list of antecedents. Note that a proof checker for the ordinary resolution calculus is sufficient for extended resolution proofs, too, as all definitional clauses produced by the extension rule can be added initially, and then only applications of the resolution rule are required to be checked.

The ASCII version of the trace format itself is almost identical to the DIMACS format and since the traces generated by EBDDRES are quite large we also have a compact binary version, comparable to the one used by MINISAT. Currently the translation from the default ASCII format into the binary format is only possible through an external tool. Due to this current limitation we were not able to generate binary traces where the ASCII trace was of size 1GB or more.

For the experiments we used a cluster of Pentium IV 3.0 GHz PCs with 2GB main memory running Debian Sarge Linux. The time limit was set to 1000 seconds, the memory limit to 1GB main memory and no size limit on the generated traces was imposed. Besides the *pigeon hole* instances (ph*) we used combinatorial benchmarks from [22, 30], more specifically *mutilated checker board* (mutcb*) and *Urquhart* formulae (urq*), FPGA routing (fpga*), and one suite of structural instances from [31], which represent equivalence checking problems for adder circuits (add*). The latter suite of

Table 1. Comparison of trace generation with MINISAT and with EBDDRES

	MINISAT			EBDDRES											
	solve resources		trace size	solve resources		trace gen.	trace size		bdd nodes	recursive bdd and-steps					trace chk
	sec	MB	MB	sec	MB	sec	ASCII MB	binary MB	$\times 10^3$	all $\times 10^3$	triv. $\times 10^3$	lines $\times 10^3$	red. $\times 10^3$	core $\times 10^3$	sec
ph7	0	0	0	0	0	0	1	0	3	20	10	10	0	10	0
ph8	0	4	1	0	3	0	3	1	15	67	34	33	0	33	0
ph9	6	4	11	0	3	0	3	1	8	90	45	45	0	45	0
ph10	44	4	63	1	17	1	30	10	136	538	270	269	1	268	1
ph11	887	6	929	1	13	1	21	8	35	670	335	334	1	333	1
ph12	*	-	-	2	28	1	33	12	31	1150	575	574	1	573	1
ph13	*	-	-	10	102	8	260	92	850	5230	2615	2614	2	2612	8
ph14	*	-	-	10	111	7	204	74	166	6554	3278	3276	2	3274	7
mutcb9	0	4	0	0	5	0	5	2	27	141	71	70	22	48	0
mutcb10	0	4	1	0	8	0	11	4	58	311	156	155	49	106	0
mutcb11	1	4	4	1	17	1	31	10	153	818	409	409	129	280	1
mutcb12	8	4	22	2	33	2	69	22	320	1763	882	881	282	600	2
mutcb13	113	5	244	6	102	5	181	61	817	4453	2227	2226	707	1519	5
mutcb14	490	8	972	14	250	10	393	132	1694	9446	4724	4723	1515	3208	10
mutcb15	*	-	-	36	498	25	1009	*	4191	23119	11560	11559	3690	7868	25
mutcb16	*	-	-	-	*	-	-	-	-	-	-	-	-	-	-
urq35	96	4	218	2	28	1	37	13	24	1216	608	608	0	608	1
urq45	*	-	-	-	*	-	-	-	-	-	-	-	-	-	-
fpga108	0	0	0	6	47	4	135	47	186	4087	2044	2043	3	2040	4
fpga109	0	0	0	3	44	2	70	24	88	2218	1109	1109	1	1108	2
fpga1211	0	0	0	54	874	38	1214	*	1312	33783	16892	16891	41	16850	38
add16	0	0	0	0	4	0	6	2	30	100	51	50	1	49	0
add32	0	0	0	1	9	1	24	8	122	445	223	222	4	217	1
add64	0	4	0	12	146	9	338	112	1393	5892	2948	2944	19	2925	9
add128	0	4	0	-	*	-	-	-	-	-	-	-	-	-	-

The first column lists the name of the instance. Columns 2-4 contain the data for MINISAT, first the time taken to solve the instance including the time to produce the trace, then the memory used, and in column 4 the size of the generated trace. The data for EBDDRES takes up the rest of the table. It is split into a more general part in columns 5-9 on the left. The right part provides more detailed statistics in columns 10-15. The first column in the general part of EBDDRES shows the time taken to solve the instance with EBDDRES including the time to generate and dump the trace. The latter is shown separately in column 7. The memory used by EBDDRES, column 6, is linearly related to the number of generated BDD nodes in column 10 and the number of generated cache lines in column 13. The number of recursive steps of the *BDD-and* operation occurs in column 11. Among these steps many trivial base cases occur (column 12) and the number of cache lines in column 13 is simply the number of non trivial steps. Among the cache lines several redundant lines occur (column 14) in which the result is equal to one of the arguments. The core consists of irredundant cache lines necessary for the proof. Their number is listed in the next to last column (column 15). The last column (column 16) shows the time taken by the trace checker to validate the proof generated by EBDDRES. The * denotes either *time out* (>1000 seconds) or *out of memory* (>1GB main memory).

benchmarks is supposed to be very easy for BDDs with variable quantification, which is not implemented in EBDDRES. Starting with ZCHAFF [4], these benchmarks also became easy for search based solvers.

As expected, the experimental data in Tab. 1 shows that even our simplistic approach in conjoining BDDs to solve SAT, is able to outperform MINISAT on certain hard combinatorial instances in the ph* and mutcb* family. In contrast to the simplified exposition in Sec. 3 a tree shaped computation turns out to be more efficient for almost all benchmarks. Only for one benchmark family (mutcb*) we used the linear combination. The results of Sec. 3 transfer to the more general case easily.

For comparison we used the latest version 1.14 of MINISAT, with proof generation capabilities. MINISAT in essence was the fastest SAT solver in the SAT'05 SAT solver competition. MINISAT in combination with the preprocessor SATELITE [32] was even faster in the competition, but we could not use SATELITE, because it cannot generate proofs. The binary trace format of EBDDRES (see column 9 of Tab. 1) is comparable— though not identical—to the trace format of MINISAT. EBDDRES was able to produce smaller traces for certain instances. Since for EBDDRES the traces grow almost linear in the number of steps, we expect much smaller traces for more sophisticated BDD-based SAT approaches.

Similar to related approaches, EBDDRES does not use dynamic variable ordering but relies on the choice of a good initial static order instead. We experimented with various static ordering algorithms. Only for the add* family of benchmarks it turns out that the variable order generated by our implementation of the FORCE algorithm of [33] yields better results. For all other families we used the original ordering. It is given by the order of the variable indices in the DIMACS file. Improvements on running times and trace size will most likely not be possible from better variable ordering algorithms. But we expect an improvement through clustering of BDDs and elimination of variables through existential quantification.

In general, whether BDD-based methods or SAT solvers behave superior turned out to be highly problem dependent, as other empirical studies also suggest [18, 19, 22].

5 Conclusion and Future Work

Resolution proofs are used in many practical applications for proof checking, debugging, core extraction, abstraction refinement, and interpolation. This paper presents and evaluates a practical method to obtain extended resolution proofs for conjoining BDDs in SAT solving. Our results enable the use of BDDs for these purposes instead—or in combination with—already established methods based on DPLL with clause learning.

As future work the ideas presented in this paper need to be extended to other BDD operations besides conjunction, particularly to existential quantification. We also conjecture that equivalence reasoning and, more generally, Gaussian elimination over GF(2), can easily be handled in the same way.

Finally we want to thank Eugene Goldberg for very fruitful discussions about the connection between extended resolution and BDDs.

References

1. M. Davis and H. Putnam. A computing procedure for quantification theory. *JACM*, 7, 1960.
2. M. Davis, G. Logemann, and D. Loveland. A machine program for theorem-proving. *Communications of the ACM*, 5(7), 1962.
3. J. P. Marques-Silva and K. A. Sakallah. GRASP — a new search algorithm for satisfiability. In *Proc. ICCAD'96*.
4. M. Moskewicz, C. Madigan, Y. Zhao, L. Zhang, and S. Malik. Chaff: Engineering an efficient SAT solver. In *Proc. DAC'01*.

5. E. Goldberg and Y. Novikov. BerkMin: A fast and robust SAT-solver. In *Proc. DATE'02.*
6. N. Eén and N. Sörensson. An extensible SAT-solver. In *Proc. SAT'03.*
7. A. Biere, A. Cimatti, E. Clarke, and Y. Zhu. Symbolic model checking without BDDs. In *Proc. TACAS'99.*
8. M. Velev and R. Bryant. Effective use of boolean satisfiability procedures in the formal verification of superscalar and VLIW microprocessors. *J. Symb. Comput.*, 35(2), 2003.
9. I. Shlyakhter, R. Seater, D. Jackson, M. Sridharan, and M. Taghdiri. Debugging overconstrained declarative models using unsatisfiable cores. In *Proc. ASE'03.*
10. C. Sinz, A. Kaiser, and W. Küchlin. Formal methods for the validation of automotive product configuration data. *AI EDAM*, 17(1), 2003.
11. K. McMillan and N. Amla. Automatic abstraction without counterexamples. In *Proc. TACAS'03.*
12. Y. Xie and A. Aiken. Scalable error detection using boolean satisfiability. In *Proc. POPL'05.*
13. K. McMillan. Interpolation and SAT-based model checking. In *Proc. CAV'03*, volume 2725 of *LNCS.*
14. L. Zhang and S. Malik. Validating SAT solvers using an independent resolution-based checker: Practical implementations and other applications. In *Proc. DATE'03.*
15. D. Motter and I. Markov. A compressed breath-first search for satisfiability. In *ALENEX'02.*
16. J. Franco, M. Kouril, J. Schlipf, J. Ward, S. Weaver, M. Dransfield, and W. Fleet. SBSAT: a state–based, BDD–based satisfiability solver. In *Proc. SAT'03.*
17. R. Damiano and J. Kukula. Checking satisfiability of a conjunction of BDDs. In *DAC'03.*
18. J. Huang and A. Darwiche. Toward good elimination orders for symbolic SAT solving. In *Proc. ICTAI'04.*
19. G. Pan and M. Vardi. Search vs. symbolic techniques in satisfiability solving. In *SAT'04.*
20. H.-S. Jin, M. Awedh, and F. Somenzi. CirCUs: A hybrid satisfiability solver. In *Proc. SAT'04.*
21. R. Bryant. Graph-based algorithms for Boolean function manipulation. *IEEE Trans. on Comp.*, 35(8), 1986.
22. T. E. Uribe and M. E. Stickel. Ordered binary decision diagrams and the Davis-Putnam procedure. In *Proc. Intl. Conf. on Constr. in Comp. Logics*, volume 845 of *LNCS*, 1994.
23. J. F. Groote and H. Zantema. Resolution and binary decision diagrams cannot simulate each other polynomially. *Discrete Applied Mathematics*, 130(2), 2003.
24. A. Urquhart. The complexity of propositional proofs. *Bulletin of the EATCS*, 64, 1998.
25. J. A. Robinson. A machine-oriented logic based on the resolution principle. *JACM*, 12, 1965.
26. G. Tseitin. On the complexity of derivation in propositional calculus. In *Studies in Constructive Mathematics and Mathematical Logic*, 1970.
27. S. Cook. A short proof of the pigeon hole principle using extended resolution. *SIGACT News*, 8(4), 1976.
28. A. Haken. The intractability of resolution. *Theoretical Comp. Science*, 39, 1985.
29. P. Beame, H Kautz, and A Sabharwal. Towards understanding and harnessing the potential of clause learning. *J. Artif. Intell. Res. (JAIR)*, 22, 2004.
30. F. Aloul, A. Ramani, I. Markov, and K. Sakallah. Solving difficult instances of boolean satisfiability in the presence of symmetry. *IEEE Trans. on Comp. Aided Design*, 22(9), 2003.
31. D. Plaisted, A. Biere, and Y. Zhu. A satisfiability procedure for quantified boolean formulae. *Discrete Applied Mathematics*, 130(2), 2003.
32. N. Eén and A. Biere. Effective preprocessing in SAT through variable and clause elimination. In *Proc. SAT'05*, volume 3569 of *LNCS.*
33. F. Aloul, I. Markov, and K. Sakallah. FORCE: A fast and easy–to–implement variable-ordering heuristic. In *ACM Great Lakes Symp. on VLSI*, 2003.

A   Example Extended Resolution Proof

To further illustrate proof generation and avoidance of tautological clauses, we give an example showing construction of part (ER-2a) of the ER proof. Let $S = (C, D)$ be an ordered clause set with $C = (ab)$ and $D = (\bar{a}b)$. The defining equations for the BDDs for C and D then are (using variable ordering $b > a$)

$$c \leftrightarrow (b \; ? \; n_1 : c_0) \qquad\qquad d \leftrightarrow (b \; ? \; n_1 : d_0)$$
$$c_0 \leftrightarrow (a \; ? \; n_1 : n_0) \qquad\qquad d_0 \leftrightarrow (a \; ? \; n_0 : n_1)$$

and the resulting BDD h_2 denoting the conjunction of C and D is given by $h_2 \leftrightarrow (b \; ? \; n_1 : n_0)$. So the first part of our ER proof consists of introducing five new variables c, c_0, d, d_0 and h_2 together with their defining clauses by five applications of the extension rule. Then the resolution part follows. The last recursive step to derive $c \wedge d \to h_2$ looks like this:

$$
\begin{array}{c}
\vdots \\
\cfrac{
  \cfrac{
    \cfrac{(\bar{c}bc_0) \quad (\bar{c_0}\bar{d_0}n_0)}{(\bar{c}b\bar{d_0}n_0)}
  }{(h_2 b \bar{n}_0) \qquad (\bar{c}\bar{d}bn_0)}
  \bigg/ (\bar{c}\bar{d}h_2 b)
  \qquad
  \cfrac{
    \cfrac{(\bar{n}_1\bar{n}_1 n_1) \quad (\bar{c}bn_1)}{(\bar{c}b\bar{n}_1 n_1)} \quad (\bar{d}bn_1)
  }{(\bar{c}\bar{d}bn_1) \qquad (h_2 b \bar{n}_1)}
  \bigg/ (\bar{c}\bar{d}h_2 \bar{b})
}{(\bar{c}\bar{d}h_2)}
\end{array}
$$

When we apply the rules stated above to avoid tautological clauses (three times rule (R1)), the whole proof for $c \wedge d \to h_2$ reduces to the following:

$$
\begin{array}{c}
\cfrac{
  \cfrac{
    \cfrac{(\bar{c}bc_0) \quad \cfrac{(\bar{c_0}an_0) \quad (\bar{d_0}\bar{a}n_0)}{(\bar{c_0}\bar{d_0}n_0)}}{(\bar{c}b\bar{d_0}n_0)}
  }{(h_2 b \bar{n}_0) \qquad (\bar{c}\bar{d}bn_0)}
}{(\bar{c}\bar{d}h_2 b)}
\qquad
\cfrac{(\bar{c}bn_1) \quad (h_2 b \bar{n}_1)}{(\bar{c}h_2 \bar{b})}
\end{array}
$$
$$(\bar{c}\bar{d}h_2)$$

Optimal Difference Systems of Sets with Multipliers

Vladimir D. Tonchev and Hao Wang

Department of Mathematical Sciences,
Michigan Technological University,
Houghton, MI 49931-1295, USA
tonchev@mtu.edu, hwang@mtu.edu

Abstract. Difference Systems of Sets (DSS) are combinatorial structures that are used in code synchronization. A DSS is *optimal* if the associated code has minimum redundancy for the given block length n, alphabet size q, and error-correcting capacity ρ. An algorithm is described for finding optimal DSS that admit prescribed symmetries defined by automorphisms of the cyclic group of order n, together with optimal DSS found by this algorithm.

Keywords: code synchronization, comma-free code, cyclic difference set, multiplier.

1 Introduction

Difference systems of sets are used for the construction of comma-free codes to resolve the code synchronization problem. We consider the following situation of transmitting data over a channel, where the data being sent is thought as a stream of symbols from a finite alphabet $F_q = \{0, 1, ..., q-1\}$. The data stream consists of consecutive messages, each message being a sequence of n consecutive symbols:

$$\ldots \underbrace{x_1 \ldots x_n}\, \underbrace{y_1 \ldots y_n} \ldots,$$

where $\mathbf{x} = x_1...x_n$ and $\mathbf{y} = y_1...y_n$ are two consecutive messages. The code synchronization problem is how to decode a message correctly without any additional information concerning the position of word synchronization points (i.e., the position of commas separating words).

A q-ary code of length n is a subset of the set F_q^n of all words of length n over $F_q = \{0, 1, ..., q-1\}$. If q is a prime power, we often identify F_q with a finite field of order q. A *linear* q-ary code (q a prime power), is a linear subspace of F_q^n.

Define the *ith overlap* of $\mathbf{x} = x_1...x_n$ and $\mathbf{y} = y_1...y_n$ as

$$T_i(\mathbf{x}, \mathbf{y}) = x_{i+1}...x_n y_1...y_i, \ \ 1 \le i \le n-1.$$

It is clear that a code $C \subseteq F_q^n$ can resolve the synchronization problem if any overlap of two codewords from C, which may be identical, is not a codeword. A *comma-free* code is a code that satisfies this requirement. The *comma-free index* $\rho(C)$ of a code $C \subseteq F_q^n$ is defined as

$$\rho(C) = \min d(z, T_i(x, y)),$$

D. Grigoriev, J. Harrison, and E.A. Hirsch (Eds.): CSR 2006, LNCS 3967, pp. 612–618, 2006.

where the minimum is taken over all $x, y, z \in C$ and all $i = 1, ..., n - 1$, and d is the Hamming distance between words in F_q^n. The comma-free index $\rho(C)$ allows one to distinguish a code word from an overlap of two code words provided that at most $\lfloor \rho(C)/2 \rfloor$ errors have occurred in the given code word [5].

A **difference system of sets** (DSS) with parameters $(n, \tau_0, \ldots, \tau_{q-1}, \rho)$ is a collection of q disjoint subsets $Q_i \subseteq \{1, \ldots, n\}$, $|Q_i| = \tau_i$, $0 \leq i \leq q - 1$, such that the multi-set

$$M = \{a - b \pmod{n} \mid a \in Q_i, \, b \in Q_j, \, i \neq j\} \tag{1}$$

contains every number i, $(1 \leq i \leq n - 1)$, at least ρ times. A DSS is **perfect** if every number i, $(1 \leq i \leq n - 1)$ appears exactly ρ times in the multi-set of differences (1). A DSS is **regular** if all subsets Q_i are of the same size: $\tau_0 = \tau_1 = \ldots = \tau_{q-1} = m$. We use the notation (n, m, q, ρ) for a regular DSS on n points with q subsets of size m. DSS were introduced in [7], [8] as a tool for the construction of comma-free codes with prescribed comma-free index obtained as cosets of linear codes. Since the zero vector belongs to any linear code, the comma-free index of a linear code is zero. However, it is possible to find codes with comma-free index $\rho > 0$ being cosets of linear codes by utilizing difference systems of sets. Given a DSS, $\{Q_0, \ldots, Q_{q-1}\}$, with parameters $(n, \tau_0, \ldots, \tau_{q-1}, \rho)$, one can define a linear q-ary code $C \subseteq F_q^n$ of dimension $n - r$ and redundancy r, where

$$r = \sum_{i=0}^{q-1} |Q_i|, \tag{2}$$

whose information positions are indexed by the numbers not contained in any of the sets $Q_0, \ldots, Q_{q-1}$, and having all r redundancy symbols equal to zero. Replacing in each vector $x \in C$ the positions indexed by Q_i with the symbol i ($0 \leq i \leq q - 1$), yields a coset C' of C with comma-free index at least ρ [7].

The redundancy of a code counts the number of positions added for error detection and correction. Thus, it is desirable that the redundancy is as small as possible.

Let $r_q(n, \rho)$ denote the minimum redundancy (2) of a DSS with parameters n, q, and ρ. A DSS is called *optimal* if it has minimum redundancy for the given parameters. Levenshtein [7] proved the following lower bound on $r_q(n, \rho)$:

$$r_q(n, \rho) \geq \sqrt{\frac{q\rho(n - 1)}{q - 1}}, \tag{3}$$

with equality if and only if the DSS is perfect and regular.

Wang [13] proved a tighter lower bound on $r_q(n, \rho)$ as follows.

Theorem 1.

$$r_q(n, \rho) \geq \begin{cases} \sqrt{\frac{q\rho(n-1)}{q-1}} + 1 \; if \; \sqrt{\frac{q\rho(n-1)}{q-1}} \; is \; a \; square\text{-}free \; integer; \\ \sqrt{\frac{q\rho(n-1)}{q-1}} \qquad otherwise, \end{cases} \tag{4}$$

with equality if and only if the DSS is perfect and the difference between the sizes of any two distinct Q_i and Q_j is not greater than one.

A **cyclic** (n, k, λ) **difference set** $D = \{x_1, x_2, ..., x_k\}$ is a collection of k residues modulo n, such that for any residue $\alpha \not\equiv 0 \pmod{n}$ the congruence

$$x_i - x_j \equiv \alpha \pmod{n}$$

has exactly λ solution pairs (x_i, x_j) with $x_i, x_j \in D$.

If $D = \{x_1, x_2, ..., x_k\}$ is a cyclic difference set with parameters (n, k, λ), then the collection of all singletons, $\{\{x_1\}, ..., \{x_k\}\}$, is a perfect regular DSS with parameters $(n, m = 1, q = k, \rho = \lambda)$. Hence, difference systems of sets are a generalization of cyclic difference sets. Various constructions of DSS obtained by using general partitions of cyclic difference sets are given in Tonchev [10], [11], and Mutoh and Tonchev [9]. Cummings [4] gave another construction method for DSS and two sufficient conditions for a systematic code to be comma-free. Although these papers give explicit constructions of several infinite families of optimal DSS, no general method is known for finding optimal DSS and the value of $r_q(n, \rho)$ for given parameters n, q, ρ, except in the case when $q = 2$ and $\rho = 1$ or 2.

In [12], we described a general algorithm for finding optimal DSS with given parameters n, q, ρ based on searching through all partitions of Z_n into q disjoint parts. Since this algorithm tests all (up to some obvious isomorphisms) possibilities, the expense of computation increases exponentially with linear increase of the values of the parameters.

In this paper, we present a restricted algorithm for finding optimal DSS that admit prescribed symmetries defined by automorphisms of the cyclic group of order n, known as *multipliers* (Section 2). Examples of optimal DSS found by this algorithm are given in Section 3.

2 An Algorithm for Finding DSS Fixed by a Multiplier

If D is a cyclic difference set and s is an integer, the set $D + s = \{x + s \pmod{n} \mid x \in D\}$ is called a cyclic **shift** of D.

Suppose t and n are relatively prime. If $tD = D + s$ for some integer s, then t is called a **multiplier** of D.

Multipliers play an important role in the theory of cyclic difference sets. It is known that if D is a cyclic (n, k, λ) difference set then every prime number $t > \lambda$ that divides $k - \lambda$ and does not divide n is a multiplier of D [6].

In this Section, we describe an algorithm for finding optimal DSS obtained as q-partitions that are fixed by some multiplier.

For any pair of positive relatively prime integers n and p with $p < n$, the map

$$f_p : \alpha \longrightarrow p\alpha \pmod{n}, \ \alpha \in Z_n \tag{5}$$

partitions the group Z_n into pairwise disjoint sets called *orbits* of Z_n under f_p.

If a DSS with parameter n is preserved by a map f_p, where p is a positive integer such that $p < n$ and $\gcd(p, n) = 1$, then p is called a **multiplier** of the DSS.

Example 1. Let $n = 19$ and $p = 7$. The positive integers modulo 19 are partitioned into the following seven orbits under the multiplier $p = 7$:

$$\{0\}, \{1, 7, 11\}, \{2, 14, 3\}, \{4, 9, 6\}, \{5, 16, 17\}, \{8, 18, 12\}, \{10, 13, 15\}. \tag{6}$$

The three sets $\{1, 7, 11\}, \{2, 14, 3\}, \{8, 18, 12\}$ form a perfect and optimal DSS with $q = 3, \rho = 3$, and $r = 9$. Another optimal DSS with the same parameters is $\{1, 2, 4\}$, $\{7, 14, 9\}, \{11, 3, 6\}$. Both DSS are preserved by the map

$$f_7 : \alpha \longrightarrow 7\alpha \pmod{19}.$$

Thus, $p = 7$ is a multiplier of these two DSS.

Let $S \subset \mathbb{Z}_n$ be a cyclic (n, k, λ) difference set and p be a multiplier of S, that is, S is preserved by the map $f_p : i \longrightarrow pi \pmod{n}$. Then f_p partitions S into disjoint orbits. Our algorithm 2.1 is designed to find optimal DSS being a union of such orbits. The external function GENORBITS(S, n, p) generates all orbits partitioning S under the map f_p. The return value of GENORBITS(S, n, p) are the number of orbits, stored in $size$, and an array of all orbits, where $orbits[i]$ is the ith orbit for $0 \leq i \leq size - 1$.

This algorithm takes a set S and two integers n and p as inputs, where S is preserved under the map f_p. The algorithm tests all possible partitions such that $\bigcup_{i=0}^{q-1} Q_i = \bigcup_{j=0}^{size-1} orbits[j]$, to find an optimal DSS.

Algorithm 2.1. SEARCHDSS(S, n, p, q)

```
procedure RECSEARCH(t)
  if t = size
    then  { if {Q_i} is a DSS
          {   then return (1)
    else for i ← 0 to q − 1
    do  { Q_i ← Q_i ∪ orbits[t]
        { if RECSEARCH(t + 1) = 1
        {   then return (1)

main
  size, orbits ← GENORBITS(S, n, p)
  if size < q
    then return (0)
  Q_0 ← orbits[0]
  return (RECSEARCH(1))
```

A slight modification of the algorithm 2.1 applies to partitions of $\mathbb{Z}_n$ preserved by a given multiplier. Note that the set of all nonzero integers modulo n is a trivial cyclic $(n, n-1, n-2)$ difference set. Instead of checking all partitions of $\mathbb{Z}_n$, algorithm 2.2 tests all cases where Q_i is one orbit or a union of several orbits under some map f_p. Now the input parameters are n, $S = \mathbb{Z}_n$ and a positive integer p, where n and p are relatively prime and $p < n$.

Some optimal DSS found by this algorithm are given in next section. The source code of a program written in Mathematica implementing this algorithm is available at www.math.mtu.edu/~hwang.

Algorithm 2.2. SEARCHDSS2(S, n, p, q)

procedure RECSEARCH2(t)
 if $\{Q_i\}$ is a DSS
  then return (1)
 for $i \leftarrow 0$ **to** $q - 1$
  do $\begin{cases} Q_i \leftarrow Q_i \bigcup orbits[t] \\ \textbf{if } \text{RECSEARCH2}(t+1) = 1 \\ \quad \textbf{then return } (1) \end{cases}$

main
 $size, orbits \leftarrow$ GENORBITS(S, n, p)
 if $size < q$
  then return (0)
 for $i \leftarrow 0$ **to** $size - q$
  do $\begin{cases} Q_0 \leftarrow orbits[i] \\ \textbf{return } (\text{RECSEARCH2}(i+1)) \end{cases}$

3   Some Optimal DSS with Multipliers

As an illustration of our algorithm, we list below optimal DSS found by using multipliers.

The first few examples are DSS obtained from cyclic difference sets fixed by a multiplier. We used the list of cyclic difference sets given in [1]. Note that if $\lambda = 1$, the only perfect DSS is the collection of singletons $Q_0 = \{x_1\}, \ldots, Q_{k-1} = \{x_k\}$.

- $n = 15, q = 3, \rho = 2, p = 2, PG(3, 2)$
  $\{0;1,2,4,8;5,10\}$
- $n = 31, q = 3, \rho = 5, p = 2, QR$
  $\{1, 2, 4, 8, 16; 5, 10, 20, 9, 18; 7, 14, 28, 25, 19\}$
- $n = 43, q = 3, \rho = 7, p = 11, QR$
  $\{1, 11, 35, 41, 21, 16, 4; 6, 23, 38, 31, 40, 10, 24;9, 13, 14, 25, 17, 15, 36\}$
- $n = 85, q = 3, \rho = 2, p = 2, PG(3, 4)$
  $\{0;1, 2, 4, 8, 16, 32, 64, 43, 7, 14, 28, 56, 27, 54, 23, 46;17, 34, 68, 51\}$
- $n = 156, q = 3, \rho = 2, p = 5, PG(3, 5)$
  $\{0, 1, 5, 25, 125, 11, 55, 119, 127, 28, 140, 76, 68, 46, 74, 58, 134, 86, 118, 122, 142, 87, 123, 147, 111; 13, 65, 91, 143, 117; 39\}$
- $n = 103, q = 3, \rho = 17, p = 13, QR$
  $\{1, 13, 66, 34, 30, 81, 23, 93, 76, 61, 72, 9, 14, 79, 100, 64, 8;$
  $2, 26, 29, 68, 60, 59, 46, 83, 49, 19, 41, 18, 28, 55, 97, 25, 16;$
  $4, 52, 58, 33, \ 17, 15, 92, 63, 98, 38, 82, 36, 56, 7, 91, 50, 32\}$
- $n = 127, q = 3, \rho = 21, p = 2, QR$
  $\{1, 2, 4, 8, 16, 32, 64, 19, 38, 76, 25, 50, 100, 73, 47, 94, 61, 122, 117, 107, 87; 9, 18, 36,$
  $72, 17, 34, 68, 11, 22, 44, 88, 49, 98, 69, 21, 42, 84, 41, 82, 37, 74; 13, 26, 52, 104, 81, 35,$
  $70, 15, 30, 60, 120, 113, 99, 71, \ 31, 62, 124, 121, 115, 103, 79\}$

- $n = 127, q = 9, \rho = 28, p = 2, QR$
  {1, 2, 4, 8, 16, 32, 64; 9, 18, 36, 72, 17, 34, 68; 11, 22, 44, 88, 49, 98, 69; 13, 26, 52, 104, 81, 35, 70; 15, 30, 60, 120, 113, 99, 71; 19, 38, 76, 25, 50, 100, 73; 21, 42, 84, 41, 82, 37, 74; 31, 62, 124, 121, 115, 103, 79; 47, 94, 61, 122, 117, 107, 87}
- $n = 585, q = 3, \rho = 2, p = 2, PG(3, 8)$
  {0; 1, 2, 4, 8, 16, 32, 64, 128, 256, 512, 439, 293, 5, 10, 20, 40, 80, 160, 320, 55, 110, 220, 440, 295, 17, 34, 68, 136, 272, 544, 503, 421, 257, 514, 443, 301, 39, 78, 156, 312, 43, 86, 172, 344, 103, 206, 412, 239, 478, 371, 157, 314, 123, 246, 492, 399, 213, 426, 267, 534, 483, 381, 177, 354; 65, 130, 260, 520, 455, 325, 195, 390}
- $n = 151, q = 5, \rho = 30, p = 2, QR$
  {1, 2, 4, 8, 16, 32, 64, 128, 105, 59, 118, 85, 19, 38, 76;
  5, 10, 20, 40, 80, 9, 18, 36, 72, 144, 137, 123, 95, 39, 78;
  11, 22, 44, 88, 25, 50, 100, 49, 98, 45, 90, 29, 58, 116, 81;
  17, 34, 68, 136, 121, 91, 31, 62, 124, 97, 43, 86, 21, 42, 84;
  37, 74, 148, 145, 139, 127, 103, 55, 110, 69, 138, 125, 99, 47, 94}
- $n = 199, q = 5, \rho = 30, p = 2, QR$
  {1, 5, 25, 125, 28, 140, 103, 117, 187, 139, 98, 92, 62, 111, 157, 188, 144, 123, 18, 90, 52, 61, 106, 132, 63, 116, 182, 114, 172, 64, 121, 8, 40;
  2, 10, 50, 51, 56, 81, 7, 35, 175, 79, 196, 184, 124, 23, 115, 177, 89, 47, 36, 180, 104, 122, 13, 65, 126, 33, 165, 29, 145, 128, 43, 16, 80;
  4, 20, 100, 102, 112, 162, 14, 70, 151, 158, 193, 169, 49, 46, 31, 155, 178, 94, 72, 161, 9, 45, 26, 130, 53, 66, 131, 58, 91, 57, 86, 32, 160}

The next examples are optimal DSS found by partitioning the whole cyclic group $\mathbb{Z}_n$ (or the trivial $(n, n - 1, n - 2)$ cyclic difference set) with a given multiplier p.

- $n = 31, q = 4, \rho = 6, p = 2$:
  {0; 1, 2, 4, 8, 16; 5, 10, 20, 9, 18; 7, 14, 28, 25, 19}
- $n = 43, q = 4, \rho = 8, p = 11$:
  { 0; 2, 22, 27, 39, 42, 32, 8; 3, 33, 19, 37, 20, 5, 12; 7, 34, 30, 29, 18, 26, 28 }
- $n = 127, q = 2, \rho = 16, p = 2$:
  { 0, 1, 2, 4, 8, 16, 32, 64, 3, 6, 12, 24, 48, 96, 65, 9, 18, 36, 72, 17, 34, 68, 19, 38, 76, 25, 50, 100, 73, 21, 42, 84, 41, 82, 37, 74; 5, 10, 20, 40, 80, 33, 66, 7, 14, 28, 56, 112, 97, 67, 11, 22, 44, 88, 49, 98, 69, 13, 26, 52, 104, 81, 35, 70, 47, 94, 61, 122, 117, 107, 87, 55, 110, 93, 59, 118, 109, 91 }

Acknowledgments

This research was supported by NSF Grant CCR-0310632.

References

1. L. D. Baumert, *Cyclic Difference Sets*, Springer-Verlag, 1971.
2. D. J. Clague, *New classes of synchronous codes*, IEEE Trans. on Electronic Computers **EC-16**, 290-298.
3. H. C. Crick, J. S. Griffith, and L. E. Orgel, *Codes without commas*, Proc. Nat. Acad. Sci. 43 (1957), 416-421.
4. L. J. Cumming, *A Family of Circular Systematic Comma-Free Codes*, (to appear).

5. S. W. Golomb, B. Gordon, and L. R. Welch, *Comma-free code*, Canad. J. Math., vol. 10, no. 2, pp. 202-209, 1958.

6. M. Hall, Jr., "Combinatorial Theory", Second Ed., Wiley, New York 1986.

7. V. I. Levenshtein, *One method of constructing quasilinear codes providing synchronization in the presence of errors*, Problems of Information Transmission, vol. 7, No. 3 (1971), 215-222.

8. V. I. Levenshtein, *Combinatorial problems motivated by comma-free codes*, J. of Combinatorial Designs, vol. 12, issue 3 (2004), 184-196.

9. Y. Mutoh, and V. D. Tonchev, *Difference Systems of Sets and Cyclotomy*, Discrete Math., to appear.

10. V. D. Tonchev, *Difference systems of sets and code synchronization*, Rendiconti del Seminario Matematico di Messina, Series II, vol. 9 (2003), 217-226.

11. V. D. Tonchev, *Partitions of difference sets and code synchronization*, Finite Fields and their Applications, Finite Fields and Their Appl. 11 (2005), 601-621.

12. V. D. Tonchev, H. Wang, *An algorithm for optimal difference systems of sets*, J. Discrete Optimization, accepted.

13. H. Wang, *A New Bound for Difference Systems of Sets*, J. of Combinatorial Mathematics and Combinatorial Computing, (to appear).

Authentication Mechanism Using One-Time Password for 802.11 Wireless LAN

Binod Vaidya[1], SangDuck Lee[2], Jae-Kyun Han[3], and SeungJo Han[2]

[1] Dept. of Electronics & Computer Eng., Tribhuvan Univ., Nepal
bvaidya@ioe.edu.np
[2] Dept. of Information & Communication Eng., Chosun Univ., Korea
dandylsd@hanmail.net, sjbhan@chosun.ac.kr
[3] Korea National Open Univ., Korea
jiguin@knou.ac.kr

Abstract. Wireless local area networks (WLANs) are quickly becoming ubiquitous in our every day life. The increasing demand for ubiquitous services imposes more security threats to communications due to open mediums in wireless networks. The further widespread deployment of WLANs, however, depends on whether secure networking can be achieved. We propose a simple scheme for implementing authentication based on the One-Time Password (OTP) mechanism. The authentication protocol is proposed to solve the weak authentication and security flaw problem of the WEP in 802.11 WLAN. Further we have simulated the implementation of proposed scheme and EAP-OTP and analyzed the performance in terms of different performance metrics such as response time and authentication delay.

1 Introduction

Wireless local area networks (WLANs) are quickly becoming ubiquitous in our every day life. Users are adopting the technology to save the time, cost, and mess of running wire in providing high speed network access. Specifically, the IEEE 802.11 [1] WLAN, popularly known as Wireless Fidelity (Wi-Fi) has grown steadily in popularity since its inception and is now well positioned to complement much more complex and costly technologies such as the Third Generation (3G).

The freedom and mobility that WLANs promise also present some serious security challenges. WLANs are not limited by network jacks nor are they limited by geography. WLANs provide unprecedented flexibility in that an area not originally intended as a collaborative workspace can accommodate a large number of wireless clients. The further widespread deployment of WLANs, however, depends on whether secure networking can be achieved.

2 Security in Wireless LAN

In the IEEE 802.11 Wireless LANs, [2] security measures such as Wired Equivalent Privacy (WEP) and shared Authentication scheme have been originally

D. Grigoriev, J. Harrison, and E.A. Hirsch (Eds.): CSR 2006, LNCS 3967, pp. 619–628, 2006.

employed. However, the new security protocols such as Wi-Fi Protected Access (WPA) and IEEE 802.11i protocols have been established that replace WEP and provide real security. [3]

2.1   Wired Equivalent Privacy

WEP offers an integrated security mechanism that is aimed at providing the same level of security experienced in wired networks. WEP is based on the RC4 symmetric stream cipher and uses static, pre-shared, 40 bit or 104 bit keys on client and access point. [4]

Several WEP key vulnerabilities are due to many reasons such as use of static WEP keys, use of 40-bit or 128-bit RC4 keys for the WEP and vulnerable 24-bit Initialization Vector (IV). [5]

2.2   Authentication

The current Wireless LAN standard includes two forms of authentication: open system and shared key. The open system authentication is a null authentication process where the client knowing Service Set Identifier (SSID) always successfully authenticates with the AP. The second authentication method utilizes a shared key with a challenge and a response. Actually, shared-key authentication uses WEP encryption and therefore can be used only on products that implement WEP.

2.3   Wi-Fi Protected Access

The WPA protocol, backed by the Wi-Fi Alliance, is considered as a temporary solution to the weakness of WEP. WPA is designed to secure all versions of 802.11 devices, including multi-band and multi-mode. Its authentication is based on the 802.1x authentication protocol that was developed for wired networks, as well as Extensible Authentication Protocol (EAP). WPA is forward-compatible with the 802.11i specification.

WPA offers two primary security enhancements over WEP: Enhanced data encryption through Temporal Key Integrity Protocol (TKIP) and Enterprise-level user authentication via IEEE 802.1x. [6]

2.4   IEEE 802.11i Protocol

The 802.11i is a standard for WLANs that provides improved encryption for networks that use the IEEE 802.11 standards. To reach that purpose, the IEEE 802.11i Task Group has defined a new protocol called, Robust Security Network (RSN). [7] The 802.11i standard was officially ratified by the IEEE in June 2004. As anticipated by most Wi-Fi manufacturers, the IEEE 802.11i protocol is the ultimate security mechanism for IEEE 802.11 WLANs, which is known as WPA version 2 (WPA2). [6]

The 802.11i protocol consists of three protocols, organized into two layers. On the bottom layer are TKIP and CCMP (Counter Mode with Cipher Block

Chaining Message Authentication Code Protocol) - an AES (Advanced Encryption Standard) based protocol. [8] While TKIP is optional in 802.11i, CCMP is mandatory for anyone implementing 802.11i. Above the TKIP and CCMP is the 802.1x protocol, which provides user-level authentication and encryption key distribution. However, AES requires a dedicated chip, and this may mean hardware upgrades for most existing Wi-Fi networks.

3   One Time Password System

One form of attack on network computing systems is eavesdropping on network connections to obtain authentication information such as the login IDs and password of legitimate users. Once this information is captured, it can be used at a later time to gain access to the system. One-time Password (OTP) system [9] designed to counter this type of attack, called a replay attack.

A sequence of OTP produced by applying the secure hash function multiple times gives to the output of the initial step (called S). That is, the first OTP to be used is produced by passing S through the secure hash function a number of times (N) specified by the user. The next OTP to be used is generated by passing S through the secure hash function N-1 times. An eavesdropper who has monitored the transmission of OTP would not be able to generate the next required password because doing so would mean inverting the hash function.

The OTP improves security by limiting the danger of eavesdropping and replay attacks that have been used against simple password system. The use of the OTP system only provides protections against passive eavesdropping and replay attacks. It does not provide for the privacy of transmitted data, and it does not provide protection against active attacks. The success of the OTP system to protect host system is dependent on the non-invert (one-way) of the secure hash functions used.

4   Design of Authentication Scheme

There is a desire to increase the security and performance of authentication mechanisms due to strong user authentication, protection of user credentials, mutual authentication and generation of session keys. As demand for strong user authentication grows, OTP-based authentications tend to become more common.

We propose a simple scheme for implementing authentication mechanism based on the hash function. The authentication protocol is proposed to solve the weak authentication and security flaw problem of WEP in 802.11 WLAN.

4.1   Overviews

In this scheme, the AP receives the password which is processed from the client and the server's original password exclusive-OR (XOR) the stream bits, in order to defend malice attack. The server examines client's password, stream bits and transmits new stream bits to the client for next authentication by privacy-key cryptography. The AP computes simple hash function, and compares the hash value between the client and the server. If the result is true, the AP sends authentication success frame to the client. Then the client responds for update client's database. Then the client sends One-time Password (OTP) to the AP. The key exchange between the client and the AP generates a session key. Finally, the client refreshes secret key of the WEP in IEEE 802.11, to defend replay attack.

4.2 Rationale

In order to prevent client and AP in this protocol from several attack methods. The techniques include per-packet authentication and encryption, etc are adopted. This scheme can be used for defending malice attack by verification of the client's password and defending replay attack by refreshing secret key.

4.3 Working Principle

There are four phases in proposed authentication scheme, which are as follows:

1. Phase 0: Discovery phase;
2. Phase 1: OTP Authentication phase;
3. Phase 2: Secure key exchange phase; and
4. Phase 3: Refreshing OTP phase.

The conversation phases and relationship between the parties is shown in Fig 1.

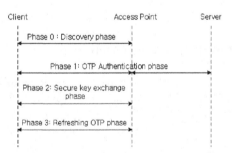

Fig. 1. Conversation Overviews

Phase 0: Discovery Phase

In the discovery phase (phase 0), the client and AP locate each other and discover each other's capabilities. IEEE 802.11 provides integrated discovery support utilizing Beacon frames, allowing the client (known as the station or STA) to determine the Media Access Control (MAC) address and capabilities of Access Point or AP.

Phase 1: OTP Authentication Phase

The authentication phase (phase 1) begins once the client and AP discover each other. This phase always includes OTP authentication. In this phase, the client authenticates itself as a legal member for a particular AP. The steps are shown as the follows:

Step 1: The AP send Request message.

Step 2: The client will then send $\{ID_C\|Y_C\|H(ID_C, A)\}$. Where ID_C is client's identity; $Y_C = a_C^R \bmod q$; R_C is random number produced by a client; $H()$ is one way hash function; $A = PW_C \oplus SB$; PW_C is client's password; and SB is stream bits generated by the server to share with the client.

Step 3: After receiving $\{ID_C\|Y_C\|H(ID_C, A)\}$ from the client, the AP sends $\{ID_A\|ID_C\}$ to the server. Where ID_A is AP's Identity.

Step 4: Depending on ID_C, the server examines PW_C, SB, and client's privacy key (K_C) from user's database, and generates a new stream bits (SB_N) and A. The ID_C and SB_N are encrypted with K_C producing $E_{K_C}(ID_C\|SB_N)$. Then $E_{K_C}(ID_C\|SB_N)$, ID_C, and A are encrypted by server with symmetric cryptosystem to get $E_{AS}(ID_C\|A\|E_{K_C}(ID_C\|SB_N))$. Then $\{ID_A\|E_{AS}(ID_C\|A\|E_{K_C}(ID_C\|SB_N))\}$ is send to the AP.

Step 5: Upon receiving $\{ID_A\|E_{AS}(ID_C\|A\|E_{K_C}(ID_C\|SB_N))\}$ from the server, then the AP computes and verifies $H(ID_C, A)$ between the client and the server. If true, it will send authentication success frame $\{ID_C\|Y_A\|E_{K_C}(ID_C\| SB_N)\}$ to the client. Where $Y_A = a_A^R \bmod q$; R_A is random number produced by AP. Y_A will be used for key exchange in later steps.

Step 6: The client accomplishes authentication procedure, then it sends $\{ID_C \|H(SB_N)\}$ to the server through the AP to update client's database in the server.

Step 7: The client and the AP apply Y_A or Y_C to generate a session key (K), then the client sends $\{ID_C\|E_K(otp\|ctr\|ID_C)\}$ to the AP. Where otp is the client's one time password shared with the AP; and ctr is the counter of otp.

Step 8: The AP responds with $\{ID_C\|H(otp, ctr)\}$ to the client.

It should be noted the counter is a positive integer and will decrease by one once the otp is changed. When it decreases to zero, the client should reinitialize his otp.

Phase 2: Secure Key Exchange Phase

The Secure key exchange phase (phase 2), begins after the completion of OTP authentication. In practice, the phase 2 is used to negotiate a new secret key between the client and the AP. Similar to the authentication phase, a MAC is added to each packet and then is encrypted together, and the receiver should check the MAC. The steps of the protocol are as shown in the following:

Step 1: The client sends $\{ID_C\|E_K(otp_{I+1}\|H(otp_I, ctr, ID_C))\}$ to AP. It should be noted that $H(opt_{I+1})$ is equal to otp_I and K is session key by OTP authentication phase (phase 1).

Step 2: The AP checks the MAC and ID_C by checking if $H(opt_{I+1})$ is equal to otp_I. If this is true, the client is a legal member. It will replace otpI by opt_{I+1} and decrease the counter by 1. And AP transmits $H(otp_I, ctr)$ to client.

Step 3: The client checks Hash value and decreases the counter by 1.

By these three steps, the client and the AP can achieve the goal of mutual authentication. In this phase, both the client and the AP have the same new secret key of WEP just for this session.

Phase 3: Refreshing OTP Phases

When the counter decreases to zero, the client should change its One-time Password, otherwise the AP has the right to prohibit the client from using its services. The steps of the protocol are shown in the following:

Step 1: The client sends its $\{ID_C \| E_K(otp_{I+1} \| otp_N \| ctr_N, H(otp_{I+1}, otp_N, ctr_N, ID_C))\}$ to AP, where $H(otp_{I+1})$ is equal to otp_I, otp_N is its new One-time Password, ctr_N is a new counter for this new password (otp_N) and K is session key by authentication phase.

Step 2: The AP verifies the MAC and otp_{I+1} (check $H(otp_{I+1})$ is equal to otp_I). If this is true, the client is a legal member, and AP replaces the otp_I by otp_N and resets the ctr to ctr_N. Then the AP sends $\{ID_C \| H(otp_I, ctr)\}$ to the client.

Step 3: The client checks his ID and the Hash value.

4.4   Security Analysis

The proposed authentication mechanism is analyzed:

Malice attack Protection - In this scheme, the AP receives the password which is processed from the client and the server's original password exclusive-OR (XOR) the stream bits, in order to defend malice attack.

Mutual Authentication - This scheme provides mutual authentication. Strong mutual authentication can prevent user from being impersonated.

Replay Protection - This scheme is inherently designed to avoid replay attacks by choosing One-time Password is that it refreshes secret key of the WEP in IEEE 802.11 anytime. Because of refresh OTP, the lifetime of the share key can be extended.

Confidentiality - Each the element of the key exchange is protected by a shared secret that provides high confidentiality.

5   Related Works

Lamport [10] proposed a One-time password/hash-chaining technique, which has been used in many applications. S/KEY [11] is an authentication system that uses one-time passwords. The S/KEY system is designed to provide users with one-time passwords which can be used to control user access to remote hosts. One-time Password (OTP) system [9] designed to provide protections against passive eavesdropping and replay attacks.

The IETF RFC 3748 [12] has defined that with Extensible Authentication Protocol (EAP), OTP method can be used. Further the Internet draft [13] defines the One-time Password (OTP) method, which provide one-way authentication but not key generation. As a result, the OTP method is only appropriate for use on networks where physical security can be assumed. This method is not suitable for using in wireless IP networks, or over the Internet, unless the EAP conversation is protected.

6   Simulation Model

In order to evaluate the performance of the proposed authentication scheme in Wireless LAN, we have simulated several scenarios in wireless network using OPNET Modeler, [14] which a discrete event-driven simulator tool capable of modeling both wireless and wireline network.

6.1   Scenario

For analyzing the effect of proposed authentication scheme in WLAN, we have designed experimental model.

The experimental model for evaluating proposed authentication mechanism in Wireless LAN is depicted in Fig. 2. In this model, it consists of wireless clients, Access Point (AP), authentication server and main FTP server.

The main FTP server is capable of providing file transfer service. File transfer service can be provided to wireless clients in a wireless network. The wireless network consists of various wireless users using IEEE 802.11b devices. For the experimental purpose, two scenarios have been designed. First one with the implementation of proposed authentication scheme in wireless network whereas second with implementation of EAP-OTP in wireless network. The Network Analyzer is used to capture network statistics and the measurements were collected from the server.

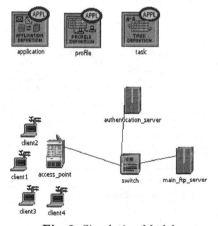

Fig. 2. Simulation Model

6.2   Assumptions

We have implemented the proposed authentication mechanism and the EAP-OTP in our experiment.

For the wireless users, we have considered IEEE 802.11b as our WLAN protocol. And the transmission speed is 11 Mbps between the AP and the clients.

For the experimental evaluation, we have assumed the various performance metrics in order to analyze the effect of proposed authentication scheme and the EAP-OTP. We have considered following performance metrics:

- Response Time: the total time required for traffic to travel between two points. It includes connection establishment, security negotiation time as well as the actual data transfer time.
- Authentication Delay: the time involved in an authentication phase of a given protocol.

6.3   Results and Analysis

In order to investigate the performance of proposed and EAP-OTP authentication schemes in wireless 802.11 network, we have analyzed different aspects of experimental results obtained. Particularly we have investigated the impact of proposed authentication scheme and EAP-OTP in Wireless IP network on the response time and authentication delay.

Fig. 3 illustrates the mean response time of FTP traffic for different numbers of wireless clients for two scenarios - first scenario is proposed authentication scheme and second is EAP-OTP. It can be seen that the response time in second scenarios has slightly lower than that of first one.

The Authentication delays for different numbers of wireless clients in both scenarios - proposed scheme and EAP-OTP are depicted in Fig. 4. On analyzing two scenarios of the experiment, it can be seen that the Authentication delay for proposed is reduced than that for EAP-OTP.

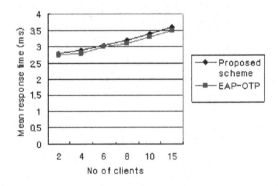

Fig. 3. Response Time

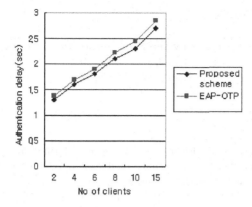

Fig. 4. Authentication Delay

7   Conclusion and Future Work

In this paper we have shown the major security measures in wireless LAN and drawback of the WEP in IEEE 802.11b. We propose authentication scheme for Wireless IP network. Proposed authentication scheme has advantage that it makes the simpler computational complexity with sufficient security. The concept of refresh password is proposed to renew the secret key of the WEP in IEEE 802.11. Moreover, it will be convenient for a user to register once to a server for roaming every AP.

We have simulated the experimental scenarios for implementation of proposed scheme and for implementation of EAP-OTP. We have analyzed the results of the experimental analysis of the authentication schemes used in Wireless IP network. Performance analysis of two different authentication methods over wireless IP network was investigated in order to view its impact on performance measurements such as response time and authentication delay. In terms of authentication delay, proposed authentication scheme shows better performance.

As the properties of authentication protocols are extremely subtle, informal reasoning might not be sufficient. Thus the use of rigorous and formal analytical techniques is desirable for understanding the capabilities of the authentication scheme. The most popular approach to analyze authentication protocols is through the use of logic. The one of known logics is the BAN logic [15] [16], named after its inventors M. Burrows, M. Abadi, and R. Needham. The BAN logic maps initial knowledge states and assumptions into formulas in the logic. The predefined postulates, initial states, assumptions and protocols steps lead to reasoning into final beliefs that can then be used to verify the design objective. In the future work, we will formulate a formal analysis of authentication mechanism using one-time password protocol for IEEE 802.11 wireless network using BAN logic.

Acknowledgement

This research was supported by the Program for the Training of Graduate Students in Regional Innovation which was conducted by the Ministry of Commerce Industry and Energy of the Korean Government.

References

1. Institute of Electrical and Electronics Engineers: Information technology - Telecommunications and information exchange between systems - Local and metropolitan area networks - Specific Requirements Part 11 - Wireless LAN Medium Access Control (MAC) and Physical Layer (PHY) Specifications, IEEE Standard 802.11-2003, 2003
2. M. Gast: 802.11 Wireless Networks - The Definitive Guide, O'Reilly, Dec. 2002
3. B. Potter, and B. Fleck: 802.11 Security, O'Reilly, Dec. 2002
4. W. A. Arbaugh, N. Shankar, and Y.C. Justin Wan: Your 802.11 wireless network has no clothes, IEEE Wireless Communications, Vol. 9, Dec. 2002, pp 44-51
5. J. R. Walker: Unsafe at any key size - an analysis of the WEP encapsulation, 802.11 Security Papers at NetSys.com, Oct. 2000.
   http://www.netsys.com/library/papers/walker-2000-10-27.pdf
6. J. Edney, and W. A. Arbaugh: Real 802.11 Security - Wi-Fi Protected Access and 802.11i, Addison Wesley, July 2003
7. P. Cox: Robust Security Network - The future of wireless security, System Experts Corporation.
   http://www.systemexperts.com/win2k/SecureWorldExpo-RSN.ppt
8. K.H Baek, S. W. Smith, and D. Kotz: A Survey of WPA and 802.11i RSN Authentication Protocols, Dartmouth College Computer Science, Technical Report TR2004-524, Nov 2004
9. N. Haller, C.Metz, P. Nesser, and M. Straw; A One-Time Password System, IETF RFC 2289, Feb 1998
10. L. Lamport: Password Authentication with insecure communication, Communications of the ACM, Vol. 24 No. 11, Nov. 1981, pp 770-722
11. B. Aboba, L. Blunk, J. Vollbrecht, J. Carlson, and H. Levkowetz: Extensible Authentication Protocol (EAP), IETF RFC 3748, June 2004
12. N. Haller: The S/KEY One-Time Password System, Proc. of the Symposium on Network and Distributed Systems Security, Internet Society, CA, USA, Feb. 1994
13. L. Blunk, J. Vollbrecht, and B. Aboba: The One Time Password (OTP) and Generic Token Card Authentication Protocols, Internet draft <draft-ietf-eap-otp-00.txt>
14. OPNET Modeler Simulation Software, http://www.opnet.com.
15. M. Burrows, M. Abadi and R. Needham: A logic of authentication, ACM Transactions on Computer Systems, 8(1), Feb. 1990, pp 18-36.
16. M. Abadi, and M. Tuttle: A semantics for a logic of authentication, Proc. of the Tenth Annual ACM Symposium on Principles of Distributed Computing, Aug. 1991, pp. 201-216.

Optimizing Personalized Retrieval System Based on Web Ranking

Hao-ming Wang[1,3], Ye Guo[2], and Bo-qin Feng[1]

[1] School of Electronic and Information Engineering, Xi'an Jiaotong University,
Xi'an, Shaanxi 710049, P.R. China
{wanghm, bqfeng}@mail.xjtu.edu.cn
[2] School of Information, Xi'an University of Finance & Economics,
Xi'an, Shaanxi 710061, P.R. China
guoyexinxi@126.com
[3] School of I & C, Swiss Federal Institute of Technology(EPFL),
1015 Lausanne, Switzerland

Abstract. This paper drew up a personalized recommender system model combined the text categorization with the pagerank. The document or the page was considered in two sides: the content of the document and the domain it belonged to. The features were extracted in order to form the feature vector, which would be used in computing the difference between the documents or keywords with the user's interests and the given domain. It set up the structure of four block levels in information management of a website. The link information was downloaded in the domain block level, which is the top level of the structure. In the host block level, the links were divided into two parts, the inter-link and the intra-link. All links were setup with different weights. The stationary eigenvector of the link matrix was calculated. The final order of documents was determined by the vector distance and the eigenvector of the link matrix.

1 Introduction

Applying the peer-to-peer architectural paradigm to Web search engines has recently become a subject of intensive research. Whereas proposals have been made for the decomposition of content-based retrieval techniques, such as classical text-based vector space retrieval or latent semantic indexing, it is much less clear of how to decompose the computation for ranking methods based on the link structure of the Web.

As a user, in order to find, collect and maintenance the information, which maybe useful for the specific aims, s/he have to pay more time, money and attention on the retrieval course.

There are many search engines, such as Yahoo, Google, etc. to help the user to search and collect the information from the Internet. The features of the Internet, such as mass, semi-structure, have become drawbacks in using the information widely in Internet.[1]

D. Grigoriev, J. Harrison, and E.A. Hirsch (Eds.): CSR 2006, LNCS 3967, pp. 629–640, 2006.

In order to make the information retrieval system more efficient, the artificial intelligence (AI) technologies have been suggested to use. The IR models can be divided into two items, the one is based on the large scales machine learning, and another is based on the intelligent personalization.

For the user interested in the special domain, for example, the computer science or architectonics, it may be a good way to support the results by using intelligent personalization.

This paper describes the architecture of information retrieval model based on the intelligent personalization. The aim is to provide the information related to the given domain. The model receives the request of user, interprets it, selects and filters the information from Internet and local database according to the profile of the user. The user profile is maintained according to the feedback of the user. The pagerank values of documents are computed before they are stored.

This paper is organized as follows: Section 2 introduces the basic concept of TFIDF and the Pagerank. Section 3 describes the architecture of the new system. Section 4 discusses the process of re-ranking. Section 5 introduces the query process. Section 6 discusses the shortcoming and the room for improvement of the system, and section 7 gives the conclusion.

2 Basic Concept

2.1 TFIDF and Text Categorization

TFIDF (Term Frequency / Inverse Document Frequency) is the most common weighting method used to describe documents in the Vector Space Model (VSM), particularly in IR problems. Regarding text categorization, this weighting function has been particularly related to two important machine learning methods: kNN (k-nearest neighbor) and SVM(Support Vector Machine). The TFIDF function weights each vector component (each of them relating to a word of the vocabulary) of each document on the following basis.

Assuming vector $\tilde{d} = (d^{(1)}, d^{(2)}, ..., d^{|F|})$ represents the document d in a vector space. It is obviously that documents with similar content have similar vector. Each dimension of the vector space represents a word selected by the feature selection.

The values of the vector elements $d^{(i)}(i \in (0, |F|))$ for a document d are calculated as a combination of the statistics $TF(w, d)$ and $DF(w)$ (document frequency).

$TF(w, d)$ is the number of word occurred in document d. $DF(w)$ is the number of documents in which the word w occurred at least once time. The $IDF(w)$ can be calculated as

$$IDF(w) = log \frac{N_{all}}{DF(w)}.$$

Where N_{all} is the total number of documents. Obviously, the $IDF(w)$ were low if w occurred in many documents and it were the highest if w occurred in only one.

The value $d^{(i)}$ of feature w_i for the document d is then calculated as

$$d^{(i)} = TF(w_i, d) \times IDF(w_i).$$

$d^{(i)}$ is called the weight of word w_i in document d. [9-12]

The TFIDF algorithm learns a class model by combining document vectors into a prototype vector $\tilde{C}$ for every class $C \in \varsigma$. Prototype vectors are generated by adding the document vectors of all documents in the class.

$$\tilde{C} = \sum_{d \in C} \tilde{d}.$$

This model can be used to classify a new document d'. d' can be represented by a vector $\tilde{d}'$. And the cosine distance between the prototype vector of each class and $\tilde{d}'$ is calculated. The d' is belonged to the class with which the cosine distance has the highest value.

$$H_{TFIDF}(d') = argmax cos(\tilde{d}', \tilde{C}).$$

Where $H_{TFIDF}(d')$ is the category to which the algorithm assigns document d'.

$$H_{TFIDF}(d') = argmax(\frac{\tilde{d}' \cdot \tilde{C}}{\|\tilde{d}'\| \cdot \|\tilde{C}\|}) = argmax(\frac{\sum_{i=1}^{|F|}[d'^{(i)} \cdot C^{(i)}]}{\sqrt{\sum_{i=1}^{|F|}[d'^{(i)}]^2} \cdot \sqrt{\sum_{i=1}^{|F|}[C^{(i)}]^2}}).$$

2.2 Pagerank

For a search engine, after finding all documents using the query terms, or related to the query terms by semantic meaning, the result, which maybe a large number of web pages, should be managed in order to make this list clearly. Many search engines sort this list by some ranking criterion. One popular way to create this ranking is to exploit the additional information inherent in the web due to its hyper linking structure. Thus, link analysis becomes the means to ranking. One successful and well publicized link-based ranking system is PageRank, the ranking system used by the Google search engine.

The Google search engine is based on the popular PageRank algorithm first introduced by Brin and Page in Ref.[6]. The algorithm can be described as:

Let u be the web page. Then let F_u be the set of pages u points to and B_u be the set of pages that point to u. Let N_u be the number of links from u and let c be a factor used for normalization (so that the total rank of all web pages is constant).

$$R(u) = c \sum_{v \in B_u} \frac{R(v)}{N_v}.$$

This algorithm displays that the pagerank of page u comes from the pages that point to it, and u also transfers its pagerank to the pages which it points to.

632   H.-m. Wang, Y. Guo, and B.-q. Feng

Considering the pages and the links as a graph $G = P(Page, Link)$, we can describe the graph by using the adjacency matrix. The entries of the matrix, for example p_{ij}, can be defined as:

$$p_{ij} = \begin{cases} 1 & Link \quad i \rightarrow j \quad exists \\ 0 & Otherwise. \end{cases}$$

Here $i, j \in (1, n)$ and n is a number of web pages. Because the total probability from one page to others can be considered 1, the rows, which correspond to pages with a non-zero number of out-links $deg(i) > 0$, can be made row-stochastic (row entries non-negative and sum to 1) by setting $p_{ij} = p_{ij}/deg(i)$. That means if the page u has m out-links, the probability of following each of out-links is $1/m$. We assume all the m out-links from page u have the similar probability. Actually, there is difference among them, and one link maybe more important than others. We can assume the probability list,

$$\{w_1, w_2, \cdots, w_m; \sum_{i=1}^{m} w_i = 1\}.$$

which can promise the pagerank more precise.

If we considered the property of the adjacency matrix, we could find the adjacency matrix correspond to a Markov chain.

According to the Chapman-Kolmogorov Equations, for the Markov chains, we can get

$$p_{ij}^{n+m} = \sum_{k=0}^{\infty} p_{ik}^n p_{kj}^m \quad (n, m \geq 0, \forall i, \forall j)$$

If we let $P^{(n)}$ denote the matrix of $n - step$ transition probabilities p_{ij}^n, then we can asserts that

$$P^{(n+m)} = P^{(n)} \cdot P^{(m)};$$

$$P^{(2)} = P^{(1)} \cdot P^{(1)} = P \cdot P = P^2;$$

$$P^{(n)} = P^{(n-1+1)} = P^{(n-1)} \cdot P^{(1)} = P^{n-1} \cdot P = P^n.$$

That is, the $n - step$ transition matrix can be obtained by multiplying the matrix P by itself n times.

The case discussed above is ideal. For a real adjacency matrix P, in fact, there are many special pages without any out-link from them, which are called *dangling page*. Any other pages can reach the dangling page in $n(n \geq 1)$ steps, but it is impossible to get out. The dangling page is called absorbing state. In the adjacency matrix, the row, corresponding to the dangling page is all zeros. Thus, the matrix P is not a row-stochastic. It should be deal with in order to meet the requirement of the row-stochastic.

One way to overcome this difficulty is to slightly change the transition matrix P. We can replace the rows, all of the zeros, with $v = (1/n)e^T$, where e^T is the row vector of all $1s$ and n is the number of pages of P contains. The P will be changed to $P' = P + d \cdot v^T$. Where

$$d = \begin{cases} 1 & if \quad deg(i) = 0 \\ 0 & Otherwise. \end{cases}$$

is the dangling page indictor. If there were a page without any out-link from it, we could assume it can link to every other pages in P with the same probability. After that there is not row with all 0s in matrix P'. P' is row-stochastic. [26]

Because P' corresponds to the stochastic transition matrix over the graph G, Pagerank can be viewed as the stationary probability distribution over pages induced by a random walk on the web. The pagerank can be defined as a limiting solution of the iterative process:

$$x_j^{(k+1)} = \sum_i P'_{ij} x_i^{(k)} = \sum_{i \to j} x_i^{(k)} / deg(i).$$

Because of the existing of zero entries in the matrix P', it cannot be insure the existence of the stationary vector. The problem comes from that the P' may be reducible.

In order to solve the problem, P' should be modified by adding the connection between every pair of pages.

$$Q = P'' = cP' + (1 - c)ev^T, \quad e = (1, 1, \cdots, 1)^T.$$

Where c is called dangling factor, and $c \in (0, 1)$. In most of the references, the c is set $[0.85, 1)$.

After that, it can be consider that all of the pages are connected (Strong connection). From one of the pages, the random surfer can reach every other page in the web. The Q is irreducible. For $Q_{ii}^{(k)} > 0, (i, k \in [1, n])$, the Q is aperiodic.

Above all, the matrix Q is row-stochastic, irreducible and aperiodic. The Perron-Frobenius theorem guarantees the equation $x^{(k+1)} = Q^T x^{(k)}$ (for the eigensystem $Q^T x = x$) converges to the principal eigenvector with eigenvalue 1, and there is a real, positive, and the biggest eigenvector.

3   Architecture of the System

The goal of this system is to help the users to find the information on Internet easily and quickly. It requests the system should process the information resource independently, gather the information the users are interested in, filter the repeating one, wipe off the useless one, and store them to the local database. This system should be feasible, friendly, adaptable, and transplantable.

The system would not take the place of Yahoo or Google, it will be an entrance of personalized search engine. With the interaction between the user and the system, it will gather the personalized information of the user. Fig.1 shows

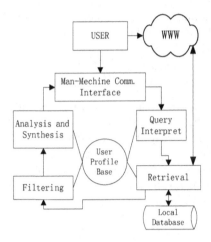

Fig. 1. The architecture of the system

the architecture of the system. The architecture is composed of five components, local database, and a user profile base. The architecture of the system is described in details as follows:[4]

(1) Man-Machine Communication component (MM): this component serves as the communication interface between the user and the system. The user inputs the keywords or other kinds of the requests to the system and receives the results from the interface.

(2) Query Interpret component (QI): this component enhances the user's query based on the user profile. Due to the difference of the knowledge and the domain, not all of the users can express his/her request exactly. As an information retrieval system, it should decide what kinds of sources should to be queried, how to modify the query expression the user has submitted in order to utilize the underlying search engines more, and how to order the feedback results.

(3) Retrieval component (RE): this component sends the requests and gets the information from Internet and the local database. There are two expectations in information retrieval, the *precision* and the *recall*. The precision shows the degree of the feedback results case to the user's needs. The recall shows the percentage of the feedback results to the total records, which are related to the user's needs. The results from the Internet maybe more general and the results from the local database maybe more accurate. In this system, the output list will be combined by the two kinds of results.

(4) Filtering component (FI): this component filters the raw data from the RE based on the user profile. The feedback results from the Internet are the raw data according to the keywords and query expression. They cannot meet the user's needs satisfactorily. This component filters the results according to the domains or the subjects the user interested. The related data of user's feature and the domain have been stored in the User Profile Base.

(5) Analysis & Synthesis component (AS): this component uses the filtered information to enhance decision making, uses data mining techniques and the user profile to analyze and synthesize the information retrieved.

(6) User Profile Base (UPB): a knowledge base for the users. There are two kinds of forms.

- The one is the holistic user profiles for all the users of a system; this conceptual profile base can either be distributed across the system or stored in a central location, the holistic user profile consists of a personal profile, a functional area profile, a current project profile, an organizational environment profile, and a client-type profile.
- Another is the storing feature or the visiting model of the specifically kind of the user. The initial data are set by manual. When the user visits the Internet, the historical pages are recorded and downloaded. The system extracts the information from the pages, extracts the class keywords, and constructs the vector of every page. The distance between the page vector and the domain vector is calculated. According to the distance, the user's personalize model is built. This model will be refined based on the feedbacks the user interested.

(7) Local Database (LD): it stores the data, which have been downloaded from the Internet according to the historical pages. Most of the knowledge on Internet are non-structure or semi-structure. They are different from the data stored in the local relational database. Most of the non-structure and semi-structure data are organized as the natural language model. Those data should be converted before they are stored to the relational database, after that, they can be shared and utilized effectively.

4 Re-ranking the Pages

In this section, we introduce two kinds of calculation used in this system. The first is the similarity computing of the vector, and the second is the pagerank re-computing.

4.1 Extracting Keywords

In our system, we should select the keywords from the given domain. In the traditional algorithm of text categorization or recommender system, all the terms (words) are considered, and the importance of each term is decided by the numbers of it appeared in the documents. But actually, some terms in a given domain maybe more important than the others.

The Ref.[9][15] introduces a new weighting method based on statistical estimation of the importance of a word for a specific categorization problem. This method also has the benefit to make feature selection implicit, since useless features for the categorization problem considered get a very small weight. Extensive experiments reported in the paper shows that this new weighting method improves significantly the classification accuracy as measured on many categorization tasks.

In our system, the method can be described as selecting the $top-n$ keywords from the paper set, which have been categorized in manual, by calculating the weight of every word.

We donated the keywords list with the vector:

$$\widetilde{D_i} = \{(k_j, w_j), j \in (1, m)\}, i \in (1, n)$$

Where n is the numbers of domains, m is the numbers of keywords in a given domain, (k_j, w_j) is the keyword and its weight in the given domain, and $\widetilde{D_i}$ is the vector of the domain D_i. The detail can be found in the Ref.[16].

Similarly, we denote the document d with the vector $\tilde{d}$:

$$\tilde{d_i} = \{(k_j, w_j), j \in (1, m)\}, i \in (1, n)$$

Where n is the numbers of documents, m is the numbers of keywords in a given domain, (k_j, w_j) is the keyword and its weight in the given document, and $\tilde{d_i}$ is the vector of the document d_i.

In the next section, we will replace the domain D, the document d with the corresponding vector $\widetilde{D}$ and $\tilde{d}$.

4.2   Downloading Links

According the log file of the server, we can get the visiting queue. In this section, we will download the links between the pages in domain level. In Fig.2, we divide the information management into 4 block levels. They are, 1st: pages; 2nd: Directories. Such as $http://liawww.epfl.ch/Research$; 3rd: Host. Such as $http://liawww.epfl.ch$; 4th: Domain. Such as $http://www.epfl.ch$.

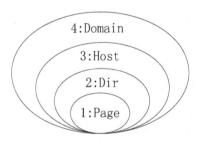

Fig. 2. Levels of Information Management

The Ref.[14] points out that in order to get the best performance, the block level should be set at host. We download the links in domain level in order to calculate the pagerank in host level. We should set up a crawl to download the links. The Ref.[17] showed how to set up and manage the crawls.

4.3   Setting Up the Link Matrix

In section 2.2, we introduce the calculating of pagerank, which is used by Google.

In the algorithm, the entry $p_{ij} = 1$ when the link $i \rightarrow j$ *exists*. The Ref.[5][7-8][13-14]drew the methods to improve the effects of calculating.

In our system, we consider the host as a block level. The out-links can be divided two parts: the intra-links and the inter-links. The former one mean that the pages linked to and from are belonged to the same host. Otherwise they are belonged to the latter one.

The weight of two kinds out-link are different. We assume the total weight of all out-link pages is 1. All the intra-links share the 3/4 weight, and all inter-links share the 1/4 weight. The link matrix P will be changed to:

$$
p_{ij} = \begin{cases} \frac{3}{4*dig(intra)} & i, j \in H_m; \\ \frac{1}{4*dig(inter)} & i \in H_m, j \in H_n, m \neq n; \\ 0 & Otherwise. \end{cases}
$$

The P should be dealt with just fellow the steps discussed above to P''(or Q) in order to promise $Q^{(k)}$ converge and have a real, positive and the biggest eigenvector.

4.4   Re-ranking

In the local database, there are: (1)the pages downloaded from the website according to the log file, (2) the vectors of the pages and the domains, (3)the relating link information of the website. There are two lists, (1)the order of the pages, which related to the difference between domain's and page's vector , and (2)the importance (or probability) of the pages in the given domain.

In this section, we combine the list(1) and (2) in order to adjust the order of the pages stored in local database.

$$
score(d) = \alpha * sim(\widetilde{d}, \widetilde{D}) + (1 - \alpha) * PR(d).
$$

Where d is the page stored in local database, $\alpha \in (0, 1)$ is the adjust factor, $PR(d)$ is the Pagerank of d, and $sim(\widetilde{d}, \widetilde{D})$ is the difference between the page d and the domain $\widetilde{D}$.

The result of the computation is a new order list of the pages, which will be stored in local database, too.

5   Query Process

In this section, we introduce the query process.

(1) Extracting the domain vectors

We use the Open Directory Cluster as the reference. We assume there are m domains. For each domain, we select n papers related to the domain in order to obtain the vector $\widetilde{D_i}(i \in (1, m))$ of given domain. Obviously, the work should be done once only.

(2) Login

This system will be an entrance of personalized search engine. If the user wanted to apply the service, he should register and submit the basic information of himself. The system will create a new user profile in order to record the habit of the new user. If the use wanted not to register, he could visit the Internet freely.

(3) Submitting the keywords

Just as the description in Fig.1, the MM component receives the keywords k and modifies them according to the user's profile. It will do nothing when the user entered the system in the first time. The QI component will submit the keywords to the search engine directly. But in the other times, the QI component will compute the two kinds of difference:

- between k and the domain vector $\widetilde{D_i}$. The m nearest neighbors have the most probability of the keywords belonged to. Assuming the domain list is $D_{ca} = \{\widetilde{D_{k1}}, \widetilde{D_{k2}}, \cdots \widetilde{D_{kl}}, (kl \leq m)\}$.
- between k and the document vectors $\widetilde{d_i}$, $\widetilde{d_i} \in \widetilde{D_i}(\widetilde{D_i} \in D_{ca})$.

The RE component will list all the $top - n$ pages of every domain $\widetilde{D_i}(\widetilde{D_i} \in D_{ca})$ stored in local database. Meanwhile, the RE will send the keywords to the search engine.

All of the results (pages, documents) will be order by the FI component in order to output. The user can select the pages he was interested in, and the log file will record it simultaneity.

(4) Maintaining the Use Profile and Local database

After the user logout, the system will download the link information of the domain (Show in Fig.2) and the pages, which the user has visited. All of the computation will be done by the AS component, and the local database and user profile will be maintained.

6   Shortcoming and Room for Improvement

There are two sides need to be improved, the feature extraction and the pagerank computation.

In our system, we use the traditional algorithm to extract the features of the document and the domain. The feature is the word appeared in the document. Just as we know, in a given document. There are many words have a more importance or have a bigger weight than others do. In this algorithm, we do not consider this condition.

We simulate the pagerank algorithm just as the Google did. Different from the real computation, we centralize the pagerank among the domain, the 4th block level. We hardly consider the influence from and to other domains. It makes the warp from the real one. This model may not respond to the real importance from the link view.

For the pagerank computing, two questions are put forward in Ref.[5]. One is that a page may correlative with the given subject, but might not contain the keywords of query. It makes the page failed to be select by the query. The other question is that some websites might contain a lot of hyperlinks or had a high pagerank, but it might be less correlative with the keywords. The Google claimed that the second problem had been solved. But we have not found the report about how to do that in recent papers.

For the SVM method, it is effective when the pages have one kind of schema. There are many kinds of mass, business information in Internet, and they are arranged in another way different from those pages of science and technology. For those pages, another method should be developed, which includes the information analysis, storing and issuing, too.

7   Conclusion

This paper discussed the disadvantages, which could not provide the personalized service, of the current search engines. It drew up a new personalization model, which combine the text categorization with pagerank computation.

The keywords of the given domain and the documents were extracted in order to form the vectors, which will be used in text categorization. The links of the domain, which in the top of the block level of a website, were downloaded. The model changed the weight of the link in order to distinguish the intra-links and the inter-links of a host. The pagerank was computed, and it was combined with vector difference to form the order list of the page in the given domain.

In the last paragraph, it pointed out that there were some problems should be solved in order to make the retrieval results more veraciously.

Acknowledgements

This work was supported by project 2004F06 of Nature Science Research of Shaanxi Province, P.R. China.

References

1. WANG Ji-Cheng, et al. State of the Art of Information Retrieval on the Web. Journal of Computer Research & Development. Vol.382001(2)pp.187-193.
2. James Kempf. Evolving the Internet Addressing Architecture. Proceedings of the 2004 International Symposium on Applications and the Internet (SAINT'04).
3. D.Rafiei, A.Mendelzon. What is this page known for Computing web page reputations. In 9th International World Wide Web Conference, Amsterdam, Netherlands, May 2000.
4. Neal G. Shaw, et al. A comprehensive agent-based architecture for intelligent information retrieval in a distributed heterogeneous environment. Decision Support Systems 32 (2002) pp.401-415.
5. R.Lempel, S.Moran. The stochastic approach for link-structure analysis (SALSA) and the TKC effect. In 9th International World Wide Web Conference, Amsterdam, Netherlands, May 2000.

6. L. Page, S. Brin, R. Motwani, T. Winograd. The PageRank Citation Ranking: Bringing Order to the Web. Stanford Digital Libraries Working Paper, 1998.
7. T. Haveliwala. Efficient computation of PageRank. Technical report, Computer Science Department, Stanford University, 1999.
8. Wenpu Xing, Ali Ghorbani. "Weighted PageRank Algorithm," cnsr, pp. 305-314, Second Annual Conference on Communication Networks and Services Research (CNSR'04), 2004.
9. Pascal Soucy,Guy W. Mineau. Beyond TFIDF Weighting for Text Categorization in the Vector Space Model. In proceeding of IJCAI-05. pp.1136-1141.
10. Gongde Guo, Hui Wang, et. al . An kNN Model-Based Approach and Its Application in Text Categorization. In Proceeding of CICLing 2004: 559-570.
11. Ralf Steinberger, Bruno Pouliquen, Johan Hagman. Cross-Lingual Document Similarity Calculation Using the Multilingual Thesaurus EUROVOC. In Proceedings of Computational Linguistics and Intelligent Text Processing (CICLing 2002). pp. 415-424.
12. LI-PING JING, HOU-KUAN HUANG. Improved feature selection approach tfidf in text mining - Machine. In Proceedings of the First International Conference of Machine learning and Cybernetics, 2002. pp.944-946.
13. Monica Bianchini, Marco Gori, Franco Scarselli. Inside Pagerank. ACM Transactions on Internet Technology. Vol.(5), No.1,2005(2). pp. 92-128.
14. Xue-Mei Jiang, Gui-Rong Xue, Wen-Guan Song, Hua-Jun Zeng, Zheng Chen, Wei-Ying Ma. Exploiting PageRank at Different Block Level. In Proceedings of WISE 2004. pp. 241-252.
15. B. Arslan, F. Ricci, N. Mirzadeh, A. Venturini. A dynamic approach to feature weighting. In Proceedings of Data Mining 2002 Conference.
16. Thorsten Joachims. A Probabilistic Analysis of the Rocchio Algorithm with TFIDF for Text Categorization. In Proceedings of the Fourteenth International Conference on Machine Learning table of contents(ICML). 1997. pp.143 - 151.
17. http://www.searchtools.com/robots/robot-code.html

Instruction Selection for ARM/Thumb Processors Based on a Multi-objective Ant Algorithm*

Shengning Wu and Sikun Li

School of Computer Science,
National University of Defense Technology,
410073, Changsha, China
kanvard@yahoo.com.cn, lisikun@263.net.cn

Abstract. In the embedded domain, not only performance, but also memory and energy are important concerns. A dual instruction set ARM processor, which supports a reduced Thumb instruction set with a smaller instruction length in addition to a full instruction set, provides an opportunity for a flexible tradeoff between these requirements. For a given program, typically the Thumb code is smaller than the ARM code, but slower than the latter, because a program compiled into the Thumb instruction set executes a larger number of instructions than the same program compiled into the ARM instruction set. Motivated by this observation, we propose a new Multi-objective Ant Colony Optimization (MOACO) algorithm that can be used to enable a flexible tradeoff between the code size and execution time of a program by using the two instruction sets selectively for different parts of a program. Our proposed approach determines the instruction set to be used for each function using a subset selection technique, and the execution time is the total one based on the profiling analyses of the dynamic behavior of a program. The experimental results show that our proposed technique can be effectively used to make the tradeoff between a program's code size and execution time and can provide much flexibility in code generation for dual instruction set processors in general.

1   Introduction

Embedded systems often have some conflicting requirements such as high performance, small volume, low energy consumption, and low price, etc. Subsequently, compilers for embedded processors have to meet multiple conflicting objectives for the generated embedded code.

A dual instruction set processor such as ARM/Thumb offers a challenge in code generation while allows an opportunity for a tradeoff between the code size and execution time of a program. The ARM processor core [1] is a leading processor design for the embedded domain. In addition to support the 32 bit ARM full

---

* This research is supported partially by National Natural Science Foundation of China (No. 90207019) and National 863 Development Plan of China (No. 2002AA1Z1480).

D. Grigoriev, J. Harrison, and E.A. Hirsch (Eds.): CSR 2006, LNCS 3967, pp. 641–651, 2006.

instruction set, these processors also support a 16 bit Thumb reduced instruction set. A Thumb instruction has a smaller bit width than an ARM instruction, but a single full instruction can perform more operations than a single reduced instruction. By using the Thumb instruction set it is possible to obtain significant reductions in code size in comparison to the corresponding ARM code, but at the expense of a significant increase in the number of instructions executed by the program. In general, a program compiled into the Thumb instruction set runs slower than its ARM instruction set counterpart.

This motivates us to exploit the tradeoff between the code size and the execution time of a program when generating code for embedded systems. Some work has been done around the code generation for dual instruction set processors. However, most related research effort has focused on only one optimization objective, lacking the flexibility of making the tradeoff among multiple conflicting objectives.

Computing multi-objective optimal solutions is more complex than that for a single objective problem and is computationally intractable for those NP-hard code optimization problems. In practice, we are usually satisfied with "good" solutions. In the last 20 years, a new kind of approximate algorithm has emerged which can provide high quality solutions to combinatorial problems in short computation time. These methods are nowadays commonly called as meta-heuristics. In addition to single-solution search algorithms such as Simulated Annealing (SA) [2], Tabu search (TS) [3], there is a growing interest in population-based meta-heuristics. Those meta-heuristics include evolutionary algorithms (EA: Genetic Algorithms (GA) [4], Evolution Strategies (ES)[5], Genetic Programming (GP)[6], etc.), and so on. The ability of these population-based algorithms to find multiple optimal solutions in one single simulation run makes them unique in solving multi-objective optimization problems.

Ant Colony Optimization (ACO) is a meta-heuristic inspired by the shortest path searching behavior of various ant species. It has been applied successfully to solve various combinatorial optimization problems (for an overview see Dorigo and Di Caro [7]). Recently, some researchers have designed ACO algorithms to deal with multi-objective problems (MOACO algorithms) (see [8] for an overview) [9][10][11].

In this paper, we propose a simple MOACO algorithm as a baseline to show the effectiveness of MOACO algorithms to solve the complex multi-objective optimization problem, that is, to optimize simultaneously the code size and execution time of a program for a dual instruction set processor. Note that the goal is not to design powerful ACO algorithms that often use several pheromone matrixes or ant colonies [8][9][10][11]. We will design these algorithms to get better results in the future. Another contribution of this paper is to stimulate much more research on the problem of code generation for embedded systems satisfying simultaneously conflicting objectives such as performance, size and energy. Although these objectives are considered in the embedded domain, there is so little research to treat them as multi-objective optimization problems.

The rest of this paper is organized as follows. In the next section, we give an overview of related work. In section 3, we give a formal description of our problem and then detail a new MOACO algorithm for the selection of ARM and Thumb instructions. Section 4 presents experimental results, with conclusions left for section 5.

2 Relate Works

Halambi et al. [12] developed a method to reduce the code size significantly by generating mixed instruction set code. The technique groups consecutive instructions that can be translated into the reduced instruction, and decides whether to actually translate them based on the estimation of the size of the resulting code.

Krishnaswamy and Gupta [13] proposed several heuristics for a function-level coarse-grained approach, with the emphasis on enhancing the instruction cache performance in terms of execution time and energy consumption. In addition, they provided a fine-grained approach, which is shown to be only comparable to the coarse-grained approach due to the overhead of mode switch instructions.

Sheayun Lee, et al. [14] proposed an approach to determine the instruction set to be used for each basic block of a program using a path-based profitability analysis, so that the execution time of a program is minimized while the code size constraint is satisfied.

Largely, all the techniques mentioned above belong to single objective optimization problems, and could not meet the flexibility requirement of the code generation for embedded processors.

3 Compilation for ARM/Thumb Processors

This section details the proposed techniques for compiling a given program for ARM/Thumb processors. We give a formal description of the problem in section 3.1. Section 3.2 details our proposed MOACO algorithm.

3.1 Problem Description

A program can be represented by a function call graph $G = <V, E>$, where V is the set of functions and E is a set of ordered pairs of elements of V called edges representing the control flow of the program. We assume that the number of functions of a program is n, $V = \{v_i | i = 1, 2, \cdots, n\}$, and $E = \{e_{ij} = <v_i, v_j> |$ there is a control flow from v_i to $v_j\}$. Let $s_{v_i}^{ARM}$ and $s_{v_i}^{Thumb}$ denote the code size of a function v_i compiled into the ARM and Thumb instruction sets respectively. Let $t_{v_i}^{ARM}$ and $t_{v_i}^{Thumb}$ denote the execution time of v_i compiled into the ARM and Thumb instruction sets respectively.

We associate a decision variable x_i with each v_i,

$$x_i = \begin{cases} 1 & \text{if } v_i \text{ compiled into ARM instruction set} \\ 0 & \text{if } v_i \text{ compiled into Thumb instruction set} \end{cases} \qquad (1)$$

Now we estimate the total code size and total execution time of a program. First, we calculate the code size. Let $overhead_i^s$ be the size cost of mode switch instructions between v_i and the parents of it. Then the total code size of a program can be obtained by summing the size of each function ARM its associated size overhead:

$$Size_{program} = \sum_i (s_{v_i}^{ARM} \cdot x_i + s_{v_i}^{Thumb} \cdot (1 - x_i) + overhead_i^s) \qquad (2)$$

$$overhead_i^s = c_s \cdot (\bigcup_j ((x_j + x_i) \bmod 2)), \exists e_{ji} \in E \qquad (3)$$

where c_s denotes the size cost of instructions for a single mode switch.

Second, we calculate the total execution time of a program. We assume that the number of parents of v_i is k, $v_{i_1}, v_{i_2}, \cdots, v_{i_k}$ are the parents of v_i, the times v_i is called by its parents are $c_{i_1}, c_{i_2}, \cdots, c_{i_k}$ respectively, and that the decision variables associated with its parents are $x_{i_1}, x_{i_2}, x_{i_k}$ respectively. Then the total execution time of a program can be obtained by summing the execution time of each function and its associated time overhead:

$$Time_{program} = \sum_i (\sum_{j=1}^k ((t_{v_i}^{ARM} \cdot x_i + t_{v_i}^{Thumb}(1 - x_i) + overhead_{i_j}^t) \cdot c_{i_j})) \quad (4)$$

$$overhead_{i_j}^t = c_t \cdot ((x_{i_j} + x_i) \bmod 2) \qquad (5)$$

where $overhead_{i_j}^t$ is the execution time overhead of mode switch instructions between the parent v_{i_j} and v_i, c_t denotes the execution time cost of instructions for a single mode switch.

In summary, the problem can be formulated as:

$$
\begin{aligned}
&\text{minimize} \quad Size_{program}, \\
&\text{minimize} \quad Time_{program}, \\
&MinSize \leq Size_{program} \leq MaxSize, \\
&MinTime \leq Time_{program} \leq MaxTime
\end{aligned}
\qquad (6)
$$

Note that although we are to solve a multi-objective optimization problem, if we consider the mode switching overhead the coefficients of problem constraints and objective functions are computed dynamically, because the instruction set assignment of a function affects the assignment of its adjacent functions.

3.2 Our Elitist Multi-objective Ant Algorithm for ARM

Given a program, we can obtain its function call graph $G = <V, E>$, where V is a function set whose elements are now called nodes, i.e., functions. Elements of E are edges. Then we associate a 0-1 variable with each node, and when the variable is 1 then we compile the corresponding function to ARM code, 0 for Thumb code. This is a SS problem: the set of objects is the set of functions; the goal is to find the subset of functions that must be compiled to ARM code so that the performance and code size are optimal. We refer to such a problem as an ARM-SS problem.

Informally, our new MOARM-Ant-SS (Multi-objective ARM Ant-SS) algorithm works as follows: each of ants successively generates a subset (i.e., a feasible solution to the ARM-SS) through the repeated selection of nodes. All subsets are initially empty. The first node is chosen randomly; subsequent nodes are chosen within a set Candidates that contains all feasible nodes with respect to the nodes the ant has chosen so far using a probabilistic state transition rule. Once all ants have constructed a feasible subset, all current solutions are compared with a non-dominated set of solutions since the beginning of the run. The best solutions of the current set and old elite set are selected into the new elite set and worse ones are removed. Only the ants whose solutions are the best in the elite set can modify pheromone information by applying a global updating rule: some pheromone is evaporated, and some is added.

Pheromone and Heuristic Factor Definition. Pheromone and heuristic information are two important factors that influence ants' decision making procession. When ant k who has already selected the subset of nodes S_k selects a node $o_i \in Candidates$,

1. The pheromone factor $\tau(o_i, S_k)$ evaluates the learned desirability of adding node o_i to subset S_k, corresponding to the quantity of pheromone on the considered node, i.e., $\tau(o_i, S_k) = \tau(o_i)$.
2. The heuristic factor $\eta(o_i, S_k)$ evaluates the promise of node o_i based on information local to the ant, i.e., the subset $S_k$it has built so far. We prefer the node (function) which consumes more execution time and less code size to be selected (compiled to ARM code). So we need time factor $\eta_{time}(o_i, S_k) = t_i \Big/ \sum_i t_i$ and code size factor $\eta_{size}(o_i, S_k) = \frac{1}{size_i/\sum_i size_i}$.

MOARM-Ant-SS State Transition Rule. In MOARM-Ant-SS the state transition rule is as follows: ant k who has already selected the subset of nodes S_k selects node $o_i \in Candidates$ by applying the rule given by Eq. (7)

$$o = \begin{cases} \arg \max_{o_i \in Candidates} \{[\tau(o_i, S_k)]^\alpha [\eta_{time}(o_i, S_k)]^{\lambda\beta} [\eta_{size}(o_i, S_k)]^{(1-\lambda)\beta}\} & \text{if } q \leq q_0 \\ X \end{cases}$$

$$(7)$$

where $\lambda = i/nbAnts$, nbAnts is the number of ants, α and β are two parameters that determine the relative importance of pheromone and heuristic factors; q is a random number uniformly distributed in $[0 .. 1]$, q_0 is a parameter, and X is a random variable selected according to the probability distribution given in Eq. (8). The parameter q_0 determines the relative importance of exploitation versus exploration: every time ant k having subset S_k has to choose the next node o_i, it samples a random number $0 \leq q \leq 1$. If $q \leq q_0$ then best nodes (according to

Eq. (7)) are chosen (exploitation), otherwise a node is chosen according to Eq. (8) (biased exploration).

$$p(o_i, S_k) = \frac{[\tau(o_i, S_k)]^\alpha \cdot [\eta_{size}(o_i, S_k)]^{\lambda\beta} \cdot [\eta_{time}(o_i, S_k)]^{(1-\lambda)\beta}}{\sum_{o_i \in Candidates}[\tau(o_i, S_k)]^\alpha \cdot [\eta_{size}(o_i, S_k)]^{\lambda\beta} \cdot [\eta_{time}(o_i, S_k)]^{(1-\lambda)\beta}}$$

$$(8)$$

Pheromone Updating Rule. MOARM-Ant-SS globally updates the pheromone trail with each solution of the elite set since the beginning run. Pheromone evaporation is simulated by multiplying the quantity of pheromone on each pheromone node by a persistence rate ρ. Then a quantity δ_τ of pheromone is added. The pheromone updating rule is given by Eq. (9)

$$\tau(o) \leftarrow \tau(o) \cdot \rho + \sum_j \delta_{\tau_j}(o) \qquad (9)$$

$$\delta_{\tau_j}(o) = \begin{cases} c/(T_j \cdot SIZE_j) & \text{if } o \in \text{subset constructed by elitest ant } j \\ 0 & \text{otherwise} \end{cases} \qquad (10)$$

$0 \leq \eta \leq 1$ is the pheromone persistence rate, ant j is the one who has constructed one of the global non-dominated subsets, c is a constant,T_j is the execution time of the corresponding program, and $SIZE_j$ the code size. Less time and space consumed, more pheromone added.

The algorithm follows the MAX-MIN Ant system [15]: we explicitly impose lower and upper bounds τ_{min} and τ_{max} on pheromone trails (with $0 < \tau_{min} < \tau_{max}$). These bounds restrict differences between the pheromone on nodes, which encourages wider exploration.

The pseudo-code of MOARM-Ant-SS algorithm is as follows.

MOARM-Ant-SS Algorithm
Input: an ARM-SS problem(S, f_{time}, f_{size}),
       a set of parameters $\{\alpha, \beta, \rho, \tau_{min}, \tau_{max}, q_0, nbAnts\}$.
Output: a feasible near-optimal subset of nodes S'.
1. Initialize the pheromone trail $\tau(o)$ associated with each $o \in S$ to τ_{max}, $EliteSet = \Phi$.
2. Repeat
3.   for each ant $k \in \{1, \ldots, nbAnts\}$, construct a subset S_k as follows:
4.   randomly choose a first node $o_i \in S$
5. $S_k \leftarrow \{o_i\}$
6. $Candidates \leftarrow \{o_j \in S|(S_k \cup \{o_j\})$ satisfying constraints $\}$
7.   while $Candidates \neq \emptyset$ do
8.   choose an node $o_i \in Candidates$ with state transition rule
9. $S_k \leftarrow S_k \cup \{o_i\}$
10.   remove o_i from $Candidates$
11.   remove from $Candidates$ every object $o_j$such that
12. $(S_k \cup \{o_j\})$violating any constraint
13.   end while
14.   end for

15. $CurrentSet = \{S_1, \cdots, S_{nbAnts}\}$
16.   select the non-dominated solutions for$EliteSet$from
        $(CurrentSet \cup EliteSet)$
17.   for each $o \in S$, update the pheromone trail $\tau(o)$ as follows:
18. $\tau(o) \leftarrow \tau(o) \cdot \rho + \delta_\tau(o)$
19.   if $\tau(o) < \tau_{\min}$then$\tau(o) \leftarrow \tau_{\min}$
20.   if $\tau(o) > \tau_{\max}$then$\tau(o) \leftarrow \tau_{\max}$
21.   end for
22. Until maximum number of cycles reached or acceptable solution found
23. return the best solution found since the beginning

4   Implementation and Results

In this section, we describe our implementation of the proposed technique targeted for the ARM7TDMI dual instruction set processor, present the experimental results to show the effectiveness of the approach and give some discussions.

We retargeted MachineSUIF [16] compiler to ARM architectures. HALT, the Harvard Atom-Like Tool, which works by instrumentation, is used to study program behavior and the performance of computer hardware. We extended HALT to profile the benchmarks, and use it to gather the execution count of each function of a program. The code size is estimated directly from the instruction sequence of each function, compiled into ARM and Thumb instructions respectively. The execution time of each function is approximated by the instruction count of that function.

To validate the effectiveness of our approach, we implemented the proposed technique and performed a set of experiments. The benchmark programs are from the SNU-RT benchmark suite [17], which is a collection of application programs for real-time embedded systems. The benchmarks have no library calls, and all the functions needed are implemented in the source code.

To compare our proposed techniques with heuristic methods, we applied our selective code transformation to the set of benchmarks, generating different versions of code for each program. We assume that $f_1, f_2, \cdots, f_n$ are the functions of a benchmark program, and n is the number of the functions. Let $S_{f_i}^{ARM}$ and $S_{f_i}^{Thumb}$ denote the code size of f_i compiled into the ARM and Thumb instruction sets respectively. Then we can get the minimal and maximal code size of the whole program:$MinTotal = \sum_i \min\{S_{f_i}^{ARM}, S_{f_i}^{Thumb}\}$, $MaxTotal = \sum_i \max\{S_{f_i}^{ARM}, S_{f_i}^{Thumb}\}$. Total code size budget$size = MaxTotal - MinTotal$. We set the code size budget to 0, 10% , 20% , $\cdots$, 100% of $size$, then add these to $MinTotal$ to get different code size constraints.

Figure 1 summarizes the experimental results for our Ant algorithm. For each application, we display average results over 10 trials. In all experiments, the numeric parameters are set to the following values: $nbAnts$=30, α=1, β=3, q_0=0.9, ρ=0.99, $\tau_{\min}$=0.01, and $\tau_{\max}$=6. The results verify that there exists a clear tradeoff between the code size and performance of programs, and that our

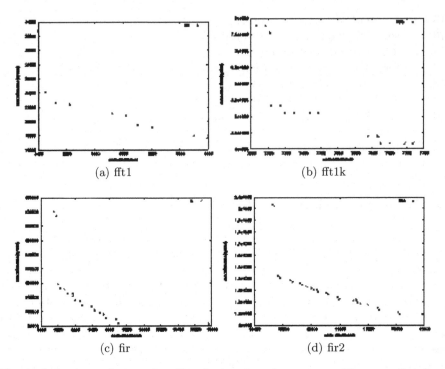

(a) fft1   (b) fft1k

(c) fir   (d) fir2

Fig. 1. Code size and execution time for benchmark programs. The execution time decreases while code size increases gradually.

approach can effectively exploit this tradeoff. With the increase of the code size, the execution time decreases and the phenomenon differs from one benchmark to another. For all the benchmark programs, the execution time decreased significantly even for a modest increase in the code size. This is because the *fabs* functions are executed heavily and are transformed to ARM code at early stage.

For the *fft1* benchmark program, based on the heuristic method, when size budget arrives at 10% of *size* (corresponding to size constraint of 8458.3 Bytes), *fabs* function can be transformed into ARM code. When size budget arrives at 60% of *size* (size constraint of 8759.8 Bytes), sin function is added into the set of functions that can be transformed into ARM code. When size budget is *size* × 90%(size constraint of 8940.7 Bytes), *fft1* function added. And when given total *size* budget, all functions can be transformed into ARM code. All these changes can be seen apparently from Figure 1 as several obvious decrease in execution time. Relative to the heuristic method, our techniques can make better tradeoff between code size and execution time. For example, given size budget of *size* × 10%, *log* function can also be add to ARM set, in addition to *fabs* function.

The tradeoff between code size and execution time of benchmark *fft1k* is similar to *fft1* benchmark, except that the execution time reduction of *fft1k* is not remarkable at first, due to *fabs* function not transformed into ARM code firstly.

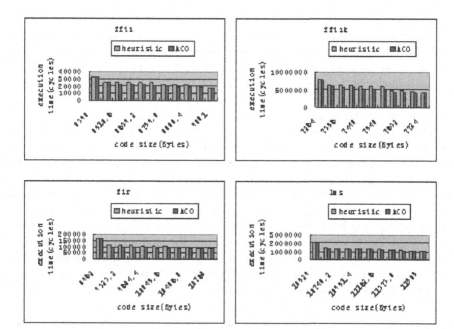

Fig. 2. Compared to the heuristic method, proposed Ant algorithm can find solutions consuming less execution time given the same code size constraints

The execution time of benchmark *lms* deceases smoothly with the increase of code size, because the functions of it are added gradually to the set of ARM code with the increase of code size.

Similar to benchmark *lms*, the execution time of benchmark *fir* decreases almost linearly with the increase of code size. But when size budget arrives at $size \times 50\%$, almost all functions (excluding *main* function) are added to the set of ARM code. For different code size budget $(0, 10\%, 20\%, \cdots, 100\%$ of $size)$, we compared our techniques with a simple heuristic method that favor the functions called more heavily. For all the benchmark programs, given the same code size constraint, our proposed techniques can often find solutions requiring less execution time than the heuristic method does. The relative maximal reduction in execution time of our algorithm compared to the heuristic method is 16.8%, and the average reduction in execution time is 9.2%.

5 Summary

Code generation for embedded systems is complex having to meet multiple conflicting requirements. Rapid growing applications of meta-heuristics to NP-hard combinatorial optimization problems inspire us to introduce meta-heuristics to compiler optimization for embedded domain.

We have presented an MOACO algorithm to enable a flexible tradeoff between code size and execution time of a program by generating mixed instruction set code for a dual instruction set processor. We have also proposed an approach to estimate the total execution time of a program by profiling analyses of the dynamic behavior of the program. The proposed techniques have been implemented for the ARM/Thumb dual instruction set processor, and their effectiveness has been demonstrated by experiments on a set of benchmark programs.

There are several research directions in the future. First, compared to the coarse grained function level instruction selection, basic block level or even instruction level selection algorithms may get better results. Second, we will deal with hybrid approaches by combining an ACO algorithm with a local search heuristic. And more powerful MOACO algorithms will be designed which should give better results. Third, we will also consider other objectives (energy, etc.) and design approaches to find multi-objective Pareto-optimal set to provide much more flexibility in compiler optimization for embedded domain. Finally, we will also focus on developing an efficient register allocation algorithm for dual instruction set processors, since the quality of the code produced by a compiler is always influenced by the phase coupling of the compiler.

Acknowledgement

Thanks to Christine Solnon for the help with subset selection problems.

References

1. S. Furber. ARM system Architecture. Publisher: Addison Wesley Longman, 1996.
2. S. Kirkpatrick, C. D. Gelatt, and M. P. Vecchi. Optimization by Simulated Annealing. Science, 220(4598):671-680, May 1983.
3. F. Glover. Tabu Search-Part I. ORSA Journal of Computing, 1(3):190-206, 1989.
4. J. H. Holland. Adaptation in Natural and Artificial Systems. Michigan Press University, 1975.
5. K. Schmitt. Using Gene Deletion and Gene Duplication in Evolution Strategies. GECCO, 919-920, 2005.
6. D. Jackson. Evolving Defence Strategies by Genetic Programming. EuroGP, 281-290, 2005.
7. M. Dorigo and G. Di Caro. The Ant Colony Optimization Meta-heutistic. In D. Corne, M. Dorigo, and F. Glover, editors, New Ideas in Optimization, pages 11-32, London, 1999. McGraw-Hill.
8. C. G. Martínez, O. Cordón, and F. Herrera. An Empirical Analysis of Multiple Objective Ant Colony Optimization Algorithms for the Bi-criteria TSP. In M. Dorigo, M. Birattari, C. Blum, et al., editors, Ant Colony, Optimization and Swarm Intelligence: 4th International Workshop, ANTS 2004, Brussels, Belgium, September 5-8, 2004. LNCS, Volume 3172 / 2004. Springer.
9. S. Iredi, D. Merkle, and M. Middendorf. Bi-Criterion Optimization with Multi Colony Ant Algorithms. E. Zitzler et al., editors, EMO 2001, LNCS 1993, pages 359-372, 2001. Springer.

10. M. Guntsch and M. Middendorf. Solving Multi-criteria Optimization Problems with Population-Based ACO. C. M. Fonseca et al., et al., EMO 2003, LNCS 2632, pages 464-478, 2003. Springer.
11. K. Doerner, W. J. Gutjahr, R. F. Hartl, C. Strauss and C. Stummer. Pareto Ant Colony Optimization: A Metaheuristic Approach to Multiobjective Portfolio Selection. Annals of Operations Research, 131, 79-99, 2004.
12. A. Halambi, A. Shrivastava, P. Biswas, N. Dutt, and A. Nicolau. An efficient Compiler Technique for Code Size Reduction Using Reduced Bit-width ISAs. In Proceedings of the DATE, France, 2002.
13. A. Krishnaswamy and R. Gupa. Profile Guided Selection of ARM and Thumb Instructions. In Proceedings of LCTES/SCOPES, Germany, 2002.
14. S. Lee, J. Lee, S. L. Min, J. Hiser, and J. W. Davidson. Code Generation for a Dual Instruction Set Processor Based on Selective Code Transformation. SCOPES 2003, LNCS 2826:33-48, 2003. Springer.
15. T. Stutzle, and H. Hoos. Improvements on the Ant System: Introducing the *MAX-MIN* Ant System. In G. Smith, N. Steele and R. Albrecht, editors, Proceedings of the Artificial Neural Nets and Genetic Algorithms, 245–249, 1997
16. M. D. Smith and G. Holloway. An introduction to Machine-SUIF and Its Portable Libraries for Analysis and Optimization. The Machine-SUIF documentation set. Harvard University, 2002.
17. SNU Real-Time Benchmark Suite. http://archi.snu.ac.kr/realtime/benchmark.

A New Flow Control Algorithm for High Speed Computer Network

Haoran Zhang[1] and Peng Sun[2]

[1] College of Information Science and Engineering, Zhejiang Normal University,
Jinhua, Zhejiang, China
hylt@zjnu.cn
[2] Yunnan Power Company
yizhili000@163.com

Abstract. Developing an effective flow control algorithm to avoid congestion is a hot topic in computer network society. This paper gives a mathematical model for general network at first, then discrete control theory is proposed as a key tool to design a new flow control algorithm for congestion avoidance in high speed network, the proposed algorithm assures the stability of network system. The simulation results show that the proposed method can adjust the sending rate and queue level in buffer rapidly and effectively. Moreover the method is easy to implement and apply to high speed computer network.

1 Introduction

High-speed computer networks are generally store-and-forward backbone networks consisting of switching nodes and communication links based on a certain topology. All the links and all the nodes are characterized by their own capacities for packet transmission and packet storing, respectively. A node which reaches its maximum storing capacity due to the saturation of its processors or one or more of its outgoing transmission links is called congested. Some of the packets, arriving at a congested node, cannot be accepted and have to be retransmitted at a later instance. This would lead to a deterioration of the network's throughput and delay performance or even the worst situation–network collapse. Therefore, congestion control is an important problem arising from the networks management.The flow control scheme can adjust the sending packets rate in source host, which can effectively avoid the traffic congestion. Therefore, the proper flow control scheme is a direct way to be executed.

Many complex systems such as computer network can be expounded by control theory. Hence, an increasing amount of research is devoted to merging control theory into flow control. The first introduction that applied control theory to flow control appeared in ATM network. Fendick, Bonomi and Yin et al. proposed ABR traffic control from 1992 to 1995 [1,2,3]. In these schemes, if the queue length in a switch is greater than a threshold, then a binary digit is set in the control management cell. However, they suffer serious problems of stability, exhibit oscillatory dynamics, and require large amount of buffer in order to avoid

D. Grigoriev, J. Harrison, and E.A. Hirsch (Eds.): CSR 2006, LNCS 3967, pp. 652–659, 2006.

cell loss. As a consequence, explicit rate algorithms have been largely considered and investigated. See Jain(1996) for an excellent survey[5]. Most of the existing explicit rate schemes lack of two fundamental parts in the feedback control design:(1) the analysis of the closed-loop network dynamics; (2) the interaction with VBR traffic. In the paper of Jain et al., an explicit rate algorithm was proposed, which basically computes input rates dividing the measured available bandwidth by the number of active connections [4]. In Zhao et al. [6], the control design problem is formulated as a standard disturbance rejection problem where the available bandwidth acts as a disturbance for the system. In Basar and Srikant [7], the problem is formulated as a stochastic control problem where the disturbance is modeled as an autoregressive process. The node has to estimate this process using recursive least squares. In mascolo [8], Smith's principle is exploited to derive a controller in case a FIFO buffering is maintained at output links.

For all of the above works, the designed control algorithm is based on continuous time system theory. But as we know, the discrete time control is effective in computer control system and easy to put into execution. Hence, this paper will design a discrete control algorithm to satisfy practical control requirements.

2 Mathematical Model of General Network

In this section, we will develop the mathematical model of a general network that employs a store and forward packet switching service, that is, packets enter the network from the source edge nodes, are then stored and forwarded along a sequence of intermediate nodes and communication links, finally reaching their destination nodes. Figure 1 depicts a store and forward packet switching network.

In Fig.1, $S_i(i = 0, \ldots, n)$ denotes source node, $D_i(i = 0, \ldots, n)$ denotes destination node, B denotes bottleneck node, $u_{0i}(i = 0, \ldots, n)$ denotes sending rate, $q_i(i = 0, \ldots, n)$ denotes buffer queue level.

One approach to the buffer size is to use variable-size buffers. The advantage here is better memory utilization, at the price of more complicated buffer management. Another possibility is to dedicate a single large circular buffer per

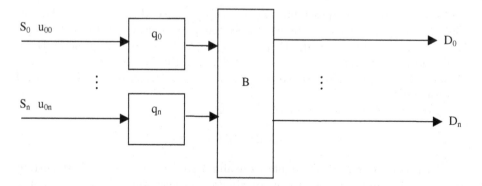

Fig. 1. Store and forward packet switching network

connection. This system also makes good use of memory, provided that all connections are heavily loaded but is poor if some connections are lightly loaded. As easy of buffer management, it make brief of problem. Moreover, per-flow buffering is also auspicated for Internet routers in order to ensure QoS [9, 10]. For these considerations, we adopt the second buffers management. We assume all connections are heavily loaded and divided available bandwidth in average.

From above discussion, we can analyze this problem with a connection case. Consider a flow reaches destination node from anyone of n source nodes along a sequence of several intermediate nodes. The nodes in the path of the connection are numbered $1, 2, 3, \cdots, n$, and the source node is numbered 0. The time taken to get service at each node is finite and deterministic. The source sending rate is denoted by $u_{0i}(i = 1, 2, \ldots, n)$ and the intermediate nodes service rate is $u_i(i = 1, 2, \ldots, n)$. We define

$$u_b = \min(u_i \mid 0 \leq k \leq n) \tag{1}$$

to be the bottleneck service rate.

We will hence work in discrete time mode, so continuous time parameter t is replaced by the step index k. One step corresponds to one round-trip (RTT). In bottleneck node, the service rate is close [11], hence we consider $u_b(k) = u_b(k + 1)$ as a constant. At the time marked 'NOW', which is the end of the kth epoch, all the packets sent in epoch k-1 have been acknowledged. So the only unacknowledged packets are those sent in the kth epoch itself, and this is the same as the number of outstanding packets. This can be approximated by the sending rate $u_0(k)$ multiplied by the sending interval RTT(k). Therefore we have model of this problem:

$$q_b(k + 1) = q_b(k) + u_0(k)\text{RTT}(k) - u_b(k)\text{RTT}(k) \tag{2}$$

$q_b(k)$ is buffer queue length at kth epoch.

3   Design Discrete Controller

Consider the input is the sending rate of source node and the output is the bottleneck queue level. Figure 2 depicts the system control principle. In Fig.2, $S_i(i = 0, \ldots, n)$ denotes source node, $D_i(i = 0, \ldots, n)$ denotes destination node, B denotes bottleneck node, $u_{0i}(i = 0, \ldots, n)$ denotes sending rate, $q_i(i = 0, \ldots, n)$ denotes buffer queue level.controller$i(i = 0, \ldots, n)$ will be designed to control network system.

Taking the Z transformation of (2), we get

$$zq_b(z) = q_b(z) + u_0(z)\text{RTT}(z) - u_b(z)\text{RTT}(z) \tag{3}$$

For each connection, TCP maintains available, RTT, that is the best current estimate of round-trip time to the destination in question. When a segment is sent, a timer is started, both to see how long the acknowledgment takes and

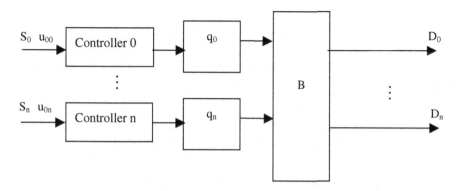

Fig. 2. The system control principle

to trigger a retransmission if it takes too long. If the acknowledgment gets back before the timer expires, TCP measures how long the acknowledgment took, Say, M. It then updates RTT according to the formula [12]:

$$\text{RTT}(k) = \alpha \text{RTT}(k-1) + (1-\alpha)M \tag{4}$$

where is a smoothing factor that determines how much weight is given to the old value. Typically $\alpha = 7/8$ [12].

Taking the Z transformation of (4), we get

$$\text{RTT}(z) = \alpha z^{-1} \text{RTT}(z) + (1-\alpha)M \tag{5}$$

Substituting (5)into (3), we obtain

$$(z-1)q_b(z) = [u_0(z) - u_b(z)]\frac{M}{8 - 7z^{-1}} \tag{6}$$

Substituting $u_0(z) = u_b(z) + \lambda(z)$ into (6), we get

$$(z-1)q_b(z) = \lambda(z)\frac{M}{8 - 7z^{-1}} \tag{7}$$

Let $\lambda(z) = f(z) \times R(z)(f(z)$ is undetermined Z rational fraction function, $R(z)$ is step function), transform (7) as :

$$(z-1)q_b(z) = \frac{Mf(z)R(z)}{8 - 7z^{-1}} \tag{8}$$

We can get system transfer function:

$$G(z) = \frac{q_b(z)}{R(z)} = \frac{Mf(z)}{(z-1)(8 - 7z^{-1})} \tag{9}$$

Let

$$f(z) = \frac{c(z-1)(8z-7)}{(z+a)(z+b)} \tag{10}$$

Then

$$G(z) = \frac{Mcz}{(z+a)(z+b)} \tag{11}$$

M is a available constant, c is constant parameter. According to liner control theory, the system is stable only if $|a| < 1, |b| < 1$.

The system has high transient performance only if a, b value is around original point. Transient performance means that the controller can track output variation in shortest time.

According to Z transformation theorem, the system steady step error is:

$$q_b(\infty) = \lim_{z \to 1}(z-1)q_b(z) = \lim_{z \to 1}\frac{Mcz^2}{(z+a)(z+b)} = \frac{4}{5}Q \tag{12}$$

Q is distributive maximum buffer in per connection and buffer has margin with 4/5 representation, which does not discard packets when burden flow occur.

We get

$$c = \frac{4Q(1+a)(1+b)}{5M} \tag{13}$$

Due to

$$\lambda(z) = \frac{c(z-1)(8z-7)}{(z+a)(z+b)} \times \frac{z}{z-1} \tag{14}$$

$(z/(z-1))$ is Z transformation of step function)

Then

$$u_0(z) = \frac{c(z-1)(8z-7)}{(z+a)(z+b)} \times \frac{z}{z-1} + u_b(z) \tag{15}$$

Simplifying (15), we obtain

$$u_0(z) = \frac{cz(8z-7)}{(z+a)(z+b)} + u_b(z) \tag{16}$$

Factorizing (16), we obtain

$$u_0(z) = u_b(z) + cz(\frac{a_1}{z+a} + \frac{a_2}{z+b}) \tag{17}$$

So

$$a_1 = \frac{-7-8a}{b-a} \tag{18}$$

$$a_2 = \frac{-7-8b}{a-b} \tag{19}$$

Substituting (13), (18), (19) into (17), we obtain

$$u_0(k) = u_b(k) + \frac{4Q(1+a)(1+b)}{5M} + [\frac{-7-8a}{b-a}(-a)^k + \frac{-7-8b}{a-b}(-b)^k] \tag{20}$$

$u_0(k)$ is control algorithm of sending rate.

4   Numerical Experiments

Variant graphical chart correspond with variant system parameters (see as tab.1).

The simulation results are as follows:

Figure 3 is the simulation results of $a = -0.2, b = -0.1, k = 0.8$; Fig.4 is the simulation results of $a = -0.5, b = -0.1, k = 0.8$.

From simulation results, we can see that the nearer to original point of a, b, the shorter of the settling time, so a, b should be chosen carefully. It is shown from the simulation results that high system performance is guaranteed even if the parameters vary, due to the effect of the control law. The queue length in the buffer is little affected by the varying parameters and the queue length can be quickly adjusted to a given value. The settling time depends on a, b.

Table 1. The table of variant system parameters

line shape	$u_b(k)$(packets/ms)	M(ms)	Q(packets)
solid line	14.5	10	1000
Thick dotted line	20	7.5	1000
thin dotted line	27	5	1000

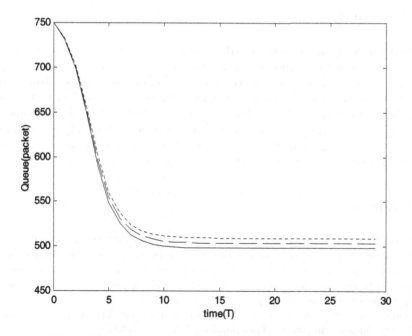

Fig. 3. The simulation results of $a = -0.2, b = -0.1, k = 0.8$

658 H. Zhang and P. Sun

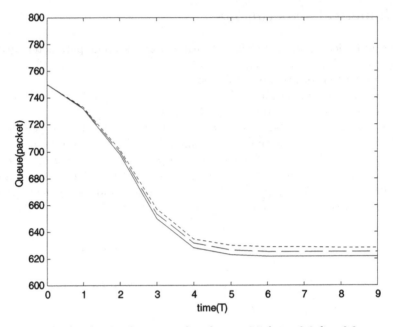

Fig. 4. The simulation results of $a = -0.5, b = -0.1, k = 0.8$

5 Conclusion

With network communication rapid development, the requirement of network performance also rises time after time. Congestion control plays an important role in network performance. This paper proposes a new flow control scheme based on discrete control theory and analysis stability and feasibility of the system. It is shown from the simulation results that this control algorithm can adjust the sending rate and queue in buffer rapidly and effectively. Moreover the queue length in the buffer is little affected by the varying parameters and the queue length can be quickly adjusted to a given value.

References

1. Fendick, K.W., Rodrigues, M.A., Weiss, A.:Analysis of a rate-based feedback control strategy for long haul data transport. Performance Evaluation, **16**(1992)67-84
2. Benmohamed, L., Wang, Y.T.: A control-theoretic ABR explicit rate algorithm for ATM switches with per-VC queuing. Proceedings of the IEEE Infocom'98, S. Francisco, CA, USA(1998)
3. Yin, N., Hluchyj, M.G.: On closed-loop rate control for ATM cell relay networks. Proceedings of infocom 94, Los Alamitos, CA, USA (1994)
4. Jain, R., Kalyanaraman, S., Goyal, R., Fahmy, S., Viswanathan, R.: The ERICA switch algorithm for ABR traffic management in ATM networks, Part I: Description. Available at http://www.cis.ohio-state.edu/~jain/papers.html(1996)

5. Jain, R.: Congestion control and traffic management in ATM networks: Resent advances and a survey. Computer Networks and ISDN System, **28**(1996)1723-1738
6. Zhao, Y., Li, S.Q., Sigarto, S.: A linear dynamic model for design of stable explicit-rate ABR control schemes. Proceeding of IEEE infocom97, Kobe, Japan(1997)
7. Altaman, E., Basar, T., Srikant, R.: Robust rate control for ABR sources. Proceedings of infocom'98, San Francisco, CA, USA(1998)
8. Saverio M.: Congestion control in high-speed communication networks using the Smith principle. Automatics, **35**(1999)1921-1935
9. Stoica, I., Shenker, S., Zhang, H.: Core-stateless fair queueing: achieving approximately fair bandwidth allocation in high-speed networks. Proceedings of sigcomm98 in ACM computer communication review, **28**(1998)118-130
10. Suter, B., Lakshman, T.V., Stiliadis, D., Choudhury, A. K.: Design considerations for supporting TCP with per-flow queueing. IEEE proceeding of infocom 98, S. Francisco, CA, USA(1998)
11. Keshav,S.: A Control Theoretic Approach to Flow Control.SIGCOMM'91(1991)
12. Andrew S.T.: Computer Networks, 3rd ed., Englewood Cliffs, Prentice Hall(1995)

Nonlinear Systems Modeling and Control Using Support Vector Machine Technique

Haoran Zhang and Xiaodong Wang

College of Information Science and Engineering,
Zhejiang Normal University, Jinhua, Zhejiang, China
{hylt, wxd}@zjnu.cn

Abstract. This paper firstly provides an short introduction to least square support vector machine (LSSVM), a new class of kernel-based techniques introduced in statistical learning theory and structural risk minimization, then designs a training algorithm for LSSVM, and uses LSSVM to model and control nonlinear systems. Simulation experiments are performed and indicate that the proposed method provides satisfactory performance with excellent generalization property and achieves superior modeling performance to the conventional method based on neural networks, at same time achieves favourable control performance.

1 Introduction

In the last decade, neural networks have proven to be a powerful methodology in a wide range of fields and applications [1, 2]. Reliable training methods have been de-veloped mainly thanks to interdisciplinary studies and insights from several fields including statistics, systems and control theory, signal processing, information theory and others. Despite many of these advances, there still remain a number of weak points such as: difficult to choose the number of hidden units, overfitting problem, existence of many local minima solutions, and so on. In order to overcome those hard problems, Major breakthroughs are obtained at this point with a new class of neural networks called support vector machine (SVM), developed within the area of statistical learning theory and structural risk minimization [3, 4]. SVM has many advantages, such as nonexistence of curse of dimensionality, possessing good generalization per-formance and so on. As an interesting variant of the standard support vector machines, least squares support vector machines (LSSVM) have been proposed by Suykens and Van-dewalle[5, 6] for solving pattern recognition and nonlinear function estimation problems. Standard SVM formulation is modified in the sense of ridge regression and taking equality instead of inequality constraints in the problem formulation. As a result one solves a linear system instead of a QP problem, so LSSVM is easy to training. In this paper we will discuss modeling and controlling nonlinear system with LSSVM.

This paper is organized as follows: In section 2 we give a short introduction to LSSVM regression. A Training algorithm is designed for LSSVM in section 3. In section 4 we present the modeling-control scheme with LSSVM. In section 5

D. Grigoriev, J. Harrison, and E.A. Hirsch (Eds.): CSR 2006, LNCS 3967, pp. 660–669, 2006.

we give a numerical experiments to check the performance of proposed modeling and control method. Finally In section 6, a conclusion is drawn.

2 An Overview of LSSVM Regression

Usually a typical regression problem is defined as follows: Suppose a set of observation data generated from an unknown probability distribution$P(x,y)$,$X = \{(x_1,y_1),\ldots,(x_l,y_l)\}$, with $x_i \subset R^n, y_i \subset R$, and a class of functions $F = \{f \mid f : R^n \rightarrow R\}$, then the basic problem is to find function $f \subset F$ that minimizes a risk functional:

$$R[f] = \int L(y - f(x))dP(x,y) \tag{1}$$

$L(\cdot)$ is a loss function, it indicates how differences between y and $f(x)$ should be penalized. As $P(x,y)$ is unknown one cannot evaluate $R[f]$ directly but only compute the empirical risk:

$$R_{\text{emp}} = \frac{1}{l} \sum_{i=1}^{l} L(y_i - f(x_i)) \tag{2}$$

and try to bound the risk $R[f]$ by $R_{\text{emp}} + R_{\text{gen}}$ where R_{gen} is the metric of complexity to function used to approximate the data given [4].

For a nonlinear regression problem, firstly we may transfer it into a linear regression problem, this can be achieved by a nonlinear map from input space into a high dimensional feature space and constructing a linear regression function there. That is:

$$f(x) = w^T \phi(x) + b \tag{3}$$

We would like to find the function with the following risk function[4]:

$$R_{\text{reg}} = \frac{1}{2}\|w\|^2 + C \times R_{\text{emp}}[f] \tag{4}$$

Here $\|w\|^2$ is a term which characterizes the function complexity of $f(\cdot)$. We introduce a square loss function, and get empirical risk :

$$R_{\text{emp}} = \frac{1}{2} \sum_{i=1}^{l} (y_i - f(x_i))^2 \tag{5}$$

C is a constant determine the trade-off between empirical risk and model complexity. Minimizing the (4) captures the main insight of statistical learning theory, stating that in order to obtain a small expected risk, one need to control both training error and model complexity, i.e. explain the data with a simple model. According to Structural Risk Minimization, the regression problem (3) can transfer to the following optimal problem [6]:

$$\min_{w,b,e} J(w,e) = \frac{1}{2}w^T w + \frac{C}{2} \sum_{i=1}^{l} e_i^2 \tag{6}$$

subject to constraints:

$$y_i = w^T \phi(x_i) + b + e_i, i = 1, \ldots, l$$

Using Lagrange function and dual theory, we can get dual optimization problem of (6):

$$\begin{bmatrix} 0 & e1^T \\ e1 & Q + C^{-1}I \end{bmatrix} \begin{bmatrix} b \\ \alpha \end{bmatrix} = \begin{bmatrix} 0 \\ y \end{bmatrix} \tag{7}$$

Where

$$y = (y_1, \ldots, y_l)^T, e1 = (1, \cdots, 1)^T, \alpha = (\alpha_1, \ldots, \alpha_l)^T$$

$$Q_{ij} = \phi(x_i)^T \phi(x_j) = k(x_i, x_j), i, j = 1, \ldots, l$$

The nonlinear regression function (the output of LSSVM) can be formulated by:

$$y(x) = \sum_{i=1}^{l} \alpha_i k(x, x_i) + b \tag{8}$$

$k(x, x_i)$ is a symmetric function which satisfies Mercer conditions. Some useful kernels are as following:
(1) polynomial kernel of order p: $k(x, x^*) = (1 + x^T x^*)^p$
(2) RBF-kernel: $k(x, x^*) = \exp(-\frac{\|x - x^*\|_2^2}{2\sigma^2})$
(3) hyperbolic kernel: $k(x, x^*) = \tanh(\beta x^T x^* + \kappa)$

3   A Training Algorithm for LSSVM

In this section we will give a simple training algorithm for LSSVM, Before describing the algorithm in detail, the special case of the Sherman-Woodbury Formula [7] in linear algebra, which is the key formula in the algorithm, is first introduced.

If G is a matrix partitioned as a bordered matrix as shown below:

$$G = \begin{bmatrix} A & u \\ u^T & \alpha \end{bmatrix}$$

Where A is a $n \times n$ matrix, u is a $n \times 1$ column vector, and α is scalar, then

$$G^{-1} = \begin{bmatrix} B & q \\ q^T & z \end{bmatrix} \tag{9}$$

Where

$$B = A^{-1} + zA^{-1}uu^T A^{-1}$$

$$q = -zA^{-1}u$$

$$z = \frac{1}{\alpha - u^T A^{-1}u}$$

Now suppose a new input training vector (x_{new}, y_{new}) is available, Equation (7) becomes:

$$\begin{bmatrix} 0 & e1^T & v_1 \\ e1 & Q + C^{-1}I & V_2 \\ v_1 & V_2^T & v_3 \end{bmatrix} \begin{bmatrix} b \\ \alpha \\ \alpha_{new} \end{bmatrix} = \begin{bmatrix} 0 \\ y \\ y_{new} \end{bmatrix}$$

Where $v_1 = k(x_1, x_{new})$, $V_2 = [k(x_2, x_{new}), \ldots, k(x_n, x_{new})]^T$ is a $(n-1) \times 1$ column vector, $v_3 = k(x_{new}, x_{new}) + 1/C$.

It is important to notice that the square matrix on the left side now can be partitioned into a bordered matrix as follows:

$$A_{new} = \begin{bmatrix} 0 & e1^T & v_1 \\ e1 & Q + C^{-1}I & V_2 \\ v_1 & V_2^T & v_3 \end{bmatrix} = \begin{bmatrix} A_{old} & s \\ s^T & \beta \end{bmatrix}$$

Where

$$A_{old} = \begin{bmatrix} 0 & e1^T \\ e1 & Q + C^{-1}I \end{bmatrix}$$

$$s = \begin{bmatrix} v_1 & V_2^T \end{bmatrix}^T$$

$$\beta = v_2$$

Applying formula (9), the inverse of the new square matrix A_{new} can be expressed in terms of the inverse of the old square matrix A_{old} and the column vector s , as depicted below:

$$A_{new}^{-1} = \begin{bmatrix} B & q \\ q^T & Q + C^{-1}I \end{bmatrix} \tag{10}$$

Where

$$B = A_{old}^{-1} + z A_{old}^{-1} s s^T A_{old}^{-1}$$

$$q = -z A_{old}^{-1} s$$

$$z = \frac{1}{\beta - s^T A_{old}^{-1} s}$$

In other words, the inverse of the new square matrix A_{new} can be computed via the above formulas efficiently, without any matrix inversion. Once obtaining the inverse of A_{new} , the new optimum b, α, α_{new} are given by:

$$\begin{bmatrix} b \\ \alpha \\ \alpha_{new} \end{bmatrix} = A_{new}^{-1} \begin{bmatrix} 0 \\ y \\ y_{new} \end{bmatrix} = \begin{bmatrix} B & q \\ q^T & z \end{bmatrix} \begin{bmatrix} 0 \\ y \\ y_{new} \end{bmatrix} \tag{11}$$

A summary of this training algorithm, namely the algorithm for finding parameters value b and α can be summarized by:

(1) Initialization: $n = 2$;
(2) Calculate A_2^{-1}, b, α;

(3) Get new data (x_{new}, y_{new}) and compute A_{new}^{-1} with formula(10);

(4) Compute b, α, α_{new} with formula (11), then $\alpha = \begin{bmatrix} \alpha \\ \alpha_{new} \end{bmatrix}$;

(5) If there are more input training samples, then $A_{old}^{-1} = A_{new}^{-1}$, goto step(3), otherwise, output b, α , quit.

4   Modeling and Control Framework with LSSVM

In this section we will use the LSSVM regression algorithm described previously to: (1) model a nonlinear dynamical system (design of the modeling LSSVM-block), (2) generate the control input to the nonlinear plant (design of the control-block). Here we use a typical modeling + control framework[8], as sketched in the Fig.1.

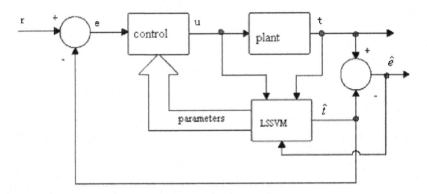

Fig. 1. A modeling-control framework with LSSVM

Where, t is the actual output of a plant, $\hat{t}$ is the output of the LSSVM. $\hat{e}$ is the error between the output of the plant and the output of the LSSVM, it is the identification error, and depends on the approximation capability of the LSSVM itself. r is the reference input of the plant. e is the control error. u is the control signal to the plant. The control aim is to minimize error e ,i.e. make the output of plant t to follow reference input r.

Note that the actual plant is suppose to be unknown, we must design LSSVM block to model the plant's dynamical behavior according to it's input-output data sets. In general, the output of a discrete time system at the time $k + 1$ can be represented as a function of n previous samples of the output itself and m previous samples of the input signal (NARMAX model). The relationship is:

$$t(k+1) = f(t(k), \ldots, t(k-n+1), u(k), \ldots, u(k-m+1)) = f(x(k))$$

$$x(k) = [t(k), \ldots, t(k-n+1), u(k), \ldots, u(k-m+1)]^T$$

Where $x \in R^{m+n}$. At the time $k + 1$, the LSSVM gives an estimate of $t(k + 1)$, called $\hat{t}(k + 1)$:

$$\hat{t}(k+1) = \hat{f}(x(k), w)$$

In order to design the LSSVM block (namely solving the parameter vector w), firstly we must build a training set Z, as follow:

$$[t(n-1), \ldots, t(0), u(n-1), \ldots, u(n-m), t(n)] = (x_1, t_1)$$

$$[t(n), \ldots, t(1), u(n), \ldots, u(n-m+1), t(n+1)] = (x_2, t_2)$$

$$[t(n+1), \ldots, t(2), u(n+1), \ldots, u(n-m+2), t(n+2)] = (x_3, t_3)$$

$$\cdot$$
$$\cdot$$
$$\cdot$$

and so on.

We can use the above data set to train LSSVM. Due to its good generalization performance, the trained LSSVM can characterize the dynamical behavior of the unknown plant.

Then we begin to design the control-block. The job of the controller is to provide the control signal u in such a way to minimize the control error e on the basis of the LSSVM-block. The idea is quite simple and based on the following criteria [8]. A cost J measuring the quadratic error between the reference and the output of the LSSVM is defined as:

$$J(k+1) = \frac{1}{2}e(k+1)^2 = \frac{1}{2}[r(k+1) - \hat{t}(k+1)]^2$$

The goal is to minimize J by finding a simple control law for u. Here we adopt gradient descent algorithm, as follow:

$$u(k+1) = u(k) - \mu \frac{\partial J(k+1)}{\partial u(k)}$$

where μ is the control descent step to be properly set in order to avoid undesired oscillations. If a RBF kernel is supposed to be used, according to formula (8), it is possible to obtain:

$$\hat{t}(k+1) = \sum_{i=1}^{l} \alpha_i \exp(-\frac{\|x(k) - x_i\|_2^2}{2\sigma^2}) + b$$

$$J(k+1) = \frac{1}{2}[r(k+1) - \hat{t}(k+1)]^2$$

$$= \frac{1}{2}[r(k+1) - \sum_{i=1}^{l} \alpha_i \exp(-\frac{\|x(k) - x_i\|_2^2}{2\sigma^2}) - b]^2$$

$$\frac{\partial J(k+1)}{\partial u(k)} = \frac{\partial J(k+1)}{\partial \hat{t}(k+1)} \times \frac{\partial \hat{t}(k+1)}{\partial u(k)}$$

$$= \frac{1}{\sigma^2} e(k+1) \sum_{i=1}^{l} \alpha_i (u(k) - x_{i,n+1}) \exp(-\frac{\|x(k) - x_i\|_2^2}{2\sigma^2})$$

$$u(k+1) = u(k) - \mu \frac{\partial J(k+1)}{\partial u(k)}$$

$$= u(k) - \frac{\mu}{\sigma^2} e(k+1) \sum_{i=1}^{l} \alpha_i (u(k) - x_{i,n+1}) \exp(-\frac{\|x(k) - x_i\|_2^2}{2\sigma^2}) \qquad (12)$$

where $x_{i,n+1}$ denotes the component $n+1$ of the vector x_i.

In general, nonlinear control systems like the one described in this section theoretically analyzing stability is a difficult thing, experientially as long as the following conditions hold true: (1) the LSSVM-block converges to the actual physical systems; (2) the control law stabilizes the LSSVM-block, we can say the controlled system is stable, so we can choose suitable parameter μ to stabilize controlled system by computer simulation analysis.

5   Numerical Results

In this section we will take a concrete example to illustrate proposed control method. The plant under consideration is a spring-mass-damper system with a hardening spring:

$$\ddot{y}(t) + \dot{y}(t) + y(t) + y^3(t) = u(t)$$

we begin by generating 100 data dot $(u(k), y(k))$. The data set is then split into two portions: one for training and one for testing.

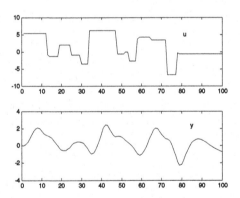

Fig. 2. Training and testing data set

Since it is a second order plant, we will use two past outputs and two past control signals as regressors, namely:

$$x(k) = [y(k), y(k-1), u(k), u(k-1)]^T$$

$$t(k) = y(k+1)$$

Then we can train LSSVM-block by $\{x(k), t(k)\}$. We give out LSSVM's design parameter as follow: $\sigma = 3, C = 150$. In order to show the performance of LSSVM, we use both LSSVM and RBF neural network to model the above plant. The following table illustrates training and testing error of both LSSVM and NN:

As shown in Tab.1, we can see that the performance of LSSVM is superior to the performance of RBF NN method, LSSVM method possesses the best generalization ability.

Figure 3 illustrate LSSVM's training samples and testing samples fitting quality in the case of 50 training samples, 50 testing samples, we can see that trained LSSVM-block can exactly model the plant's dynamical behavior.

Table 1. The simulation error of LSSVM and RBF NN

training sets	testing sets	training err of LSSVM	testing err of LSSVM	training err of RBF NN	testing err RBF NN
20	20	0.0406	0.1494	6.5143e-015	0.3305
30	30	0.0391	0.1050	5.7547e-014	0.3880
40	40	0.0478	0.0882	4.7945e-014	0.1094
50	50	0.0470	0.0698	9.0175e-014	0.1635
60	40	0.0440	0.0654	1.6850e-013	0.1021
70	30	0.0419	0.0975	3.0721e-013	0.1454

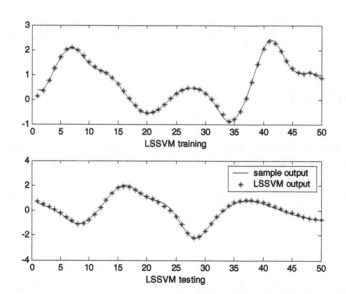

Fig. 3. LSSVM training and testing results

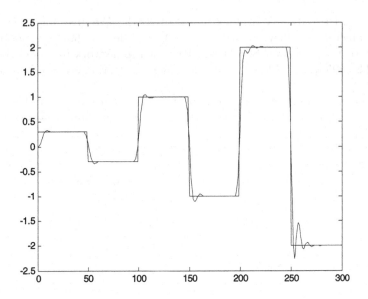

Fig. 4. Reference input and controlled plant output

After the LSSVM-block has been designed, we set up the controller by formula (12). We get parameter $\mu = 0.7$, and carry out numerical experiment. The simulation result is as follow:

In the Fig.4, the segmenting straight line is reference output, the other line is the actual output of plant. From the obtained simulation results, we deduce that the method based on LSSVM can model and control a nonlinear unknown system efficiently.

6   Conclusion

In this paper we introduce the use of least square support vector machines for solving nonlinear systems' modeling and control problems. An introduction to LSSVM is given at first, then gives its training algorithm, and uses it to build a modeling-control framework to control a nonlinear system, the numerical experiment has shown the efficiency of the LSSVM based modeling and control method.

References

1. Cherkassky, V., Mulier, F.: Learning from data: concepts, theory and methods. John Wiley and Sons, New-York(1998)
2. Sjoberg, J., Zhang, Q., Ljung, L., Benveniste A., Delyon, B.: Nonlinear black-box model-ing in system identification: a unified overview. Automatica, **31**(1995) 1691-1724

3. Vapnik, V.: the nature of statistical learning theory. Springer-Verlag, New-York(1995).
4. Vapnik, V.: Statistical learning theory. John Wiley and Sons, New-York(1998)
5. Suykens, J. A. K., Vandewalle, J.: Least squares support vector machine classifiers. Neu-ral Processing Letters, **9**(1999)293-300
6. Suykens, J. A. K., Vangestel, T., Demoor, B.: Least Squares Support Vector Machines. World Scientific, Singapore(2002)
7. Cowen, C. C.: Linear Algebra for Engineering and Science. West Pickle Press, West Lafayette, Indiana(1996)
8. Davide, A., Andrea B., Luca T.: SVM Performance Assessment for the Control of Injec-tion Moulding Processes and Plasticating Extrusion. Special Issue of the International Journal of Systems Science on Support Vector and Kernel Methods,**9**(2002)723-735

Fast Motif Search in Protein Sequence Databases

Elena Zheleva[1] and Abdullah N. Arslan[2]

[1] University of Maryland, Department of Computer Science,
College Park, MD 20742, USA
elena@cs.umd.edu
[2] University of Vermont, Department of Computer Science,
Burlington, VT 05405, USA
aarslan@cs.uvm.edu

Abstract. Regular expression pattern matching is widely used in computational biology. Searching through a database of sequences for a motif (a simple regular expression), or its variations is an important interactive process which requires fast motif-matching algorithms. In this paper, we explore and evaluate various representations of the database of sequences using suffix trees for two types of query problems for a given regular expression: 1) Find the first match, and 2) Find all matches. Answering Problem 1 increases the level and effectiveness of interactive motif exploration. We propose a framework in which Problem 1 can be solved in a faster manner than existing solutions while not slowing down the solution of Problem 2. We apply several heuristics both at the level of suffix tree creation resulting in modified tree representations, and at the regular expression matching level in which we search subtrees in a given predefined order by simulating a deterministic finite automaton that we create from the given regular expression. The focus of our work is to develop a method for faster retrieval of PROSITE motif (a restricted regular expression) matches from a protein sequence database. We show empirically the effectiveness of our solution using several real protein data sets.

Keywords: regular expression matching, motif search, suffix tree, PROSITE pattern, heuristic, preprocessing.

1   Introduction

Motifs are short amino acid sequence patterns which describe families and domains of proteins. PROSITE motifs are represented in the form of regular expressions that do not use Kleene closure. The use of motifs to determine the function(s) of proteins is one of the most important tools of sequence analysis [5]. The number of protein sequences stored in public databases such as SWISS-PROT [4] has been growing immensely and searching for a motif match in such databases can be very slow (many minutes to hours) if we exhaustively search the database [3]. When we explore a motif we may need to repeat the search process for many variations of the description of the motif to see

D. Grigoriev, J. Harrison, and E.A. Hirsch (Eds.): CSR 2006, LNCS 3967, pp. 670–681, 2006.

the effect of modifications on the motif and the matching sequences interactively. This requires finding the matching sequences very fast (in seconds) if there is one.

In this paper we develop a method which finds all the matches. Our emphasis is finding the first match very quickly to improve the response time, and the interactivity. We create an ordered index of the protein sequences which allows us to answer queries about an existing match faster than previous solutions, and to answer queries about all matches as fast as other existing solutions. We show empirically the effectiveness of our solution using several real protein data sets.

Bieganski et al. [3] create a generalized suffix tree from a given database of protein sequences, and then use this tree to find matches to a given motif. They create an equivalent deterministic finite automaton from a given regular expression. They combine the depth-first traversal with the simulation of the automaton on the labels of the tree edges to find matches (strings accepted by the automaton). They implement this method into a tool, namely *MotifExplorer*. They show experimentally that the performance of *MotifExplorer* is superior to grep, the regular expression pattern matching utility of UNIX, which essentially creates a non-deterministic finite automaton and searches for the pattern in the entire database sequentially. They report that on average *MotifExplorer* is 16 times faster. Their results show that in 95% of the motif queries, a generalized suffix tree proves to be a better choice than an unprocessed sequence database.

In this paper we modify the suffix-tree approach in [3]. We study the effectiveness of several heuristics for speeding up the motif search. We recognize the importance of subtree leaf count and use it in pattern matching with suffix trees. First, we consider the sibling nodes in descending order of their corresponding subtree leaf-counts. Secondly and separately, we consider the sibling nodes in descending order of label-lengths on the incoming edges. We study combinations of these two approaches. We also consider a height and branching factor constrained tree in which nodes satisfy certain constraints on leaf-counts. We preprocess the generalized suffix tree to obtain such a tree, which we call a weight-balanced semi-suffix tree. A similar tree is useful in answering certain dictionary queries [2]. We use these trees in developing several heuristic algorithms. We implement, and test them on real protein sequences, and motifs. We report which heuristics yield a speed-up on answering the decision problem using these heuristics. We note that using the leaf-counts ordering is the most effective factor in the improved results. We observe that in 88.4% of the cases in our tests, our heuristic-based methods perform much better than the method presented in [3].

We include pointers for basic definitions and properties of regular expressions, finite automata, and suffix trees in Section 2. In Section 3 we first summarize related work on protein motif searching that uses generalized suffix trees, and then we explain in Subsection 3.1 various orderings we study for suffix trees. We present our heuristic algorithms on ordered trees, and results in Section 4. We conclude in Section 5.

2   Background

Given a text T, and regular expression R, a classical solution for finding in T matches for R is to construct an equivalent finite automaton (FA) M for R and simulate M on symbols in T, and report whenever a final state in M is entered [1].

Instead of simulating M on the sequences in a given database we use the approach proposed by Bieganski et. al [3] and we simulate M on the labels of the suffix tree that we create from the database.

Suffix trees are data structures that are useful for various problems which involve finding substrings in a given string. For example finding a pattern (substring) of length m takes $O(m)$ time. Gusfield [8] provides a comprehensive study and analysis of suffix trees.

A suffix tree for a string s is a rooted directed tree in which each path between the root and a leaf represents a unique suffix of a given input string. Each non-leaf (or internal) node, except the root, has more than one children, and each edge is labelled with a non-empty substring of S. All edges out of a node have distinct first characters of their labels.

From the definition, we can see that a suffix tree on s will contain exactly n leaves. A character $\$ \notin \Sigma(s)$ is usually appended to s to guarantee that a suffix tree exists for every string and that each suffix will end in a leaf, and not in the middle of an edge. This adds an extra suffix to the tree - $\$$. The *label of a tree node* v is the in-order concatenation of the edge labels on the path from the root of the tree to v. The *string − depth* of v is the number of characters in that label. A path may end in the middle of an edge. Every suffix is denoted by $s[i..n]$ where i is the position in the string where the suffix starts. The label

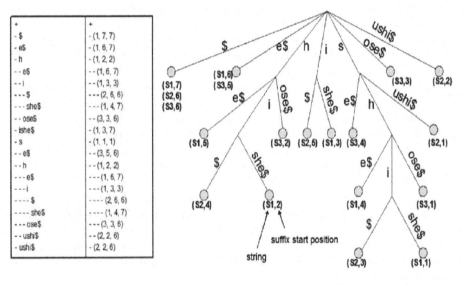

Fig. 1. The generalized suffix tree of the strings $S_1 = shishe$, $S_2 = sushi$, $S_3 = shose$

length can be restrained by the so called *edge-label compression* in which every edge label is represented by the pair (x, y), where x is the starting position of the suffix in the string and y is the ending position of the suffix. This allows a suffix tree to be created in $O(n)$ space.

Sometimes we would like to represent a set of strings rather than one string by a suffix tree. In such a case, we would need a *generalized suffix tree (GST)*. From here on in this paper, a suffix tree would imply a generalized suffix tree. A generalized suffix tree for strings $S_1, S_2, ..., S_n$ can be built in $O(| S_1 | + | S_2 | + ... + | S_n |)$ time using Ukkonen's algorithm [11]. Figure 1 shows a suffix tree for the strings *shishe, sushi, shose*. The table in the figure shows the edge-label compression. There are two differences in the representation of a generalized suffix tree as compared to a suffix tree to reflect the extra information that it contains. First of all, each edge label is represented by three numbers (i, x, y) where i is the index of the sequence S_i under consideration, and (x, y) are the beginning and ending positions of the edge label in S_i. The second difference is that each leaf node contains a multiple of distinct pairs (S_i, j) of numbers where S_i is a string whose suffix $S_i[j..n]$ is represented by the path from the root to the leaf, and where $n = |S_i|$. We will refer to the set of sequences as the *database* of sequences.

3   Protein Motif Matching

Protein sequences are sequences from the alphabet of 20 amino acids. Proteins are grouped into families and domains depending on their functional attributes and evolutionary ancestry. The proteins which belong to a certain family usually have some sequence similarities which have been conserved during evolution. A conserved region is like a signature for a protein family or domain, and it distinguishes the members of one family from all other unrelated proteins. These signatures are useful for finding the function of an unknown, newly sequenced protein, especially when alignment techniques do not produce any results. PROSITE is an established database which contains information about such signatures [5], and it is connected with SWISS-PROT which is a database for protein annotation [4]. There are different ways to describe the conserved regions in protein sequences, and protein *motifs* are such pattern descriptors. The use of protein sequence patterns (or motifs) to determine the function(s) of proteins is one of the most important tools of sequence analysis [5].

Typically, motifs span 10 to 30 amino acid residues. Motifs are restricted regular expressions (RRE) which do not contain the Kleene closure operator *. The symbol 'x' is used for a position where any amino acid is accepted. Ambiguities are indicated by listing the acceptable amino acids for a given position, between square brackets '[ ]', or by listing between a pair of curly brackets '{ } ' the amino acids that are not accepted at a given position. Each element in a pattern is separated from its neighbor by a '-'. Repetition of an element of the pattern can be indicated by following that element with a numerical value or, if it is a gap ('x'), by a numerical range between parentheses. For example, x(3) corresponds to x-x-x, and x(2,4) corresponds to x-x or x-x-x or x-x-x-x. When a pattern is restricted to either the N- or C-terminal of a sequence, that pattern either starts with a '<' symbol or respectively ends with a '>' symbol.

For example, a so-called P-loop motif is represented in PROSITE as [AG]-x(4)-G-K-[ST] (PROSITE entry PS00017), which means that the first position of the motif can be occupied by either Alanine or Glycine, the second, third, fourth, and fifth positions can be occupied by any amino acid residue, and the sixth and seventh positions have to be Glycine and Lysine, respectively, followed by either Serine or Threonine.

We take a protein motif and compare it to a set of protein sequences. We convert the motif to a regular expression which in turn we convert to an equivalent deterministic FA (DFA). We consider in the motif matching frame the following two problems:

Problem 1. Given a set of n protein sequences $S = \{S_1, S_2, ..., S_n\}$, and a motif M, if M matches a substring in any $S_i \in S$ then return that substring, otherwise, return $NULL$.

Problem 2. Given a set of n protein sequences $S = \{S_1, S_2, ..., S_n\}$, and a motif M, return all substrings of $S_i \in S$ for i=[1..n] which match M.

Currently, the motif matching supported by the PROSITE database website [5] does not preprocess the protein sequence database. It runs a *Perl* script called ps_scan against the sequences in the database SWISS-PROT. The script provides a very comprehensive output [6].

Bieganski et al. [3] were the first to propose matching the PROSITE motifs against a preprocessed database, in particular a sequence database represented as a generalized suffix tree. Their tool *MotifExplorer* represents the motif as a

DFA:

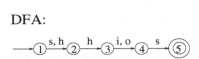

Char	Transition	Match
$	No	
e	No	
h	Yes	
- e	No	
- i	No	
- o	No	
i	No	
s	Yes	
- e	No	
- h	Yes	
- - e	No	
- - i	Yes	
- - - $	No	
- - - s	Yes	s h i s
h	No	
- - o	Yes	
s	Yes	s h o s
e	No	
- u	No	
o	No	
u	No	

Fig. 2. An example of applying *TreeFindMatches* to the suffix tree of the strings $S_1 = shishe$, $S_2 = sushi$, $S_3 = shose$ for regular expression $(s|h)h(i|o)s$

DFA and simulates it on a preprocessed protein sequence database. We modify and use their approach.

First we develop an algorithm for Problem 2. We create a GST with root r from the sequence database. We construct a DFA M whose start state is q_0 from a given motif. Our pattern matching algorithm $TreeFindMatches(M, r, q_0)$ solves Problem 2 by simulating the automaton M on the labels of the tree edges in depth-first manner starting at state q_0. In order to solve Problem 1, we use a modified version of the algorithm that stops whenever it finds a match (if there is one). Figure 2 shows applying $TreeFindMatches$ to the tree in Figure 1 for regular expression $(s|h)h(i|o)s$. The first column shows the parts of the tree in the order that they are traversed, the second column shows whether there exists a transition in the finite automaton for each symbol under consideration. The third column displays all the matches found.

We study motif search not only in an ordinary generalized suffix tree as in [3] but also in other similar data structures which are ordered by various parameters. We study the effect of these data structures, and as we report in Section 4 compared to the method used in [3] we achieve much faster results in 88.4% of the cases in our tests in answering Problem 1.

3.1   In Ordered Suffix Trees

We first create a generalized suffix tree T from the sequences in a given database. Then we preprocess T to create a new tree S using various criteria: ordering by leaf count, ordering by incoming edge-label length, bounding the balance of the tree. We explain the motivation for each of these trees. In Section 4 we report the effectiveness of these trees.

Ordered by Leaf Count. Let $p_{v,i}$ be the probability that a match will be found in subtree T_i of node v. After this node has been reached in the pattern matching process, the expected number of branches (i.e. edges) out of v that will be explored in the pattern matching is $b_{exp} = \sum_{i=1}^{n} i p_{v,i}$. Since subtrees are always explored left to right, if we order the subtrees in decreasing order of their associated probabilities, we will minimize b_{exp} for v when looking for one match. When we look for all matches, the branch search will be exhaustive independent of the processing order of the subtrees.

We assume that a given pattern appears in every sequence suffix with equal probability. This is a realistic assumption if input patterns are uniformly selected from the universe of suffixes in the database. Under this assumption, the leaf count of subtree T_i can be used in approximating the probability $p_{v,i}$. We compute and associate $p_{v,i}$ with every subtree root node v using the formula $p_{v,i} = \frac{n_{v,i}}{n_v}$ where $n_{v,i}$ is the number of leaves in T_i, and n_v is the number of leaves in the tree rooted at v. We sort the children of each node in the tree by their leaf count in breadth-first manner. If there is a partial match in the label of the first edge, the chances that there is a full match in the subtree are the highest. Otherwise, we eliminate the biggest subtree from the pattern matching process.

Ordered by Incoming Edge Label Length. Nodes in suffix trees distinguish between string depth (corresponding number of characters on the path from the root to that node) and depth (number of nodes on that path). All children of a node have the same tree depth but they may have different string depths. When we order the tree by leaf count, there is a hidden assumption that the string depth of all the subtree roots are the same. However, this is often not the case because edges coming out of a node can have different label lengths. Exploring a subtree of many short edges first (short in term of edge label length) may be more costly than matching one with longer edges. Therefore, for subtrees with the same probability of finding a match, we would like to explore the one with the longest edge first. We preprocess T by sorting the nodes at each level in descending order in their incoming edge label length. The preprocessed tree is represented by S. We also try sorting the nodes at each level in ascending order in their incoming edge label length. We note that short-label length may indicate a large leaf-count. Therefore, there is a tradeoff between the potentials of descending leaf-count heuristic, and descending label-length heuristic.

Weight-Balanced Semi-suffix Trees. Leaf count is central to the idea of creating weight-balanced trees [10]. We preprocess a suffix tree T to create a new suffix tree S with bounded branching factor and height as shown in Figure 3 (Arslan [2] creates a similar tree for a different problem). We call the new tree $semi - suffix$ since it is based on a suffix tree but a few edges out of a node can start with the same character. Subtrees T_i are created recursively based on an observation that there exists a node v in any generalized suffix tree T such that $\frac{n}{2|\Sigma|} \le n_v \le \frac{n}{2}$ for alphabet Σ [2]. Let N_{p_i} denote the number of words with prefix p_i. We create subtrees T_i in increasing order in i. The subtree T_i is selected such that the root of the subtree T_i has leaf-count N_{p_i} that satisfies $\frac{N_i}{|2\Sigma|} \le N_{p_i} \le \frac{N_i}{2}$ where $N_i = n - \sum_{j=1}^{i-1} N_{p_j}$. If there are k subtrees created this way then N_i is

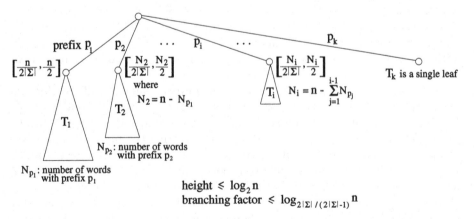

Fig. 3. The resulting weight-balanced semi-suffix tree. The intervals next to the nodes are the range of leaf-counts for these nodes. Σ denotes the alphabet for the suffixes.

also the total number of leaves in subtrees from T_i to T_k, i.e. $N_i = \sum_{j=i}^{k} N_{p_j}$ for all i, $1 < i \leq k$, and $N_1 = n$. The following are true for the tree S: 1) The height h is $\leq \log_2 n$, 2) The branching factor b is $\leq \log_{2|\Sigma|/(|2\Sigma|-1)} n$.

4 Implementation and Tests

For building the generalized suffix tree, we have used a suffix tree library *libstree* version 0.4 [9] which builds a tree using Ukkonen's algorithm and represents the edges coming out of a node as a linked list. Each edge has references to its right sibling, to its source node and destination node. Each node has a reference to its incoming edge and to the edge which connects it with its leftmost child. This allows depth-first iteration, breadth-first iteration, and bottom-up iteration of the nodes in the tree. Based on the tree produced by the library, we have created the ordered trees discussed in the previous section.

For building the DFA, we have used a regular expression parsing program written by Amer Gerzic [7]. We have modified the pattern matching to suit our needs for finding matches in a suffix tree according to our regular expression matching algorithm *TreeFindMatches*.

We aim to test whether ordering the tree will affect the performance of the pattern matching search. We create a GST from a sample of protein sequences, and preprocess the GST to create 8 types of suffix trees. For each motif, we create an DFA and simulate it on each tree, and count the number of characters processed in the simulation. Under the assumption that the tree can fit in main memory, the character count is a good estimator for the performance of the tree. For the current state of the SWISS-PROT 43.6 database, a suffix tree on the database is approximately 1Gb.

We create and use in our tests the following types of trees:

1. Original - generalized suffix tree with no ordering of nodes.
2. Weight-balanced - weight-balanced semi-suffix tree (see Figure 3).
3. Asc. Label (Desc. Leaf) - GST in which the subtrees of each node are in ascending order of incoming edge label length (conflict resolution: highest leaf count first).
4. Desc. Leaf (Asc. Label) - GST in which the subtrees of each node are in descending order of leaf count (conflict resolution: shortest incoming-edge labels first).
5. Desc. Leaf (Desc. Label) - GST in which the subtrees of each node are in descending order of leaf count (conflict resolution: longest incoming-edge labels first).
6. Desc. (Leaf*Label) - GST in which the subtrees of each node are in descending order of (incoming edge label length)*(leaf count).
7. Desc. Label - GST in which the subtrees of each node are in descending order of incoming-edge label length (no conflict resolution).
8. Asc. Label - GST in which the subtrees of each node are in ascending order of incoming-edge label length (no conflict resolution).

We consider 20 samples of randomly selected protein sequences from SWISS-PROT 43.6. Each sample contains 30 sequences. We run 730 of the PROSITE motifs (all the PROSITE motifs except those that contain x(n,m), {} or >) against the unordered and ordered suffix trees created from the samples. The average length of the 600 protein sequences is 340 characters (amino acids) (More details about the parameters used in the tests, and the test results can be obtained by contacting the authors).

We find the part that the pattern matching will cover for each tree by counting the number of characters processed (simulated on the DFA) until a match is found. When there is no match, the ordering of the leaves does not reduce the number of characters read, and the characters processed are the same as in Problem 2 (finding all matches). Therefore, we do not include in the discussion the cases when there is no match.

Figure 4 shows the distribution of the number of characters processed for all cases when a match is found (Problem 1 is answered) in our tests. For example we see in the figure that in more than 70 cases, approximately 23 characters are processed when we use the weight-balanced tree. Most of the 730 motifs that we tested do not have matches in our samples since we built the tree from a fraction of the whole database. In 88.4% of the cases in which there is a match, the weight-balanced tree (Tree 2) performs better than the original tree (Tree 1). We designed our tests such that each case represents a sample in which a certain motif has a match. Figure 5 shows the distribution of the speedup of the weight-balanced tree over the original tree for all cases. In the figure we number bins

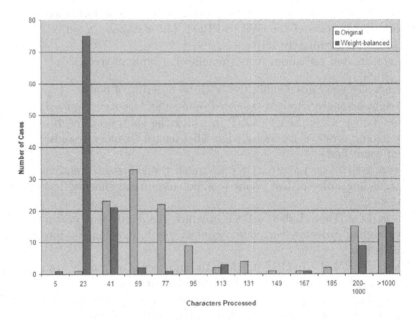

Fig. 4. The distribution of number of characters processed in answering Problem 1: Weight-balanced tree versus original tree

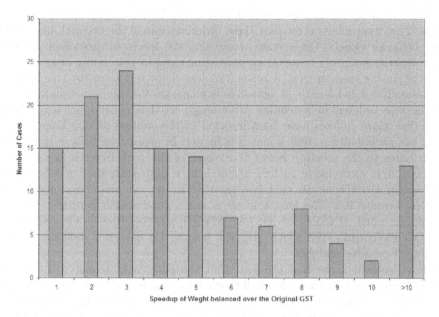

Fig. 5. Histogram of speedup of weight-balanced tree over original tree

such that bin i represents a speedup factor (i.e. the factor by which our method is faster over the ordinary suffix tree method) between factors $i - 1$ and i. Bin 2 corresponds to a speedup factor between 1 and 2. All the cases when the weight-balanced tree is faster are included in the bins 2 and higher. In 11.6% of the cases our heuristic algorithm did not help with the performance. They are at bin 1 indicating a slowdown (speedup factor is between 0 and 1). The smallest speedup factor we observed in our tests was 0.14.

Empirically, trees 2, 3, 4, 5 and 6 have very similar behavior. To save space we do not include histograms for the performance of each tree. They are very much similar to figures 4 and 5. In all of them, the decreasing order of subtree leaf count plays a role which proves that this is the important factor in the ordering. The median for the number of characters processed is 62 for the original tree. It is 16 for trees 2, 3, 4, 5, 6. The average number of characters processed for all ordered and unordered trees is between 393.6 and 406.9 characters in answering Problem 1 when there is a match. In most of the cases the answer was found very quickly. For 11.6% of the motifs, it took very long to answer Problem 1 (observed worst-case for the number of processed characters in these cases: 2070 for unprocessed tree, and 4241 for an ordered tree). They increased the averages.

The leaf count heuristic aims to minimize the time to find a match by directing the search to a subtree with the highest leaf-count where the probability for a match is the highest under a uniform distribution of matches. One may argue that the time required to find a match should increase as the leaf-counts increase. Although this may be factor, it is insignificant for most of the cases as our experiments shows the superior performance of the leaf-count heuristic.

Ordering only by edge-label length (Trees 7 and 8) does not yield to better results. The experiments show that these orderings mimic the original, unordered tree behavior closely. On average, traversing the leaves in ascending order of label lengths is slightly better than processing them in descending order of label lengths. In the median case, it reads 30% less characters than the tree ordered in descending label length. A plausible explanation for the better performance of the tree ordered in ascending label length is that short label length is a sign that more suffixes have been inserted in the subtree of that edge, therefore, the probability to find a match is higher. In general we believe that what contributes to the speedup is not the length of the path but the fact that on average, it is more likely to find a match in a path with shorter labels/high leaf count. For the motifs that had matches in our samples, the motif length was on average 9.6. The length alone is not a good search heuristic. For example, in motif [DNSTAGC]-[GSTAPIMVQH]-x-x-G-[DE]-S-G-[GS]-[SAPHV]-[LIVMFYWH]-[LIVMFYSTANQH], the length, 12, does not indicate a large variety of possible matches.

For Problem 2, we find the part that the pattern matching will cover in each tree by counting the number of characters read until all matches are found. Our results confirm the results of Bieganski et al. [3] that using a generalized suffix tree can speed up the search for motif matches when compared to an unprocessed database. Using weight-balanced semi-suffix tree compared to an unprocessed generalized suffix tree slowed down answering Problem 2 by 1% on average in our tests. We explain this "delay" with the higher branching factor, and common prefixes on the edge-labels in the weight-balanced semi-suffix tree. Ordering the nodes by leaf count or label length does not affect the solution of Problem 2 as compared to the original tree. Therefore, we do not observe a speedup when solving Problem 2 as compared to an unordered tree.

The effectiveness of leaf-count as an estimator for probability of a match under uniform distribution suggests the following branch-and-bound algorithm: Start at the root with an empty priority queue. For each node v visited calculate for each branch from v a key which is the number of leaves that are reached from this branch divided by the number of leaves at the subtree rooted at v, and enqueue this number. Continue next at the node with the largest key, until a match is found. This algorithm branches to the node where the match probability is the highest, and the probabilities are updated dynamically as the search proceeds.

5   Conclusion

In this paper, we explore and evaluate alternative representations of a database of sequences for two types of query problems for a given regular expression: 1) Find the first match (if there is one), and 2) Find all the matches. We propose a framework in which both problems can be addressed. We represent the database of sequences as a generalized suffix tree, and obtain new data structures based on this tree. We show empirically that when solving the first problem, we can achieve a speedup for most of the test cases after ordering the nodes in the generalized suffix tree in a meaningful manner. The experiments suggest that the most important factor is to process the subtrees in the decreasing order of leaf count at each node level. This framework addresses the second problem in a similar manner as the existing suffix-tree method [3]. The framework allows probabilistic analysis which could be a topic of future work.

References

[1] A. Aho. Algorithms for finding patterns in strings. *In J. van Leeuwen (Ed.), Handbook of Theoretical Computer Science: Algorithms and Complexity, Elsevier Science Publishers B.V.*, 5:255–300, 1990.

[2] A. N. Arslan. Efficient approximate dictionary look-up for long words over small alphabets. In *Jose R. Correa, Alejandro Hevia, Marcos A. Kiwi (Eds.). Lecture Notes in Computer Science 3887, pp. 118-129, ISBN 3-540-32755*, Valdivia, Chile, March 20-24 2006. LATIN 2006: Theoretical Informatics, 7th Latin American Symposium, Springer.

[3] P. Bieganski, J. Riedl, J. V. Carlis, and E. F. Retzel. Motif explorer - a tool for interactive exploration of aminoacid sequence motifs. In *Proceedings of Pacific Symposium on Biocomputing*, 1996.

[4] B. Boeckmann, A. Bairoch, R. Apweiler, M.-C. Blatter, A. Estreicher, E. Gasteiger, M.J. Martin, K. Michoud, I. Phan C. O'Donovan, S. Pilbout, and M. Schneider. The swiss-prot protein knowledgebase and its supplement trembl in 2003. *Nucleic Acids Research*, 31:365–370, 2003.

[5] L. Falquet, M. Pagni, P. Bucher, N. Hulo, C.J. Sigrist, K. Hofmann, and A. Bairoch. The prosite database, its status in 2002. *Nucleic Acids Research*, 30:235–238, 2002.

[6] A. Gattiker, E. Gasteiger, and A. Bairoch. Scanprosite: a reference implementation of a prosite scanning tool. *Applied Bioinformatics*, 1:107–108, 2002.

[7] A. Gerzic. Write your own regular expression parser. http://www.codeguru.com/ CppCppcpp_mfc/parsing/article.php/c4093/, 2003.

[8] D. Gusfield. *Algorithms on strings, trees, and sequences: computer science and computational biology.* Cambridge University Press, 1997.

[9] C. Kreibich. libstree - a generic suffix tree library. http://www.cl.cam.ac.uk/ ~cpk25/libstree/, 2004.

[10] K. Mehlhorn. *Data Structures and Algorithms: Sorting and Searching.* Springer-Verlag, 1977.

[11] E. Ukkonen. On-line construction of suffix-trees. *Algorithmica*, 14:249–260, 1995.

Author Index

684   Author Index

Lecture Notes in Computer Science

For information about Vols. 1–3882

please contact your bookseller or Springer